COMBINATORICS AND PROBABILITY

COMBINATORICS AND PROBABILITY

Celebrating Béla Bollobás's 60th birthday

GRAHAM BRIGHTWELL
London School of Economics

IMRE LEADER
University of Cambridge

ALEXANDER SCOTT
University of Oxford

ANDREW THOMASON
University of Cambridge

CAMBRIDGE UNIVERSITY PRESS
Cambridge, New York, Melbourne, Madrid, Cape Town, Singapore, São Paulo

Cambridge University Press
The Edinburgh Building, Cambridge CB2 2RU, UK

Published in the United States of America by Cambridge University Press, New York

www.cambridge.org
Information on this title: www.cambridge.org/9780521872072

© Cambridge University Press 2007

This publication is in copyright. Subject to statutory exception
and to the provisions of relevant collective licensing agreements,
no reproduction of any part may take place without
the written permission of Cambridge University Press.

First published 2007

Printed in the United Kingdom at the University Press, Cambridge

A catalogue record for this publication is available from the British Library

ISBN-13 978-0-521-87207-2 hardback

Cambridge University Press has no responsibility for the persistence or accuracy of URLs for external or third-party internet websites referred to in this publication, and does not guarantee that any content on such websites is, or will remain, accurate or appropriate.

Contents

Preface	*page* ix
Foreword	xi
Photographs	xx
Contributors	xxv

Measures of Pseudorandomness for Finite Sequences: Minimal Values   1
N. ALON, Y. KOHAYAKAWA, C. MAUDUIT, C. G. MOREIRA
AND V. RÖDL

MaxCut in H-Free Graphs   31
NOGA ALON, MICHAEL KRIVELEVICH AND BENNY SUDAKOV

A Tale of Three Couplings: Poisson–Dirichlet and GEM Approximations for Random Permutations   51
RICHARD ARRATIA, A. D. BARBOUR AND SIMON TAVARÉ

Positional Games   83
JÓZSEF BECK

Degree Distribution of Competition-Induced Preferential Attachment Graphs   132
N. BERGER, C. BORGS, J. T. CHAYES, R. M. D'SOUZA
AND R. D. KLEINBERG

On Two Conjectures on Packing of Graphs   157
BÉLA BOLLOBÁS, ALEXANDR KOSTOCHKA AND KITTIKORN
NAKPRASIT

Approximate Counting and Quantum Computation   171
M. BORDEWICH, M. FREEDMAN, L. LOVÁSZ
AND D. WELSH

Absence of Zeros for the Chromatic Polynomial on Bounded Degree Graphs   189
CHRISTIAN BORGS

Duality in Infinite Graphs   201
HENNING BRUHN AND REINHARD DIESTEL

Homomorphism-Homogeneous Relational Structures   217
PETER J. CAMERON AND JAROSLAV NEŠETŘIL

A Spectral Turán Theorem   231
FAN CHUNG

Automorphism Groups of Metacirculant Graphs of Order a Product of Two Distinct Primes   245
EDWARD DOBSON

On the Number of Hamiltonian Cycles in a Tournament   271
EHUD FRIEDGUT AND JEFF KAHN

The Game of JumbleG   285
ALAN FRIEZE, MICHAEL KRIVELEVICH, OLEG PIKHURKO AND TIBOR SZABÓ

2-Bases of Quadruples   297
ZOLTÁN FÜREDI AND GYULA O. H. KATONA

On Triple Systems with Independent Neighbourhoods   309
ZOLTÁN FÜREDI, OLEG PIKHURKO AND MIKLÓS SIMONOVITS

Quasirandomness, Counting and Regularity for 3-Uniform Hypergraphs   329
W. T. GOWERS

Triangle-Free Hypergraphs   371
ERVIN GYŐRI

Odd Independent Transversals are Odd   379
PENNY HAXELL AND TIBOR SZABÓ

The First Eigenvalue of Random Graphs   399
SVANTE JANSON

On the Number of Monochromatic Solutions of $x + y = z^2$ 413
AYMAN KHALFALAH AND ENDRE SZEMERÉDI

Rapid Steiner Symmetrization of Most of a Convex Body and the Slicing Problem   429
B. KLARTAG AND V. MILMAN

A Note on Bipartite Graphs Without 2k-Cycles   445
ASSAF NAOR AND JACQUES VERSTRAËTE

Book Ramsey Numbers and Quasi-Randomness   451
V. NIKIFOROV, C. C. ROUSSEAU AND R. H. SCHELP

Homomorphism and Dimension   461
PATRICE OSSONA DE MENDEZ AND PIERRE ROSENSTIEHL

The distance of a permutation from a subgroup of S_n 473
RICHARD G.E. PINCH

On Dimensions of a Random Solid Diagram   481
BORIS PITTEL

The Small Giant Component in Scale-Free Random Graphs   505
OLIVER RIORDAN

A Dirac-Type Theorem for 3-Uniform Hypergraphs   547
VOJTĚCH RÖDL, ANDRZEJ RUCIŃSKI AND ENDRE SZEMERÉDI

On Dependency Graphs and the Lattice Gas   571
ALEXANDER D. SCOTT AND ALAN D. SOKAL

Solving Sparse Random Instances of Max Cut and Max 2-CSP in Linear Expected Time   599
ALEXANDER D. SCOTT AND GREGORY B. SORKIN

Preface

This volume brings together research in Combinatorics, Probability and Analysis that was presented at the conference entitled "Combinatorics in Cambridge", held in Cambridge during August 4th–7th 2003. The meeting was organized to mark the 60th birthday of Béla Bollobás on August 3rd. The esteem in which Béla is held by mathematicians throughout the world is attested to both by the exceptional academic standard of the conference and by the high quality of the contributions to this volume.

The mathematical activities of the conference took place in the newly built Centre for Mathematical Sciences, home to the Department of Pure Mathematics and Mathematical Statistics in the University of Cambridge. Accommodation was provided in Trinity College, where the memorable conference banquet was held also. The generous support of the London Mathematical Society, the Heilbronn Fund of Trinity College, the British Combinatorial Committee and DPMMS is gratefully acknowledged. The smooth day-to-day running of the conference owed much to the serenity and good humour of Tricia Simmons, of the University of Memphis.

The editors warmly thank all those who have contributed to this volume, both authors and referees, and indeed we thank all those whose participation in the conference served to make it so successful. Furthermore, since each of us has known the privilege that it is to be one of Béla's research students, we take this opportunity to thank him warmly too.

The illustration on the front cover is of the Fountain in Great Court, Trinity College, Cambridge, and was drawn by Gabriella Bollobás.

Foreword

The thirty-one papers in this volume describe recent research in combinatorics, convex geometry, probability theory, statistical physics and theoretical computer science, and they form the proceedings of the conference "Combinatorics in Cambridge", held in Cambridge during August 4th–7th 2003.

The conference was timed to celebrate the 60th birthday of Béla Bollobás. Speakers were invited with the aim of representing the many areas, both geographical and mathematical, in which Béla has worked. The conference format comprised thirty-five half-hour lectures, with no parallel sessions; this arrangement proved very effective and was well received by the participants, but the time constraints did mean that not all those whom the organizers would have wished to speak were able to do so. We were delighted that so many excellent mathematicians agreed to speak at the meeting: that they did so is a tribute to the very high regard in which Béla is held by the worldwide research community, and the quality of those who participated despite being unable to speak is an equal tribute. We are therefore very grateful to all those participants who have contributed to this volume (and, simultaneously, to the journal *Combinatorics, Probability and Computing*).

The lectures themselves took place in the newly built Centre for Mathematical Sciences, which offered a superb environment both for the lectures and for other mathematical discussions. Memorable non-mathematical features of the conference were the punting afternoon, the piano recital given by Lindesay Dixon and, most unforgettable of all, the conference banquet, of which more later.

It is necessary to say a few words here about Béla Bollobás himself, his life and his work. It is difficult, though, to do so without adopting a valedictory tone which would be completely out of place. Those who know Béla will see at once how a slap on the back and a "well done, old chap" would not fit the bill, and indeed might invite injury. At the age of sixty, Béla is able to look back on nearly five decades of mathematical research of the highest standard, on well over three hundred research papers (with more than a hundred co-authors), on half a dozen textbooks and on around forty research students. But it nevertheless remains the case that the conference was not about celebrating achievements that are safely in the past.

Béla is still, as he always has been, a man of enormous energy, and he remains as active as ever: with his army of research students and his countless visitors and collaborators, his enthusiasm is undiminished, and he still seeks out fresh avenues of research. Now is not the time to sum up a career, for who can say what is yet to come? As for the past, even though that, manifestly, was what lay behind the conference, no serious attempt will be made in this foreword to do it justice. The papers in this volume are testimony enough to a man who has, in Endre Szemerédi's words, "played a fundamental role in the development of modern discrete mathematics".

So the remarks that follow are in no way intended to set out Béla's research, nor indeed are they meant to encapsulate his life or his character. Béla is unquestionably a very distinctive man: he is immensely gifted, he has a very powerful personality, and he has an unusual history spanning several countries – it is then no surprise that so many legends, so many myths and stories seem to surround him. Rather than paint a rounded picture, we offer instead what are in effect a few snapshots of aspects that some might find new, or at least (we hope) interesting.

Béla is of Pecheneg ancestry on his father's side, and he combines Hungarian and Austrian roots on his mother's side. His paternal grandfather fell on hard times, but Béla's father was talented and determined to make a good living as a doctor. As a boy, he used to earn a little money by giving lessons to his classmates and later, when he wished to study medicine at university, Béla's grandfather moved the family to Budapest. Béla's father subsequently became a renowned ear, nose and throat surgeon. He set up his own home in Dunaharaszti just outside Budapest, and this was where Béla's family were living at the time of his birth on 3rd August 1943.

The years of Béla's boyhood were troubled times in Hungary, and many whose lives had been shipwrecked in the prevailing political wind found sanctuary in the Bollobás home. For Béla himself, the habitués provided a wonderful opportunity: from them he learnt languages, history, fencing (from a general), and how to head a ball (from Kocsis, inside right of the great Hungarian football team of the '50s). Not long afterwards these people were mostly banished from the capital, but the sense of an enlarged family remained, made yet larger by the Hungarian attitude to servants, who are seen as part of the extended family. In this home, with its ever open door and its welcoming table, can be traced the source of Béla's own hospitality, and especially of his treatment of his research students, whom he always regards as part of his family.

At the age of 14 Béla started to attend the Czukor utca school in Budapest, which was then considered to be the best in Hungary, and Béla's father moved the family into the capital. Béla was already developing a reputation for mathematical talent, enough to bring him an invitation to tea at the Golf Hotel to meet Paul Erdős and his mother. Of course, Erdős set Béla a problem, which Béla solved by the end of the meal, and so began a great collaboration and a great friendship. Erdős would always visit Béla's house when he was in Hungary.

The first International Olympiad took place in 1959, when Béla was still only 15. He competed with great success in each of the first three Olympiads, winning the last of them by a clear margin. At the age of 18 he entered the Eötvös Loránd University. His "junior" year, 1963–64, however, was spent in Cambridge, at Trinity College, on a visit arranged by Erdős with his friend Davenport, and here Béla enjoyed supervision by Cassels, Swinnerton-Dyer, C.T.C Wall and John Polkinghorne. In this context, supervision means the personal style of teaching involving two undergraduates bringing the work they have done in, say, Functional Analysis, to a supervisor for an hour's discussion: some years later, Béla was to become a supervisor himself, as we shall describe shortly.

On graduating in 1966 he was immediately given a job at the Mathematical Institute of the Hungarian Academy of Sciences (now the Rényi Institute). Whilst there he wrote a PhD in geometry, under the formal direction of Fejes Tóth but informally directed by Erdős. The academic year 1967–68 was spent at the Lomonosov University in Moscow with Gelfand. In February 1969 he arrived to spend half a year at Christchurch College, Oxford at the invitation of Coulson. Aside from mathematics, he found time to earn a Blue as a modern pentathlete and to get his oar rowing for his college. In the summer, facing the prospect of returning to Hungary and so losing the opportunity to travel in the West, Béla decided not to go back. Instead, he travelled to the buffer state of Yugoslavia, where he had arranged a tryst with Gabriella. They were married in a church in Novi Sad, and then returned, via Italy, to England. During his earlier visit to Trinity College, Béla had been told he could return to write a PhD there, and so in October 1969 he registered as a research student in topology under the supervision of Frank Adams. Within just a year, before Adams actually arrived in Cambridge from Manchester, Béla wrote a dissertation in functional analysis and won a Research Fellowship at Trinity (in his spare time picking up a fencing Blue for Cambridge).

This brings the story up to the modern era, so to speak, regarding which we shall be much briefer. In 1971 Béla joined the staff of the Department of Pure Mathematics and Mathematical Statistics in Cambridge (where he stayed till his resignation in 1996) and at the same time he became a Fellow of Trinity College (a position he still holds). From 1981 until 1995 he also held a Chair at Louisiana State University. In 1995 he was made Jabie Hardin Chair of Excellence in Graph Theory and Combinatorics at the University of Memphis. From 2006 he will hold a Senior Research Fellowship at Trinity College.

The basis of "Combinatorics in Cambridge" was a celebration of Béla's research. Most conference participants would know little about his work as a teacher, but this side of his life has been equally effective and distinctive, and indeed it has had an impact on his research, insofar as many of his research students were first influenced by Béla through being taught by him as undergraduates. No Cambridge mathematics student who attended a Bollobás lecture will forget it, and those students who were in Béla's own college, Trinity, will be indelibly marked by his supervisions. The conference learnt a little of why this is true from the speeches made at the banquet by Tim Gowers and Imre Leader.

Tim explained, "In my first term at Cambridge I took four courses, one of which Béla gave. And while I cannot even remember who the other three lecturers were, I will never forget Béla's course, and in particular the first lecture. The course in question was an introduction to analysis, which, I need hardly say, he gave without notes. He began with a proof of the isoperimetric inequality in the plane. His argument was beautiful and satisfying, but just as we were congratulating ourselves on following it he suddenly asked us where he had gone wrong. We were all completely stumped. The answer was that he had taken for granted that there *was* a maximum area that could be achieved, a fact that he would not be able to prove until we had done some analysis. With this example, Béla had aroused our interest and got us thinking."

"Not unreasonably, he assumed that we would carry on thinking and being interested. Several people here will have desperately tried to make themselves inconspicuous when Béla decides to ask a question of his audience. If no answer is forthcoming from the audience in general, he turns to individuals he knows, and of course it doesn't occur to him that they might not be able to supply the answer... I am told that on one occasion he said to an unfortunate undergraduate who was wearing a Pembroke T-shirt, 'Come on, even people from Pembroke should know that.'"

To accompany the lectures, example sheets, of the lecturer's own devising, are handed out, and it is the questions on these sheets that the students work on in their own time and discuss with the supervisor. Imre Leader recalls Béla's example sheets. "He has an inexhaustible supply of *nice* questions. They are not arbitrary or ugly. They are hard and interesting. They are the sort of question that, when you read them, you want to work on them. They are *unusual* questions, questions that make you sit up and say 'what?', questions that get to the heart of the matter. It is quite amazing that his questions are so often both hard and beautiful. Some he has collected; most he has invented."

"He did it in all subjects: not just Graph Theory, Combinatorics, Functional Analysis (his own areas), but also Metric Spaces, Topological Spaces, Ring Theory, Measure Theory, and so on. I once complained to him that there were no hard problems in Linear Algebra; the next day, there arrived in my pigeonhole a list of six gorgeous ones! And the same happened with Complex Variable. In most subjects, his questions are much better than those of the experts in the field. Actually, very often a hard question on another lecturer's sheet can be traced back to a Bollobás question!"

As a supervisor, Béla demanded high standards, and he would often set his own questions when he found the lecturer's too feeble. Imre continued: "Once, Keith Carne discovered that the problem that Béla had set him had only been solved (in a research paper) six months previously. He mentioned this to Béla, who said 'But you have a huge advantage: you know that the result is true'. In supervisions, Béla always lets you think. And, if needs be, sweat. You knew you were doing badly when he'd suddenly ask you what sports you played, or otherwise changed the subject. A sign that you were doing even worse was that he'd make himself a cup of coffee and offer you a dry biscuit."

Tim observed, "As we all know, Béla has views about everything, not always easy to guess in advance but always passionately held." Béla's ever forthcoming advice can sometimes be delivered indirectly, though, and Tim recalled the time in his third year when he had to choose which courses to do. "I had the idea that it would be good to do a mixture of pure and applied mathematics, so that I would not be too narrowly focused and would understand some of the great scientific theories of the early twentieth century. I put this to Béla, and he responded enthusiastically: 'That is an excellent idea. But of course you will do 90 percent pure.' I got the message, did 100 percent pure and to this day do not understand relativity or quantum mechanics."

When considering Béla's impact on the wider mathematical world, there are three aspects of his work that demand attention. The first is, of course, his own research. His creativity, energy and technical power have yielded a continuous stream of excellent results, meaning in particular that he has for decades been in constant demand as a conference speaker. His first paper, published with Erdős in 1962, gave two proofs of a conjecture of Pósa; they each found a proof independently (Béla while he was still at the Czukor utca school). In a 1965 paper he proved the fundamental theorem about strongly saturated hypergraphs, and for many years this remained one of the few non-trivial exact results about hypergraphs. Many significant papers appeared in the 1970s, principally about extremal graph and hypergraph theory – for example with Erdős he established the logarithmic speed of the Erdős-Stone theorem – but papers in functional analysis also form a major component of Béla's output during this period.

Up till this point random graphs feature little in his work, but the situation changes dramatically in the 1980s, with a flow of deep and seminal results on this subject. The most newsworthy was, no doubt, Béla's resolution of the old question of the chromatic number of random graphs, but there are many other memorable ones, such as for example the $n^{2/3}$-sized window illuminating the old 'double jump' of Erdős and Rényi, the threshold for the appearance of arbitrary subgraphs, the configuration model for graphs with a given degree sequence, and the existence of Hamiltonian cycles in regular graphs of high degree. The ideas impinge not just on other areas of mathematics such as analysis and probability, but are of significance in other disciplines, including computer science and physics. It is probably for his work on random graphs that Béla is most widely known.

Random combinatorial models continue to be a dominant theme in his subsequent work, with applications to percolation and to web graphs becoming prominent, but at the same time he has proved major results in extremal graph theory, isoperimetric theory and set systems. Some examples of these are the systematic study of discrete isoperimetric inequalities (with Leader), of judicious partitions (with Scott) and of hereditary properties (with several students and with Simonovits), the discovery of a general graph polynomial (with Riordan) and the interlace polynomial (with Arratia and Sorkin), hard results about scale-free graph models (with Janson and Riordan), progress on random geometric graphs (with Balister, Sarkar and Walters), and (with Balister, Riordan and Walters) developments in bootstrap and Voronoi percolation. Far from taking it easy as he enters his seventh decade, Béla appears to be as productive as ever, if not more so. The quantity, the breadth and the quality of his research output over the years is remarkable.

Béla's books make up the second of the three aspects of his work that must be mentioned. Here we mean principally his textbooks rather than the several volumes that he has edited, though not even Béla's academic editing life has been routine: he and Gabi befriended Littlewood in his later years, and Béla subsequently became Littlewood's literary executor, editing the famous *Littlewood's Miscellany*. Béla's own book, *Extremal Graph Theory*, published in 1978, was the first book to treat the subject in such depth and detail, and it remains an indispensible reference to this day, notwithstanding the developments that have been made since it was written. *Graph Theory* was published as a graduate text, although it was intended to accompany the undergraduate course Béla developed at Cambridge: as a student text it was unrivalled in depth, and in its recent rebirth as *Modern Graph Theory* it forms a fine handbook for even the professional combinatorialist. His *Combinatorics* and *Linear Analysis* are more compact, and very readable, treatments of these areas and are popular texts. In each of these books, the material Béla includes has been carefully selected and the proofs have all been polished (and, not infrequently, improved) from the original. When combined with the author's manifest command of his subject and enthusiasm for it, the result is a very appealing read. As for *Random Graphs*, it appeared at a critical moment in the mid-1980s, when this important subject was undergoing a huge surge in development. With its precise and thorough exposition of the fundamentals, including much due to Béla himself, the book achieved canonical status and has been hugely influential.

The third significant contribution that Béla has made to mathematics is through his supervision of research students. It was mentioned earlier that he has now had about forty students, and not all are combinatorialists: Keith Ball, Keith Carne, Tim Gowers, Jonathan Partington and Charles Read constitute a proud legacy in analysis. Whereas forty is not such a large number in disciplines where research is a team activity, or where supervision is delegated to subordinates, Béla has always given his students close personal attention, and lots of it. Though he demands much of his students he enjoys working with them, and most of them are familiar with that combination of fine hospitality and fine mathematics that is the life of a Bollobás student.

Both Tim Gowers and Imre Leader recounted some first-hand details at the banquet. Imre said, "Béla always takes a huge interest in his research students. Many of his students collaborate with him. He is *always* visitable. At his house, one works, with kitchen breaks for coffee with him and Gabi, or else Gabi may bring coffee and biscuits in to the room where we are working. It is an environment incredibly suited for work. Béla has amazing drive. He is quick to think up new questions, but always nice ones. He has an uncanny ability to see in advance if a method will work. And he has immense power: he can often force a solution through in a way that others cannot. He is also very good at knowing which example to look at (to get insight into a general problem). One might imagine that, when collaborating with Béla, he would be happy if you go away and work on the problem. But, if the stage we've reached is fun, he *wants* to do it together. 'You pig', he'll say, if you have gone away and done it yourself. That is the point – Béla is *fun* to do maths with."

The porcine epithet came up in Tim's recollections too, when it was applied to a friend of his. "Later, Béla apologised and explained that 'pig' was not a strong insult in Hungarian – one shudders to think what is." Tim gave a further non-apocryphal account. "One cannot really succeed as a mathematician without being obsessed with the subject to some extent. Of course, Béla shows magnificently that it is possible to combine that obsession with a rich and varied life, but there must be times when mathematics comes first. Several years ago he made this point very clearly to two of his students as he drove them down the A1 to London. Eager not to waste a valuable opportunity to talk mathematics, he began to explain to them a problem in graph theory. It was a little complicated, so he demonstrated what he was talking about on the windscreen – with both hands. History does not tell us, but knowing Béla it seems likely that his speed at the time was not entirely negligible. Apparently, even with the demonstration the two students found the problem hard to take in."

Tim spoke of what he had himself had learned from Béla. Basic lessons included "... there are certain things that One Should Not Do. I have learnt, either directly or second-hand, that no civilized person wears shorts on an aeroplane, that it is illbred to answer the telephone without immediately identifying oneself, and that it is absolutely not done to supervise in the morning." More importantly, "... when I had just started as a research student, I mentioned to Béla an interesting problem I had just heard about. His immediate response was, 'Why don't you solve it?' I'd like to be able to report that I did, but unfortunately I can't. Nevertheless, in just five words Béla had got across two important messages: first, that one should not be afraid of unsolved problems, and second, that one should be greedy for opportunities to make mathematical progress." And on a more personal note, he concluded, "There is one other extraordinary side of Béla that makes him a very unusual research supervisor, and that is the close personal interest he takes in his students... it is the sort of love that a father has for his children. I don't think anybody can have been a student of Béla's without being very aware of this analogy."

Béla's affection for his research students lay behind a hugely enjoyable party given by Gabi in their garden on the afternoon of Saturday August 2nd, preceding the conference. On a superbly sunny afternoon, most of Béla's students from the last thirty years (plus a couple of "honorary students"), along with spouses and children, came together to wish him well on his birthday. It was enlightening for students from different generations, who in many cases had never met before, to compare notes: for example, whereas the earliest students lived in the understanding that Béla, as a pentathlete, might kill you in any of three different ways, it was felt by the latest generations that this nowadays represented a low probability event.

The climax of the afternoon came with the presentation of the gift. It was in line with Béla's great interest in historical dates, and it related to that most momentous decade of Hungarian history, the 1950s – to be precise, to that high water mark on 25th November 1953, when Béla was just ten years old. It was gratifying that Béla, who for years has grilled his students on whether they know this or that historical date, could not immediately recall the significance of this one. It was, of course, the day that Hungary beat England 6–3 at Wembley. For those who don't know about football, this event marked the first time England had been beaten at home since Trinity College invented the game a hundred years earlier (as Béla would have it). So we presented Béla with an original match-day programme, full of stories and pictures of his boyhood heroes, together with a copy of the next day's *Times*, written in praise of the "Magic Magyars". Such is the power of the Bollobás legend that we half expected to see Béla's own name in the list of those who played that day (it wasn't).

The conference banquet itself, as alluded to earlier, will live long in the memories of those who were there. To understand why, the reader needs to be aware that the

Wednesday of that conference week was the hottest day of modern times; indeed, trains from Cambridge to London had been cancelled because the rails had buckled in the heat, and the air conditioning equipment in the conference lecture hall had died in the struggle. It was expected that the temperature would exceed one hundred degrees for the first time ever in England, though that in fact did not happen till the following Sunday. But on this sweltering evening, a banquet was held in Trinity College to celebrate Béla's birthday and, true to character, Béla insisted his friends dress for this formal occasion. It has already been said what a tribute it is to Béla that so many good mathematicians participated in the conference: now we saw the true measure of that tribute, for, somehow or other, scores of them managed to beg or borrow a black dinner jacket, and to wear it cheerfully in the steaming heat. Moved by compassion at the sight of such loyalty and noble fortitude, the presiding Fellow at the dinner, Professor Glynn, told the company that they might remove their jackets before sitting down to eat. But it cannot often have happened that two hundred people sat down to a sumptuous banquet and finished it weighing less than when they started.

At the end of the dinner, short speeches were given by Tim Gowers, by Gyula Katona, who illuminated Béla's early years, and by Imre Leader. Some of what they said has already been mentioned. What has not yet been mentioned, though, is that each speaker paid tribute to Gabriella. In a style in keeping with the end-of-dinner spirit, Tim commented, "Of course, Gabi makes an enormous contribution to the family atmosphere. One of the great joys of knowing Béla is that one gets to know her as well, though she is so glamorous that it is not easy to think of her as a mother. Like Béla, she is free with her advice, which is not always the same as his but carries equal conviction. I have to confess that I have not yet followed her recent advice to me to develop the muscles in my upper body, but if in the future you notice changes of that kind, you will know why." The various tributes were brief and individual, but the response they received showed that no-one was in any doubt as to the profound importance of Gabriella to Béla's story.

That is enough by way of an introduction to this volume. What remains is the mathematics, which we trust will be found a fitting tribute to the first sixty years of Béla's life.

Andrew Thomason

Contributors

for address details, see the appropriate papers

Noga Alon, *Tel Aviv University*
Richard Arratia, *University of Southern California*
A.D. Barbour, *Universität Zürich*
József Beck, *Rutgers University*
N. Berger, *Microsoft Research*
Béla Bollobás, *University of Memphis and Trinity College, Cambridge*
M. Bordewich, *University of Oxford*
Christian Borgs, *Microsoft Research*
Henning Bruhn, *Universität Hamburg*
Peter J. Cameron, *Queen Mary, University of London*
J. T. Chayes, *Microsoft Research*
Fan Chung, *University of California, San Diego*
Reinhard Diestel, *Universität Hamburg*
Edward Dobson, *Mississippi State University*
R. M. D'Souza, *Microsoft Research*
M. Freedman, *Microsoft Research*
Ehud Friedgut, *Hebrew University*
Alan Frieze, *Carnegie Mellon University*
Zoltán Füredi, *Alfréd Rényi Institute of Mathematics and University of Illinois at Urbana-Champaign*
W.T. Gowers, *University of Cambridge*
Ervin Győri, *Alfréd Rényi Institute of Mathematics*
Penny Haxell, *University of Waterloo*
Svante Janson, *Uppsala University*
Jeff Kahn, *Rutgers University*
Gyula O.H. Katona, *Alfréd Rényi Institute of Mathematics*
Ayman Khalfalah, *Rutgers University*
B. Klartag, *Tel Aviv University*

R. D. Kleinberg, *M.I.T.*
Y. Kohayakawa, *Universidade de São Paulo*
Alexandr Kostochka, *University of Illinois at Urbana-Champaign and Insititute of Mathematics, Novosibirsk*
Michael Krivelevich, *Tel Aviv University*
L. Lovász, *Microsoft Research*
C. Mauduit, *Institut de Mathématiques de Luminy, CNRS*
V. Milman, *Tel Aviv University*
C. G. Moreira, *IMPA, Rio de Janeiro*
Kittikorn Nakprasit, *University of Illinois at Urbana-Champaign*
Assaf Naor, *Microsoft Research*
Jaroslav Nešetřil, *Charles University*
V. Nikiforov, *University of Memphis*
Patrice Ossona de Mendez, *Centre d'Analyse et de Mathématiques Sociales, CNRS*
Oleg Pikhurko, *Carnegie Mellon University*
Richard G.E. Pinch, *Cheltenham*
Boris Pittel, *Ohio State University*
Oliver Riordan, *University of Cambridge*
V. Rödl, *Emory University*
Pierre Rosenstiehl, *Centre d'Analyse et de Mathématiques Sociales, EHESS*
C. C. Rousseau, *University of Memphis*
Andrzej Ruciński, *Adam Mickiewicz University*
R. H. Schelp, *University of Memphis*
Alexander D. Scott, *University College London*
Miklós Simonovits, *Alfréd Rényi Institute of Mathematics*
Alan D. Sokal, *New York University*
Gregory B. Sorkin, *IBM T.J. Watson Research Center*
Benny Sudakov, *Princeton University*
Endre Szemerédi, *Rutgers University*
Tibor Szabó, *ETH Zürich*
Simon Tavaré, *University of Southern California and University of Cambridge*
Jacques Verstraëte, *Microsoft Research*
D. Welsh, *University of Oxford*

Measures of Pseudorandomness for Finite Sequences: Minimal Values

N. ALON[1],[†] Y. KOHAYAKAWA[2],[‡] C. MAUDUIT[3],
C. G. MOREIRA[4,§] and V. RÖDL[5,¶,∥]

[1] Raymond and Beverly Sackler Faculty of Exact Sciences,
Tel Aviv University, Tel Aviv 69978, Israel
(e-mail: noga@math.tau.ac.il)

[2] Instituto de Matemática e Estatística, Universidade de São Paulo,
Rua do Matão 1010, 05508-090 São Paulo, Brazil
(e-mail: yoshi@ime.usp.br)

[3] Institut de Mathématiques de Luminy, CNRS-UPR9016, 163 av. de Luminy,
case 907, F-13288, Marseille Cedex 9, France
(e-mail: mauduit@iml.univ-mrs.fr)

[4] IMPA, Estrada Dona Castorina 110, 22460-320 Rio de Janeiro, RJ, Brazil
(e-mail: gugu@impa.br)

[5] Department of Mathematics and Computer Science, Emory University, Atlanta, GA 30322, USA
(e-mail: rodl@mathcs.emory.edu)

Received 19 March 2004; revised 30 June 2005

For Béla Bollobás on his 60th birthday

Mauduit and Sárközy introduced and studied certain numerical parameters associated to finite binary sequences $E_N \in \{-1, 1\}^N$ in order to measure their 'level of randomness'. Two of these parameters are the *normality measure* $\mathcal{N}(E_N)$ and the *correlation measure* $C_k(E_N)$ *of order k*, which focus on different combinatorial aspects of E_N. In their work, amongst others, Mauduit and Sárközy investigated the minimal possible value of these parameters.

[†] Supported by a USA–Israel BSF grant, by a grant from the Israel Science Foundation and by the Hermann Minkowski Minerva Center for Geometry at Tel Aviv University.
[‡] Partially supported by FAPESP and CNPq through ProNEx projects (Proc. CNPq 664107/1997-4 and Proc. FAPESP 2003/09925-5) and by CNPq (Proc. 306334/2004-6 and 479882/2004-5).
[§] Partially supported by MCT/CNPq through a ProNEx project (Proc. CNPq 662416/1996-1) and by CNPq (Proc. 300647/95-2).
[¶] Partially supported by NSF grant 0300529.
[∥] Part of this work was done at IMPA, whose hospitality the authors gratefully acknowledge. This research was partially supported by IM-AGIMB/IMPA. The authors gratefully acknowledge the support of a CNPq/NSF cooperative grant (910064/99-7, 0072064) and the Brazil/France Agreement in Mathematics (Proc. CNPq 69-0014/01-5 and 69-0140/03-7).

In this paper, we continue the work in this direction and prove a lower bound for the correlation measure $C_k(E_N)$ (k even) for arbitrary sequences E_N, establishing one of their conjectures. We also give an algebraic construction for a sequence E_N with small normality measure $\mathcal{N}(E_N)$.

1. Introduction and statement of results

In a series of papers, Mauduit and Sárközy studied finite pseudorandom binary sequences $E_N = (e_1, \ldots, e_N) \in \{-1, 1\}^N$. In particular, they investigated in [11] certain 'measures of pseudorandomness', to be defined shortly. We restrict ourselves to the Mauduit–Sárközy parameters directly relevant to the present note, and refer the reader to [10] and [11] for detailed discussions concerning the definitions below, related measures, and further related literature.

Let $k \in \mathbb{N}$, $M \in \mathbb{N}$, and $X \in \{-1, 1\}^k$ be given. Also, let $D = \{d_1, \ldots, d_k\}$, where the d_i are integers with $1 \leqslant d_1 < \cdots < d_k \leqslant N - M + 1$. Below, we write $\operatorname{card} S$ for the cardinality of a set S, and if S is a set of numbers, then we write $\sum S$ for the sum $\sum_{s \in S} s$. We let

$$T(E_N, M, X) = \operatorname{card}\{n : 0 \leqslant n < M, \ n + k \leqslant N, \text{ and } (e_{n+1}, e_{n+2}, \ldots, e_{n+k}) = X\} \quad (1.1)$$

and

$$\begin{aligned} V(E_N, M, D) &= \sum \{e_{n+d_1} e_{n+d_2} \cdots e_{n+d_k} : 0 \leqslant n < M\} \\ &= \sum_{0 \leqslant n < M} \prod_{1 \leqslant i \leqslant k} e_{n+d_i} = \sum_{0 \leqslant n < M} \prod_{d \in D} e_{n+d}. \end{aligned} \quad (1.2)$$

In words, $T(E_N, M, X)$ is the number of occurrences of the pattern X in E_N, counting only those occurrences whose first symbol is among the first M elements of E_N. On the other hand, one may think of the quantity $V(E_N, M, D)$ as the 'correlation' among k length M segments of E_N 'relatively positioned' according to $D = \{d_1, \ldots, d_k\}$.

The *normality measure* of E_N is defined as

$$\mathcal{N}(E_N) = \max_k \max_X \max_M \left| T(E_N, M, X) - \frac{M}{2^k} \right|, \quad (1.3)$$

where the maxima are taken over all $1 \leqslant k \leqslant \log_2 N$, $X \in \{-1, 1\}^k$, and $0 < M \leqslant N + 1 - k$. The *correlation measure of order k* of E_N is defined as

$$C_k(E_N) = \max\{|V(E_N, M, D)| : M \text{ and } D \text{ such that } M - 1 + d_k \leqslant N\}. \quad (1.4)$$

In what follows, we shall sometimes make use of terms commonly used in the area of combinatorics on words. In particular, sequences will sometimes be referred to as *words*. Moreover, a word u *occurs* in a word w if w contains u as a 'contiguous segment' (that is, $w = tuv$, where t is a 'prefix' of w and v is a 'suffix' of w).

In Section 1.1 we shall state and discuss our results concerning the correlation measure C_k, while in Section 1.2 we shall state and discuss our results on the normality measure $\mathcal{N}$.

1.1. Typical and minimal values of correlation

In [4], Cassaigne, Mauduit and Sárközy studied, amongst others, the typical value of $C_k(E_N)$ for random binary sequences E_N, with all the 2^N sequences in $\{-1,1\}^N$ equiprobable, and the minimal possible value for $C_k(E_N)$. The investigation of the typical value of $C_k(E_N)$ is continued in [2], where Theorems A and B below are proved. (In what follows, we write log for the natural logarithm.)

Theorem A. *Let $0 < \varepsilon_0 < 1/16$ be fixed and let $\varepsilon_1 = \varepsilon_1(N) = (\log\log N)/\log N$. There is a constant $N_0 = N_0(\varepsilon_0)$ such that if $N \geqslant N_0$, then, with probability at least $1 - \varepsilon_0$, we have*

$$\frac{2}{5}\sqrt{N\log\binom{N}{k}} < C_k(E_N) < \sqrt{(2+\varepsilon_1)N\log\left(N\binom{N}{k}\right)}$$
$$< \sqrt{(3+\varepsilon_0)N\log\binom{N}{k}} < \frac{7}{4}\sqrt{N\log\binom{N}{k}} \qquad (1.5)$$

for every integer k with $2 \leqslant k \leqslant N/4$.

Note that Theorem A establishes the typical order of magnitude of $C_k(E_N)$ for a wide range of k, including values of k proportional to N. The next result tells us that $C_k(E_N)$ is concentrated in the case in which k is small.

Theorem B. *For any fixed constant $\varepsilon > 0$ and any integer function $k = k(N)$ with $2 \leqslant k \leqslant \log N - \log\log N$, there is a function $\Gamma(k,N)$ and a constant N_0 for which the following holds. If $N \geqslant N_0$, then the probability that*

$$1 - \varepsilon < \frac{C_k(E_N)}{\Gamma(k,N)} < 1 + \varepsilon \qquad (1.6)$$

holds is at least $1 - \varepsilon$.

Clearly, Theorem A tells us that $\Gamma(k,N)$ is of order $\sqrt{N\log\binom{N}{k}}$. Let us now turn to the minimal possible value of the parameter $C_k(E_N)$. In [4], the following result is proved.

Theorem C. *For all k and $N \in \mathbb{N}$ with $2 \leqslant k \leqslant N$, we have*
(i) $\min\{C_k(E_N) : E_N \in \{-1,1\}^N\} = 1$ *if k is odd,*
(ii) $\min\{C_k(E_N) : E_N \in \{-1,1\}^N\} \geqslant \log_2(N/k)$ *if k is even.*

Theorem C(i) follows simply from the observation that the alternating sequence $E_N = (1, -1, 1, -1, \ldots)$ is such that $C_k(E_N) = 1$ for odd k. Owing to Theorem C(i), when concerned with minimal values of $C_k(E_N)$, we are only interested in even k. In [4], it is conjectured that for any even $k \geqslant 2$ there is a constant $c > 0$ such that for $N \to \infty$ we have

$$\min\{C_k(E_N) : E_N \in \{-1,1\}^N\} \gg N^c, \qquad (1.7)$$

which would be a considerable strengthening of Theorem C(ii). In this paper, we prove the conjecture above in a more general form. We shall prove the following result.

Theorem 1.1. *If k and N are natural numbers with k even and $2 \leqslant k \leqslant N$, then*

$$C_k(E_N) > \sqrt{\frac{1}{2}\left\lfloor\frac{N}{k+1}\right\rfloor} \tag{1.8}$$

for any $E_N \in \{-1,1\}^N$.

The lower bound given in (1.8) decreases as k increases. One may ask whether, in fact, $C_{2k}(E_N) \geqslant c\sqrt{kN}$ for some absolute constant $c > 0$, or at least $C_{2k}(E_N) \geqslant c\sqrt{N}$ for some absolute constant $c > 0$. The results below (and the results in Section 2.3) are partial answers in this direction.

It turns out that if we look at the maximum of $C_2(E_N), C_4(E_N), \ldots, C_k(E_N)$ (with k again even), then a lower bound of order $\sqrt{kN}$ may indeed be proved.

Theorem 1.2. *There is an absolute constant $c > 0$ for which the following holds. For any positive integers ℓ and N with $\ell \leqslant N/3$, we have*

$$\max\{C_2(E_N), C_4(E_N), \ldots, C_{2\ell}(E_N)\} \geqslant c\sqrt{\ell N} \tag{1.9}$$

for all $E_N \in \{-1,1\}^N$.

In view of Theorem A, the lower bound in Theorem 1.2 is best possible apart from a multiplicative factor of $O(\sqrt{\log(N/2\ell)})$, for all $\ell \leqslant N/8$.

One may also prove lower bounds of the form $c\sqrt{N}$ for some absolute constant $c > 0$ if one considers correlations of two consecutive even orders $2k - 2$ and $2k$ (with k not too large).

Theorem 1.3. *Let positive integers k and N with $2 \leqslant k \leqslant \sqrt{N/6}$ be given. If N is large enough, then*

$$\max\{C_{2k-2}(E_N), C_{2k}(E_N)\} \geqslant \sqrt{\frac{1}{2}\left\lfloor\frac{N}{3}\right\rfloor} \tag{1.10}$$

for any $E_N \in \{-1,1\}^N$.

Some further results are stated and proved in Section 2.3 (see Theorems 2.7, 2.9, and 2.10).

1.2. Typical and minimal values of normality

We now turn to the normality measure $\mathcal{N}(E_N)$. In [2], the following result is proved.

Theorem D. *For any given $\varepsilon > 0$ there exist N_0 and $\delta > 0$ such that if $N \geqslant N_0$, then*

$$\delta\sqrt{N} < \mathcal{N}(E_N) < \frac{1}{\delta}\sqrt{N} \tag{1.11}$$

with probability at least $1 - \varepsilon$.

Here, we shall give an explicit construction for sequences $E_N \in \{-1,1\}^N$ with $\mathcal{N}(E_N)$ small. Theorem D tells us that, typically, $\mathcal{N}(E_N)$ is of order $\sqrt{N}$. We shall exhibit a sequence E_N with $\mathcal{N}(E_N) = O(N^{1/3}(\log N)^{2/3})$.

Theorem 1.4. *For any sufficiently large N, there exists a sequence $E_N \in \{-1,1\}^N$ with*
$$\mathcal{N}(E_N) \leq 3N^{1/3}(\log N)^{2/3}. \tag{1.12}$$

A simple argument shows that $\mathcal{N}(E_N) \geq (1/2 + o(1))\log_2 N$ for any $E_N \in \{-1,1\}^N$ (see Proposition 3.1). In view of Theorem 1.4, we have
$$\left(\frac{1}{2} + o(1)\right)\log_2 N \leq \min_{E_N \in \{-1,1\}^N} \mathcal{N}(E_N) \leq 3N^{1/3}(\log N)^{2/3} \tag{1.13}$$
for all large enough N. It would be interesting to close the rather wide gap in (1.13).

The construction of the sequence $E_N \in \{-1,1\}^N$ in Theorem 1.4 may be generalized to larger alphabets Σ, as long as the cardinality of Σ is a power of a prime (see Section 3.3). Finally, we remark that one of the ingredients in the proof of (1.12) for our sequence E_N allows one to give a short proof of the celebrated Pólya–Vinogradov inequality on incomplete character sums (see Section 3.4), which is somewhat simpler than the known proofs.

2. The minimum of the correlation measure

2.1. Auxiliary lemmas from linear algebra

The proof of Theorem 1.1 that we give in Section 2.2 is based on the following elementary lemma from linear algebra (see, *e.g.*, [1, Lemma 9.1] or [5, Lemma 7]), whose proof we include for completeness.

Lemma 2.1. *For any symmetric matrix $\mathbf{A} = (A_{ij})_{1 \leq i,j \leq n}$, we have*
$$\mathrm{rk}(\mathbf{A}) \geq \frac{(\mathrm{tr}(\mathbf{A}))^2}{\mathrm{tr}(\mathbf{A}^2)} = \frac{\left(\sum_{1 \leq i \leq n} A_{ii}\right)^2}{\sum_{1 \leq i,j \leq n} A_{ij}^2}. \tag{2.1}$$

Consequently, if $A_{ii} = 1$ for all i and $|A_{ij}| \leq \varepsilon$ for all $i \neq j$, then
$$\mathrm{rk}(\mathbf{A}) \geq \frac{n}{1 + \varepsilon^2(n-1)}. \tag{2.2}$$

In particular, if $\varepsilon = \sqrt{1/n}$, then $\mathrm{rk}(\mathbf{A}) \geq n/2$.

Proof. Let $r = \mathrm{rk}(\mathbf{A})$. Then $\mathbf{A}$ has exactly r nonzero eigenvalues, say, $\lambda_1, \ldots, \lambda_r$. By the Cauchy–Schwarz inequality, we have
$$(\mathrm{tr}(\mathbf{A}))^2 = (\lambda_1 + \cdots + \lambda_r)^2 \leq r(\lambda_1^2 + \cdots + \lambda_r^2) = r\,\mathrm{tr}(\mathbf{A}^2),$$
and it now suffices to notice that, because $\mathbf{A}$ is symmetric, we have
$$\mathrm{tr}(\mathbf{A}^2) = \sum_{1 \leq i \leq n}\left(\sum_{1 \leq j \leq n} A_{ij} A_{ji}\right) = \sum_{1 \leq i,j \leq n} A_{ij}^2,$$
as required. Inequality (2.2) follows immediately from (2.1). □

The next lemma, due to the first author [1], improves Lemma 2.1 for larger values of ε.

Lemma 2.2. *Let $\mathbf{A} = (A_{ij})_{1 \leq i,j \leq n}$ be an $n \times n$ real matrix with $A_{ii} = 1$ for all i and $|A_{ij}| \leq \varepsilon$ for all $i \neq j$, where $\sqrt{1/n} \leq \varepsilon \leq 1/2$. Then*

$$\mathrm{rk}(\mathbf{A}) \geq \frac{1}{100\varepsilon^2 \log(1/\varepsilon)} \log n. \tag{2.3}$$

If $\mathbf{A}$ is symmetric, then (2.3) holds with the constant $1/100$ replaced by $1/50$.

For completeness, we give the proof of Lemma 2.2. We shall need the following auxiliary lemma [1].

Lemma 2.3. *Let $\mathbf{A} = (A_{i,j})$ be an $n \times n$ matrix of rank d, and let $P(x)$ be an arbitrary polynomial of degree k. Then the rank of the $n \times n$ matrix $(P(A_{i,j}))$ is at most $\binom{k+d}{k}$. Moreover, if $P(x) = x^k$, then the rank of $(P(A_{i,j})) = (A_{i,j}^k)$ is at most $\binom{k+d-1}{k}$.*

Proof. Let $\mathbf{v}_1 = (v_{1,j})_{j=1}^n$, $\mathbf{v}_2 = (v_{2,j})_{j=1}^n$, ..., $\mathbf{v}_d = (v_{d,j})_{j=1}^n$ be a basis of the row space of $\mathbf{A}$. Then the vectors $(v_{1,j}^{k_1} v_{2,j}^{k_2} \cdots v_{d,j}^{k_d})_{j=1}^n$, where $k_1, k_2, \ldots, k_d$ range over all nonnegative integers whose sum is at most k, span the row space of the matrix $(P(A_{i,j}))$. If $P(x) = x^k$, then it suffices to take all these vectors corresponding to $k_1, k_2, \ldots, k_d$ whose sum is precisely k. $\square$

Proof of Lemma 2.2. Let us first note that the nonsymmetric case follows from the symmetric case: if $\mathbf{A}$ is not symmetric, it suffices to consider the symmetric matrix $(\mathbf{A}^T + \mathbf{A})/2$, whose rank is at most twice the rank of $\mathbf{A}$. We therefore suppose that $\mathbf{A}$ is symmetric, and proceed to prove (2.3) with the constant $1/100$ replaced by $1/50$.

Let $\delta = 1/16$. Consider first the case in which $\varepsilon \leq 1/n^\delta$. In this case, let $m = \lfloor 1/\varepsilon^2 \rfloor$, and let $\mathbf{A}'$ be the submatrix of $\mathbf{A}$ consisting of the, say, first m rows and first m columns of $\mathbf{A}$. By the choice of m, we have that $1/\sqrt{m} \geq \varepsilon$, and hence Lemma 2.1 applies to $\mathbf{A}'$, and we deduce that $\mathrm{rk}(\mathbf{A}) \geq \mathrm{rk}(\mathbf{A}') \geq m/2$. It now suffices to check that, because $\varepsilon \leq \min\{1/2, 1/n^\delta\}$ and $\delta = 1/16$, we have

$$\frac{1}{2}m \geq \frac{3}{8\varepsilon^2} = \frac{3}{2^7 \delta \varepsilon^2} > \frac{1}{50\varepsilon^2 \log(1/\varepsilon)} \log n, \tag{2.4}$$

and we are done in this case. We now suppose that $1/n^\delta \leq \varepsilon \leq 1/2$. In this case, we let

$$k = \left\lfloor \frac{\log n}{2 \log(1/\varepsilon)} \right\rfloor \geq \left\lfloor \frac{1}{2\delta} \right\rfloor = 8, \tag{2.5}$$

and let $m = \lfloor 1/\varepsilon^{2k} \rfloor$. Note that, then, we have $m \leq n$. We again let $\mathbf{A}'$ be the submatrix of $\mathbf{A}$ consisting of the first m rows and first m columns of $\mathbf{A}$. We now have

$$\varepsilon^k \leq \frac{1}{\sqrt{m}}. \tag{2.6}$$

Let $\mathbf{A}''$ be the matrix obtained from $\mathbf{A}'$ by raising all its entries to the kth power. Because of (2.6) and the hypothesis on the entries of $\mathbf{A}$, Lemma 2.1 applies and tells us that

$$\mathrm{rk}(\mathbf{A}'') \geqslant \frac{1}{2}m = \frac{1}{2}\left\lfloor \frac{1}{\varepsilon^{2k}} \right\rfloor \geqslant \frac{0.49}{\varepsilon^{2k}}, \tag{2.7}$$

where the last inequality follows easily from the fact that $\varepsilon \leqslant 1/2$ and $k \geqslant 8$ (see (2.5)). We now observe that Lemma 2.3 tells us that

$$\mathrm{rk}(\mathbf{A}'') \leqslant \binom{k + \mathrm{rk}(\mathbf{A}')}{k} \leqslant \left(\frac{\mathrm{e}(k + \mathrm{rk}(\mathbf{A}'))}{k}\right)^k. \tag{2.8}$$

Putting together (2.7) and (2.8), we get

$$\mathrm{rk}(\mathbf{A}) \geqslant \mathrm{rk}(\mathbf{A}') \geqslant \frac{k}{\varepsilon^2}\left(\frac{0.49^{1/k}}{\mathrm{e}} - \varepsilon^2\right), \tag{2.9}$$

which, because $0.49^{1/8}/\mathrm{e} \geqslant 1/3$ and $\varepsilon^2 \leqslant 1/4$, implies that $\mathrm{rk}(\mathbf{A}) \geqslant k/12\varepsilon^2$. Therefore, we have

$$\mathrm{rk}(\mathbf{A}) > \frac{1}{50\varepsilon^2 \log(1/\varepsilon)} \log n, \tag{2.10}$$

and we are done. □

2.2. Proof of the lower bounds for correlation

We shall prove Theorems 1.1 and 1.2 in this section. These results will be deduced from suitable applications of Lemmas 2.1 and 2.2; to describe these applications, we first need to introduce some notation.

Let $E_N = (e_i)_{1 \leqslant i \leqslant N} \in \{-1, 1\}^N$ be given. Let a positive integer $M \leqslant N$ be fixed and set $N' = N - M + 1$. Moreover, fix a family $\mathscr{L}$ of subsets of $[N']$. We now define a vector $\mathbf{v}_L = (v_{L,i})_{0 \leqslant i < M} \in \{-1, 1\}^M$ for all $L \in \mathscr{L}$, letting

$$v_{L,i} = \prod_{x \in L} e_{i+x} \tag{2.11}$$

for all $0 \leqslant i < M$ (note that $1 \leqslant i + x \leqslant M - 1 + N' = N$ for any x in (2.11)). Let us now define an $\mathscr{L} \times \mathscr{L}$ matrix $\mathbf{A} = (A_{L,L'})_{L,L' \in \mathscr{L}}$, putting

$$A_{L,L'} = \frac{1}{M}\langle \mathbf{v}_L, \mathbf{v}_{L'}\rangle = \frac{1}{M} \sum_{0 \leqslant i < M} v_{L,i} v_{L',i} \tag{2.12}$$

for all $L, L' \in \mathscr{L}$. Clearly, the diagonal entries of $\mathbf{A}$ are all 1. Suppose now that $L \neq L'$. Then

$$A_{L,L'} = \frac{1}{M}\langle \mathbf{v}_L, \mathbf{v}_{L'}\rangle = \frac{1}{M} \sum_{0 \leqslant i < M} \left(\prod_{x \in L} e_{i+x}\right)\left(\prod_{y \in L'} e_{i+y}\right)$$

$$= \frac{1}{M} \sum_{0 \leqslant i < M} \prod_{z \in L \triangle L'} e_{i+z}, \tag{2.13}$$

where we write $L \triangle L'$ for the symmetric difference of the sets L and L'. Let $\mathscr{L}^\triangle = \{L \triangle L' : L, L' \in \mathscr{L}, L \neq L'\}$ and let K be the set of the cardinalities of the members of $\mathscr{L}^{\triangle}$, that is, $K = \{|S| : S \in \mathscr{L}^{\triangle}\}$. It follows from (2.13) and the definition of $C_k(E_N)$ that

$$\max\{C_k(E_N) : k \in K\} \geq M \max\{|A_{L,L'}| : L, L' \in \mathscr{L}, L \neq L'\}. \tag{2.14}$$

Lemma 2.1 and (2.14) imply the following result.

Lemma 2.4. *We have*

$$\max\{C_k(E_N) : k \in K\} > \sqrt{M - \frac{M^2}{|\mathscr{L}|}}. \tag{2.15}$$

Proof. Let $\mathbf{B} = (\mathbf{v}_L^T)_{L \in \mathscr{L}}$ be the $|\mathscr{L}| \times M$ matrix with rows $\mathbf{v}_L^T$ ($L \in \mathscr{L}$). Observing that $\mathbf{A} = M^{-1}\mathbf{B}\mathbf{B}^T$, we see that $\mathbf{A}$ has rank at most M. Combining this with the lower bound for the rank of $\mathbf{A}$ given by Lemma 2.1, we get

$$M \geq \mathrm{rk}(\mathbf{A}) > \frac{|\mathscr{L}|}{1 + \varepsilon^2|\mathscr{L}|}, \tag{2.16}$$

where $\varepsilon = \max\{|A_{L,L'}| : L, L' \in \mathscr{L}, L \neq L'\}$. It follows from (2.16) that

$$\varepsilon > \sqrt{\frac{1}{M} - \frac{1}{|\mathscr{L}|}}. \tag{2.17}$$

Inequality (2.15) follows from (2.14) on multiplying (2.17) by M. □

We are now ready to prove Theorem 1.1.

Proof of Theorem 1.1. Let k, N, and E_N be as in the statement of Theorem 1.1. Set $\ell = k/2$ and $M = \lfloor N/(k+1) \rfloor$ and, as above, let $N' = N - M + 1$. We take for $\mathscr{L} \subset \mathscr{P}([N'])$ a set system of $t = \lfloor N'/\ell \rfloor$ pairwise disjoint ℓ-element subsets $L_1, \ldots, L_t$ of $[N']$. Note that

$$|\mathscr{L}| = t = \left\lfloor \frac{N - \lfloor N/(k+1) \rfloor + 1}{k/2} \right\rfloor \geq \left\lfloor \frac{2N}{k+1} \right\rfloor \geq 2M. \tag{2.18}$$

Therefore, it follows from (2.15) and (2.18) that

$$C_k(E_N) > \sqrt{M - \frac{M^2}{|\mathscr{L}|}} \geq \sqrt{M - \frac{M}{2}} = \sqrt{\frac{1}{2}\left\lfloor \frac{N}{k+1} \right\rfloor}, \tag{2.19}$$

as required. □

Lemma 2.4 was deduced from an application of Lemma 2.1 to the matrix $\mathbf{A} = (A_{L,L'})$; the next lemma will be obtained from an application of Lemma 2.2 to $\mathbf{A}$.

Lemma 2.5. *If $2M \leq |\mathscr{L}| < e^{50M}$, then*

$$\max\{C_k(E_N) : k \in K\} \geq \min\left\{\frac{1}{2}M, \sqrt{\frac{1}{50}M(\log|\mathscr{L}|) \Big/ \log\frac{50M}{\log|\mathscr{L}|}}\right\}. \tag{2.20}$$

Proof. Let $\varepsilon = \max\{|A_{L,L'}| : L, L' \in \mathscr{L}, L \neq L'\}$. Inequality (2.2) and the fact that $\mathrm{rk}(\mathbf{A}) \leqslant M$, coupled with $M \leqslant |\mathscr{L}|/2$, give that

$$\varepsilon^2 > \frac{1}{M} - \frac{1}{|\mathscr{L}|} \geqslant \frac{1}{|\mathscr{L}|}, \tag{2.21}$$

and hence $\varepsilon > \sqrt{1/|\mathscr{L}|}$. If $\varepsilon > 1/2$, then (2.20) follows immediately (recall (2.14)). Therefore, we may suppose that $\sqrt{1/|\mathscr{L}|} \leqslant \varepsilon \leqslant 1/2$, and hence we may apply Lemma 2.2 to the symmetric matrix $\mathbf{A}$. Combining the fact that $\mathbf{A}$ has rank at most M with Lemma 2.2, we obtain that

$$M \geqslant \mathrm{rk}(\mathbf{A}) \geqslant \frac{1}{50\varepsilon^2 \log(1/\varepsilon)} \log |\mathscr{L}|, \tag{2.22}$$

whence

$$\varepsilon^2 \log \frac{1}{\varepsilon} \geqslant \frac{1}{50M} \log |\mathscr{L}|. \tag{2.23}$$

Using that $1/\varepsilon \geqslant \log 1/\varepsilon$, we have from (2.23) that

$$\varepsilon \geqslant \varepsilon^2 \log \frac{1}{\varepsilon} \geqslant \frac{1}{50M} \log |\mathscr{L}|. \tag{2.24}$$

Plugging (2.24) into (2.23), we get

$$\varepsilon^2 \log \frac{50M}{\log |\mathscr{L}|} \geqslant \varepsilon^2 \log \frac{1}{\varepsilon} \geqslant \frac{1}{50M} \log |\mathscr{L}|, \tag{2.25}$$

and hence

$$\varepsilon \geqslant \sqrt{\frac{\log |\mathscr{L}|}{50M} \bigg/ \log \frac{50M}{\log |\mathscr{L}|}}. \tag{2.26}$$

Inequality (2.20) follows easily from (2.14), (2.26), and the definition of ε. $\square$

We shall now deduce Theorem 1.2 from Lemma 2.5.

Proof of Theorem 1.2. Let ℓ and N with $\ell \leqslant N/3$ be given. Let $M = \lfloor N/3 \rfloor$, and set $N' = N - M + 1 \geqslant 2N/3$. We take for $\mathscr{L}$ the set system of all ℓ-element subsets of $[N']$. Then, clearly, $\mathscr{L}^\triangle = \{L \triangle L' : L, L' \in \mathscr{L}, L \neq L'\}$ is the family of non-empty subsets of $[N']$ of even cardinality not greater than 2ℓ. Hence, $K = \{|S| : S \in \mathscr{L}^\triangle\} = \{2, 4, \ldots, 2\ell\}$. Moreover,

$$|\mathscr{L}| = \binom{N'}{\ell} \geqslant N' \geqslant \frac{2N}{3} \geqslant 2M, \tag{2.27}$$

and, as $M = \lfloor N/3 \rfloor \geqslant N/5$ because $N \geqslant 3$, we have

$$|\mathscr{L}| \leqslant 2^N = (2^{N/M})^M \leqslant 2^{5M} < e^{50M}. \tag{2.28}$$

Inequalities (2.27) and (2.28) tell us that Lemma 2.5 may be applied. We deduce from Lemma 2.5 that

$$\max\{C_2(E_N), C_4(E_N), \ldots, C_{2\ell}(E_N)\} \geqslant \min\left\{\frac{1}{2}M, \sqrt{\frac{1}{50}M(\log |\mathscr{L}|) \bigg/ \log \frac{50M}{\log |\mathscr{L}|}}\right\}. \tag{2.29}$$

If the minimum on the right-hand side of (2.29) is achieved by $M/2 = \lfloor N/3 \rfloor/2$, then we are already done; suppose therefore that the minimum is given by the other term. Observe that

$$\frac{1}{50} M (\log |\mathscr{L}|) \Big/ \log \frac{50M}{\log |\mathscr{L}|} \geq \frac{1}{50} \left\lfloor \frac{N}{3} \right\rfloor (\log |\mathscr{L}|) \Big/ \log \frac{50N/3}{\log |\mathscr{L}|}, \qquad (2.30)$$

and, moreover,

$$|\mathscr{L}| = \binom{N'}{\ell} \geq \left(\frac{2N}{3\ell}\right)^{\ell}, \qquad (2.31)$$

so that

$$\log |\mathscr{L}| \geq \ell \log \frac{2N}{3\ell}. \qquad (2.32)$$

By (2.30) and (2.32), it suffices to show that

$$\frac{1}{150} N \ell \left(\log \frac{2N}{3\ell}\right) \Big/ \log \frac{50N/3}{\ell \log(2N/3\ell)} \geq c' N \ell \qquad (2.33)$$

for some absolute constant $c' > 0$. Routine calculations show that a suitable constant $c' > 0$ will do in (2.33). We only give a sketch: suppose first that $1 \leq \ell = o(N)$. In this case, it is simple to check that the left-hand side of (2.33) is in fact

$$\left(\frac{1}{150} + o(1)\right) N \ell. \qquad (2.34)$$

Suppose now that $c'' N \leq \ell \leq N/3$. In this case, the left-hand side of (2.33) is at least

$$\frac{1}{150} N \ell (\log 2) \Big/ \log \frac{50/3}{c'' \log 2}, \qquad (2.35)$$

and (2.33) follows for some small enough $c' > 0$. $\qquad \square$

2.3. Some further lower bounds for correlation
In this section, we deduce some further consequences of Lemmas 2.4 and 2.5, using other families $\mathscr{L}$.

2.3.1. Projective plane bounds. We shall prove Theorem 1.3 (see Section 1.1) by making use of systems of sets derived from projective planes. Recall that Theorem 1.3 tells us that, for any $2 \leq k \leq \sqrt{N/6}$ and any $E_N \in \{-1,1\}^N$, at least one of $C_{2k-2}(E_N)$ and $C_{2k}(E_N)$ is $\geq c\sqrt{N}$, for some absolute constant $c > 0$. (We shall not try to obtain the best value of c in what follows.) We shall use the following fact.

Lemma 2.6. *Let positive integers k and n with $k \leq (1/2)\sqrt{n}$ be given. If n is large enough, then there is a family $\mathscr{L}$ of k-element subsets of $[n]$ with $|\mathscr{L}| = n$ and such that $|L \cap L'| \leq 1$ for all distinct L and $L' \in \mathscr{L}$.*

One may prove Lemma 2.6 by considering suitable projective planes on m points, with m only slightly larger than n: one may first delete $m - n$ points from the plane at random, to obtain a system with n points and $\geq n$ 'lines' of cardinality only slightly smaller than $\sqrt{n}$, and then one may remove some points from these 'lines' to turn them into k-element sets. (The constant $1/2$ in the upper bound for k in Lemma 2.6 may in fact be replaced by any constant < 1.)

Proof of Theorem 1.3. Let k and N as in the statement of the theorem be given. Let $M = \lfloor N/3 \rfloor$ and

$$N' = N - M + 1 \geq \frac{2}{3}N \geq 2M. \tag{2.36}$$

Observe that $k \leq \sqrt{N/6} = (1/2)\sqrt{2N/3} \leq (1/2)\sqrt{N'}$. We now use that N is supposed to be large and invoke Lemma 2.6, to obtain a family $\mathscr{L}$ of k-element subsets of $[N']$ with $|\mathscr{L}| = N'$ and $|L \cap L'| \leq 1$ for any two distinct L and $L' \in \mathscr{L}$.

By (2.36), we have

$$M - \frac{M^2}{|\mathscr{L}|} \geq \frac{1}{2}M = \frac{1}{2}\left\lfloor \frac{N}{3} \right\rfloor. \tag{2.37}$$

Moreover, $|L \triangle L'| \in \{2k-2, 2k\}$ for all distinct L and $L' \in \mathscr{L}$. Inequality (1.10) follows from (2.15). □

If a projective plane of order k exists, then one may give a lower bound of order $\sqrt{N}$ for $C_{2k}(E_N)$.

Theorem 2.7. *For any constant $1/\sqrt{2} < \alpha < 1$, there is a constant $c = c(\alpha) > 0$ for which the following holds. Given any $\varepsilon > 0$, there is N_0 such that if $N \geq N_0$ and k is a power of a prime and $|k - \alpha\sqrt{N}| \leq \varepsilon\sqrt{N}$, then*

$$C_{2k}(E_N) \geq c\sqrt{N} \tag{2.38}$$

for any $E_N \in \{-1, 1\}^N$.

Proof. We only give a sketch of the proof. Let k be a large prime power as in the statement of our result, and set

$$N' = k^2 + k + 1 \quad \text{and} \quad M = N - N' + 1. \tag{2.39}$$

Using that $k = (\alpha + o(1))\sqrt{N}$, we have

$$N' = (\alpha^2 + o(1))N \quad \text{and} \quad M = (1 - \alpha^2 + o(1))N. \tag{2.40}$$

We now use that k is a prime power, and let $\mathscr{L}$ be the family of lines of a projective plane with point set $[N']$. Clearly, every member of $\mathscr{L}$ has $k+1$ elements and

$$|\mathscr{L}| = N' = (\alpha^2 + o(1))N. \tag{2.41}$$

We shall now apply Lemma 2.4. By (2.40), we have

$$M - \frac{M^2}{|\mathscr{L}|} = (1 - \alpha^2 + o(1))N - \frac{(1 - \alpha^2 + o(1))^2 N^2}{(\alpha^2 + o(1))N}$$

$$= \left(1 - \frac{1 - \alpha^2 + o(1)}{\alpha^2 + o(1)}\right)(1 - \alpha^2 + o(1))N$$

$$= (1 + o(1))\left(2 - \frac{1}{\alpha^2}\right)(1 - \alpha^2)N. \tag{2.42}$$

Clearly, $|L \triangle L'| = 2k$ for all distinct L and $L' \in \mathscr{L}$. Therefore, inequality (2.15) in Lemma 2.4, together with the hypothesis that $1/\sqrt{2} < \alpha < 1$, imply the desired result. □

The proof of Theorem 2.7 above is based on Lemma 2.4; one may use Lemma 2.5 instead, which would give a somewhat different value for the constant c in (2.38). A bound of the form (2.38) for k of order N may also be proved in the case in which there exists a $4k \times 4k$ Hadamard matrix. Indeed, it suffices to consider such a matrix as the incidence matrix of a system $\mathscr{L}$ of $2k$-element subsets of a $4k$-element set; the system $\mathscr{L}$ would then have the property that all pairwise symmetric differences of its members are of cardinality $2k$.

The condition that k should be a power of a prime in Theorem 2.7 may be removed by making use of Vinogradov's three primes theorem (to be more precise, we use a strengthening of that result). The key observation is the following.

Lemma 2.8. *For any $\varepsilon > 0$, there is an integer k_0 for which the following holds. If $k \geq k_0$ is an odd integer, then there there is a family $\mathscr{L}$ of $(k+3)$-element subsets of $[n]$, where $|n - k^2/3| \leq \varepsilon k^2$, such that $||\mathscr{L}| - n/3| \leq \varepsilon n$ and*

$$|L \triangle L'| = 2k \tag{2.43}$$

for all distinct L and $L' \in \mathscr{L}$. If $k \geq k_0$ is even, then there is a family $\mathscr{L}$ of $(k+4)$-element subsets of $[n]$, where $|n - k^2/4| \leq \varepsilon k^2$, such that $||\mathscr{L}| - n/4| \leq \varepsilon n$ and (2.43) holds for all distinct L and $L' \in \mathscr{L}$.

Proof. We give a sketch of the proof. Let $\varepsilon > 0$ be fixed and suppose first that k is a large odd integer.

We use a strengthening of Vinogradov's theorem, according to which any large enough odd integer k may be written as a sum of three primes p_1, p_2, and p_3 that satisfy $p_i = (1/3 + o(1))k$, where $o(1) \to 0$ as $k \to \infty$ (an old theorem of Haselgrove [8] implies this result). Let $\mathscr{L}_1$, $\mathscr{L}_2$, and $\mathscr{L}_3$ be projective planes of order p_1, p_2, and p_3, respectively, and suppose that $p_1 \leq p_2$ and p_3. We take the $\mathscr{L}_i$ on pairwise disjoint point sets X_i and let $X = X_1 \cup X_2 \cup X_3$. Clearly, $n = |X| = 3(1/3 + o(1))^2 k^2 = (1/3 + o(1))k^2$. Let the lines of $\mathscr{L}_i$ be $L_1^{(i)}, \ldots, L_{n_i}^{(i)}$, where $n_i = p_i^2 + p_i + 1 = (1/3 + o(1))^2 k^2 = (1/3 + o(1))n$. We let $\mathscr{L}$ be the set system on X given by

$$\mathscr{L} = \{L_j^{(1)} \cup L_j^{(2)} \cup L_j^{(3)} : 1 \leq j \leq n_1\}. \tag{2.44}$$

The members of $\mathscr{L}$ are therefore $(k+3)$-element subsets of X, with $|L \triangle L'| = 2(p_1 + p_2 + p_3) = 2k$ for all distinct L and $L' \in \mathscr{L}$, and the case in which k is a large odd integer follows.

For even k, it suffices to let $p_4 = (1/4 + o(1))k$ be an odd prime (whose existence follows from the prime number theorem) and apply Haselgrove's result to $k - p_4$, and then construct $\mathscr{L}$ as the union of 4 suitable projective planes. We omit the details. $\square$

Lemmas 2.4 and 2.8 imply the following result.

Theorem 2.9. *For all $\varepsilon > 0$, there are constants $c > 0$, k_0, and N_0 for which the following hold.*

(i) If $k \geqslant k_0$ is an odd integer with

$$\left(\frac{3}{2} + \varepsilon\right)\sqrt{N} \leqslant k \leqslant (\sqrt{3} - \varepsilon)\sqrt{N}, \tag{2.45}$$

then $C_{2k}(E_N) \geqslant c\sqrt{N}$ for all $E_N \in \{-1, 1\}^N$ as long as $N \geqslant N_0$.

(ii) If $k \geqslant k_0$ is an even integer with

$$\left(\frac{4}{5}\sqrt{5} + \varepsilon\right)\sqrt{N} \leqslant k \leqslant (2 - \varepsilon)\sqrt{N}, \tag{2.46}$$

then $C_{2k}(E_N) \geqslant c\sqrt{N}$ for all $E_N \in \{-1, 1\}^N$ as long as $N \geqslant N_0$.

We omit the proof of Theorem 2.9. We only remark that it suffices to take for $\mathscr{L}$ in Lemma 2.4 the systems given by Lemma 2.8. One may prove results similar to Theorem 2.9 for other ranges of k of order $\sqrt{N}$ using the method above: one simply proves variants of Lemma 2.8 by writing k as the sum of h nearly equal primes, for other values of h.

We close by making the following remark. In the discussion above, we have used the family of lines in projective planes; it is easy to check that one may also use hyperplanes in projective d-spaces for other values of d, to obtain lower bounds for $C_{2k}(E_N)$ for any k of order $N^{1-1/d}$, in certain ranges (as in Theorem 2.9). Furthermore, for any k of order N, again in certain ranges, we may use Hadamard matrices arising from quadratic residues modulo primes to prove lower bounds of order $\sqrt{N}$ for $C_{2k}(E_N)$. We omit the details.

2.3.2. A variant of Theorem 1.2. In this section, we shall prove a result similar in nature to Theorem 1.2.

Theorem 2.10. *There is an absolute constant $c > 0$ for which the following holds. For any positive integers ℓ and N with $\ell \leqslant N/25$ and N large enough, we have*

$$\max\{C_{2\ell+2}(E_N), C_{2\ell+4}(E_N), \ldots, C_{4\ell}(E_N)\} \geqslant c\sqrt{\ell N} \tag{2.47}$$

for all $E_N \in \{-1, 1\}^N$.

The proof of Theorem 2.10 is based on the following lemma.

Lemma 2.11. Let $1 \leqslant \ell \leqslant n/9e$. Then there is a system $\mathscr{L}$ of 2ℓ-element subsets of $[n]$ with

$$|\mathscr{L}| \geqslant \frac{1}{2}\binom{n/9e}{\ell} \tag{2.48}$$

and

$$|L \triangle L'| \geqslant 2\ell + 2 \tag{2.49}$$

for all distinct L and $L' \in \mathscr{L}$.

Proof. We give a sketch of the proof. Comparing with a geometric series, one may check that, say,

$$\sum_{\ell \leqslant j \leqslant 2\ell}\binom{2\ell}{j}\binom{n-2\ell}{2\ell-j} \leqslant 2\binom{2\ell}{\ell}\binom{n-2\ell}{\ell}. \tag{2.50}$$

Let $\mathscr{L}$ be a maximal family of 2ℓ-element subsets of $[n]$, with any two of its members satisfying (2.49) for all distinct L and $L' \in \mathscr{L}$. Then, clearly,

$$2\binom{2\ell}{\ell}\binom{n-2\ell}{\ell}|\mathscr{L}| \geqslant |\mathscr{L}|\sum_{\ell \leqslant j \leqslant 2\ell}\binom{2\ell}{j}\binom{n-2\ell}{2\ell-j} \geqslant \binom{n}{2\ell}. \tag{2.51}$$

Therefore,

$$\begin{aligned}|\mathscr{L}| &\geqslant \frac{1}{2}\binom{n}{2\ell}\bigg/\binom{2\ell}{\ell}\binom{n-2\ell}{\ell} = \frac{(n)_\ell (n-\ell)_\ell (\ell!)^2}{2(2\ell)_\ell (n-2\ell)_\ell (2\ell)!} \\ &\geqslant \frac{(n-\ell)_\ell}{2(2\ell)_\ell 4^\ell} \geqslant \frac{1}{2}\left(\frac{n-\ell}{8\ell}\right)^\ell \geqslant \frac{1}{2}\left(\frac{n}{9\ell}\right)^\ell \geqslant \frac{1}{2}\binom{n/9e}{\ell}, \end{aligned} \tag{2.52}$$

as required. $\square$

Proof of Theorem 2.10. This follows from Lemmas 2.5 and 2.11; we shall only give a sketch of the proof, because the argument is simple and very similar to the argument given in the proof of Theorem 1.2. Let ℓ and N be as given in the statement of Theorem 2.10. The case in which $\ell = 1$ is covered by Theorem 1.1 (in fact, the case in which ℓ is bounded follows from that result). Therefore, we suppose $\ell \geqslant 2$. Let $M = \lfloor N/50 \rfloor$ and $N' = N - M + 1$. Let $\mathscr{L}$ be a family of 2ℓ-element subsets of $[N']$ of maximal cardinality satisfying (2.49) for all distinct L and $L' \in \mathscr{L}$. By Lemma 2.11, we have

$$2M \leqslant \frac{1}{2}\left(\frac{N'/9e}{2}\right)^2 \leqslant \frac{1}{2}\binom{N'/9e}{\ell} \leqslant |\mathscr{L}| \leqslant 2^{N'} < e^{50M} \tag{2.53}$$

for all large enough N. Therefore, Lemma 2.5 applies and we deduce that, for all $E_N \in \{-1, 1\}^N$, we have

$$\max\{C_{2\ell+2}(E_N), C_{2\ell+4}(E_N), \ldots, C_{4\ell}(E_N)\}$$

$$\geqslant \min\left\{\frac{1}{2}M, \sqrt{\frac{1}{50}M(\log|\mathscr{L}|)}\bigg/\log\frac{50M}{\log|\mathscr{L}|}\right\}. \tag{2.54}$$

If the minimum on the right-hand side of (2.54) is achieved by $M/2$, we are done. In the other case, we may check that (2.47) follows for a suitable absolute constant $c > 0$ by, say, analysing the cases $1 \leqslant \ell = o(N)$ and $\ell \geqslant c'N$ separately (see the proof of Theorem 1.2). □

2.4. Bounds from coding theory

We observe that one may prove lower bounds for the parameter $C_k(E_N)$ by invoking upper bounds for the size of codes with a given minimum distance (bounds in the range that we are interested in are given in [9, p. 565] (see also [12])). For simplicity, let us take the case in which $k = 2$. A sequence with $C_2(E_N)$ small gives rise to a large number of nearly orthogonal $\{-1, 1\}$-vectors of a given length: it suffices to consider all the $N - M + 1$ segments of E_N of length M, where we take $M = (\alpha + o(1))N$ for a suitable positive constant α. From the fact that $C_2(E_N)$ is small, we may deduce that these $N - M + 1$ vectors are pairwise nearly orthogonal. Therefore, these binary vectors have pairwise Hamming distance at least $M/2 - \Delta$, for some small $\Delta > 0$. On the other hand, bounds from the theory of error-correcting codes give us lower bounds for Δ, because we have a family of $N - M + 1$ such vectors. The bounds one deduces with this approach are somewhat weaker than the bounds obtained above.

However, we mention that the argument above applies in a more general setting. For $E_N \in \{-1, 1\}^N$, let

$$\widetilde{C}_k(E_N) = \max\{V(E_N, M, D) : M \text{ and } D \text{ with } M - 1 + d_k \leqslant N\}, \quad (2.55)$$

where $D = \{d_1, \ldots, d_k\}$ is as in Section 1; the only difference between $C_k(E_N)$ and $\widetilde{C}_k(E_N)$ is that, in the definition of $\widetilde{C}_k(E_N)$, we do not take $V(E_N, M, D)$ in absolute value (cf. (1.4) and (2.55)). Clearly, $\widetilde{C}_k(E_N) \leqslant C_k(E_N)$. The argument from coding theory briefly sketched in the previous paragraph applies to $\widetilde{C}_k(E_N)$ as well.

3. The minimum of the normality measure

3.1. Remarks on $\min \mathcal{N}(E_N)$

We start with two observations on $\mathcal{N}(E_N)$. Put

$$\mathcal{N}_k(E_N) = \max_X \max_M \left| T(E_N, M, X) - \frac{M}{2^k} \right|, \quad (3.1)$$

where the maxima are taken over all $X \in \{-1, 1\}^k$ and $0 < M \leqslant N + 1 - k$. Note that, then, we have $\mathcal{N}(E_N) = \max\{\mathcal{N}_k(E_N) : k \leqslant \log_2 N\}$.

Proposition 3.1.

(i) We have $\min_{E_N} \mathcal{N}_k(E_N) = 1 - 2^{-k}$ for any $k \geqslant 1$ and any $N \geqslant 2^k$.
(ii) We have

$$\min_{E_N} \mathcal{N}(E_N) \geqslant \left(\frac{1}{2} + o(1)\right) \log_2 N. \quad (3.2)$$

Proof. To prove (i), we simply consider powers of appropriate de Bruijn sequences [6]. More precisely, we take a circular sequence in which every member of $\{-1,1\}^k$ occurs exactly once, open it up (turning it into a linear sequence), and repeat it an appropriate number of times. The fact that $\mathcal{N}_k(E_N) \geq 1 - 2^{-k}$ for this sequence E_N may be seen by taking $M = 1$ in (3.1) with X the prefix of E_N of length k. We leave the other inequality for the reader.

Let us now prove (ii). If a sequence $E_N \in \{-1,1\}^N$ contains no segment of length $k = \lfloor \log_2 N - \log_2 \log_2 N \rfloor$ of repeated 1s, then

$$\mathcal{N}_k(E_N) \geq \frac{N-k+1}{2^k} = (1+o(1))\frac{N}{2^k} \geq (1+o(1))\log_2 N, \tag{3.3}$$

as required. Suppose now that $E_N = (e_i)_{1 \leq i \leq N}$ does contain such a segment, say, $(e_{M_0}, \ldots, e_{M_0+k-1}) = (1, \ldots, 1)$. Fix $\ell = \ell(N) \to \infty$ as $N \to \infty$ with $\ell = o(k)$, and let X_ℓ be the sequence of ℓ consecutive 1s. Let $M_1 = M_0 + k - \ell$, and note that then

$$T(E_N, M_1, X_\ell) - T(E_N, M_0, X_\ell) = M_1 - M_0 + 1 = k - \ell + 1 = (1+o(1))k. \tag{3.4}$$

Therefore

$$\left(T(E_N, M_1, X_\ell) - \frac{M_1}{2^\ell}\right) - \left(T(E_N, M_0, X_\ell) - \frac{M_0}{2^\ell}\right)$$
$$= (1+o(1))k - (M_1 - M_0)2^{-\ell} = (1+o(1))k. \tag{3.5}$$

It follows from (3.5) that for some $M_0 \leq M^* \leq M_1$ we have

$$\left| T(E_N, M^*, X_\ell) - \frac{M^*}{2^\ell} \right| \geq \left(\frac{1}{2} + o(1)\right)k = \left(\frac{1}{2} + o(1)\right)\log_2 N. \tag{3.6}$$

Therefore, $\mathcal{N}(E_N) \geq \mathcal{N}_\ell(E_N) \geq (1/2 + o(1))\log_2 N$, as required. □

We suspect that the logarithmic lower bound in Proposition 3.1(ii) is far from the truth.

Problem 3.2. *Is there an absolute constant $\alpha > 0$ for which we have*

$$\min_{E_N} \mathcal{N}(E_N) > N^\alpha$$

for all large enough N?

3.2. A sequence E_N with small $\mathcal{N}(E_N)$

Our aim in this section is to prove Theorem 1.4. We start by describing the construction of E_N.

Let s be a positive integer and let $\mathbb{F}_{2^s} = \mathrm{GF}(2^s)$ be the finite field with 2^s elements. Fix a primitive element $x \in \mathbb{F}_{2^s}^*$, and let $m = |\mathbb{F}_{2^s}^*| = 2^s - 1$. We consider $\mathbb{F}_{2^s}$ as a vector space over $\mathbb{F}_2$, and fix a nonzero linear functional

$$b: \mathbb{F}_{2^s} \to \mathbb{F}_2. \tag{3.7}$$

We now let

$$\widetilde{E}_m = (b(x), b(x^2), \ldots, b(x^m)) \in \mathbb{F}_2^m = \{0,1\}^m \tag{3.8}$$

and let
$$E_m = ((-1)^{b(x)}, (-1)^{b(x^2)}, \ldots, (-1)^{b(x^m)}) \in \{-1, 1\}^m. \tag{3.9}$$

Finally, set
$$E_N = E_m^q = E_m \cdots E_m \quad (q \text{ factors}), \tag{3.10}$$

where E_m^q denotes the concatenation of q copies of E_m; clearly, E_N has length $N = qm$.

Theorem 3.3. *Let $s \geqslant 2$. With E_N as defined in (3.10), we have*
$$\mathcal{N}(E_N) \leqslant q + 2(\log_2(m-1))\sqrt{m}. \tag{3.11}$$

Theorem 1.4 will be deduced from Theorem 3.3 in Section 3.2.2 below. Let us now give a rough outline of the proof of Theorem 3.3. Essentially all the work will concern the sequence E_m defined above.

In what follows, we shall first prove that any reasonably long segment of E_m has small 'discrepancy'; we shall show that the entries of segments of E_m of length k add up to $O((\log k)\sqrt{m})$ (see Corollary 3.7). We shall then show two results concerning the number of occurrences of (short) words in E_m. We shall first show that all the words of length $k \leqslant s$ (except for the word $(0,\ldots,0)$) occur exactly the same number of times in E_m (see Lemma 3.8). We shall then prove a similar fact for *segments* of E_m, although for segments the conclusion will be weaker (see Lemma 3.10). Theorem 3.3 will then be deduced from these facts in Section 3.2.2.

3.2.1. Auxiliary lemmas. We start with a well-known lemma concerning the 'discrepancy' of matrices whose rows have uniformly bounded norm and, pairwise, have nonpositive inner product (see, e.g., [7, Theorem 15.2] for a similar statement).

Lemma 3.4. *Let $H = (h_{ij})_{1 \leqslant i,j \leqslant M}$ be an M by M real matrix and let $\mathbf{v}_i$ be the ith row of H $(1 \leqslant i \leqslant M)$. Let $A, B \subset [M]$ be given, and suppose that*
$$\|\mathbf{v}_a\| \leqslant \sqrt{m} \tag{3.12}$$

for all $a \in A$ and
$$\langle \mathbf{v}_a, \mathbf{v}_{a'} \rangle = \sum_{1 \leqslant b \leqslant M} h_{ab} h_{a'b} \leqslant 0 \tag{3.13}$$

for all $a \neq a'$ with $a, a' \in A$. Then
$$\left| \sum_{a \in A, b \in B} h_{ab} \right| \leqslant \sqrt{m|A||B|}. \tag{3.14}$$

Proof. Let $\mathbf{1}_B \in \{0,1\}^M$ be the characteristic vector of B. By the Cauchy–Schwarz inequality, we have
$$\left| \sum_{A,B} h_{ab} \right| = \left| \left\langle \sum_{a \in A} \mathbf{v}_a, \mathbf{1}_B \right\rangle \right| \leqslant \left\| \sum_{a \in A} \mathbf{v}_a \right\| \sqrt{|B|}. \tag{3.15}$$

From (3.12) and (3.13), we have

$$\left\| \sum_{a \in A} \mathbf{v}_a \right\|^2 = \sum_{a \in A} \|\mathbf{v}_a\|^2 + \sum_{a \in A} \sum_{a' \neq a \in A} \langle \mathbf{v}_a, \mathbf{v}_{a'} \rangle \leqslant m|A|. \tag{3.16}$$

Plugging (3.16) into (3.15), we have

$$\left| \sum_{A,B} h_{ab} \right| \leqslant \sqrt{m|A||B|},$$

as required. $\square$

We now define a matrix $\mathbf{E}$ from E_m; we shall apply Lemma 3.4 to $\mathbf{E}$ to deduce the discrepancy property we seek for E_m. Let

$$\mathbf{E} = (E_{ij})_{1 \leqslant i,j \leqslant m} = \begin{bmatrix} (-1)^{b(x)} & (-1)^{b(x^2)} & \cdots & (-1)^{b(x^m)} \\ (-1)^{b(x^2)} & (-1)^{b(x^3)} & \cdots & (-1)^{b(x)} \\ \vdots & \vdots & \ddots & \vdots \\ (-1)^{b(x^m)} & (-1)^{b(x)} & \cdots & (-1)^{b(x^{m-1})} \end{bmatrix}. \tag{3.17}$$

Note that $\mathbf{E}$ is an $m \times m$ circulant, symmetric $\{-1,1\}$-matrix whose first row is E_m. For convenience, let $\mathbf{e}_i = (E_{ij})_{1 \leqslant j \leqslant m}$ ($1 \leqslant i \leqslant m$) denote the ith row of $\mathbf{E}$. Moreover, if $\mathbf{v} = (v_j)_{1 \leqslant j \leqslant m}$ and $\mathbf{w} = (w_j)_{1 \leqslant j \leqslant m}$ are two real m-vectors, let $\mathbf{v} \circ \mathbf{w}$ denote the m-vector $(v_j w_j)_{1 \leqslant j \leqslant m}$.

Lemma 3.5. *The following hold for $\mathbf{E}$.*
(i) *Every row of $\mathbf{E}$ adds up to -1, that is, $\sum_{1 \leqslant j \leqslant m} E_{ij} = -1$ for all $1 \leqslant i \leqslant m$.*
(ii) *For all $i \neq i'$ ($1 \leqslant i, i' \leqslant m$), we have $\mathbf{e}_i \circ \mathbf{e}_{i'} = \mathbf{e}_{i''}$ for some $1 \leqslant i'' \leqslant m$.*
(iii) *The matrix $\mathbf{E}$ satisfies*

$$\mathbf{E}\mathbf{E}^T = -\mathbf{J} + (m+1)\mathbf{I}, \tag{3.18}$$

where $\mathbf{J}$ is the $m \times m$ matrix with all entries 1 and $\mathbf{I}$ is the $m \times m$ identity matrix.
(iv) *For all A and $B \subset [m]$, we have*

$$\left| \sum_{a \in A, b \in B} E_{ab} \right| \leqslant \sqrt{m|A||B|}. \tag{3.19}$$

Proof. Since $b \colon \mathbb{F}_{2^s} \to \mathbb{F}_2$ is a nonzero linear functional, $b^{-1}(0)$ is a hyperplane in $\mathbb{F}_{2^s}$ and hence has cardinality 2^{s-1}. Given that $\mathbb{F}_{2^s}^* = \{x^j : 1 \leqslant j \leqslant m = 2^s - 1\}$, we conclude that $b(x^j) = 1$ for 2^{s-1} values of j with $1 \leqslant j \leqslant m$ and $b(x^j) = 0$ for all the other $2^{s-1} - 1$ values of j with $1 \leqslant j \leqslant m$. Therefore, statement (i) follows. Let now $1 \leqslant i < i' \leqslant m$ be fixed. Then

$$\begin{aligned} \mathbf{e}_i \circ \mathbf{e}_{i'} &= ((-1)^{b(x^i) + b(x^{i'})}, (-1)^{b(x^{i+1}) + b(x^{i'+1})}, \ldots, (-1)^{b(x^{i-1}) + b(x^{i'-1})}) \\ &= ((-1)^{b(x^i + x^{i'})}, (-1)^{b(x^{i+1} + x^{i'+1})}, \ldots, (-1)^{b(x^{i-1} + x^{i'-1})}) \\ &= ((-1)^{b(x^i(1 + x^{i'-i}))}, (-1)^{b(x^{i+1}(1 + x^{i'-i}))}, \ldots, (-1)^{b(x^{i-1}(1 + x^{i'-i}))}). \end{aligned} \tag{3.20}$$

However, as $0 < i' - i < m$, we have $1 + x^{i'-i} \neq 0$, and hence $1 + x^{i-i'} = x^k$ for some $1 \leq k \leq m$. Therefore, we have from (3.20) that

$$\mathbf{e}_i \circ \mathbf{e}_{i'} = ((-1)^{b(x^{i+k})}, (-1)^{b(x^{i+1+k})}, \ldots, (-1)^{b(x^{i-1+k})}) = \mathbf{e}_{i+k} \quad (3.21)$$

(naturally, the index of $\mathbf{e}_{i+k}$ is modulo m). Equation (3.21) proves (ii).

Equation (3.18) is an immediate consequence of (i) and (ii), and hence (iii) is clear. Finally, for (iv), it suffices to notice that $\|\mathbf{e}_i\| = \sqrt{m}$ for all i and that, from the above discussion, $\langle \mathbf{e}_i, \mathbf{e}_{i'} \rangle = -1 < 0$ for all $i \neq i'$. Therefore, Lemma 3.4 applies and (3.19) follows. $\square$

Lemma 3.5(iv) tells us that 'rectangles' in the matrix $\mathbf{E}$ have small discrepancy (in the sense of (3.19)). We shall now deduce a similar result for 'triangles' in $\mathbf{E}$, which will later be used to show that *segments* of E_m have small discrepancy.

Lemma 3.6. *Let A and $B \subset [m]$ be given and suppose $A = \{a_1, \ldots, a_t\}$, $B = \{b_1, \ldots, b_t\}$, where $a_1 < \cdots < a_t$ and $b_1 < \cdots < b_t$. The following assertions hold for the matrix $\mathbf{E} = (E_{ij})_{1 \leq i,j \leq m}$.*

(i) *We have*

$$\left| \sum_{i+j \leq t+1} E_{a_i b_j} \right| \leq (t \log_2 t + 1)\sqrt{m}. \quad (3.22)$$

(ii) *Similarly,*

$$\left| \sum_{i+j \geq t+1} E_{a_i b_j} \right| \leq (t \log_2 t + 1)\sqrt{m}. \quad (3.23)$$

Proof. Inequality (3.22) follows from Lemma 3.5(iv), by induction on t. Note first that (3.22) holds for $t = 1$. Now suppose that $t > 1$ and that (3.22) holds for smaller values of t. By the triangle inequality, we have

$$\left| \sum_{i+j \leq t+1} E_{a_i b_j} \right| \leq \left| \sum_{(i,j) \in S} E_{a_i b_j} \right| + \left| \sum_{(i,j) \in T_1} E_{a_i b_j} \right| + \left| \sum_{(i,j) \in T_2} E_{a_i b_j} \right|, \quad (3.24)$$

where $S = \{(i,j) : i, j \leq \lceil t/2 \rceil\}$, $T_1 = \{(i,j) : i \leq \lceil t/2 \rceil, j > \lceil t/2 \rceil\}$, and $T_2 = \{(i,j) : j \leq \lceil t/2 \rceil, i > \lceil t/2 \rceil\}$. We now estimate the three terms on the right-hand side of (3.24) by using (3.19) and the induction hypothesis twice. We have

$$\left| \sum_{i+j \leq t+1} E_{a_i b_j} \right| \leq \left\lceil \frac{t}{2} \right\rceil \sqrt{m} + 2\left(\left\lfloor \frac{t}{2} \right\rfloor \log_2 \left\lfloor \frac{t}{2} \right\rfloor + 1\right)\sqrt{m}$$

$$\leq \left\lceil \frac{t}{2} \right\rceil \sqrt{m} + (t(\log_2 t - 1) + 2)\sqrt{m}$$

$$\leq (t \log_2 t + 1)\sqrt{m} + \left\lceil \frac{t}{2} \right\rceil \sqrt{m} - (t-1)\sqrt{m}$$

$$\leq (t \log_2 t + 1)\sqrt{m}, \quad (3.25)$$

which completes the induction step, and (i) is proved. The proof of assertion (ii) is similar, and hence it is omitted. □

We shall now show that segments of E_m have small discrepancy, in the sense that they have the same number of 1s as -1s, up to a small error. We observe that Corollary 3.7 also considers segments of E_m that 'wrap around' the end of E_m; equivalently, that result considers E_m as a circular sequence.

Corollary 3.7. *For any $1 \leq r \leq m$ and $2 \leq k \leq m$, we have*

$$\left| \sum_{0 \leq i < k} (-1)^{b(x^{r+i})} \right| \leq \left(\log_2 k + \left(1 - \frac{1}{k}\right) \log_2(k-1) + \frac{2}{k} \right) \sqrt{m}. \quad (3.26)$$

In particular, for all $1 \leq r \leq m$ and $2 \leq k \leq m$, we have

$$\left| \sum_{0 \leq i < k} (-1)^{b(x^{r+i})} \right| \leq 2(\log_2 k)\sqrt{m}. \quad (3.27)$$

Proof. Note that

$$\left(1 - \frac{1}{k}\right) \log_2(k-1) + \frac{2}{k} \leq \log_2 k \quad (3.28)$$

if and only if

$$(k-1)^{1-1/k} 2^{2/k} \leq k \quad (3.29)$$

if and only if

$$\left(1 - \frac{1}{k}\right)^k \leq \frac{1}{4}(k-1), \quad (3.30)$$

which holds if $k \geq 3$. Therefore, (3.27) follows directly from (3.26) for $3 \leq k \leq m$. If $k = 2$, then (3.27) holds by inspection. To prove (3.26), we apply Lemma 3.6(i) and (ii). For the application of (i), we consider the sets $A = \{1, 2, \ldots, k\}$, and $B = \{r, r+1, \ldots, r+k-1\}$, whereas for the application of (ii) we consider $A' = \{2, 3, \ldots, k\}$ and $B' = \{r-k+1, r-k+2, \ldots, r-1\}$. Taking into account that $\mathbf{E} = (E_{ij})$ is circulant, we deduce that

$$k \sum_{0 \leq i < k} (-1)^{b(x^{r+i})} = \sum \{E_{ab} : a \in A, b \in B, a+b \leq k+r\}$$

$$+ \sum \{E_{a'b'} : a' \in A', b' \in B', a'+b' \geq r+1\} \quad (3.31)$$

(see Figure 1). Therefore, by the triangle inequality, we have

$$k \left| \sum_{0 \leq i < k} (-1)^{b(x^{r+i})} \right| \leq \left| \sum \{E_{ab} : a \in A, b \in B, a+b \leq k+r\} \right|$$

$$+ \left| \sum \{E_{a'b'} : a' \in A', b' \in B', a'+b' \geq r+1\} \right|$$

$$\leq (k \log_2 k + 1)\sqrt{m} + ((k-1)\log_2(k-1) + 1)\sqrt{m}, \quad (3.32)$$

and (3.26) follows on dividing (3.32) by k. □

$$\begin{array}{ccccc}
& E_{1,r} & E_{1,r+1} & \cdots & E_{1,r+k-1} \\
& E_{2,r-1} & E_{2,r} & & \reflectbox{$\ddots$} \\
\reflectbox{$\ddots$} & \vdots & \vdots & \reflectbox{$\ddots$} & \\
E_{k,r-k+1} & \cdots & E_{k,r-1} & E_{k,r} &
\end{array}$$

Figure 1. Portion of the matrix **E** to which Lemma 3.6(i) and (ii) are applied. Note that $E_{1,r} = E_{2,r-1} = \cdots = E_{k,r-k+1} = (-1)^{b(x^r)}$, $E_{1,r+1} = E_{2,r} = \cdots = E_{k,r-k+2} = (-1)^{b(x^{r+1})}$, etc.

The next lemma states that the number of occurrences of shorter words in $\widetilde{E}_m$ is basically equal to the expectation of this number in the case of the random sequence of length m. To state this precisely, we introduce some notation. Let $1 \leqslant k \leqslant s$ be fixed. For all $1 \leqslant r \leqslant m$, let $\widetilde{E}_m^{(r)}$ denote the segment of $\widetilde{E}_m$ of length k starting at its rth letter, i.e.,

$$\widetilde{E}_m^{(r)} = (b(x^r), b(x^{r+1}), \ldots, b(x^{r+k-1})) \tag{3.33}$$

($\widetilde{E}_m$ is considered as a cyclic sequence). Now, for all $X \in \{0,1\}^k$, let $f_X = f_X(\widetilde{E}_m)$ denote the number of occurrences of X as a segment in $\widetilde{E}_m$, where we consider $\widetilde{E}_m$ as a cyclic sequence; that is,

$$f_X = \text{card}\{r : 1 \leqslant r \leqslant m \text{ and } \widetilde{E}_m^{(r)} = X\}. \tag{3.34}$$

Lemma 3.8. *For all $1 \leqslant k \leqslant s$, we have*

$$f_X = f_X(\widetilde{E}_m) = \begin{cases} (m+1)2^{-k} - 1 = 2^{s-k} - 1 & \text{if } X = (0,\ldots,0) \in \{0,1\}^k, \\ (m+1)2^{-k} = 2^{s-k} & \text{otherwise.} \end{cases} \tag{3.35}$$

Proof. Let $1 \leqslant r \leqslant m$ and $\delta = (\delta_i)_{1 \leqslant i \leqslant k} \in \{0,1\}^k$ be given. Note that

$$\langle \delta, \widetilde{E}_m^{(r)} \rangle = \sum_{1 \leqslant i \leqslant k} \delta_i b(x^{r+i-1}) = b\left(x^r \sum_{1 \leqslant i \leqslant k} \delta_i x^{i-1}\right). \tag{3.36}$$

We shall now use the fact that x does not satisfy a polynomial over $\mathbb{F}_2$ of degree less than s (indeed, if $p(x) = 0$ for a polynomial p over $\mathbb{F}_2$ of degree t, then a standard argument shows that $1, x, \ldots, x^{t-1}$ spans $\mathbb{F}_{2^s}$ as a vector space over $\mathbb{F}_2$ and hence $\deg(p) = t \geqslant s$). We use this fact in (3.36): as $k \leqslant s$, we see that $\sum_{1 \leqslant i \leqslant k} \delta_i x^{i-1} \neq 0$ as long as $\delta \neq (0, \ldots, 0)$, and hence this sum is x^t for some $1 \leqslant t \leqslant m$ independent of r. Therefore, $\langle \delta, \widetilde{E}_m^{(r)} \rangle = b(x^{r+t})$, and we have

$$\sum_{1 \leqslant r \leqslant m} (-1)^{\langle \delta, \widetilde{E}_m^{(r)} \rangle} = \sum_{1 \leqslant r \leqslant m} (-1)^{b(x^{r+t})} = -1 \tag{3.37}$$

by Lemma 3.5(i), since we have in (3.37) above the sum of the entries of the $(t+1)$st row of **E**. If $\delta = (0, \ldots, 0)$, then clearly the sum in (3.37) is m.

Let us now observe that the left-hand side of (3.37) may also be written as
$$\sum_X (-1)^{\langle \delta, X \rangle} f_X, \tag{3.38}$$
where the sum is over all $X \in \{0,1\}^k$. Therefore, we have established a system of 2^k linear equation for the f_X ($X \in \{0,1\}^k$):
$$\sum_{X \in \{0,1\}^k} (-1)^{\langle \delta, X \rangle} f_X = \begin{cases} m & \text{if } \delta = (0, \ldots, 0) \in \{0,1\}^k, \\ -1 & \text{otherwise.} \end{cases} \tag{3.39}$$

The matrix associated to the system of equations (3.39) is the $2^k \times 2^k$ Hadamard matrix $\mathbf{H}_k = [(-1)^{\langle \delta, X \rangle}]_{\delta, X \in \{0,1\}^k}$. For convenience, let $\mathbf{f} = (f_X)_{X \in \{0,1\}^k}$ and let $\mathbf{g} = (g_\delta)_{\delta \in \{0,1\}^k}$, where $g_\delta = m$ if $\delta = (0, \ldots, 0)$ and $g_\delta = -1$ otherwise. Then (3.39) may be written as
$$\mathbf{H}_k \mathbf{f} = \mathbf{g}. \tag{3.40}$$

Now, since
$$\sum_{\delta \in \{0,1\}^k} (-1)^{\langle \delta, X \rangle} (-1)^{\langle \delta, Y \rangle} = \sum_{\delta \in \{0,1\}^k} (-1)^{\langle \delta, X \triangle Y \rangle} = 0 \tag{3.41}$$
if $X \neq Y$, we have
$$\mathbf{H}_k^T \mathbf{H}_k = 2^k \mathbf{I}, \tag{3.42}$$
where, naturally, $\mathbf{I}$ is the $2^k \times 2^k$ identity matrix. Therefore, from (3.40) and (3.42) we have
$$2^k \mathbf{f} = \mathbf{H}_k^T \mathbf{H}_k \mathbf{f} = \mathbf{H}_k^T \mathbf{g}. \tag{3.43}$$
The last product in (3.43) may be computed explicitly, and one obtains that
$$\mathbf{H}_k^T \mathbf{g} = \begin{bmatrix} m - 2^k + 1 \\ m + 1 \\ \vdots \\ m + 1 \end{bmatrix}, \tag{3.44}$$
where the entry $m - 2^k + 1$ corresponds to $X = (0, \ldots, 0)$. Equation (3.35) now follows from (3.43) and (3.44). $\square$

Setting $k = s$ in Lemma 3.8, we see that words of length s occur in $\widetilde{E}_m$ at most once. Since every occurrence of a word of length at least s gives us an occurrence of its prefix of length s, we conclude that words longer than s occur no more than once in $\widetilde{E}_m$. We thus have the following corollary to Lemma 3.8, to be used later in the proof of Theorem 3.3.

Corollary 3.9. *Suppose $\ell \geqslant s = \log_2(m+1)$. Any $Y \in \{0,1\}^\ell$ occurs at most once in $\widetilde{E}_m$, even considering $\widetilde{E}_m$ as a cyclic sequence; that is,*
$$\mathrm{card}\{r : 1 \leqslant r \leqslant m \text{ and } (b(x^r), \ldots, b(x^{r+\ell-1})) = Y\} \leqslant 1. \tag{3.45}$$

As it turns out, not only has $\widetilde{E}_m$ the property that shorter words occur evenly in it (as shows Lemma 3.8), but $\widetilde{E}_m$ has this property on its longer *segments* (in a weaker sense):

Measures of Pseudorandomness 23

for $k \leqslant s = \log_2(m+1)$, every k-letter word $X \in \{0,1\}^k$ occurs roughly $n2^{-k}$ times in any segment of $\widetilde{E}_m$ of length n, as long as n is reasonably large.

To make the above statement precise, we introduce some notation. Let $1 \leqslant r \leqslant m$ and $1 \leqslant n \leqslant m$ be given. Let $\widetilde{E}_m^{(r,n)}$ be the segment of $\widetilde{E}_m$ of length n starting at the rth letter of $\widetilde{E}_m$, that is, set

$$\widetilde{E}_m^{(r,n)} = (b(x^r), b(x^{r+1}), \ldots, b(x^{r+n-1})). \tag{3.46}$$

Now let $1 \leqslant k \leqslant s$. We shall be interested in the segments $\widetilde{E}_m^{(t,k)}$ of length k of $\widetilde{E}_m$, for $r \leqslant t < r+n$. For $X \in \{0,1\}^k$, set

$$f_X = f_X(\widetilde{E}_m^{(r,n)}) = \mathrm{card}\{t : r \leqslant t < r+n \text{ and } \widetilde{E}_m^{(t,k)} = X\}. \tag{3.47}$$

In what follows, we write $O_1(a)$ for any term b such that $|b| \leqslant a$. We are now ready to state our lemma on the frequency of words in segments of $\widetilde{E}_m$.

Lemma 3.10. *For any $1 \leqslant r \leqslant m$, $2 \leqslant n \leqslant m$, and $1 \leqslant k \leqslant s$, we have*

$$f_X = f_X(\widetilde{E}_m^{(r,n)}) = n2^{-k} + O_1\bigl(2(\log_2 n)\sqrt{m}\bigr) \tag{3.48}$$

for all $X \in \{0,1\}^k$.

The proof of Lemma 3.10 will be similar to the proof of Lemma 3.8, except that we shall now make use of Corollary 3.7, instead of using the fact that the sum of the entries of the whole sequence E_m is -1.

Proof of Lemma 3.10. Let $\delta = (\delta_i)_{1 \leqslant i \leqslant k} \in \{0,1\}^k$ be fixed. As before, we have

$$\langle \delta, \widetilde{E}_m^{(t,k)} \rangle = \sum_{1 \leqslant i \leqslant k} \delta_i b(x^{t+i-1}) = b\left(x^t \sum_{1 \leqslant i \leqslant k} \delta_i x^{i-1}\right) = b(x^{t+u}), \tag{3.49}$$

for some $1 \leqslant u \leqslant m$ independent of t. Therefore, by Corollary 3.7,

$$\left| \sum_{X \in \{0,1\}^k} (-1)^{\langle \delta, X \rangle} f_X \right| = \left| \sum_{r \leqslant t < r+n} (-1)^{\langle \delta, \widetilde{E}_m^{(t,k)} \rangle} \right|$$

$$= \left| \sum_{r \leqslant t < r+n} (-1)^{b(x^{t+u})} \right| \leqslant 2(\log_2 n)\sqrt{m}. \tag{3.50}$$

As before, let $\mathbf{H}_k$ be the $2^k \times 2^k$ Hadamard matrix $[(-1)^{\langle \delta, X \rangle}]_{\delta, X \in \{0,1\}^k}$, and let $\mathbf{f} = (f_X)_{X \in \{0,1\}^k}$. If $\mathbf{g} = \mathbf{H}_k \mathbf{f}$ and $\mathbf{g} = (g_\delta)_{\delta \in \{0,1\}^k}$, then (3.50) implies that

$$g_\delta = \begin{cases} n & \text{if } \delta = (0,\ldots,0), \\ O_1\bigl(2(\log_2 n)\sqrt{m}\bigr) & \text{otherwise.} \end{cases} \tag{3.51}$$

Using that $\mathbf{H}_k^T \mathbf{H}_k = 2^k \mathbf{I}$, we have

$$\mathbf{f} = 2^{-k} \mathbf{H}_k^T \mathbf{H}_k \mathbf{f} = 2^{-k} \mathbf{H}_k^T \mathbf{g}. \tag{3.52}$$

One may easily observe that the entries of $\mathbf{H}_k^T \mathbf{g}$ are all equal to

$$n + O_1\bigl(2^{k+1}(\log_2 n)\sqrt{m}\bigr). \tag{3.53}$$

The asserted conclusion (3.48) follows from (3.52) and (3.53). □

3.2.2. Proof of Theorems 1.4 and 3.3. We shall prove Theorem 3.3 using Lemmas 3.8 and 3.10 and Corollary 3.9, whereas we shall deduce Theorem 1.4 from Theorem 3.3 by making a suitable choice for q and m in the construction of E_N. Let us start with the proof of Theorem 3.3.

Proof of Theorem 3.3. Let E_N be as defined in (3.10), and let $X \in \{-1, 1\}^k$ with $1 \leq k \leq \log_2 N$ be given. Let $1 \leq M \leq N - k + 1$ and let us compute $T(E_N, M, X)$; our aim is to compare $T(E_N, M, X)$ and $M2^{-k}$.

We first suppose $k \leq s$, so that we may apply Lemmas 3.8 and 3.10.

Let $M = \alpha m + \beta$, where α and β are integers with $0 \leq \beta < m$. Clearly, $0 \leq \alpha \leq q$. We use the following notation below, for conciseness: if P is some property, then $[P] = 0$ if P is false and $[P] = 1$ if P is true.

By definition (1.1), we have $T(E_m, \beta, X) \leq \beta$. Suppose for a moment that $\beta \geq 2$. Then, by Lemma 3.10 applied with $r = 1$ and $n = \beta \geq 2$, we have

$$\begin{aligned}T(E_m, \beta, X) &\leq \beta 2^{-k} + 2(\log_2 \beta)\sqrt{m} \\ &\leq \beta 2^{-k} + 2(\log_2(m-1))\sqrt{m}.\end{aligned} \tag{3.54}$$

As $m = 2^s - 1 \geq 3$, the upper bound (3.54) for $T(E_m, \beta, X)$ does hold for $\beta = 0$ and $\beta = 1$ as well. Lemma 3.8 tells us that $T(E_m, m, X) \leq (m+1)2^{-k} - [X = \mathbf{1}]$ (note that the 'exceptional' sequence in (3.35), which concerns $\widetilde{E}_m \in \{0, 1\}^m$, is the zero sequence $\mathbf{0} \in \{0, 1\}^k$, which translates to the all 1 sequence $\mathbf{1} \in \{-1, 1\}^k$ when considering $E_m \in \{-1, 1\}^m$). We conclude from this and (3.54) that

$$\begin{aligned}T(E_N, M, X) &= \alpha T(E_m, m, X) + T(E_m, \beta, X) \\ &\leq \alpha(m2^{-k} + 2^{-k} - [X = \mathbf{1}]) + \beta 2^{-k} + 2(\log_2(m-1))\sqrt{m} \\ &= \alpha m 2^{-k} + \beta 2^{-k} + \alpha(2^{-k} - [X = \mathbf{1}]) + 2(\log_2(m-1))\sqrt{m} \\ &\leq M 2^{-k} + q + 2(\log_2(m-1))\sqrt{m}.\end{aligned} \tag{3.55}$$

Similarly, by Lemmas 3.8 and 3.10, we have

$$\begin{aligned}T(E_N, M, X) &= \alpha T(E_m, m, X) + T(E_m, \beta, X) \\ &\geq \alpha(m2^{-k} + 2^{-k} - [X = \mathbf{1}]) + \beta 2^{-k} - 2(\log_2(m-1))\sqrt{m} \\ &= \alpha m 2^{-k} + \beta 2^{-k} + \alpha(2^{-k} - [X = \mathbf{1}]) - 2(\log_2(m-1))\sqrt{m} \\ &\geq M 2^{-k} - q - 2(\log_2(m-1))\sqrt{m}.\end{aligned} \tag{3.56}$$

From (3.55) and (3.56), we have

$$\left| T(E_N, M, X) - \frac{M}{2^k} \right| \leq q + 2(\log_2(m-1))\sqrt{m}. \tag{3.57}$$

We have thus completed the analysis for the case in which $k \leq s$. Suppose now that $k > s$. Recall that Corollary 3.9 tells us that, in this case, X occurs in E_m at most once, that is,

$T(E_m, m, X) \leq 1$ and hence $0 \leq T(E_N, M, X) \leq q$. Note also that

$$0 \leq \frac{M}{2^k} \leq \frac{N}{2^{s+1}} = \frac{N}{2(m+1)} < \frac{1}{2}q. \tag{3.58}$$

Therefore,

$$\left| T(E_N, M, X) - \frac{M}{2^k} \right| \leq q. \tag{3.59}$$

Inequality (3.11) follows from (3.57) and (3.59). □

We shall now prove Theorem 1.4.

Proof of Theorem 1.4. Let an integer N be given. In what follows, we may suppose that N is suitably large for our inequalities to hold. We start by choosing an integer s so that $m = 2^s - 1$ satisfies

$$\frac{14}{17} \left(\frac{N}{\log_2 N} \right)^{2/3} \leq m \leq \frac{5}{3} \left(\frac{N}{\log_2 N} \right)^{2/3}. \tag{3.60}$$

We now let

$$q = \left\lfloor \frac{11}{9} N^{1/3} (\log_2 N)^{2/3} \right\rfloor, \tag{3.61}$$

set $N' = qm$, and consider $E_{N'} = E_m \cdots E_m = E_m^q$. We have

$$N' = qm \geq \left\lfloor \frac{11}{9} N^{1/3} (\log_2 N)^{2/3} \right\rfloor \times \frac{14}{17} \left(\frac{N}{\log_2 N} \right)^{2/3} \geq N \tag{3.62}$$

for all large enough N. We let E_N be the prefix of $E_{N'}$ of length N.

We claim that E_N satisfies (1.12). Clearly, it suffices to show that $E_{N'}$ is such that $\mathcal{N}(E_{N'}) \leq 3N^{1/3} (\log_2 N)^{2/3}$. To prove this last inequality, we simply show that the right-hand side of (3.11) is at most $3N^{1/3}(\log_2 N)^{2/3}$.

We have

$$\log_2(m-1) < \log_2 \left(\frac{5}{3} \left(\frac{N}{\log_2 N} \right)^{2/3} \right) < \frac{2}{3} \log_2 N. \tag{3.63}$$

Moreover,

$$\sqrt{m} \leq \left(\frac{5}{3} \left(\frac{N}{\log_2 N} \right)^{2/3} \right)^{1/2} < \frac{31}{24} \left(\frac{N}{\log_2 N} \right)^{1/3} \tag{3.64}$$

for all large enough N. Therefore,

$$q + 2(\log_2(m-1))\sqrt{m} < \frac{11}{9} N^{1/3} (\log_2 N)^{2/3} + \frac{31}{18} (\log_2 N) \left(\frac{N}{\log_2 N} \right)^{1/3}$$
$$< 3N^{1/3} (\log_2 N)^{2/3}, \tag{3.65}$$

implying that the right-hand side of (3.11) is at most $3N^{1/3}(\log_2 N)^{2/3}$, as required. □

We close with a remark concerning some recent work of Carpi and de Luca [3], generalizing de Bruijn sequences [6]. Those authors have proved a number of interesting results on *uniform words*: words w such that for any two words u and v of the same length, the number of occurrences of u and v in w differ by at most 1. It would be interesting to see whether their constructions could be used to obtain words with small normality measure.

3.3. Larger alphabets

We now sketch a generalization of the construction in Section 3.2 to alphabets of cardinality larger than 2. As it turns out, the construction generalizes easily to alphabets of cardinality that are powers of primes.

Let s be a positive integer and q a power of a prime, and let $\mathbb{F}_{q^s} = \mathrm{GF}(q^s)$ be the finite field with q^s elements. Fix a primitive element $x \in \mathbb{F}_{q^s}^*$, and let $m = |\mathbb{F}_{q^s}^*| = q^s - 1$. We consider $\mathbb{F}_{q^s}$ as a vector space over $\mathbb{F}_q$, and fix a nonzero linear functional

$$b: \mathbb{F}_{q^s} \to \mathbb{F}_q. \tag{3.66}$$

Let $\psi: \mathbb{F}_q \to S^1 \subset \mathbb{C}$ be an additive character with $\mathrm{card}\{\psi(y): y \in \mathbb{F}_q\} = q$ (that is, we take ψ injective), and put

$$\widetilde{E}_m = (b(x), b(x^2), \ldots, b(x^m)) \in \mathbb{F}_q^m \tag{3.67}$$

and

$$E_m = (\psi(b(x)), \psi(b(x^2)), \ldots, \psi(b(x^m))) \in (S^1)^m. \tag{3.68}$$

Finally, set

$$E_N = E_m^\ell = E_m \cdots E_m \quad (\ell \text{ factors}), \tag{3.69}$$

where E_m^ℓ denotes the concatenation of ℓ copies of E_m; clearly, E_N has length $N = \ell m$. The sequence E_N, considered as a word over the q-letter alphabet

$$\Sigma_q = \{\psi(y): y \in \mathbb{F}_q\}, \tag{3.70}$$

is such that

$$\mathcal{N}^{(q)}(E_N) = O(N^{1/3}(\log N)^{2/3}), \tag{3.71}$$

where

$$\mathcal{N}^{(q)}(E_N) = \max_k \max_X \max_M \left| T(E_N, M, X) - \frac{M}{q^k} \right|, \tag{3.72}$$

and the maxima are taken over all $1 \leq k \leq \log_q N$, $X \in \Sigma_q^k$, and $0 < M \leq N + 1 - k$.

Let us sketch the proof of (3.71). This time, we let

$$\mathbf{E} = (E_{ij})_{1 \leq i,j \leq m} = \begin{bmatrix} \psi(b(x)) & \psi(b(x^2)) & \cdots & \psi(b(x^m)) \\ \psi(b(x^2)) & \psi(b(x^3)) & \cdots & \psi(b(x)) \\ \vdots & \vdots & \ddots & \vdots \\ \psi(b(x^m)) & \psi(b(x)) & \cdots & \psi(b(x^{m-1})) \end{bmatrix}. \tag{3.73}$$

Then $\mathbf{E}$ is an $m \times m$ circulant, complex matrix whose first row is E_m. Again, let $\mathbf{e}_i = (E_{ij})_{1 \leq j \leq m}$ ($1 \leq i \leq m$) denote the ith row of $\mathbf{E}$. Moreover, if $\mathbf{v} = (v_j)_{1 \leq j \leq m}$ and $\mathbf{w} =$

$(w_j)_{1 \leqslant j \leqslant m}$ are two complex m-vectors, let $\mathbf{v} \circ \mathbf{w}$ denote the m-vector $(v_j \overline{w_j})_{1 \leqslant j \leqslant m}$, where $\overline{z}$ denotes the complex conjugate of $z \in \mathbb{C}$.

It turns out that Lemma 3.5 generalizes to the the matrix $\mathbf{E}$ defined in (3.73), in the following way.

Lemma 3.11. *The following hold for $\mathbf{E}$.*

(i) *Every row of $\mathbf{E}$ adds up to -1, that is, $\sum_{1 \leqslant j \leqslant m} E_{ij} = -1$ for all $1 \leqslant i \leqslant m$.*
(ii) *For all $i \neq i'$ ($1 \leqslant i, i' \leqslant m$), we have $\mathbf{e}_i \circ \mathbf{e}_{i'} = \mathbf{e}_{i''}$ for some $1 \leqslant i'' \leqslant m$.*
(iii) *The matrix $\mathbf{E}$ satisfies*

$$\mathbf{E}\mathbf{E}^* = -\mathbf{J} + (m+1)\mathbf{I}, \tag{3.74}$$

where $\mathbf{E}^*$ is the adjoint of $\mathbf{E}$.
(iv) *For all A and $B \subset [m]$, we have*

$$\left| \sum_{a \in A, b \in B} E_{ab} \right| \leqslant \sqrt{m|A||B|}. \tag{3.75}$$

Lemma 3.11(i)–(iii) may be checked easily. For Lemma 3.11(iv), one observes that Lemma 3.4 may be generalized in a natural way to complex matrices, with exactly the same proof.

Lemma 3.12. *Let $H = (h_{ij})_{1 \leqslant i,j \leqslant M}$ be an M by M complex matrix and let $\mathbf{v}_i$ be the ith row of H ($1 \leqslant i \leqslant M$). Let $A, B \subset [M]$ be given, and suppose that*

$$\|\mathbf{v}_a\| = \sqrt{\sum_{1 \leqslant j \leqslant m} |h_{aj}|^2} \leqslant \sqrt{m} \tag{3.76}$$

for all $a \in A$ and

$$\langle \mathbf{v}_a, \mathbf{v}_{a'} \rangle = \sum_{1 \leqslant b \leqslant m} h_{ab} \overline{h_{a'b}} \leqslant 0 \tag{3.77}$$

for all $a \neq a'$ with $a, a' \in A$. Then

$$\left| \sum_{a \in A, b \in B} h_{ab} \right| \leqslant \sqrt{m|A||B|}. \tag{3.78}$$

To prove Lemma 3.11(iv), one applies Lemma 3.12 to the matrix $\mathbf{E}$ given in (3.73). The remainder of the argument is as before, with some small changes. The $2^k \times 2^k$ Hadamard matrix $\mathbf{H}_k = [(-1)^{\langle \delta, X \rangle}]_{\delta, X \in \{0,1\}^k}$ that occurs later in the proof should be replaced by the $q^k \times q^k$ matrix $\mathbf{H}_k = [\psi(\langle \delta, X \rangle)]_{\delta, X}$, where δ and X vary over $\mathbb{F}_q^k$, which is a unitary matrix, up to a multiplicative constant: $\mathbf{H}_k \mathbf{H}_k^* = mI$. We omit the details.

3.4. The Pólya–Vinogradov inequality

Let p be a prime and let $\chi: \mathbb{F}_p = \mathbb{Z}/p\mathbb{Z} \to S^1 \subset \mathbb{C}$ be a multiplicative character, where, as usual, $\chi(0) = 0$. With the methods in Section 3.2.1 (and Lemma 3.12 above) one may easily prove the celebrated Pólya–Vinogradov inequality, in the following form.

Theorem 3.13. *For all integers r and $2 \leqslant k \leqslant p$, we have*
$$\left| \sum_{0 \leqslant h < k} \chi(r+h) \right| \leqslant 2(\log_2 k)\sqrt{p-1}. \tag{3.79}$$

We give an outline of the proof of Theorem 3.13. This time, we let $\mathbf{E} = (e_{ij})_{i,j} = (\chi(i-j))_{0 \leqslant i,j < p}$. Note that $\mathbf{E}$ is circulant: $e_{00} = e_{11} = e_{22} = \cdots$, $e_{01} = e_{12} = e_{23} = \cdots$, $e_{10} = e_{21} = e_{32} \cdots$, etc. The rows $\mathbf{v}_i$ ($0 \leqslant i < p$) of $\mathbf{E}$ have Euclidean norm $\sqrt{p-1}$. Moreover, one may check that
$$\langle \mathbf{v}_i, \mathbf{v}_{i'} \rangle = -1 \tag{3.80}$$
for all $i \neq i'$. Indeed,
$$\langle \mathbf{v}_i, \mathbf{v}_{i'} \rangle = \sum_{0 \leqslant j < p} \chi(i-j)\overline{\chi(i'-j)} = \sum_{0 \leqslant j < p,\, j \neq i, i'} \chi\left(\frac{i-j}{i'-j}\right)$$
$$= \sum_{0 \leqslant j < p,\, j \neq i, i'} \chi\left(1 - \frac{i'-i}{i'-j}\right). \tag{3.81}$$

As j varies over $\mathbb{F}_p \setminus \{i, i'\}$, the argument $1 - (i'-i)/(i'-j)$ of χ in the last term in (3.81) varies over $\mathbb{F}_p \setminus \{0, 1\}$. Since $\chi(1) = 1$ and $\sum_{0 \leqslant j < p} \chi(j) = 0$, we conclude from (3.81) that (3.80) does indeed hold.

Therefore, by Lemma 3.12, we have
$$\left| \sum_{a \in A, b \in B} \chi(a-b) \right| \leqslant \sqrt{(p-1)|A||B|} \tag{3.82}$$
for all A and $B \subset \{0, \ldots, p-1\}$. Theorem 3.13 now follows from (3.82) in the same way that (3.27) follows from (3.19) (and the fact that $\mathbf{E}$ is circulant).

Acknowledgements

The authors are grateful to Eduardo Tengan and Norihide Tokushige for their careful reading of this paper and for their many comments. The authors are also most pleased to thank the referee for his or her very meticulous work.

References

[1] Alon, N. (2003) Problems and results in extremal combinatorics I. In *EuroComb'01* (Barcelona), *Discrete Math.* **273** 31–53.
[2] Alon, N., Kohayakawa, Y., Mauduit, C., Moreira, C. G. and Rödl, V. Measures of pseudorandomness for finite sequences: typical values. In preparation.
[3] Carpi, A. and de Luca, A. (2004) Uniform words. *Adv. Appl. Math.* **32** 485–522.
[4] Cassaigne, J., Mauduit, C. and Sárközy, A. (2002) On finite pseudorandom binary sequences VII: The measures of pseudorandomness. *Acta Arith.* **103** 97–118.
[5] Codenotti, B., Pudlák, P. and Resta, G. (2000) Some structural properties of low-rank matrices related to computational complexity. In *Selected Papers in Honor of Manuel Blum* (Hong Kong, 1998), *Theoret. Comput. Sci.* **235** 89–107.

[6] de Bruijn, N. G. (1946) A combinatorial problem. *Nederl. Akad. Wetensch., Proc.* **49** 758–764 (*Indagationes Math.* **8** (1946) 461–467).

[7] Erdős, P. and Spencer, J. (1974) *Probabilistic Methods in Combinatorics*, Vol. 17 of *Probability and Mathematical Statistics*, Academic Press, New York/London.

[8] Haselgrove, C. B. (1951) Some theorems in the analytic theory of numbers. *J. London Math. Soc.* **26** 273–277.

[9] MacWilliams, F. J. and Sloane, N. J. A. (1977) *The Theory of Error-Correcting Codes II*, Vol. 16 of *North-Holland Mathematical Library*, North-Holland, Amsterdam.

[10] Mauduit, C. (2002) Finite and infinite pseudorandom binary words. In *WORDS* (Rouen, 1999), *Theoret. Comput. Sci.* **273** 249–261.

[11] Mauduit, C. and Sárközy, A. (1997) On finite pseudorandom binary sequences I: Measure of pseudorandomness, the Legendre symbol. *Acta Arith.* **82** 365–377.

[12] Tietäväinen, A. (1980) Bounds for binary codes just outside the Plotkin range. *Inform. and Control* **47** 85–93.

MaxCut in *H*-Free Graphs

NOGA ALON,[1,†] MICHAEL KRIVELEVICH[2,‡] and BENNY SUDAKOV[3,§]

[1] Schools of Mathematics and Computer Science, Raymond and Beverly Sackler Faculty of Exact Sciences,
Tel Aviv University, Tel Aviv 69978, Israel
(e-mail: nogaa@post.tau.ac.il)

[2] Department of Mathematics, Raymond and Beverly Sackler Faculty of Exact Sciences,
Tel Aviv University, Tel Aviv 69978, Israel
(e-mail: krivelev@post.tau.ac.il)

[3] Department of Mathematics, Princeton University, Princeton, NJ 08544, USA
(e-mail: bsudakov@math.princeton.edu)

Received 21 October 2003; revised 6 April 2005

For Béla Bollobás on his 60th birthday

For a graph G, let $f(G)$ denote the maximum number of edges in a cut of G. For an integer m and for a fixed graph H, let $f(m,H)$ denote the minimum possible cardinality of $f(G)$, as G ranges over all graphs on m edges that contain no copy of H. In this paper we study this function for various graphs H. In particular we show that for any graph H obtained by connecting a single vertex to all vertices of a fixed nontrivial forest, there is a $c(H) > 0$ such that $f(m, H) \geq \frac{m}{2} + c(H)m^{4/5}$, and that this is tight up to the value of $c(H)$. We also prove that for any even cycle C_{2k} there is a $c(k) > 0$ such that $f(m, C_{2k}) \geq \frac{m}{2} + c(k)m^{(2k+1)/(2k+2)}$, and that this is tight, up to the value of $c(k)$, for $2k \in \{4, 6, 10\}$. The proofs combine combinatorial, probabilistic and spectral techniques.

1. Introduction

All graphs considered here are finite, undirected and have no loops and no parallel edges, unless otherwise specified. For a graph G, let $f(G)$ denote the maximum number of edges in a cut of G, that is, the maximum number of edges in a bipartite subgraph of G. For a positive integer m let $f(m)$ denote the minimum value of $f(G)$, as G ranges over all

---

[†] Research supported in part by a USA–Israel BSF grant, by a grant from the Israel Science Foundation and by the Hermann Minkowski Minerva Center for Geometry at Tel Aviv University.
[‡] Research supported in part by a USA–Israel BSF grant and by a grant from the Israel Science Foundation.
[§] Research supported in part by NSF grants DMS-0355497, DMS-0106589, and by an Alfred P. Sloan fellowship.

graphs with m edges. Thus, $f(m)$ is the largest integer f such that *any* graph with m edges contains a bipartite subgraph with at least $f(m)$ edges. Edwards [12, 13] proved that for every m,

$$f(m) \geq \frac{m}{2} + \frac{-1 + \sqrt{8m+1}}{8},$$

and noticed that this is tight in infinitely many cases, whenever $m = \binom{k}{2}$ for some integer k.

Answering a question of Erdős, it is shown in [2] that the limsup of the difference

$$f(m) - \left(\frac{m}{2} + \frac{-1 + \sqrt{8m+1}}{8}\right)$$

tends to infinity as m tends to infinity. In fact, as shown in [2], there are two absolute positive constants c_1, c_2 such that

$$f(m) \geq \frac{m}{2} + \sqrt{m/8} + c_1 m^{1/4}$$

for infinitely many values of m, and

$$f(m) \leq \frac{m}{2} + \sqrt{m/8} + c_2 m^{1/4}$$

for all m. More information on $f(m)$, including a determination of its precise value for many additional values of m, appears in [4] and [9].

The situation is more complicated if we consider only H-free graphs G, that is, graphs G that contain no copy of a fixed, given graph H. Let $f(m, H)$ denote the minimum possible cardinality of $f(G)$, as G ranges over all H-free graphs on m edges. Similarly, for a set $\mathcal{H}$ of graphs, let $f(m, \mathcal{H})$ denote the minimum possible cardinality of $f(G)$, as G ranges over all graphs on m edges that contain no member of $\mathcal{H}$. It is not difficult to show (see, *e.g.*, [3]) that for every fixed graph H there are some $\epsilon = \epsilon(H) > 0$ and $c = c(H) > 0$ such that $f(m, H) \geq \frac{m}{2} + cm^{1/2+\epsilon}$ for all m, but the problem of estimating the error term more precisely is not easy, even for relatively simple graphs H. The case $H = K_3$ in which $f(m, K_3)$ is the minimum possible size of the maximum cut in a triangle-free graph with m edges, has been studied extensively. After a series of papers by various researchers ([14], [20], [21]), the first author proved in [2] that $f(m, K_3) = \frac{m}{2} + \Theta(m^{4/5})$, that is, there are $c_1, c_2 > 0$ such that

$$\frac{m}{2} + c_1 m^{4/5} \leq f(m, K_3) \leq \frac{m}{2} + c_2 m^{4/5} \quad (1.1)$$

for all m.

In a recent joint paper with Bollobás [3], we studied the value of $f(m, \mathcal{H}_r)$ for the families $\mathcal{H}_r = \{C_3, \ldots, C_{r-1}\}$ of all cycles of length less than r. Here $f(m, \mathcal{H}_r)$ is the minimum possible size of a maximum cut in a graph with m edges and girth at least r. The problem of estimating $f(m, \mathcal{H}_r)$ had in fact been considered much earlier by Erdős in [14]. He conjectured that for every $r \geq 4$ there exists a constant $c_r > 0$ such that, for every $\epsilon > 0$,

$$\frac{m}{2} + m^{c_r - \epsilon} \leq f(m, \mathcal{H}_r) \leq \frac{m}{2} + m^{c_r + \epsilon}$$

provided $m > m(\epsilon)$. He also mentioned that together with Lovász they proved that

$$\frac{m}{2} + c_2 m^{c''_r} \leqslant f(m, \mathcal{H}_r) \leqslant \frac{m}{2} + c_1 m^{c'_r},$$

where $1/2 < c''_r < c'_r < 1$ for all $r > 3$, and c''_r (as well as c'_r) tend to one as r tends to infinity.

In [3] it is proved that for every $r \geqslant 4$ there is a $c(r) > 0$ such that

$$f(m, \mathcal{H}_r) \geqslant \frac{m}{2} + c(r) m^{\frac{r}{r+1}}$$

for all m. It is further shown that this is tight, up to the value of $c(r)$, for $r = 5$ (note that it is tight for $r = 4$ as well, by (1.1)). The authors of [3] conjecture that this is in fact tight for all $r \geqslant 4$.

In the present paper we study the value of $f(m, H)$ for several additional graphs H. Although we are unable to determine the asymptotic behaviour of the error term of this function (that is, the behaviour of $f(m, H) - m/2$) up to a constant factor for general graphs H, we can do it for infinitely many graphs H. Our main new results are the following.

Theorem 1.1. *Let H be a tree with $h > 1$ edges.*
(i) *If h is even, then $f(m, H) \geqslant \frac{m}{2} + \frac{m}{2h-2}$.*
(ii) *If h is odd, then $f(m, H) \geqslant \frac{m}{2} + \frac{m}{2h}$.*
Both bounds hold as equalities if $\binom{h}{2}$ divides m.

Theorem 1.2. *For every odd integer $r > 3$ there is a $c(r) > 0$ such that*

$$f(m, C_{r-1}) \geqslant \frac{m}{2} + c(r) m^{r/(r+1)}$$

for all m. This is tight, up to the value of $c(r)$, for $r \in \{5, 7, 11\}$.

Note that the lower bounds in the last theorem strengthens the lower bound for $f(m, \mathcal{H}_r)$, for odd r. It seems plausible that the above estimate holds for even values of r as well, but this remains open for all $r > 4$.

Theorem 1.3. *Let H be a graph obtained by connecting a single vertex to all vertices of a fixed nontrivial forest. Then there is a $c = c(H) > 0$ such that*

$$f(m, H) \geqslant \frac{m}{2} + cm^{4/5}$$

for all m. This is tight (up to the value of c) for each such H.

Theorem 1.4. *Let $K_{t,s}$ denote the complete bipartite graph with classes of vertices of size t and s.*
(i) *For each $s \geqslant 2$ there is a $c(s) > 0$ such that*

$$f(m, K_{2,s}) \geqslant \frac{m}{2} + c(s) m^{5/6},$$

for all m, and this is tight up to the value of $c(s)$.

(ii) *For each $s \geqslant 3$ there is a $c'(s) > 0$ such that*

$$f(m, K_{3,s}) \geqslant \frac{m}{2} + c'(s)m^{4/5},$$

for all m, and this is tight up to the value of $c'(s)$.

We can handle some additional forbidden graphs H, as well as certain families of graphs $\mathcal{H}$. Our proofs combine combinatorial, probabilistic and spectral techniques, including extensions of ideas that appear in [21], [2], [3]. Throughout the paper, we omit all floor and ceiling signs whenever these are not crucial, to simplify the presentation.

The rest of the paper is organized as follows. In Section 2 we present the (simple) proof of Theorem 1.1. In Section 3 we prove an extension of a lemma established by Shearer in [21]. This lemma shows that graphs in which every edge lies in a relatively small number of triangles have large cuts. Combining this lemma with some additional ideas we prove Theorems 1.2, 1.3 and 1.4 in Section 4. The final Section 5 contains some concluding remarks and open problems.

2. Trees

In this section we prove Theorem 1.1. We need the following simple lemma proved in [14].

Lemma 2.1. *Let G be a graph with m edges and chromatic number at most h. If h is even, then $f(G) \geqslant \frac{m}{2} + \frac{m}{2h-2}$, and if h is odd, then $f(G) \geqslant \frac{m}{2} + \frac{m}{2h}$.*

The proof is simple: split the set of vertices of G into h independent sets, partition these randomly into a part consisting of $\lfloor h/2 \rfloor$ of these sets and its complement, and compute the expected number of edges in the cut obtained, to get the desired result.

Proof of Theorem 1.1. Let G be an H-free graph with m edges, where H is a tree with h edges. It is not difficult to see that G is $(h-1)$-degenerate, that is, any subgraph of it contains a vertex of degree at most $h-1$. Indeed, otherwise there is a subgraph G' of G in which all degrees are at least h. It is easy and well known that each such subgraph contains a copy of each tree with h edges, contradicting H-freeness. Therefore, G is indeed $(h-1)$-degenerate, and hence h-colourable. By Lemma 2.1, it follows that if h is even then $f(G) \geqslant \frac{m}{2} + \frac{m}{2h-2}$, and if h is odd, then $f(G) \geqslant \frac{m}{2} + \frac{m}{2h}$. This proves Theorem 1.1(i),(ii). The fact that if $\binom{h}{2}$ divides m then the result is tight follows by letting G be the vertex-disjoint union of $m/\binom{h}{2}$ cliques, each of size h. Then G is H-free, has m edges, and

$$f(G) = \left\lfloor \frac{h^2}{4} \right\rfloor \frac{m}{\binom{h}{2}},$$

which is $\frac{m}{2} + \frac{m}{2h-2}$ for even h and $\frac{m}{2} + \frac{m}{2h}$ for odd h. □

3. Cuts, degrees, and codegrees

In [21] Shearer proved that if $G = (V, E)$ is a triangle-free graph with n vertices and m edges, and if $d_1, d_2, \ldots, d_n$ are the degrees of its vertices, then

$$f(G) \geq \frac{m}{2} + \frac{1}{8\sqrt{2}} \sum_{i=1}^{n} \sqrt{d_i}. \tag{3.1}$$

Here we extend his argument and prove a similar result for graphs in which most edges do not lie in too many triangles.

Let $G = (V, E)$ be a graph, and consider the following randomized procedure for obtaining a cut of G. Let $h : V \mapsto \{0, 1\}$ be a random function obtained by picking, for each $v \in V$ randomly and independently, the value of $h(v) \in \{0, 1\}$, where both choices are equally likely. Call a vertex $v \in V$ *stable* if it has more neighbours u satisfying $h(u) \neq h(v)$ than neighbours w satisfying $h(w) = h(v)$, otherwise call it *active*. Let $h' : V \mapsto \{0, 1\}$ be the random function obtained from h as follows. For each $u \in V$, if u is active, then choose randomly $h'(u) \in \{0, 1\}$, where both choices are equally likely and all choices are independent. Otherwise, define $h'(u) = h(u)$. Finally, define $V_0 = (h')^{-1}(0)$ and $V_1 = (h')^{-1}(1)$.

This produces a cut (V_0, V_1) consisting of all edges of G that connect a vertex of V_0 and a vertex of V_1. Although in some cases (for example, when G is a complete graph), the expected number of edges in this cut is far smaller than $m/2$, it turns out that when the graph is relatively sparse, the expected size of the cut is large. This is shown by considering the probability that for a given edge uv, $h'(u) \neq h'(v)$. Clearly,

$$\Pr[h'(u) \neq h'(v)] = \frac{1}{2} \Pr[h'(u) \neq h'(v)|\ h(u) = h(v)] + \frac{1}{2} \Pr[h'(u) \neq h'(v)|\ h(u) \neq h(v)]. \tag{3.2}$$

In order to estimate the conditional probability $\Pr[h'(u) \neq h'(v)|h(u) = h(v)]$ note that if at least one of the vertices u or v is active, the probability that $h'(u) \neq h'(v)$ is precisely $1/2$, whereas if they are both stable then $h'(u)$ cannot differ from $h'(v)$. Thus

$$\Pr[h'(u) \neq h'(v)|\ h(u) = h(v)] = \frac{1}{2} - \frac{1}{2} \Pr[u \text{ and } v \text{ are stable}|\ h(u) = h(v)].$$

Similarly, if $h(u) \neq h(v)$ and both these vertices are stable, then certainly $h'(u) \neq h'(v)$, whereas if at least one of them is active, then the probability that $h'(u) \neq h'(v)$ is $1/2$. Thus

$$\Pr[h'(u) \neq h'(v)|\ h(u) \neq h(v)]$$
$$= \Pr[u \text{ and } v \text{ are stable}|\ h(u) \neq h(v)] + \frac{1}{2}(1 - \Pr[u \text{ and } v \text{ are stable}|\ h(u) \neq h(v)])$$
$$= \frac{1}{2} + \frac{1}{2} \Pr[u \text{ and } v \text{ are stable}|\ h(u) \neq h(v)].$$

Substituting in (3.2) we conclude that

$$\Pr[h'(u) \neq h'(v)]$$
$$= \frac{1}{2} + \frac{1}{4}(\Pr[u \text{ and } v \text{ are stable}|\ h(u) \neq h(v)] - \Pr[u \text{ and } v \text{ are stable}|\ h(u) = h(v)]). \tag{3.3}$$

The main remaining task is thus to estimate the difference between these last two conditional probabilities. Intuitively, if u and v do not have too many common neighbours, then the conditional probability of u and v to be stable is smaller when $h(u) = h(v)$ than when $h(u) \neq h(v)$, since in the first case each of them already has at least one vertex with the same h value as itself. In what follows we show that this is indeed the case.

Lemma 3.1. *There are two absolute positive constants c_1, c_2 such that the following holds. Let $G = (V, E)$ be a graph, and let $h, h' : V \mapsto \{0,1\}$ be the two random functions defined by the randomized procedure described above. Let u, v be two adjacent vertices of G, and suppose they have k common neighbours. Let the degree of u in G be $a + k + 1$ and let the degree of v in G be $b + k + 1$. Suppose, further, that $a \geq k$ and $b \geq k$. Then the probability $\Pr[h'(u) \neq h'(v)]$ that the edge uv lies in the random cut produced by the procedure satisfies*

$$\Pr[h'(u) \neq h'(v)] \geq \frac{1}{2} + \frac{c_1}{\sqrt{a+k+1}} + \frac{c_1}{\sqrt{b+k+1}} - \frac{c_2 k}{\sqrt{ab}}.$$

Proof. For convenience we assume that a, b, k are all even. The computation for the other possible cases is similar. For any two integers $m \geq r \geq 0$ let

$$B(m, r) = \frac{1}{2^m} \sum_{i=0}^{r} \binom{m}{i}$$

denote the probability that at most r of m random coin flips are heads. For $\epsilon_1, \epsilon_2 \in \{0, 1\}$, put

$$P_{\epsilon_1, \epsilon_2} = \Pr[u \text{ and } v \text{ are stable} \mid h(u) = \epsilon_1, h(v) = \epsilon_2].$$

By (3.3) it suffices to show that

$$P_{0,1} + P_{1,0} - P_{0,0} - P_{1,1} \geq \frac{4c_1}{\sqrt{a+k+1}} + \frac{4c_1}{\sqrt{b+k+1}} - \frac{4c_2 k}{\sqrt{ab}}.$$

Let K denote the set of common neighbours of u and v, let A denote the set of neighbours of u which are not adjacent to v, and let B denote the set of neighbours of v which are not adjacent to u. By the assumptions of the lemma $|K| = k$, $|A| = a$, $|B| = b$. Note that the stability of u and v is determined by the h-values of the vertices in $\{u, v\} \cup K \cup A \cup B$. Thus, for example, to compute $P_{0,1}$ note that the probability that the h-value of exactly $k/2 + \delta$ of the members of K is 0 is given by

$$\frac{1}{2^k} \binom{k}{k/2 + \delta}.$$

In this case, and assuming $h(u) = 0$ and $h(v) = 1$, both u and v will be stable if and only if the h-value of at most $a/2 - \delta$ members of A is 0, and the h-value of at most $b/2 + \delta$ members of B is one. Hence

$$P_{0,1} = \frac{1}{2^k} \sum_{-k/2 \leq \delta \leq k/2} \binom{k}{k/2 + \delta} B\left(a, \frac{a}{2} - \delta\right) B\left(b, \frac{b}{2} + \delta\right).$$

The computation of the other terms $P_{\epsilon_1,\epsilon_2}$ is similar. Thus we have

$$P_{0,1} + P_{1,0} - P_{0,0} - P_{1,1}$$
$$= \frac{1}{2^k} \sum_{-k/2 \leq \delta \leq k/2} \binom{k}{k/2+\delta} \left[B\left(a, \frac{a}{2}-\delta\right) B\left(b, \frac{b}{2}+\delta\right) + B\left(a, \frac{a}{2}+\delta\right) B\left(b, \frac{b}{2}-\delta\right) \right.$$
$$\left. - B\left(a, \frac{a}{2}-\delta-1\right) B\left(b, \frac{b}{2}-\delta-1\right) - B\left(a, \frac{a}{2}+\delta-1\right) B\left(b, \frac{b}{2}+\delta-1\right) \right]$$
$$= \frac{1}{2^k} \sum_{-k/2 \leq \delta \leq k/2} \binom{k}{k/2+\delta} [\Delta_1(\delta) + \Delta_2(\delta)],$$

where

$$\Delta_1 = \Delta_1(\delta) = \frac{1}{2^a} \binom{a}{a/2-\delta} B\left(b, \frac{b}{2}+\delta\right) + B\left(a, \frac{a}{2}-\delta-1\right) \frac{1}{2^b} \binom{b}{b/2+\delta}$$
$$+ \frac{1}{2^a} \binom{a}{a/2+\delta} B\left(b, \frac{b}{2}-\delta\right) + B\left(a, \frac{a}{2}+\delta-1\right) \frac{1}{2^b} \binom{b}{b/2-\delta},$$

and

$$\Delta_2 = \Delta_2(\delta) = B\left(a, \frac{a}{2}-\delta-1\right) B\left(b, \frac{b}{2}+\delta-1\right) + B\left(a, \frac{a}{2}+\delta-1\right) B\left(b, \frac{b}{2}-\delta-1\right)$$
$$- B\left(a, \frac{a}{2}-\delta-1\right) B\left(b, \frac{b}{2}-\delta-1\right) - B\left(a, \frac{a}{2}+\delta-1\right) B\left(b, \frac{b}{2}+\delta-1\right).$$

However, since

$$B\left(b, \frac{b}{2}+\delta\right) + B\left(b, \frac{b}{2}-\delta\right) = 1 + \frac{\binom{b}{b/2+\delta}}{2^b},$$

and

$$B\left(a, \frac{a}{2}-\delta-1\right) + B\left(a, \frac{a}{2}+\delta-1\right) = 1 - \frac{\binom{a}{a/2-\delta}}{2^a},$$

it follows that

$$\Delta_1 = \frac{\binom{a}{a/2-\delta}}{2^a} + \frac{\binom{a}{a/2-\delta}}{2^a} \cdot \frac{\binom{b}{b/2+\delta}}{2^b} + \frac{\binom{b}{b/2+\delta}}{2^b} - \frac{\binom{b}{b/2+\delta}}{2^b} \cdot \frac{\binom{a}{a/2-\delta}}{2^a} = \frac{\binom{a}{a/2-\delta}}{2^a} + \frac{\binom{b}{b/2+\delta}}{2^b}.$$

Therefore

$$\frac{1}{2^k} \sum_{-k/2 \leq \delta \leq k/2} \binom{k}{k/2+\delta} \Delta_1 = \frac{\binom{k+a}{(k+a)/2}}{2^{k+a}} + \frac{\binom{k+b}{(k+b)/2}}{2^{k+b}} = \Theta\left(\frac{1}{\sqrt{k+a+1}} + \frac{1}{\sqrt{k+b+1}}\right).$$

In addition, for $\delta = 0$, $\Delta_2 = 0$, whereas for $\delta > 0$,

$$\Delta_2 = B\left(a, \frac{a}{2}-\delta-1\right) \frac{1}{2^b} \sum_{j=b/2-\delta}^{b/2+\delta-1} \binom{b}{j} - B\left(a, \frac{a}{2}+\delta-1\right) \frac{1}{2^b} \sum_{j=b/2-\delta}^{b/2+\delta-1} \binom{b}{j}$$
$$= -\frac{1}{2^a} \sum_{i=a/2-\delta}^{a/2+\delta-1} \binom{a}{i} \cdot \frac{1}{2^b} \sum_{j=b/2-\delta}^{b/2+\delta-1} \binom{b}{j}.$$

For $\delta < 0$ the result is obtained from the one above by replacing each δ by $-\delta$.

For each fixed i, $\frac{1}{2^a}\binom{a}{i} \leq O(1/\sqrt{a})$ holds uniformly in i, a, and a similar estimate holds for $\frac{1}{2^b}\binom{b}{j}$. Also note that

$$\frac{1}{2^k} \sum_{-k/2 \leq \delta \leq k/2} \binom{k}{k/2+\delta} \delta^2 = \frac{k}{4},$$

since the expression on the left-hand side is the variance of the binomially distributed random variable with parameters k and $1/2$. Therefore

$$\frac{1}{2^k} \sum_{-k/2 \leq \delta \leq k/2} \binom{k}{k/2+\delta} \Delta_2 \leq O\left(\frac{1}{\sqrt{ab}}\right) \left(\frac{1}{2^k} \sum_{-k/2 \leq \delta \leq k/2} \binom{k}{k/2+\delta} \delta^2 \right) = O\left(\frac{k}{\sqrt{ab}}\right).$$

This completes the proof. □

In order to apply the last lemma, we need the following simple facts, which appear, in some versions, already in [12], [13], and [2].

Lemma 3.2.
(i) There is a positive constant c such that for every graph $G = (V, E)$ with n vertices, m edges, and positive minimum degree, $f(G) \geq \frac{m}{2} + cn$.
(ii) Let $G = (V, E)$ be a graph with m edges, suppose $U \subset V$ and let G' be the induced subgraph of G on U. If G' has m' edges, then $f(G) \geq f(G') + \frac{m-m'}{2}$.

Proof. A simple proof of part (i) is to take a random linear order $v_1, v_2, \ldots, v_n$ on the vertices of G, and to define a cut (A, B) by starting with $A = B = \emptyset$ and placing each vertex v_i, in its turn, either in A or in B, trying to maximize the number of edges in the bipartite graph spanned by the vertex classes A and B. In each such addition, the number of edges added to the bipartite graph is clearly at least half the number of edges connecting v_i to the previous vertices $\{v_1, \ldots, v_{i-1}\}$, and hence at the end of the process we have at least $\frac{m}{2}$ edges with ends at A and B. Moreover, if v_i has an odd number of neighbours among the vertices $\{v_1, \ldots, v_{i-1}\}$, then the number of edges added to the cut while inserting v_i exceeds half the number of edges connecting it to the previous vertices by at least $1/2$. As the probability that a given vertex has an odd number of neighbours preceding it in the randomly chosen order is bounded away from zero, the assertion of part (i) follows.

The proof of part (ii) is even simpler. Given a partition $U = A \cup B$, $A \cap B = \emptyset$, of U such that the number of edges of G between A and B is $f(G')$, add the vertices in $V - U$ one by one, where each vertex in its turn is added to A or to B, trying to maximize the number of edges in the bipartite graph spanned by these vertex classes. This clearly produces a cut in G that contains all the $f(G')$ edges of the initial cut of G', and at least half of all other edges, implying the assertion of (ii). □

We can now prove the following lemma, which will be one of the main tools used in the next section.

Lemma 3.3. *There exists an absolute positive constant ϵ such that for every positive constant C there is a $\delta = \delta(C) > 0$ with the following property. Let $G = (V, E)$ be a graph with n vertices (with positive degrees), m edges, and degree sequence $d_1, d_2, \ldots, d_n$. Suppose, further, that the induced subgraph on any set of $d \geqslant C$ vertices, all of which have a common neighbour, contains at most $\epsilon d^{3/2}$ edges. Then*

$$f(G) \geqslant \frac{m}{2} + \delta \sum_{i=1}^{n} \sqrt{d_i}.$$

Proof. As long as there is a vertex of degree smaller than C in G, delete it. If during this process we delete more than $\frac{1}{4C} \sum_{i=1}^{n} \sqrt{d_i}$ vertices, then the desired result, with $\delta = \frac{c}{4C}$, where c is the constant from Lemma 3.2(i), follows from the assertion of that lemma. Therefore we may assume that the process terminates after at most $\frac{1}{4C} \sum_{i=1}^{n} \sqrt{d_i}$ such deletion steps. It thus terminates with an induced subgraph G' on n' vertices in which all degrees are at least C. Let $d'_1, d'_2, \ldots, d'_{n'}$ be the degree sequence of G', and let m' denote its total number of edges. Note that G' is obtained from G by deleting fewer than $C \frac{1}{4C} \sum_{i=1}^{n} \sqrt{d_i} = \frac{1}{4} \sum_{i=1}^{n} \sqrt{d_i}$ edges. Since each deletion of an edge decreases the sum of roots of the degrees by at most 2, we conclude that

$$\sum_{i=1}^{n'} \sqrt{d'_i} \geqslant \frac{1}{2} \sum_{i=1}^{n} \sqrt{d_i} \qquad (3.4)$$

Let V' be the set of vertices of G', and consider the randomized procedure for obtaining a cut of G' by choosing the random functions $h, h' : V' \mapsto \{0, 1\}$ as described in the beginning of the section. Lemma 3.1 enables us to estimate the probability that an edge uv of G' for which the degrees of u and v in G' are $d(u), d(v)$, respectively, and

the number of common neighbours of u and v is at most $\min\left(\frac{d(u)}{2}, \frac{d(v)}{2}\right)$ (3.5)

lies in the cut produced in this way. We first show that the number of edges uv for which the condition (3.5) is violated is not too large. Indeed, assign each edge violating this condition to its end with the smaller degree (or to either of them in case of equality). Let u be a vertex of degree d in G', and consider the set of all edges uv assigned to u. For each such edge, the degree of v in the subgraph induced on the neighbours of u in G' is at least $d/2$. Since $d \geqslant C$, there are at most $\epsilon d^{3/2}$ edges in this subgraph, and hence the number of edges uv assigned to u is at most $4\epsilon \sqrt{d}$. Summing over all vertices we conclude that the number of edges of G' that violate (3.5) is at most

$$4\epsilon \sum_{i=1}^{n'} \sqrt{d'_i}. \qquad (3.6)$$

Every other edge uv of G' satisfies the assumptions of Lemma 3.1. Therefore, the probability that it lies in our cut is at least

$$\frac{1}{2} + \frac{c_1}{\sqrt{d(u)}} + \frac{c_1}{\sqrt{d(v)}} - \frac{2c_2 k}{\sqrt{d(u)d(v)}},$$

where $d(u), d(v)$ are the degrees of u and v in G', and k is the number of their common neighbours in G'. By linearity of expectation, and assuming, for a moment, that all edges of G' satisfy (3.5), we conclude, from Lemma 3.1, that the expected size of the cut is at least

$$\frac{m'}{2} + c_1 \sum_{i=1}^{n'} \sqrt{d'_i} - \Delta,$$

where Δ is the sum of contributions of the last term (the term $\frac{2c_2 k}{\sqrt{d(u)d(v)}}$) over all edges uv of G'. In reality, however, not all edges of G' satisfy (3.5); for each edge that does not satisfy it we do not add the corresponding term from the lemma. Thus we lose, for each such edge, at most 1, and as the number of these edges is bounded by (3.6), and we may assume, for example, that $\epsilon < \frac{c_1}{8}$, we still conclude that the expected size of our cut is at least

$$\frac{m'}{2} + \frac{c_1}{2} \sum_{i=1}^{n'} \sqrt{d'_i} - \Delta \qquad (3.7)$$

where Δ is now the sum of contributions of the $\frac{2c_2 k}{\sqrt{d(u)d(v)}}$-term over all edges uv of G' that satisfy (3.5). To bound Δ, as before, assign each edge uv, as above, to its end with the smaller degree (or to any of them in case of equality). Let u be a vertex of degree d in G', and consider the set of all edges uv assigned in this way to u. For each such edge, the degree of v in the subgraph induced on the neighbours of u in G' is k. In addition $\frac{1}{\sqrt{d(u)d(v)}} \leq \frac{1}{d(u)}$. Since $d(u) \geq C$, there are at most $\epsilon d(u)^{3/2}$ edges in this induced subgraph, and hence the total contribution of the $\frac{2c_2 k}{\sqrt{d(u)d(v)}}$-terms assigned to u is at most $4c_2 \epsilon \sqrt{d(u)}$. Summing over all vertices we conclude that

$$\Delta \leq 4c_2 \epsilon \sum_{i=1}^{n'} \sqrt{d'_i}.$$

This, together with (3.7), implies that

$$f(G') \geq \frac{m'}{2} + \left(\frac{c_1}{2} - 4c_2 \epsilon\right) \sum_{i=1}^{n'} \sqrt{d'_i}.$$

By the last inequality, (3.4) and part (ii) of Lemma 3.2, and by choosing $\epsilon < \frac{c_1}{16c_2}$ the desired result follows. □

In what follows, it will be convenient to use the following variant of the last lemma as well.

Lemma 3.4. *There exist two absolute positive constants ϵ and δ such that the following holds. Let $G = (V, E)$ be a graph with n vertices, m edges, and degree sequence $d_1, d_2, \ldots, d_n$. Suppose, further, that for each i the induced subgraph on all the d_i neighbours of vertex* number i contains at most $\epsilon d^{3/2}$ edges. Then

$$f(G) \geq \frac{m}{2} + \delta \sum_{i=1}^{n} \sqrt{d_i}.$$

This lemma differs from Lemma 3.3 by the fact that here δ is an absolute constant, but more crucially, by the fact that we assume that the induced subgraph on the full neighbourhood of each vertex is sparse (and not that the induced subgraph on any large subset of this neighbourhood is sparse). Note that since ϵ is small this means that if G satisfies the assumptions, then the induced subgraph on the neighbourhood of each low-degree vertex of G (if there are any) contains no edges at all. The proof of this lemma is essentially identical to the proof of Lemma 3.3, without the extra complication of producing the graph G'. Here we apply our randomized procedure to get a cut of the original graph G, and estimate its expected size as before using the fact that each neighbourhood of size d spans at most $\epsilon d^{3/2}$ edges. By choosing, for example,

$$\epsilon < \min\left(\frac{c_1}{8}, \frac{c_1}{16c_2}\right) \quad \text{and} \quad \delta = \frac{c_1}{4},$$

where c_1, c_2 are the constants from Lemma 3.1, and by repeating the arguments in the previous proof, we obtain the desired result.

4. *H*-free graphs

In this section we present the proofs of Theorems 1.2, 1.3 and 1.4. We need several known results. The first one is a simple, well-known upper bound for the size of the maximum cut in a regular graph in terms of its eigenvalues. See, *e.g.*, [2] for a proof.

Lemma 4.1. *Let $G = (V, E)$ be a d-regular graph with n vertices and $m = nd/2$ edges, and let $\lambda_1 \geq \lambda_2 \geq \ldots \geq \lambda_n$ be the eigenvalues of (the adjacency matrix of) G. Then*

$$f(G) \leq (d - \lambda_n)n/4 = \frac{m}{2} - \lambda_n n/4.$$

(Note that since the trace of the adjacency matrix of G is zero, $\lambda_n \leq 0$.)
The second result is due to Bondy and Simonovits [10].

Lemma 4.2. *Let $k \geq 2$ be an integer and let G be a graph on n vertices. If G contains no cycle of length $2k$, then the number of edges of G is at most $100kn^{1+1/k}$.*

We also need the following result of Kővári, T. Sós and Turán [17].

Lemma 4.3. *Let $K_{t,s}$ denote, as in Theorem 1.4, the complete bipartite graph with $t + s$ vertices and ts edges. For every $s \geq t$, every graph with n vertices that contains no copy of $K_{t,s}$ has at most*

$$\frac{1}{2}(s-1)^{1/t}n^{2-1/t} + \frac{1}{2}(t-1)n$$

edges.

4.1. Even cycles

Proof of Theorem 1.2. Let $r - 1 = 2k \geq 4$ be an even integer, and let $G = (V, E)$ be a C_{2k}-free graphs with n vertices and m edges.

Define $D = bm^{2/(r+1)}$, where $b = b(r) > 1$ will be chosen later. We consider two possible cases depending on the existence of dense subgraphs in G.

Case 1. G is $(D-1)$-degenerate, that is, it contains no subgraph with minimum degree at least D. In this case, as is well known, there exists a labelling $v_1, \ldots, v_n$ of the vertices of G so that for every i, the number of neighbours v_j of v_i with $j < i$ is strictly smaller than D. (To see this, let u be a vertex of minimum degree in G, define $v_n = u$, delete it from G and repeat the process.) Let $d^+(v_i)$ denote the number of neighbours v_j of v_i with $j < i$ and let $d(v_i)$ denote the total degree of v_i. Then

$$\sum_{i=1}^{n} \sqrt{d(v_i)} \geq \sum_{i=1}^{n} \sqrt{d^+(v_i)} > \frac{\sum_{i=1}^{n} d^+(v_i)}{\sqrt{D}} = \frac{m}{\sqrt{D}} \geq \frac{m^{r/(r+1)}}{\sqrt{b}}.$$

Note that the neighbourhood of a vertex of G cannot contain a path of length $2k - 2$, and hence the number of edges induced on any subset of cardinality d in it is smaller than kd, which is smaller than $\epsilon d^{3/2}$ for all $d > (k/\epsilon)^2 = (\frac{r-1}{2\epsilon})^2$. Therefore, by Lemma 3.3,

$$f(G) \geq \frac{m}{2} + \delta \sum_{i=1}^{n} \sqrt{d(v_i)} \geq \frac{m}{2} + \delta \frac{m^{r/(r+1)}}{\sqrt{b}},$$

where $\delta = \delta(r)$, as needed.

Case 2. G contains a subgraph G' with minimum degree at least D. But in this case, the number of vertices of G' is $N \leq 2m^{(r-1)/(r+1)}/b$ and as the minimum degree is at least $bm^{2/(r+1)} \geq \frac{b^{(r+1)/(r-1)}}{2^{2/(r-1)}} N^{2/(r-1)}$ this is impossible, in view of Lemma 4.2, for a sufficiently large chosen value of $b = b(r)$.

This completes the proof of the required lower bound for $f(m, C_{r-1})$ for all odd $r > 3$. The fact that the error term is tight, up to the value of $c(r)$, for $r \in \{5, 7, 11\}$ is proved using Lemma 4.1. An (n, d, λ)-graph is a d-regular graph $G = (V, E)$ on n vertices, such that the absolute value of every eigenvalue of the adjacency matrix of G, besides the largest one, is at most λ. The properties of such graphs in which λ is much smaller than d have been studied extensively. It is known that in this case the graph has some strong pseudo-random properties; see, e.g., [8], Chapter 9.2 or [18] and their references. The extremal graphs we use here are, indeed, appropriate (n, d, λ)-graphs.

The Erdős–Rényi graph G, constructed in [15], is the polarity graph of a finite projective plane of order p. This graph is a C_4-free (n, d, λ)-graph, where $n = p^2 + p + 1$, $d = p + 1$ and $\lambda = \sqrt{p}$, and it exists for every prime power p. It has, in fact, $p + 1$ loops, which we omit. See, e.g., [3] for the proof of these facts. Let m denote the number of edges of this graph. By Lemma 4.1 its maximum cut is of size at most

$$\frac{m}{2} + \frac{n\sqrt{p}}{4} \leq \frac{m}{2} + O(m^{5/6}).$$

Therefore, the error term in Theorem 1.2 is tight for $r - 1 = 4$.

For every q which is an odd power of 2, the incidence graph of the generalized 4-gon has a polarity. The corresponding polarity graph is a $(q+1)$-regular graph with

$n = q^3 + q^2 + q + 1$ vertices. See [11], [19] for more details. This graph contains no cycle of length 6 and it is not difficult to compute its eigenvalues (they can be derived, for example, from the eigenvalues of the corresponding incidence graph, given in [23]; see also [6]). Indeed, all the eigenvalues, besides the trivial one (which is $q + 1$) are either 0 or $\sqrt{2q}$ or $-\sqrt{2q}$. Let m denote the number of edges of this graph. By Lemma 4.1 we conclude that

$$f(G) \leqslant \frac{m}{2} + \frac{n\sqrt{2q}}{4} \leqslant \frac{m}{2} + O(m^{7/8}),$$

showing that the assertion of the theorem is tight for $r - 1 = 6$ as well.

For every q which is an odd power of 3, the incidence graph of the generalized 6-gon has a polarity. The corresponding polarity graph is a $(q + 1)$-regular graph with $q^5 + q^4 + \cdots + q + 1$ vertices. See [11], [19] for more details. This graph contains no cycle of length 10 and its eigenvalues can be easily derived from the the eigenvalues of the corresponding incidence graph, given in [23]; see also [6]. All the eigenvalues, besides the trivial one, are either $\sqrt{3q}$ or $-\sqrt{3q}$ or $\sqrt{q}$ or $-\sqrt{q}$. Using Lemma 4.1 this shows that the assertion of Theorem 1.2 is tight for $r - 1 = 10$ as well. □

Remark. Using similar reasoning, we can show that for any graph H which is the union of an arbitrary number of cycles of length 4, all having a single common point,

$$f(m, H) \geqslant \frac{m}{2} + c(H)m^{5/6},$$

and this is tight, up to the value of $c(H)$. To do so we have to combine the previous proof with the fact that the number of edges in any H-free graph on n vertices is at most $c(H)n^{3/2}$, a result that follows from the main result of [16] (see also [5]). A similar result holds for certain other graphs H; we omit the details.

4.2. Forests with a common neighbour

Proof of Theorem 1.3. The proof is similar to one of the proofs in [2] (see also [3]). Let H be the graph obtained from a forest F with at least one edge, by adding an additional vertex and by connecting it to all vertices of F. Let $G = (V, E)$ be an H-free graph with n vertices and m edges. Define $D = m^{2/5}$ and proceed as before, by considering two possible cases.

Case 1. G contains no subgraph with minimum degree at least D. In this case, we proceed as in the previous proof. Here, too, the induced subgraph of G on any set of common neighbours of a vertex can span only a linear number of edges, as it contains no copy of F. Thus we can apply, again, Lemma 3.3 and conclude, as in the proof of Theorem 1.2, that in this case

$$f(G) \geqslant \frac{m}{2} + \Omega\left(\frac{m}{\sqrt{D}}\right) \geqslant \frac{m}{2} + \Omega(m^{4/5}),$$

as needed.

Case 2. There exists a subset W of x vertices of G such that the induced subgraph G' of G on W has minimum degree at least D. We first prove that in this case G' (and hence G) contains an induced subgraph G'' on a set W' of vertices of G, with at least $xD/4$

edges, which is r-colourable for $r = O(\frac{x}{D})$. To see this, let R be a random subset of at most $2x/D$ vertices of G' obtained by picking, with repetitions, $2x/D$ vertices of G', each chosen randomly with uniform probability. Let u be a fixed vertex of G'. The probability that u does not have a neighbour in R is

$$\left(1 - \frac{d_{G'}(u)}{x}\right)^{2x/D} < \exp\{-(D/x)(2x/D)\} < 1/4,$$

where $d_{G'}(u)$ denotes the degree of u in G'. It follows that for every fixed edge uv of G', the probability that both u and v have neighbours in R is at least $1/2$. Let W' be the set of all vertices of W that have a neighbour in R, and let G'' be the induced subgraph of G on W'. By linearity of expectation, the expected number of edges of G'' is at least half the number of edges of G', that is, at least $xD/4$. Hence there exists a set R of at most $2x/D$ vertices of G' so that the corresponding graph G'' has at least $xD/4$ edges. Fix such an R and note that as the neighbourhood of each vertex of G contains no copy of the forest F, the induced subgraph on it is colourable by at most $|F|$ colours, implying that indeed the induced subgraph G'' is $r = O(x/D)$-colourable.

By Lemma 2.1 it follows that $f(G'')$ exceeds half the number of edges of this subgraph by at least

$$\Omega\left(\frac{xD}{4} \cdot \frac{1}{2r}\right) = \Omega(D^2) = \Omega(m^{4/5}).$$

This implies, by Lemma 3.2(ii), that

$$f(G) \geq \frac{m}{2} + \Omega(m^{4/5}),$$

as needed. The fact that this is tight up to the constant in the Ω-notation follows from the spectral properties of the graph constructed in [1]. This is a triangle-free (n, d, λ)-graph with $d = \Theta(n^{2/3})$ and $\lambda = \Theta(n^{1/3})$. As it is triangle-free, it contains no copy of H, and its spectral properties imply, by Lemma 4.1, that if m is the number of its edges then

$$f(G) \leq \frac{m}{2} + O(n^{4/3}) = \frac{m}{2} + O(m^{4/5}). \qquad \square$$

4.3. Complete bipartite graphs

Proof of Theorem 1.4(i) (sketch). The proof is similar to the previous ones. Let G be a $K_{2,s}$-free graph with m edges, where $s \geq 2$. If G is D-degenerate, for $D = bm^{1/3}$ where $b = b(s) > 0$ will be chosen later, then the desired result follows. Indeed, as before, Lemma 3.3 implies that in this case

$$f(G) \geq \frac{m}{2} + \Omega\left(\frac{m}{\sqrt{D}}\right) = \frac{m}{2} + \Omega(m^{5/6}).$$

Otherwise we get, for the right choice of $b = b(s)$, a contradiction to Lemma 4.3. The Erdős–Rényi graph shows that the result is tight. $\qquad \square$

Proof of Theorem 1.4(ii) (sketch). The proof of this case is similar to the previous ones as well, but contains two extra twists. Let $G = (V, E)$ be a $K_{3,s}$-free graph with m edges, and n vertices (of positive degrees), where $s \geq 3$. We may and will assume that m is sufficiently large. If $n \geq \frac{m^{4/5}}{2}$, the desired result follows from Lemma 3.2(i). Thus we may assume that $n < \frac{m^{4/5}}{2}$. As long as there is a vertex of degree smaller than $m^{1/5}$ in G, omit it. This process terminates after deleting fewer than $m^{1/5}n < \frac{m}{2}$ edges, and thus we obtain an induced subgraph G' of G with at least $m/2$ edges and minimum degree at least $m^{1/5}$. Let V' denote the set of vertices of G'. Note that the induced subgraph on the neighbourhood of any vertex of degree d of G' contains no copy of $K_{2,s}$, and hence contains fewer than $sd^{3/2}$ edges, by Lemma 4.3.

Let $\eta > 0$ be a small fixed real, to be chosen later, and consider a random subset V'' of V' obtained by picking each vertex of V' randomly and independently, with probability η. Let G'' be the induced subgraph of G' (and hence of G) on V''. If a vertex has degree d in G', then its expected degree in G'' (assuming it is a vertex of G'') is ηd. The expected number of edges in its neighbourhood is at most $\eta^2 s d^{3/2}$. Since all degrees are large, it follows that with high probability, every vertex of degree d in G' that lies in V'' has degree at least, say, $\eta d/2$ in G''. Similarly, for every pair of vertices with codegree exceeding, say, $\eta \sqrt{d}$, the number of their common neighbours in V'' is highly concentrated around its expectation. This implies that with high probability the number of edges in every neighbourhood of G'' is at most, say, $2\eta^2 s d^{3/2}$. If η is sufficiently small as a function of s and ϵ, where ϵ is the number from Lemma 3.4, we can ensure that the assumptions of Lemma 3.4 hold for G''. Moreover, this graph has m' edges, where $m' \geq \eta^2 m/4 = \Omega(m)$, and it is, of course, $K_{3,s}$-free.

We can now proceed as in the previous proofs. If G'' is $bm^{2/5}$-degenerate, then by Lemma 3.4 $f(G'')$ exceeds half the number of its edges by at least $\Omega(m^{4/5})$ and the desired result follows, in this case, from Lemma 3.2(ii). Otherwise, it contains a subgraph with minimum degree $bm^{2/5}$, and hence at most $2m^{3/5}/b$ vertices, but this is impossible, for a sufficiently large $b = b(s)$, in view of Lemma 4.3 and the fact that the graph contains no copy of $K_{3,s}$. This completes the proof of the lower bound for $f(m, K_{3,s})$.

The fact that the result is tight follows from Lemma 4.1 and the spectral properties of the projective norm graphs constructed in [7] (see [6] or [22] for a computation of their eigenvalues). For any prime p and for appropriate choice of the parameters, this construction gives a $K_{3,3}$-free (n, d, λ)-graph with $n = p^3 - p^2$, $d = p^2 - 1$ and $\lambda = p$. $\square$

5. Concluding remarks and open problems

(1) Closely related to the MaxCut problem is the so-called *judicious partition problem*, where the task is to find a partition $V = V_1 \cup V_2$ such that both parts V_1 and V_2 span the smallest possible number of edges. Formally, for a graph $G = (V, E)$ we define

$$g(G) = \min_{V = V_1 \cup V_2} \max\{e(V_1), e(V_2)\},$$

where as usual $e(V_i)$ denotes the number of edges of G spanned by V_i. Bounding $g(G)$ from above immediately supplies a lower bound for $f(G)$: $f(G) \geq m - 2g(G)$. In the other direction, in our joint paper with Bollobás [3] we obtained a general result, connecting the size of an optimal bipartite cut with the best value of a judicious partition in it. We proved that if a graph $G = (V, E)$ with m edges has a bipartite cut of size $\frac{m}{2} + \delta$, then there exists a partition $V = V_1 \cup V_2$ such that both parts V_1, V_2 span at most $\frac{m}{4} - (1 - o(1))\frac{\delta}{2} + O(\sqrt{m})$ edges for the case $\delta = o(m)$, and at most $(\frac{1}{4} - \Omega(1))m$ edges for $\delta = \Omega(m)$. This result immediately enables us to extend Theorems 1.1, 1.2, 1.3 and 1.4 to the corresponding ones for judicious partitioning.

(2) The main technical part of Section 3 is given by Lemma 3.1 and Lemma 3.3, extending the result of Shearer in [21]. We can give an alternative proof of a variant of Lemma 3.3, which gives essentially the same result, but is (possibly) somewhat more natural. Here is the argument for the triangle-free case. The same reasoning can be extended to the more general case considered in Section 3 as well.

Theorem 5.1. *There exists an absolute constant $c_1 > 0$ such that every triangle-free graph G with vertex set $V(G) = [n]$ and degree sequence $(d_1, \ldots, d_n)$ has a cut with at least $dn/4 + c_1 \sum_{i=1}^{n} \sqrt{d_i}$ edges, where $d = \frac{1}{n} \sum_{i=1}^{n} d_i$ is the average degree of G.*

Proof. The theorem is an easy consequence of the following lemma.

Lemma 5.2. *Let G be as above. Then there exist disjoint subsets $A, B \subset [n]$ such that $e(A, B) - e(A) - e(B) \geq c_2 \sum_{i=1}^{n} \sqrt{d_i}$, where $c_2 > 0$ is an absolute constant.*

To derive Theorem 5.1, set $c_1 = c_2/2$, apply Lemma 5.2, and then add vertices from $V(G) \setminus (A \cup B)$ vertex by vertex to A or to B, each time adding a vertex to the set where it has fewer neighbours, breaking ties arbitrarily. The obtained cut clearly has at least $dn/4 + c_1 \sum_{i=1}^{n} \sqrt{d_i}$ edges. (See also Lemma 3.2(ii).)

Proof of Lemma 5.2. Let

$$\psi(d) = \max\left\{k : \sum_{i=k}^{d} \binom{d}{i} \left(\frac{1}{4}\right)^i \left(\frac{3}{4}\right)^{d-i} \geq \frac{1}{3}\right\}. \tag{5.1}$$

By de Moivre–Laplace $\psi(d) \geq d/4 + c\sqrt{d}$, for some absolute constant $c > 0$. Now, for $1 \leq i \leq n$, set

$$p_i = \sum_{j=\psi(d_i)}^{d_i} \binom{d_i}{j} \left(\frac{1}{4}\right)^j \left(\frac{3}{4}\right)^{d_i-j}, \tag{5.2}$$

and observe that $p_i \geq 1/3$ by the definition (5.1) of $\psi(d)$.

Form disjoint subsets A, B using the following procedure.

- Form A by taking each $i \in [n]$ into A independently and with probability $1/4$.
- Given A, define

$$B_0 = \{i \notin A : d(i, A) \geq \psi(d_i)\}.$$

- For each $i \in B_0$, include i in B independently and with probability $1/(3p_i)$.

Let X, Y, Z be random variables, counting the number of edges between A and B, inside A, and inside B, respectively. We estimate the means of X, Y, Z. Obviously, each $e \in E(G)$

is in A with probability $1/16$, and therefore

$$\mathbb{E}[Y] = |E(G)|/16 = dn/32. \tag{5.3}$$

Let $i \in [n]$. Then

$$\Pr[i \in B_0] = \Pr[(i \notin A) \text{ and } (d(i, A) \geq \psi(d_i))] = \frac{3}{4} \cdot p_i,$$

and therefore $\Pr[i \in B] = \Pr[i \in B_0]\Pr[i \in B|i \in B_0] = 1/4$. Since the degree of each $i \in B$ to A is at least $\psi(d_i)$, we get

$$\mathbb{E}[X] \geq \sum_{i=1}^{n} \frac{\psi(d_i)}{4} \geq \sum_{i=1}^{n} \left(\frac{d_i}{16} + \frac{c\sqrt{d_i}}{4} \right) = \frac{dn}{16} + \frac{c}{4} \sum_{i=1}^{n} \sqrt{d_i}. \tag{5.4}$$

Now let $e = (i, j) \in E(G)$. Since G is triangle-free, i and j do not have common neighbours. It follows that

$$\Pr[i, j \in B_0] = \Pr[(i, j \notin A) \text{ and } (d(i, A) \geq \psi(d_i)) \text{ and } (d(j, A) \geq \psi(d_j))]$$

equals

$$\frac{9}{16} \left(\sum_{k=\psi(d_i)}^{d_i-1} \binom{d_i-1}{k} \left(\frac{1}{4}\right)^k \left(\frac{3}{4}\right)^{d_i-1-k} \right) \left(\sum_{k=\psi(d_j)}^{d_j-1} \binom{d_j-1}{k} \left(\frac{1}{4}\right)^k \left(\frac{3}{4}\right)^{d_j-1-k} \right) < \frac{9}{16} p_i p_j,$$

and thus $\Pr[(i, j) \in B] < 1/16$. Summing up, we obtain

$$\mathbb{E}[Z] < \frac{1}{16} |E(G)| = \frac{dn}{32}. \tag{5.5}$$

Using linearity of expectation and estimates (5.3), (5.4), (5.5), we derive

$$\mathbb{E}[X - Y - Z] \geq \frac{c}{4} \sum_{i=1}^{n} \sqrt{d_i}.$$

Set $c_2 = c/4$. Choose A, B for which $X - Y - Z \geq c_2 \sum_{i=1}^{n} \sqrt{d_i}$. The lemma and hence the theorem follow. $\square$

(3) When some edges of the graph are contained in too many triangles, the assertion of Lemma 3.3 may cease to hold. As an example, consider the following graph $G = (V, E)$ constructed by Delsarte and Goethals, and by Turyn. Let q be a prime power and let V be the elements of the two-dimensional vector space over $GF(q)$. Thus G has $n = q^2$ vertices. Partition the $q + 1$ lines through the origin of the space into two sets P and N, where $|P| = k$. Two vertices x and y of the graph G are adjacent if and only if $x - y$ is parallel to a line in P. It is easy to check that G is $d = k(q - 1)$-regular, and hence its number of edges, m, is $kq^2(q - 1)/2 = \Theta(kq^3)$. Moreover, this is a strongly regular graph and its smallest eigenvalue is $-k$ (see, e.g., [18]). It follows that $f(G)$ exceeds half the number of edges by at most $O(kq^2)$ (this can, in fact, be shown without using the spectral bound as well, since the graph is an edge-disjoint union of cliques of size q, and the maximum cut in such a clique exceeds half the number of its edges by only $O(q)$). The number $O(kq^2)$ is (much) smaller than $n\sqrt{d} = \Theta(q^{5/2}k^{1/2})$ whenever k is (much) smaller than q. Therefore,

in each such case, the conclusion of Lemma 3.3 does not hold. Note that the number of common neighbours of each pair of adjacent vertices of G is $q-2+(k-1)(k-2)$. This is (of course) bigger than $\epsilon\sqrt{d}$.

(4) The authors of [3] conjecture that, for any fixed graph H, there is an $\epsilon(H) > 0$ such that
$$f(m,H) \geq \frac{m}{2} + \Omega(m^{3/4+\epsilon}).$$
It clearly suffices to prove this conjecture for complete graphs H. A related plausible conjecture is that for every fixed graph H there is a constant $c(H)$ such that
$$f(m,H) = \frac{m}{2} + \Theta(m^{c(H)}).$$
It would be nice to prove this conjecture, and possibly even to determine the value of $c(H)$ for every graph H. This seems difficult.

(5) We conjecture that the assertion of Theorem 1.2 is tight, up to the value of $c(r)$, for all odd $r > 3$. Even if this is true, however, a proof will not be easy, as it would imply that the maximum number of edges of a C_{r-1}-free graph on n vertices is $\Theta(n^{1+2/(r-1)})$ for all odd $r > 3$. This is known only for $r - 1 \in \{4, 6, 10\}$, despite a considerable amount of effort to prove it for other values.

(6) It seems plausible to conjecture that the assertion of Theorem 1.4 can be extended for bigger values of t, as follows:
$$f(m,K_{t,s}) \geq \frac{m}{2} + c(s)m^{\frac{3t-1}{4t-2}},$$
for all $s \geq t \geq 2$. By the statement of the theorem this holds (and is tight) for $t \in \{2,3\}$. If true for larger values of t, this is tight at least for all $s \geq (t-1)! + 1$, as shown by the projective norm graphs.

References

[1] Alon, N. (1994) Explicit Ramsey graphs and orthonormal labelings. *Electronic J. Combin.* **1** R12, 8pp.
[2] Alon, N. (1996) Bipartite subgraphs. *Combinatorica* **16** 301–311.
[3] Alon, N., Bollobás, B., Krivelevich, M. and Sudakov, B. (2003) Maximum cuts and judicious partitions in graphs without short cycles. *J. Combin. Theory Ser. B* **88** 329–346.
[4] Alon, N. and Halperin, E. (1998) Bipartite subgraphs of integer weighted graphs. *Discrete Math.* **181** 19–29.
[5] Alon, N., Krivelevich, M. and Sudakov, B. (2003) Turán numbers of bipartite graphs and related Ramsey-type questions. *Combin. Probab. Comput.* **12** 477–494.
[6] Alon, N. and Rödl, V. (2005) Sharp bounds for some multicolor Ramsey numbers. *Combinatorica* **25** 125–141.
[7] Alon, N., Rónyai, L. and Szabó, T. (1999) Norm-graphs: variations and applications. *J. Combin. Theory Ser. B* **76** 280–290.
[8] Alon, N. and Spencer, J. (2000) *The Probabilistic Method*, second edn, Wiley, New York.

[9] Bollobás, B. and Scott, A. D. (2002) Better bounds for max cut. In *Contemporary Combinatorics* (B. Bollobás, ed.), Bolyai Society Mathematical Studies, Springer, pp. 185–246.
[10] Bondy, A. and Simonovits, M. (1974) Cycles of even length in graphs. *J. Combin. Theory Ser. B* **16** 97–105.
[11] Brouwer, A. E., Cohen, A. M. and Neumaier, A. (1989) *Distance-Regular Graphs*, Springer, Berlin.
[12] Edwards, C. S. (1973) Some extremal properties of bipartite subgraphs. *Canadian J. Math.* **3** 475–485.
[13] Edwards, C. S. (1975) An improved lower bound for the number of edges in a largest bipartite subgraph. In *Proc. 2nd Czechoslovak Symposium on Graph Theory*, Prague, pp. 167–181.
[14] Erdős, P. (1979) Problems and results in graph theory and combinatorial analysis. In *Graph Theory and Related Topics* (Proc. Conf. Waterloo, 1977), Academic Press, New York, pp. 153–163.
[15] Erdős, P. and Rényi, A. (1962) On a problem in the theory of graphs (in Hungarian). *Publ. Math. Inst. Hungar. Acad. Sci.* **7** 215–235.
[16] Füredi, Z. (1991) On a Turán type problem of Erdős. *Combinatorica* **11** 75–79.
[17] Kővári, T., Sós, V. T. and Turán, P. (1954) On a problem of K. Zarankiewicz. *Colloquium Math.* **3** 50–57.
[18] Krivelevich, M. and Sudakov, B. Pseudo-random graphs. To appear.
[19] Lazebnik, F., Ustimenko, V. A. and Woldar, A. J. (1999) Polarities and 2k-cycle-free graphs. *Discrete Math.* **197/198** 503–513.
[20] Poljak, S. and Tuza, Zs. (1994) Bipartite subgraphs of triangle-free graphs. *SIAM J. Discrete Math.* **7** 307–313.
[21] Shearer, J. (1992) A note on bipartite subgraphs of triangle-free graphs. *Random Struct. Alg.* **3** 223–226.
[22] Szabó, T. (2003) On the spectrum of projective norm-graphs. *Inform. Process. Lett.* **86** 71–74.
[23] Tanner, R. M. (1984) Explicit concentrators from generalized N-gons. *SIAM J. Algebraic Discrete Methods* **5** 287–293.

A Tale of Three Couplings: Poisson–Dirichlet and GEM Approximations for Random Permutations

RICHARD ARRATIA,[1] A. D. BARBOUR[2†] and SIMON TAVARÉ[3‡]

[1]Department of Mathematics, University of Southern California, Los Angeles, CA 90089, USA
(e-mail: rarratia@math.usc.edu)

[2]Angewandte Mathematik, Universität Zürich, Winterthurerstrasse 190,
CH-8057, Zürich, Switzerland
(e-mail: a.d.barbour@math.unizh.ch)

[3]Department of Mathematics, University of Southern California, Los Angeles, CA 90089, USA
and
Department of Applied Mathematics and Theoretical Physics,
University of Cambridge, Cambridge, UK
(e-mail: stavare@usc.edu)

Received 31 January 2005; revised 7 June 2005

For Béla Bollobás on his 60th birthday

For a random permutation of n objects, as $n \to \infty$, the process giving the proportion of elements in the longest cycle, the second-longest cycle, and so on, converges in distribution to the Poisson–Dirichlet process with parameter 1. This was proved in 1977 by Kingman and by Vershik and Schmidt. For soft reasons, this is equivalent to the statement that the random permutations and the Poisson–Dirichlet process can be coupled so that zero is the limit of the expected ℓ_1 distance between the process of cycle length proportions and the Poisson–Dirichlet process. We investigate how rapid this metric convergence can be, and in doing so, give two new proofs of the distributional convergence.

One of the couplings we consider has an analogue for the prime factorizations of a uniformly distributed random integer, and these couplings rely on the 'scale-invariant spacing lemma' for the scale-invariant Poisson processes, proved in this paper.

1. Introduction

Pick a permutation of n objects, with all $n!$ possibilities equally likely. Write $L_1^{(n)}$ for the length of the longest cycle, $L_2^{(n)}$ for the length of the second-longest cycle, and so on, with $L_j^{(n)} = 0$ if the permutation has fewer than j cycles. Kingman [18] and Vershik and

---

[†] Supported in part by Schweizerischer Nationalfondsprojekte Nos. 20-61753.00 and 20-67909.02.
[‡] Holder of a Royal Society–Wolfson Research Merit Award.

Shmidt [22] proved that these lengths, taken as proportions of n, converge in distribution, to a process $(L_1, L_2, \ldots)$ now known as the Poisson–Dirichlet process (with parameter 1):

$$n^{-1}\left(L_1^{(n)}, L_2^{(n)}, \ldots\right) \Rightarrow (L_1, L_2, \ldots). \tag{1.1}$$

A condition equivalent to (1.1) is that for every fixed k, the distribution of the random vector $n^{-1}(L_1^{(n)}, L_2^{(n)}, \ldots, L_k^{(n)})$ converges to that of $(L_1, L_2, \ldots, L_k)$.

The convergence in (1.1) is, by definition, based on the product topology for $\mathbb{R}^\infty$, and states that for every bounded continuous $f : \mathbb{R}^\infty \to \mathbb{R}$, we have

$$\lim_{n \to \infty} \mathbb{E} f\left(n^{-1}\left(L_1^{(n)}, L_2^{(n)}, \ldots\right)\right) = \mathbb{E} f((L_1, L_2, \ldots)).$$

Both processes $n^{-1}(L_1^{(n)}, L_2^{(n)}, \ldots)$ and $n^{-1}(L_1^{(n)}, L_2^{(n)}, \ldots)$ take values in the simplex $\Delta := \{(x_1, x_2, \ldots) : x_1 + x_2 + \cdots = 1 \text{ and } \forall i, x_i \geqslant 0\}$, and the ℓ_1 distance is a bounded metric on Δ which induces the same topology as the restriction of the product topology to $\Delta \subset \mathbb{R}^\infty$. It then follows from Strassen's theorem, together with the the Kantorovich–Rubinstein theorem (see Chapter 11 of Dudley [10],) that the distributional convergence claimed in (1.1) is *equivalent* to the following statement about couplings: 'The processes $(L_1^{(n)}, L_2^{(n)}, \ldots)$ and $(L_1, L_2, \ldots)$ can be coupled, that is, simultaneously constructed on a single probability space, so that

$$\mathbb{E} \sum_{i \geqslant 1} |L_i^{(n)}/n - L_i| \to 0 \tag{1.2}$$

as $n \to \infty$.'

In this paper, we investigate how rapidly the left side of (1.2) can converge to zero. Rather than divide the cycle lengths by n, we prefer to multiply the Poisson–Dirichlet components by n, so the quantity of interest is n times the expected ℓ_1 distance involved in (1.2). Thus, for any given coupling we consider

$$d_{\text{PD}}(n) := \mathbb{E} \sum_{i \geqslant 1} |L_i^{(n)} - n L_i| \tag{1.3}$$

$$= n \mathbb{E} \left\| n^{-1} \left(L_1^{(n)}, L_2^{(n)}, \ldots\right) - (L_1, L_2, \ldots) \right\|_1.$$

Any coupling in which $d_{\text{PD}}(n) = o(n)$ as $n \to \infty$ serves as a proof of the distributional convergence in (1.1). We give three different couplings that do so, and hence three different proofs of (1.1). The first coupling is the 'natural' coupling involved in the usual proof of (1.1), and what is new is our analysis of the expected ℓ_1 distance for this coupling. Our second and third couplings are novel, and thus provide two new proofs of (1.1). We consider the convergence in (1.1) to be so fundamental that *every* explicit coupling that achieves (1.2) would be interesting, and this, rather than a quest to minimize the distance $d_{\text{PD}}(n)$, is our primary motivation. In particular, fourth and further couplings would be interesting, even if they offer no improvement on the metric bound.

We give a lower bound in Theorem 2.2 stating that for *any* coupling,

$$\liminf_{n \to \infty} (\log n)^{-1} d_{\text{PD}}(n) \geqslant 1/4.$$

In Section 5, we analyse the obvious natural coupling, based on the canonical cycle notation for a permutation, to show that it achieves

$$\limsup_{n\to\infty} (\log n)^{-1} d_{\mathrm{PD}}(n) \leqslant \mathbb{E}|Q|,$$

where Q is a random variable defined in (5.12). Simulation shows that $\mathbb{E}|Q| \doteq 0.39357 \pm 0.00001$. In Section 6 we consider a second coupling, which achieves

$$\limsup_{n\to\infty} (\log n)^{-1} d_{\mathrm{PD}}(n) \leqslant 1/3.$$

(We believe, but do not attempt to prove, that this coupling has $\lim_{n\to\infty}(\log n)^{-1} d_{\mathrm{PD}}(n) = 1/3$.) In Section 8 we consider a third coupling, which achieves

$$\limsup_{n\to\infty} (\log n)^{-1} d_{\mathrm{PD}}(n) \leqslant 1/4,$$

and hence by comparison with the lower bound, this third coupling satisfies

$$\lim_{n\to\infty} (\log n)^{-1} d_{\mathrm{PD}}(n) = 1/4. \tag{1.4}$$

In Section 9 we give the analogous result for θ-biased permutations.

1.1. Age-ordered cycles

The age-ordering of the cycles of a permutation is described in detail and from first principles in Section 4.1. For an introductory overview, it suffices to say that starting from the list of cycle lengths $(L_1^{(n)}, L_2^{(n)}, \ldots)$, one can apply auxiliary randomization (a size-biased permutation) to re-order the list to get an 'age-ordered' list $(A_1^{(n)}, A_2^{(n)}, \ldots)$, with $L_i^{(n)} = i$th-largest of $A_1^{(n)}, A_2^{(n)}, \ldots$. Furthermore, this age-ordered list, rescaled by n, has a distributional limit $(A_1, A_2, \ldots)$, called the GEM process, which has a remarkably simple structure.

In each of our three couplings, we simultaneously construct the age-ordered cycle lengths $(A_1^{(n)}, A_2^{(n)}, \ldots)$ and the scaled GEM $(nA_1, nA_2, \ldots)$, aiming to make them close to one another. Then we construct the Poisson–Dirichlet $(L_1, L_2, \ldots)$ from the GEM $(A_1, A_2, \ldots)$ by applying RANK, that is, we take L_i to be the ith-largest of $A_1, A_2, \ldots$. Lemma 3.2 below shows that RANK can only reduce the l_1-distance, so that

$$\sum_{i\geqslant 1} |L_i^{(n)} - nL_i| \leqslant \sum_{i\geqslant 1} |A_i^{(n)} - nA_i|, \tag{1.5}$$

and taking expectations then shows that

$$d_{\mathrm{PD}}(n) \leqslant d_{\mathrm{GEM}}(n) := \mathbb{E}\left\{\sum_{i\leqslant 1} |A_i^{(n)} - nA_i|\right\}. \tag{1.6}$$

The second and third couplings are rather different from the first. For coupling 1, we show that

$$\lim_{n\to\infty} (\log n)^{-1} d_{\mathrm{GEM}}(n) = \mathbb{E}|Q|,$$

so that, by Lemma 3.2, $\limsup_{n\to\infty}(\log n)^{-1} d_{\mathrm{PD}}(n) \leqslant \mathbb{E}|Q|$. In contrast, for coupling 2, we show that $\liminf_{n\to\infty}(\log n)^{-1} d_{\mathrm{PD}}(n) \leqslant 1/3$, but do not get an estimate for $d_{\mathrm{GEM}}(n)$;

rather, we show that, for some permutation σ of $\mathbb{N}$, $\mathbb{E}\sum_{i\geqslant 1}|A_i^{(n)} - nA_{\sigma(i)}|$ is small, and then apply Lemma 3.2. We use a similar strategy with coupling 3, matching a rearrangement of the GEM random variables, and then applying Lemma 3.2 to deduce an upper bound on $d_{\mathrm{PD}}(n)$. This gives $\limsup_{n\to\infty}(\log n)^{-1} d_{\mathrm{PD}}(n) \leqslant 1/4$, and comparison with the lower bound from Theorem 2.1 then yields (1.4).

We believe that there is an essential difference between $d_{\mathrm{PD}}(n)$ and $d_{\mathrm{GEM}}(n)$. The idea is that if one starts with a close coupling of $n^{-1}(L_1^{(n)}, L_2^{(n)}, \dots)$ with $(L_1, L_2, \dots)$, and applies a size-biased permutation to each process, then, since the sizes are slightly different, it is *not* possible to use exactly the same permutation for size-biasing. The following conjecture uses a constant strictly greater than $1/4$, to assert that there is an essential difference between $d_{\mathrm{PD}}(n)$ and $d_{\mathrm{GEM}}(n)$; it uses the particular constant $\mathbb{E}|Q|$ for the simple reason that our first coupling achieves this constant.

Conjecture ($100). Any coupling of the age-ordered cycle lengths with the GEM has $\liminf_{n\to\infty}(\log n)^{-1} d_{\mathrm{GEM}}(n) \geqslant \mathbb{E}|Q|$, where Q is specified by (5.12).

1.2. Comparison with prime factorization of an integer

Consider a random positive integer chosen uniformly from 1 to n, and let $P_i(n)$ be the size of its ith-largest prime factor, with $P_i(n) = 1$ if i is greater than the number of factors of the random integer. Billingsley [7] proved in 1972 that

$$(\log n)^{-1}(\log P_1(n), \log P_2(n), \dots) \Rightarrow (L_1, L_2, \dots),$$

with the same Poisson–Dirichlet limit as in (1.1).

As in (1.3), one can consider the ℓ_1 distance to the limit, but this time scaling up by $\log n$ rather than by n. Arratia [2] gave a coupling, similar to our third coupling for permutations, which has

$$\mathbb{E}\sum |\log P_i(n) - (\log n)L_i| = O(\log \log n). \tag{1.7}$$

A conjecture from [2], that $O(1)$ can be achieved in place of $O(\log \log n)$, remains open, with a $100 prize offered for its resolution. Our third coupling for permutations, as well as the coupling in [2], relies on the 'scale-invariant spacing lemma' for the scale-invariant Poisson processes, proved in Section 7 in this paper.

2. A lower bound for all couplings

A lower bound asymptotic to $(1/4)\log n$ for $\mathbb{E}\sum_{i\geqslant 1}|L_i^{(n)} - nL_i|$ can be derived from the fact that the intensity $\mu(a,b)$, defined by

$$\mu(a,b) = \mathbb{E}\sum_{i\geqslant 1} \mathbb{1}(L_i \in (a,b)) \quad \text{for } 0 < a < b \leqslant 1,$$

is given by the explicit expression $\mu(a,b) = \int_a^b x^{-1} dx$. An alternate statement of this property of the Poisson–Dirichlet process, from [12] and [17], is

$$\mathbb{E}\sum_{j\geqslant 1} \phi(L_j) = \int_0^1 \phi(x)\frac{dx}{x}, \tag{2.1}$$

for any function ϕ that makes the integral absolutely convergent. One can also deduce (2.1) directly from the distributional convergence in (1.1), together with the fact that $\mathbb{E}C_i^{(n)} = 1/i$ for $1 \leqslant i \leqslant n$, where $C_i^{(n)}$ is the number of cycles of length i.

Theorem 2.1. *Let $L_i^{(n)}$ denote the size of the ith-largest cycle of a random permutation, and let L_i be the ith coordinate of the Poisson–Dirichlet process with parameter 1. Uniformly over all couplings of these two processes,*

$$\liminf_{n\to\infty} (\log n)^{-1} \mathbb{E} \sum_{i \geqslant 1} |L_i^{(n)} - nL_i| \geqslant \frac{1}{4}. \tag{2.2}$$

Proof. The factor $1/4$ arises as the long term average of the sawtooth function $d(x, \mathbb{Z}) = |x - \lfloor x + .5 \rfloor|$. Since each $L_i^{(n)} \in \mathbb{Z}$, any coupling has

$$\mathbb{E} \sum_{i \geqslant 1} |L_i^{(n)} - nL_i| \geqslant \mathbb{E} \sum_{i \geqslant 1} d(nL_i, \mathbb{Z}).$$

Using (2.1), we see that

$$\mathbb{E} \sum_{i \geqslant 1} d(nL_i, \mathbb{Z}) = \int_0^1 d(nx, \mathbb{Z}) x^{-1} dx = \int_0^n d(x, \mathbb{Z}) x^{-1} dx \sim \frac{1}{4} \log n.$$

The asymptotic in the last step comes from the fact that $d(x, \mathbb{Z}) = x$ if $0 < x \leqslant 1/2$, so that $\int_0^{1/2} d(x, \mathbb{Z}) x^{-1} dx = 1/2$, while for $k = 1, 2, \ldots,$

$$d(x, \mathbb{Z}) = \begin{cases} k - x, & \text{if } x \in [k - 1/2, k], \\ x - k, & \text{if } x \in [k, k + 1/2], \end{cases}$$

so that the contribution from 'one sawtooth' is

$$\int_{k-.5}^{k+.5} \frac{d(x, \mathbb{Z})}{x} dx = \int_{k-.5}^{k} \frac{k-x}{x} dx + \int_{k}^{k+.5} \frac{x-k}{x} dx$$

$$= k \left(\int_{k-.5}^{k} \frac{dx}{x} - \int_{k}^{k+.5} \frac{dx}{x} \right)$$

$$= k \log(k^2/(k^2 - 1/4)) = -k \log(1 - 1/(4k^2)).$$

Hence

$$\int_0^1 d(nx, \mathbb{Z}) x^{-1} dx = \frac{1}{2} + \sum_{k=1}^{n-1} -k \log\left(1 - \frac{1}{4k^2}\right) + \int_{n-.5}^n \frac{n-x}{x} dx, \tag{2.3}$$

and, since $-k \log(1 - \frac{1}{4k^2}) \sim 1/(4k)$, the sum, as well as the entire right side of (2.3), is asymptotic to $(1/4) \log n$. □

3. Lemmas for sorting and ℓ_1 distances

There is a well-known 'rearrangement inequality' stating that, for any nonincreasing sequences $(l_i, 1 \leqslant i \leqslant k)$ and $(m_i, 1 \leqslant i \leqslant k)$ and permutations ρ and σ, $\sum_1^k l_i m_i \geqslant \sum_1^k l_{\rho(i)} m_{\sigma(i)}$. Lemma 3.1 treats differences rather than products, and we provide a simple direct proof, very similar to the proof of the rearrangement inequality. Lemma 3.2, an extension of Lemma 3.1, implies that the RANK function is a contraction for the ℓ_1 distance.

Lemma 3.1 may be viewed as a special case of the result (1.1.14) in [20], or Exercises 11.8.1–2 in [10]. Recall that for any random variables X, Y, with cumulative distribution functions F, G, the Wasserstein distance, $\inf_{\text{couplings}} \mathbb{E}|X - Y|$, equals $\int_{-\infty}^{\infty} |F(t) - G(t)| \, dt$, and this infimum is achieved by constructing both X and Y using their quantiles applied to a single uniform [0,1] random variable. Dudley [10, p. 342] traces the history back to Gini in 1914 and Dall'Aglio [8]. The special case of Lemma 3.1 is that of empirical distributions: X is chosen uniformly from the multiset $\{l_1, \ldots, l_k\}$ and Y is chosen uniformly from the multiset $\{m_1, \ldots, m_k\}$. But in a sense this is circular, since proofs of the result on $\inf_{\text{couplings}} \mathbb{E}|X - Y|$ start with the discrete case, and then take limits to get the general case.

Lemma 3.1. *If $l_1 \geq l_2 \geq \cdots \geq l_k$ and $m_1 \geq m_2 \geq \cdots \geq m_k$, then for any permutations ρ, σ on $\{1, 2, \ldots, k\}$,*

$$\sum_{1}^{k} |l_i - m_i| \leq \sum_{1}^{k} |l_{\rho(i)} - m_{\sigma(i)}|. \tag{3.1}$$

Proof. Without loss of generality, the permutation σ can be taken to be the identity with $\sigma(i) = i$ for $i = 1$ to k. Thus our goal is to show that the 'score' $s(\rho) := \sum_1^k |l_{\rho(i)} - m_i|$ is minimized by taking ρ to be the identity permutation.

Any time there is a reversal, i.e., $i < j$ with $l_{\rho(i)} > l_{\rho(j)}$, we can apply the transposition $(i \ j)$ without increasing the score. To check this, write $a = m_i, A = m_j, b = l_{\rho(j)}, B = l_{\rho(i)}$, so that $a \leq A$ and $b \leq B$. We have $s(\rho) - s(\rho \circ (i \ j)) = |a - B| + |A - b| - (|a - b| + |A - B|) \geq 0$, which can be verified by considering cases such as $a < A < b < B$, which yields zero, $a < b < A < B$, which yields $2|A - b|$, or $a < b < B < A$, which yields $2(B - b)$. An arbitrary permutation ρ can be transformed into the identity in a finite number of such steps. □

Lemma 3.2. *If $l_1 \geq l_2 \geq \cdots$ and $m_1 \geq m_2 \geq \ldots$ then, for any permutations ρ, σ on $\mathbb{N}$,*

$$\sum_{i \geq 1} |l_i - m_i| \leq \sum_{i \geq 1} |l_{\rho(i)} - m_{\sigma(i)}|.$$

Hence the map RANK is a contraction on ℓ_1.

Proof. As in the previous lemma, without loss of generality σ can be taken to be the identity permutation, so our goal is to show, for an arbitrary permutation of $\mathbb{N}$, that $\sum_{i \geq 1} |l_i - m_i| \leq \sum_{i \geq 1} |l_{\rho(i)} - m_i|$. This in turn is equivalent to showing that, for arbitrary $k \geq 1$,

$$\sum_{i=1}^{k} |l_i - m_i| \leq \sum_{i \geq 1} |l_{\rho(i)} - m_i|. \tag{3.2}$$

Fix k. The set of ordered pairs $\{(\rho(1), 1), \ldots, (\rho(k), k), (1, \rho^{-1}(1)), \ldots, (k, \rho^{-1}(k))\}$ is a matching between two sets, say A and B, with

$$k \leq n = |A| = |B| \leq 2k.$$

Note that $\{1, 2, \ldots, k\} \subset A \cap B$. Label the elements of these sets so that $A = \{i_1, i_2, \ldots, i_n\}$ and $B = \{j_1, j_2, \ldots, j_n\}$ with $i_1 < i_2 < \cdots < i_n$ and $j_1 < j_2 < \cdots < j_n$ (and hence $1 = i_1 = j_1, \ldots, k = i_k = j_k$). The matching of A and B can be expressed as $\{(1, \tau(1)), \ldots, (n, \tau(n))\}$ for a permutation τ of $\{1, \ldots, n\}$. We have, writing for example $i(a) \equiv i_a$,

$$\sum_{a=1}^{k} |l_a - m_a| = \sum_{a=1}^{k} |l_{i(a)} - m_{j(a)}| \leq \sum_{a=1}^{n} |l_{i(a)} - m_{j(a)}|$$

$$\leq \sum_{a=1}^{n} |l_a - m_{\tau(a)}| \leq \sum_{i \geq 1} |l_{\rho(i)} - m_i|.$$

In the above, the middle inequality is justified by Lemma 3.1 applied to the permutation τ, while the first and third inequalities are simply by inserting additional nonnegative terms. □

4. Canonical cycle notation and the GEM

The easy path to understanding the structure of $(L_1^{(n)}, L_2^{(n)}, \ldots)$, which are the cycle lengths taken in decreasing order, is to consider first a more elaborate construction: the cycle lengths in the order produced by writing out the canonical cycle notation for a permutation on $[n] := \{1, 2, \ldots, n\}$.

4.1. Age order of the cycles

A permutation $\rho \in \mathscr{S}_n$ can be written as an (ordered) product of cycles in the following way: start the first cycle with the integer 1, followed by its image $\rho(1)$, the image of $\rho(1)$ and so on. Once this cycle is completed, the second cycle starts with the smallest unused integer followed by its images, and so on. For example, the permutation $\rho \in \mathscr{S}_{10}$ given by

$$\rho = \begin{pmatrix} 1 & 2 & 3 & 4 & 5 & 6 & 7 & 8 & 9 & 10 \\ 9 & 2 & 7 & 3 & 6 & 4 & 5 & 8 & 10 & 1 \end{pmatrix} \quad (4.1)$$

is decomposed as

$$\rho = (1\ 9\ 10)(2)(3\ 7\ 5\ 6\ 4)(8), \quad (4.2)$$

a permutation with two singleton cycles (or fixed points), one cycle of length 3, and one of length 5. This example with $n = 10$ has cycle lengths in *decreasing* order given by $(L_1^{(n)}, L_2^{(n)}, \ldots) = (5, 3, 1, 1, 0, 0, \ldots)$, while the cycle lengths in canonical, or *age* order are $(A_1^{(n)}, A_2^{(n)}, \ldots) = (3, 1, 5, 1, 0, 0, \ldots)$.

The distribution now known as the GEM arose in population biology, in [14, 11, 19]. Just as the Poisson–Dirichlet distribution (with parameter 1) may be characterized as the limit occurring in (1.1), the GEM distribution with parameter 1 may be characterized as the distributional limit for the proportions of elements in the cycles taken in age order:

$$n^{-1}\left(A_1^{(n)}, A_2^{(n)}, \ldots\right) \Rightarrow (A_1, A_2, \ldots). \tag{4.3}$$

Conditional on the values $(L_i^{(n)}, i \geq 1)$, the joint distribution of $(A_i^{(n)}, i \geq 1)$ is obtained by size-biasing. The size-biased order derived from any finite collection $\{l_1, \ldots, l_m\} \subset \mathbb{R}_+$ with sum L is the random permutation σ of $\{1, 2, \ldots, m\}$ obtained from the relations

$$\mathbb{P}[\sigma(1) = j] = l_j/L;$$

$$\mathbb{P}[\sigma(r+1) = j \mid \sigma(1) = j_1, \ldots, \sigma(r) = j_r] = l_j \left\{L - \sum_{i=1}^r l_{j_i}\right\}^{-1},$$

$0 \leq r \leq m - 1$. For an infinite collection $\{l_1, l_2, \ldots\}$ with sum $L < \infty$, the same construction can be used to define the size-biased order σ of $\mathbb{N}$, but if $L = \infty$, the algorithm fails. However, the size-biased order can be more generally defined as a random ordering of $\{l_j, j \geq 1\}$ with the property that any *finite* collection of them are in their size-biased order. Such a random order can be realized by the following simple construction. Assign to each l_j the value $\lambda_j = l_j^{-1} S_j$, where $(S_j, j \geq 1)$ are independent standard exponential random variables, and arrange the l_j in increasing order of λ_j. If the sets $\{l_j, j \geq 1\} \cap (M^{-1}, M)$ are finite for each $M \in \mathbb{N}$, the l_j can be indexed by $i \in \mathbb{Z}$ in increasing order of λ_j. The size-biased ordering of a random sequence $\{L_j, j \geq 1\}$ is obtained in the same way, by first sampling the multiset consisting of its values, and then using the independent S_js to determine their order.

4.2. Coupling to independent discrete random variables

The probabilistic structure found in writing the canonical cycle notation for a randomly chosen permutation involves independent Bernoulli $(1/i)$ random variables, for $i = 1, 2, \ldots$. We call this structure the *Feller coupling*, and trace it back to Feller (1945) and Rényi (1962); see also Arratia, Barbour and Tavaré (1992). In writing the canonical cycle notation for a random $\rho \in \mathscr{S}_n$, say with $n = 10$, one always starts with '(1', and then makes a ten-way choice, between '(1)(2', '(1 2',..., and '(1 10'. One continues with a nine-way choice, an eight-way choice,..., a two-way choice, and finally a one-way choice.

Say that D_i, chosen from 1 to i, is used to make the i-way choice – take item D_i from the i available choices listed in increasing order; the example (4.2) has $(D_{10}, D_9, \ldots, D_2, D_1) = (9, 9, 1, 1, 5, 3, 3, 2, 1, 1)$. Clearly, the map constructing canonical cycle notation,

$$(D_1, D_2, \ldots, D_n) \mapsto \rho,$$

from $[1] \times [2] \times \cdots \times [n]$ to $\mathscr{S}_n$, is one-to-one, and hence a bijection. This is the natural coupling; starting with $D_1, D_2, D_3, \ldots$ independent, and with no further randomization, we produce random permutations, simultaneously for $n = 1, 2, \ldots$, distributed uniformly over $\mathscr{S}_n$.

Define ξ_i to be the indicator function

$$\xi_i = \mathbb{1}\{D_i = 1\} = \mathbb{1}\{\text{end cycle when there is an } i\text{-way choice}\}. \tag{4.4}$$

Thus
$$\mathbb{P}[\xi_i = 1] = \frac{1}{i}, \qquad \mathbb{P}[\xi_i = 0] = 1 - \frac{1}{i}, \qquad i \geq 1,$$
and
$$\xi_1, \xi_2, \xi_3, \ldots, \quad \text{are independent.}$$

The sequence $\xi_1 \xi_2 \xi_3 \ldots \xi_n$ read from ξ_n down to ξ_1 determines the list of cycle lengths in age order. The length $A_1^{(n)}$ of the first cycle is the waiting time to the first of $\xi_n, \xi_{n-1}, \ldots$ to take the value 1, the length $A_2^{(n)}$ of the next cycle is the waiting time to the next 1, and so on.

We may view this in terms of 'an artificial 1 in position $n+1$', and consider the sequence $\xi_1 \xi_2 \xi_3 \ldots \xi_n 1$ of length $n+1$ that begins and ends with 1. Every i-spacing in $1\xi_2 \xi_3 \ldots \xi_n 1$, that is, every pattern $10^{i-1}1$ of two ones separated by $i-1$ zeros, corresponds to a cycle of length i. The size of the rightmost spacing in $1\xi_2\xi_3\ldots\xi_n 1$ gives the size of the first cycle in canonical cycle notation. The example (4.2) with $n = 10$ has $(D_1, \ldots, D_n) = (1, 1, 2, 3, 3, 5, 1, 1, 9, 9)$, so $\xi_1 \xi_2 \xi_3 \ldots \xi_n 1 = 11000011001$. We read the spacings from right to left to see that the cycle lengths in age order are $(A_1^{(n)}, A_2^{(n)}, \ldots, A_4^{(n)}) = (3, 1, 5, 1)$, with $(A_1^{(n)}, A_2^{(n)}, \ldots) = (3, 1, 5, 1, 0, 0, \ldots)$.

The single set of realizations $(D_i, i \geq 1)$ enables one to generate a coupled set of realizations of uniform random permutations $\sigma(n)$ for each $n \geq 1$, for which, for each fixed j, the random variables

$$C_j^{(n)} := \sum_{r=1}^{n-j-1} \mathbb{1}\{\xi_r = 1, \xi_{r+1} = \cdots = \xi_{r+j-1} = 0, \xi_{r+j} = 1\}$$
$$+ \mathbb{1}\{\xi_{n-j} = 1, \xi_{n-j+1} = \cdots = \xi_n = 0\},$$

the numbers of cycles of lengths j in $\sigma(n)$, converge a.s. as $n \to \infty$ to random variables $C_j^{(\infty)}$. The joint distribution of $\{C_j^{(\infty)}, j \geq 1\}$ is that of independent Poisson random variables with means $1/j$: see Arratia and Tavaré [6] and Diaconis and Pitman [9].

Now let $M_0 = 1$ and $M_i = \min\{j > m_{i-1}, \xi_j = 1\}$, $i \geq 1$; set $M_i^{(n)} = M_i \wedge n$. Then the differences $B_i^{(n)} = M_i^{(n)} - M_{i-1}^{(n)}$, $i \geq 1$, form the cycle lengths of $\sigma(n)$ in reverse age (size-biased) order, followed by an infinite string of zeros, and we also have $B_i^{(n)} \to B_i := M_i - M_{i-1}$ a.s. for each i. A random vector with the same distribution as $(B_i^{(n)}, i \geq 1)$ can equally be realized by sampling $(C_j^{(n)}, 1 \leq j \leq n)$, and then arranging the $C_j^{(n)}$ elements of size j, $1 \leq j \leq n$, together in reversed size-biased order, again filling out with zeros. Now define the events

$$E_{1m}^{(n)} := \{C_j^{(n^2)} = C_j^{(\infty)}, 1 \leq j \leq m\}, \qquad E_{2m}^{(n)} := \{B_j^{(n^2)} = B_j, 1 \leq j \leq n\}$$

and

$$E_{3m}^{(n)} := \left\{\min_{j>n} B_j^{(n^2)} \leq m\right\}.$$

Then simple calculation shows that $\lim_{n \to \infty} \mathbb{P}[E_{lm}^{(n)}] = 0$ for each fixed m, $l = 1, 2, 3$. However, all elements of the sequence $(B_i, i \geq 1)$ of size at most m are realized in the correct order on the set $E^{(n)}_{1m} \cap E^{(n)}_{2m} \cap E^{(n)}_{3m}$, by taking $(C^{(n^2)}_j, 1 \leq j \leq n^2)$, performing the size-biased ordering to get $(B^{(n^2)}_j, 1 \leq j \leq n)$, and then taking the elements of size at most m in $(B^{(n^2)}_j, 1 \leq j \leq n)$; since, on this set, we could equally well have begun with $(C^{(\infty)}_j, 1 \leq j \leq m)$ in place of $(C^{(n^2)}_j, 1 \leq j \leq m)$, it follows that the infinite sequence $(B_i, i \geq 1)$ is itself in reversed size-biased order. Hence, to construct the age-ordered distributions, we just need to sample independent Poisson random variables with means $1/j, j \geq 1$, and reverse size-bias the outcome, to obtain the sequence $(B_i, i \geq 1)$, and from it to deduce the $(\xi_j, j \geq 1)$.

5. The first coupling: using canonical cycle notation

A simple calculation shows that $A^{(n)}_1$ is uniformly distributed over $\{1, 2, \ldots, n\}$ – it might be constructed from a uniform $[0,1]$ random variable U using the ceiling function, $A^{(n)}_1 = \lceil nU \rceil$. Similarly, $A^{(n)}_2$ is uniform on 1 to $n - A^{(n)}_1$, meaning that conditional on $n - A^{(n)}_1 = m$, if $m > 0$ then $A^{(n)}_2$ is uniformly distributed over $\{1, 2, \ldots, m\}$, and otherwise $A^{(n)}_2$ is zero. We could take $A^{(n)}_2 = \lceil (n - A^{(n)}_1)U' \rceil$ with U, U' i.i.d. Likewise $A^{(n)}_3$ is uniform on 1 to $n - A^{(n)}_1 - A^{(n)}_2$, and so on.

The easy way to view this is in terms of the number of elements *not* yet used after k cycles have been taken:

$$N_0 = n; \qquad N_k \equiv N^{(n)}_k := n - A^{(n)}_1 - \cdots - A^{(n)}_k \quad \text{for } k \geq 1. \tag{5.1}$$

Then, conditional on $N_0 = n, N_1 = n_1, \ldots, N_k = n_k$, if $n_k > 0$ then N_{k+1} is uniformly distributed over $\{0, 1, \ldots, n_k - 1\}$, while if $n_k = 0$ then $N_{k+1} = 0$. There is an obvious coupling which achieves this: take $U_1, U_2, \ldots$ to be i.i.d. uniform $[0,1]$ random variables and from them construct the process $A^{(n)}_1, A^{(n)}_2, \ldots$ via

$$N^{(n)}_0 = n, \qquad N^{(n)}_k := \lfloor N^{(n)}_{k-1} U_k \rfloor \quad \text{for } k \geq 1, \tag{5.2}$$

$$A^{(n)}_k := N^{(n)}_{k-1} - N^{(n)}_k \quad \text{for } k \geq 1. \tag{5.3}$$

This yields the process $(A^{(n)}_1, A^{(n)}_2, \ldots)$ distributed exactly as the process of cycle lengths in age order, for a random permutation.

Define

$$S_0 = 1, \qquad S_k := U_1 U_2 \cdots U_k, \qquad A_k := S_{k-1} - S_k, \quad \text{for } k \geq 1. \tag{5.4}$$

Since $\lfloor nx \rfloor / n \to x$ for all x, by induction on $k \geq 0$ we see that $N^{(n)}_k / n \to S_k$ as $n \to \infty$, and by differencing, that $A^{(n)}_k / n \to A_k$. This proves, using only elementary analysis, that

$$n^{-1}(A^{(n)}_1, A^{(n)}_2, \ldots) \to (A_1, A_2, \ldots) \tag{5.5}$$

under the coupling (5.3), for *all* realizations of the uniform random variables U_i; the weak convergence in (4.3) naturally follows.

With or without coupling, (5.4) is a simple description of the GEM(1) limit $(A_1, A_2, \ldots)$; it is due to Ignatov [15, 16] and others; see [5, p. 119] for further history.

5.1. Analysis of the expected ℓ_1 distance

The goal of this section is to explain and prove the following theorem, stating that the expected ℓ_1 distance for the coupling above is asymptotic to a certain constant times $\log n$ as $n \to \infty$.

Theorem 5.1. *For the process of cycle lengths in age order, of a permutation chosen uniformly at random from $\mathscr{S}_n$, in comparison with the GEM process (5.3), scaled up by a factor of n, the expected ℓ_1 distance (1.6), for the coupling given by (5.2)–(5.4), has the following asymptotic value as $n \to \infty$:*

$$d_{\text{GEM}}(n) \sim \mathbb{E}|Q| \log n, \qquad (5.6)$$

where the distribution of Q is given by (5.12).

To start the proof, let us write $t_k(n) = \mathbb{E}|T_k(n)|$ for the kth term of (1.6), the scaled-up expected ℓ_1 distance, for the coupling given by (5.2)–(5.4):

$$t_k(n) := \mathbb{E}|T_k(n)| \quad \text{where} \quad T_k(n) := nA_k - A_k^{(n)}. \qquad (5.7)$$

We use curly braces to denote the fractional part function, $\{x\} := x - \lfloor x \rfloor$, so that $\lfloor x \rfloor = x - \{x\}$.

Lemma 5.2. *For $k \geqslant 0$,*

$$N_k^{(n)} - nS_k = -\sum_{j=1}^{k} \{N_{j-1}^{(n)} U_j\} U_{j+1} \cdots U_k. \qquad (5.8)$$

Proof. For $k = 0$, both sides are zero. For $k \geqslant 1$,

$$\begin{aligned} N_k^{(n)} - nS_k &= \lfloor N_{k-1}^{(n)} U_k \rfloor - nS_{k-1} U_k \\ &= N_{k-1}^{(n)} U_k - \{N_{k-1}^{(n)} U_k\} - nS_{k-1} U_k \\ &= -\{N_{k-1}^{(n)} U_k\} + U_k \left(N_{k-1}^{(n)} - nS_{k-1} \right). \end{aligned} \qquad (5.9)$$

Using induction to expand $(N_{k-1}^{(n)} - nS_{k-1})$, we get (5.8). □

We use the notation $\overline{U} := 1 - U$ for the complement of a uniformly distributed variable. Starting from (5.7), and using (5.9) to get the third line below,

$$\begin{aligned} T_k(n) &= (nS_{k-1} - nS_k) - \left(N_{k-1}^{(n)} - N_k^{(n)} \right) \\ &= \left(N_k^{(n)} - nS_k \right) - \left(N_{k-1}^{(n)} - nS_{k-1} \right) \\ &= -\{N_{k-1}^{(n)} U_k\} - \left(N_{k-1}^{(n)} - nS_{k-1} \right) \overline{U}_k \\ &= -\{N_{k-1}^{(n)} U_k\} + \overline{U}_k \sum_{j=1}^{k-1} \{N_{k-j-1}^{(n)} U_{k-j}\} \prod_{l=k-j+1}^{k-1} U_l, \end{aligned} \qquad (5.10)$$

an empty product being taken to be 1.

Now let $W_1, W_2, \ldots$ and $V_1, V_2, \ldots$ be mutually independent uniform random variables, and consider the random variables

$$Q_k = -V_0 + \sum_{j=1}^{k-1} \left\{ V_j \prod_{l=1}^{j} W_l \right\} \qquad (5.11)$$

and

$$Q = -V_0 + \sum_{j \geq 1} \left\{ V_j \prod_{l=1}^{j} W_l \right\}. \qquad (5.12)$$

Since $\mathbb{E}|Q - Q_k| = 2^{-k}$ is summable, $Q_k \to Q$ almost surely and in L_1. Clearly $\mathbb{E}Q = 0$, but for $\mathbb{E}|Q|$ we have no closed form expression.

(**Heuristic.** Imagine k large, but n much larger, say larger even than e^k, so that nS_k is large. We claim that T_k is close in distribution to Q_k, and hence close to Q. Superficially, this does not appear to be so – in the fourth line of the display (5.10), there are k independent uniform variables, while in (5.11), there are $2k - 1$ independent uniforms. However, when N is large, $\{NU\}$ is approximately independent of U, in the sense that the low order bits of U give the approximate value of U, while higher order bits of U control the value of $\{NU\}$.)

To carry out this approximation, start with the i.i.d. uniform [0,1] variables $W_1, W_2, \ldots$ and $V_1, V_2, \ldots$. Note that Q_k from (5.11) is exactly equal in distribution to

$$Q'_k = -\overline{V}_k + \overline{W}_k \sum_{j=1}^{k-1} \left\{ V_{k-j} \prod_{l=k-j+1}^{k-1} W_l \right\}. \qquad (5.13)$$

Given $n \geq 1$, we *define*

$$N_0 := n; \qquad N_k := \lfloor N_{k-1} W_k \rfloor, \quad k \geq 1,$$

and

$$U_k := (N_k + V_k)/N_{k-1} \quad \text{if } N_{k-1} \geq 1; \qquad U_k := V_k \quad \text{if } N_{k-1} = 0.$$

It is then an elementary calculation to *check* that we have satisfied (5.2); *i.e.*, we check that $N_k = \lfloor N_{k-1} U_k \rfloor$, that U_k is uniform, and that $U_1, U_2, \ldots$ are independent. It is also immediate that

$$\{N_{k-1} U_k\} = V_k \quad \text{if } N_{k-1} \geq 1, \qquad (5.14)$$

and that

$$|U_k - W_k| \leq 1/\max(1, N_{k-1}). \qquad (5.15)$$

Combining this coupling with (5.10), we have

$$T_k = -\overline{V}_k + \overline{U}_k \sum_{j=1}^{k-1} \left\{ V_{k-j} \prod_{l=k-j+1}^{k-1} U_l \right\}, \qquad (5.16)$$

which matches formula (5.13) for Q'_k, except that Q'_k has the variables W which are independent of the Vs, while T_k has the variables U which are dependent on the Vs. We use (5.15) in Lemma 5.3 to show that $\mathbb{E}|T_k - Q'_k|$ is $O(n^{-\delta})$ small, uniformly in $k \leq (1-\varepsilon)\log n$. This, together with $\mathbb{E}|Q'_k| = \mathbb{E}|Q_k| \to \mathbb{E}|Q|$, implies that as $n \to \infty$,

$$\sum_{k \leq (1-\varepsilon)\log n} \mathbb{E}|T_k| \sim \sum_{k \leq (1-\varepsilon)\log n} \mathbb{E}|Q_k| \sim (1-\varepsilon)\mathbb{E}|Q|\log n.$$

Combining this with estimates for the remaining ranges, and arguing with fixed but arbitrarily small $\varepsilon > 0$, the final outcome is that the expected ℓ_1 distance given by (1.6), for this particular coupling, is asymptotic to $\mathbb{E}|Q|\log n$, as was stated formally in Theorem 5.1.

For the range $(1-\varepsilon)\log n < k \leq (1+\varepsilon)\log n$, it suffices to use (5.10) to see that for all k, n, $\mathbb{E}|T_k| < 2$; we have $2\varepsilon \log n$ terms in this range, so the net contribution, at most $4\varepsilon \log n$, is made small relative to $\log n$ by taking ε small.

For $k \geq k_0 := (1+\varepsilon)\log n$, large deviation theory easily shows that $\sum_{k \geq k_0} \mathbb{E}T_k \to 0$. Here are the details. Consider a 'good event' G_k, defined by

$$G_k = \left\{ \frac{\log S_k}{k} < (-1 + \varepsilon/2) \right\}.$$

As in the proof of Lemma 5.3, standard large deviation theory gives the upper bound $\mathbb{P}(G_k^c) \leq e^{-\delta k}$, where $\delta = \delta(\varepsilon) > 0$. Using (5.2) and (5.4), we have $0 \leq N_k^{(n)} \leq nS_k$ so that $|T_k| \leq nS_k$, and on the good event this gives $|T_k| \leq \exp(\log n + (-1 + \varepsilon/2)k)$. For $k = k_0$, this upper bound is no more than $n^{-\varepsilon/2 + \varepsilon^2/2}$, and goes down exponentially fast with increasing k, so the sum over all $k \geq k_0$ is comparable to the bound at k_0. For the contribution from the bad event, observe first that from (5.10), we have $|T_k| \leq k$ always, and $\sum_{k \geq k_0} k e^{-\delta k} = O(e^{-\delta k_0/2}) = O(n^{-\delta/2})$.

The proof of Theorem 5.1 is completed by the following lemma.

Lemma 5.3. *Fix $\varepsilon > 0$, and let $k_0 \equiv k_0(n) := \lfloor (1-\varepsilon)\log n \rfloor$. There exists a $\delta > 0$, so that uniformly in $1 \leq k \leq k_0$,*

$$\mathbb{E}|T_k - Q'_k| = O(n^{-\delta}). \tag{5.17}$$

Proof. Consider a 'good event' G_n, defined by

$$G_n = \left\{ \frac{\log S_{k_0}}{k_0} > -(1 + \varepsilon/3) \right\}.$$

Since $-\log S_k$ is the sum of k i.i.d. exponentially distributed random variables with mean 1, large deviation theory gives an upper bound of the form $\mathbb{P}(G_n^c) \leq e^{-I(\varepsilon/3)k_0} \leq n^{-\delta_1}$, where $\delta_1 = (1-\varepsilon)I(\varepsilon/3) > 0$.

For the contribution from the bad event, observe first that from (5.10) and (5.13) we have $|T_k - Q'_k| \leq 1 + k$ always, so with respect to this contribution (5.17) is satisfied.

Assume that the good event occurs. For $\varepsilon \in (0, 1)$, we have the bound $(1 + \varepsilon/3)(1-\varepsilon) < 1 - \varepsilon/2$, and hence $nS_{k_0} > n^{\varepsilon/2}$. Since N_k is obtained from nS_k using k applications of the floor function, we have $N_k > n^{\varepsilon/2} - k_0$ for every $k \leq k_0$. Finally, consider the difference of Q'_k from (5.13) and the coupling expression (5.16) for T_k. One converts Q'_k to T_k by replacing factors W_i by U_i, in expressions that are products of factors in $(0, 1)$, so that such a replacement causes a change not exceeding $|U_i - W_i|$ in absolute value, and by (5.15) this is at most $1/N_{k_0} < 1/(n^{\varepsilon/2} - k_0)$. There are $1 + 2 + \cdots + k - 1$ places to make such a replacement, and hence for every $k \leqslant k_0$, on the good event

$$|T_k - Q'_k| \leqslant k_0^2/N_{k_0} = O(n^{-\varepsilon/4}). \qquad \square$$

6. The second coupling: to exploit $\mathbb{E}|U - U'| = 1/3$

Our second coupling achieves

$$d_{\mathrm{PD}}(n) \sim \frac{1}{3} \log n,$$

with the constant $1/3$ arising as the expected value of the difference of two independent uniform $(0,1)$ random variables. To do this, we consider the 'scale-invariant' Poisson process $\mathscr{X}$ on $(0, \infty)$ with intensity dx/x, restricted to $(0, n)$, and for each $i = 2, 3, \ldots$, we *throw away* any points representing a second, third, or further arrival of $\mathscr{X}$ in an interval $(i - 1, i]$. The 'first arrival' in $(i - 1, i]$, if there is one, corresponds to $\xi_i = 1$ in the Feller coupling.

For large i, the 'first arrival' in $(i - 1, i]$ is approximately at $i - U$, where U is uniform, and if the Feller process has adjacent arrivals at $i < j$, corresponding to a cycle of length $j - i$, there is a matching spacing in $\mathscr{X}$ of length approximately $(j - U') - (i - U) = (j - i) + (U - U')$; with $\mathbb{E}|U - U'| = 1/3$, this accounts for a $(1/3) \log n$ contribution to $d_{\mathrm{PD}}(n)$. The act of restoring 'extra' arrivals in an interval $(i - 1, i]$ can be viewed as splitting a small piece, of length at most 1, from a spacing already accounted for as $(j - U') - (i - U)$, and matching these additional small pieces with zeros in the age-ordered cycle process – since $A_j^{(n)} = 0$ for j greater than the number of cycles of the random permutation. Then Lemma 3.2 is applied.

6.1. Details for this coupling

In the point process $\mathscr{X}$, for any $0 < a < b < n$, the number $\mathscr{X}(a, b]$ of points in the interval $(a, b]$ is Poisson with mean $\mathbb{E}\mathscr{X}(a, b] = \int_a^b dx/x = \log(b/a)$, and disjoint intervals have independent counts. If we label the points of $\mathscr{X}$ in decreasing order, so that

$$n > X_1 > X_2 > \cdots,$$

then the sequence $X_1, X_2, X_3 \ldots$ is distributed exactly as $nU_1, nU_1U_2, nU_1U_2U_3, \ldots$, which are the points in (5.4), scaled up by n. Thus the GEM, scaled up by n, is

$$(nA_1, nA_2, nA_3, \ldots) = (n - X_1, X_1 - X_2, X_2 - X_3, \ldots). \tag{6.1}$$

We extend of course by taking $L_i :=$ ith-largest of the $A_1, A_2, \ldots$.

Let

$$N_i = \mathscr{X}(i - 1, i], \quad i = 2, 3, \ldots \tag{6.2}$$

so that N_i is Poisson with mean

$$\lambda_i := \mathbb{E}N_i = \log(i/(i-1)) = 1/i + 1/(2i^2) + 1/(3i^3) + \cdots,$$

and $\mathbb{P}(N_i = 0) = 1 - (1/i)$. Taking $N_1 = \infty$, we define

$$\xi_i := \min(1, N_i), \quad i = 1, 2, 3, \ldots, \tag{6.3}$$

thereby coupling the scale-invariant Poisson process $\mathscr{X}$ with the Bernoulli process $\xi_1, \xi_2, \ldots$ of the Feller coupling (4.4). The paragraph following (4.4) gives the age-ordered cycle lengths $A_1^{(n)}, A_2^{(n)}, \ldots$ in terms of spacings between successive ones in $1\xi_n\xi_{n-1}\cdots\xi_3\xi_2 1$. In detail, write

$$K := \xi_1 + \xi_2 + \cdots + \xi_n$$

so that K represents the number of cycles in our random cycle structure. Define $M_0 \equiv n+1$ and for $j = 1$ to K, let M_j be the index of the jth 1 in $\xi_n \cdots \xi_2 \xi_1$, so that $M_K = 1$ always,

$$A_j^{(n)} = M_{j-1} - M_j, \quad j = 1 \text{ to } K, \tag{6.4}$$

and $A_j^{(n)} = 0$ for $j > K$. Of course we take $L_i^{(n)} := i\text{th-largest of } A_1^{(n)}, A_2^{(n)}, \ldots$. At this stage, we have defined our second coupling; it only remains to estimate $d_{\text{PD}}(n)$.

The Poisson process $\mathscr{X}$, restricted to $(i-1, i]$, can be constructed from the count N_i, together with an i.i.d. sequence $X_{i,1}, X_{i,2}, \ldots$ of locations in $(i-1, i]$ distributed with density $f_i(x) := 1/(x\lambda_i)$. This density can be expressed as a mixture of the uniform density on $(i-1, i]$, taken with weight $1/(i\lambda_i) = 1/(1 + 1/(2i) + 1/(3i^2) + \cdots$, and some other density on $(i-1, i]$, taken with weight $d(i) := 1 - 1/(i\lambda_i) = 1/(2i) + O(i^{-2})$. Thus we can take V_i uniform on $(0,1)$, so that

$$\mathbb{E}|(i - V_i) - X_{i,1}| < d(i), \tag{6.5}$$

for every $i \geq 2$. For the interval $(0,1]$, there is an infinite number of arrivals (but not in i.i.d. locations); we may label these in decreasing order as $X_{1,1} > X_{1,2} > \cdots > 0$. Note that $X_{1,1}$ is uniformly distributed in $(0,1)$, so we may define $d(1) = 0$ and consider (6.5) to also hold for $i = 1$. Finally, note that the $\xi_1, \xi_2, \ldots$ from (6.3) and the $V_1, V_2, \ldots$ from (6.5) can be taken to be mutually independent.

The Poisson process $\mathscr{X}$ restricted to $(0, n]$ is realized as the random set of points

$$\{X_{i,j} : 1 \leq i \leq n, \ 1 \leq j \leq N_i\}. \tag{6.6}$$

We get a subset $\hat{\mathscr{X}}$ of $\mathscr{X}$ by taking only those $X_{i,j}$ with $N_i > 0$ and $j = 1$, so that the cardinality of $\hat{S}$ is K. Label these points as $\hat{X}_j$ with $n \geq \hat{X}_1 > \hat{X}_2 > \cdots > \hat{X}_K$, with $\hat{X}_K \in (0, 1]$. For $j = 1$ to K, let M_j be the index of the jth 1 in $\xi_n \cdots \xi_2 \xi_1$, so that $\hat{X}_j = X_{M_j, 1} \in (M_j - 1, M_j]$ and $M_K = 1$ always. If the spacings construction (6.1) were applied to $\hat{\mathscr{X}}$ in place of $\mathscr{X}$, the ℓ_1 distance to the age-ordered list of cycle lengths, that is, the ℓ_1 distance between $(n - \hat{X}_1, \hat{X}_1 - \hat{X}_2, \ldots, \hat{X}_{K-1} - \hat{X}_K, 0, 0, \ldots)$ and $(A_1^{(n)}, A_2^{(n)}, \ldots, A_K^{(n)}, 0, 0, \ldots)$, would be

$$D = |(n - \hat{X}_1) - (n + 1 - M_1)| + \sum_{j=2}^{K} |(\hat{X}_{j-1} - \hat{X}_j) - (M_{j-1} - M_j)|. \tag{6.7}$$

Looking back at (6.5), we write $U_j := V_{M_j}$ so that

$$|\hat{X}_j - (M_j - U_j)| \leq d(M_j) \tag{6.8}$$

and hence

$$\left| D - \sum_{j=2}^{K} |U_{j-1} - U_j| \right| \leq 2 + 2\sum_{j=2}^{K} d(M_j). \tag{6.9}$$

Since $\mathbb{E}K = 1 + 1/2 + \cdots + 1/n \sim \log n$, and $\lim_{i \to \infty} d(i) = 0$, and always $M_j \geq j$, it follows first that the expectation of the left side of (6.9) is $o(\log n)$; then since $\mathbb{E}\sum_{2}^{K} |U_{j-1} - U_j| \sim \mathbb{E}|U - U'|\log n = (1/3)\log n$, it follows that $\mathbb{E}D \sim (1/3)\log n$.

What is the effect of replacing $\mathscr{X}$ with $\hat{\mathscr{X}}$ in the spacings construction? There is an easy bound, provided that we consider the spacings arranged from large to small. Namely, the ℓ_1 distance between $\text{RANK}(n - \hat{X}_1, \hat{X}_1 - \hat{X}_2, \ldots)$ and $n(L_1, L_2, \ldots) = \text{RANK}(n - X_1, X_1 - X_2, \ldots)$ is at most

$$2\sum_{j=2}^{n}(N_i - 1)^+ + 2. \tag{6.10}$$

(For the first sum above: any second, third, or further arrival in $(i-1, i]$ serves to split off a piece of a spacing involving the first arrival, but the split off piece has length at most 1, so the effect on the ℓ_1 distance of the ranked lists of deleting one of these extra arrivals is at most $1 + 1$, using Lemma 3.2. The second term in (6.10), 2, comes from splitting the interval $(0, \hat{X}_K)$ into an infinite number of subpieces.) Since $\mathbb{E}N_i \sim 1/i$, the expectation of the positive part of $(N_i - 1)$ is $O(1/i^2)$, and the expected ℓ_1 distance in (6.10) is $O(1)$ as $n \to \infty$.

Now comes a subtle point: the rearrangement inequality in Lemma 3.2 implies that the ℓ_1 distance between $\text{RANK}(n - \hat{X}_1, \hat{X}_1 - \hat{X}_2, \ldots)$ and $(L_1^{(n)}, L_2^{(n)}, \ldots) = \text{RANK}(A_1^{(n)}, A_2^{(n)}, \ldots)$ is *at most* D, and hence has mean asymptotically *at most* $(1/3)\log n$. We believe that there exists a matching lower bound, asymptotic to $(1/3)\log n$, but in view of the third coupling, it does not seem worth pursuing this.

The net result of these arguments is that we have proved

Theorem 6.1. *The second coupling, given by* (6.1)–(6.4), *achieves*

$$\limsup_{n \to \infty} (\log n)^{-1} d_{\text{PD}}(n) \leq 1/3.$$

7. The scale-invariant spacings lemma

The following 'scale-invariant spacings lemma' was first presented in [1], an unpublished manuscript, and was used in a coupling for prime factors with the Poisson–Dirichlet process in [2]; see also [4]. It will be used in our third coupling, in Section 8.

Start with the Poisson process $\mathscr{P}$ with intensity $\theta/x\,dx$. Consider the process $\mathscr{Y}$ with a point for each spacing in $\mathscr{P}$. To be precise, the points of $\mathscr{P}$ can with probability one be labelled $X_i \in (0, \infty)$ for $i \in \mathbb{Z}$ so that

$$\cdots < X_2 < X_1 < 1 < X_0 < X_{-1} < \cdots, \tag{7.1}$$

Three Couplings for Cycle Structure 67

with $\lim_{i \to \infty} X_i = 0, \lim_{i \to -\infty} X_i = \infty$. The spacings are the points $Y_i := X_i - X_{i+1}$ for $-\infty < i < \infty$, and

$$\mathscr{Y} := \sum_{i \in \mathbb{Z}} \delta_{Y_i} \qquad (7.2)$$

is a random counting measure on $(0, \infty)$; here $\delta_{Y_i}(A) = 1$ if $Y_i \in A$ and $= 0$ otherwise.

Lemma 7.1 (The scale-invariant spacing lemma). *For any $\theta > 0$, the random measures $\mathscr{P}$ and $\mathscr{Y}$ have the same distribution.*

Proof. Start with a Poisson process $\mathscr{R}$ on $(0, \infty)^2$ having points (W, Y) with intensity $\theta \exp(-wy) \, dy \, dw$. The intensity for the projection π_1 on the w coordinate is $\theta/w \, dw$, and similarly for the y-projection π_2; i.e., each projection is a copy of the process $\mathscr{P}$. Label the points of $\mathscr{R}$ as (W_i, Y_i) in decreasing order of their w-coordinates, say with $W_{-1} > 1 > W_0$, so that $\cdots > W_{-1} > W_0 > W_1 > W_2 > \cdots$. Define, for $-\infty < j < \infty$,

$$X_j := \sum_{-\infty < i \leqslant j} Y_i, \qquad (7.3)$$

and let $\mathscr{X}$ be the process with these points. Since the spacings of $\mathscr{X}$ are by construction the points Y_i of a process which has the same distribution as $\mathscr{P}$, the goal is to show that $\mathscr{X}$ also has the same distribution as $\mathscr{P}$. We do this by calculating, for $k = 0, 1, 2, \ldots$, for $0 < x_0 < x_1 < \cdots < x_k$, the intensity for the process $\mathscr{X}$ to have points at x_i for $i = 0$ to k and no points in (x_i, x_{i+1}) for $i = 0$ to $k-1$.

Consider, for $c > 0$, the restriction of $\mathscr{R}$ to $(c, \infty) \times (0, \infty)$. The intensity function of the y-projection of this process is

$$f_c(y) = \int_{w > c} \theta \exp(-wy) \, dw = \theta \exp(-cy)/y.$$

For the case $c = 1$, this intensity function arises in the study of the Poisson–Dirichlet process [17], and the sum of the y-coordinates has the Gamma distribution with parameter θ and density $g(x; \theta) = x^{\theta-1} e^{-x}/\Gamma(\theta)$. For general $c > 0$, since $f_c(y) = c f_1(cy)$, the process with parameter c is the same as the process with parameter 1, rescaled by dividing each y-coordinate by c. In particular, the sum S_c of the y-coordinates of the points in this process has density function $g(x; \theta, c) = c(xc)^{\theta-1} e^{-xc}/\Gamma(\theta)$.

For $\mathscr{X}$ to have a point at x, there must be some value c such that $S_c = x$. Taking w_0 to be the infimum of such c, the process $\mathscr{R}$ must have a point on the line $w = w_0$ (not necessarily the point labelled (W_0, Y_0)), and $\mathscr{R}$ must have x for the sum of the y-coordinates of the points in $(w_0, \infty) \times (0, \infty)$. Thus the intensity function for the pair (x, w_0) is $(\theta/w_0) g(x; \theta, w_0)$. Integrating out w_0 yields

$$\int_{w > 0} \frac{\theta}{w} \frac{w(xw)^{\theta-1} e^{-xw}}{\Gamma(\theta)} \, dw = \frac{\theta}{x}$$

which shows that $\mathscr{X}$ has the same intensity as the Poisson process $\mathscr{P}$.

For a Poisson process on $(0, \infty)$ with intensity $f(x) \, dx$, for $0 < a < b$, the intensity function to have two consecutive points at a, b is $f(a) f(b) \exp(-\int_a^b f(x) \, dx)$; and for the process $\mathscr{P}$ this simplifies to $(\theta/a)(\theta/b)(a/b)^\theta = \theta^2 a^{\theta-1} b^{-\theta-1}$. Similarly, the intensity function for $\mathscr{P}$ to have three consecutive points at $a < b < c$ simplifies to $\theta^3 a^{\theta-1} b^{-1} c^{-\theta-1}$, and for four consecutive points at $a < b_1 < b_2 < c$ the expression simplifies to $\theta^4 a^{\theta-1} b_1^{-1} b_2^{-1} c^{-\theta-1}$. Since the case with two points does not illustrate the full pattern, we show below the details of computing the intensity for $\mathscr{X}$ to have three consecutive points at $a = x_0$, $b = x_0 + y_0$, $c = x_0 + y_0 + y_1$, to check that it agrees with the intensity for $\mathscr{P}$.

The configuration of three consecutive points specified for $\mathscr{X}$ above requires that for some $0 < w_1 < w_0$, $\mathscr{R}$ has points at (w_0, y_0) and (w_1, y_1), with no points in the strip $(w_1, w_0) \times (0, \infty)$, and with the sum S_{w_0} of y-coordinates of points in $(w_0, \infty) \times (0, \infty)$ equal to x_0. The intensity function for this, with respect to $dx_0 \, dy_0 \, dy_1 \, dw_0 \, dw_1$, is a product with four factors:

$$\theta \exp(-w_0 y_0) \cdot \theta \exp(-w_1 y_1) \cdot (w_1/w_0)^\theta \cdot w_0 (x_0 w_0)^{\theta-1} \exp(-x_0 w_0)/\Gamma(\theta),$$

which reduces to

$$\theta^2 x_0^{\theta-1} \Gamma(\theta)^{-1} w_1^\theta \exp(-w_1 y_1) \exp(-w_0 (x_0 + y_0)).$$

Integrating over $0 < w_1 < w_0$ yields

$$\frac{\theta^2 x_0^{\theta-1}}{\Gamma(\theta)} \int_0^\infty dw_1 \, w_1^\theta e^{-w_1 y_1} \int_{w_1}^\infty dw_0 \, e^{-w_0(x_0+y_0)}$$

$$= \frac{\theta^2 x_0^{\theta-1}}{\Gamma(\theta)} \int_0^\infty dw_1 \, w_1^\theta e^{-w_1 y_1} \frac{e^{-w_1(x_0+y_0)}}{x_0 + y_0}$$

$$= \frac{\theta^2 x_0^{\theta-1}}{(x_0+y_0)\Gamma(\theta)} \int_0^\infty dw_1 \left(\frac{w_1(x_0+y_0+y_1)}{x_0+y_0+y_1}\right)^\theta e^{-w_1(x_0+y_0+y_1)}$$

$$= \frac{\theta^2 \Gamma(\theta+1)}{\Gamma(\theta)} x_0^{\theta-1} \frac{1}{x_0+y_0} (x_0+y_0+y_1)^{-\theta-1}$$

$$= \theta^3 a^{\theta-1} b^{-1} c^{-\theta-1}$$

as desired.

For the general case, to calculate the intensity for $k+1$ given points to be consecutive points of $\mathscr{X}$, take $b_0 < b_1 < \cdots < b_k$ and set

$$y_i = b_{i+1} - b_i, 0 \leqslant i \leqslant k-1; \qquad a = b_0; \qquad c = b_k.$$

The intensity for $\mathscr{R}$ to have points at (w_j, y_j) for $j = 0$ to $k-1$, with $0 < w_{k-1} < \cdots < w_1 < w_0$, to have no points in $\cup_1^{k-1}(w_{j-1}, w_j) \times (0, \infty)$, and to satisfy $S_{w_0} = a$, is

$$\prod_{j=0}^{k-1} \theta e^{-w_j y_j} (w_{k-1}/w_0)^\theta w_0 (aw_0)^{\theta-1} e^{-x_0 w_0}/\Gamma(\theta).$$

Integrating over $w_{k-1} < w_{k-2} < \cdots < w_0$, the innermost integral is still

$$\int_{w_1}^\infty dw_0 \exp(-w_0(x_0+y_0)) = \int_{w_1}^\infty dw_0 \exp(-w_0 b_1),$$

which produces the 'b_1^{-1}' factor along with the function $\exp(-w_1 b_1)$. This combines with the already present factor $\exp(-w_1 y_1)$, so the next integration is $\int_{w_2}^\infty dw_1 \exp(-w_1 b_2)$, which produces the '$b_2^{-1}$' factor. Continuing to integrate, the final result is

$$\theta^k \Gamma(\theta+1)\Gamma(\theta)^{-1} a^{\theta-1} \left\{ \prod_{j=1}^{k-1} b_j^{-1} \right\} c^{-\theta-1} = \theta^{k+1} a^{\theta-1} \left\{ \prod_{j=1}^{k-1} b_j^{-1} \right\} c^{-\theta-1},$$

as required. □

8. The third coupling: to exploit $\mathbb{E}|U - 1/2| = 1/4$

8.1. Motivating and defining the coupling

As in the proof of Lemma 7.1, start with the Poisson process $\mathscr{R}$ on $(0,\infty)^2$ with intensity $\theta \exp(-wy)\, dy\, dw$, now for $\theta = 1$. The joint intensity $e^{-wy}\, dw\, dy$ may be factored as the marginal intensity $(1/y)\, dy$ for Y, times the conditional intensity $y e^{-wy}\, dw$ for W given $Y = y$. Hence the Y values form the Poisson process on $(0,\infty)$ with intensity dy/y, and given a Y arrival at y, its 'label' W is exponentially distributed with mean $1/y$. So we construct our realization of $\mathscr{R}$ by first sampling the Y-process. If its points are labelled in any way as $(Y_j^*, j \geq 1)$, attach to each Y_j^* the associated $W_j^* = S_j^*/Y_j^*$, where $(S_j^*, j \geq 1)$ is an independent sequence of independent standard exponential random variables.

We now re-label, dropping the *s. Instead of labelling as at (7.3), we now label so that

$$\cdots > W_2 > W_1 > W_0 > W_{-1} > W_{-2} > \cdots, \tag{8.1}$$

and we define $X_i = \sum_{j \geq i} Y_j$, so that, by Lemma 7.1, both of the sets $\{X_i, i \in \mathbb{Z}\}$ and $\{Y_i, i \in \mathbb{Z}\}$ are distributed as the scale-invariant Poisson process on $(0,\infty)$ with intensity dx/x.

The event that there is a point of X at any integer n, or that $Y_i = Y_j$ for some $i \neq j$, has probability 0, and we remove all such outcomes from our probability space! Translate the indexing so that X_1 is the largest X_i before n. Thus

$$0 < \cdots < X_2 < X_1 < n < X_0 < X_{-1} < X_{-2} < \cdots < \infty, \tag{8.2}$$

the spacings are indexed with

$$Y_i := X_i - X_{i+1} \in (0,\infty), \quad \text{for } i \in \mathbb{Z},$$

and

$$X_1 = \sum_{j \geq 1} Y_j < n \quad \text{and} \quad X_0 = \sum_{j \geq 0} Y_j > n. \tag{8.3}$$

The GEM, scaled up by a factor of n, is constructed from the subintervals of $(0,n)$ with boundaries $X_1, X_2, \ldots$; that is,

$$(nA_1, nA_2, nA_3 \ldots) = \left(\left(n - \sum_{i \geq 1} Y_i \right), Y_1, Y_2, \ldots \right).$$

The Poisson–Dirichlet is formed by applying the function RANK, so that L_i is the ith-largest of $A_1, A_2, \ldots$, for $i \geq 1$.

We next define a deterministic step function $f : (0, \infty) \to \mathbb{Z}_+$ to be applied to the points $Y_i, i \in \mathbb{Z}$. The function f is defined, using the Euler constant γ, by

$$\begin{aligned} (0, \exp(-\gamma)) &\mapsto 0, \\ [\exp(-\gamma), \exp(-\gamma + 1)) &\mapsto 1, \\ [\exp(-\gamma + 1), \exp(-\gamma + 1 + 1/2)) &\mapsto 2, \end{aligned} \quad (8.4)$$

and so on. For $k = 1, 2, \ldots$, the map f takes an interval of $x^{-1} dx$ measure $1/k$ onto k, so that the number Z_k of times that k appears in the multiset $\{f(Y_i), i \in \mathbb{Z}\}$ is Poisson with $\mathbb{E}Z_k = 1/k$, with $Z_1, Z_2, \ldots$ independent. Also, the interval $(0, e^{-\gamma})$, which is mapped to zero, has infinite dx/x measure, so the multiset $\{f(Y_i), i \in \mathbb{Z}\}$ has an infinite number of copies of zero.

A small table of approximate values shows that starting the step boundaries with offset $\exp(-\gamma)$ gives us a function which is very close to 'round to the nearest integer':

$$\begin{aligned} f^{-1}(0) &\doteq (0, 0.561459483566885), \\ f^{-1}(1) &\doteq [0.561459483566885, 1.52620511159586), \\ f^{-1}(2) &\doteq [1.52620511159586, 2.51628683093936), \\ f^{-1}(3) &\doteq [2.51628683093936, 3.51176116633948), \\ f^{-1}(10) &\doteq [9.50437851808436, 10.5039627325698), \\ f^{-1}(100) &\doteq [99.5004187539487, 100.500414587371), \\ f^{-1}(1000) &\doteq [999.500041687504, 1000.50004164584), \\ f^{-1}(10000) &\doteq [9999.50000416701, 10000.5000041666). \end{aligned}$$

The sequence Y_i is exactly in a size-biased permutation: this is clear from our construction. The sequence $f(Y_i)$ is not exactly in a size-biased permutation, although it comes close. *If* the sequence $f(Y_i)$ *were* in a size-biased permutation, it *would have* all the zeros coming first, followed by positive integers tending from small to large, and the indicator function of the set

$$\left\{ 1 + \sum_{j > i} f(Y_j) : i \in \mathbb{Z} \right\} \quad (8.5)$$

would be distributed exactly as the variables $\xi_1, \xi_2, \ldots$ for the Feller coupling in Section 4.2, because

$$(Z_1, Z_2, \ldots) =_d \left(C_1^{(\infty)}, C_2^{(\infty)}, \ldots \right),$$

and the values of each are then arranged in (reversed) size-biased order. (A history of this result, tracing it back to Rényi and Ignatov [15, 16], is given in Section 2.1 of [2].)

In order to have a size-biased permutation of the multiset $\{f(Y_i), i \in \mathbb{Z}\}$ which is close to the identity permutation on $\mathbb{Z}$, we simply re-use the exponential random variables S_j^*,

Table 1.

i	8	7	6	5	4	3	2	1	0	−1
Y_i	0.125	0.25	0.5	2.8	6.6	0.1	1.6	7.8	20.2	2.3
$f(Y_i)$	0	0	0	3	7	0	2	8	20	2
$X_i = \sum_{j:j \geq i} Y_j$	0.25	0.5	1.0	3.8	10.4	10.5	12.1	19.9	40.1	42.4
$\sum_{j:j \geq i} f(Y_j)$	0	0	0	3	10	10	12	20	40	42

defining new W-coordinates $\widetilde{W}_j^* = S_j^*/f(Y_j^*)$, and hence, in the labelling (8.2),

$$\widetilde{W}_i = \frac{S_i}{f(Y_i)} = W_i \frac{Y_i}{f(Y_i)}. \tag{8.6}$$

Since the S_j^* are i.i.d. standard exponential, the points $f(Y_i)$, taken in order of decreasing tilde labels $\widetilde{W}_i$, are in a size-biased permutation, tending from small to large.

Use the $f(Y_i)$, taken in order of their tilde labels, to form the sequence $\xi_1, \xi_2, \ldots$ for the Feller coupling, and construct the age-ordered cycle lengths $A_1^{(n)}, A_2^{(n)}, \ldots$. Finally, we take $L_i^{(n)} :=$ ith-largest of the $A_1^{(n)}, A_2^{(n)}, \ldots$.

To summarize: we have defined a coupling, with the $L_i, L_i^{(n)}, A_i, A_i^{(n)}$ for $i \geq 1$ all realized together. It only remains to estimate the ℓ_1 distances $d_{\text{PD}}(n)$ and $d_{\text{GEM}}(n)$!

A technicality. To be careful, whenever $Y_i < e^{-\gamma}$, so that $f(Y_i) = 0$, we take $\widetilde{W}_i = \infty$. With probability 1, the non-infinite values among the $\widetilde{W}_i$ are all distinct, and on this good event, there is a permutation σ of the integers, with the property that for distinct i, j, $\sigma(i) > \sigma(j)$ if and only $\widetilde{W}_i > \widetilde{W}_j$ or ($\widetilde{W}_i = \widetilde{W}_j = \infty$ and $Y_i < Y_j < e^{-\gamma}$). Modulo translation, this permutation is unique. The Feller coupling variables $\xi_1, \xi_2, \ldots$ are defined by

$$\xi_i = 1 \text{ if and only if } i \in \left\{ 1 + \sum_{\sigma(j) > \sigma(k)} f(Y_j) : k \in \mathbb{Z} \right\}. \tag{8.7}$$

8.2. An example

Suppose that the sequence (Y_i), taking the index i decreasing through $\mathbb{Z}$, has negative powers of two, until the partial sum is exactly 1, followed by the values $2.8, 6.6, 0.1, 1.6, 7.8, 20.2, 2.3$ and 5.4. We take $20 \leq n \leq 40$.

Note that in (8.3), the placement of the origin, $i = 0$, depends on n via $X_1 < n < X_0$. Thus, in case $20 \leq n \leq 40$ the last three columns, starting with $7.8, 20.2$ and 2.3, are labelled with $i = 1, 0, -1$; if $n = 41$ they are labelled with $i = 2, 1, 0$; and if $n = 19$ they are labelled with $i = 0, -1, -2$.

The example, with $n = 35$. The scaled-up GEM has first component $nA_1 = n - X_1 = 35 - 19.9 = 15.1$, $nA_2 = Y_1 = 7.8$, $nA_3 = Y_2 = 1.6$, $nA_4 = Y_3 = 0.1$, $nA_5 = Y_4 = 6.6$, $nA_6 = 2.8$, $nA_7 = 0.5, \ldots$

The example, with minimal differences in the size-biased permutations. This is the most common situation, in which the labels $\widetilde{W}$ defined in (8.6), excluding cases where $f(Y_i) = 0$, have the same relative order as in (8.1). In this example, the Feller coupling, as at (8.5), is based on 1 plus the partial sums of the sequence

$$\ldots, 0, 0, 0, 3, 7, 0, 2, 8, 20, 2, \ldots,$$

so that $\xi_1 \xi_2 \ldots$ has ones in positions indexed by

$$1, 4, 11, 13, 21, 41, 43, \ldots. \tag{8.8}$$

We have $A_1^{(n)} = 15$ (the spacing from 35+1 down to 21), $A_2^{(n)} = 8 = f(Y_1)$, $A_3^{(n)} = 2 = f(Y_2)$, $A_4^{(n)} = 7 = f(Y_4)$ (notice that the value $f(Y_3) = 0$ does not serve as the amount of spacing between two successive ones in $\xi_1 \xi_2 \cdots \xi_n 1$) and then $A_5^{(n)} = 3 = f(Y_5)$ and $A_j^{(n)} = 0$ for any $j > 5$. Using the rearrangement Lemma 3.2 to give an upper bound on $d_{\text{PD}}(n)$, we will match $nA_j = Y_{j-1}$ with $A_j^{(n)} = f(Y_{j-1})$ for $j = 2, 3$ and we will match $nA_j = Y_{j-1}$ with $A_{j-1}^{(n)} = f(Y_{j-1})$ for $j \geqslant 5$, so the 'gap' at $j = 4$ contributes nothing to $d_{\text{PD}}(n)$. This example shows why our third coupling may be good for $d_{\text{PD}}(n)$, but not good enough for $d_{\text{GEM}}(n)$ – because when $X_i = x$ and $X_i - X_{i+1} < e^{-\gamma}$ we will then usually have to match $A_j^{(n)} = f(Y_j)$ with $nA_j = Y_{j-1}$ for all $j > i$, for a contribution to $d_{\text{GEM}}(n)$ that is order of x, and $\int_1^n x e^{-\gamma}/x \, dx/x$ is order of $\log n$.

The example, with dislocation in the size-biased permutations. Note that since $Y_2 = 1.6$ is left of $Y_1 = 7.8$, we have of necessity that $W_2 > W_1$. But in (8.6), $\widetilde{W}_2$ is obtained from W_2 by multiplying by the factor $Y_2/f(Y_2) = 1.6/2$, while $\widetilde{W}_1$ is obtained from W_1 by multiplying by the somewhat larger factor $Y_1/f(Y_1) = 7.8/8$, so if W_1 was only slightly larger than W_1, we will have $\widetilde{W}_2 < \widetilde{W}_1$, and $8 = f(Y_1)$ will come left of $2 = f(Y_2)$. Suppose that this indeed happens. The Feller coupling, as at (8.5), is based on 1 plus the partial sums of the sequence $\ldots, 0, 0, 0, 3, 7, 0, 8, 2, 20, 2, \ldots$, so that $\xi_1 \xi_2 \ldots$ has ones in positions indexed by $1, 4, 11, 19, 21, 41, 43, \ldots$. We have $A_1^{(n)} = 15$, just as in (8.8), but $A_2^{(n)} = 2 = f(Y_2)$ and $A_3^{(n)} = 8 = f(Y_1)$, in contrast with (8.8), and this contributes a large amount to $d_{\text{GEM}}(n)$. Using the rearrangement Lemma 3.2 to give an upper bound on $d_{\text{PD}}(n)$, we will match $nA_3 = Y_2$ with $A_2^{(n)} = f(Y_2)$ and $nA_2 = Y_1$ with $A_3^{(n)} = f(Y_1)$, so the extra difference in the size-biased permutation does not cause an increase in the upper bound on $d_{\text{PD}}(n)$.

8.3. The principal differences

Our scaled-up GEM is viewed as a list of interval lengths subdividing $(0, n)$, and the Feller process as a list of interval lengths subdividing $[1, n+1]$; in both cases, the sum of the lengths is exactly n. The scaled-up GEM has as its components $nA_1 = n - \sum_{i \geqslant 1} Y_i$ and $Y_1, Y_2, \ldots$. A natural fit to the age-ordered cycle lengths would be that using $(n+1) - (1 + \sum_{i \geqslant 1} f(Y_i))$ and $f(Y_1), f(Y_2), \ldots$, giving a distance of at most $T_1 + T_2$, where

$$T_1 := \sum_{i \geqslant 1} |Y_i - f(Y_i)| \quad \text{and} \quad T_2 = \left| \sum_{i \geqslant 1} (Y_i - f(Y_i)) \right|.$$

Although this match may not quite work, because the $\widetilde{W}$-ordering is not exactly the same as the original, the essence of the proof is nonetheless to show that the error term T_1 is indeed the dominant contribution to $d_{\mathrm{PD}}(n)$. Our first two lemmas concern the asymptotics of T_1 and T_2.

Lemma 8.1.
$$T_1 := \sum_{i \geq 1} |Y_i - f(Y_i)| \quad \text{has} \quad \mathbb{E}T_1 \sim \frac{1}{4} \log n. \tag{8.9}$$

Proof. In the proof of Theorem 2.1, we calculated $\mathbb{E} \sum_{i \geq 1} |nL_i - g(nL_i)| \sim (1/4) \log n$, where g is the function 'round to the nearest integer', that is, $g(x) = \lfloor x + .5 \rfloor$. This sum differs from that in (8.9) in two ways. First, since our coupling takes $nL_i = i$th-largest of $(n - \sum_{i \geq 1} Y_i), Y_1, Y_2, \ldots$, the first of those differences is no longer present. Secondly, the function g is replaced by f, and the table above makes it clear that $f(x) - g(x) \in \{-1, 0, 1\}$ for all $x > 0$. Write h_i for the harmonic sum $h_i = 1 + 1/2 + \cdots + 1/i$. From the expansion $h_i = \gamma + \log i + 1/(2i) - 1/(12i^2) + O(i^{-4})$ we have

$$\exp(-\gamma + h_i) = i \exp\left(\frac{1}{2i} - \frac{1}{12i^2} + \cdots\right) = i + .5 + \frac{1}{24i} + O(i^{-2}) \tag{8.10}$$

so the set difference $f^{-1}(i) \setminus g^{-1}(i)$ is an interval of length $\sim 1/(24i)$. Hence

$$\mathbb{E} \sum_{i: Y_i < n} |f(Y_i) - g(Y_i)| = \int_0^n |f(x) - g(x)| \, dx/x = O(1); \tag{8.11}$$

note that if we were integrating dx rather than dx/x, this error would be order of $\log n$ and unacceptable. $\square$

Lemma 8.2.
$$T_2 := \left| \sum_{i \geq 1} (Y_i - f(Y_i)) \right| \quad \text{has} \quad \mathbb{E}T_2 = O(\sqrt{\log n}). \tag{8.12}$$

Proof. We will use Lemma 8.1 to approximate f by g, the function which rounds to the nearest integer. Start by considering

$$H_n := \sum_{i: 1/2 \leq Y_i < n} (Y_i - g(Y_i)).$$

This is compound Poisson, with values in $[-.5, .5]$, and total Poisson intensity $\int_{1/2}^n dx/x = \log(2n)$. The calculation

$$\int_{i-1/2}^{i+1/2} (y - g(y)) dy/y = \int_{i-1/2}^{i+1/2} (y - i) dy/y$$
$$= 1 - i \log((i+1/2)/(i-1/2)) = O(i^{-2})$$

shows that $\mathbb{E}H_n = O(1)$ as $n \to \infty$, and the bound $|y - g(y)| \leq 1/2$ shows that $\mathrm{Var}(H_n) = \int_{1/2}^n (y - g(y))^2 dy/y \leq (1/4) \log(2n)$; combined, these give $\mathbb{E}H_n^2 \leq c \log n$, for some constant

$c < \infty$ and for all n. This yields $\mathbb{E}|H_n| \leqslant \sqrt{\mathbb{E}H_n^2} = O(\sqrt{\log n})$. If we change H_n to

$$H'_n := \sum_{i:0<Y_i<n}(Y_i - g(Y_i)),$$

then $H_n - H'_n = \sum_{i:0<Y_i<1/2} Y_i$. The sum is positive a.s., with expectation $1/2$, so $\mathbb{E}|H_n - H'_n| = 1/2$, and hence $\mathbb{E}|H'_n| = O(\sqrt{\log n})$.

Next, from (8.11) in the proof of Lemma 8.1, there is an $O(1)$ error in replacing g by the function f, so now we have

$$\mathbb{E}\left|\sum_{i:Y_i<n}(Y_i - f(Y_i))\right| = O(\sqrt{\log n}).$$

Finally, the function $f(x) - x$ is bounded, and

$$\mathbb{E}\left\{\sum_{i\leqslant 0}\mathbb{1}\{Y_i < n\}\right\} \leqslant 1 + \sum_{i<0}\mathbb{1}\{X_{i-1} - X_i < n\}$$

$$\leqslant 1 + \int_n^\infty \frac{n\,dx}{x(n+x)} \leqslant 2,$$

so that thus

$$\mathbb{E}\left|\sum_{i\geqslant 1}(Y_i - f(Y_i))\right| = O(1) + \mathbb{E}\left|\sum_{i:Y_i<n}(Y_i - f(Y_i))\right| = O(\sqrt{\log n}). \qquad \square$$

8.4. Analysis of the coupling

The remainder of the argument concerns the effect of switching the order from the W-ordering to the $\widetilde{W}$-ordering. To do so, a number of preliminaries are needed. To start with, define

$$f_\infty(w) := \mathbb{P}\left[\sum_{i:W_i>w} Y_i > 1\right].$$

Then we have the following lemma.

Lemma 8.3. *For all $w > 0$,*

$$f_\infty(w) \leqslant \min\{1, 2(1+w)e^{-w}\}.$$

Proof. For any bounded function $g : \mathbb{R}_+ \to \mathbb{R}_+$, let

$$(Tg)(w) := w\int_w^\infty x^{-2}e^{-x}\,dx + w\int_w^\infty x^{-1}\int_0^\infty e^{-xz}g(x(1-z))\,dz\,dx,$$

and, for all $w \geqslant 0$, set $f_0(w) = 0$ and

$$f_n(w) := \mathbb{P}\left[\sum_{i=J(w)}^{J(w)+n} Y_i > 1\right], \quad n \geqslant 1,$$

where $J(w) := \min\{i : W_i > w\}$. Then it is clear that $f_\infty(w) \uparrow f_n(w)$ as $n \to \infty$, and, because

$$\mathbb{P}\left[\sum_{i=J(x)}^{J(x)+n} Y_i > 1 - z\right] = f_n(x(1-z)), \quad x > 0, \quad 0 \leqslant z \leqslant 1,$$

by scaling, it also follows that $f_{n+1} = Tf_n$ for all $n \geqslant 0$. Clearly, $f_0(w) \leqslant K(1+w)e^{-w}$ for all $K, w > 0$, and $f_n(w) \leqslant 1$ for all n and w. Then, if $f_n(w) \leqslant K(1+w)e^{-w}$ for all $w > 0$, it follows from the definition of T that

$$f_{n+1}(w) \leqslant w \int_w^\infty x^{-2} e^{-x} \, dx + w \int_w^\infty x^{-1} \int_0^\infty e^{-xz} K(1 + x(1-z)) e^{-x(1-z)} \, dz \, dx$$

$$= w \int_w^\infty e^{-x} \left\{\frac{1}{x^2} + \frac{K}{x} + \frac{K}{2}\right\}$$

$$\leqslant e^{-w} + Ke^{-w} + \tfrac{1}{2} Kwe^{-w}.$$

Thus it follows that $f_{n+1}(w) \leqslant K(1+w)e^{-w}$ for all w such that $Kw/2 \geqslant 1$. However, $f_{n+1}(w) \leqslant 1 \leqslant K(1+w)e^{-w}$ for all $w < 2/K$ provided that $(K+2)e^{-2/K} \geqslant 1$, true for $K = 2$. Hence, for $K = 2$, it follows by induction that $f_n(w) \leqslant \min\{1, 2(1+w)e^{-w}\}$ for all n, proving the lemma. □

Now, if $W_0^1 := \sup\{w : \sum_{i: W_i > w} Y_i > 1\}$, it follows that

$$\{W_0^1 > w\} = \left\{\sum_{i: W_i > w} Y_i > 1\right\},$$

and hence that

$$\mathbb{E} W_0^1 = \int_0^\infty \mathbb{P}[W_0^1 > w] \, dw \leqslant 2 \int_0^\infty (1+w)e^{-w} \, dw = 4.$$

Hence we have proved the following corollary.

Corollary 8.4.

$$\mathbb{E} W_0 \leqslant 4n^{-1}.$$

We now turn to consideration of the differences between the values $\{f(Y_i), i \geqslant 1\} = \{f(Y_i), W_i > W_0\}$ and $\{f(Y_i), \widetilde{W}_i > W_0\}$. The latter set is a 'left-hand segment' of the infinite reverse size-biased multiset $\{f(Y_i), i \in \mathbb{Z}\}$, and is close to the coupled reverse age-ordered cycles. The two sets differ because, for some $i < 0$, $\widetilde{W}_i > W_0$, and, for some $i > 0$, $\widetilde{W}_i < W_0$; the effects of these two exchanges are treated in Lemmas 8.5 and 8.6, respectively. We begin by defining

$$c_1 := \min_{y: f(y) \geqslant 1} \{y^{-1} f(y)\}, \quad c_2 := \max_{y > 0} \{y^{-1} f(y)\}.$$

Then, for $y = (y_1, y_2, \ldots)$ an increasing sequence of positive reals and $u, v > 0$, we let $A(u, v; y)$ denote the event

$$\{(W_0, Y_0) = (u, v)\} \cap \{\pi_2 \mathscr{R}\{(u, \infty) \times \mathbb{R}_+\} = y\}.$$

Lemma 8.5. *Defining*

$$\eta_1 := \sum_{i \geq 1} \{f(Y_{-i})\mathbb{1}\{\widetilde{W}_{-i} > W_0\}\},$$

we have $\mathbb{E}\eta_1 \leq c_2/c_1$.

Proof. We use the fact that, conditional on $A(w_0, y_0; y)$, we have

$$\mathscr{L}(W_{-i}, i \geq 1) = \mathscr{L}\left(w_0 \prod_{j=1}^{i} U_j, i \geq 1\right),$$

where $U_1, U_2, \ldots$ are independent and uniformly distributed on $(0, 1)$. Hence

$$\mathbb{E}\{f(Y_{-i})\mathbb{1}\{\widetilde{W}_{-i} > W_0\}\}$$

$$= \mathbb{E}\left(f(Y_{-i})\mathbb{1}\left\{W_0\prod_{j=1}^{i}U_j(Y_{-i}/f(Y_{-i})) > W_0\right\}\right)$$

$$\leq \mathbb{E}\left\{f(Y_{-i})\mathbb{1}\{U_i(Y_{-i}/f(Y_{-i})) > 1\}\prod_{j=1}^{i-1}\mathbb{1}\{U_j > c_1\}\right\}$$

$$\leq \mathbb{E}\{f(Y_{-i})(Y_{-i} - f(Y_{-i}))_+/Y_{-i}\}(1-c_1)^{i-1} \leq c_2(1-c_1)^{i-1},$$

and the lemma follows. □

Lemma 8.6. *Defining*

$$\eta_2 := \sum_{i \geq 1}\{f(Y_i\mathbb{1}\{\widetilde{W}_i < W_0\}\},$$

we have $\mathbb{E}\eta_2 \leq 4c_2$.

Proof. Conditional on $A(w_0, y_0; y)$, the point set $\{W_i, i \geq 1\}$ has the same distribution as the set $\{\widehat{W}_j, j \geq 1\}$, where $\widehat{W}_j := w_0 + y_j^{-1}E_j$ and $E_1, E_2, \ldots$ are independent standard exponential random variables. Hence

$$\sum_{i \geq 1}\mathbb{E}\{f(Y_i)\mathbb{1}\{\widetilde{W}_i < W_0\} \mid A(w_0, y_0; y)\}$$

$$= \sum_{j \geq 1} f(y_j)\mathbb{P}[(y_j/f(y_j))(w_0 + y_j^{-1}E_j) < w_0].$$

But it is easy to see that

$$f(y_j)\mathbb{P}[(y_j/f(y_j))(w_0 + y_j^{-1}E_j) < w_0]$$
$$= f(y_j)\mathbb{P}[E_j < (f(y_j) - y_j)_+ w_0] \leq f(y_j)w_0,$$

so that

$$\sum_{j \geq 1} f(y_j)\mathbb{P}[(y_j/f(y_j))(w_0 + y_j^{-1}E_j) < w_0] \leq w_0\sum_{j \geq 1}f(y_j) \leq w_0c_2\sum_{j \geq 1}y_j \leq w_0c_2n.$$

The lemma now follows from Corollary 8.4. □

Corollary 8.7. *Enumerate the set* $\{f(Y_i) : i \in \mathbb{Z} \setminus \{0\}, \widetilde{W}_i > W_0\}$ *as* $\{F_j, j \geq 1\}$. *Then there is a matching* σ *of the set* $\{F_j, j \geq 1\}$ *and the set* $\{Y_i, i \geq 1\}$ *such that*

$$\mathbb{E}\left|\sum_{i\geq 1}(Y_i - F_{\sigma(i)})\right| = \mathbb{E}\left|\left(n - \sum_{i\geq 1} Y_i\right) - \left(n - \sum_{j\geq 1} F_j\right)\right| = O(\sqrt{\log n})$$

and

$$\limsup_{n\to\infty} \{\log n\}^{-1} \mathbb{E}\left\{\sum_{i\geq 1}|Y_i - F_{\sigma(i)}|\right\} \leq \frac{1}{4}.$$

Proof. Match Y_i with $f(Y_i)$ for all $i \geq 1$ such that also $\widetilde{W}_i > W_0$, and use Lemmas 8.5 and 8.6 to control the remainder, which is therefore of order $O(1)$. This implies that the first expectation is bounded by $T_2 + O(1)$ and the second by $T_1 + O(1)$, and Lemmas 8.1 and 8.2 complete the bound. □

Since the $f(Y_i)$ in the $\widetilde{W}_i$-ordering are in size-biased order, the set $\{F_j, j \geq 1\}$ consists of a 'left-hand segment' from the reverse size-biased order, and is close to being that for which the sum is closest to n from below. If $f(Y_0)$ were added to the collection, and the difference between Y_i and $f(Y_i)$ were temporarily neglected, the sum would be the first to exceed n. Thus, from Corollary 8.7, we would be more or less finished if we always had $f(Y_0)$ as next element after the $\{F_j, j \geq 1\}$ in the $\widetilde{W}_i$-ordering. This need not quite be the case, and the next two lemmas control the possible error made when completing the approximation.

We first consider the case in which $f(Y_0) > Y_0$, and so $\widetilde{W}_0 < W_0$. Here, the main concern is that there may be indices $i < 0$ such that $\widetilde{W}_0 < \widetilde{W}_i < W_0$, so that the corresponding $f(Y_i)$ would be taken before $f(Y_0)$. The possible contribution from indices $i > 0$ satisfying $\widetilde{W}_0 < \widetilde{W}_i < W_0$ is already more than covered by Lemma 8.6.

Lemma 8.8. *If* $f(y_0) > y_0$, *then*

$$\sum_{i\geq 1}\mathbb{E}\{(f(Y_{-i}) \wedge f(Y_0))\mathbb{1}\{\widetilde{W}_0 < \widetilde{W}_{-i} \leq W_0\} \mid A(w_0, y_0; y)\} \leq c_2^2/c_1.$$

Proof. We argue much as for Lemma 8.5, obtaining

$$\mathbb{E}\{(f(Y_{-i}) \wedge f(y_0))\mathbb{1}\{\widetilde{W}_0 < \widetilde{W}_{-i} \leq W_0\} \mid A(w_0, y_0; y)\}$$

$$\leq \mathbb{E}\left\{(f(Y_{-i}) \wedge f(y_0))\mathbb{1}\{(y_0/f(y_0)) < U_i(Y_{-i}/f(Y_{-i}) \leq 1\}\right.$$

$$\left.\prod_{j=1}^{i-1}\mathbb{1}\{U_j > c_1/c_2\} \,\bigg|\, A(w_0, y_0; y)\right\}$$

$$\leq f(y_0)\{c_2(f(y_0) - y_0)_+/f(y_0)\}(1 - c_1/c_2)^{i-1}$$

$$\leq c_2(1 - c_1/c_2)^{i-1}.$$

The lemma now follows immediately. □

For the complementary case, in which $f(Y_0) < Y_0$ and so $\widetilde{W}_0 > W_0$, we are principally concerned about indices $i > 0$ such that $W_0 < \widetilde{W}_i < \widetilde{W}_0$, so that $f(Y_0)$ would be taken before the corresponding $f(Y_i)$. The possible contribution from indices $i < 0$ satisfying $W_0 < \widetilde{W}_i < \widetilde{W}_0$ is taken care of by Lemma 8.5.

Lemma 8.9. *If $f(y_0) < y_0$, then*

$$\sum_{i \geq 1} \mathbb{E}\{f(Y_0)\mathbb{1}\{W_0 < \widetilde{W}_i < \widetilde{W}_0\} \mid A(w_0, y_0; y)\} \leq 4c_2.$$

Proof. The argument here is like that of Lemma 8.6. We start by computing

$$\mathbb{E}\{f(Y_0)\mathbb{1}\{W_0 < \widehat{W}_j(y_j/f(y_j)) < \widetilde{W}_0\} \mid A(w_0, y_0; y)\}$$
$$= \mathbb{E}\{f(y_0)\mathbb{1}\{w_0 < (w_0 + y_j^{-1}E_j)(y_j/f(y_j)) < w_0(y_0/f(y_0))\}\}$$
$$\leq f(y_0)\mathbb{P}\left[w_0\left\{1 - \frac{y_j}{f(y_j)}\right\} < \frac{E_j}{f(y_j)} < w_0\left\{\frac{y_0}{f(y_0)} - \frac{y_j}{f(y_j)}\right\}\right]$$
$$\leq f(y_0)f(y_j)w_0\{(y_0/f(y_0)) - 1\} \leq w_0 f(y_j).$$

Once again, adding over j, this yields

$$\sum_{i \geq 1} \mathbb{E}\{f(Y_0)\mathbb{1}\{W_0 < \widetilde{W}_i < \widetilde{W}_0\} \mid A(w_0, y_0; y)\} \leq w_0 \sum_{j \geq 1} f(y_j),$$

and the lemma follows from Corollary 8.4. □

Using these two lemmas, the main theorem can be proved. Let $\tau(i)$ denote the index of the ith nonzero element of the set $\{f(Y_l), l \in \mathbb{Z}\}$ in decreasing $\widetilde{W}$-ordering, and let I_n be such that

$$\sum_{i=1}^{I_n} f(Y_{\tau(i)}) \leq n < \sum_{i=1}^{I_n+1} f(Y_{\tau(i)}).$$

Write $Y_0^{(1)} = n - \sum_{i \geq 1} Y_i$ and $Y_i^{(1)} = Y_i$, $i \geq 1$, and then $Y_0^{(2)} = n - \sum_{i=1}^{I_n} f(Y_{\tau(i)})$ and $Y_i^{(2)} = f(Y_{\tau(i)})$, $1 \leq i \leq I_n$, with $Y_i^{(2)} = 0$ for $i > I_n$. These are the scaled GEM and age-ordered cycle lengths to be matched, as realized in our coupling.

Theorem 8.10. *There is a matching ρ of $\{Y_i^{(1)}, i \geq 0\}$ and $\{Y_i^{(2)}, i \geq 0\}$ such that*

$$\mathbb{E}\left\{\sum_{i \geq 0} |Y_i^{(1)} - Y_{\rho(i)}^{(2)}|\right\} \leq \frac{1}{4} \log n + o(\log n).$$

Hence, by comparison with the lower bound from Theorem 2.1, the coupling defined in Section 8.1 achieves

$$d_{\mathrm{PD}}(n) \sim \frac{1}{4} \log n.$$

Proof. The proof consists mainly of finding upper bounds for the possible error in particular matchings, in a number of particular cases.

To start with, consider the case where $f(Y_0) > Y_0$, so that $\widetilde{W}_0 < W_0$, and the set of lengths $\{F_j, j \geq 1\}$, the values of $f(Y_l)$ for which $\widetilde{W}_l > W_0$, typically needs augmenting in order to have total length n. Let F_{-l}, $1 \leq l \leq L$, be the values $f(Y_{-i})$ for those $i \geq 1$ for which $\widetilde{W}_0 < \widetilde{W}_{-i} < W_0$, taken in $\widetilde{W}$-order, writing L for their total number. Define Q_l, $l \geq 0$, to be the undershoot $n - \sum_{j \geq 1} F_j - \sum_{s=1}^{l} F_{-s}$ at stage l, when all of the $\{F_j, j \geq 1\}$ and the first l of the F_{-s} have been taken. Clearly, once the undershoot $Q_l \leq 0$, no more elements are taken.

The first sub-case is when the undershoot $Q_0 \leq 0$. Here, the F_j already match the Y_i as given by Corollary 8.7, and there is additional mismatch only because of the unmatched element $Y_0^{(1)} = n - \sum_{i \geq 1} Y_i$ and the piece of length $\sum_{j \geq 1} F_j - n = -Q_0 \geq 0$ which has to be removed from the F_js; hence the error can be kept to at most

$$\left(\sum_{j \geq 1} F_j - n \right) + \left(n - \sum_{i \geq 1} Y_i \right) + \sum_{i \geq 1} |Y_i - F_{\sigma(i)}| \leq \left| \sum_{j \geq 1} F_j - \sum_{i \geq 1} Y_i \right| + \sum_{i \geq 1} |Y_i - F_{\sigma(i)}|. \quad (8.13)$$

We next consider the sub-case in which $0 < Q_0 \leq f(Y_0)$. Here, we begin by taking the successive elements F_{-l}, $1 \leq l \leq L$, and, when they are exhausted, any remaining interval is then more than covered by the element $f(Y_0)$, Q_L being matched with $Y_0^{(1)}$. Usually, F_{-l} is matched with 0, at a cost of F_{-l}. However, if one of the F_{-l} is big enough to itself cover the whole remaining interval, i.e., $F_{-l} \geq Q_{l-1}$, it is used to do so, with Q_{l-1} matched with $Y_0^{(1)}$, and no more are then needed; this happens in particular if $F_{-l} \geq f(Y_0)$. (The possible effect of elements arising in the $\widetilde{W}$-interval $(\widetilde{W}_0, W_0)$ from $f(Y_i)$ with $i \geq 1$ is controlled by Lemma 8.6, and can introduce an extra error of no more than $2\eta_2$.) We can then bound the error additional to $\sum_{i \geq 1} |Y_i - F_{\sigma(i)}|$ by the expression

$$\sum_{l=1}^{L} \mathbb{1}\{Q_{l-1} > 0\} \left[\mathbb{1}\{F_{-l} < f(Y_0)\}(F_{-l}\mathbb{1}\{F_{-l} < Q_{l-1}\} + R_{1,l}\mathbb{1}\{F_{-l} \geq Q_{l-1}\}) \right.$$
$$\left. + \mathbb{1}\{F_{-l} \geq f(Y_0)\}R_{1,l} \right] + R_{1,L+1}\mathbb{1}\{Q_L > 0\},$$

where, for $1 \leq l \leq L+1$, the error $R_{1,l}$ in matching Q_{l-1} and $Y_0^{(1)}$ is

$$R_{1,l} := \left| \left[n - \sum_{j \geq 1} F_j - \sum_{s=1}^{l-1} F_{-s} \right] - \left[n - \sum_{i \geq 1} Y_i \right] \right| \leq \left| \sum_{j \geq 1} F_j - \sum_{i \geq 1} Y_i \right| + \sum_{s=1}^{l-1} F_{-s}.$$

In this sum, since $Q_0 > 0$, there is exactly one of the $R_{1,l}$, and some or all of those of the F_{-l} that are smaller than $f(Y_0)$. Hence, in this sub-case, the total error is at most

$$\left| \sum_{j \geq 1} F_j - \sum_{i \geq 1} Y_i \right| + \sum_{i \geq 1} |Y_i - F_{\sigma(i)}|$$
$$+ 2 \sum_{i \geq 1} f(Y_{-i})\mathbb{1}\{f(Y_{-i}) < f(Y_0)\}\mathbb{1}\{\widetilde{W}_0 < \widetilde{W}_{-i} \leq W_0\} + 2\eta_2. \quad (8.14)$$

Within the case where $f(Y_0) > Y_0$, there now remains only the possibility that $Q_0 > f(Y_0)$. Here, the previous procedure can be used to match, but using $n' = f(Y_0) + \sum_{j \geq 1} F_j$ in place of n throughout. This leaves an interval of length at most $n - n'$ unmatched.

However, we have

$$n - n' \leq \sum_{i \geq 0} Y_i - \sum_{j \geq 1} F_j - f(Y_0) = Y_0 - f(Y_0) + \sum_{i \geq 1} Y_i - \sum_{j \geq 1} F_j$$

$$\leq 1 + \left| \sum_{j \geq 1} F_j - \sum_{i \geq 1} Y_i \right|, \tag{8.15}$$

to be added to the error in the previous sub-case.

The case in which $f(Y_0) < Y_0$, so that $W_0 < \widetilde{W}_0$, is argued in rather similar fashion. Here, we shall denote the set $\{f(Y_i), W_0 < \widetilde{W}_i < \widetilde{W}_0\} \subset \{F_j, j \geq 1\}$ by $\{F_j, j \in R\}$. The matching of Corollary 8.7 is not quite a matching for our coupling, if R is not empty. To modify the matching to become one, we consider sub-cases. First, if $f(Y_0) \geq n - \sum_{j \notin R} F_j$, then $n - \sum_{i \geq 1} Y_i$ is matched to $n - \sum_{j \notin R} F_j$ instead of to $n - \sum_{j \geq 1} F_j$, and the elements $\{F_j, j \in R\}$ are missing in the new matching, so that there is an extra error of at most $2 \sum_{j \in R} F_j$. Note also that, under these circumstances, $\sum_{j \in R} F_j \leq f(Y_0)$, so that the extra error is at most

$$2 \sum_{j \in R} (F_j \wedge f(Y_0)). \tag{8.16}$$

The next sub-case has $n - \sum_{j \geq 1} F_j \leq f(Y_0) < n - \sum_{j \notin R} F_j$, in which case we can match $n - \sum_{i \geq 1} Y_i$ with $f(Y_0)$; some of the $\{F_j, j \in R\}$ are again missing. The former match differs from the original by at most

$$f(Y_0) - \left(n - \sum_{j \geq 1} F_j\right) \leq \left(n - \sum_{j \notin R} F_j\right) - \left(n - \sum_{j \geq 1} F_j\right) = \sum_{j \in R} F_j,$$

again leading to an upper bound of $2 \sum_{j \in R} F_j$ for the extra error; and an alternative matching with error at most $2f(Y_0)$ could also be achieved by matching $f(Y_0)$ with 0, so that (8.16) is a bound for the extra error in this sub-case, too. In the final sub-case, in which $f(Y_0) < n - \sum_{j \geq 1} F_j$, we again match $n - \sum_{i \geq 1} Y_i$ to $f(Y_0)$ and the pieces making up the undershoot $n - \sum_{j \geq 1} F_j - f(Y_0)$ with 0, leading to an error of at most

$$\left| n - \sum_{i \geq 1} Y_i - f(Y_0) \right| + \left(n - \sum_{j \geq 1} F_j - f(Y_0) \right)$$

$$\leq \left(\sum_{i \geq 1} Y_i + Y_0 - n \right) + |f(Y_0) - Y_0| + n - \sum_{j \geq 1} F_j - f(Y_0)$$

$$\leq 2 + \sum_{i \geq 1} Y_i - \sum_{j \geq 1} F_j,$$

to replace the original error of $|\sum_{i \geq 1} Y_i - \sum_{j \geq 1} F_j|$ in matching $n - \sum_{i \geq 1} Y_i$ to $n - \sum_{j \geq 1} F_j$; thus the increase is here at most 2. Taking expectations, it follows that the overall bound in the case $f(Y_0) < Y_0$ is at most

$$\left| \sum_{j \geq 1} F_j - \sum_{i \geq 1} Y_i \right| + \sum_{i \geq 1} |Y_i - F_{\sigma(i)}| + 2 \sum_{i \geq 1} f(Y_0) \mathbb{1}\{W_0 < \widetilde{W}_i < \widetilde{W}_0\} + 2. \tag{8.17}$$

The conclusion of the theorem now follows from the bounds (8.13)–(8.17), Lemmas 8.6–8.9 and Corollary 8.7. $\square$

9. θ-biased permutations

To derive a lower bound for $\mathbb{E} \sum_{i \geq 1} |L_i^{(n)} - nL_i|$, we use the intensity measure $\mu(dx) = \theta(1-x)^{\theta-1}x^{-1}\,dx$ for $0 < x < 1$, called the 'frequency spectrum' in Ewens [12], corresponding to the Poisson–Dirichlet distribution with parameter θ. We obtain the following result.

Theorem 9.1. *For any $\theta > 0$, let $L_i^{(n)}$ denote the size of the ith-largest component of the Ewens sampling formula, and let L_i be the ith coordinate of the Poisson–Dirichlet process with parameter θ. Uniformly over all couplings of these two processes,*

$$\liminf_{n \to \infty} (\log n)^{-1} \mathbb{E} \sum_{i \geq 1} |L_i^{(n)} - nL_i| \geq \frac{1}{4}\theta. \tag{9.1}$$

Proof. As in Theorem 2.1, any coupling has

$$\mathbb{E} \sum_{i \geq 1} |L_i^{(n)} - nL_i| \geq \int_{(0,1]} d(nx, \mathbb{Z})\mu(dx)$$

$$= \int_0^1 d(nx, \mathbb{Z})(1-x)^{\theta-1}\frac{\theta}{x}\,dx = \int_0^n d(x, \mathbb{Z})\left(1 - \frac{x}{n}\right)^{\theta-1}\frac{\theta}{x}\,dx$$

$$= \theta \int_0^n d(x, \mathbb{Z}) x^{-1}\,dx + \theta \int_0^n d(x, \mathbb{Z})\left(\left(1 - \frac{x}{n}\right)^{\theta-1} - 1\right)\frac{1}{x}\,dx$$

$$\sim \frac{\theta}{4} \log n,$$

this last following from the monotone convergence theorem and (2.3). $\square$

Theorem 9.2. *For any $\theta > 0$, the coupling of Section 8.1, using exactly the function f given by (8.4), but with the scale-invariant Poisson processes taken to have intensity $\theta/x\,dx$, achieves*

$$d_{\mathrm{PD}}(n) \sim \frac{\theta}{4} \log n.$$

Proof. Every consideration in Section 8, with the factor θ inserted into the intensity for the scale-invariant Poisson process, goes through exactly as it did in the special case $\theta = 1$. $\square$

Acknowledgement

The authors thank Ludger Rüschendorf for the reference to the alternative proof of Lemma 3.1.

References

[1] Arratia, R. (1996) Independence of prime factors: total variation and Wasserstein metrics, insertions and deletions, and the Poisson–Dirichlet process. Preprint.

[2] Arratia, R. (2002) On the amount of dependence in the prime factorization of a uniform random integer. In *Contemporary Combinatorics* (B. Bollobás, ed.), Vol. 10 of *Bolyai Society Mathematical Studies*, pp. 29–91.

[3] Arratia, R., Barbour, A. D. and Tavaré, S. (1992) Poisson process approximations for the Ewens Sampling Formula. *Ann. Appl. Probab.* **2** 519–535.

[4] Arratia, R., Barbour, A. D. and Tavaré, S. (1999) The Poisson–Dirichlet distribution and the scale invariant Poisson process. *Combin. Probab. Comput.* **8** 407–416.

[5] Arratia, R., Barbour, A. D. and Tavaré, S. (2003) *Logarithmic Combinatorial Structures: A Probabilistic Approach*, Vol. 1 of *EMS Monographs in Mathematics*, European Mathematical Society Publishing House, Zürich.

[6] Arratia, R. and Tavaré, S. (1992) The cycle structure of random permutations. *Ann. Probab.* **20** 1567–1591.

[7] Billingsley, P. (1972) On the distribution of large prime divisors. *Periodica Mathematica Hungarica* **2** 283–289.

[8] Dall'Aglio, G. (1956) Sugli estremi dei momenti delle funzioni di ripartizione doppia. *Ann. Scuola Norm. Sup. Pisa* **10** 35–74.

[9] Diaconis, P. and Pitman, J. W. (1986) Permutations, record values and random measures. Unpublished lecture notes, Statistics Department, University of California, Berkeley.

[10] Dudley, R. M. (1989) *Real Analysis and Probability*, Wadsworth and Brooks/Cole, Pacific Grove, CA.

[11] Engen, S. (1975) A note on the geometric series as a species frequency model. *Biometrika* **62** 97–699.

[12] Ewens, W. J. (1972) The sampling theory of selectively neutral alleles. *Theoretical Population Biology* **3** 87–112.

[13] Feller, W. (1945) The fundamental limit theorems in probability. *Bull. Amer. Math. Soc.* **51** 800–832.

[14] Griffiths, R. C. (1979) On the distribution of allele frequencies in a diffusion model. *Theoretical Population Biology* **15** 140–158.

[15] Ignatov, T. (1981) Point processes generated by order statistics and their applications. In *Point Processes and Queuing Problems* (P. Bartfái and J. Tomkó, ed.), North-Holland, pp. 109–116.

[16] Ignatov, T. (1982) A constant arising in the asymptotic theory of symmetric groups, and Poisson–Dirichlet measures. *Theory Probab. Appl.* **27** 136–147.

[17] Kingman, J. F. C. (1975) Random discrete distributions. *J. Royal Statist. Soc. Ser. B* **37** 1–22.

[18] Kingman, J. F. C. (1977) The population structure associated with the Ewens sampling formula. *Theoretical Population Biology* **11** 274–283.

[19] McCloskey, J. W. (1965) A model for the distribution of individuals by species in an environment. PhD thesis, Michigan State University.

[20] Rachev, S. T. and Rüschendorf, L. (1998) *Mass Transportation Problems, Part I: Theory*, Springer, New York.

[21] Rényi, A. (1962) On the outliers of a series of observations. *A Magyar Tudományos Akadémia Matematikai és Fizikai Tudományok Osztályának Közleményei*, **12** 105–121. Reprinted in *Selected papers of Alfréd Rényi*, Vol. 3, Akadémiai Kiadó (1976), pp. 50–65.

[22] Vershik, A. M. and Shmidt, A. A. (1977) Limit measures arising in the asymptotic theory of symmetric groups I. *Theory Probab. Appl.* **22** 70–85.

Positional Games

JÓZSEF BECK

Department of Mathematics, Rutgers University,
Busch Campus, Hill Center, New Brunswick, New Jersey 08903, USA
(e-mail: jbeck@math.rutgers.edu)

Received 10 March 2004; revised 7 August 2004

For Béla Bollobás on his 60th birthday

1. Informal introduction

This paper is partly an overview of the subject (Sections 1–4), and partly a research paper containing proofs for new results (Sections 5–7).

Every 'theory' of games concentrates on one particular aspect of games, and pretty much ignores the rest.

(1) Traditional *game theory* (J. von Neumann, J. Nash, *etc.*) focuses on the 'lack of complete information' (games like Poker). Its main result is a *minimax* theorem about *mixed* strategies ('random choice'), and it is basically linear algebra.

Games of 'complete information', like Chess, Go, Checkers, Nim, and Tic-Tac-Toe, are (almost completely) ignored by the traditional theory.

(2) One existing 'theory' for games of complete information is the 'theory of Nim-like games' (Berlekamp, Conway, Guy, *etc.*), which focuses on 'sum games', and it is basically algebra ('addition theory').

(3) In this paper I am talking about something completely different: I discuss a 'fake probability theory' of 'Tic-Tac-Toe-like games'. Note that 'Tic-Tac-Toe-like games' are officially called *positional games*.

The basic challenge of games of complete information is the 'combinatorial chaos'. To analyse a position (say in Chess), one has to examine all of its options, and all the options of these options, and all the options of the options of these options, and so on. The exhaustive search through the 'game tree' takes an *enormous* amount of time (usually more than the age of the universe). A (desperate) attempt to make up for the lack of time is to study the *random walk* on the game tree, that is, to study the *randomized game* where both players play randomly.

The *extremely surprising part* is that the probabilistic analysis of the randomized game can often be *converted* into optimal strategies via potential arguments. It is basically a game-theoretic adaptation of the so-called *probabilistic method* in combinatorics ('Erdős theory') applied to hopelessly complicated Tic-Tac-Toe-like games, where the 'usual shortcuts' (pairing strategy, *etc.*) fail miserably.

I consider it a 'mathematical paradox' for the following reason. Game theory is about *perfect* players, and I find it shocking that a play between *random generators* ('dumb players') has anything to do with a play between perfect players! 'Poker and randomness' is a natural combination: *mixed* strategy (*i.e.*, random choice among deterministic strategies) is necessary to make up for the lack of complete information (see, *e.g.*, the role of 'bluffing' in Poker). On the other hand, 'Tic-Tac-Toe and randomness' sounds like a mismatch. To explain the connection between 'Tic-Tac-Toe' and 'randomness' requires a long and detailed analysis.

First note that the connection is not 'trivial', in the sense that an optimal strategy is *never* a 'random play'. (In fact, a 'random play' usually leads to a quick, catastrophic defeat.) For games of 'complete information' the optimal strategies are *always* deterministic ('pure'). The 'fake probability theory' is employed to *find* a deterministic optimal strategy. This is where the connection is: the 'fake probability theory' is *motivated* by traditional probability theory, but eventually it is *derandomized* by *potential arguments*. In other words, I eventually get rid of probability theory completely, but the intermediate 'probabilistic step' is an absolutely crucial, inevitable part of the understanding process.

The 'fake probability theory' consists of the following main chapters:

 (i) game-theoretic first moment;
 (ii) game-theoretic second and higher moments;
(iii) game-theoretic 'independence'.

By using these tools I could find the *exact* solution of infinitely many natural games, thought to be completely hopeless before, like some 'clique games' and some 'subspace' versions of 'multi-dimensional Tic-Tac-Toe'.

Before describing the results in detail, first I want to talk about positional games (*i.e.*, 'Tic-Tac-Toe-like games') in general.

Chess, Tic-Tac-Toe, and Hex are among the most well-known games of complete information with no chance move. What is common in these apparently very different games? Well, in each game that player wins who achieves a 'winning configuration' first. A 'winning configuration' in Tic-Tac-Toe is a '3-in-a-row', in Hex it is a 'connecting chain of hexagons', and in Chess it is a 'capture of the opponent's King' (called a *checkmate*).

The objective of other well-known games of complete information like Checkers and Go is more complicated. In Checkers the goal is to be the first player to either capture all of the opponent's pieces ('checkers') or make it impossible for the opponent to make a move. The capture of a single piece ('jumping over') is a 'mini-win configuration', and similarly, an arrangement where the opponent cannot make a move is a 'winning configuration'.

In Go the goal is to capture as many stones of the opponent as possible ('capturing' means to 'surround a set of opponent's stones by a connected set').

These games are clearly very different, but the basic question is always the same: Which player can achieve a winning configuration *first*?

The bad news is that nobody knows *how* to achieve a winning configuration *first*, except by exhaustive case study. There is no general theorem whatsoever answering the question of 'how'. 'Strategy stealing' gives a partial answer to 'when', but doesn't say a word about 'how'. (Note that 'doing it first' means 'competition', a key characteristic of all game-playing.)

But if we ignore 'doing it first', then an even more fundamental question arises: What can a player achieve by his own moves against an adversary? Which configurations are achievable (but not necessarily first)? Or the (equivalent) 'complementary' question: What are the impossible configurations?

The main objective of 'positional game theory' is to 'draw the line' between achievable and impossible configurations in the class of 'Tic-Tac-Toe-like games'; for simplicity I ignore 'doing it first'.

Why 'Tic-Tac-Toe-like games'? Well, 'Tic-Tac-Toe-like games' are the *simplest case* in the sense that they are *static* games. Unlike Chess, Go, and Checkers, where the players repeatedly *relocate* or even *remove* pieces ('*dynamic* games') from the board, in Tic-Tac-Toe and Hex the players make permanent marks on the board, and 'relocating' or 'removing' is illegal. (Chess is particularly complicated. There are 6 types of pieces: King, Queen, Bishop, Knight, Rook, Pawn, and each one has its own set of rules of 'how to move the piece'. The 'instructions' for playing Tic-Tac-Toe are just a couple of lines, but the 'instructions' for playing Chess are several pages long.) The 'relative' simplicity of Tic-Tac-Toe-like games makes this class an ideal candidate for a mathematical theory.

What exactly are these 'Tic-Tac-Toe-like' games? Well, they are simply 'generalized Tic-Tac-Toe'. Nobody knows what 'generalized Chess' or 'generalized Go' are supposed to mean, but I think (almost) everybody would agree on what 'generalized Tic-Tac-Toe' should mean. In Tic-Tac-Toe the 'board' is a $3 \times 3 = 9$ element set, and there are 8 'winning triplets'. Similarly, one can play 'generalized Tic-Tac-Toe' on an arbitrary finite hypergraph, where the hyperedges are called 'winning sets', the union set is the 'board', the players alternately occupy elements of the 'board', and the winner is the player who occupies a whole 'winning set' first. This is what I call the *strong game*.

Unfortunately nobody knows *how* to win a strong game. For example, the $4 \times 4 \times 4 = 4^3$ Tic-Tac-Toe is a first-player's win, but the winning strategy is extremely complicated ('the size of a phone-book', according to O. Patashnik). The $5 \times 5 \times 5 = 5^3$ version is expected to be a draw, but no one can prove it. 'To occupy a winning line *first*', the common goal of these games, is an extremely difficult task. We know nothing beyond 'strategy stealing'. 'Strategy stealing' refers to the *existence argument* that 'if a draw is impossible, then the game is a first-player's win'. Strategy stealing does not give any hint about how to actually win. One way to do it is to search *all* strategies, but it is absurd: the total number is a doubly exponential function of the board size. The exhaustive search through all positions (searching the 'game tree', or the 'position graph'), called *backward labelling*, is more efficient, but still requires exponential time ('hard').

As I said before, nobody knows how to win a strong game. I had much more luck with *weak games*, the class of games where 'doing it first' is ignored. In a weak game, played on an arbitrary finite hypergraph, the two players are called 'Maker' and 'Breaker' (alternative names are 'Builder' and 'Blocker'). To win a strong game, a player has to 'build and block' at the *same* time. In a weak game these two jobs are separated, which makes the analysis somewhat *easier* (but not *easy*: Hex is clearly equivalent to a weak game, but it doesn't help to find an explicit first-player's winning strategy).

In a weak game, Maker's goal is to occupy a whole winning set, but not necessarily first. On the other hand, Breaker's goal is simply to prevent Maker from achieving his goal; Breaker does *not* want to build.

There is a trivial one-sided implication between strong and weak games (of course, played on the same hypergraph): if Breaker can mark every winning set (*i.e.*, wins the weak game), then the *same* strategy gives a draw in the strong game; I call it a *blocking draw*.

What I have been discussing so far was the 'achievement' version. The *reverse game* ('avoidance' version) is equally interesting (or even more interesting). In a reverse strong game the player who occupies a whole winning set first is the *loser*.

For example, consider the (K_n, K_q) clique game: the 'board' is a complete graph K_n on n vertices, the players alternately select edges from K_n, and the goal is a copy of K_q ('clique'). An almost trivial case study shows that the first player wins the strong (K_5, K_3) clique game in his 4th move, or before (the goal is a 'triangle'). The reverse 'triangle' game is much more exciting (and difficult). A very complicated case study (involving a computer) shows that the second player wins the (K_6, K_3) reverse strong game (the winning strategy is far from simple).

The general definition of the *reverse weak game* goes as follows. As usual, it is played on an arbitrary finite hypergraph. One player is a kind of 'anti-builder': he wants to avoid occupying a whole winning set – I call him *Avoider*. The other player is a kind of 'anti-blocker': he wants to force the reluctant Avoider to build a winning set – 'anti-builder' is officially called *Forcer*.

The best way to illustrate the striking connection between 'positional games' and 'random structures' is to study the weak and reverse weak versions of the (K_n, K_q) clique game. If $q = q(n)$ is 'very small' in terms of n, then Maker (or Forcer) can easily win. On the other hand, if $q = q(n)$ is 'not so small' in terms of n, then Breaker (or Avoider) can easily win. Where is the *game-theoretic breaking point*?

Well, for 'small' ns I don't know the answer, but for 'large' ns I know the *exact* value of the breaking point! Indeed, assume that n is sufficiently large, say $n > 10^{100}$. If we take the 'lower integral part'

$$q = \lfloor 2\log_2 n - 2\log_2 \log_2 n + 2\log_2 e - 3 \rfloor$$

(base 2 logarithm), then Maker (or Forcer) wins. On the other hand, if we take the 'upper integral part'

$$q = \lceil 2\log_2 n - 2\log_2 \log_2 n + 2\log_2 e - 3 \rceil,$$

then Breaker (or Avoider) wins.

(I have to admit, I 'cheated' a little bit: if $2\log_2 n - 2\log_2 \log_2 n + 2\log_2 e - 3$ is 'very close' to an integer, then I don't know who wins. But for the overwhelming majority of

ns the expression above is *not* too close to an integer. So for the overwhelming majority of ns I know exactly.)

First of all, I find it very surprising that the weak clique game and the *reverse* weak clique game have *exactly* the same breaking point. I feel this contradicts 'common sense'. I would expect that an eager Maker in the 'straight' game has a good chance of building a *bigger* clique than a reluctant Avoider in the reverse version, but this 'natural' expectation turned out to be wrong. I cannot give any *a priori* reason why the two breaking points coincide. All that I can say is that the highly technical proof of the 'straight' case (40 pages) can be easily adapted (*e.g., maximum* is replaced by *minimum*) to yield the same breaking point for the reverse game, but this is hardly the answer we are looking for.

The reader is probably wondering about the mysterious expression

$$2\log_2 n - 2\log_2 \log_2 n + 2\log_2 e - 3.$$

What is this? Well, an expert in the theory of random graphs immediately recognizes that $2\log_2 n - 2\log_2 \log_2 n + 2\log_2 e - 3$ is exactly 2 less than the 'clique number of the symmetric random graph $\mathbf{G}(n, 1/2)$' (1/2 is the edge probability).

A combination of the first and second moment methods (standard probability theory) shows that the clique number $\omega(\mathbf{G}(n, 1/2))$ of the random graph has a very strong concentration. Typically it is concentrated on a *single* integer with probability $\to 1$ as $n \to \infty$ (and even in the worst case there are at most two values). Indeed, the expected number of q-cliques in $\mathbf{G}(n, 1/2)$ equals

$$f(q) = f_n(q) = \binom{n}{q} 2^{-\binom{q}{2}}.$$

The function $f(q)$ drops under 1 around $q \approx 2\log_2 n$. The 'real solution' of the equation $f(q) = 1$ is

$$q = 2\log_2 n - 2\log_2 \log_2 n + 2\log_2 e - 1 + o(1), \tag{1.1}$$

which is exactly 2 more than the game-theoretic breaking point

$$q = 2\log_2 n - 2\log_2 \log_2 n + 2\log_2 e - 3 + o(1), \tag{1.2}$$

mentioned above.

The strong concentration of the clique number of the random graph is not as terribly surprising as it seems at first sight. Indeed, $f(q)$ is a very *rapidly* changing function:

$$\frac{f(q)}{f(q+1)} = \frac{q+1}{n-q} 2^q = n^{1+o(1)}$$

if $q \approx 2\log_2 n$. On an intuitive level it is explained by the fact that if q switches to $q+1$, then $\binom{q}{2}$ switches to $\binom{q+1}{2} = \binom{q}{2} + q$, which is a 'square root size increase'.

Is there a 'reasonable clique game' for which the breaking point is exactly (1.1), that is, the clique number of the random graph? The answer is 'yes', and the game is a *Picker–Chooser game*. To motivate the Picker–Chooser game, note that the *alternating* Tic-Tac-Toe-like play splits the board into two equal (or almost equal) parts. But there are many other ways to divide the board into two equal parts. The 'I-cut-you-choose' way (motivated by how a couple shares a single piece of cake after dinner) goes as follows: in each move, Picker picks two previously unselected points of the board, Chooser chooses one of them, and the other one goes back to Picker. In the *Picker–Chooser* game Picker is the 'builder' (*i.e.*, he wants to occupy a whole winning set) and Chooser is the 'blocker' (*i.e.*, his goal is to mark every winning set).

When Chooser is the 'builder' and Picker is the 'blocker', then I call it the *Chooser–Picker* game.

The proof of the theorem that (1.1) is the exact value of the breaking point for the (K_n, K_q) Picker–Chooser clique game is based on the concepts of

(a) game-theoretic first moment; and
(b) game-theoretic second moment.

The proof is far from trivial, but not so terribly difficult either (because Picker has so much control of the game). It is a perfect 'stepping stone' before conquering the much more challenging weak and reverse weak, and also the Chooser–Picker versions. These three clique games all have the *same* breaking point, namely (1.2). What is (1.2)?

Well, (1.2) is the 'real solution' of the equation $f(q) = \binom{n}{2}/2$. The intuitive meaning of the equation $f(q) = \binom{n}{2}/2$ is that the overwhelming majority of the edges of the random graph are covered by exactly one copy of K_q. In other words, the random graph may have a large number of copies of K_q, but they are well spread (un-crowded); in fact, there is room enough to be typically pairwise edge-disjoint. This suggests the following *heuristic argument*. Assume that we are at a 'last stage' of a clique game where Maker (in the weak game) has a large number of 'almost complete' K_qs: 'almost complete' in the sense that, in each one, all but *two edges* are occupied by Maker, and these 'extremely dangerous' K_qs are pairwise edge-disjoint. Then Breaker can still escape from losing: he can block these disjoint unoccupied edge-pairs by a simple *pairing strategy*! It is exactly the pairing strategy that distinguishes the Picker–Chooser game from the rest of the bunch. Indeed, in each of the weak, reverse weak, and Chooser–Picker games the 'blocker' can easily *win* the *disjoint game*, where the winning sets are disjoint, and contain at least two elements each, by employing a pairing strategy. In sharp contrast, Chooser always *loses* a 'sufficiently large' disjoint game in the Picker–Chooser version (more precisely: if there are at least 2^n disjoint n-element winning sets, then Picker wins the Picker–Chooser game).

This is the best *intuitive explanation* that I could come up with to understand breaking point (1.2). This intuition *requests* the 'random graph heuristic', that is, to (artificially) introduce a random structure in order to understand a deterministic game of complete information.

But the connection is much deeper than that. To *prove* that (1.2) is the exact value of the game-theoretic breaking point, one *requires* a *fake probability theory* (or at least this is how I could do it). The main steps of the proof are:

(1) game-theoretic first moment;
(2) game-theoretic higher moments ('sliding potentials'); and
(3) game-theoretic *independence*.

Developing (1)–(3) is a long and difficult task. The word 'fake' in *fake probability theory* refers to the fact that, when I actually define an optimal strategy, the 'probabilistic part' completely disappears. It is a 'metamorphosis': as a caterpillar turns into a butterfly, the 'probabilistic arguments' are converted into (deterministic) 'potential arguments'.

Note that 'potential arguments' are widely used in puzzles ('one-player games'). A well-known example is Conway's *Solitaire Army* puzzle: arrange men behind a line, and then, by playing 'jump and remove' horizontally or vertically, move a man as far across the line as possible. Conway's beautiful 'golden ratio' proof, a striking potential argument, shows that it is *impossible* to send a man forward 5 (4 is possible). Conway's result is from the early 1960s. It was shown very recently that if 'jumping a man diagonally' is permitted, then '5' is replaced by '9'; in other words, it is impossible to send a man forward 9, but 8 is possible. (The idea is the same, but the details are substantially more complicated.)

It is quite natural to use potential arguments to describe *impossible configurations* (like Conway did). It is more surprising that potential arguments are equally useful to describe *achievable configurations* ('Maker's win') as well. But the biggest surprise is that the 'Maker's win criteria' and the 'Breaker's win criteria' often coincide, yielding *exact* solutions of several seemingly hopeless games. However, there is a basic difference: Conway's argument works for small values like '5', but my arguments give sharp results for 'large values' of the parameters only; I call it a 'game-theoretic law of large numbers'.

These *exact solutions* all depend on the concept of 'game-theoretic independence'. What is 'game-theoretic independence'? Well, there is a 'trivial' and a 'nontrivial' interpretation of game-theoretic *independence*.

The 'trivial' (but still very useful) interpretation is about *disjoint* games. Consider a sequence of hypergraphs such that, in each one, Breaker (as the second player) can mark every winning set. If the hypergraphs are pairwise disjoint (in the strong sense that the 'boards' are disjoint), then of course Breaker can block the union hypergraph as well. Disjointness guarantees that, in any component, either player can play 'independently' from the rest of the components. For example, the concept of *pairing strategy* is based on this simple observation.

In the 'nontrivial' interpretation, the initial game does *not* fall apart into disjoint components. Instead Maker can *force* that eventually, in a much later stage of the play, the family of unblocked (yet) hyperedges *does* fall apart into much smaller (disjoint) components. This is how Maker can eventually finish the job blocking the whole initial hypergraph, namely 'blocking componentwise' in the 'small' components.

A convincing 'probabilistic intuition' behind the 'nontrivial' version is the well-known *Lovász Local Lemma* [12]. The LLL is a remarkable probabilistic *sieve* argument to prove the *existence* of certain very complicated structures that we are unable to construct directly (see [12] and [1]). To be precise, let $E_1, E_2, \ldots, E_s$ denote events in a probability space. In the applications, the E_is are 'bad' events, and we want to avoid all of them, that is, we wish to show that $\text{Prob}(\cup_{i=1}^{s} E_i) < 1$. A trivial way to guarantee this is to assume $\sum_{i=1}^{s} \text{Prob}(E_i) < 1$. A completely different way is when $E_1, E_2, \ldots, E_s$ are *mutually independent* and all $\text{Prob}(E_i) < 1$. Indeed, then $\text{Prob}(\cup_{i=1}^{s} E_i) = 1 - \prod_{i=1}^{s}(1 - \text{Prob}(E_i)) < 1$. The LLL applies in the very important case when we don't have mutual independence, but 'independence dominates' in the following precise sense: each one of these events is *independent* of all but a *small number* of other events; if $\text{Prob}(E_i) \leq p < 1$ holds uniformly, then a good choice for 'small number' is $(4p)^{-1}$.

A typical application of the LLL goes as follows (see [12]).

Erdős–Lovász theorem. *Let $\mathscr{F} = \{A_1, A_2, A_3, \ldots\}$ be an n-uniform hypergraph. Suppose that each A_i intersects at most 2^{n-3} other $A_j \in \mathscr{F}$ ('local size'). Then there is a 2-colouring of the 'board' $V = \bigcup_i A_i$ such that no $A_i \in \mathscr{F}$ is monochromatic.*

The conclusion means that there exists a *drawing terminal position* (well, *almost* true: in a drawing terminal position the two colour-classes are of *equal* size). The very surprising message of the Erdős–Lovász theorem is that the 'global size' of hypergraph $\mathscr{F}$ is irrelevant (it can even be infinite!): all that matters is the 'local size'.

Of course, the existence of a single (or even many) drawing terminal position does *not* guarantee at all the existence of a *drawing strategy*. But perhaps it is still true that 'under the Erdős–Lovász condition (or under some slightly weaker local condition), Breaker (or Avoider, or Picker) has a *blocking draw strategy*, i.e., he can block every winning set in the weak (or reverse weak, or Chooser–Picker) game on $\mathscr{F}$'.

This is a wonderful problem; I call it the 'game-theoretic LLL conjecture'. Unfortunately the conjecture is still open in general, in spite of all my efforts trying to prove it during the last 10 years.

Open Problem 1. *Prove the 'game-theoretic LLL conjecture'.*

I could prove, however, several partial results, which led to interesting applications. In fact, I have developed 3 different techniques, which all have their own range of applications.

A very important special case, when the conjecture is 'nearly proved', is the class of *almost disjoint hypergraphs*: where any two hyperedges have at most one common point. This is certainly the case for 'lines', the winning sets of the n^d Tic-Tac-Toe.

What can I prove about the n^d Tic-Tac-Toe? Well, I proved that it is a draw game even if the dimension d is as large as $d = c_1 n^2 / \log n$, that is, nearly quadratic in terms of (the winning size) n. What is more, the 'draw' is a stronger *blocking draw*: the second player can mark every winning line (if they play as long as the whole board is occupied).

Note that this bound is nearly best possible: if $d > c_2 n^2$ then the second player *cannot* force a blocking draw.

How come that for the clique game I could find the *exact* value of the breaking point, but for the multi-dimensional Tic-Tac-Toe I couldn't even find the asymptotic truth (because of a factor of $\log n$ in the denominator)? Well, the answer is somewhat technical. The winning lines in the multi-dimensional n^d Tic-Tac-Toe form an extremely *irregular* hypergraph: the maximum degree is *much* larger than the average degree. This is why I cannot apply my 'blocking criteria' directly to the 'n^d hypergraph'. First I have to employ a 'truncation procedure' to bring the maximum degree close to the average degree, and the price that I pay for the 'degree reduction' is the loss of a factor of $\log n$.

However, if I consider the n^d *torus* Tic-Tac-Toe, then the corresponding hypergraph becomes perfectly uniform (the torus is a *group*). For example, every point of the n^d torus Tic-Tac-Toe has $(3^d - 1)/2$ winning lines passing through it. For the n^d torus Tic-Tac-Toe

I can prove asymptotically sharp thresholds – in fact, very sharp results in the quantitative sense that the 'error' is 'logarithmically small'.

A *winning line* in the n^d Tic-Tac-Toe is 'n points on a line forming an arithmetic progression'. This motivates the 'arithmetic progression game': the board is the interval $1, 2, \ldots, N$, and the goal is to build an n-term arithmetic progression. The corresponding hypergraph is 'nearly regular', and this is why I can prove asymptotically very sharp thresholds: the breaking point for n is $\log_2 N + O(\log_2 \log_2 N)$; again the 'error' is 'logarithmically small'.

Let us return to the n^d torus Tic-Tac-Toe. If 'winning line' is replaced by 'winning plane' (or 'winning subspace of dimension ≥ 2' in general), then I can determine the *exact* value of the game-theoretic threshold, just like in the clique game. For example, a *winning plane* is an $n \times n$ lattice in the n^d torus. This is another 'rapidly changing structure': if n switches to $n+1$, then $n \times n$ switches to $(n+1) \times (n+1)$, which is again a 'square root size increase' just like in the case of the 'clique'. This formal similarity to the clique game explains why I had a chance of finding the *exact* value of the game-theoretic breaking point (the actual proof is rather different).

Let me summarize the meaning of 'game-theoretic independence'. It is about Tic-Tac-Toe-like games for which the 'local size' is much smaller than the 'global size'. Even if the game starts out as a coherent entity, either player can force it to develop into 'small' *composites* in the sense that every component-size is less than the 'local size'. A sort of intuitive explanation behind it is the LLL, which is a sophisticated application of *statistical* independence.

The exact solution of the clique game and the 'subspace game' of dimension ≥ 2 played on the n^d torus are both difficult and long. They will be published somewhere else. In the rest of this paper I focus on some simple but still very interesting applications of the basic technique. (The only exception is Section 7, which is a little more complicated.) From now on I stop the informal style, and switch to a rigorous treatment.

2. Strong and weak games

I first recall a few concrete games which are won by the first player to complete some kind of winning configuration. These examples motivate the general definitions.

Tic-Tac-Toe (or Noughts and Crosses). The game board is a big square which is partitioned into $3 \times 3 = 9$ congruent small squares. Whoever moves first puts a nought in one of the nine small squares. The opponent puts a cross into any other small square, and then they alternate nought and cross in the remaining empty squares until one player wins by getting three of his own squares in a line. If neither player gets 3-in-a-row the play is a draw. There are eight winning triplets: three horizontal, three vertical and two diagonal lines (of slope 1 and -1).

Note that Tic-Tac-Toe is a '3-in-a-row' game on a 3×3 board. The 2-dimensional generalization is the 'n-in-a-row' game on an $n \times n$ board; we simply call it the n^2 game. The multi-dimensional generalization is n^d Tic-Tac-Toe, or simply the n^d game.

n^d game. Two players alternately put their marks in the cells of a d-dimensional cube of size $n \times \cdots \times n = n^d$. The winner is the player who occupies a full-length-line *first*, i.e., who has n of his marks in a line *first*. More precisely, the board V of the n^d game is the set of d-tuples

$$V = \{\mathbf{a} = (a_1, a_2, \ldots, a_d) : 1 \leqslant a_j \leqslant n \text{ for each } 1 \leqslant j \leqslant d\}.$$

The winning sets of the n^d game are the n-in-a-line sets, that is, the n-element sequences

$$\left(\mathbf{a}^{(1)}, \mathbf{a}^{(2)}, \ldots, \mathbf{a}^{(n)}\right)$$

of the board V such that, for each j, the sequence $a_j^{(1)}, a_j^{(2)}, \ldots, a_j^{(n)}$ composed of the jth coordinates is either $1, 2, 3, \ldots, n$ ('increasing'), or $n, n-1, n-2, \ldots, 1$ ('decreasing'), or a *constant*. In other words, the winning sets are exactly the n-in-a-lines on the n^d board. If neither player gets n-in-a-line the play is a draw.

Note that in higher dimensions most of the winning lines are some kind of diagonal. The special case $n = 3, d = 2$ gives ordinary Tic-Tac-Toe.

The total number of winning lines in the n^d game is $((n+2)^d - n^d)/2$. Indeed, for each j, the sequence $a_j^{(1)}, a_j^{(2)}, \ldots, a_j^{(n)}$ composed of the jth coordinates is either *increasing* (i.e., $1, 2, 3, \ldots, n$), or *decreasing* (i.e., $n, n-1, n-2, \ldots, 1$), or a *constant* $c \in \{1, 2, \ldots, n\}$. Since for each coordinate we have $(n+2)$ possibilities, this gives $(n+2)^d$. But we have to subtract n^d since in a line at least one coordinate must change. Finally, we have to divide by 2 because every line has two orientations.

An alternative geometric way of getting $((n+2)^d - n^d)/2$ goes as follows. Imagine the board n^d is surrounded by an additional layer of cells, one cell thick. This new object is a cube

$$(n+2) \times (n+2) \times \cdots \times (n+2) = (n+2)^d.$$

It is easy to see that every winning line of the n^d-board extends to a uniquely determined *pair* of cells in the new surface layer. So the total number of lines is

$$((n+2)^d - n^d)/2. \tag{2.1}$$

A far-reaching generalization of n^d Tic-Tac-Toe is the class of *strong positional games* (see [9], [10], [13], [16]).

Strong positional games. Let $(V, \mathscr{F})$ be an arbitrary finite hypergraph. Here V is a finite set, called the *board* of the game, and $\mathscr{F}$ is an *arbitrary* family of subsets of V, called the family of *winning sets*. The two players, First Player and Second Player, alternately occupy previously unoccupied elements ('points') of board V. That player wins who occupies all the elements of some winning set $A \in \mathscr{F}$ *first*; otherwise the play ends in a draw.

Sometimes we just give the family $\mathscr{F}$ of winning sets. Then the board V is the union $\bigcup_{A \in \mathscr{F}} A$ of all winning sets.

First I recall Zermelo's classical theorem. Note that the old theorems are lettered and the new main theorems are numbered (see Sections 5–7).

Theorem A. *Every finite perfect-information 2-player game is* determined, *which means that either*

(a) *First Player has a winning strategy, or*
(b) *Second Player has a winning strategy, or*
(c) *both of them have a drawing strategy.*

Remarks. These three alternatives are what one calls the three *outcomes of a game*. Of course every play has three possible outcomes (First Player wins, Second Player wins, or a draw), but the outcome of a particular play is not necessarily the 'same' as the *outcome of the game*. For example, Second Player can easily lose in ordinary Tic-Tac-Toe (if First Player opens in the centre, and Second Player replies on the side), even if Tic-Tac-Toe is a draw game (*i.e.*, belongs to case (c)).

Proof. The *existence* proof of Zermelo's theorem is a simple application of De Morgan's law. Indeed, there are three alternatives only: either

(a) First Player (= **I**) has a winning strategy: $\exists x_1 \forall y_1 \exists x_2 \forall y_2 \cdots$ such that **I** wins; or
(b) Second Player (= **II**) has a winning strategy: $\forall x_1 \exists y_1 \forall x_2 \exists y_2 \cdots$ such that **II** wins;
(c′) or the negation of (a)∨(b):

$$\neg((\exists x_1 \forall y_1 \exists x_2 \forall y_2 \cdots \textbf{ I wins}) \vee (\forall x_1 \exists y_1 \forall x_2 \exists y_2 \cdots \textbf{ II wins})),$$

which, by De Morgan's law, is equivalent to

$$(\forall x_1 \exists y_1 \forall x_2 \exists y_2 \cdots \textbf{ I loses or draw}) \wedge (\exists x_1 \forall y_1 \exists x_2 \forall y_2 \cdots \textbf{ II loses or draw}).$$

So the third alternative is that both players have a drawing strategy, *i.e.*, alternative (c). This completes the existence proof of Theorem A. □

Every game can be visualized as a 'tree of all possible plays' (called a game tree). The exhaustive search of the game tree gives a *constructive* proof of Zermelo's theorem. It provides an *explicit* winning (or drawing) strategy. The bad news is that exhaustive search is usually impractical. From a complexity point of view a better way to do the same thing is to work with the (usually smaller) position-graph; then 'exhaustive search' is called 'backward labelling'. Unfortunately even 'backward labelling' is impractical: it takes 'exponential time'.

Strategy stealing argument. It is well known that whoever plays first in a strong positional game can force at least a draw. In other words, for strong games alternative (b) cannot hold. Heuristically this is obvious because strong positional games are symmetric, and First Player has the advantage of the first move (note that an extra move in a strong game does not harm a player). The precise proof of this remarkable result is just a little bit more complicated.

Theorem B (strategy stealing). *Let* $(V, \mathcal{F})$ *be an arbitrary finite hypergraph. Then, playing the strong positional game on* $(V, \mathcal{F})$, *First Player can force at least a draw, i.e., a draw or possibly a win.*

Unfortunately the 'strategy stealing argument' doesn't supply any *explicit* strategy.

Open Problem 2. *Find an explicit First Player's drawing strategy in every n^d game.*

When a draw is impossible: Ramsey theory. When can First Player win in a strong positional game? A partial answer (sufficient condition) is the following: First Player has a *winning strategy* in a strong game when a draw is impossible.

Of course this condition is not necessary. For example, the full branches of a binary tree with n levels form an n-uniform family of 2^{n-1} winning sets such that First Player has an easy win (the players take the vertices of the tree). But this strong game has plenty of drawing terminal positions (*e.g.*, all vertices of degree one are red, and the rest of the vertices are all blue).

A drawing terminal position in a strong game $(V, \mathcal{F})$ means a 'halving' 2-colouring of the board V such that no winning set $A \in \mathcal{F}$ is monochromatic. A slightly more general concept is when we allow *arbitrary* 2-colourings (*i.e.*, the two colour classes may have very different sizes).

The *chromatic number* $\chi(\mathcal{F})$ of hypergraph $\mathcal{F}$ is the least integer $r \geq 2$ such that the elements of the board V can be coloured with r colours yielding no monochromatic $A \in \mathcal{F}$. If the chromatic number of $\mathcal{F}$ is bigger than 2, then a draw is impossible. In this case First Player's (at least) drawing strategy in Theorem B is actually a winning strategy (see, *e.g.*, [15]).

Theorem C (Ramsey criterion). *Suppose that the board V is finite, and the family $\mathcal{F}$ of winning sets has chromatic number at least three. Then First Player has a winning strategy in the strong positional game on $(V, \mathcal{F})$.*

Theorem C doesn't say a word about *how* to win. Unfortunately nobody knows how to win. The only general advice I know is 'try exhaustive search'! Theorem C describes a subclass of strong positional games with the remarkable property that one can easily diagnose the winner without being able to say how one wins.

A fundamental result in Ramsey theory is the Hales–Jewett theorem [16]. This is basically the combinatorial content of van der Waerden's famous theorem on arithmetic progressions. It states that there is a (least) *finite* threshold number $h(n)$ such that the family of all 'n-in-a-lines' in the n^d game (*i.e.*, the family of winning sets) has chromatic number ≥ 3 if $d \geq h(n)$. We call $h(n)$ the Hales–Jewett threshold. By the Ramsey criterion (Theorem C), First Player has a winning strategy in the n^d game if $d \geq h(n)$.

Open Problem 3. *Find an explicit First Player's winning strategy in the n^d game when $d \geq h(n)$.*

Unfortunately our present knowledge on the Hales–Jewett threshold number $h(n)$ is rather poor. The best-known upper bound on $h(n)$ is a breakthrough result of Shelah [18] from 1988. It is a primitive recursive function (the *supertower* function), which is much-much better than the original van der Waerden–Hales–Jewett threshold (the original argument gave the 'notorious' Ackermann function: see, *e.g.*, [15]). Unfortunately Shelah's bound is still enormous for 'pedestrian combinatorics'. For example, it is an open problem whether $h(n)$ is less than the 'plain' *tower* function $\text{tower}_n(2)$, where $\text{tower}_k(x)$ denotes the k-fold iteration of the exponential function: $\text{tower}_1(x) = 2^x$ and for $k \geq 2$, $\text{tower}_k(x) = 2^{\text{tower}_{k-1}(x)}$ (so, *e.g.*, $\text{tower}_2(x) = 2^{2^x}$). To get an idea about the order of magnitude of Shelah's bound, the reader is recommended to study the *supertower* function defined as follows. Let $\text{tower}(k; x) = \text{tower}_k(x)$, and

$$\text{supertower}(n) = \text{tower}\Big(\text{tower}\big(\cdots \text{tower}(\text{tower}(2;2);2) \cdots ;2\big);2\Big),$$

where the right-hand side is an n-fold iteration.

In the other direction, Hales and Jewett [16] proved the linear lower bound $h(n) \geq n$ by an explicit construction. A couple of years ago I could improve this linear bound to a nearly quadratic bound $h(n) > c_0 \cdot n^2 / \log n$ where $c_0 > 0$ is an absolute constant (this was the first improvement since 1963). Of course there remains an enormous discrepancy between the (nearly) quadratic lower bound and Shelah's supertower upper bound, and we don't have the slightest idea which one is closer to the truth.

It seems highly unlikely that the game-theoretic breaking point for the n^d game is anywhere close to the Hales–Jewett threshold number $h(n)$, but no method is known for handling this problem.

I mention two 'plausible conjectures'.

Open Problem 4. *Is it true that if the n^d game is a First Player's win then the n^{d+1} game is also a win?*

A good reason why Open Problem 4 is so difficult is that one can construct a finite hypergraph (*i.e.*, a strong game) which is a First Player's win, but adding an extra winning set turns it into a draw game. An even more sophisticated 'extra set paradox' is the following. One can construct a finite hypergraph which is a draw, but it has an induced sub-hypergraph (*i.e.*, all winning sets in a subset of the board) which is a First Player's win. One can even construct an example where the hypergraph is uniform.

I challenge the reader to find such examples.

Open Problem 5. *Is it true that if the n^d game is a draw then the $(n+1)^d$ game is also a draw?*

Note that Golomb and Hales [14] proved the weaker result that 'if the n^d game is a draw then the $(n+2)^d$ game is also a draw'. I am sure the reader is wondering 'Why is it easier from n to $n+2$ than from n to $n+1$?' Well, a good reason is that the n^d game is the 'interior' of the $(n+2)^d$ game, explaining why it is more natural to switch from n to $n+2$. The lack of any simple connection between the n^d and $(n+1)^d$ games raises the possibility of an occasional *nonmonotonicity* here; that is, it may be, for some n and d, that the n^d game is a draw, the $(n+1)^d$ game is a First Player's win, and the $(n+2)^d$ game is a draw again.

I believe (but cannot prove) that *nonmonotonicity* never happens.

Weak game: Maker–Breaker version. As we have seen in Theorem B, Second Player has no chance of winning a strong game against a perfect First Player. Then, why does he not just concentrate on preventing First Player from building a winning set, and simply ignore his own desire for building? We thus can name him Breaker, and the other one Maker. (I usually assume that Maker is the first player, and Breaker is the second player.)

In general, on the same hypergraph $(V, \mathcal{F})$, one can play the 'symmetric' strong game, and also the 'asymmetric' *weak game*: the *Maker–Breaker version*. Here Maker's aim is to claim every element of a winning set $A \in \mathcal{F}$ (but not necessarily first), and Breaker's aim is simply to prevent Maker from doing so (*i.e.*, Breaker's goal is to put his mark in every winning set). The winner is the one who achieves his goal; so a draw is impossible by definition.

If First Player can force a win in the strong game on $(V, \mathcal{F})$, then of course the *same* play gives him, as Maker, a win in the weak (Maker–Breaker) version (on the same hypergraph). But the converse is not true. It is possible that Maker, as the first player, has a winning strategy in the weak version, while Second Player can force a draw in the strong version. This happens for example in the standard 3^2 Tic-Tac-Toe: the strong 3^2 game is a draw, but the weak version is a win for Maker (as the first player). This is why one can call a Maker's winning strategy in the weak game a 'weak win strategy'. Breaker's winning strategy in the weak game can be called a 'blocking draw strategy'. I know a lot about the complementary concepts of 'weak win' and 'blocking draw', but I know (almost) nothing about a 'strong win' (*i.e.*, a win in the strong game) except Theorem C.

While playing the strong version on a hypergraph, both players have their own threats, and either of them, fending off the other's, may build his own winning set. Therefore, a play is a delicate balancing between threats and counter-threats and can be of very intricate structure even if the hypergraph itself is simple.

The weak version is usually somewhat simpler. Maker doesn't have to waste valuable moves fending off his opponent's threats. Maker can simply concentrate on his own goal of building. For example, the Maker–Breaker version of Open Problem 4 is trivial: Maker uses the winning strategy within, for example, a d-dimensional subcube of n^{d+1}.

However, 'easier' doesn't mean that Maker–Breaker games are 'easy'. For example, the well-known and notoriously difficult Hex is equivalent to a Maker–Breaker game, but it doesn't help to 'break' the game. In Hex the board is a rhombus of hexagons of size $n \times n$ (the standard size is $n = 11$). The two players, White (the first player) and Black (the second player) take the two pairs of opposite sides of the board. The players alternately put their pieces on unoccupied hexagons (White has white pieces, and Black has black pieces). White wins if his pieces connect his opposite sides of the board, and Black wins if his pieces connect the other pair. Observe that Hex is *not* a strong game: the winning sets for White and Black are mostly different (the only common winning sets are the chains connecting opposite corner points). Let Hex(+) denote the weak (Maker–Breaker) game in which the board is the $n \times n$ Hex board, Maker = White, Breaker = Black, and the winning sets are the connecting chains of White. We claim that Hex and Hex(+) are equivalent. First note that a draw in Hex is impossible. Indeed, in order to prevent the opponent from making a connecting chain, one must build a 'river', which itself contains a chain connecting the *other* pair of opposite sides (this fact seems plausible, but the precise proof is not completely trivial). This means that Breaker's goal in Hex(+) (*i.e.*, 'blocking') is equivalent to Black's goal in Hex (*i.e.*, 'building first'). Here 'equivalent' means that Breaker has a winning strategy in Hex(+) if and only if Black has a winning strategy in Hex. Since a draw is impossible, Hex and Hex(+) are equivalent.

By the strategy stealing argument White has a winning strategy in Hex. This remarkable result was discovered by J. Nash in the 1940s (perhaps the first application of strategy stealing). The fact that Hex is equivalent to a *weak game* doesn't make it 'easy': we don't have the slightest clue how to actually find an *explicit* winning strategy. It remains a long-standing open problem.

3. Potential functions

Positions with limited potential: Solitaire Army. *Solitaire Army* is a particular case of the class of *Solitaire* puzzles (*Peg Solitaire* in America). These puzzles are not 'games': a 'game' requires 2 players, and here there is only one player. The reason why I discuss this puzzle is to demonstrate the potential function technique on a simple (and very elegant) example.

The common feature of the Solitaire puzzles is that each one is played with a board and men (pegs), the board contains a number of holes each of which can hold one man. Each move consists of a jump by one man over one or more other men, the men jumped over being removed from the board. Each move therefore reduces the number of men on the board.

The Solitaire Army is played on the infinite plane and the holes are in the lattice points. (see pp. 715–7 in Vol. 2 of *Winning Ways* [11]). The permitted move is to jump a man horizontally or vertically but not diagonally. Let us draw a horizontal line across the infinite board and start with all men behind this line. Assume this line is the horizontal axis, so all men are in the lower half-plane. How many men do we need to send one man forward 1, 2, 3, 4, or 5 holes into the upper half-plane?

Obviously two men are needed to send a man forward one hole. After playing around for a few minutes everyone can see how four men are enough to send a man forward two holes, and how eight men are enough to send a man forward three holes. It takes a couple of hours to discover how twenty men are enough to send a man forward four holes.

But the really surprising result is the case of five holes: it is *impossible* to send a man forward five holes into the upper half-plane. This striking result was discovered by Conway in 1961.

Conway's 'potential proof' goes as follows. We assign a weight to each hole subject to the condition that if H1, H2, H3 are *any* three consecutive holes in a row or in a column, and $w(H1), w(H2), w(H3)$ are the corresponding weights, then $w(H1) + w(H2) \geqslant w(H3)$. We can evaluate a position by the sum of the weights of those holes which are occupied by men – this sum is called the *value* of the position,

The meaning of inequality $w(H1) + w(H2) \geqslant w(H3)$ is very simple. The effect of a move where a man in H1 jumps over another man in H2 and arrives at H3 is that we replace men with weights $w(H1)$ and $w(H2)$ by a man with weight $w(H3)$. Since $w(H1) + w(H2) \geq w(H3)$, this change cannot be an increase in the value of the new position.

Inequality $w(H1) + w(H2) \geq w(H3)$ guarantees that *no play is possible from an initial position to a target position if the target position has a higher value.*

Let w be a positive number which satisfies $w + w^2 = 1$: w equals the golden ratio $\frac{\sqrt{5}-1}{2}$. Now Conway's potential argument goes as follows. Assume that one succeeded in sending a man 5 holes forward into the upper half-plane by starting from a configuration of a finite number of men in the lower half-plane. Write 1 where the man stands 5 holes forward into the upper half-plane, and extend it in the following way:

$$
\begin{array}{c}
1 \\
w \\
w^2 \\
w^3 \\
w^4 \\
\cdots \; w^9 \; w^8 \; w^7 \; w^6 \; w^5 \; w^6 \; w^7 \; w^8 \; w^9 \; \cdots \\
\cdots \; w^{10} \; w^9 \; w^8 \; w^7 \; w^6 \; w^7 \; w^8 \; w^9 \; w^{10} \; \cdots \\
\cdots \; w^{11} \; w^{10} \; w^9 \; w^8 \; w^7 \; w^8 \; w^9 \; w^{10} \; w^{11} \; \cdots
\end{array}
$$
$$\ldots\ldots\ldots\ldots\ldots\ldots\ldots\ldots\ldots\ldots\ldots\ldots\ldots\ldots\ldots$$

The value of the top line of the lower half-plane is

$$w^5 + 2w^6 + 2w^7 + 2w^8 + \cdots = w^5 + 2\frac{w^6}{1-w} = w^5 + 2\frac{w^6}{w^2} = w^5 + 2w^4 = w^3 + w^4 = w^2.$$

So the value of the whole lower half-plane is

$$w^2(1 + w + w^2 + w^3 + \cdots) = w^2 \frac{1}{1-w} = w^2 \frac{1}{w^2} = 1,$$

which is exactly the value of the target position. So no finite number of men in the lower half-plane will suffice to send a man forward five holes into the upper half-plane.

One can even show that eight men are in fact *needed* to send a man forward three holes, and similarly, twenty men are *needed* to send a man forward four holes (see Beasley's excellent little book [3]). Note that the 'doubling pattern' breaks for four holes; this is the warning sign for the big surprise coming in the case of 'five holes forward'.

We challenge the reader to solve the following exercise.

Exercise. We generalize Solitaire Army in such a way that 'to jump a man diagonally' is permitted. Show that it is impossible to send a man forward 9 holes.

Erdős–Selfridge theorem. The Erdős–Selfridge theorem [13] is the pioneering application of the 'potential technique' for *2-player games*. It gives a sufficient condition for a blocking draw. The result was proved in 1973, and since then it has continued to play a central role in the developments of positional games. It changed the outlook of the subject: it shifted the emphasis from Ramsey theory and matching theory to the probabilistic method.

Theorem D (Erdős–Selfridge theorem). *Let $\mathcal{F}$ be an n-uniform hypergraph, and assume that $|\mathcal{F}| + \text{MaxDeg}(\mathcal{F}) < 2^n$, where $\text{MaxDeg}(\mathcal{F})$ denotes the maximum degree of hypergraph $\mathcal{F}$. Then playing on $\mathcal{F}$ Second Player can force a blocking draw.*

Remark. A well-known result of Erdős states that, under the slightly weaker condition $|\mathcal{F}| < 2^{n-1}$,

(a) there is a proper 2-colouring; and what is slightly more,
(b) there is a *drawing terminal position*.

Both (a) and (b) can be proved by the 'counting argument'. Of course, the Erdős–Selfridge theorem implies (b) (and (b) implies (a)). The reason why I include the proofs of (a) and (b) is to demonstrate a key idea: the dual roles of 'randomization' and 'derandomization' in positional game theory.

The proof of (a) goes as follows. Let N denote the size of the union set (board) V of hypergraph $\mathcal{F}$. A simple counting shows that under the condition $|\mathcal{F}| < 2^{n-1}$ there exists a proper 2-colouring. Indeed, there are 2^N 2-colourings of board V, and for every single winning set $A \in \mathcal{F}$ there exist 2^{N-n+1} 'bad' 2-colourings which are monochromatic on A. By hypothesis $2^N - |\mathcal{F}|2^{N-n+1} > 0$, which implies that throwing out all 'bad' 2-colourings, there must remain at least one proper 2-colouring (*i.e.*, where no $A \in \mathcal{F}$ is monochromatic).

To prove (b) we have to find a drawing terminal position (*i.e.*, a 2-colouring of the board by 'colours' X and O such that the two colour classes have the same size, and each winning set contains both marks). For notational simplicity assume that N is even. The idea is exactly the same as that of (a), except that we restrict ourselves to the $\binom{N}{N/2}$ terminal positions (*i.e.*, 'halving' 2-colourings), instead of the 2^N (arbitrary) 2-colourings. The analogue of $2^N - |\mathcal{F}|2^{N-n+1} > 0$ is the following requirement: $\binom{N}{N/2} - 2|\mathcal{F}|\binom{N-n}{N/2} > 0$. This holds because

$$\frac{\binom{N-n}{N/2}}{\binom{N}{N/2}} = \frac{N/2}{N} \frac{(N/2)-1}{N-1} \frac{(N/2)-2}{N-2} \cdots \frac{(N/2)-n+1}{N-n+1} \leqslant 2^{-n},$$

and (b) follows.

The previous argument can be stated in the following slightly different form: the *average number* ('expected value') of winning sets completely occupied by either player is precisely

$$2|\mathcal{F}|\frac{\binom{N-n}{N/2}}{\binom{N}{N/2}}, \quad \text{which is less than 1.}$$

Since the *minimum* is less or equal to the *average, there must exist* at least one *drawing* terminal position (*i.e.*, no player owns a whole winning set). This kind of 'counting argument', due to Erdős, was the forerunner of the probabilistic method (see, *e.g.*, [1]).

The Erdős–Selfridge theorem can be viewed as a 'derandomization' of the first moment method. Indeed, if a drawing position exists, then there is a (slight) chance of a drawing strategy. The message of the Erdős–Selfridge theorem is that the existing drawing terminal position, proved by the first moment method, can be *upgraded* to a drawing strategy (in fact to a *blocking draw strategy*).

Finally, we prove the theorem itself.

Proof of Theorem D. Let $\mathscr{F} = \{A_1, A_2, \ldots, A_M\}$ where $M < 2^{n-1}$. Assume we are at the stage of the play where First Player already occupied $x_1, x_2, \ldots, x_i$, and Second Player occupied $y_1, y_2, \ldots, y_{i-1}$. The question is how to choose Second Player's next point y_i. Those winning sets which contain at least one y_j ($j \leq i-1$) are 'harmless' – we call them 'dead sets'. The winning sets which are not 'dead' are called 'survivors'. The survivors have a chance of being completely occupied by First Player at the end of the play, so they each represent some 'danger'. What is the total danger of the *whole* position? We evaluate the given position by the following expression, called the 'danger function': $D_i = \sum_{s \in S_i} 2^{-u_s}$ where u_s is the number of unoccupied elements of the 'survivor' A_s ($s \in S_i = $ 'index set of the survivors'), and index i indicates that we are at the stage of choosing Second Player's ith point y_i. A natural choice for y_i is to minimize the danger D_{i+1} at the next stage. How to do that? Well, the simple linear structure of the danger function D_i gives an easy answer to this question. Indeed, let y_i and x_{i+1} denote the next two moves. What is the effect of these two points on D_i? How do we get D_{i+1} from D_i? Clearly y_i 'kills' all the survivors $A_s \ni y_i$, which means we have to subtract the sum

$$\sum_{s \in S_i:\; y_i \in A_s} 2^{-u_s}$$

from D_i. On the other hand, x_{i+1} doubles the danger of each survivor $A_s \ni x_{i+1}$, that is, we have to add the sum $\sum_{s \in S_i:\; x_{i+1} \in A_s} 2^{-u_s}$ back to D_i. Warning: if some survivor A_s contains both y_i and x_{i+1}, then we do not have to give the corresponding term 2^{-u_s} back because that A_s was previously 'killed' by y_i.

The natural choice for y_i is the unoccupied z for which $\sum_{s \in S_i:\; z \in A_s} 2^{-u_s}$ attains its maximum. Then what we subtract is at least as large as what we add back:

$$D_{i+1} \leq D_i - \sum_{s \in S_i:\; y_i \in A_s} 2^{-u_s} + \sum_{s \in S_i:\; x_{i+1} \in A_s} 2^{-u_s}$$

$$\leq D_i - \sum_{s \in S_i:\; y_i \in A_s} 2^{-u_s} + \sum_{s \in S_i:\; y_i \in A_s} 2^{-u_s} = D_i.$$

In other words, Second Player can force the *decreasing property* $D_1 \geq D_2 \geq \cdots \geq D_{\text{last}}$ of the danger function.

Second Player's ultimate goal is to prevent First Player from completely occupying some $A_j \in \mathscr{F}$, that is, to avoid $u_j = 0$. If $u_j = 0$ for some j then $D_{\text{last}} \geq 2^{-u_j} = 1$. By hypothesis

$$D_{\text{start}} = D_1 = \sum_{A:\; x_1 \in A \in \mathscr{F}} 2^{-n+1} + \sum_{A:\; x_1 \notin A \in \mathscr{F}} 2^{-n} \leq (|\mathscr{F}| + \text{MaxDeg}(\mathscr{F})) 2^{-n} < 1,$$

so by the *decreasing property* of the danger function, $D_{\text{last}} < 1$. This is exactly what we were talking about in the Solitaire Army puzzle: *no play is possible from an initial position if the target position has a higher value than the initial position*. This completes the proof of the Erdős–Selfridge theorem. □

Remarks. (1) Exactly the same proof gives the following blocking criterion in the general, not necessarily uniform, case. If

$$\sum_{A \in \mathcal{F}} 2^{-|A|} + \max_{x \in V} \sum_{A \in \mathcal{F}:\ x \in A} 2^{-|A|} < 1,$$

then Second Player can block every winning set $A \in \mathcal{F}$.

(2) If $\mathcal{F}$ is n-uniform, then multiplying the 'danger' 2^{-u_s} of a 'survivor' by 2^n, the renormalized danger becomes 2^{n-u_s}. The exponent, $n - u_s$, is the number of First Player's marks in a survivor (i.e., Second Player-free) set (u_s denotes the number of unoccupied points). This means the following *power-of-two scoring system*. A winning set containing an O (Second Player's mark) scores zero, a blank winning set scores 1, a set with a single X (First Player's mark) and no O scores 2, a set with two Xs and no O scores 4, a set with three Xs and no O scores 8, and so on (i.e., the 'values' are integers rather than small fractions). Occupying a whole n-element winning set scores 2^n, i.e., due to the renormalization, the 'target value' becomes 2^n (instead of 1).

It is just a matter of taste which scoring system one prefers: the first one, where the scores were negative powers of 2 and the target value was 1, or the second one, where the scores were positive powers of 2 and the target value was 2^n.

(3) Theorem D is sharp. The full branches of a binary tree with n levels form an n-uniform family of 2^{n-1} winning sets such that First Player can occupy a full branch in n moves (the players occupy vertices of the binary tree).

Another *extremal system* is the following. The board V is a $(2n-1)$-element set $\{w, x_1, x_2, \ldots, x_{n-1}, y_1, y_2, \ldots, y_{n-1}\}$. The family of winning sets consists of all possible n-element subsets A of V with the following two properties: (1) $w \in A$, (2) A contains exactly one point from each pair $\{x_i, y_i\}$, $i = 1, 2, \ldots, n - 1$. The number of winning sets is 2^{n-1}, and again First Player can occupy a winning set in n moves.

Note that very recently my PhD students, Klay Kruczek and Eric Sundberg, could find, for every n, an 'extremal system' for Theorem D (i.e., an n-uniform hypergraph of size 2^{n-1}, where First Player wins) such that it takes *more* than n moves for First Player to occupy a whole winning set (if Second Player plays rationally). In fact, Second Player can postpone his inevitable loss as long as $2n - 3$ moves. They named their tricky construction a 'tumbleweed'.

We are very far from being able to classify all 'extremal systems' for Theorem D. (Why is it so difficult? I don't know!)

Theorem D gives a quick proof of the fact that the 4^2 game is a draw, in fact Second Player can force a blocking draw. Indeed, in the 4^2 game there are 10 winning sets, the maximum degree is 3, and $3 + 10 < 2^4 = 16$, so Theorem D applies. The same argument works for all n^2-games with $n \geq 4$.

A very clever and complicated 'pairing' of S. Golomb shows that the 8^3 game has a pairing strategy draw (see *Winning Ways* [11], Vol. 2, pp. 677–8). By using Theorem D one can easily show that the 8^3 game has a blocking draw. Indeed, in the 8^3 game there are $(10^3 - 8^3)/2 = 244$ winning sets, the maximum degree is $2^3 - 1 = 7$, and $244 + 7 < 2^8 = 256$, so Theorem D applies. The same argument works for all n^3 games with $n \geq 8$. (Of course a blocking draw is not necessarily a pairing strategy draw.)

Hales–Jewett conjecture. The n^d-hypergraph is *almost disjoint*: the winning sets are 'lines' (more precisely: finite subsets of straight lines), and any two straight lines have at most one point in common. In general, a hypergraph is called *almost disjoint* if any two winning sets have at most one point in common. If an almost disjoint hypergraph has a pairing strategy draw, then there are at least twice as many points as winning sets (indeed, two sets cannot share the same point-pair). Hales and Jewett conjectured [16] that for the subclass of n^d-games the 'converse' holds. They conjectured that the n^d-game has a draw, in fact a pairing strategy draw, if there are at least twice as many points (*i.e.*, 'cells') as winning lines. This is the *Hales–Jewett conjecture*. It has two natural interpretations. The stronger form is as follows.

Open Problem 6. *Is it true that if the n^d game has at least twice as many points as lines, then it is a pairing strategy draw?*

In view of the well-known Marriage Lemma (Hall's theorem) a *necessary and sufficient* condition for the existence of a draw-forcing pairing is that for every subfamily of winning lines the union set has at least twice as many points as the number of lines in the subfamily. (We actually use the Bigamy Corollary of the Marriage Lemma: every man needs 2 wives.) Consequently, what Open Problem 6 really says is that *the point/line ratio attains its minimum for the family of all lines, and for any proper subfamily the ratio is greater or equal*. This *Ratio Conjecture* is very compelling, not only when a pairing strategy exists, but in general for arbitrary n^d-game. The Ratio Conjecture was formulated in Patashnik [17].

Erdős–Selfridge [13] and *Winning Ways* [11] both gave the following broader interpretation of the Hales–Jewett conjecture; I call it the 'weaker form'.

Open Problem 7. *Is it true that if the n^d game has at least twice as many points as lines, then it is a draw (but not necessarily a pairing strategy draw)?*

Note that Open Problem 7 is nearly solved: I proved that it holds true for all dimensions $d \geq 33$. More precisely, for all $d \geq 16$ with n even, and for all $d \geq 33$ with n odd. Asymptotically I can prove much more: the n^d game can have far fewer points than lines, in fact the point/line ratio can tend to zero, and Second Player can still force a blocking draw.

By formula (2.1) the number of winning lines is $((n+2)^d - n^d)/2$, so the stronger (weaker) Hales–Jewett conjecture states that the n^d game has a pairing strategy draw (has a draw) if $n^d \geq (n+2)^d - n^d$, or equivalently, if $2 \geq ((n+2)/n)^d$. Since $((n+2)/n)^d = (1 + \frac{2}{n})^d \approx e^{2d/n}$ if n is large, the conjecture is *asymptotically* equivalent to the threshold

$$n \geq \frac{2d}{\log 2} = 2.8854d. \tag{3.1}$$

We can get pretty close to this threshold: in an unpublished paper R. Schroeppel (Arizona) proved that the n^d game has a pairing strategy draw if $n \geq 3d$, $d \geq 2$ *even*, and if $n \geq 3d - 1$, $d \geq 3$ *odd*. Schroeppel's new idea is to involve the concept of 'fractional matching'.

Is Theorem D powerful enough to settle the weaker Hales–Jewett conjecture? Unfortunately the answer is 'no'. Indeed, the Erdős–Selfridge theorem applies to the n^d-hypergraph if

$$\frac{(n+2)^d - n^d}{2} + \{(3^d - 1)/2 \text{ or } 2^d - 1\} < 2^n,$$

depending on the parity of n. This means Second Player can force a blocking draw if $n > \text{const} \cdot d \cdot \log d$. This 'superlinear' result clearly falls short of the 'linear' Hales–Jewett conjecture (see (3.1)).

Unfortunately a pairing strategy 'beats' the Erdős–Selfridge theorem for all n^d-games (except for the single case of 4^2). If the Erdős–Selfridge theorem yields mediocre results for the n^d games, then what was the point of introducing it? Well, the importance of the Erdős–Selfridge theorem is five-fold.

(1) *Combining* the Erdős–Selfridge power-of-two scoring system with a pairing strategy I could nearly solve the weaker Hales–Jewett conjecture, and could in fact go far beyond that. I call the technique developed for this purpose 'BigGame SmallGame Decomposition'.
(2) By using the Erdős–Selfridge scoring system in *both* the 'BigGame' and 'SmallGame', I can get an asymptotically *better* blocking draw criterion.
(3) A straightforward adaptation of the Erdős–Selfridge argument gives a weak win criterion, which is the best-known result for almost disjoint hypergraphs.
(4) By developing a 'higher moment version' of the 'linear' (4), I proved the best-known weak win criterion for arbitrary (not necessarily almost disjoint) hypergraphs.
(5) Finally, for the *clique game* the Erdős–Selfridge theorem gives an excellent upper bound, very close to the 'truth' (the 'gap' is at most 2).

Here I discuss (5).

Blocking draws in the clique game. (5) is about the 'straight' clique game (K_n, K_q): the board is K_n, i.e., the players claim *edges* of K_n and the goal is to have a copy of K_q first. The well-known Erdős–Szekeres theorem ('Ramsey theory') states that, given any 2-colouring of the edges of the complete graph K_n with $n \geq \binom{2q-2}{q-1}$ vertices, there is always a monochromatic copy of K_q. Since every play has a winner, strategy stealing applies (see Theorem C), and yields a First Player's winning strategy. In other words, if $n = \binom{2q-2}{q-1} \approx 4^{q-1}/\sqrt{\pi q}$, or equivalently, if $q = (\frac{1}{2} + o(1))\log_2 n$, then First Player can force a win in the 'straight' clique game (K_n, K_q). Of course, it does *not* say a word about what First Player's winning strategy actually looks like. To find an explicit winning strategy remains a big open problem.

In the other direction, the Erdős–Selfridge theorem implies a blocking draw if

$$\binom{n}{q} < 2^{\binom{q}{2}-1}. \tag{3.2}$$

This is equivalent to

$$q \geqslant 2\log_2 n - 2\log_2\log_2 n + c_0 + o(1) \quad \text{where} \quad c_0 = 2\log_2 e - 1. \tag{3.3}$$

The deduction of (3.3) from (3.2) is a routine calculation by using Stirling's formula and taking binary logarithm of both sides.

The reason why I gave the very precise form (3.3) instead of simply writing $(2 + o(1))\log_2 n$ was that (3.3) comes *very close* to the truth! By using a more sophisticated argument I can show that the *exact* threshold for a blocking draw is actually 2 less than (3.3): if

$$q > 2\log_2 n - 2\log_2\log_2 n + 2\log_2 e - 3 + o(1), \tag{3.4}$$

then Second Player can still block every K_q; on the other hand, if

$$q < 2\log_2 n - 2\log_2\log_2 n + 2\log_2 e - 3 + o(1), \tag{3.5}$$

then First Player can completely occupy a K_q (but not necessarily first – this is called a weak win), implying that a blocking draw is impossible.

To prove (3.5) I needed an 'advanced weak win criterion'. The precise form of this criterion is rather complicated. Instead I just give an oversimplified heuristic form as follows: If $\mathscr{F}$ is m-uniform, the average degree is 'substantially larger' than 2^m, and the average degree is 'substantially larger' than the 'typical pair-degree', then First Player can force a weak win.

The clique-hypergraph $\mathscr{F}_{n,q}$ is degree-regular, so the maximum degree equals the average degree. For many degree-regular hypergraphs, including the clique-hypergraph, the weak win criterion and the game-theoretic LLL Conjecture (see Open Problem 1) nearly complement each other, which leads to asymptotically very good results.

A common feature of the proofs of the 'advanced criteria' is to define an appropriate auxiliary hypergraph of 'big sets'. The 'advanced weak win criterion' is a 'game-theoretic higher moment method' ('higher moment' means 'big sets'), and the proof uses the technique of 'sliding potentials' (see [7]). 'Sliding potentials' is a technique to divided the whole play into several phases, and in each phase one uses a new potential function.

The best blocking draw criteria are proved by using the 'BigGame SmallGame Decomposition' (see [4], [5], [8]). It means dividing the board into two non-interacting subgames (with disjoint subboards), and use the *big game* (played on a family of big sets) to control the *small game*: a family of 'dangerous' winning sets which are almost completely occupied by the opponent. In the simpler version, the small game is handled by a pairing strategy; a nice illustration of this is Section 7. In the more advanced versions, the small game is handled by an Erdős–Selfridge-type scoring system.

The Erdős–Selfridge theorem itself hardly ever gives the best quantitative result. To get the exact solution of an interesting game like the 'clique game' (or the 'subspace game') one adapts the proof technique ('power-of-two scoring system') rather than using the Erdős–Selfridge theorem itself.

4. Weak win criteria

The primary objective of game-playing is to win ('winning is everything' as they say in the USA). Unfortunately we don't know how to force a strong win (a 'win in a strong game'). The next best thing is a weak win (a 'Maker's win in a weak game'). In this section I discuss weak win criteria, and in the next section I show several applications.

The Ramsey criterion (Theorem C) for a strong win was a pure *existence* argument (via strategy stealing), which left the question of 'How to actually win?' completely unanswered. Weak games are different: there is an *explicit* version of the Ramsey criterion, which is a simple 'copycat' pairing strategy.

Theorem E (weak win by Ramsey). *Let $(W, \mathcal{G})$ be a hypergraph such that it contains two disjoint copies of a hypergraph with chromatic number ≥ 3: let $(V, \mathcal{F})$ be a finite hypergraph of chromatic number at least 3, let $(V', \mathcal{F}')$ be a disjoint copy of $(V, \mathcal{F})$, let $W \supseteq V \cup V'$ and $\mathcal{G} \supseteq \mathcal{F} \cup \mathcal{F}'$. Then Maker has an explicit winning strategy in the weak game on $(W, \mathcal{G})$, no matter whether Maker is the first or second player.*

Proof. Let $f : V \to V'$ be the isomorphism between $(V, \mathcal{F})$ and $(V', \mathcal{F}')$. Maker wins by using the following copycat pairing strategy: if Breaker's last move was $x \in V$ or $x' \in V'$, then Maker's next move is $f(x) \in V'$ or $f^{-1}(x') \in V$ (unless it was occupied by Maker before; then Maker's next move is arbitrary). Since the chromatic number of $(V, \mathcal{F})$ is larger than two, one of the two players will completely occupy a winning set. If this player is Maker, we are done. If Breaker occupies some $A \in \mathcal{F}$, then Maker occupies $f(A) \in \mathcal{F}'$, and we are done again. $\square$

Theorem E is *folklore*; for an interesting infinite version, see [2].

The first application of Theorem E is the weak n^d-game where $d \geq h(n) + 1$. The condition guarantees that the board contains two disjoint copies of $n^{h(n)}$. The copycat pairing of Theorem E provides an *explicit* winning strategy for Maker (whether Maker is the first or second player).

Next consider the weak clique game (K_n, K_q) where $n \geq 2\binom{2q-2}{q-1}$. The condition guarantees that the board contains two disjoint copies of K_m with $m = \binom{2q-2}{q-1}$ (Erdős–Szekeres threshold). The copycat pairing of Theorem E supplies an *explicit* winning strategy for Maker if $q = \frac{1}{2}\log_2 n$ (and again whether Maker is the first or second player).

Theorem E was a Ramsey-type weak win criterion. The next criterion is completely different: it is a 'weak win counterpart' to the Erdős–Selfridge theorem.

Theorem F (Erdős–Selfridge-type weak win criterion). *Assume that $\mathcal{F}$ is n-uniform. Let V denote the union set (board), and assume that, fixing any two distinct points of V, there* are no more than $\Delta_2 = \Delta_2(\mathscr{F})$ winning sets $A \in \mathscr{F}$ containing both points; we call $\Delta_2(\mathscr{F})$ the *pair-degree*. If

$$\frac{|\mathscr{F}|}{|V|} > 2^{n-3} \cdot \Delta_2,$$

then First Player has a weak win in $(V, \mathscr{F})$.

In particular, for almost disjoint $\mathscr{F}$, the condition simplifies to $|\mathscr{F}| > 2^{n-3}|V|$.

Remark. If $\mathscr{F}$ is n-uniform, then $\frac{|\mathscr{F}|}{|V|}$ is $\frac{1}{n}$ times the *average degree*. In other words, the hypothesis of Theorem F is a simple 'density condition'.

Proof of Theorem F. I basically repeat the proof of the Erdős–Selfridge theorem. Assume we are at the stage of the play where First Player already occupied $x_1, x_2, \ldots, x_i$, and Second Player occupied $y_1, y_2, \ldots, y_i$. The question is how to choose First Player's next point x_{i+1}. Those winning sets which contain at least one y_j ($j \leq i$) are 'useless' for First Player. We call them 'dead sets'. The winning sets which are not 'dead' (yet) are called 'survivors'. The survivors have a chance of being completely occupied by First Player ('weak win'). What is the total 'chance' of the position? We evaluate the given position by the following 'chance function': $C_i = \sum_{s \in S_i} 2^{n-u_s}$ where u_s is the number of unoccupied points of the survivor A_s ($s \in S_i$ = 'index set of the survivors', and index i indicates that we are at the stage of choosing the $(i+1)$st point x_{i+1} of First Player. Note that the 'chance function' can be much greater than 1 (it is not a *probability*), but it is always nonnegative.

A natural choice for x_{i+1} is to maximize the 'chance' C_{i+1} at the next stage. Let x_{i+1} and y_{i+1} denote the next moves of the two players. What is their effect on C_{i+1}? Well, first x_{i+1} doubles the chances for each survivor $A_s \ni x_{i+1}$, that is, we have to add the sum $\sum_{s \in S_i: \; x_{i+1} \in A_s} 2^{n-u_s}$ to C_i.

On the other hand, y_{i+1} 'kills' all the survivors $A_s \ni y_{i+1}$, which means we have to subtract the sum

$$\sum_{s \in S_i: \; y_{i+1} \in A_s} 2^{n-u_s}$$

from C_i.

Warning: we have to make a correction to those survivors A_s which contain both x_{i+1} and y_{i+1}. These survivors A_s were 'doubled' first and killed second. So what we have subtract from C_i is not

$$\sum_{s \in S_i: \; \{x_{i+1}, y_{i+1}\} \subset A_s} 2^{n-u_s}$$

but the twice-as-large

$$\sum_{s \in S_i: \; \{x_{i+1}, y_{i+1}\} \subset A_s} 2^{n-u_s+1}.$$

It follows that

$$C_{i+1} = C_i + \sum_{s \in S_i:\, x_{i+1} \in A_s} 2^{n-u_s} - \sum_{s \in S_i:\, y_{i+1} \in A_s} 2^{n-u_s} - \sum_{s \in S_i:\, \{x_{i+1}, y_{i+1}\} \subset A_s} 2^{n-u_s}. \quad (4.1)$$

Now the natural choice for x_{i+1} is the unoccupied z for which $\sum_{s \in S_i:\, z \in A_s} 2^{n-u_s}$ attains its maximum. Then clearly

$$C_{i+1} \geq C_i - \sum_{s \in S_i:\, \{x_{i+1}, y_{i+1}\} \subset A_s} 2^{n-u_s}.$$

We trivially have

$$\sum_{s \in S_i:\, \{x_{i+1}, y_{i+1}\} \subset A_s} 2^{n-u_s} \leq \Delta_2 2^{n-2}.$$

Indeed, there are at most Δ_2 winning sets A_s containing the given two points $\{x_{i+1}, y_{i+1}\}$, and $2^{n-u_s} \leq 2^{n-2}$, since x_{i+1} and y_{i+1} were definitely unoccupied points at the previous stage.

Therefore,

$$C_{i+1} \geq C_i - \Delta_2 2^{n-2}.$$

What happens at the end? Let ℓ denote the number of stages, i.e., the ℓth stage is the last one. Clearly $\ell = |V|/2$. Inequality $C_\ell = C_{\text{last}} > 0$ means that Second Player could not 'kill' (block) all the winning sets. Indeed, at the last stage all points are occupied, so $C_\ell = C_{\text{last}} > 0$ means that First Player was able to completely occupy a winning set, meaning a weak win.

So all what we have to check is that $C_\ell = C_{\text{last}} > 0$. But this is trivial. Indeed, $C_{\text{start}} = C_0 = |\mathscr{F}|$, so we have

$$C_{\text{last}} \geq |\mathscr{F}| - \frac{|V|}{2} \Delta_2 2^{n-2}.$$

It follows that $C_{\text{last}} > 0$ if $|\mathscr{F}| > 2^{n-3}|V|\Delta_2$, which implies that at the end of the play First Player was able to completely occupy a winning set.

Finally note that if $\mathscr{F}$ is almost disjoint, then fixing any two points of the board V, there is at most one winning set containing both of them. So $\Delta_2 = \Delta_2(\mathscr{F}) = 1$. □

Next I apply the two 'first moment criteria' (the Erdős–Selfridge theorem and its 'weak win counterpart') to the three basic 'Ramsey-type games': n^d Tic-Tac-Toe, the clique game, and the van der Waerden game. One obtains the following quantitative results.

(1) n^d Tic-Tac-Toe. If $d < c_1 \cdot n/\log n$ then either player can block every winning line; on the other hand, if $d > (\log 2) n^2 / 2$ then First Player can occupy a whole winning line (weak win).

(2) (K_n, K_q) clique game. If $q > 2\log_2 n$ then either player can block every copy of K_q; on the other hand, if $q < c_2 \sqrt{\log n}$ then First Player can occupy a whole K_q (weak win).

(3) (N, n) van der Waerden game ('arithmetic progression game'). If $n > 2\log_2 N$ then either player can block every n-term AP; on the other hand, if $n = (1 - o(1))\log_2 N$ then First Player can completely occupy an n-term AP (weak win).

Observe that the 'first moment criteria' are not powerful enough to determine the true order of magnitude in either case. (To get the true order one needs 'higher moments'.)

First I prove (1), then (3), and finally (2).

Let us apply Theorem F to the n^d game. The corresponding hypergraph is almost disjoint, so First Player has a weak win in the n^d game if

$$\frac{(n+2)^d - n^d}{2} > 2^{n-3} n^d,$$

which is equivalent to

$$\left(1 + \frac{2}{n}\right)^d > 2^{n-2} + 1. \tag{4.2}$$

Inequality (4.2) holds for the 3^3, 4^4, 5^7, 6^{10}, 7^{14}, 8^{19}, 9^{25}, 10^{31}, ... games, and guarantees a weak win. (Exhaustive search upgrades the games 3^3 and 4^3 to a strong win.)

The previous list is complemented by the following list of small blocking draw games: 4^2, 8^3, 11^4, 14^5, 18^6, 20^7, and so on. They are all pairing strategy draws with the exception of 4^2. (Potential function technique begins to 'beat' a pairing strategy only in higher dimensions, like $d \geq 16$.) There is a *big* gap between the two lists.

Note that inequality (4.2) holds for $n < \sqrt{2d/\log 2}$, or equivalently, if $d > (\log 2) \cdot n^2/2$. In the other direction, the Erdős–Selfridge theorem guarantees a blocking draw if $N > c_0 d \log d$, which is equivalent to $d < c_1 n/\log n$ (this is pretty weak: even a pairing strategy can beat it). Which one is closer to the truth: the 'quadratic' weak win result or the (nearly) 'linear' strong draw result (via E-S)? Well, the true order turns out to be 'quadratic'. I can prove, by roughly 'squaring' the Erdős–Selfridge bound, that if $d < c_2 n^2/\log n$ then the n^d game is a blocking draw.

This proves that, if n is a fixed large integer, then the game-theoretic breaking point between a weak win and a blocking draw in the n^d-game occurs when (the dimension) d is roughly around n^2 (there is an 'error factor' of $\log n$). Note that in this range the point/line ratio tends to zero, so a pairing strategy doesn't have a chance.

The second weak win criterion (Theorem F) gives a 'close-to-optimal' bound for a weak win in the n^d game (dimension d is 'around' *quadratic* in terms of n), but the right order of magnitude remains an unsolved problem. I don't know how to get rid of a factor of $\log n$.

A more impressive example is the 'weak van der Waerden game': the board is the interval $[N] = \{1, 2, \ldots, N\}$, and the players, Maker and Breaker, alternately take previously unselected integers from $[N]$. Maker wins if he can occupy a whole n-term arithmetic progression; otherwise Breaker wins.

An elementary calculation shows that the total number of arithmetic progressions of length n in $[N]$ is $\geq N(\lfloor \frac{N}{n-1} \rfloor - 1)/2$. The corresponding hypergraph is not almost disjoint, but the pair-degree is very small (in terms of N): the pair-degree is $\leq \binom{n}{2}$. Indeed, fixing two integers $1 \leqslant x < y \leqslant N$ and two indices $0 \leqslant i < j \leqslant n-1$, the equations $x = a + id$ and $y = a + jd$ uniquely determine the arithmetic progression $a, a+d, \ldots, a+(n-1)d$ of length n. Theorem F applies if $\lfloor \frac{N}{n-1} \rfloor > 2^{n-2}\binom{n}{2} + 1$. For $n = 4, 5, 6$ this gives the weak win thresholds 76, 325, 1206. Asymptotically, for large n, we get the weak win threshold $\approx 2^{n-3}n^3$. It follows that Maker wins if $n \leqslant (1 + o(1))\log_2 N$.

This result can be complemented by the following (see [5]): Breaker wins the weak van der Waerden game if $n \geqslant (1 + o(1))\log_2 N$. (This beats Erdős–Selfridge by a factor of two – see (3) above.) The game-theoretic breaking point of the weak van der Waerden game is therefore $(1 + o(1))\log_2 N$.

The van der Waerden game is a 'victory' for Theorem F, but how about the weak clique game (K_n, K_q)? How close can one get to the truth by applying Theorem F to the weak clique game (K_n, K_q)? Well, Theorem F implies a weak win if

$$\binom{n}{q} > 2^{\binom{q}{2}-3}\binom{n}{2}\Delta_2.$$

For this particular family of winning sets the pair-degree Δ_2 is at most $\binom{n}{q-3}$. Indeed, two distinct edges determine at least 3 different vertices. Therefore, it suffices to check the inequality

$$\binom{n}{q} > 2^{\binom{q}{2}-3}\binom{n}{2}\binom{n}{q-3},$$

which yields $n \geqslant 2^{q^2/2}$, or in terms of n, $q < \sqrt{2\log_2 n}$.

Well, this is a very *disappointing* result. It is asymptotically much weaker than the 'Ramsey-type bound' $q = \frac{1}{2}\log_2 n$ obtained by combining Theorem E and the Erdős–Szekeres upper bound. The reason why Theorem F gave such a poor result in the weak clique game is that the 'clique-hypergraph' is very far from being almost disjoint. Indeed, two different cliques can share a lot of edges; in fact two different K_qs can have as many as $\binom{q-1}{2}$ common edges.

On the other hand, the reason why Theorem F gave such a good result for the van der Waerden game was that the corresponding hypergraph is (1) 'nearly' almost disjoint; and (2) 'nearly' degree-regular.

My techniques work best for degree-regular hypergraphs. The n^d-hypergraph is highly degree-irregular: this is why I don't know the right order of magnitude of the game-theoretic breaking point.

Next I study the clique game for k-graphs where $k \geqslant 3$, which corresponds to *higher Ramsey numbers*. This means the straightforward generalization of complete graphs, where the board is a complete k-uniform hypergraph instead of K_n, and of course the players claim k-sets instead of edges. Even if the case $k = 2$ gave disappointing result, for $k \geqslant 4$ we will have a pleasant surprise: for the first time we will be able to separate a (higher) *Ramsey number* from the corresponding *weak game number*.

Clique games on k-graphs where $k \geqslant 3$. For every natural number n write $[n] = \{1, \ldots, n\}$. If S is a set, then let $\binom{S}{k}$ denote the family of all k-element subsets of S. Then $\binom{[n]}{2}$ can be interpreted as a complete graph with n vertices, i.e., $\binom{[n]}{2} = K_n$.

Let $2 \leq k \leq q < n$. The board of the game is $\binom{[n]}{k}$, and the family of winning sets consists of all possible $\binom{S}{k}$ where $S \in \binom{[n]}{q}$. The family of winning sets is a $\binom{q}{k}$-uniform hypergraph of size $\binom{n}{q}$. We consider the weak (*i.e.*, Maker–Breaker) game on this particular hypergraph, and call it the *weak* $(\binom{n}{k}, \binom{q}{k})$ *clique game*.

The general form of Ramsey's theorem states that for every $k \geq 2$ and for every $q \geq k$, there is a least finite threshold number $R_k(q)$ such that the family of winning sets of the $(\binom{n}{k}, \binom{q}{k})$ clique game has chromatic number at least 3 if $n \geq R_k(q)$. If $n \geq R_k(q)$ then by Theorem C First Player has a strong win in the $(\binom{n}{k}, \binom{q}{k})$ clique game (but we don't know what the winning strategy looks like), and if $n \geq 2R_k(q)$ then by Theorem E First Player has an explicit ('copycat') weak win in the $(\binom{n}{k}, \binom{q}{k})$ clique game.

What do we know about the size of the higher Ramsey number $R_k(q)$? Well, it is huge! Indeed, let $\text{tower}_k(x)$ denote the k-fold iteration of the exponential function: $\text{tower}_1(x) = 2^x$ and for $k \geq 2$, $\text{tower}_k(x) = 2^{\text{tower}_{k-1}(x)}$. So $\text{tower}_2(x) = 2^{2^x}$, $\text{tower}_3(x) = 2^{2^{2^x}}$, and so on. We call the index k in $\text{tower}_k(x)$ the *height* (it is in fact the height of the exponential expression).

For graphs (*i.e.*, $k = 2$) it is well known that

$$2^{q/2} < R_2(q) < 4^q.$$

For k-graphs with $k \geq 3$,

$$2^{q^2/6} < R_3(q) < 2^{2^{4q}}, \tag{4.3}$$

$$2^{2^{q^2/6}} < R_4(q) < 2^{2^{2^{4q}}}, \tag{4.4}$$

and in general,

$$\text{tower}_{k-2}(q^2/6) < R_k(q) < \text{tower}_{k-1}(4q). \tag{4.5}$$

The last two bounds are due to Erdős, Hajnal and Rado (see, *e.g.*, [15]).

First let $k = 3$. If $q = c_1 \cdot \log \log n$, then by (4.3), First Player has a strong win in the $(\binom{n}{3}, \binom{q}{3})$ clique game, but the winning strategy is not known. On the other hand, by Theorem E under the same condition $q = c_1 \cdot \log \log n$ Maker has an *explicit* weak win strategy by playing copycat. Similarly, if $q = c_2 \cdot \log \log \log n$ then by (4.4) First Player has a strong win in the $(\binom{n}{4}, \binom{q}{4})$ clique game (the strategy is unknown), and under the same condition Maker has an *explicit* weak win strategy by playing copycat, and so on.

What happens if one replaces Theorem E with Theorem F? The Erdős–Selfridge-type weak win criterion (Theorem F) applies to the weak $(\binom{n}{k}, \binom{q}{k})$ clique game when

$$\binom{n}{q} > 2^{\binom{q}{k}-3} \binom{n}{k} \Delta_2.$$

For this particular family of winning sets the pair-degree Δ_2 satisfies the obvious inequality $\Delta_2 \leq \binom{n}{q-k-1}$. Indeed, two distinct k-sets cover at least $k+1$ points.

This leads to the inequality

$$\binom{n}{q} > 2^{\binom{q}{k}-3} \binom{n}{k} \binom{n}{q-k-1},$$

which means $n \geq 2^{q^k/k!}$, or in terms of n, $q \leq (k!\log_2 n)^{1/k}$. Therefore, if

$$q = c_k(\log n)^{\frac{1}{k}}$$

then Maker has a weak win in the $(\binom{n}{k}, \binom{q}{k})$ clique game. This means that for k-graphs with $k \geq 4$ the weak Ramsey criterion (Theorem E) definitely *fails to give the true order of magnitude of the breaking point for a weak win* in the clique game. Indeed, by (4.5) the Ramsey theory threshold $R_k(q)$ is greater than the tower function tower$_{k-2}(q^2/6)$ of height $k-2$ (increasing with k). On the other hand, Maker can force a weak win around $n = 2^{q^k/k!}$, which has *constant height* independent of k. It is trivial to see that tower$_{k-2}(q^2/6)$ is asymptotically *much* larger than $n = 2^{q^k/k!}$ if $k \geq 4$; in fact they have completely different asymptotic behaviours.

By using more sophisticated arguments, I can find the *exact* game-theoretic breaking points for 'almost all' clique games ($k \geq 2$).

5. Beyond Ramsey theory

In Section 4 we learned that the clique game for k-graphs, where $k \geq 4$, is an example where the potential function technique (Theorem F) 'beats' Ramsey theory (Theorem E) in a dramatic fashion. Indeed, the weak game number is *much* smaller than the corresponding Ramsey number.

A different kind of example is a game where Ramsey theory (Theorem E) fails to give anything, *i.e.*, there is *no* Ramsey phenomenon whatsoever, but Theorem F still gives an interesting result. This kind of game is the unrestricted n-in-a-row game with *arbitrary slopes*: see subsection (2) below.

But first I recall the well-known unrestricted 5-in-a-row game, and in general, the unrestricted n-in-a-row game.

(1) Unrestricted means that it is played on an *infinite* chessboard (infinite in every direction). In the unrestricted 5-in-a-row game the players alternately occupy little squares of an infinite chessboard, First Player marks his squares by X, and Second Player marks his squares by O. That player wins who first gets 5 consecutive marks of his own in a row horizontally, or vertically, or diagonally (of slope 1 or −1). The unrestricted n-in-a-row differs in only one aspect: the winning size is n instead of 5.

Unrestricted n-in-a-row on the plane is a *semi-infinite* game (in fact, a semi-infinite strong positional game): the board is infinite but the winning sets are all finite. Since the board is infinite, we have to define the *length* of the game. We assume that the players take turns until either of them wins in a finite number of moves, or until they have taken their nth turns for every natural number n. In other words, the length of a play is at most ω, where ω denotes (as usual) the first countable (infinite) ordinal number. Semi-infinite strong (and weak) positional games are all *determined*, i.e., Theorem A holds for semi-infinite games. Indeed, basically the same proof works: if player P has no winning strategy, then the opponent, player Q, can always make a next move so that player P still has no winning strategy, and this is exactly how player Q can force a draw. The point is that the winning sets are finite, so if a player wins, he wins in a finite number of moves – this is why player P cannot win and player Q forces a draw.

112   *J. Beck*

The unrestricted 4-in-a-row game is an easy First Player's win. The unrestricted 5-in-a-row game is conjectured to be a First Player's win, too, but there is no proof. For 'reasonable board size', say 20 by 20, there is a First Player's winning strategy (computer-assisted proof; huge case study), but no one knows how to extend it to the whole plane ('extra set paradox'). In the other direction, the unrestricted 8-in-a-row is known to be a draw game (this is the best that we know). Here I include a simple pairing strategy proof of the slightly weaker fact that the unrestricted 9-in-a-row game is a draw: see p. 677 in [11]. Indeed, cover the infinite chessboard with disjoint copies of the following 8 by 8 matrix:

$$n = 8 : \begin{bmatrix} \backslash & \backslash & - & - & | & | & / & / \\ - & - & \backslash & \backslash & / & / & | & | \\ - & - & / & / & \backslash & \backslash & | & | \\ / & / & | & | & - & - & \backslash & \backslash \\ \backslash & \backslash & | & | & - & - & / & / \\ | & | & \backslash & \backslash & / & / & - & - \\ | & | & / & / & \backslash & \backslash & - & - \\ / & / & - & - & | & | & \backslash & \backslash \end{bmatrix}.$$

What this 8 by 8 matrix represents is a direction-marking of the $4 \cdot 8 = 32$ 'torus-lines' of the 8×8 torus. The direction marks $-$, $|$, $\backslash$, and $/$ mean (respectively) 'horizontal', 'vertical', 'diagonal of slope -1', and 'diagonal of slope 1'. Each one of the 32 torus-lines contains 2 marks of its own direction. The periodic extension of the 8 by 8 matrix over the whole plane gives a pairing strategy draw for the unrestricted 9-in-a-row game. Either player responds to the opponent's last move by taking the *nearest* similarly marked square in the direction indicated by the mark in the opponent's last move square.

(2) Unrestricted *n*-in-a-row with arbitrary slopes. Consider the following generalization of the unrestricted *n*-in-a-row: the players occupy lattice points instead of 'little squares', and *n*-in-a-row means consecutive lattice points on *any* straight line. The point is that *every* rational slope is allowed, not just the four Tic-Tac-Toe directions.

We challenge the reader to solve the following exercise.

Exercise. Show that there is no 'Ramsey phenomenon' here: it is possible to 2-colour the set of integer lattice points in the plane in such a way that every set of 100 consecutive lattice points on a straight line contains both colours.

Of course '100' just means a 'large constant'. The correct value is actually 4 (this elegant result is due to A. Dumitrescu and R. Radoicic), but it is much easier to prove 100.

Even if the 'Ramsey phenomenon' fails here, I can still prove a 'game theorem': Maker can have *n*-in-a-row for *every* finite *n*. The proof is a straightforward application of Theorem F. Consider an $m \times m$ square lattice, in fact assume that the lower-left corner is the origin ($m = m(n)$ will be specified later). The set $(k + aj, l + j)$ $j = 0, 1, \ldots, n-1$ gives *n consecutive* lattice points on a line *inside* the $m \times m$ square if $0 \leqslant k \leqslant m/2$, $0 \leqslant l \leqslant m/2$, and $1 \leqslant a \leqslant \frac{m}{2n}$; we call it a (k, l, a)-set. Let $\mathscr{F}$ denote the family of all (k, l, a)-sets. $\mathscr{F}$ is an *n*-uniform hypergraph. Since $0 \leqslant k \leqslant m/2$, $0 \leqslant l \leqslant m/2$, and $1 \leqslant a \leqslant \frac{m}{2n}$, we have

$|\mathscr{F}| \geqslant m^3/8n$. Since the 'board' $|V| = m^2$, and the pair-degree of $\mathscr{F}$ is at most n, Theorem F applies if

$$\frac{m^3}{8n} > 2^{n-3} \cdot n \cdot m^2. \tag{5.1}$$

We need a simple remark here: of course Maker cannot force Breaker to stay inside the $m \times m$ square, but if Breaker moves outside of the $m \times m$ square, that clearly just helps Maker to occupy an n-in-a-row faster.

By choosing $m = m(n) = n^2 \cdot 2^n + 1$, inequality (5.1) holds, and implies a Maker's weak win.

Theorem 1. *Playing 'unrestricted n-in-a-row with arbitrary slopes', Maker can occupy an n-in-a-row in less than $n^4 \cdot 4^n$ moves ($n \geqslant 1$ is arbitrary).* □

Another example of the same type is the *weak all-subset game*.

(3) Weak all-subset game. In the clique game the players take edges, that is, *pairs* of the vertex set. Here the players take *arbitrary subsets* of a 'ground set'. More precisely, there are two players, Maker and Breaker, who alternately take previously unselected subsets of the n-element set $[n] = \{1, 2, \ldots, n\}$. First Player (Maker) wins if he has all 2^{100} subsets of some 100-element subset of $[n]$; otherwise Second Player (Breaker) wins.

Can Maker win if n is sufficiently large? Well, one thing is very clear: Maker's opening move *has to be* the 'empty set'.

Observe that there is no 'Ramsey phenomenon' here. Indeed, colour the subsets depending on the parity of the size: 'even' means red and 'odd' means blue. This particular 2-colouring kills any chance of an 'all-subset Ramsey theorem'. But there is a 'game theorem': We show that Maker (First Player) has a winning strategy if n is sufficiently large.

How do we prove the 'game theorem'? Well, it seems a natural idea to use Theorem F. Unfortunately direct application does *not* work. Indeed, $|\mathscr{F}| = \binom{n}{100}$ is actually *less* than $|V| = \binom{n}{0} + \binom{n}{1} + \binom{n}{2} + \cdots + \binom{n}{100}$, so Theorem F cannot give anything.

To get around this difficulty I use a technical trick: I prove the following statement.

Stronger Statement. *There is a 101-element subset S of $[n]$ such that Maker (First Player) can have all at-most-100-element subsets of S, assuming n is sufficiently large.*

In fact, I can prove it for arbitrary k (not just for 100), and then the Stronger Statement goes as follows: *There is a $k + 1$-element subset S of $[n]$ such that Maker (First Player) can have all at-most-k-element subsets of S, assuming n is sufficiently large depending on k.*

The reason to switch to the Stronger Statement is that now $|\mathscr{F}| = \binom{n}{101}$ is roughly n-times greater than $|V| = \binom{n}{0} + \binom{n}{1} + \binom{n}{2} + \cdots + \binom{n}{100}$. This makes it possible to *repeat the proof* of Theorem F. Note that direct application cannot work for the simple reason that the corresponding hypergraph is 'very degree-irregular', and it is 'bad' to work with the same pair-degree during the whole course of the play.

Proof. The proof of the Stronger Statement goes as follows. As usual, in each move Maker makes the 'best choice': he chooses a subset of 'maximum value', using the usual power-of-two scoring system, and if there are two *comparable* subsets, say U and W with $U \subset W$, having the same 'absolute maximum', then Maker always chooses the *smaller set*, i.e., U instead of W. (Note that if $U \subset W$, then the 'value' of U is *always* at least as large as the 'value' of W.) Maker's opening move is of course the 'empty set'. Let $U^{(i)}$ be an arbitrary move of Maker: $U^{(i)}$ is an i-element subset ($0 \leqslant i \leqslant k$). Let W be Breaker's next move right after $U^{(i)}$. We know that W cannot be a subset of $U^{(i)}$ – indeed, otherwise Maker would prefer W instead of $U^{(i)}$ – so $|U^{(i)} \cup W| \geqslant i+1$. There are at most $\binom{n}{k+1-(i+1)}$ $(k+1)$-element subsets S containing *both* $U^{(i)}$ and W: this is the 'actual pair-degree'. Note that for $i = 0, 1, 2, \ldots, k$ Maker chooses an i-element set at most $\binom{n}{i}$ times, so by repeating the proof of Theorem F one obtains

$$C_{\text{last}} \geqslant C_{\text{start}} - \sum_{i=0}^{k} \binom{n}{i}\binom{n}{k+1-(i+1)} 2^{2^{k+1}-3}, \tag{5.2}$$

where C_j is the 'chance function' at the jth stage of the play. Clearly $C_{\text{start}} = C_0 = \binom{n}{k+1}$, so (5.2) is equivalent to

$$C_{\text{last}} \geqslant \binom{n}{k+1} - \sum_{i=0}^{k} \binom{n}{i}\binom{n}{k+1-(i+1)} 2^{2^{k+1}-3}. \tag{5.3}$$

A trivial calculation shows that

$$\binom{n}{k+1} > \sum_{i=0}^{k} \binom{n}{i}\binom{n}{k+1-(i+1)} 2^{2^{k+1}-3} \quad \text{if} \quad n > (k+1)! 2^{2^{k+1}}. \tag{5.4}$$

By (5.3) and (5.4),

$$C_{\text{last}} > 0 \quad \text{if} \quad n > (k+1)! 2^{2^{k+1}}. \tag{5.5}$$

If the 'chance function' is not zero at the end of a play, then Breaker could *not* block every 'winning set', so there must exist a 'winning set' completely occupied by Maker. In other words, there must exist a $(k+1)$-element subset S of $[n]$ such that Maker occupied all at-most-k-element subsets of S. This proves the Stronger Statement. □

The last step is trivial: throwing out an arbitrary integer from S we obtain a k-element subset S' of $[n]$ such that Maker occupied all 2^k subsets of S'. This solves the 'all-subset game'.

Note that inequality

$$n > (k+1)! 2^{2^{k+1}}$$

holds (see (5.5)) if $k = (1+o(1))\log_2 \log_2 n$.

In the other direction, we show that the choice $k = (1+o(1))\log_2 \log_2 n$ is impossible. Indeed, the Erdős–Selfridge criterion applies if

$$\binom{n}{k} < 2^{2^k - 1}.$$

This holds for $k = (1+o(1))\log_2 \log_2 n$.

Theorem 2. *The game-theoretic breaking point for the weak all-subset game is $(1 + o(1))\log_2\log_2 n$, that is, the iterated binary logarithm of n.* □

Note that the power set, the goal of the all-subset game, is a 'very rapidly changing configuration': $2^{n+1} = 2 \times 2^n$, that is, changing the value of n by one doubles the size of the power set. This is why such a crude approach can still give a satisfying solution.

The Ramsey criteria (Theorems C and E) prove that the 'game world' is *at least* as large as the 'Ramsey world'. The 'unrestricted n-in-a-row with arbitrary slopes' and the all-subset game were the first examples to demonstrate that the 'game world' is actually bigger than the 'Ramsey world'. Another example is the following.

(4) Goal: 2-coloured AP. Let $N \geq W(n)$ where $W(n)$ is the van der Waerden threshold number for n-term arithmetic progressions (APs), and consider a play in the twice-as-long interval $[2N] = \{1, 2, \ldots, 2N\}$. If Second Player follows the 'copycat strategy': for First Player's x he replies by $2N + 1 - x$, then of course at the end of the play Second Player will occupy an n-term AP, and at the same time First Player will occupy an n-term AP. In other words, if Second Player is Mr. Red, and First Player is Mr. Blue, then Mr. Red can force the appearance of *both* a monochromatic *red* and a monochromatic *blue* n-term AP.

How about if Mr. Red's goal is an *arbitrary 2-coloured n-term AP*? Of course Ramsey theory cannot help here. But again the potential function technique works very well!

The precise definition of the '2-coloured goal game' goes as follows. Fix an arbitrary Red–Blue sequence of length 100: (say) $R, B, R, R, R, B, B, R, B, \ldots$. We call it the 'goal sequence'. Mr. Red and Mr. Blue alternately take new integers from the interval $[N] = \{1, 2, \ldots, N\}$, and colour them with their own colours. Mr. Red wins if at the end of the play there is an arithmetic progression of length 100 which is coloured exactly like the given Red–Blue 'goal sequence'; otherwise Mr. Blue wins. I call this the *goal sequence game*. Is it true that Mr. Red has a winning strategy in the goal sequence game if N is sufficiently large?

The answer is 'yes'. As in the case of the all-subset game, one cannot directly apply Theorem F. Instead one has to repeat the whole proof. Of course 100 can be replaced by an arbitrary finite number.

Theorem 3. *Consider an arbitrary Red–Blue 'goal sequence' S of length n. Assume that Mr. Red and Mr. Blue play the goal sequence game on the interval $[N] = \{1, 2, \ldots, N\}$, where the 'goal sequence' is S. If $N \geq n^3 \cdot 2^{n-2}$ then at the end of the play Mr. Red can force the appearance of an arithmetic progression of length n which is coloured exactly the same way as the given Red–Blue 'goal sequence' S.*

Proof. For simplicity assume that Mr. Red is the first player. Assume that we are in the middle of a play: so far Mr. Red has coloured integers $x_1, x_2, \ldots, x_i$ ('red points') and Mr. Blue has coloured integers $y_1, y_2, \ldots, y_i$ ('blue points'). This defines a partial 2-colouring of $[N]$.

Let $\mathcal{F} = \mathcal{F}(N,n)$ be the family of all n-term APs in $[N]$; clearly $|\mathcal{F}| > N^2/4(n-1)$. The partial 2-colouring of $[N]$ defines a partial 2-colouring of every n-term AP $A \in \mathcal{F}$. We introduce the following natural adaptation of the power-of-two scoring system: if the partial 2-colouring of an n-term AP $A \in \mathcal{F}$ contradicts the given 'goal sequence' S, then A has *zero value*, and we call it a 'dead set'. The rest of the elements of $\mathcal{F}$ are called 'survivors'; the partial 2-colouring of a 'survivor' $A \in \mathcal{F}$ has to be consistent with the given 'goal sequence' S. The *value* of a 'survivor' $A \in \mathcal{F}$ is 2^j if j points of A were coloured in the partial 2-colouring.

We define the 'chance function' in the standard way:

$$C_i = \sum_{A \in \mathcal{F}} \text{value}_i(A).$$

As usual, we study how the consecutive moves x_{i+1} and y_{i+1} affect the 'chance function'.

For an arbitrary n-term AP $A \in \mathcal{F}$, 2-colour the elements of A copying the given 'goal sequence' S. Then let $\text{red}_S(A)$ denote the set of red elements of A, and let $\text{blue}_S(A)$ denote the set of blue elements of A.

We have

$$C_{i+1} = C_i + \sum_{\substack{A \in \mathcal{F}: \\ x_{i+1} \in \text{red}_S(A)}} \text{value}_i(A) - \sum_{\substack{A \in \mathcal{F}: \\ x_{i+1} \in \text{blue}_S(A)}} \text{value}_i(A)$$

$$+ \sum_{\substack{A \in \mathcal{F}: \\ y_{i+1} \in \text{blue}_S(A)}} \text{value}_i(A) - \sum_{\substack{A \in \mathcal{F}: \\ y_{i+1} \in \text{red}_S(A)}} \text{value}_i(A)$$

$$- \sum_{\substack{A \in \mathcal{F}: \\ \{x_{i+1},y_{i+1}\} \in \text{red}_S(A)}} \text{value}_i(A) - \sum_{\substack{A \in \mathcal{F}: \\ \{x_{i+1},y_{i+1}\} \in \text{blue}_S(A)}} \text{value}_i(A)$$

$$+ \sum_{\substack{A \in \mathcal{F}: \\ x_{i+1} \in \text{red}_S(A) \\ y_{i+1} \in \text{blue}_S(A)}} \text{value}_i(A) + \sum_{\substack{A \in \mathcal{F}: \\ x_{i+1} \in \text{blue}_S(A) \\ y_{i+1} \in \text{red}_S(A)}} \text{value}_i(A).$$

How should Mr. Red choose his next move x_{i+1}? Well, Mr. Red's optimal move is to compute the numerical value of the function

$$f(u) = \sum_{\substack{A \in \mathcal{F}: \\ u \in \text{red}_S(A)}} \text{value}_i(A) - \sum_{\substack{A \in \mathcal{F}: \\ u \in \text{blue}_S(A)}} \text{value}_i(A)$$

for every unoccupied integer $u \in [N]$, and to choose that $u = x_{i+1}$ for which the *maximum* is attained. This choice implies

$$C_{i+1} \geq C_i - \binom{n}{2} 2^{n-2}.$$

Indeed, there are at most $\binom{n}{2}$ n-term arithmetic progressions containing both x_{i+1} and y_{i+1}. Therefore,

$$C_{\text{last}} \geq C_{\text{start}} - \frac{N}{2} \binom{n}{2} 2^{n-2}.$$

Since $C_{\text{start}} = C_0 = |\mathcal{F}| \geq N^2/4(n-1)$, it suffices to guarantee the inequality

$$\frac{N^2}{4(n-1)} > \frac{N}{2}\binom{n}{2}2^{n-2},$$

which is equivalent to $N > n(n-1)^2 \cdot 2^{n-2}$. Under this condition $C_{\text{last}} > 0$, that is, at the end of the play there must exist an n-term AP which is 2-coloured by Mr. Red and Mr. Blue exactly the same way as the 'goal sequence' S. Theorem 3 follows. □

We leave the 'clique version' of Theorem 3 to the reader in the form of an exercise.

Exercise. Fix an arbitrary 2-coloured clique $K_{100}(\text{red}; \text{blue})$ of 100 vertices ('2-coloured goal graph'). Consider the following game: Mr. Red and Mr. Blue alternately take edges of a complete graph K_n on n vertices, and colour them with their own colours. Mr. Red wins if at the end of the play there is an isomorphic copy of $K_{100}(\text{red}; \text{blue})$; otherwise Mr. Blue wins. Show that Mr. Red has a winning strategy if n is sufficiently large.

6. Bending the rules: (2 : 2) and biased versions

(1 : 1) versus (2 : 2): copycat breakdown. Every game that we have been studying so far was played in the usual (1 : 1) way: each player occupied one new point per move. A minor change is to consider the (2 : 2) play instead of the (1 : 1) play: the players still alternate, but either player occupies *two* new points per move. The game remains *fair*, but there is a surprising difference: copycat pairing does *not* work any longer. I illustrate this 'breakdown' on two examples. I begin with the ordinary (1 : 1) versions.

(1) Consider the following arithmetic progression game. The board is $\mathbb{Z}$, the set of all integers. The two players alternate, and each takes one new integer per move. First Player wins if at some point, at an even stage of the play (*i.e.*, right after Second Player's move), his longest arithmetic progression (AP) is longer than Second Player's longest AP by at least *one*. If this never happens Second Player wins.

A simple copycat pairing yields that Second Player wins. Indeed, Second Player chooses a 'half-integer' (say) $1/2$ as the 'centre', and reflects his opponent's moves: if First Player's last move was m, then Second Player replies by $1 - m$. □

The second example is a clique version of game (1).

(2) The board is an infinite complete graph K_∞. The two players alternately take new edges, one edge per move. First Player wins if at some stage of the play his largest clique is larger than Second Player's largest clique by at least *two* (vertices). If this never happens Second Player wins.

Again Second Player wins by a copycat pairing. Indeed, for simplicity assume that the vertex set is $\{1, 2, 3, \ldots\} \cup \{-1, -2, -3, \ldots\}$, then Second Player makes the following natural pairing: $i \leftrightarrow (-i)$ ($i = 1, 2, 3, \ldots$). Every edge $\{i, j\}$ has its 'mirror image' $\{-i, -j\}$; the 'mirror image' edge is a different edge, except for the 'fixpoint edges' $\{i, -i\}$. Second

Player's strategy is to take the 'mirror image' of First Player's last move; if First Player's last move was a 'fixpoint edge' then Second Player replies arbitrarily.

Let K_q denote First Player's largest clique at an arbitrary even stage (*i.e.*, right after Second Player's move). It is easy to see that K_q cannot contain two 'fixpoint edges' (we recommend the reader to draw a picture).

Removing an endpoint of a possible 'fixpoint edge' from K_q leads to a 'fixpoint-edge-free' subclique K_{q-1}. The 'mirror image' of K_{q-1} must belong to Second Player, so his largest clique has at least $q-1$ vertices. $\square$

How about the (2 : 2) version of games (1) and (2)? Copycat pairing clearly breaks down, which leads to the following open problem.

Open Problem 8.
(a) *Consider the* (2 : 2) *version of the infinite arithmetic progression game* (1). *Can Second Player guarantee that, during the* whole *course of the play, his longest AP and First Player's longest AP differ in length by at most* 100?
(b) *Consider the* (2 : 2) *version of the infinite clique game* (2). *Can Second Player guarantee that, during the* whole *course of the play, his largest clique and First Player's largest clique differ in vertex size by at most* 100?

Of course, 100 simply means 'large constant'; any absolute constant would do. The point in Open Problem 8 is to control the opponent's 'lead' during the *whole course* of the play. How about if we want 'balance' at the *end* only? To understand the problem, I study the following finite (2 : 2) version of clique game (2).

(3) The board is a finite complete graph K_n ($n \geq 3$). The two players alternately take new edges, two edges per move. First Player wins if at the end of the play his largest clique is larger than Second Player's largest clique by at least *three* (vertices); otherwise Second Player wins.

By using the *strategy stealing* argument I show that Second Player has a winning strategy in the (2 : 2)-type game (3). This is an indirect proof: *assume* that First Player has a winning strategy Str, and I will derive a contradiction.

I begin with a simple observation. If one slightly changes the rules so that Second Player may 'pass' when he wishes to do so (*i.e.*, to skip a move, or just to take one edge), then of course strategy Str still guarantees for First Player a lead of ≥ 3 at the end of the play.

Second Player can 'steal' and use Str as follows. Let $\{1, 2, 3, \ldots, n\}$ be the vertex set, and without loss of generality we can assume that First Player's opening move by Str is either

Case 1: $\{1, 2\}$ and $\{3, 4\}$; or
Case 2: $\{1, 2\}$ and $\{1, 3\}$.

In Case 1 Second Player deletes 1 and 3, and adds $1'$ and $3'$ to the vertex set. Second Player plays on the *virtual board* K'_n, where the new vertex set is $\{1', 2, 3', 4, 5, \ldots, n\}$, by using strategy Str with the slight change that '1' means the new vertex $1'$ and '3' means $3'$. Second Player's opening move in K'_n is of course $\{1', 2\}$ and $\{3', 4\}$. Suppose that

First Player follows his strategy Str, and Second Player keeps replying to the opponent's moves by Str (as we explained above). At the end of the play, let K_s denote Second Player's largest clique, and let K_f denote First Player's largest clique in K'_n. By hypothesis $s \geq f + 3$. This holds for the virtual game on K'_n. In the *real* game on K_n vertices 1 and 3 are added back and 1' and 3' are removed. It follows that at the end of this play in the real game, First Player's largest clique has at most $f + 2$ vertices, and Second Player's largest clique has at least $s - 2$ vertices. Since First Player uses Str, his clique in the real game is bigger than Second Player's largest clique by at least 3, implying $f + 2 \geq (s - 2) + 3$. This means $f + 1 \geq s$, which contradicts the inequality $s \geq f + 3$ obtained from the virtual game.

Note that Case 2 is even simpler, since Second Player has to remove only one vertex, namely 1, instead of two, and add 1' to the vertex set in the virtual game.

In both cases we obtain a contradiction. This proves that First Player cannot have a winning strategy in the (2 : 2) game (3), so it is Second Player who has a winning strategy. □

As usual with the strategy stealing argument, we don't know any explicit Second Player's winning strategy.

What happens for the finite (2 : 2) version of game (1)? I have to admit, I don't know. The proof for game (3) fails to work for APs, since removing an integer from an AP breaks it into two shorter pieces (the worst case is to have two 'halves').

After the (2 : 2) game, which is still fair, let's move to a real biased game, and consider (say) the (2 : 1)-type game where Second Player is the 'underdog', *i.e.*, Second Player takes only one per move. In general, if player P takes p points per move, and the opponent, player Q, takes q points per move from a given hypergraph, and (say) $p < q$, then we call player P the *underdog* and player Q the *topdog*.

(4) For simplicity assume that the players play a (2 : 1) game on an n-uniform hypergraph $\mathscr{F}$. Moreover, assume that $\mathscr{F}$ has at least 3^n pairwise disjoint winning sets. Then First Player, who takes 2 points per move, always wins the strong game: First Player can always occupy a whole winning set *first*.

Proof. The proof is rather simple. Again I use strategy stealing. There are two cases: either

Case 1: playing the (2 : 1) strong game on $\mathscr{F}$, Second Player has a drawing strategy; or
Case 2: playing the (2 : 1) strong game on $\mathscr{F}$, First Player has a winning strategy.

Of course in Case 2 we are done. In Case 1 let Str denote Second Player's drawing strategy. First Player is going to use Str; to explain how, consider a play (First Player marks X and Second Player marks O)

X X O X X O X X O X X O X X O X X O X X O X X O X X O X X O

and decompose it into blocks as follows (the first block has 7 marks, the rest have 6 each):

X X O X X O X– X O X X O X– X O X X O X– X O X X O X– X O X X O X– ⋯

First Player's first 4 Xs are arbitrary ('free'), his 5th X is a reply to the 2 Os of Second Player by using Str. The next 3 Xs of First Player are arbitrary ('free'), his 9th X is a reply to the last 2 Os of Second Player by using Str. The next 3 Xs of First Player are arbitrary ('free'), his 13th X is a reply to the last 2 Os of Second Player by using Str, and so on. This means the play is divided into blocks of length 6; the first block is exceptional: it has 7 marks. Each block has one X which is determined by Str (replying to the last two Os), 3 'free' Xs which are arbitrary, and 2 Os. By using the 'reply' Xs, First Player can prevent Second Player from occupying a winning set first.

On the other hand, by using his 3 'free' Xs against the opponent 2 Os (per block), First Player can definitely occupy a whole winning set. Indeed, (3 : 2) is biased in favour of First Player, and there are at least 3^n pairwise disjoint n-element winning sets. Actually, First Player will occupy a whole n-element winning set in n phases. In phase 1, by using his free Xs, First Player marks 3^n disjoint winning sets, one X per set. Second Player can block at most 2/3 of them (by putting an O), so at the end of phase 1 there are $\geqslant 3^{n-1}$ n-sets which are O-free and have one X each. Similarly, at the end of phase 2 there are $\geqslant 3^{n-2}$ n-sets which are O-free and have 2 Xs each. Repeating this argument, at the end of phase n there exists $\geqslant 3^{n-n} = 1$ n-set which is O-free and has n Xs, i.e., this winning set is completely occupied by First Player. This shows that

(i) Second Player cannot occupy a winning set first (since First Player uses Str); and
(ii) First Player can occupy a whole winning set.

(i) and (ii) together give that First Player can occupy a whole winning set *first*, which contradicts the hypothesis of Case 1. This contradiction proves that Case 1 is impossible, *i.e.*, First Player can force a strong win. □

Again I don't know any explicit First Player's strong-win strategy in the (2 : 1)-type game (4).

This means that, playing the (2 : 1) game, the underdog *always* loses the strong game, unless the hypergraph is very small. In other words, 'doing it first' is a hopeless task for the underdog. The relevant question is 'Can the underdog do it or not?', the 'reasonable' goal is a weak win, and the relevant game is the weak game.

Weak win for the underdog. If there is a 'density theorem', then of course the underdog can achieve a weak win by simply showing up. For example, in the (2 : 1) version of the (N, n) van der Waerden game, at the end of a play the underdog has $N/3$ integers from the interval $1, 2, \ldots, N$. By the deep Szemerédi–Gowers theorem, *every* $N/3$-element subset of the interval $1, 2, \ldots, N$ contains an n-term AP if N is larger than a five-times iterated exponential function of n.

Unfortunately we don't know any *effective* density version of the Hales–Jewett theorem, and the density version of the Ramsey theorem is simply *not true*. What can we do then? Can large *chromatic number* help us out here? Is it true that, if the chromatic number of the hypergraph is sufficiently large, then playing the (2 : 1) game on the hypergraph, the underdog can achieve a weak win? Unfortunately, the answer is 'no', even if the underdog is the first player, i.e., the (1 : 2) game. The counter-example is due to F. Galvin: for every $r \geqslant 3$ there are uniform hypergraphs with chromatic number $\geqslant r$ such that, playing the (1 : 2) game on the hypergraph, the second player can prevent the first from occupying a whole winning set. Galvin's construction is very complicated.

Since large chromatic number does not necessarily guarantee the underdog's weak win, Ramsey theory becomes 'useless' for biased games. In other words, there is no underdog analogue of Theorem E. But what actually *does* work here is the potential function technique. One can easily formulate and prove the biased version of Theorem F (weak win criterion), and also the biased version of Theorem D (Erdős–Selfridge theorem). These two generalizations (see Theorems G–H [6] below) are the basic tools to handle biased games.

Theorem G (biased weak win). *If*

$$\sum_{A \in \mathscr{F}} \left(\frac{p+q}{p}\right)^{-|A|} > p^2 \cdot q^2 \cdot (p+q)^{-3} \cdot \Delta_2(\mathscr{F}) \cdot |V(\mathscr{F})|,$$

where $\Delta_2(\mathscr{F})$ is the pair-degree of hypergraph $\mathscr{F}$, and $V(\mathscr{F})$ is the board, then First Player can completely occupy a winning set $A \in \mathscr{F}$ in the biased $(p : q)$ play on $\mathscr{F}$ (First Player takes p new points and Second Player takes q new points per move).

Applying the biased criterion to a fair game. I give a surprising application of Theorem G: I apply it to a *fair* tournament game! We recall that a 'tournament' means a 'directed complete graph' such that every edge of a complete graph is directed by one of the two possible orientations; it represents a tennis tournament where any two players played with each other, and an arrow points from the winner to the loser.

Fix an arbitrary tournament T_k on k vertices. The two players are Red and Blue, who alternately take new edges of a complete graph K_n, and for each new edge choose one of the two possible orientations ('arrow'). Either player colours his arrow with his own colour. At the end of a play, the players create a 2-coloured tournament on n vertices. Red wins if there is a red copy of T_k; otherwise Blue wins. Is it true that, if n is sufficiently large compared to k, then Red always has a winning strategy?

The answer is 'yes', and I prove it by a simple application of the biased weak win criterion Theorem G. This is rather surprising, since the tournament game itself is *fair*. The idea is to associate with the fair tournament game a biased (1 : 3) hypergraph game! To understand the application, we recommend the reader to draw a picture. The board $V = V(\mathscr{F})$ of the biased (1 : 3) hypergraph game is the set of $2\binom{n}{2}$ arrows of $K_n(\uparrow\downarrow)$, where $K_n(\uparrow\downarrow)$ means that every edge of the complete graph K_n shows up with both orientations. The winning sets $A \in \mathscr{F}$ are the arrow-sets of all possible copies of T_k in $K_n(\uparrow\downarrow)$. So $\mathscr{F}$ is a $\binom{k}{2}$-uniform hypergraph, and trivially $|\mathscr{F}| \geqslant \binom{n}{k}$. If the mth move of Red and Blue are, respectively, $i_1 \to j_1$ and $i_2 \to j_2$, then these two moves (arrows) automatically exclude the extra arrows $j_1 \to i_1$ and $j_2 \to i_2$ from $K_n(\uparrow\downarrow)$ for the rest of the play. (There may be some coincidence among i_1, j_1, i_2, j_2, but it does not make any difference.) This means 1 arrow for Maker, and 3 arrows for Breaker in the hypergraph game on $\mathcal{F}$, which explains how the biased (1 : 3) play enters the story.

All that is left is to apply Theorem G to the (1 : 3) hypergraph game on $\mathcal{F}$. If

$$\sum_{A \in \mathcal{F}} 4^{-\binom{k}{2}} > \frac{9}{64} \cdot \Delta_2(\mathcal{F}) \cdot |V(\mathcal{F})|,$$

where $\Delta_2(\mathcal{F})$ is the pair-degree and $V(\mathcal{F})$ is the board, then Red can force a win in the tournament game. We have the trivial equality $|V(\mathcal{F})| = 2\binom{n}{2}$, and the less trivial inequality $|\Delta_2(\mathcal{F})| \leqslant \binom{n-3}{k-3} \cdot k!$. Indeed, a tournament T_k *cannot* contain parallel arrows (*i.e.*, both orientations of an edge), so a pair of red and blue arrows contained by a copy of T_k must span at least 3 vertices, and there are at most $\binom{n-3}{k-3} \cdot k!$ ways to extend an unparallel arrow-pair to a copy of T_k.

Combining these facts, in view of Theorem G, it suffices to check that

$$\binom{n}{k} > \frac{9}{64} \cdot 4^{\binom{k}{2}} \cdot \binom{n-3}{k-3} \cdot k! \cdot 2\binom{n}{2}.$$

This inequality trivially follows if $n \geqslant c_0 \cdot k^{k+3} \cdot 4^{\binom{k}{2}}$. The threshold $n_0(k) = c_0 \cdot k^{k+3} \cdot 4^{\binom{k}{2}}$ works uniformly for all goal tournaments T_k on k vertices. $\square$

We conclude this section with the biased version of the Erdős–Selfridge theorem (see [6]).

Theorem H (biased Erdős–Selfridge). *If*

$$\sum_{A \in \mathcal{F}} (1+q)^{-|A|/p} < 1, \quad \text{or} \quad < \frac{1}{1+q},$$

then First Player (or Second Player) can block every winning set $A \in \mathcal{F}$ in the biased $(q : p)$ (or $(p : q)$) play on $\mathcal{F}$.

7. An illustration of the BigGame SmallGame Decomposition

A major technical difficulty in studying the n^d game comes from the highly irregular nature of the n^d hypergraph. For example, if n is odd, the maximum degree (attained in the centre) is $(3^d - 1)/2$, which is much larger than the average degree:

$$\text{AverDeg}(n^d\text{-hypergraph}) = \frac{n \cdot \text{family size}}{\text{board size}} = \frac{n((n+2)^d - n^d)/2}{n^d} \approx \frac{n}{2}(e^{\frac{2d}{n}} - 1).$$

Therefore, the average degree of the n^d-hypergraph is about (very roughly) the nth root of the maximum degree. (Basically the same holds for n even; then the maximum degree is the smaller but still exponential $2^d - 1$.)

On the other hand, if we switch to the n^d *torus*, and consider the family of all 'torus-lines' (instead of all 'geometric lines' in the n^d hypercube), then we obtain a *degree-regular* hypergraph. This is the family of winning sets of the n^d *torus game*. The 8^2 torus game already came up in Section 6: we proved that this particular torus game was a pairing strategy draw, and the given explicit pairing implied that the 'unrestricted 9-in-a-row' was a draw game.

The n^d torus is an abelian group, implying that the n^d torus-hypergraph is translation-invariant (any two points 'look the same'). The n^d torus-hypergraph is degree-regular: every point has degree $(3^d - 1)/2$ (which is by the way the same as the degree of the centre in the n^d hypergraph when n is odd). So the total number of winning sets (torus-lines) is $(3^d - 1)n^{d-1}/2$. Of course, the board size remains n^d. We owe the reader a formal definition of the concept of 'torus-line'. A torus-line L is formally defined by a point $P \in L$ and a vector $\mathbf{v} = (a_1, \ldots, a_d)$ where each coordinate a_i is either 0, or +1, or −1 ($1 \leq i \leq d$). The n points of line L are $P + k\mathbf{v}$ (mod n) where $k = 0, 1, \ldots, n - 1$.

A peculiarity of this new 'line-concept' is that two different torus-lines may have more than one point in common! We recommend the reader to study the 4^2 torus game.

We show that this 'pair-intersection' cannot happen when n is *odd*, and if n is *even* then it can happen only under very special circumstances.

Statement. Any two different torus-lines have at most *one* common point if n is *odd*, and at most *two* common points if n is *even*. In the second case the distance between the two common points along either torus-line is always $n/2$.

Proof. The proof of this elementary fact goes as follows. Let L_1 and L_2 be two different torus-lines with (at least) two common points P and Q. Then there exist $k, l, \mathbf{v}, \mathbf{w}$ with $1 \leq k, l \leq n - 1$, $\mathbf{v} = (a_1, \ldots, a_d)$, $\mathbf{w} = (b_1, \ldots, b_d)$ (where a_i and b_i are either 0, or +1, or −1 ($1 \leq i \leq d$)), $\mathbf{v} \neq \pm\mathbf{w}$, such that $Q \equiv P + k\mathbf{v} \equiv P + l\mathbf{w}$ (mod n). It follows that $k\mathbf{v} \equiv l\mathbf{w}$ (mod n), or equivalently, $ka_i \equiv lb_i$ (mod n) for every $i = 1, 2, \ldots, d$. Since a_i and b_i are either 0, or +1, or −1 ($1 \leq i \leq d$)), the only solution is $k = l = n/2$ (no solution if n is *odd*). □

What can we say about the two-dimensional n^2 torus game? Well, we know everything. From Section 6 we know that the 8^2 torus game is a pairing strategy draw, and this is a sharp result (since the point/line ratio of the 7^2 torus is $7^2/4 \cdot 7 = 7/4$: less than 2).

The Erdős–Selfridge theorem applies if $4n + 4 < 2^n$, which gives that the n^2 torus game is a blocking draw for every $n \geq 5$. The 4^2 torus game is also a draw (mid-size 'case study'), but I don't know any elegant proof. On the other hand, the 3^2 torus game is an easy First Player's win.

Next consider the three-dimensional n^3 torus game. The Erdős–Selfridge theorem applies if $13n^2 + 13 < 2^n$, which gives that the n^3 torus game is a blocking draw for every $n \geq 11$. I am convinced that the 10^3 torus game is also a draw, but I don't know how to prove it.

How about the four-dimensional n^4 torus game? The Erdős–Selfridge theorem applies if $40n^3 + 40 < 2^n$, which gives that the n^4 torus game is a blocking draw for every $n \geq 18$. This $n = 18$ can be improved to $n = 15$ by using a new method called 'BigGame SmallGame Decomposition'. This method was developed for proving the best known draw-criterion for the n^d game (multi-dimensional Tic-Tac-Toe). That application is rather complicated, and gives 'good' results in relatively large dimensions only (for example, by 'beating' any pairing strategy for $d \geq 33$; see the Hales–Jewett conjecture in Section 3).

The following proof is perhaps the best illustration of this method on a (relatively) simple low-dimensional example.

Theorem 4. *The 15^4 torus game is a blocking draw.*

Proof. Since $n = 15$ is odd, the corresponding hypergraph is almost disjoint. This fact will be used repeatedly during the whole proof.

The basic idea is a *decomposition* of the 15^4 torus game into *two non-interacting games* with disjoint boards; I call them the *big game* and the *small game*. This is in Breaker's mind only; Maker does not know anything about the decomposition. Whenever Maker picks a point from the *big board* (board of the big game) then Breaker responds in the big board; whenever Maker picks a point from the *small board* (board of the small game) then Breaker responds in the small board (that is, Breaker follows the 'same board rule').

Non-interacting games means that playing the big game Breaker has to make his moves without any knowledge of the action in the small game, and similarly, playing the small game Breaker has to make his moves without any knowledge of the action in the big game, even if, of course, Breaker *does* know everything ('game of perfect information'). In other words, we *assume* that Breaker is 'schizophrenic': he has two 'personalities'; one for the big game and one for the small game, and the two personalities know nothing about each other ('Iron Curtain principle'). This assumption, at first sight weird, is crucial to the proof.

The *small game* deals with the winning sets (torus-lines) that are *dangerous*, where Maker is close to winning in the sense that all but 2 points are occupied by Maker. The *small board* is a kind of 'emergency room': in the small game, Breaker's goal is to block the 2-element unoccupied parts of the 'dangerous' winning sets by using a trivial pairing strategy. The small game is Breaker's last chance to block the 'dangerous' sets before it is too late.

Breaker's goal in the *big game* is to prevent too complex winning-set-configurations ('forbidden configurations') from graduating into the *small game*. For example, a pairing strategy works in the small game *only* if the 2-element unoccupied parts of the dangerous winning sets remain *pairwise disjoint*. The big game has to enforce, among many other things, this 'disjointness property'. This is how the big game guarantees that Breaker's pairing strategy in the small game will indeed work. This is why we must fully understand the small game first. The big game plays a auxiliary role only: to keep the small game under control.

In the *big game* Breaker uses the Erdős–Selfridge power-of-two scoring system, and *not* the Erdős–Selfridge theorem itself; see Lemma 5 below. The key numerical fact is that the number of *big sets* depends primarily on the maximum degree $D = (3^4 - 1)/2 = 40$ (rather than on the much larger total number $40 \cdot 15^3$ of winning sets). This is why the number of big sets is not too large. This is how Breaker can force – by using the power-of-two scoring system – the small game to remain 'trivial'; so simple that even a pairing strategy can block every 'dangerous' winning set.

The board of the big game is going to shrink during a play (more precisely: the *unoccupied part* of the big game is going to shrink). The reason for the 'shrinking' is that the 2-element 'emergency sets' are constantly removed from the big board, and added to the small board. Consequently, the small board – which, by definition, is exactly the *complement* of the big board – keeps growing during a play. At the beginning of a play the big board is equal to the whole board $V = n^d$-torus (*i.e.*, the small game is 'not born yet'). Let $V_{\text{BIG}}(i)$ and $V_{\text{small}}(i) = V \setminus V_{\text{BIG}}(i)$ denote the big board and the small board after Maker's ith move and before Breaker's ith move. Then

$$V = V_{\text{BIG}}(0) \supseteq V_{\text{BIG}}(1) \supseteq V_{\text{BIG}}(2) \supseteq V_{\text{BIG}}(3) \supseteq \cdots,$$
$$\emptyset = V_{\text{small}}(0) \subseteq V_{\text{small}}(1) \subseteq V_{\text{small}}(2) \subseteq V_{\text{small}}(3) \subseteq \cdots.$$

The big game is played on the family of *big sets*, and the small game is played on the family of *emergency sets*. What are the big sets, and what are the emergency sets? Well, we already explained, at least in a heuristic way, what the emergency sets are. After the long 'heuristic introduction' it is time now to introduce the precise definitions.

Let $x_1, x_2, \ldots, x_i, \ldots$ and $y_1, y_2, \ldots, y_i, \ldots$ denote, respectively, the points of Maker and Breaker in a particular play. At the beginning, when the board of the small game is empty (*i.e.*, the small game is 'not born yet'), Breaker chooses his points $y_1, y_2, y_3, \ldots$ according to an Erdős–Selfridge type scoring system applied to the family $\mathscr{B}$ of big sets ($\mathscr{B}$ will be defined later). The hypergraph of the game is the family of all 'torus-lines of the 15^4 torus'; we denote it by $\mathscr{F}$. $\mathscr{F}$ is a 15-uniform hypergraph, and $|\mathscr{F}| = 40 \cdot 15^3$. In the course of a play in the big game a 15-element winning set $A \in \mathscr{F}$ (torus-line) is called 'dead' when it contains a mark of Breaker. Note that dead winning sets (*i.e.*, elements of $\mathscr{F}$) no longer represent any danger (they are marked by Breaker, so Maker cannot completely occupy them). At any time in the big game, the elements of $\mathscr{F}$ which are not yet dead are called 'survivors'. A survivor $A \in \mathscr{F}$ becomes 'dangerous' when Maker occupies its 13th point (all 13 points have to be in the big board); then the unoccupied 2-element part of this dangerous $A \in \mathscr{F}$ becomes an 'emergency set'. Whenever an emergency set arises (in the big game), then it is *removed* from the board of the big game, and at the same time it is added to the board of the small game. This is why the big game is shrinking. The board of the small game is precisely the union of *all 2-element emergency sets*, and consequently, the board of the big game is precisely the complement of the union of all emergency sets.

If the (growing) family of 2-element emergency sets remains 'disjoint' (*i.e.*, the 2-element emergency sets never overlap during the whole course of a play), then Breaker can easily block them in the small game (on the small board) by using the following trivial pairing strategy: when Maker takes a member of a 2-element emergency set, then Breaker takes the other one. The big game is designed exactly to enforce, among other properties, the 'disjointness of the emergency sets'. Therefore, Breaker must prevent the appearance of any 'forbidden configuration of type 1' (to understand the definition we *strongly recommend* the reader to draw a picture).

Forbidden configuration of type 1. At some stage of the play there exist two *dangerous* sets $A_1 \in \mathscr{F}$ and $A_2 \in \mathscr{F}$ such that their 2-element emergency parts $E_1 (\subset A_1)$ and $E_2 (\subset A_2)$ have a common point. (Since $\mathscr{F}$ is almost disjoint, they cannot have more than one point in common.)

If there exists a forbidden configuration of type 1, then at some stage of the play Maker occupied $13 + 13 = 26$ points of a 'pair-union' $A_1 \cup A_2$ (where $A_1, A_2 \in \mathscr{F}$) in the big board (during the big game), and Breaker could not yet put a single mark in $A_1 \cup A_2$ in the big game (perhaps Breaker could do it in the small game, but that does not count).

Note that the total number of 'intersecting pairs' $\{A_1, A_2\}$ with $|A_1 \cap A_2| = 1$ ($A_1, A_2 \in \mathscr{F}$) is exactly

$$15^4 \binom{40}{2}. \tag{7.1}$$

Indeed, the torus has 15^4 points, and each point has the same degree 40.

Another 'potentially bad configuration', that Breaker is advised to prevent, is any 'forbidden configuration of type 2' (again we *strongly recommend* the reader to draw a picture).

Forbidden configuration of type 2. There exists a *survivor* $A_0 \in \mathscr{F}$

(a) which never graduates into a *dangerous set*; and at some stage of the play,
(b) A_0's intersection with the big board is completely occupied by Maker, or possibly empty; and at the same time,
(c) A_0's intersection with the small board is completely covered by pairwise disjoint 2-element emergency sets.

The 'danger' of (a)–(c) is obvious: since A_0 never graduates into a *dangerous set*, A_0's intersection with the small board remains 'invisible' for Breaker in the whole course of the small game, so there is a real chance that A_0 will be completely occupied by Maker.

In (c) let k denote the size of the intersection of A_0 with the small board; (a) implies that the possible values of k are $3, 4, \ldots, 15$. Accordingly we can talk about a forbidden configuration of type $(2, k)$ where $3 \leq k \leq 15$.

Let A_0 be a forbidden configuration of type $(2, k)$ ($3 \leq k \leq 15$); then there are k disjoint 2-element emergency sets $E_1, E_2, \ldots, E_k$ which cover the k-element intersection of A_0 with the small board (see (c)). Let $A_1, A_2, \ldots, A_k$ denote the *supersets* of $E_1, E_2, \ldots, E_k$: $E_i \subset A_i$ where every $A_i \in \mathscr{F}$ is a dangerous set ('almost disjointness' implies that for every E_i there is a unique A_i).

Then, at the particular stage of the play described by (b)–(c), Maker occupied at least $(15 - 2) + (15 - 3) + \cdots + (15 - k) + (15 - k - 1) + (15 - k) = (30 + 25k - k^2)/2$ points (almost disjointness) of a union set $\bigcup_{i=0}^{k} A_i$ in the big board (in the big game), and Breaker could not yet put a single mark in $\bigcup_{i=0}^{k} A_i$ in the big game.

Note that the total number of configurations $A_0, \{A_1, A_2, \ldots, A_k\}$ satisfying

(α) $A_0, A_1, \ldots, A_k \in \mathscr{F}$, and
(β) $|A_0 \cap A_i| = 1$ for $1 \leq i \leq k$, and
(γ) $A_0 \cap A_i$ with $1 \leq i \leq k$ are k distinct points,

is at most

$$(40 \cdot 15^3) \cdot \binom{15}{k} \cdot (40 - 1)^k. \tag{7.2}$$

Indeed, first choose A_0, then choose the k distinct points $A_0 \cap A_i$ with $1 \leq i \leq k$, and finally choose the k sets $A_i \in \mathscr{F}$ ($i = 1, \ldots, k$) (use almost disjointness).

Note that after an arbitrary move of Maker in the big game, the big board may shrink (because of the possible appearance of new dangerous sets; the 2-element emergency parts are removed from the big board), but the big board does *not* change after a move of Breaker.

In order to prevent all forbidden configurations (*i.e.*, type 1 and type $(2,k)$ where $k = 3, 4, \ldots, 15$), Breaker needs the following 'shrinking' variant of the Erdős–Selfridge theorem.

Lemma 5. *Let* $\mathscr{B}_1, \mathscr{B}_2, \ldots, \mathscr{B}_l$ *be a sequence of finite hypergraphs. Let* $m_1, m_2, \ldots, m_l$ *be positive integers. Consider the following 'shrinking' game. Let V be the union set of the l hypergraphs; we call V the 'initial board'. The two players, called White and Black, alternate. A 'move' of White is to take a previously unoccupied point of the board and at the same time White may remove an arbitrary unoccupied part from the board (if he wants any). A 'move' of Black is standard: he takes a previously unoccupied point of the board. After White's move the board may shrink, and the players are always forced to take the next point from the 'available board' (which is a 'decreasing' subset of the initial board). White wins, if at some stage of the play, there exist* $i \in \{1, \ldots, l\}$ *and* $B \in \mathscr{B}_i$ *such that, White has* m_i *points from B and Black has none. Otherwise Black wins.*

Assume that

$$\sum_{i=1}^{l} \bigl(|\mathscr{B}_i| + \mathrm{MaxDeg}(\mathscr{B}_i)\bigr) 2^{-m_i} < 1.$$

Then Black has a winning strategy, no matter whether Black is the first or second player.

First I explain how to apply Lemma 5 (and prove it later). To prevent the appearance of any forbidden configuration of type 1, one has to control all possible candidates (of course no one knows in advance which winning set will eventually become *dangerous*), so let

$$\mathscr{B}_1 = \left\{ A_1 \cup A_2 : \{A_1, A_2\} \in \binom{\mathscr{F}}{2},\ |A_1 \cap A_2| = 1 \right\} \quad \text{and let}\quad m_1 = 2(15-2) = 26.$$

To avoid the appearance of any forbidden configuration of type $(2,3)$, define hypergraph $\mathscr{B}_2$ as follows:

$$\left\{ A_0 \cup \cdots \cup A_3 : \{A_0, \ldots, A_3\} \in \binom{\mathscr{F}}{4},\ A_0 \cap A_i,\, 1 \leq i \leq 3,\ \text{are distinct points} \right\},$$

and let $m_2 = (15-2) + (15-3) + (15-4) + (15-3) = 48$. In general, to avoid the appearance of any forbidden configuration of type $(2, j+1)$, define hypergraph $\mathscr{B}_j$ as follows:

$$\left\{ A_0 \cup \cdots \cup A_{j+1} : \{A_0, \ldots, A_{j+1}\} \in \binom{\mathscr{F}}{j+2},\ A_0 \cap A_i,\, 1 \leq i \leq j+1,\ \text{are distinct points} \right\}, \tag{7.3}$$

and let

$$m_j = (15-2) + (15-3) + \cdots + (15-j-1) + (15-j-2) + (15-j-1) = \frac{54 + 23j - j^2}{2}. \tag{7.4}$$

For technical reasons we use definition (7.3)–(7.4) for $j = 2, 3, \ldots, 6$ only. This takes care of forbidden configurations of type $(2,k)$ with $3 \leq k \leq 7$. To prevent the appearance of any forbidden configuration of type $(2,k)$ where $8 \leq k \leq 15$, we use a *single* extra hypergraph $\mathscr{B}_7$. Let hypergraph $\mathscr{B}_7$ be defined as

$$\left\{ A_0 \cup A_1 \cup \cdots \cup A_8 : \{A_0, A_1, \ldots, A_8\} \in \binom{\mathscr{F}}{9}, A_0 \cap A_i, 1 \leq i \leq 8, \text{ are distinct points} \right\}, \tag{7.5}$$

and let

$$m_7 = (15-2) + (15-3) + (15-4) + (15-5) + (15-6) + (15-7) + (15-8) + (15-9) = 76. \tag{7.6}$$

In the definition of m_7 we didn't include the number of marks of Maker in A_0: this is how we can deal with all types $(2,k)$ where $8 \leq k \leq 15$ *at once*. Indeed, any forbidden configuration of type $(2,k)$ with $8 \leq k \leq 15$ contains a *subconfiguration* which is 'covered' by $\mathscr{B}_7$ and m_7 (we strongly recommend the reader to draw a picture).

Next we estimate the sum

$$\sum_{i=1}^{7} (|\mathscr{B}_i| + \text{MaxDeg}(\mathscr{B}_i)) 2^{-m_i}.$$

By (7.1)–(7.2) we have

$$|\mathscr{B}_1| = 15^4 \binom{40}{2}, \text{ and } |\mathscr{B}_j| = (40 \cdot 15^3) \cdot \binom{15}{j+1} \cdot (40-1)^{j+1} \text{ for } 2 \leq j \leq 7.$$

Since the 15^4-torus is a group, the hypergraphs $\mathscr{B}_i$ ($1 \leq i \leq 7$) are all degree-regular; every point has the same degree, namely the average degree (this observation simplifies the calculations). In $\mathscr{B}_1$ every set has the same size $2 \cdot 15 - 1 = 29$, and for each $j = 2, \ldots, 7$ every $B \in \mathscr{B}_j$ has size $\geq (15-2) + (15-3) + \cdots + (15-j-1) + (15-j-2) + (15-j-1) = (54 + 23j - j^2)/2$. It follows that

$$|\mathscr{B}_1| + \text{MaxDeg}(\mathscr{B}_1) = 15^4 \cdot \binom{40}{2}\left(1 + \frac{29}{15^4}\right),$$

and for $j = 2, \ldots, 7$,

$$|\mathscr{B}_j| + \text{MaxDeg}(\mathscr{B}_j) \leq (40 \cdot 15^3) \cdot \binom{15}{j+1} \cdot (40-1)^{j+1}\left(1 + \frac{54 + 23j - j^2}{2 \cdot 15^4}\right).$$

Therefore, easy calculations give

$$\sum_{i=1}^{7} (|\mathscr{B}_i| + \text{MaxDeg}(\mathscr{B}_i)) 2^{-m_i} \leq 15^4 \cdot \binom{40}{2}\left(1 + \frac{29}{15^4}\right) 2^{-26}$$

$$+ \sum_{j=2}^{6} (40 \cdot 15^3) \cdot \binom{15}{j+1} \cdot (40-1)^{j+1}\left(1 + \frac{54 + 23j - j^2}{2 \cdot 15^4}\right) 2^{-(54+23j-j^2)/2}$$

$$+ (40 \cdot 15^3) \cdot \binom{15}{8} \cdot (40-1)^8\left(1 + \frac{54 + 23 \cdot 8 - 8^2}{2 \cdot 15^4}\right) 2^{-76} < \frac{9}{10} < 1.$$

Now at last we are ready to define the *big sets*, i.e., the forbidden configurations.

Let $\mathscr{B} = \mathscr{B}_1 \cup \mathscr{B}_2 \cup \cdots \cup \mathscr{B}_7$ be the family of big sets. Applying Lemma 5 (Maker is 'White' and Breaker is 'Black') we obtain that, in the big game, played on the family of big sets, Breaker can prevent the appearance of any forbidden configuration of type 1 or type $(2,k)$ with $3 \leqslant k \leqslant 15$. We claim that, combining the 'Lemma 5 strategy' with the trivial pairing strategy in the family of emergency sets (small game), Breaker can block hypergraph $\mathscr{F}$, that is, Breaker can put his mark in every torus-line of the 15^4 torus game. Indeed, an arbitrary $A \in \mathscr{F}$ (torus-line)

(1) either eventually becomes *dangerous*; then its 2-element emergency set will be blocked by Breaker in the small board (in the small game) by the trivial pairing strategy ('disjointness of the emergence sets' is enforced by preventing forbidden configurations of type 1),

(2) or A never becomes *dangerous*; then it will be blocked by Breaker in the big board: indeed, otherwise there is a forbidden configuration of type $(2,k)$ with some $k \in \{3,\ldots,15\}$.

It remains to prove Lemma 5.

Proof of Lemma 5. The proof is a standard application of the power-of-two scoring system. If $B \in \mathscr{B}_i$ is marked by Black then it scores 0. If $B \in \mathscr{B}_i$ is unmarked by Black and has w points of White, then it scores 2^{w-m_i}. The 'target value' (*i.e.*, White's win) is $\geqslant 2^0 = 1$; the 'initial value' is less than 1 (by hypothesis). Black can guarantee the usual 'decreasing property', which implies that no play is possible if the 'target value' is larger than the 'initial value'. Lemma 5 follows. □

This completes the proof of Theorem 4. □

References

[1] Alon, N. and Spencer, J. (1992) *The Probabilistic Method*, Academic Press, New York.
[2] Baumgartner, J., Galvin, F., Laver, R. and McKenzie, R. (1973) Game theoretic versions of partition relations. In *Infinite and Finite sets*, *Colloq. Math. Soc. János Bolyai*, Keszthely, Hungary, pp. 131–135.
[3] Beasley, J. D. (1992) *The Ins and Outs of Peg Solitaire*, Oxford University Press, Oxford, New York.
[4] Beck, J. (1981) On positional games. *J. Combin. Theory Ser. A* **30** 117–133.
[5] Beck, J. (1981) Van der Waerden and Ramsey type games. *Combinatorica* **2** 103–116.
[6] Beck, J. (1982) Remarks on positional games. *Acta Math. Acad. Sci. Hungarica* **40** (1–2) 65–71.
[7] Beck, J. (2002) Positional games and the second moment method. *Combinatorica* **22** (2) 169–216.
[8] Beck, J. (2004) Multi-dimensional Tic-Tac-Toe, to appear in *J. Combin. Theory Ser. A*.
[9] Beck, J. and Csirmaz, L. (1982) Variations on a game. *J. Combin. Theory Ser. A* **33** 297–315.
[10] Berge, C. (1976) Sur les jeux positionelles. *Cahiers Centre Études Rech. Opér.* **18**.
[11] Berlekamp, E. R., Conway, J. H. and Guy, R. K. (1982) *Winning Ways*, Vol. 1–2, Academic Press, London.
[12] Erdős, P. and Lovász, L. (1975) Problems and results on 3-chromatic hypergraphs and some related questions. In *Infinite and Finite Sets* (A. Hajnal et al., eds), Vol. 11 of *Colloq. Math. Soc. J. Bolyai*, North-Holland, Amsterdam, pp. 609–627.

[13] Erdős, P. and Selfridge, J. (1973) On a combinatorial game. *J. Combin. Theory Ser. A* **14** 298–301.

[14] Golomb, S. W. and Hales, A. W. (2002) Hypercube Tic-Tac-Toe. In *More Games of No Chance*, MSRI Publications, Vol. 42, pp. 167–182.

[15] Graham, R. L., Rothschild, B. L. and Spencer, J. H. (1980) *Ramsey Theory*, Wiley-Interscience Ser. in Discrete Math., New York.

[16] Hales, A. W. and Jewett, R. I. (1963) Regularity and positional games. *Trans. Amer. Math. Soc.* **106** 222–229.

[17] Patashnik, O. (1980) Qubic: $4 \times 4 \times 4$ Tic-Tac-Toe. *Mathematics Magazine* **53** Sept. 1980, 202–216.

[18] Shelah, S. (1988) Primitive recursive bounds for van der Waerden numbers. *J. Amer. Math. Soc.* **1** (3) 683–697.

Degree Distribution of Competition-Induced Preferential Attachment Graphs

N. BERGER[1], C. BORGS[1], J. T. CHAYES[1],
R. M. D'SOUZA[1] and R. D. KLEINBERG[2][†]

[1]Microsoft Research, One Microsoft Way, Redmond WA 98052, USA
(e-mail: berger@its.caltech.edu, borgs@microsoft.com, jchayes@microsoft.com, raissa@alum.mit.edu)

[2]MIT CSAIL, 77 Massachusetts Ave, Cambridge MA 02139, USA
(e-mail: rdk@math.mit.edu)

Received 25 May 2004; revised 8 February 2005

For Béla Bollobás on his 60th birthday

We introduce a family of one-dimensional geometric growth models, constructed iteratively by locally optimizing the trade-offs between two competing metrics, and show that this family is equivalent to a family of preferential attachment random graph models with upper cut-offs. This is the first explanation of how preferential attachment can arise from a more basic underlying mechanism of local competition. We rigorously determine the degree distribution for the family of random graph models, showing that it obeys a power law up to a finite threshold and decays exponentially above this threshold.

We also rigorously analyse a generalized version of our graph process, with two natural parameters, one corresponding to the cut-off and the other a 'fertility' parameter. We prove that the general model has a power-law degree distribution up to a cut-off, and establish monotonicity of the power as a function of the two parameters. Limiting cases of the general model include the standard preferential attachment model without cut-off and the uniform attachment model.

1. Introduction

1.1. Network growth models

This paper is dedicated, with great affection and admiration, to Béla Bollobás on the occasion of his 60th birthday. Two of us (C. B. and J. T. C.) are privileged to count Béla among our dearest friends. And all of us have been inspired by his pioneering work on graph processes in general, and scale-free graphs in particular. We use the opportunity of this birthday volume to provide complete proofs of results on a new graph model, first announced in [6].

[†] Supported by a Fannie and John Hertz Foundation Fellowship.

There is currently tremendous interest in understanding the mathematical structure of networks – especially as we discover the pervasiveness of network structures in natural and engineered systems. Much recent theoretical work has been motivated by measurements of real-world networks, indicating that they have certain 'scale-free' properties, such as a power-law distribution of degrees. For the Internet graph, in particular, both the graph of routers and the graph of autonomous systems (AS) seem to obey power laws [15, 16]. However, these observed power laws hold only for a limited range of degrees, presumably due to physical constraints and the finite size of the Internet.

Many random network growth models have been proposed which give rise to power-law degree distributions. Most of these models rely on a small number of basic mechanisms, mainly preferential attachment[1] [20, 4] or copying [18], extending ideas known for many years [13, 21, 23, 22] to a network context. Variants of the basic preferential attachment mechanism have also been proposed, and some of these lead to changes in the values of the exponents in the resulting power laws. For extensive reviews of work in this area, see Albert and Barabási [2], Dorogovtsev and Mendes [12], and Newman [19]; for a survey of the relatively limited amount of mathematical work see [7]. Most of this work concerns network models without reference to an underlying geometric space. Nor do most of these models allow for heterogeneity of nodes, or address physical constraints on the capacity of the nodes. Thus, while such models may be quite appropriate for geometry-free networks, such as the web graph, they do not seem to be ideally suited to the description of other observed networks, *e.g.*, the Internet graph.

In this paper, instead of assuming preferential attachment, we show that it can arise from a more basic underlying process, namely competition between opposing forces. The idea that power laws can arise from competing effects, modelled as the solution of optimization problems with complex objectives, was proposed originally by Carlson and Doyle [10]. Their 'highly optimized tolerance' (HOT) framework has reliable design as a primary objective. Fabrikant, Koutsoupias and Papadimitriou (FKP) [14] introduce an elegant network growth model with such a mechanism, which they called 'heuristically optimized trade-offs'. As in many growth models, the FKP network is grown one node at a time, with each new node choosing a previous node to which it connects. However, in contrast to the standard preferential attachment types of models, a key feature of the FKP model is the underlying geometry. The nodes are points chosen uniformly at random from some region, for example a unit square in the plane. The trade-off is between the geometric consideration that it is desirable to connect to a nearby point, and a networking consideration, that it is desirable to connect to a node that is 'central' in the network as a graph. Centrality is measured by using, for example, the graph distance to the initial node. The model has a tunable, but fixed, parameter, which determines the relative weights given to the geometric distance and the graph distance.

The suggestion that competition between two metrics could be an alternative to preferential attachment for generating power-law degree distributions represents an important paradigm shift. Though FKP introduced this paradigm for network growth,

---

[1] As Aldous [3] points out, proportional attachment may be a more appropriate name, stressing the linear dependence of the attractiveness on the degree.

and FKP networks have many interesting properties, the resulting distribution is not a power law in the standard sense [5]. Instead the overwhelming majority of the nodes are leaves (degree one), and a second substantial fraction heavily connected 'stars' (hubs), producing a node degree distribution which has clear bimodal features.[2]

Here, instead of directly producing power laws as a consequence of competition between metrics, we show that such competition can give rise to a preferential attachment mechanism, which in turn gives rise to power laws. Moreover, the power laws we generate have an upper cut-off, which is more realistic in the context of many applications.

1.2. Overview of competition-induced preferential attachment

We begin by formulating a general competition model for network growth. Let $x_0, x_1, \ldots, x_t$ be a sequence of random variables with values in some space Λ. We think of the points $x_0, x_1, \ldots, x_t$ arriving one at a time according to some stochastic process. For example, we typically take Λ to be a compact subset of $\mathbb{R}^d$, x_0 to be a given point, say the origin, and $x_1, \ldots, x_t$ to be i.i.d. uniform on Λ. The network at time t will be represented by a graph, $G(t)$, on $t+1$ vertices, labelled $0, 1, \ldots, t$, and at each time step, the new node attaches to one or several nodes in the existing network. For simplicity, here we assume that each new node connects to a single node, resulting in $G(t)$ being a tree.

Given $G(t-1)$, the new node, labelled t, attaches to that node j in the existing network that minimizes a certain cost function representing the trade-off of two competing effects, namely connection or startup cost, and routing or performance cost. The connection cost is represented by a metric, $g_{ij}(t)$, on $\{0, \ldots, t\}$ which depends on $x_0, \ldots, x_t$, but not on the current graph $G(t-1)$, while the routing cost is represented by a function, $h_j(t-1)$, on the nodes which depends on the current graph, but not on the physical locations $x_0, \ldots, x_t$ of the nodes $0, \ldots, t$. This leads to the cost function

$$c_t = \min_j [\alpha g_{tj}(t) + h_j(t-1)], \tag{1.1}$$

where α is a constant which determines the relative weighting between connection and routing costs. We think of the function $h_j(t-1)$ as measuring the centrality of the node j; for simplicity, we take it to be the hop distance along the graph $G(t-1)$ from j to the root 0.

To simplify the analysis of the random graph process, we will assume that nodes always choose to connect to a point which is closer to the root, i.e., they minimize the cost function

$$\tilde{c}_t = \min_{j: \|x_j\| < \|x_t\|} [\alpha g_{tj}(t) + h_j(t-1)], \tag{1.2}$$

where $\|\cdot\|$ is an appropriate norm.

In the original FKP model, Λ is a compact subset of $\mathbb{R}^2$, say the unit square, and the points x_i are independently uniformly distributed on Λ. The cost function is of the form

---

[2] In simulations of the FKP model, this can be clearly discerned by examining the probability distribution function (pdf); for the system sizes amenable to simulations, it is less prominent in the cumulative distribution function (cdf).

(1.1), with $g_{ij} = d_{ij}$, the Euclidean metric (modelling the cost of building the physical transmission line), and $h_j(t)$ is the hop distance along the existing network $G(t)$ from j to the root. A rigorous analysis of the degree distribution of this two-dimensional model was given in [5], and the analogous one-dimensional problem was treated in [17].

Our model is defined as follows.

Definition 1. (Border toll optimization process) Let $x_0 = 0$, and let $x_1, x_2, \ldots$ be i.i.d., uniformly at random in the unit interval $\Lambda = [0, 1]$, and let $G(t)$ be the following process. At $t = 0$, $G(t)$ consists of a single vertex 0, the root. Let $h_j(t)$ be the hop distance to 0 along $G(t)$, and let $g_{ij}(t) = n_{ij}(t)$ be the number of existing nodes between x_i and x_j at time t, which we refer to as the *jump cost* of i connecting to j. Given $G(t-1)$ at time $t-1$, a new vertex, labelled t, attaches to the node j which minimizes the cost function (1.2). Furthermore, if there are several nodes j that minimize this cost function and satisfy the constraint, we choose the one whose position x_j is nearest to x_t. The process so defined is called the *border toll optimization process* (BTOP).

As in the FKP model, the routing cost is just the hop distance to the root along the existing network. However, in our model the connection cost metric measures the number of 'borders' between two nodes: hence the name BTOP. Note the correspondence to the Internet, where the principal connection cost is related to the number of AS domains crossed – representing, *e.g.*, the overhead associated with BGP, monetary costs of peering agreements, *etc*. In order to facilitate a rigorous analysis of our model, we took the simpler cost function (1.2), so that the new node always attaches to a node to its left.

It is interesting to note that the ratio of the BTOP connection cost metric to that of the one-dimensional FKP model is just the local density of nodes: $n_{ij}/d_{ij} = \rho_{ij}$. Thus the transformation between the two models is equivalent to replacing the constant parameter α in the FKP model with a variable parameter $\alpha_{ij} = \alpha \rho_{ij}$ which changes as the network evolves in time. That α_{ij} is proportional to the local density of nodes in the network reflects a model with an increase in cost for local resources that are scarce or in high demand. Alternatively, it can be thought of as reflecting the economic advantages of being first to market.

Somewhat surprisingly, the BTOP is equivalent to a special case of the following process, which closely parallels the preferential attachment model and makes no reference to any underlying geometry.

Definition 2. (Generalized preferential attachment with fertility and aging) Let A_1, A_2 be two positive integer-valued parameters. Let $G(t)$ be the following Markov process, whose states are finite rooted trees in which each node is labelled either *fertile* or *infertile*. At time $t = 0$, $G(t)$ consists of a single fertile vertex. Given the graph at time t, the new graph is formed in two steps: first, a new vertex, labelled $t + 1$ and initialized as infertile, connects to an old vertex j with probability zero if j is infertile, and with probability

$$Pr(t+1 \to j) = \frac{\min\{d_j(t), A_2\}}{W(t)} \tag{1.3}$$

if j is fertile. Here, $d_j(t)$ is equal to 1 plus the out-degree of j, and $W(t) = \sum_j' \min\{d_j(t), A_2\}$ with the sum running over fertile vertices only. We refer to vertex $t+1$ as a child of j. If after the first step, j has more than $A_1 - 1$ infertile children, one of them, chosen uniformly at random, becomes fertile. The process so defined is called a *generalized preferential attachment process with fertility threshold A_1 and aging threshold A_2*.

The special case $A_1 = A_2$ is called the *competition-induced preferential attachment process with parameter A_1*.

The last definition is motivated by the following theorem, to be proved in Section 2. To state it, we define a graph process as a random sequence of graphs $G(0), G(1), G(2), \ldots$ on the vertex sets $\{0\}, \{0, 1\}, \{0, 1, 2\}, \ldots$, respectively.

Theorem 1.1. *As a graph process, the border toll optimization process has the same distribution as the competition-induced preferential attachment process with parameter $A = \lceil \alpha^{-1} \rceil$.*

Certain other limiting cases of the generalized preferential attachment process are worth noting. If $A_1 = 1$ and $A_2 = \infty$, we recover the standard model of preferential attachment as considered in [20, 4]. If $A_1 = 1$ and A_2 is finite, the model is equivalent to the standard model of preferential attachment with a cut-off. On the other hand, if $A_1 = A_2 = 1$, we get a uniform attachment model.

The degree distribution of our random trees is characterized by the following theorem, which asserts that almost surely (a.s.) the fraction of vertices having degree k converges to a specified limit q_k, and moreover that this limit obeys a power law for $k < A_2$, and decays exponentially above A_2.

Theorem 1.2. *Let A_1, A_2 be positive integers. Consider the generalized preferential attachment process with fertility parameter A_1 and aging parameter A_2. Let $N_0(t)$ be the number of infertile vertices at time t, and let $N_k(t)$ be the number of fertile vertices with $k - 1$ children at time t, $k \geq 1$. Then we have the following.*

(1) *There are numbers $q_k \in [0, 1]$ such that, for all $k \geq 0$,*

$$\frac{N_k(t)}{t+1} \to q_k \quad \text{a.s., as } t \to \infty. \tag{1.4}$$

(2) *There exists a number $w = w(A_1, A_2) \in [0, 2]$ such that the q_k are determined by the following equations:*

$$q_i = \left(\prod_{k=2}^{i} \frac{k-1}{k+w}\right) q_1 \quad \text{if } 1 \leq i \leq A_2, \tag{1.5}$$

$$q_i = \left(\frac{A_2}{A_2 + w}\right)^{i - A_2} q_{A_2} \quad \text{if } i > A_2, \tag{1.6}$$

$$1 = \sum_{i=0}^{\infty} q_i, \quad \text{and} \quad q_0 = \sum_{i=1}^{\infty} q_i \min\{i - 1, A_1 - 1\}.$$

(3) There are positive constants c_1 and C_1, independent of A_1 and A_2, such that

$$c_1 k^{-(w+1)} < q_k/q_1 < C_1 k^{-(w+1)} \qquad (1.7)$$

for $1 \leqslant k \leqslant A_2$.
(4) If $A_1 = A_2$, the parameter w is equal to 1, and for general A_1 and A_2, w decreases with increasing A_1, and increases with increasing A_2.

Equation (1.7) clearly defines a power-law degree distribution with exponent $\gamma = w + 1$ for $k \leqslant A_2$. Note that for measurements of the Internet the value of the exponent for the power law is $\gamma \approx 2$. In our border toll optimization model, where $A_1 = A_2$, we recover $\gamma = 2$.

The convergence claim of Theorem 1.2 is proved using a novel method which we believe is one of the main technical contributions of this work. For preferential attachment models which have been analysed in the past [1, 8, 9, 11], the convergence was established using the Azuma–Hoeffding martingale inequality. To establish the bounded-differences hypothesis required by that inequality, those proofs employed a clever coupling of the random decisions made by the various edges, such that the decisions made by an edge e only influence the decisions of subsequent edges which choose to imitate e's choices. A consequence of this coupling is that if e made a different decision, it would alter the degrees of only finitely many vertices. This in turn allows the required bounded-differences hypothesis to be established. No such approach is available for our models, because the coupling fails. The random decisions made by an edge e may influence the time at which some node v crosses the fertility or aging threshold, which thereby exerts a subtle influence on the decisions of *every* future edge, not only those which choose to imitate e.

Instead we introduce a new approach based on the second-moment method. The argument establishing the requisite second-moment upper bound is quite subtle; it depends on a computation involving the eigenvalues of a matrix describing the evolution of the degree sequence in a continuous-time version of the model.

2. Equivalence of the two models

2.1. Basic properties of the border toll optimization process

In this section we will turn to the BTOP defined in the introduction, establishing some basic properties which will enable us to prove that it is equivalent to the competition-induced preferential attachment model. In order to avoid complications we exclude the case that some of the x_is are identical, an event that has probability zero. We say that $j \in \{0, 1 \ldots, t\}$ lies to the right of $i \in \{0, 1 \ldots, t\}$ if $x_i < x_j$, and we say that j lies directly to the right of i if $x_i < x_j$ but there is no $k \in \{1, \ldots, t\}$ such that $x_i < x_k < x_j$. In a similar way, we say that j is the first vertex with a certain property to the right of i if j has that property and there exists no $k \in \{1, \ldots, t\}$ such that $x_i < x_k < x_j$ and k has the property in question. Similar notions apply with 'left' in place of 'right'.

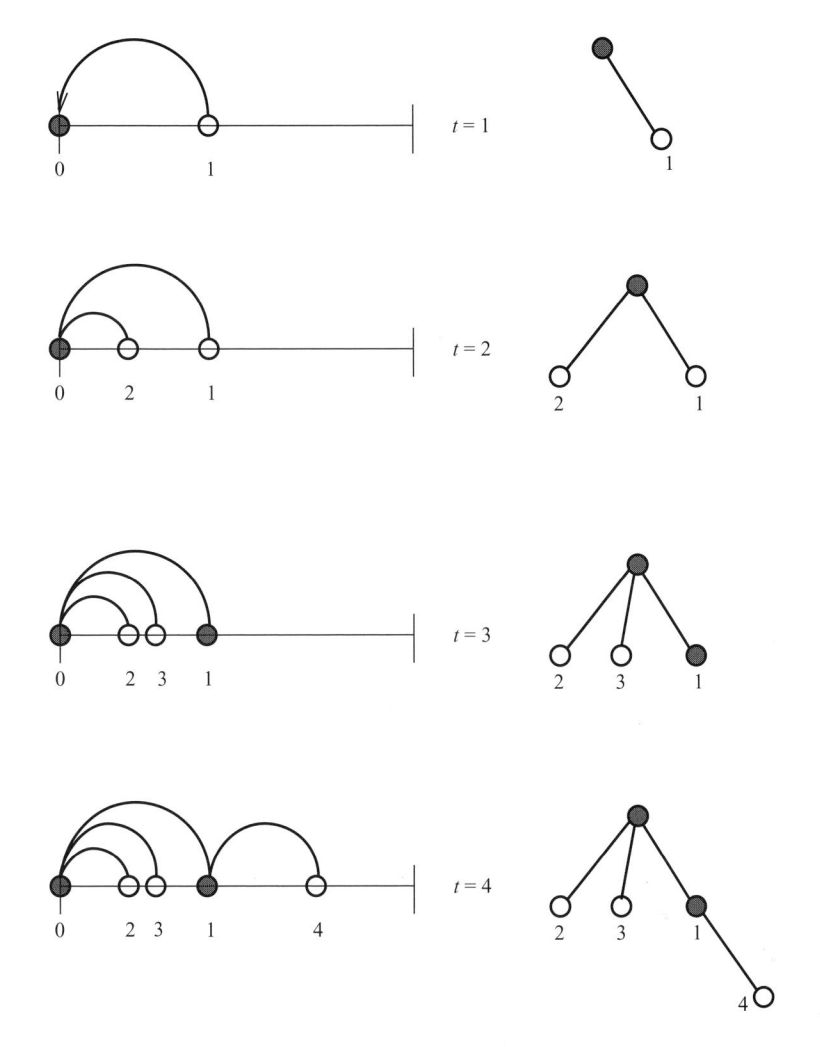

Figure 1. A sample instance of BTOP for $\alpha = 1/3, A = 3$, showing the process on the unit interval (on the left), and the resulting tree (on the right). Fertile vertices are shaded, infertile ones are not. Note that vertex 1 became fertile at $t = 3$.

Definition 3. A vertex i is called *fertile at time t* if a hypothetical new point arriving at time $t + 1$ and landing directly to the right of x_i would attach itself to the node i. Otherwise i is called *infertile at time t*.

This definition is illustrated in Figure 1.

Lemma 2.1. *Let $0 < \alpha < \infty$, let $A = \lceil \alpha^{-1} \rceil$, and let $0 < t < \infty$. Then we have the following.*

(i) *The node 0 is fertile at time t.*
(ii) *Let i be fertile at time t. If i is the rightmost fertile vertex at time t (case 1), let ℓ be the number of infertile vertices to the right of i. Otherwise (case 2), let j be the next*

fertile vertex to the right of i, and let $\ell = n_{ij}(t)$. Then $0 \leq \ell \leq A - 1$, and the ℓ infertile vertices located directly to the right of i are children of i. In case 2, if $h_j > h_i$, then j is a fertile child of i and $\ell = A - 1$. As a consequence, the hop count between two consecutive fertile vertices never increases by more than 1 as we move to the right, and if it increases by 1, there are $A - 1$ infertile vertices between the two fertile ones.

(iii) Assume that the new vertex at time $t + 1$ lands between two consecutive fertile vertices i and j, and let $\ell = n_{ij}(t)$. Then $t + 1$ becomes a child of i. If $\ell + 1 < A$, the new vertex is infertile at time $t + 1$, and the fertility of all old vertices is unchanged. If $\ell + 1 = A$ and the new vertex lies directly to the left of j, the new vertex is fertile at time $t + 1$ and the fertility of the old vertices is unchanged. If $\ell + 1 = A$ and the new vertex does not lie directly to the left of j, the new vertex is infertile at time $t + 1$, the vertex directly to the left of j becomes fertile, and the fertility of all other vertices is unchanged.

(iv) If $t + 1$ lands to the right of the rightmost fertile vertex at time t, the statements in (iii) hold with j replaced by the right endpoint of the interval $[0, 1]$, and $n_{ij}(t)$ replaced by the number of vertices to the right of i.

(v) If i is fertile at time t, it is still fertile at time $t + 1$.

(vi) If i has k children at time t, the $\ell = \min\{A - 1, k\}$ leftmost of them are infertile at time t, and any others are fertile.

Proof. The proof is straightforward but lengthy. We include the details of the argument here for completeness.

Statement (i) is trivial, statement (v) follows immediately from (iii) and (iv), and (vi) follows immediately from (ii). So we are left with (ii)–(iv). We proceed by induction on t. If (ii) holds at time t, and (iii) and (iv) hold for a new vertex arriving at time $t + 1$, (ii) clearly also holds at time $t + 1$. We therefore only have to prove that (ii) at time t implies (iii) and (iv) for a new vertex arriving at time $t + 1$.

Assume thus that (ii) holds at time t. At time $t + 1$, a new vertex arrives, and falls directly to the right of some vertex k. Let i be the nearest vertex to the left of k that was fertile at time t (if k is fertile at time t, we set $i = k$) and let j be the nearest vertex to the right of i that was fertile at time t (we assume for the moment that i is not the rightmost fertile vertex at time t), let ℓ be the number of vertices between i and j at time t.

Let us first prove that the vertex $t + 1$ connects to i. If $i = k$, this is obvious, since i is fertile at time t. We may therefore assume that $k \neq i$. For the new vertex $t + 1$, the cost of connecting to the vertex i is then equal to $\alpha(n_{ik}(t) + 1)$. Let us first compare this cost to the cost of connecting to a fertile vertex i' to the left of i. Let $i_0 = i'$, let $i_s = i$, and let $i_1, \ldots, i_{s-1}$ be the fertile vertices between i' and i, ordered from left to right. If $h_{i_{m-1}} < h_{i_m}$, we use the inductive assumption (ii) to conclude that the number of infertile vertices between i_{m-1} and i_m is equal to $A - 1$, and $h_{i_{m-1}} = h_{i_m} - 1$. A decrease of q in the hop cost is therefore accompanied by an increase in the jump cost of at least $\alpha A q \geq q$. As a consequence, it never pays to connect to a fertile vertex i' to the left of i. The cost of connecting to an infertile vertex to the left of i is even higher, since the hop count of an infertile vertex is at best equal to the hop count of the next fertile vertex to the right. We therefore only have to consider the connection cost to some of the infertile children of i. But again, the hop count is worse by 1 when compared to the hop count of i, and the jump cost is at best reduced by $(A-1)\alpha < 1$, proving that the cost of connecting to i is minimal.

To discuss the fertility of the vertices in the graph $G(t+1)$, we need to consider the arrival of a second vertex, labelled $t+2$. If $t+2$ falls to the left of $t+1$, it will face an optimization problem that has not been changed by the arrival of the vertex $t+1$, implying that the fertility of the vertices to the left of $t+1$ is unchanged. If $t+2$ falls to the right of j, the cost of connecting to j or one of the vertices to the right of j is the same as before, and the cost of connecting to a vertex to the left of j is at best equal (the cost of connecting to any vertex to the left of $t+1$ is in fact higher, because of the additional cost of jumping over the vertex $t+1$). Therefore, the vertex $t+2$ will still prefer to connect to either j or one of the vertices to the right of j, implying that the fertility of the vertices to the right of j has not changed at all. We are therefore left with analysing the case where $t+2$ falls between $t+1$ and j. Again, the vertex $t+2$ will prefer i over any vertex to the left of i (the cost analysis is the same as the one used for $t+1$ above), so we just have to compare the costs of connecting to the different vertices between i and j. If $\ell+1 < A$, this will again imply that $t+2$ connects to i; but if $\ell+1 = A$, the vertex $t+2$ will only connect to i if it does not fall to the right of the rightmost of the now $\ell+1$ vertices between i and j. If it falls to the right of this vertex, it will be as expensive to connect to the rightmost of the now $\ell+1$ vertices between i and j as it is to connect to i. Recalling our convention of connecting to the nearest vertex to the left if there is a tie in costs, this proves that now $t+2$ connects to the rightmost vertex between i and j, implying that this vertex is fertile.

The above considerations prove the fertility statements in (iii), and thus complete the proof of (iii). The case where i is the rightmost fertile vertex at time t is similar (in fact, it is slightly easier since it involves fewer cases), and leads to the proof of (iv). This completes the proof of Lemma 2.1. □

2.2. Proof of Theorem 1.1

In the BTOP, note that our cost function

$$\min_{j}[\alpha n_{tj}(t) + h_j(t-1)], \tag{2.1}$$

and hence the graph $G(t)$, only depends on the order of the vertices $x_0, \ldots, x_t$, and not on their actual positions in the interval $[0, 1]$. Let $\vec{\pi}(t)$ be the permutation of $\{0, 1, \ldots, t\}$ which orders the vertices $x_0, \ldots, x_t$ from left to right, so that

$$x_0 = x_{\pi_0(t)} < x_{\pi_1(t)} < \cdots < x_{\pi_t(t)}. \tag{2.2}$$

(Recall that the vertices $x_0, x_1, \ldots, x_t$ are pairwise distinct with probability one.) Note that $\vec{\pi}(t)$ and $\vec{\pi}(t+1)$ are related as follows: there exists $i_0 \in \{1, 2, \ldots, t+1\}$ such that

$$\pi_i(t+1) = \begin{cases} \pi_i(t) & \text{if } i < i_0, \\ t+1 & \text{if } i = i_0, \\ \pi_{i-1}(t) & \text{if } i > i_0. \end{cases} \tag{2.3}$$

Informally, the permutation $\vec{\pi}(t+1)$ is obtained by inserting the new element $t+1$ into the permutation $\vec{\pi}(t)$ in a random position i_0, where $x_{\pi_{i_0}(t)}$ is the left endpoint of the subinterval of $(0,1)$ into which x_{t+1} falls. The distribution of the random variable i_0 may be deduced as follows. Since $x_0 = 0$ and $x_1, x_2, \ldots, x_t$ are i.i.d., we know that, for all t, the permutation $\vec{\pi}(t)$ is uniformly distributed among permutations of $\{0, 1, \ldots, t\}$ which fix the element 0. This means that, conditioned on a given such permutation $\vec{\pi}(t)$, the permutation $\vec{\pi}(t+1)$ is uniformly distributed among all permutations related to $\vec{\pi}(t)$ by the transformation (2.3). In other words, i_0 is uniformly distributed in the set $\{1, 2, \ldots, t+1\}$.

With the help of Lemma 2.1, we now easily derive a description of the graph $G(t)$ which does not involve any optimization problem. To this end, let us consider a vertex i with ℓ infertile children at time t. If a new vertex falls into the interval directly to the right of i, or into one of the intervals directly to the right of an infertile child of i, it will connect to the vertex i. Since there is a total of $t+1$ intervals at time t, the probability that a vertex i with ℓ infertile children grows an offspring is $(\ell+1)/(t+1)$. By Lemma 2.1(vi), this number is equal to $\min\{A, k_i\}/(t+1)$, where $k_i - 1$ is the number of children of i. Note that fertile children do not contribute to this probability, since vertices falling into an interval directly to the right of a fertile child will connect to the child, not the parent.

Assume now that i did get a new offspring, and that it had $A-1$ infertile children at time t. Then the new vertex is either born fertile, or makes one of its infertile siblings fertile. Using the principle of deferred decisions, we may assume that with probability $1/A$ the new vertex becomes fertile, and with probability $(A-1)/A$ an old one, chosen uniformly at random among the $A-1$ candidates, becomes fertile.

We thus have shown that the solution $G(t)$ of the optimization problem (2.1) can alternatively be described by the competition-induced preferential attachment model with parameter A.

3. Convergence of the degree distribution

3.1. Overview

To characterize the behaviour of the degree distribution, we will derive a recursion which governs the evolution of the vector $\vec{N}(t)$, whose components are the number of vertices of each degree, at the time when there are t nodes in the network. The conditional expectation of $\vec{N}(t+1)$ is given by an evolution equation of the form

$$\mathbb{E}\big(\vec{N}(t+1) - \vec{N}(t) \mid \vec{N}(t)\big) = M(t)\vec{N}(t),$$

where $M(t)$ depends on t through the random variable $W(t)$ introduced in Definition 2. Owing to the randomness of the coefficient matrix $M(t)$, the analysis of this evolution equation is not straightforward. We avoid this problem by introducing a continuous-time process, with time parameter τ, which is equivalent to the original discrete-time process up to a (random) reparametrization of the time coordinate. The evolution equation for the conditional expectations in the continuous-time process involves a coefficient matrix M that is not random and does not depend on τ. We will first prove that the *expected* degree distribution in the continuous-time model converges to a scalar multiple of the eigenvector $\hat{p}$ of M associated with the largest eigenvalue w. This is followed by the much more difficult proof that the *empirical* degree distribution converges a.s. to the same limit. Finally, we translate this continuous-time result into a rigorous convergence result for the original discrete-time system.

3.2. Notation

Let A be any integer greater than or equal to $\max(A_1, A_2)$. Let $N_0(t)$ be the number of infertile vertices at (discrete) time t, and, for $k \geq 1$, let $N_k(t)$ be the number of fertile vertices with $k - 1$ children at time t. Let $\bar{N}_A(t) = N_{\geq A}(t) = \sum_{k \geq A} N_k(t)$, and $\tilde{N}_k(t) = N_k(t)$ if $k < A$. The combined attractiveness of all vertices is denoted by $W(t) = \sum_{k=1}^{A} \min\{k, A_2\} \tilde{N}_k(t)$. Let $n_k(t) = \frac{1}{t+1} N_k(t)$ and $\tilde{n}_k(t) = \frac{1}{t+1} \tilde{N}_k(t)$. Finally, the vectors $(\tilde{N}_k(t))_{k=1}^{A}$ and $(\tilde{n}_k(t))_{k=1}^{A}$ are denoted by $\tilde{N}(t)$ and $\tilde{n}(t)$ respectively. Note that the index k runs from 1 to A, not 0 to A.

3.3. Evolution of the expected value

From the definition of the generalized preferential attachment model, it is easy to derive the probabilities for the various alternatives which may happen upon the arrival of the $(t + 1)$st node.

- With probability $A_2 \tilde{N}_A(t)/W(t)$, it attaches to a node of degree $\geq A$. This increments $\tilde{N}_1$, and leaves $\tilde{N}_A$ and all $\tilde{N}_j$ with $1 < j < A$ unchanged.
- With probability $\min(A_2, k) \tilde{N}_k(t)/W(t)$, it attaches to a node of degree k, where $1 \leq k < A$. This increments $\tilde{N}_{k+1}$, decrements $\tilde{N}_k$, increments $\tilde{N}_0$ or $\tilde{N}_1$ depending on whether $k < A_1$ or $k \geq A_1$, and leaves all other $\tilde{N}_j$ with $j < A$ unchanged.

It follows that the discrete-time process $(\tilde{N}_k(t))_{k=0}^{A}$ at time t is equivalent to the state of the following continuous-time stochastic process $(\hat{N}_k(\tau))_{k=0}^{A}$ at the random stopping time $\tau = \tau_t$ of the tth event.

- With rate $A_2 \hat{N}_A(\tau)$, $\hat{N}_1$ increases by 1.
- For every $0 < k < A$, with rate $\hat{N}_k(\tau) \min(k, A_2)$, the following happens:
$$\hat{N}_k \to \hat{N}_k - 1, \quad \hat{N}_{k+1} \to \hat{N}_{k+1} + 1, \quad \hat{N}_{g(k)} \to \hat{N}_{g(k)} + 1$$
where $g(k) = 0$ for $k < A_1$ and $g(k) = 1$ otherwise.

Note that the above rules need to be modified if $A_1 = 1$. Here the birth of a child of a degree-one vertex does not change the net number of fertile degree-one vertices, N_1.

Let M be the following $A \times A$ matrix:

$$M_{i,j} = \begin{cases} -1 & \text{if } i = j = 1 < A_1, \\ -\min(j, A_2) & \text{if } 2 \leq i = j \leq A - 1, \\ \min(j, A_2) & \text{if } 2 \leq i = j + 1 \leq A, \\ \min(j, A_2) & \text{if } i = 1 \text{ and } j \geq \max(A_1, 2), \\ 0 & \text{otherwise.} \end{cases} \quad (3.1)$$

Then, for every $\tau > \sigma$, the conditional expectation of the vector $\hat{N}(\tau) = (\hat{N}_k(\tau))_{k=1}^{A}$ is given by

$$\mathbb{E}(\hat{N}(\tau) \mid \hat{N}(\sigma)) = e^{(\tau - \sigma)M} \hat{N}(\sigma). \quad (3.2)$$

It is easy to see that the matrix e^M has all positive entries, and therefore (by the Perron–Frobenius theorem) M has a unique eigenvector $\hat{p}$ of ℓ_1-norm 1 having all positive entries. Let w be the eigenvalue corresponding to $\hat{p}$. Then w is real, it has multiplicity 1, and it exceeds the real part of every other eigenvalue. Therefore, for every nonzero vector y with nonnegative entries,

$$\lim_{\tau \to \infty} e^{-\tau w} e^{\tau M} y = \langle \hat{a}, y \rangle \hat{p},$$

where $\hat{a}$ is the eigenvector of M^T corresponding to w, normalized so that $\langle \hat{a}, \hat{p} \rangle = 1$. Note that $\langle \hat{a}, y \rangle > 0$ because y is nonzero and nonnegative, and $\hat{a}$ is positive, again by Perron–Frobenius. Therefore, the vector $\mathbb{E}(e^{-\tau w} \widehat{N}(\tau))$ converges to a positive scalar multiple of $\hat{p}$, say $\lambda \hat{p}$, as $\tau \to \infty$.

In order to prove concentration for the continuous-time model, we will prove that the difference $\widehat{N}_k(\tau)/q_k - \widehat{N}_j(\tau)/q_j$ has an exponential growth rate which is at most the real part of the second eigenvalue of M, which is strictly less than w, the growth rate of the individual terms $\widehat{N}_k(\tau)/q_k$ and $\widehat{N}_j(\tau)/q_j$. From this, we will conclude that the ratio $\widehat{N}_k(\tau)/\widehat{N}_j(\tau)$ converges almost surely to q_k/q_j, for all k and j, which in turn implies the convergence of the normalized degree sequence to the vector $(q_i)_{i=0}^\infty$.

In order to prove bounds on the growth rate of the differences $\widehat{N}_k(\tau)/q_k - \widehat{N}_j(\tau)/q_j$, we will need some auxiliary bounds involving the well-known standard birth process, to be defined below.

3.4. Standard birth process

We start with the definition of the standard birth process with rate ρ. The standard birth process was first introduced by Yule in 1924 [22], and is a special case of the well-known Yule Process, defined in that paper.

Definition 4. Let $\rho > 0$ and let $\{o_n\}_{n=1}^\infty$ be independent exponential random variables so that $\mathbb{E}(o_n) = \frac{1}{\rho} n^{-1}$. For $\tau \in [0, \infty)$, let $X_\tau = \min\{n \geq 1 : \sum_{k=1}^n o_k > \tau\}$. Then X is called the *standard birth process with rate ρ*.[3]

The standard birth process is connected to our discussion through the following easy claim.

Claim 3.1. *Let $\|\widehat{N}(\tau)\| = \sum_{k=1}^A \widehat{N}_k(\tau)$. Let $T \geq 0$, let $x \geq y$, and let X be a standard birth process with rate 2. Then $\{\{X_\tau\}_{\tau \geq T} \mid X_T = x\}$ stochastically dominates $\{\{\|\widehat{N}(\tau)\|\}_{\tau \geq T} \mid \|\widehat{N}(T)\| = y\}$.*

Proof. Let us start with the observation that $\sum_{k=1}^n o_k$ is the first time τ for which $X_\tau = n + 1$. Let $\{r_n\}_{n=0}^\infty$ be i.i.d. exponential random variables with mean 1. Then $\sum_{k=1}^n o_k$ has the same distribution as $\sum_{k=0}^{n-1} r_k/(2k+2)$. The time τ_n at which the node n is born has

---

[3] The name 'standard birth process' is due to the fact that X_τ is equivalent to the following process. Start with one cell at time 0. At each time, every cell divides into two cells with rate ρ. Then X_τ is the number of cells at time τ.

the same distribution as $\sum_{k=0}^{n-1} r_k/W(k)$, where $W(k)$ denotes the combined attractiveness of all nodes at the random time τ_k. The claim now follows from the observation that $W(k) \leqslant 2k+1 \leqslant 2k+2$. □

The main purpose of this section is the proof of the following claims.

Claim 3.2. *Let X be a standard birth process with rate ρ. Then X_τ is almost surely finite for every τ. Furthermore, there exists a constant $C_s = C_s(\rho)$ such that, for every $\tau_2 > \tau_1$, $x \geqslant 1$, and $k \geqslant 1$,*

$$\mathbf{P}\big(X_{\tau_2} > kxe^{\rho(\tau_2-\tau_1)} \mid X_{\tau_1} = x\big) < \frac{C_s}{x(k-1)^2}. \quad (3.3)$$

If, in addition, $\tau_2 - \tau_1 < 1$, then

$$\mathbf{P}\big(X_{\tau_2} - X_{\tau_1} > kx[e^{\rho(\tau_2-\tau_1)} - 1] \mid X_{\tau_1} = x\big) < \frac{C_s}{x(\tau_2-\tau_1)(k-1)^2}. \quad (3.4)$$

To see the finiteness of X_τ, we need to show that $\sum_{n=1}^\infty o_n = \infty$ a.s. This follows from the following simple argument. For every k, let

$$U_k = \sum_{j=2^k+1}^{2^{k+1}} o_j.$$

For $j \in [2^k + 1, 2^{k+1}]$, with probability greater than $\frac{1}{2}$, $o_j > \frac{1}{\rho}2^{-k-2}$. Therefore, $\mathbf{P}(U_k > \frac{1}{4\rho}) > \frac{1}{2}$. The random variables $\{U_k\}_{k=1}^\infty$ are independent, and therefore $\sum_{n=1}^\infty o_n \geqslant \sum_{k=1}^\infty U_k = \infty$ almost surely.

To see (3.3) and (3.4), we use the following lemma, which is proved in [22, Section II]. Since the proof is short and simple, we choose to include it for the sake of making the exposition more self-contained.

Lemma 3.3. (Yule, 1924) *For every $\tau > 0$ and every positive integer k, $\mathbb{E}(X_\tau^k) < \infty$. Furthermore,*

$$\mathbb{E}(X_\tau) = \exp(\rho\tau), \quad (3.5)$$

and

$$\mathrm{var}(X_\tau) = \exp(2\rho\tau) - \exp(\rho\tau). \quad (3.6)$$

In particular,

$$\mathrm{var}(X_\tau) = O(\exp(2\rho\tau)), \quad (3.7)$$

and for $\tau < 1$ there exists a constant $C_v = C_v(\rho)$ so that

$$\mathrm{var}(X_\tau) \leqslant C_v\tau. \quad (3.8)$$

Proof. An equivalent description of the standard birth process is the following. Let α be an exponential variable with expected value ρ, and let $\{G_t\}$ be a Poisson point process with rate $\alpha e^{\rho t}$. Then $X_\tau = 1 + G_\tau$ has the same distribution as the standard birth process. To see this, all we need is to show that for every τ and n, the rate of the process $\{G_t\}$ at time τ conditioned on $X_\tau = n$ is ρn. Indeed,

$$\text{rate}(\tau|X_\tau = n) = \frac{\int_0^\infty \alpha e^{\rho \tau} \mathbf{P}(X_\tau = n|\alpha) \frac{1}{\rho} e^{-\alpha/\rho} d\alpha}{\int_0^\infty \mathbf{P}(X_\tau = n|\alpha) \frac{1}{\rho} e^{-\alpha/\rho} d\alpha}$$

$$= \frac{\int_0^\infty \alpha e^{\rho \tau} e^{-\frac{\alpha}{\rho}(\exp(\rho \tau) - 1)} \left(\frac{\alpha}{\rho}(\exp(\rho \tau) - 1)\right)^{(n-1)} ((n-1)!)^{-1} \frac{1}{\rho} e^{-\alpha/\rho} d\alpha}{\int_0^\infty e^{-\frac{\alpha}{\rho}(\exp(\rho \tau) - 1)} \left(\frac{\alpha}{\rho}(\exp(\rho \tau) - 1)\right)^{(n-1)} ((n-1)!)^{-1} \frac{1}{\rho} e^{-\alpha/\rho} d\alpha}$$

$$= \rho n.$$

Here the second equality follows from the fact that $X_\tau - 1$ is a Poisson variable with rate $\frac{\alpha}{\rho}(e^{\rho \tau} - 1)$, and the last equality follows by integration by parts.

From this we get that the distribution of X_τ is geometric with expected value $\exp(\rho \tau)$. To see this, we again use the fact that $X_\tau - 1$ is a Poisson variable with rate $\frac{\alpha}{\rho}(e^{\rho \tau} - 1)$, where α is an exponential variable with expectation ρ. Therefore, for every n,

$$\mathbf{P}(X_\tau = n+1) = \int_0^\infty \mathbf{P}(X_\tau - 1 = n|\alpha) \frac{1}{\rho} e^{-\alpha/\rho} d\alpha$$

$$= (n!)^{-1} \int_0^\infty e^{-\frac{\alpha}{\rho}(e^{\rho \tau} - 1)} \left(\frac{\alpha}{\rho}(e^{\rho \tau} - 1)\right)^n \frac{1}{\rho} e^{-\alpha/\rho} d\alpha$$

$$= \left(1 - e^{-\rho \tau}\right) \mathbf{P}(X_\tau = n)$$

where, again, the last step follows from integration by parts.

The relations (3.5) and (3.6) follow immediately, and (3.7) and (3.8) follow from (3.6). □

Proof of (3.3) and (3.4) in Claim 3.2. Equations (3.3) and (3.4) will follow from Chebyshev's inequality if we show that

$$\mathbb{E}(X_{\tau_2}|X_{\tau_1}) = X_{\tau_1} e^{\rho(\tau_2 - \tau_1)} \tag{3.9}$$

and

$$\text{var}(X_{\tau_2}|X_{\tau_1}) = X_{\tau_1} O\left(e^{2\rho(\tau_2 - \tau_1)}\right) \tag{3.10}$$

for $\tau_2 > \tau_1$, and

$$\text{var}(X_{\tau + \tau_1}|X_{\tau_1}) = O(\tau) \cdot X_{\tau_1} \tag{3.11}$$

for $\tau < 1$.

Equations (3.9), (3.10) and (3.11) follow from (respectively) (3.5), (3.7) and (3.8) and the fact that, conditioned on X_{τ_1}, the process $X_{\tau + \tau_1}$ is the sum of X_{τ_1} independent copies of X_τ. □

Remark. From now on we will always assume that $\rho = 2$. In particular, whenever we use the term 'standard birth process', it should be understood as 'standard birth process with rate 2'.

3.5. Concentration of the continuous-time process

In order to show concentration of the degree distribution for the continuous-time process, we will first prove the following lemma. To state it, we observe for any b with $b^\mathsf{T}\hat{p} = 0$,

$$\|b^\mathsf{T} e^{(T-\tau)M}\|_\infty \leqslant \|b\|_\infty e^{(T-\tau)v'} \tag{3.12}$$

for some $v' < w$. Without loss of generality, we may assume that $v' > w/2$. Also, for a general vector b,

$$\|b^\mathsf{T} e^{(T-\tau)M}\|_\infty \leqslant \|b\|_\infty e^{(T-\tau)w}. \tag{3.13}$$

Lemma 3.4. *Let b be a vector in $\mathbb{R}^A$ with $\|b\|_\infty \leqslant 1$. Then there exists a constant $C < \infty$, such that, for all $T > 0$,*

$$\mathrm{var}(b^\mathsf{T} \hat{N}(T)) < C \exp(2uT) \tag{3.14}$$

where $u = w$ if $b^\mathsf{T}\hat{p} \neq 0$, and $u = v'$ if $b^\mathsf{T}\hat{p} = 0$.

Proof. We use a martingale to bound the variance. Fix T, and let

$$L_\tau = \mathbb{E}(b^\mathsf{T} \hat{N}(T) \mid \hat{N}(\tau)).$$

Clearly, L_τ is a (continuous-time) martingale. By (3.2), we know that $L_\tau = b^\mathsf{T} e^{(T-\tau)M} \hat{N}(\tau)$. Let $0 < \epsilon < \exp(-10T)$ be such that $K = T/\epsilon$ is an integer number. Then, $\{U_k = L_{k\epsilon}\}_{k=0}^{K}$ is a martingale and

$$\mathrm{var}(b^\mathsf{T} \hat{N}(T)) = \sum_{k=0}^{K-1} \mathrm{var}(U_{k+1} - U_k).$$

We want to estimate the variance of $U_{k+1} - U_k$. Let $v_k = \hat{N}((k+1)\epsilon) - \hat{N}(k\epsilon)$. For two vectors $\hat{N}_1$ and $\hat{N}_2$,

$$\left(b^\mathsf{T} e^{(T-(k+1)\epsilon)M} \hat{N}_1 - b^\mathsf{T} e^{(T-(k+1)\epsilon)M} \hat{N}_2\right)^2 \leqslant \|\hat{N}_1 - \hat{N}_2\|^2 e^{2u(T-(k+1)\epsilon)},$$

where the norm $\|\cdot\|$ refers to the L^1-norm here and throughout this section, unless otherwise noted. Choose $\hat{N}(k\epsilon)$ according to its distribution, and let $\hat{N}_1$ and $\hat{N}_2$ be chosen independently, according to the distribution of $\hat{N}((k+1)\epsilon)$ conditioned on $\hat{N}(k\epsilon)$. Then

$$\mathrm{var}(U_{k+1} - U_k) = \frac{1}{2}\mathbb{E}\left[\left(b^\mathsf{T} e^{(T-(k+1)\epsilon)M}\hat{N}_1 - b^\mathsf{T} e^{(T-(k+1)\epsilon)M}\hat{N}_2\right)^2\right]$$

$$\leqslant \frac{1}{2}\mathbb{E}(\|\hat{N}_1 - \hat{N}_2\|^2)e^{2u(T-(k+1)\epsilon)}.$$

On the other hand, using the fact that, for every vector x in $\mathbb{R}^d$,

$$\left(\sum_{i=1}^{d} x_i\right)^2 \leqslant d \sum_{i=1}^{d} x_i^2,$$

we get

$$\sum_{j=1}^{A} \mathrm{var}(v_k(j)) \geqslant \frac{1}{2A}\mathbb{E}(\|\hat{N}_1 - \hat{N}_2\|^2)$$

where $v_k(j)$ is the jth component of v_k. Therefore,

$$\text{var}(U_{k+1} - U_k) \leq A \sum_{j=1}^{A} \exp[2u(T - (k+1)\epsilon)] \text{var}(v_k(j)). \quad (3.15)$$

By Claim 3.1, (3.9), and (3.11), for every $j = 1, 2, \ldots, A$,

$$\text{var}(v_k(j) \mid \widehat{N}(k\epsilon)) \leq \mathbb{E}[(v_k(j))^2 \mid \widehat{N}(k\epsilon)] \leq C_v \epsilon \|\widehat{N}(k\epsilon)\|,$$
$$|\mathbb{E}(v_k(j) \mid \widehat{N}(k\epsilon))| \leq (e^{2\epsilon} - 1)\|\widehat{N}(k\epsilon)\| \leq 4\epsilon \|\widehat{N}(k\epsilon)\|$$

and

$$\mathbb{E}[(\|\widehat{N}(k\epsilon)\|)^2] < e^{4k\epsilon}.$$

Therefore,

$$\text{var}(v_k(j)) = \mathbb{E}(\text{var}(v_k(j) \mid \widehat{N}(k\epsilon))) + \text{var}(\mathbb{E}(v_k(j) \mid \widehat{N}(k\epsilon)))$$
$$\leq C_v \epsilon \exp(wk\epsilon) + 16\epsilon^2 \exp(4k\epsilon)$$
$$< C_0 \epsilon \exp(wk\epsilon)$$

for $C_0 = C_v + 1$, by the choice of ϵ. Therefore,

$$\text{var}(b^\mathsf{T} \widehat{N}_k(T)) < A^2 C_0 \epsilon \sum_{k=0}^{K-1} \exp(wk\epsilon + 2u(T - (k+1)\epsilon))$$
$$\leq A^2 C_0 e^{2uT} \int_0^T e^{(w-2u)\tau} d\tau < C_u \exp(2uT)$$

for

$$C_u = A^2 C_0 \int_0^\infty e^{(w-2u)\tau} d\tau < \infty. \qquad \square$$

In addition, note that by (3.12) and (3.13),

$$|\mathbb{E}(b^\mathsf{T} \widehat{N}(T))| \leq e^{uT}$$

and therefore there exists C so that

$$\mathbb{E}[(b^\mathsf{T} \widehat{N}(T))^2] \leq C e^{2uT}. \quad (3.16)$$

We are now ready to state and prove the two main lemmas used to prove concentration, as follows.

Lemma 3.5. *For every $w' < w$ and every $1 \leq k \leq A$, a.s. for every τ large enough,*

$$\widehat{N}_k(\tau) > e^{w'\tau}. \quad (3.17)$$

Lemma 3.6. *There exists $v < w$ s.t. for every $1 \leq k < j \leq A$ a.s. for every τ large enough,*

$$p_j \widehat{N}_k(\tau) - p_k \widehat{N}_j(\tau) < e^{v\tau},$$

where $p_i, i = 1, \ldots, A$ are the components of the vector $\widehat{p}$.

Degree Distribution of Competition-Induced Preferential Attachment Graphs   147

The following corollary is an immediate consequence of Claim 3.2, Claim 3.1 and Lemma 3.5.

Corollary 3.7. $w \leqslant 2$.

Proof of Lemma 3.6. Choose some v strictly between v' and w in a way that $w - v < 0.25 \min(0.1, v - v', w/10)$ and let $\delta = \min(0.1, v - v', w/10)$. The vector

$$b_i = \begin{cases} p_j & \text{if } i = k, \\ -p_k & \text{if } i = j, \\ 0 & \text{otherwise} \end{cases}$$

satisfies $b^\top \hat{p} = 0$, and therefore, using (3.16) and Markov's inequality,

$$\mathbf{P}\left(p_j \widehat{N}_k(T) - p_k \widehat{N}_j(T) = b^\top \widehat{N}(T) > \frac{1}{3} e^{vT}\right) \leqslant 9Ce^{-2\delta T}. \tag{3.18}$$

Let $\{T_i\}_{i=1,2,\ldots}$ be such that $e^{2\delta T_i} = i^2$. By Borel–Cantelli, almost surely there exists i_0 such that for all $i > i_0$,

$$p_j \widehat{N}_k(T_i) - p_k \widehat{N}_j(T_i) < \frac{1}{2} e^{v T_i}. \tag{3.19}$$

Note that

$$T_i = \frac{\log i}{\delta}$$

and therefore

$$T_{i+1} - T_i = \Theta(i^{-1}). \tag{3.20}$$

We want to show that almost surely, for all T large enough,

$$p_j \widehat{N}_k(T) - p_k \widehat{N}_j(T) < e^{vT}. \tag{3.21}$$

Section 3.3 tells us that $\mathbb{E}(\|\widehat{N}(T_i)\|) = O(\exp(w T_i))$, and Lemma 3.4 tells us that $\mathrm{var}(\|\widehat{N}(T_i)\|) = O(\exp(2w T_i))$. Therefore

$$\mathbf{P}(\|\widehat{N}(T_i)\| > e^{(w+0.6\delta)T_i}) < C_l e^{-1.2\delta T_i} = C_l i^{-1.2}$$

for some constant C_l, so that, if $m(i)$ is the number of vertices arriving between T_i and T_{i+1}, then

$$\mathbf{P}\left(m(i) > \frac{1}{2} e^{v T_i}\right)$$

$$\leqslant \mathbf{P}\left(\|\widehat{N}(T_i)\| > e^{(w+0.6\delta)T_i}\right) + \mathbf{P}\left(m(i) > \frac{1}{2} e^{v T_i} \mid \|\widehat{N}(T_i)\| \leqslant e^{(w+0.6\delta)T_i}\right)$$

$$\leqslant \mathbf{P}\left(\|\widehat{N}(T_i)\| > e^{(w+0.6\delta)T_i}\right) + \mathbf{P}\left(m(i) > \frac{1}{2} e^{v T_i} \mid \|\widehat{N}(T_i)\| = e^{(w+0.6\delta)T_i}\right)$$

$$\leqslant C_l i^{-1.2} + C_s e^{-(w+0.6\delta)T_i}(T_{i+1} - T_i)^{-1}, \tag{3.22}$$

where the last inequality uses (3.4) in Claim 3.2 and the fact that

$$\frac{1}{2}e^{vT_i} > 2e^{(w+0.6\delta)T_i}(\exp(2(T_{i+1} - T_i)) - 1)$$

for i large enough.

Clearly, the first part of the right side of (3.22) is a convergent sum. We need to show that so is the second part. Remember the choice $\delta \leq w/10$. Then, using (3.20),

$$C_s e^{-(w+0.6\delta)T_i}(T_{i+1} - T_i)^{-1} = \Theta\left(i \cdot e^{-(w+0.6\delta)T_i}\right) = \Theta\left(e^{\delta T_i} \cdot e^{-(w+0.6\delta)T_i}\right)$$
$$= \Theta\left(e^{-(w-0.4\delta)T_i}\right) = O\left(e^{-9\delta T_i}\right) = O(i^{-9}).$$

Using Borel–Cantelli, we conclude that, almost surely,

$$\sum_{k=1}^{A} |\widehat{N}_k(T) - \widehat{N}_k(T_i)| < \frac{1}{2}e^{vT_i} \qquad (3.23)$$

for all k and all i large enough, and all T between T_i and T_{i+1}. Equation (3.21) follows from (3.23). □

Proof of Lemma 3.5. By Lemma 3.4, $\mathrm{var}(\widehat{N}_1(\tau)) < C_1 e^{2w\tau}$, while $\mathbb{E}(\widehat{N}_1(\tau)) > C_2 e^{w\tau}$ by Section 3.3. Therefore there exists $\rho > 0$ such that

$$\mathbf{P}\left(\widehat{N}_1(\tau) > \rho e^{w\tau}\right) > \rho. \qquad (3.24)$$

Fix some large T, and let $\tau_i = iT$. For each vertex v which is a fertile leaf at time τ_{i-1}, let ℓ_v denote the number of descendants of v (including v itself) at time τ_i which are fertile leaves. The random variables $\{\ell_v\}$ are independent, their sum is $\widehat{N}_1(\tau_i)$, and the distribution of each of them is the same as the unconditional distribution of $\widehat{N}_1(T)$. Using this fact and (3.24), we get

$$\mathbf{P}\left(\widehat{N}_1(\tau_i) > \frac{\rho^2}{2}e^{wT}\widehat{N}_1(\tau_{i-1}) \mid \widehat{N}_1(\tau_{i-1})\right) \geq 1 - e^{-\frac{1}{16}\widehat{N}_1(\tau_{i-1})} \qquad (3.25)$$

via Chernoff's bound. From (3.25), we get that almost surely there exists a constant $C_3 > 0$ such that, for all i large enough,

$$\widehat{N}_1(\tau_i) > C_3 \exp\left(i\left[wT + \log\left(\frac{\rho^2}{2}\right)\right]\right).$$

From Lemma 3.6, we may conclude that the same holds for $\widehat{N}_A(\tau_i)$, i.e., for any constant $C_4 < C_3$,

$$\widehat{N}_A(\tau_i) > C_4 \exp\left(i\left[wT + \log\left(\frac{\rho^2}{2}\right)\right]\right).$$

$\widehat{N}_A(\tau)$ is monotone increasing, and therefore there exists $C_5 > 0$ such that

$$\widehat{N}_A(\tau) > C_5 \exp\left(\tau\left[w + \frac{1}{T}\log\left(\frac{\rho^2}{2}\right)\right]\right) \qquad (3.26)$$

for all τ large enough. Using Lemma 3.6 again, we conclude that there exists $C_6 > 0$ such that

$$\widehat{N}_k(\tau) > C_6 \exp\left(\tau\left[w + \frac{1}{T}\log\left(\frac{\rho^2}{2}\right)\right]\right)$$

for all k and large enough τ. We get (3.17) by taking T so large that

$$w + \frac{1}{T}\log\left(\frac{\rho^2}{2}\right) > w'.$$

□

Proposition 3.8. *For every k and j, almost surely*

$$\lim_{t \to \infty} \frac{\widehat{N}_k(\tau)}{\widehat{N}_j(\tau)} = \frac{p_k}{p_j}. \tag{3.27}$$

Proof. This follows immediately from Lemmas 3.5 and 3.6. □

3.6. Back to discrete time

Proposition 3.9. *For the discrete-time process, and $A > \max\{A_1, A_2\}$ there exists a vector $\hat{q}$ such that, for $k \leqslant A$, we have*

$$\lim_{t \to \infty} \frac{\widetilde{N}_k(t)}{t+1} = q_k. \tag{3.28}$$

Proof. The number of infertile vertices increases at step t with probability

$$\frac{\sum_{k=1}^{A_1-1} \widetilde{N}_k(t)}{\sum_{k=1}^{A} \widetilde{N}_k(t)}$$

(their number cannot decrease). However, by (3.27), this expression tends to a limit, and therefore, using the law of large numbers,

$$\lim_{t \to \infty} \frac{N_0(t)}{t+1} = q_0 = \frac{\sum_{k=1}^{A_1-1} p_k}{\sum_{k=1}^{A} p_k}. \tag{3.29}$$

Using (3.27) once more, the proposition now follows for $k \geqslant 1$ with $q_k = (1 - q_0)p_k$. □

Note that the above proposition implies that q_k and hence p_k is independent of A if $A > k$, since the left-hand side of (3.28) does not depend on A if $A > k$. So, in particular, p_1 does not depend on A.

4. Power law with a cut-off

In the previous section, we saw that for every $A > \max\{A_1, A_2\}$, the limiting proportions up to $A - 1$ are $\lambda \hat{p}$ where $\hat{p}$ is the eigenvector corresponding to the highest eigenvalue w of the A-by-A matrix M defined in (3.1). Therefore, the components $p_1, p_2, \ldots, p_A$ of the vector $\hat{p}$ satisfy the equation

$$wp_i = -\min(i, A_2)p_i + \min(i-1, A_2)p_{i-1} \quad i \geqslant 2, \tag{4.1}$$

where the normalization is determined by $\sum_{i=1}^{A} p_i = 1$. From (4.1) we get that, for $i \leqslant A_2$,

$$p_i = \left(\prod_{k=2}^{i} \frac{k-1}{k+w}\right) p_1, \tag{4.2}$$

and for $i > A_2$,

$$p_i = \left(\frac{A_2}{A_2 + w}\right)^{i-A_2} p_{A_2}. \tag{4.3}$$

Clearly, (4.3) is exponentially decaying. There are many ways to see that (4.2) behaves like a power law with degree $1 + w$. Indeed,

$$\frac{p_i}{p_1} = \left(\prod_{k=2}^{i} \frac{k-1}{k+w}\right) = \exp\left(\sum_{k=2}^{i} \log\left(\frac{k-1}{k+w}\right)\right) \tag{4.4}$$

$$= \exp\left(\sum_{k=2}^{i} \left(\frac{-1-w}{k+w}\right) + O(1)\right) = \exp\left((-1-w)\left(\sum_{k=2}^{i}(k+w)^{-1}\right) + O(1)\right)$$

$$= \exp\left((-1-w)\left(\sum_{k=2}^{i} k^{-1}\right) + O(1)\right) = \exp\left((-1-w)\left(\sum_{k=2}^{i} \log\left(\frac{k+1}{k}\right)\right) + O(1)\right)$$

$$= \exp((-1-w)\log(i/2) + O(1)) = O(1)i^{-1-w}.$$

Note that the constants implicit in the $O(\cdot)$ symbols do not depend on A_1, A_2 or i, owing to the fact that $0 < w \leqslant 2$. Equation (4.4) can be stated in the following way.

Proposition 4.1. *There exist $0 < c < C < \infty$ such that, for every A_1, A_2 and $i \leqslant A_2$, if $w = w(A_1, A_2)$ is as in (4.1), then*

$$ci^{-1-w} \leqslant \frac{p_i}{p_1} \leqslant Ci^{-1-w}. \tag{4.5}$$

The vector $(q_1, q_2, \ldots, q_{A-1})$ is a scalar multiple of the vector $(p_1, p_2, \ldots, p_{A-1})$, so equations (1.5), (1.6), and (1.7) in Theorem 1.2 (and the comment immediately following it) are consequences of equations (4.2), (4.3), and (4.5) derived above. It remains to prove the normalization conditions

$$\sum_{i=0}^{\infty} q_i = 1 \quad \text{and} \quad q_0 = \sum_{i=1}^{\infty} q_i \min(i-1, A_1 - 1)$$

stated in Theorem 1.2. These follow from the equations

$$\sum_{i=0}^{\infty} N_i(t) = t + 1 \quad \text{and} \quad N_0(t) = \sum_{i=1}^{\infty} N_i(t) \min(i-1, A_1 - 1).$$

The first of these simply says that there are $t + 1$ vertices at time t; the second equation is proved by counting the number of infertile children of each fertile node.

5. Monotonicity properties of w

In this section we will prove that the exponent $1+w$ of the power law in Proposition 4.1 is monotonically decreasing in A_1 and monotonically increasing in A_2. For this purpose, it will be useful to define a family of matrices, parametrized by two vectors $\mathbf{y}, \mathbf{z} \in \mathbb{R}^n$, which generalizes the matrix M appearing in (3.1), whose top eigenvalue is w.

Given vectors $\mathbf{y} = (y_1, y_2, \ldots, y_n)$, $\mathbf{z} = (z_1, z_2, \ldots, z_n) \in \mathbb{R}^n$, let $M(\mathbf{y}, \mathbf{z})$ denote the n-by-n matrix whose (ij)th entry is:

$$M_{i,j}(\mathbf{y},\mathbf{z}) = \begin{cases} z_1 - y_1 & \text{if } 1 = i = j, \\ -y_j & \text{if } 2 \leqslant i = j \leqslant n, \\ y_j & \text{if } 2 \leqslant i = j+1 \leqslant n, \\ z_j & \text{if } i = 1 \text{ and } j \geqslant 2, \\ 0 & \text{otherwise.} \end{cases}$$

Thus, for instance, the matrix M defined in (3.1) is $M(\mathbf{y}, \mathbf{z})$, where $n = A$ and

$$y_j = \begin{cases} \min(j, A_2) & \text{if } 1 \leqslant j < A, \\ 0 & \text{if } j = A, \end{cases}$$

$$z_j = \begin{cases} 0 & \text{if } 1 \leqslant j < A_1, \\ \min(j, A_2) & \text{if } A_1 \leqslant j \leqslant A. \end{cases}$$

For the remainder of this section, we will assume:

$$y_i > 0 \quad \text{for } 1 \leqslant i < n, \tag{5.1}$$
$$z_i \geqslant 0 \quad \text{for } 1 \leqslant i < n, \tag{5.2}$$
$$y_n = 0, \quad z_n > 0. \tag{5.3}$$

All of these criteria will be satisfied by the matrices $M(\mathbf{y}, \mathbf{z})$ which arise in proving the desired monotonicity claim. It follows from (5.1), (5.2), and (5.3) that if we add a suitably large scalar multiple of the identity matrix to $M(\mathbf{y}, \mathbf{z})$, we obtain an irreducible matrix $M(\mathbf{y}, \mathbf{z}) + BI$ with nonnegative entries. The Perron–Frobenius theorem guarantees that $M(\mathbf{y}, \mathbf{z}) + BI$ has a positive real eigenvalue R of multiplicity 1, such that all other complex eigenvalues have modulus $\leqslant R$; consequently $M(\mathbf{y}, \mathbf{z})$ has a real eigenvalue $w = R - B$, of multiplicity 1, such that the real part of every other eigenvalue is strictly less than w.

We will study how w varies under perturbations of the parameters $\mathbf{y}, \mathbf{z}$. Let $P(\lambda, \mathbf{y}, \mathbf{z})$ be the characteristic polynomial of $M(\mathbf{y}, \mathbf{z})$, i.e.,

$$P(\lambda, \mathbf{y}, \mathbf{z}) = \det(\lambda I - M(\mathbf{y}, \mathbf{z})).$$

This is a polynomial of degree n in λ (with coefficients depending smoothly on $\mathbf{y}, \mathbf{z}$), whose largest real root $w(\mathbf{y}, \mathbf{z})$ exists and has multiplicity 1, provided $(\mathbf{y}, \mathbf{z})$ belongs to the region $V \subset \mathbb{R}^n \times \mathbb{R}^n$ determined by (5.1), (5.2), and (5.3). It follows from the implicit function theorem that $w(\mathbf{y}, \mathbf{z})$ is a smooth function of $(\mathbf{y}, \mathbf{z})$ in V, satisfying

$$\left(\frac{\partial P}{\partial y_i} + \frac{\partial w}{\partial y_i} \cdot \frac{\partial P}{\partial \lambda}\right)\bigg|_{(w,\mathbf{y},\mathbf{z})} = 0, \quad \left(\frac{\partial P}{\partial z_i} + \frac{\partial w}{\partial z_i} \cdot \frac{\partial P}{\partial \lambda}\right)\bigg|_{(w,\mathbf{y},\mathbf{z})} = 0. \tag{5.4}$$

If **x** is any vector in $\mathbb{R}^n \times \mathbb{R}^n$, and $\partial_\mathbf{x}$ is the corresponding directional derivative operator, we have from (5.4):

$$\partial_\mathbf{x} w(\mathbf{y}, \mathbf{z}) = -\frac{\partial_\mathbf{x} P(w, \mathbf{y}, \mathbf{z})}{(\partial P/\partial \lambda)|_{(w,\mathbf{y},\mathbf{z})}}. \tag{5.5}$$

We know that $(\partial P/\partial \lambda)|_{(w,\mathbf{y},\mathbf{z})} > 0$ because P is a polynomial with positive leading coefficient, w is its largest real root, and w has multiplicity 1. Thus we have established the following.

Claim 5.1. *For any vector $\mathbf{x} \in \mathbb{R}^n \times \mathbb{R}^n$, and any $(\mathbf{y}, \mathbf{z}) \in V$, put $w = w(\mathbf{y}, \mathbf{z})$. Then the directional derivatives $\partial_\mathbf{x} w(\mathbf{y}, \mathbf{z})$ and $\partial_\mathbf{x} P(w, \mathbf{y}, \mathbf{z})$ have opposite signs.*

This allows monotonicity properties of w to be deduced from calculations involving directional derivatives of P. Given the definition of $M(\mathbf{y}, \mathbf{z})$, it is straightforward to compute that

$$P(\lambda, \mathbf{y}, \mathbf{z}) = \det(\lambda I - M(\mathbf{y}, \mathbf{z})) = P_1(\lambda, y, z) - \sum_{j=2}^{n} P_j(\lambda, \mathbf{y}, \mathbf{z}), \tag{5.6}$$

where

$$P_1(\lambda, y, z) = (\lambda + y_1 - z_1) \prod_{i=2}^{n} (\lambda + y_i), \tag{5.7}$$

$$P_j(\lambda, \mathbf{y}, \mathbf{z}) = \left(\prod_{i=1}^{j-1} y_i \right) z_j \left(\prod_{i=j+1}^{n} (\lambda + y_j) \right). \tag{5.8}$$

As an easy consequence of this formula, w is strictly positive.

Lemma 5.2. *w is strictly positive.*

Proof. From (5.6)–(5.8) and the fact that $y_n = 0$, we have $P(0, \mathbf{y}, \mathbf{z}) = -P_n(0, \mathbf{y}, \mathbf{z}) = -\left(\prod_{i=1}^{n-1} y_i\right) z_n$, and this is strictly negative by (5.1) and (5.3). For sufficiently large positive λ, we know that $P(\lambda, \mathbf{y}, \mathbf{z}) > 0$ because P is a polynomial whose leading coefficient in λ is positive. By the intermediate value theorem, $P(\lambda, \mathbf{y}, \mathbf{z})$ has a strictly positive real root. $\square$

The following three lemmas encapsulate the requisite directional derivative estimates for P.

Lemma 5.3. $(\partial P/\partial z_k)|_{(w,\mathbf{y},\mathbf{z})} < 0$ *for $(\mathbf{y}, \mathbf{z}) \in V$.*

Proof. For $k > 1$,

$$\partial P/\partial z_k = -\partial P_k/\partial z_k = -\left(\prod_{i=1}^{k-1} y_i \right) \left(\prod_{i=k+1}^{n} (w + y_i) \right) < 0.$$

For $k = 1$,
$$\partial P/\partial z_1 = \partial P_1/\partial z_1 = -\prod_{i=2}^{n}(w + y_i) < 0.$$

Corollary 5.4. *w is monotonically decreasing in A_1.*

Proof. Increasing A_1 from k to $k+1$ has no effect on **y**, and its only effect on **z** is to decrease z_k from $\min(k, A_2)$ to 0. As we move in the $-z_k$ direction, the directional derivative of P is positive, so the directional derivative of w is negative by Claim 5.1. Thus w decreases as we increase A_1 from k to $k+1$. □

Lemma 5.5. *For $1 < k < n$, $(\partial P/\partial y_k)|_{(w,\mathbf{y},\mathbf{z})} < 0$ if $(\mathbf{y}, \mathbf{z}) \in V$ and $z_k = 0$.*

Proof.
$$\frac{\partial P}{\partial y_k} = \frac{\partial P_1}{\partial y_k} - \sum_{j=2}^{n} \frac{\partial P_j}{\partial y_k}$$
$$= \frac{1}{w + y_k} P_1 - \frac{1}{w + y_k} \sum_{j=2}^{k-1} P_j - \frac{1}{y_k} \sum_{j=k+1}^{n} P_j$$
$$< \frac{1}{w + y_k} P_1 - \frac{1}{w + y_k} \sum_{j=2}^{k-1} P_j - \frac{1}{w + y_k} \sum_{j=k+1}^{n} P_j$$
$$= \frac{P(w, \mathbf{y}, \mathbf{z})}{w + y_k}$$
$$= 0.$$

Lemma 5.6. *For $k > 1$, $(\partial P/\partial y_k + \partial P/\partial z_k)|_{(w,\mathbf{y},\mathbf{z})} < 0$ if $(\mathbf{y}, \mathbf{z}) \in V$ and $y_k = z_k$.*

Proof.
$$\frac{\partial P}{\partial y_k} + \frac{\partial P}{\partial z_k} = \frac{\partial P_1}{\partial y_k} - \sum_{j=2}^{n} \frac{\partial P_j}{\partial y_k} - \frac{\partial P_k}{\partial z_k}$$
$$= \frac{1}{w + y_k} P_1 - \frac{1}{w + y_k} \sum_{j=2}^{k-1} P_j - \frac{1}{y_k} \sum_{j=k+1}^{n} P_j - \frac{1}{z_k} P_k$$
$$< \frac{1}{w + y_k} P_1 - \frac{1}{w + y_k} \sum_{j=2}^{k-1} P_j - \frac{1}{w + y_k} \sum_{j=k+1}^{n} P_j - \frac{1}{w + y_k} P_k$$
$$= \frac{P(w, \mathbf{y}, \mathbf{z})}{w + y_k}$$
$$= 0.$$

Corollary 5.7. *w is monotonically increasing in A_2.*

Proof. If we change A_2 from k to $k+1$, this changes $\mathbf{y}$ into a new vector $\mathbf{y}'$ satisfying

$$y'_j - y_j = \begin{cases} 1 & \text{if } k < j < n, \\ 0 & \text{otherwise.} \end{cases}$$

It changes $\mathbf{z}$ into a new vector $\mathbf{z}'$ satisfying

$$z'_j - z_j = \begin{cases} 1 & \text{if } \max(A_1, k+1) \leqslant j \leqslant n, \\ 0 & \text{otherwise.} \end{cases}$$

Letting $\mathbf{e}_j^{(y)}$ denote a unit vector in the $+y_j$ direction, and $\mathbf{e}_j^{(z)}$ a unit vector in the $+z_j$ direction, the direction of change is expressed by the vector

$$\mathbf{x} = (\mathbf{y}', \mathbf{z}') - (\mathbf{y}, \mathbf{z}) = \left[\sum_{k+1 \leqslant j < A_1} \mathbf{e}_j^{(y)}\right] + \left[\sum_{\max(k+1, A_1) \leqslant j < n} (\mathbf{e}_j^{(y)} + \mathbf{e}_j^{(z)})\right] + \mathbf{e}_n^{(z)},$$

and $\partial_{\mathbf{x}} P$ is negative, by the preceding three lemmas. By Claim 5.1, this means w increases monotonically as we move along this path. □

Acknowledgement

We wish to thank the referee for many good suggestions that helped us improve the paper.

References

[1] Aiello, W., Chung, F. and Lu, L. (2002) Random evolution of massive graphs. In *Handbook of Massive Data Sets*, Kluwer, pp. 97–122.

[2] Albert, R. and Barabási, A.-L. (2002) Statistical mechanics of complex networks. *Rev. Mod. Phys.* **74** 47–97.

[3] Aldous, D. J. (2003) A stochastic complex network model. *Electron. Res. Announc. Amer. Math. Soc.* **9** 152–161.

[4] Barabási, A.-L. and Albert, R. (1999) Emergence of scaling in random networks. *Science* **286** 509–512.

[5] Berger, N., Bollobás, B., Borgs, C., Chayes, J. T. and Riordan, O. (2003) Degree distribution of the FKP network model. In *Proc. 30th International Colloquium on Automata, Languages and Programming*, Vol. 2719 of *Lecture Notes in Computer Science*, Springer, pp. 725–738.

[6] Berger, N., Borgs, C., Chayes, J. T., D'Souza, R. M. and Kleinberg, R. D. (2004) Competition-induced preferential attachment. In *Proc. 31st International Colloquium on Automata, Languages and Programming*. Vol. 3142 of *Lecture Notes in Computer Science*, Springer, pp. 208–221.

[7] Bollobás, B. and Riordan, O. (2003) Mathematical results on scale-free random graphs. In *Handbook of Graphs and Networks*, Wiley-VCH, Weinheim, pp. 1–34.

[8] Bollobás, B., Borgs, C., Chayes, J. T. and Riordan, O. (2003) Directed scale-free graphs. In *Proc. 14th ACM–SIAM Symposium on Discrete Algorithms*, pp. 132–139.

[9] Bollobás, B., Riordan, O., Spencer, J. and Tusnady, G. E. (2001) The degree sequence of a scale-free random graph process. *Random Struct. Alg.* **18** 279–290.

[10] Carlson, J. M. and Doyle, J. (1999) Highly optimized tolerance: a mechanism for power laws in designed systems. *Phys. Rev. E* **60** 1412.

[11] Cooper, C. and Frieze, A. M. (2001) A general model of web graphs. In *Proc. 9th European Symposium on Algorithms*, pp. 500–511.

[12] Dorogovtsev, S. N. and Mendes, J. F. F. (2002) Evolution of networks. *Adv. Phys.* **51** 1079.
[13] Eggenberger, F. and Pólya, G. (1923) Über die statistik verketteter Vorgänge. *Z. Agnew. Math. Mech.* **3** 279–289.
[14] Fabrikant, A., Koutsoupias, E. and Papadimitriou, C. H. (2002) Heuristically optimized trade-offs: a new paradigm for power laws in the internet. In *Proc. 29th International Colloquium on Automata, Languages and Programming*, Vol. 2380 of *Lecture Notes in Computer Science*, Springer, pp. 110–122.
[15] Faloutsos, M., Faloutsos, P. and Faloutsos, C. (1999) On the power-law relationships of the Internet topology. *Comput. Commun. Rev.* **29** 251.
[16] Govindan, R. and Tangmunarunkit, H. (2000) Heuristics for Internet map discovery. In *Proc. INFOCOM*, pp. 1371–1380.
[17] Kenyon, C. and Schabanel, N. Personal communication.
[18] Kumar, R., Raghavan, P., Rajagopalan, S., Sivakumar, D., Tomkins, A. and Upfal, E. (2000) Stochastic models for the web graph. In *Proc. 41st IEEE Symp. on Foundations of Computer Science*, pp. 57–65.
[19] Newman, M. E. J. (2003) The structure and function of complex networks. *SIAM Review* **45** 167–256.
[20] Price, D. J. de S. (1976) A general theory of bibliometric and other cumulative advantage processes. *J. Amer. Soc. Inform. Sci.* **27** 292–306.
[21] Simon, H. A. (1955) On a class of skew distribution functions. *Biometrika* **42** 425–440.
[22] Yule, G. U. (1924) A mathematical theory of evolution, based on the conclusions of Dr. J. C. Willis. *Philos. Trans. Roy. Soc. London, Ser. B* **213** 21–87.
[23] Zipf, G. K. (1949) *Human Behavior and the Principle of Least Effort*, Addison-Wesley, Cambridge, MA.

On Two Conjectures on Packing of Graphs

BÉLA BOLLOBÁS,[1][†] ALEXANDR KOSTOCHKA[2][‡]

and KITTIKORN NAKPRASIT[3]

[1] University of Memphis, Memphis, TN 38152, USA
and
Trinity College, Cambridge CB2 1TQ, UK
(e-mail: bollobas@msci.memphis.edu)

[2] University of Illinois, Urbana, IL 61801, USA
and
Institute of Mathematics, Novosibirsk 630090, Russia
(e-mail: kostochk@math.uiuc.edu)

[3] University of Illinois, Urbana, IL 61801, USA
(e-mail: nakprasi@math.uiuc.edu)

Received 16 January 2004; revised 1 September 2004

For Béla Bollobás on his 60th birthday

In 1978, Bollobás and Eldridge [5] made the following two conjectures.

(C1) There exists an absolute constant $c > 0$ such that, if k is a positive integer and G_1 and G_2 are graphs of order n such that $\Delta(G_1), \Delta(G_2) \leqslant n - k$ and $e(G_1), e(G_2) \leqslant ckn$, then the graphs G_1 and G_2 pack.

(C2) For all $0 < \alpha < 1/2$ and $0 < c < \sqrt{1/8}$, there exists an $n_0 = n_0(\alpha, c)$ such that, if G_1 and G_2 are graphs of order $n > n_0$ satisfying $e(G_1) \leqslant \alpha n$ and $e(G_2) \leqslant c\sqrt{n^3/\alpha}$, then the graphs G_1 and G_2 pack.

Conjecture (C2) was proved by Brandt [6]. In the present paper we disprove (C1) and prove an analogue of (C2) for $1/2 \leqslant \alpha < 1$. We also give sufficient conditions for simultaneous packings of about $\sqrt{n}/4$ sparse graphs.

1. Introduction

One of the basic notions of graph theory is that of *packing*. Two graphs, G_1 and G_2, of the same order are said to *pack* if G_1 is a subgraph of the complement $\overline{G}_2$ of G_2, or,

[†] Research supported by NSF grants DMS-9970404 and EIA-0130352, and DARPA grant F33615-01-C1900.
[‡] Research supported by NSF grant DMS-0099608 and by grants 99-01-00581 and 00-01-00916 of the Russian Foundation for Basic Research.

equivalently, G_2 is a subgraph of the complement $\overline{G}_1$ of G_2. The study of packings of graphs was started in the 1970s by Sauer and Spencer [13] and Bollobás and Eldridge [5].

In particular, Sauer and Spencer [13] proved the following result.

Theorem 1.1. *Suppose that G_1 and G_2 are graphs of order n such that $2\Delta(G_1)\Delta(G_2) < n$. Then G_1 and G_2 pack.*

The main conjecture in the area is the Bollobás–Eldridge–Catlin (BEC) conjecture (see [4, 3, 5, 10]) stating that *if G_1 and G_2 are graphs with n vertices, maximum degrees Δ_1 and Δ_2, respectively, and $(\Delta_1 + 1)(\Delta_2 + 1) \leqslant n + 1$, then G_1 and G_2 pack*. If true, this conjecture is a considerable extension of the Hajnal–Szemerédi theorem [12] on equitable colouring, which is itself an extension of the Corrádi–Hajnal theorem on equitable 3-colourings of graphs. Indeed, the Hajnal–Szemerédi theorem is the special case of the BEC conjecture when G_2 is a disjoint union of cliques of the same size [12]. The conjecture has also been proved when either $\Delta_1 \leqslant 2$ [1, 2], or $\Delta_1 = 3$ and n is huge [11]. The progress on the topic has been surveyed by Yap [16] and Wozniak [15].

The following two theorems are the main results of Bollobás and Eldridge [5].

Theorem 1.2. *Suppose that G_1 and G_2 are graphs with n vertices, $\Delta(G_1)$, $\Delta(G_2) < n - 1$, $e(G_1) + e(G_2) \leqslant 2n - 3$ and $\{G_1, G_2\}$ is not one of the following pairs: $\{2K_2, K_1 \cup K_3\}$, $\{\overline{K}_2 \cup K_3, K_2 \cup K_3\}$, $\{3K_2, \overline{K}_2 \cup K_4\}$, $\{\overline{K}_3 \cup K_3, 2K_3\}$, $\{2K_2 \cup K_3, \overline{K}_3 \cup K_4\}$, $\{\overline{K}_4 \cup K_4, K_2 \cup 2K_3\}$, $\{\overline{K}_5 \cup K_4, 3K_3\}$. Then G_1 and G_2 pack.*

Theorem 1.3. *For $0 < \alpha < 1/2$, there is an integer $n_0 = n_0(\alpha)$ such that, if G_1 and G_2 are graphs of order $n \geqslant n_0$ with $e(G_1) \leqslant \alpha n$ and $e(G_2) \leqslant \frac{1-2\alpha}{5} n^{3/2}$, then G_1 and G_2 pack.*

Let n be even, x be odd, $G_1(n)$ be a perfect matching on n vertices and $G_2(n, x)$ be the complete bipartite graph $K_{x, n-x}$. Since x is odd, the graphs $G_1(n)$ and $G_2(n, x)$ do not pack. Since $e(G_1(n)) = n/2$ and $e(G_2(n, x)) = x(n - x) < xn$, these examples show that the condition $\alpha < 1/2$ in Theorem 1.3 cannot be relaxed without imposing other restrictions on G_1 and/or G_2. However, Bollobás and Eldridge [5] could not find an example showing that the factor $(1 - 2\alpha)/5$ is close to optimal, and they were led to the following conjecture.

Conjecture 1.4. *For all $0 < \alpha < 1/2$ and $0 < c < \sqrt{1/8}$, there exists an $n_0 = n_0(\alpha, c)$ such that, if G_1 and G_2 are graphs of order $n > n_0$ satisfying $e(G_1) \leqslant \alpha n$ and $e(G_2) \leqslant c\sqrt{n^3/\alpha}$, then the graphs G_1 and G_2 pack.*

This conjecture was proved by Brandt [6] in 1995. As the main result of this paper, we prove the following extension of this theorem of Brandt to the case when G_1 has αn edges, with $1/2 \leqslant \alpha < 1$.

Theorem 1.5. *Let $1/2 \leqslant \alpha < 1$ and $c > 0$ satisfy*

$$8\alpha c^2 < 1, \qquad (1.1)$$

and put
$$\varepsilon = \frac{1}{4}\min\{1-\alpha,\ 1-8\alpha c^2\}. \tag{1.2}$$

Let G_1 and G_2 be graphs of order
$$n > (10/\varepsilon)^6, \tag{1.3}$$

such that $e(G_1) \leqslant \alpha n$, $e(G_2) \leqslant cn^{3/2}$, and $\Delta(G_2) < n-1-\frac{\sqrt{n}}{\sqrt{2\alpha(1-\alpha)}}$. Then G_1 and G_2 pack.

Observe that the only additional restriction in Theorem 1.5 is that each vertex in G_2 has at least $\frac{\sqrt{n}}{\sqrt{2\alpha(1-\alpha)}}$ non-neighbours. The example of $G_1(n)$ and $G_2(n,x)$ where x is the largest odd integer not exceeding $c\sqrt{n}$ shows that the factor $\sqrt{n}$ is unavoidable there.

The examples of a perfect matching and $G_2(n,x)$ also explain why Bollobás and Eldridge [5, p. 118] made the following conjecture.

Conjecture 1.6. *There exists an absolute constant $c > 0$ such that, if $k \geqslant 1$ and G_1 and G_2 are graphs of order n satisfying the conditions $\Delta(G_1), \Delta(G_2) \leqslant n-k$ and $e(G_1), e(G_2) \leqslant ckn$, then the graphs G_1 and G_2 pack.*

We shall disprove Conjecture 1.6; more precisely, we shall prove the following result.

Theorem 1.7. *Let k be a positive integer and q be a prime power. Then for every $n \geqslant q\frac{q^{k+1}-1}{q-1}$, there are graphs $G_1(n,k)$ and $G_2(n,q,k)$ of order n that do not pack and have the following properties:*
(a) *$G_1(n,k)$ is a forest with $n-k$ edges and maximum degree at most n/k;*
(b) *$G_2(n,q,k)$ is a $\frac{q^k-1}{q-1}$-degenerate graph with maximum degree at most $2n/q$.*

Theorem 1.7 not only disproves Conjecture 1.6, but also shows that Theorem 1.5 can not be extended even to $\alpha = 1$ without essential restrictions on the maximal degree of G_2.

The rest of the paper is organized as follows. In the next section we shall discuss properties of special enumerations of vertices in graphs; our proof of Theorem 1.5, which is to be given in Section 3, will be based on these enumerations. In Section 4 we shall make use of the proof of Theorem 1.5 to give conditions providing simultaneous packing of about $\frac{1}{4}\sqrt{n/\alpha^3}$ graphs of order n with at most αn edges each. More precisely, we shall prove the following result.

Theorem 1.8. *Let $\frac{1}{2} \leqslant \alpha < 1$,*
$$n > (50/(1-\alpha))^6, \tag{1.4}$$

and $m = \lceil \frac{1}{4}\sqrt{n/\alpha^3} \rceil$. Let $H_1, H_2, \ldots, H_m$ be graphs with n vertices and at most αn edges each. Then $H_1, H_2, \ldots, H_m$ pack.

In the final section, Section 5, we discuss counterexamples to Conjecture 1.6 and prove Theorem 1.7.

Note that the proofs of upper bounds are algorithmic, and so enable one to construct polynomial-time algorithms for packing graphs satisfying the conditions of Theorems 1.5 or 1.8.

2. Greedy and degenerate enumerations

Before embarking on the proof of Theorem 1.5, we introduce some notation and prove some auxiliary statements.

Let $v_1, v_2, \ldots, v_n$ be an enumeration of the vertices of a graph G. For $1 \leqslant i \leqslant n$, let $G(i)$ be the subgraph of G induced by the vertices $v_i, v_{i+1}, \ldots, v_n$; thus $G(1) = G$ and $G(n)$ consists of the single vertex v_n. We call $v_1, v_2, \ldots, v_n$ a *greedy enumeration* of the vertices or, somewhat loosely, a *greedy order* on G, if $d_{G(i)}(v_i) = \Delta(G(i))$ for every i, $1 \leqslant i \leqslant n$, i.e., the vertex v_i has maximal degree in $G(i)$. Similarly, the enumeration and order are *degenerate* if $d_{G(i)}(v_i) = \delta(G(i))$ for every i, $1 \leqslant i \leqslant n$, i.e., the vertex v_i has minimal degree in $G(i)$. Note that if $v_1, v_2, \ldots, v_n$ is a greedy order on G then $v_i, v_{i+1}, \ldots, v_n$ is a greedy order on $G(i)$, and an analogous assertion holds for the degenerate order. Another simple observation is that $v_1, v_2, \ldots, v_n$ is a greedy order on G if and only if it is a degenerate order on the complement $\overline{G}$. Needless to say, a graph may have numerous greedy orders and degenerate orders.

For a graph G, set

$$\varphi(G) = \sum_{v \in V(G)} \frac{1}{1 + d_G(v)}.$$

The result below is a slight extension of an inequality due to Caro [7] and Wei [14], first published in [8], implying a weak form of Turán's theorem. We formulate it in the usual way, for the complement of the graph, i.e., for finding a large independent set rather than a complete subgraph.

Theorem 2.1. *Let $v_1, v_2, \ldots, v_n$ be a greedy enumeration of the vertices of a graph G, and set $\ell = \lceil \varphi(G) \rceil$. Then the last ℓ vertices form an independent set. Equivalently, if $d_{G(i)}(v_i) \geqslant 1$ then $G(i)$ has an independent set of at least $\varphi(G)$ vertices.*

Proof. We apply induction on the number of edges of G. If there are no edges then $\varphi(G) = n$ and the entire vertex set is independent, as required. Suppose that G has $m > 0$ edges and the result holds for graphs with fewer edges. Write d for the maximal degree of G, i.e., for the degree of v_1, and let $u_1, u_2, \ldots, u_d$ be the neighbours of v_1. Then

$$\varphi(G(2)) = \varphi(G(1)) - \frac{1}{d+1} + \sum_{i=1}^{d} \left(\frac{1}{d(u_i)} - \frac{1}{d(u_i) + 1} \right)$$

$$= \varphi(G(1)) - \frac{1}{d+1} + \sum_{i=1}^{d} \frac{1}{d(u_i)(d(u_i)+1)}$$

$$\geqslant \varphi(G(1)) - \frac{1}{d+1} + d \frac{1}{d(d+1)} = \varphi(G(1)) = \varphi(G).$$

By the induction hypothesis, the last $\lceil \varphi(G(2)) \rceil \geqslant \lceil \varphi(G) \rceil = \ell$ vertices of $v_2, v_3, \ldots, v_n$ form an independent set of $G(2)$, and so of G, completing the proof. □

We shall also need the following simple but somewhat technical lemma concerning greedy orders.

Lemma 2.2. *Let α, γ and ε be positive numbers satisfying $\gamma \leqslant \alpha \leqslant 1 - 2\varepsilon$ and $k_0 \leqslant (1 - \gamma - \varepsilon/2)n - 1$ a nonnegative integer. Let $v_1, v_2, \ldots, v_n$ be an enumeration of the vertices of a graph G with m edges with the following properties:*
(i) $e(G_{k_0+1}) \leqslant m(1 - \frac{2k_0(\alpha+\varepsilon)}{n(\alpha-\gamma+\varepsilon/2)})$;
(ii) *the enumeration $v_{k_0+1}, v_{k_0+2}, \ldots, v_n$ is greedy.*
Then there is an index i, $k_0 \leqslant i \leqslant (1 - \gamma - \varepsilon/2)n$, such that

$$\Delta(G(i+1)) = d_{G(i+1)}(v_{i+1}) < \frac{2m(n-i)(\alpha+\varepsilon)}{n^2(\alpha-\gamma+\varepsilon/2)}. \quad (2.1)$$

Proof. Suppose that the assertion is false. Then for $k = \lceil (1 - \gamma - \varepsilon/2)n \rceil$ we have

$$e(G(k+1)) = e(G(k_0+1)) - \sum_{i=k_0+1}^{k} \Delta(G(i))$$

$$\leqslant m\left(1 - \frac{2k_0(\alpha+\varepsilon)}{n(\alpha-\gamma+\varepsilon/2)}\right) - \sum_{i=k_0+1}^{k} \frac{2m(n+1-i)(\alpha+\varepsilon)}{n^2(\alpha-\gamma+\varepsilon/2)}$$

$$\leqslant m - \sum_{i=1}^{k} \frac{2m(n+1-i)(\alpha+\varepsilon)}{n^2(\alpha-\gamma+\varepsilon/2)}$$

$$= m - \frac{2m(\alpha+\varepsilon)}{n^2(\alpha-\gamma+\varepsilon/2)}\left(\binom{n+1}{2} - \binom{n+1-k}{2}\right)$$

$$< m - \frac{m(\alpha+\varepsilon)}{n^2(\alpha-\gamma+\varepsilon/2)}(n^2 - (n-k)^2)$$

$$\leqslant m\left(1 - \frac{(\alpha+\varepsilon)(1-(\gamma+\varepsilon/2)^2)}{(\alpha-\gamma+\varepsilon/2)}\right) = \rho m, \quad (2.2)$$

say. To arrive at a contradiction and so complete the proof, we shall show that $\rho < 0$. To this end, set $\delta = \gamma + \varepsilon/2$, and note that

$$\rho(\alpha - \gamma + \varepsilon/2) = \delta(\delta(\alpha+\varepsilon) - 1). \quad (2.3)$$

Since, by assumption, $\delta > 0$ and

$$\delta(\alpha+\varepsilon) \leqslant (\alpha+\varepsilon/2)(\alpha+\varepsilon) < 1,$$

identity (2.3) implies that ρ is indeed negative, completing our proof. □

We shall also use the following fact observed by several authors.

Claim 2.3. *Suppose that we are packing the vertices of a graph G_1 in the reverse degenerate order into (the complement of) a graph G_2 of order N and maximal degree D_2. Suppose that* we have already packed j vertices and a vertex $w \in V(G_1)$ has x neighbours among these j vertices. If

$$j + xD_2 < N, \tag{2.4}$$

then we can also find a legal placement for w.

Proof. We cannot place w at the j vertices of G_2 that we have already used and into G_2-neighbours of the images of the x neighbours of w. However, w can be mapped into every other vertex of G_2. □

3. Proof of Theorem 1.5

Let G_1 and G_2 be graphs of order $n > (10/\varepsilon)^6$ such that $e(G_1) \leqslant \alpha n$, $e(G_2) \leqslant cn^{3/2}$, and $\Delta(G_2) < n - 1 - \frac{\sqrt{n}}{\sqrt{2\alpha(1-\alpha)}}$. Since $\alpha \geqslant 1/2$, condition (1.1) yields that $c < 1/2$. Since the greater is c, the stronger is the assertion, we may assume that

$$\frac{1}{3} < c < \frac{1}{2}. \tag{3.1}$$

Observe that, by (1.2),

$$8(\alpha + \varepsilon)c^2 < 1 - 2\varepsilon \quad \text{and} \quad \alpha + 2\varepsilon < 1. \tag{3.2}$$

Let $T_1, \ldots, T_t$ be the components of G_1 that are trees (including isolated vertices) with $v(T_1) \leqslant \cdots \leqslant v(T_t)$, where we write $v(H) = |V(H)|$ for the order of a graph H. Let $G_1^* = G_1 - T_1 - \cdots - T_t$. In other words, let G_1^* be the union of the components of G_1 containing cycles. Suppose that G_1^* has exactly γn vertices. Then it has at least γn edges and hence $\gamma \leqslant \alpha$. Since $e(G_1) \leqslant \alpha n$,

$$t \geqslant (1 - \alpha)n. \tag{3.3}$$

It is trivial to check that the following assertion holds.

Claim 3.1. *For every $1 \leqslant j < t$, we have*
(a) $\sum_{i=1}^{j} v(T_i) \leqslant \frac{1-\gamma}{1-\alpha} j$, *and*
(b) $v(T_j) \leqslant \frac{n(1-\gamma)}{t-j+1}$.

Let $w_1, w_2, \ldots, w_n$ be a degenerate order of the vertices of G_1 with the additional condition that first we list vertices in T_1, then those in T_2, and so on, and we enumerate the vertices in G_1^* only after having enumerated all vertices in $T_1, \ldots, T_t$. Let $u_1, u_2, \ldots, u_n$ be a greedy order of the vertices of G_2. Let k_0' be the maximal k such that $\deg_{G_2(k)}(u_k) > \frac{(1-\alpha)^2}{20}n$. Since $e(G_2) \geqslant \sum_{i=1}^{k_0'} \deg_{G_2(k)}(u_k)$, we have

$$k_0' < \frac{20c}{(1-\alpha)^2}\sqrt{n} \leqslant \frac{10}{(1-\alpha)^2}(0.1\varepsilon)^3 n \leqslant 0.01(1-\alpha)n. \tag{3.4}$$

Claim 3.2. *For $j = 1, \ldots, k_0'$, there is a set $U_j \subset V(G_2)$ such that*
(i) $U_j \supset \{u_1, \ldots, u_j\}$,

(ii) $|U_j| = \sum_{i=1}^{j} v(T_i)$,
(iii) *there exists a packing of* $G_1[V(T_1) \cup \cdots \cup V(T_j)]$ *and* $G_2[U_j]$.

Proof. Suppose that the claim is proved for $j' \leq j-1 \leq k'_0 - 1$. Assume that the vertices of T_j are $w_{z-y+1}, w_{z-y+2}, \ldots, w_z$. Let m be the smallest index such that $u_m \notin U_{j-1}$. By the induction assumption, $m \geq j$. Identify u_m with w_z and denote $v_0 = u_m$. To prove the claim, it is enough to find for every $i = 1, \ldots, y-1$ a vertex $v_i \in V(G_2) - U_{j-1} - \{v_0, \ldots, v_{i-1}\}$ not adjacent to the vertex $v_{i'}$, $i' < i$ that was identified with a neighbour $w_{z-i'}$ of w_i. Then we can identify v_i with w_{z-i} and continue.

Case 1: $j \leq 2c\sqrt{n}$. Then by Claim 3.1(a), $|U_{j-1} \cup \{v_0, \ldots, v_{i-1}\}| \leq \frac{j}{1-\alpha}$. Since, under conditions of the theorem, $v_{i'}$ has at least $\frac{\sqrt{n}}{\sqrt{2\alpha(1-\alpha)}}$ non-neighbours, it has a non-neighbour in $V(G_2) - U_{j-1} - \{v_0, \ldots, v_{i-1}\}$.

Case 2: $j > 2c\sqrt{n}$. Then $\deg_{G_2(j)}(v_{i'}) \leq \deg_{G_2(j)}(u_j) \leq \frac{cn^{1.5}}{j} < n/2$ and by Claim 3.1(a), $|U_{j-1} \cup \{v_0, \ldots, v_{i-1}\}| < k'_0 \frac{1-\gamma}{1-\alpha}$. By (3.4), the last expression is at most $0.01n$. Again, we can choose v_i as needed. $\square$

Let $U = U_{k'_0}$ be a set provided by the claim above. We reorder the vertices $u'_1, u'_2, \ldots, u'_n$ of G_2 as follows: first we enumerate the vertices of U in any order, and then enumerate the vertices of $G_2 - U$ in a greedy order. We will denote $k_0 = |U|$.

Claim 3.3. $\varphi(G_2 - U) = \sum_{v \in V(G_2-U)} \frac{1}{1+d_{G_2-U}(v)} \geq \frac{n}{1+2c\sqrt{n}}$.

Proof. Let $H = G_2 - U$. Since φ is convex,

$$\varphi(H) \geq \frac{v(H)}{1 + \frac{2e(H)}{v(H)}}.$$

Recall that $v(H) = n - k_0 \geq n - \frac{k'_0(1-\gamma)}{1-\alpha}$ and $e(H) \leq cn\sqrt{n} - \frac{k'_0(1-\alpha)^2}{20(1-\gamma)}n$. Thus, to prove the claim, we will verify that

$$\frac{n - \frac{k'_0}{1-\alpha}}{1 + \frac{2(cn\sqrt{n} - \frac{k'_0(1-\alpha)^2}{20}n)}{n - \frac{k'_0}{1-\alpha}}} \geq \frac{n}{1+2c\sqrt{n}}. \quad (3.5)$$

Multiplying both parts of (3.5) by the product of the denominators, opening the parentheses in the left-hand side, and cancelling n in both parts, we get

$$2cn\sqrt{n} - \frac{k'_0}{1-\alpha} - \frac{2ck'_0\sqrt{n}}{1-\alpha} \geq \frac{2n^2}{n - \frac{k'_0}{1-\alpha}}\left(c\sqrt{n} - \frac{k'_0(1-\alpha)^2}{20}\right).$$

Multiplying both parts of the last inequality by $n - \frac{k'_0}{1-\alpha}$, cancelling $2cn^2\sqrt{n}$ in both parts and dividing the rest by $\frac{-k'_0}{1-\alpha}$ we obtain that (3.5) is equivalent to

$$2cn\sqrt{n} + \left(n - \frac{k'_0}{1-\alpha}\right)(1 + 2c\sqrt{n}) \leq 0.1n^2(1-\alpha)^3,$$

which is weaker than

$$1 + 4c\sqrt{n} \leqslant 0.1n(1-\alpha)^3. \tag{3.6}$$

By (3.1), (1.2), and (1.3), inequality (3.6) holds. □

The main difficulties of packing below are: (1) packing vertices of G_2 of very high degree; (2) packing cyclic components of G_1, (3) packing big components of G_1 that are trees, and (4) finishing the packing when there is not much freedom.

Our strategy will be the following.

Step 1: Map $V(T_1 \cup \cdots \cup T_{k_0'})$ onto U.
Step 2: Find some k_1, $k_0 \leqslant k_1 \leqslant (1-\gamma-\varepsilon/2)n + \frac{1}{1-\alpha}$ so that the maximum degree of $G_2(k_1+1)$ is moderate.
Step 3: Map the vertices of G_1^* into (the complement of) $G_2(k_1+1)$.
Step 4: Map the vertices of $T_t, T_{t-1}, \ldots, T_{1+\lfloor 3n(1-\alpha)/4 \rfloor}$ into some of the remaining free vertices of G_2.
Step 5: Complete the packing by arranging the vertices of the remaining tree-components of G_1 in the rest of G_2.

Step 1 will take care of difficulty (1), Steps 2 and 3 handle (2), and at Step 4 we overcome (3).

We can complete Step 1 by Claim 3.2. Note that G_2 with the enumeration $u_1', \ldots, u_n'$ satisfies condition (ii) of Lemma 2.2 and k_0 satisfies the restrictions in this lemma. Suppose that condition (i) fails for G_2 and k_0, i.e., that

$$e(G_2 - U) > e(G_2)\left(1 - \frac{2k_0(\alpha+\varepsilon)}{n(\alpha-\gamma+\varepsilon/2)}\right).$$

Then the number $\tilde{e}(U)$ of edges in G_2 incident with U is less than

$$cn^{3/2}\frac{2k_0(\alpha+\varepsilon)}{n(\alpha-\gamma+\varepsilon/2)} < c\sqrt{n}\frac{2k_0}{\alpha-\gamma+\varepsilon/2}.$$

On the other hand, by the definition of k_0', $\tilde{e}(U) > k_0' \frac{(1-\alpha)^2}{20}n$, and by Claims 3.1 and 3.2, $k_0' \geqslant k_0 \frac{1-\alpha}{1-\gamma}$. Thus if condition (i) fails for G_2 and k_0, then

$$k_0 \frac{1-\alpha}{1-\gamma}\frac{(1-\alpha)^2}{20}n < c\sqrt{n}\frac{2k_0(1-\gamma)}{\alpha-\gamma+\varepsilon/2}$$

and hence

$$\sqrt{n} < \frac{40c(1-\gamma)}{(1-\alpha)^3(\alpha-\gamma+\varepsilon/2)} < \frac{20((1-\alpha)+(\alpha-\gamma))}{(1-\alpha)^3(\alpha-\gamma+\varepsilon/2)}$$
$$< \frac{20}{(1-\alpha)^2(\alpha-\gamma+\varepsilon/2)} + \frac{20}{(1-\alpha)^3}$$
$$\leqslant \frac{20}{(1-\alpha)^2\varepsilon/2} + \frac{20}{(1-\alpha)^3} \leqslant \frac{60}{(1-\alpha)^2\varepsilon} < \frac{15}{\varepsilon^3}.$$

This contradicts (1.3).

Therefore, G_2 with the enumeration $u_1', \ldots, u_n'$ satisfies the conditions of Lemma 2.2. This lemma implies that there is an index $k_0 \leqslant k_1 \leqslant (1-\gamma-\varepsilon/2)n$ such that the maximal degree $D = \Delta(H)$ of the graph $H = G_2(k_1 + 1)$ satisfies

$$D \leqslant \frac{2c(n - k_1)(\alpha + \varepsilon)}{\sqrt{n}(\alpha - \gamma + \varepsilon/2)}. \tag{3.7}$$

This completes Step 2. Note that the right-hand side of (3.7) is at most $\frac{4c(\alpha+\varepsilon)}{\varepsilon}\sqrt{n}$ and hence (3.7) together with (3.2) yields

$$D \leqslant \frac{4c(\alpha + \varepsilon)}{\varepsilon}\sqrt{n} \leqslant \frac{\sqrt{n}}{2c\varepsilon} \leqslant \frac{3\sqrt{n}}{2\varepsilon}. \tag{3.8}$$

Also, by Theorem 2.1 and Claim 3.3, for

$$\ell = \left\lceil \frac{n}{2c\sqrt{n}+1} \right\rceil, \tag{3.9}$$

the set $L = \{u'_{n-\ell+1}, u'_{n-\ell+2}, \ldots, u'_n\}$ of the last ℓ vertices of G_2 forms an independent set in G_2.

Now, we identify the last ℓ vertices of G_1 with vertices in L. Since L is an independent set, this identification is 'legal' so far: no edge of G_1 is identified with an edge of G_2. If $w_{n-\ell}$ is not in G_1^*, then Step 3 is done, otherwise we continue as follows. We place the vertices $w_{n-\ell}, w_{n-\ell-1}, \ldots, w_{(1-\gamma)n+1}$ one by one into the rest of G_2, the 'middle' of G_2, namely $M = V(G_2(k_1 + 1)) - L$. We show now that all these vertices can be placed into M to give us a packing of G_1^* into (the complement of) G_2.

Suppose that we have placed the vertices $w_{n-\ell}, w_{n-\ell-1}, \ldots, w_{n-j+1}$ into M, and the next vertex to be placed, w_{n-j}, has x neighbours w_h with $h > n - j$. Since $w_1, w_2, \ldots, w_n$ is a degenerate order of the vertices of G_1, the subgraph $G_1(n-j)$ has minimal degree x. Furthermore, as G_1^* has γn vertices, we find that

$$jx + 2(\gamma n - j) \leqslant 2e(G_1) \leqslant 2\alpha n,$$

and so

$$x \leqslant 2 + 2(\alpha - \gamma)n/j. \tag{3.10}$$

By Claim 2.3, we have a legal placement for w_{n-j} provided that

$$n - k_1 - j - xD > 0. \tag{3.11}$$

Thus, to complete Step 3, it suffices to check that (3.11) holds.

Suppose that (3.11) is false. Then, by (3.10) and (3.7), we have

$$n - k_1 - j \leqslant D\left(2 + \frac{2(\alpha-\gamma)n}{j}\right) < 2D + \frac{2c(\alpha + \varepsilon)(n - k_1)2(\alpha - \gamma)n}{\sqrt{n}(\alpha - \gamma + 0.5\varepsilon)j}. \tag{3.12}$$

Add j to both parts of (3.12) and divide both parts by $n - k_1$. Taking into account (3.8) and the fact that $k_1 \leqslant n(1 - \gamma - \varepsilon/2)$, we get

$$1 < \frac{2D + j}{n - k_1} + \frac{4c(\alpha + \varepsilon)(\alpha - \gamma)n}{\sqrt{n}(\alpha - \gamma + 0.5\varepsilon)j} \leqslant \frac{3\sqrt{n}/\varepsilon + j}{n(\gamma + 0.5\varepsilon)} + \frac{4c(\alpha + \varepsilon)\sqrt{n}}{j}. \tag{3.13}$$

Consider the right-hand side of (3.13) as the function $f(j)$. This is a convex function of j (when other parameters are fixed). Since $\ell < j \leqslant \gamma n$, by (3.9), it is enough to check that

$f(j) \leq 1$ for $j = \frac{n}{2c\sqrt{n}+1}$ and $j = \gamma n$. Taking (1.1) into account, we get

$$f\left(\frac{n}{2c\sqrt{n}+1}\right) \leq \frac{3\sqrt{n}/\varepsilon + 2\sqrt{n}}{n(\gamma + 0.5\varepsilon)} + \frac{4c(\alpha+\varepsilon)\sqrt{n}(1+2c\sqrt{n})}{n}$$

$$\leq \frac{\frac{3}{\varepsilon}+2}{0.5\varepsilon\sqrt{n}} + \frac{4c(\alpha+\varepsilon)}{\sqrt{n}} + 8c^2(\alpha+\varepsilon).$$

By (3.2) and (1.3), the last expression is at most

$$\left(\frac{3}{\varepsilon}+2\right)(0.1\varepsilon)^2 + \frac{(0.1\varepsilon)^3}{2c} + 1 - 2\varepsilon < \frac{3\varepsilon}{100} + \frac{\varepsilon^2}{50} + \frac{2\varepsilon^3}{1000} + 1 - 2\varepsilon < 1.$$

Now,

$$f(\gamma n) = \frac{3}{\varepsilon\sqrt{n}(\gamma+0.5\varepsilon)} + \frac{\gamma}{\gamma+0.5\varepsilon} + \frac{4c(\alpha+\varepsilon)}{\gamma\sqrt{n}}.$$

If $\gamma \geq 0.1\varepsilon^2$ then, by (3.2) and (1.3), the last expression is at most

$$\frac{6}{\varepsilon^2\sqrt{n}} + \frac{1}{1+0.5\varepsilon} + \frac{5}{c\varepsilon^2\sqrt{n}} \leq \frac{6\varepsilon}{1000} + 1 - \frac{0.5\varepsilon}{1+0.5\varepsilon} + \frac{15\varepsilon}{1000} < 1.$$

Suppose that $\gamma < 0.1\varepsilon^2$. Since $\gamma n > \ell$, we obtain by (3.9) and (3.2) that

$$f(\gamma n) \leq \frac{6}{\varepsilon^2\sqrt{n}} + \frac{0.1\varepsilon^2}{0.1\varepsilon^2+0.5\varepsilon} + \frac{4c(\alpha+\varepsilon)\sqrt{n}(1+2c\sqrt{n})}{n}$$

$$\leq \frac{6}{\varepsilon^2\sqrt{n}} + \frac{\varepsilon}{5} + \frac{4c(\alpha+\varepsilon)}{\sqrt{n}} + 8c^2(\alpha+\varepsilon)$$

$$\leq \frac{6}{\varepsilon^2\sqrt{n}} + \frac{\varepsilon}{5} + \frac{1}{2c\sqrt{n}} + (1-2\varepsilon) \leq \frac{8\varepsilon}{1000} + 1 - 1.8\varepsilon < 1.$$

This finishes Step 3.

Let G'_2 denote the subgraph of G_2 induced by the vertices not used as the images of vertices in G_1^*, and in $T_1, \ldots, T_{k'_0}$. Then by (3.4) and Claim 3.1,

$$n'_2 = |V(G'_2)| \geq (1-\gamma)n - k_0 \geq (1-\gamma)\left(n - \frac{k'_0}{1-\alpha}\right)$$

$$\geq (1-\gamma)\left(n - \frac{0.01(1-\alpha)n}{1-\alpha}\right) \geq 0.99(1-\alpha)n. \quad (3.14)$$

By the definition of k'_0, the maximum degree D' of G'_2 is at most $\frac{(1-\alpha)^2}{20}n$. Since the subgraph G'_1 of G_1 induced by $V(T_t \cup T_{t-1} \cup \cdots \cup T_{1+\lfloor 3n(1-\alpha)/4 \rfloor})$ is 1-degenerate, we can apply Claim 2.3 with $x = 1$. The claim implies that we can complete Step 4 provided

$$n'_2 > D' + |V(G'_1)|. \quad (3.15)$$

Applying (3.4), we have

$$n'_2 - |V(G'_1)| \geq \sum_{i=k'_0+1}^{\lfloor 3n(1-\alpha)/4 \rfloor} v(T_i) \geq \frac{3(1-\alpha)}{4}n - k'_0 - 1 > 0.74(1-\alpha)n - 1 > \frac{2(1-\alpha)}{3}n.$$

Taking into account that $D' \leq \frac{(1-\alpha)^2}{20}n$, we get (3.15).

Remarks. (1) Any vertex in a tree could be made the last vertex in a degenerate order. In particular, we can make the last a vertex of maximum degree.

(2) Packing each tree, we can start from identifying a vertex of the highest degree in this tree with an available vertex of the smallest degree in G_2.

Finally, let G_2'' denote the subgraph of G_2 induced by the vertices not yet used as the images of vertices in G_1. Then, as in the previous paragraph,

$$n_2'' = |V(G_2'')| > \frac{2(1-\alpha)}{3}n$$

and $\Delta(G_2'') \leq D' \leq \frac{(1-\alpha)^2}{20}n$. Let $G_1'' = T_{k_0'+1} \cup T_{k_0'+2} \cup \cdots \cup T_{\lfloor 3n(1-\alpha)/4 \rfloor}$. By Claim 3.1(b), the maximum degree D_1 of G_1'' is less than $\frac{4}{1-\alpha}$. Therefore,

$$D_1 \cdot D' \leq \frac{4}{1-\alpha} \cdot \frac{(1-\alpha)^2}{20}n = \frac{(1-\alpha)}{5}n < \frac{n_2''}{2}.$$

Thus, by Theorem 1.1, G_1'' and G_2'' pack. This proves Theorem 1.5.

4. Packing many graphs

In this section, we use Theorem 1.5 to show that one can pack many graphs if each of these graphs has at most αn edges. First, we look again into the proof of Theorem 1.5.

Lemma 4.1. Let α, c, n and G_1 and G_2 satisfy the conditions of Theorem 1.5. Let $H = G_1 \cup G_2$ be the graph with $V(H) = V(G_1) = V(G_2)$, $E(H) = E(G_1) \cup E(G_2)$ obtained by packing G_1 and G_2 as described in the proof of Theorem 1.5. Then $\Delta(H) \leq \max\{\alpha n + 0.04(1-\alpha)n, \Delta(G_2) + 2/(1-\alpha)\}$.

Proof. Suppose that the lemma is false. Then there is a vertex v with $\deg_H(v) > \max\{\alpha n, \Delta(G_2)\} + 2/(1-\alpha)$. We may assume that v is the result of identifying $w_i \in V(G_1)$ with $u_j \in V(G_2)$.

Case 1: $\deg_{G_2}(u_j) > 0.5\alpha n + 2$. If $j > k_0$, then by (3.4) and the definition of k_0',

$$\deg_{G_2}(u_j) \leq k_0' + \frac{(1-\alpha)^2}{20}n < 0.01(1-\alpha)n + \frac{(1-\alpha)^2 n}{20} < 0.04(1-\alpha)n \leq 0.02n, \quad (4.1)$$

a contradiction. Therefore, $j \leq k_0$. Hence, $w_i \in V(T_1 \cup \cdots \cup T_{k_0'})$ and $\deg_{G_1}(w_i) \leq |V(T_{k_0'})| - 1$. By Claim 3.1(b), $|V(T_{k_0'})| \leq \frac{n}{n(1-\alpha)-k_0'+1}$. In view of (3.4),

$$n(1-\alpha) - k_0' + 1 > n(1-\alpha) - 0.01n(1-\alpha) = 0.99n(1-\alpha).$$

It follows that $\deg_{G_1}(w_i) < \frac{1}{0.99(1-\alpha)}$ and the lemma holds.

Case 2: $\deg_{G_1}(w_i) > 0.5\alpha n + 2$. Since $e(G_1) = \alpha n$, there is only one vertex in G_1 with this property. Furthermore, with such a large degree, w_i is either in $V(G_1^*)$, or in $V(T_t)$. In either case, $u_j \notin U$, and by (4.1), $\deg_{G_2}(u_j) \leq 0.04(1-\alpha)n$. This proves the lemma. □

Now we are ready to prove Theorem 1.8.

Proof. Recall that $m = \lceil 0.25\sqrt{n/\alpha^3}\rceil$. We will prove by induction on k, that for $k = 1,\ldots,m$, there is a packing of $H_1,\ldots,H_k$ such that the maximal degree, $\Delta(F_k)$, of the obtained graph $F_k = H_1 \cup \cdots \cup H_k$ is at most $(1 - 0.96(1-\alpha))n + 2(k-2)/(1-\alpha)$.

For $k = 1$, the statement reduces to $\Delta(H_1) \leqslant \alpha n + 0.04n - 2/(1-\alpha)$. By (1.4), $0.04n - 2/(1-\alpha) \geqslant 0$ which proves the base case.

Suppose that the theorem is proved for some $k \leqslant m-1$. Let us check that Theorem 1.5 and Lemma 4.1 hold for our α and n, $c = e(F_k)/n^{3/2}$, $\varepsilon = 0.25(1-\alpha)$, $G_1 = H_{k+1}$, and $G_2 = F_k$. Indeed, since $k \leqslant m-1$, we have

$$e(F_k) \leqslant k\alpha n \leqslant (m-1)\alpha n < \frac{\alpha n\sqrt{n}}{4\alpha^{3/2}}$$

and hence $c \leqslant 0.25/\sqrt{\alpha}$. Therefore, $8c^2\alpha \leqslant 1/2$, which yields (1.1) and (1.2). Now, (1.3) follows from (1.4). By the inductive assumption,

$$\Delta(G_2) \leqslant (1 - 0.96(1-\alpha))n + \frac{2(k-2)}{1-\alpha} \leqslant n - \frac{2}{1-\alpha} - \left(0.96(1-\alpha)n - \frac{2(m-2)}{1-\alpha}\right)$$

$$\leqslant n - 2 - \left(0.96(1-\alpha)n - \frac{2\sqrt{n}}{4(1-\alpha)\alpha^{1.5}}\right) \leqslant n - 2 - \left(0.96(1-\alpha)n - \frac{\sqrt{2n}}{1-\alpha}\right).$$

Observe that

$$0.96(1-\alpha)n > 0.96\sqrt{n}\frac{50^3}{(1-\alpha)^2} > \frac{100\sqrt{n}}{1-\alpha},$$

and hence

$$\Delta(G_2) \leqslant n - 2 - \frac{50\sqrt{n}}{1-\alpha}.$$

Thus, the conditions of Theorem 1.5 are satisfied, and by Lemma 4.1 we can pack H_{k+1} and F_k so that the maximum degree $\Delta(F_{k+1})$ of the resulting graph $F_{k+1} = F_k \cup H_{k+1}$ exceeds $(1 - 0.96(1-\alpha))n + 2(k-2)/(1-\alpha)$ by at most $2/(1-\alpha)$. This proves the theorem. □

5. Sparse graphs that do not pack

We will construct some series of pairs of sparse graphs that do not pack. We start from a simple series and then elaborate it.

Let $G_1 = G_1(n,2)$ be a forest on n vertices whose components are stars S_1 and S_2 of degree at most $\lceil \frac{n}{2}\rceil$. By s_1 and s_2 we denote the centres of these stars.

Let $W = \{w_1, w_2, w_3\}$ and U be a set disjoint from W with $|U| = n-3$ partitioned into subsets U_1, U_2, and U_3 of about the same cardinality. We define $G_2 = G_2(n,1,2)$ as follows. Let $V_2 = V(G_2) = W \cup U$ and $E_2 = \{w_iw_j \mid 1 \leqslant i < j \leqslant 3\} \cup \bigcup_{i=1}^{3}\{uw_i, uw_{i+1} \mid u \in U_i\}$ (we sum the indices modulo 3). The graph G_2 possesses the property that every two vertices have a common neighbour and the maximum degree of G_2 is $\lceil 2n/3\rceil$. Furthermore, G_2 is 2-degenerate, *i.e.*, very sparse.

Suppose that $G_1(n,2)$ and $G_2(n,1,2)$ pack, *i.e.*, that there is an edge-disjoint placement f of the vertex set V_1 of G_1 onto V_2. Let $t_1 = f(s_1)$ and $t_2 = f(s_2)$. By the previous paragraphs, t_1 and t_2 have a common neighbour, say, t_0, in G_2. Then the vertex s_0 in G_1 with $f(s_0) = t_0$ cannot be adjacent to any of s_1 and s_2. This contradicts the definition of G_1. Thus G_1 and G_2 do not pack.

Note that this example disproves Conjecture 1.6 and shows that to extend the statement of Theorem 1.5 even to $\alpha = 1$, one needs to impose sufficiently stricter conditions on the maximum degree of G_2. The maximum of maximum degrees of G_1 and G_2 is $\lceil 2n/3 \rceil$. Below, we elaborate the above example to make this maximum less by making greater the average degree of G_2.

Let $G_1 = G_1(n,k)$ be a forest on n vertices whose k components are stars $S_1, \ldots, S_k$ of degree at most $\lceil \frac{n}{k} \rceil$. By $s_1, \ldots, s_k$ we denote the centres of these stars.

Let q be a prime power. For a nonnegative integer d, let $q_d = \frac{q^{d+1}-1}{q-1}$. In particular, $q_0 = 1$ and $q_1 = q + 1$. Suppose that $n > q^{k+1}$. To construct $G_2 = G_2(n,q,k)$, consider a k-dimensional projective space W over the field GF_q. It has q_k points and q_k hyperplanes. Let U be a set of $n - q_k$ vertices partitioned into q_k sets $U_1, \ldots, U_{q_k}$ with $|U_i| \leq \lceil \frac{n}{q_k} \rceil - 1$ for all i. Let $\{H_1, \ldots, H_{q_k}\}$ be a list of all hyperplanes in W. The graph $G_2 = G_2(n,q,k)$ has the vertex set $V_2 = W \cup U$ and the edge set

$$E_2 = \{w_1 w_2 \mid w_1 \in H_1, w_2 \in W, w_1 \neq w_2\} \cup \bigcup_{i=1}^{q_k} \{wu \mid w \in H_i, u \in U_i\}.$$

Claim 5.1. *If $n > q^{k+1}$, then*

(a) *$G_2(n,q,k)$ is q_{k-1}-degenerate,*
(b) *$|E_2| < q_{k-1} n$,*
(c) *the maximum degree of $G_2(n,q,k)$ is at most $\frac{n}{q} + q_k$.*

Proof. Order the vertices of G_2 so that first we list the vertices in U, then the vertices in $W - H_1$, and finally the points of H_1. Then every vertex v has at most q_{k-1} neighbours following v in this order. This proves (a). Note that (a) yields (b).

To check (c), observe that every vertex in U has degree q_{k-1}. Every point of a k-dimensional projective space over GF_q is contained in q_{k-1} hyperplanes. Therefore, every $w \in W$ is adjacent to at most $q_{k-1}(\lceil \frac{n}{q_k} \rceil - 1) < \frac{n}{q}$ vertices in U. Since $|W| = q_k$, this proves (c). □

Claim 5.1 implies that for fixed q and k, $G_2(n,q,k)$ has linear in n number of edges. Furthermore, if $n > q \cdot q_k$, then the maximum degree of G_2 is less than $\frac{2n}{q}$. Thus, for every k and any prime power $q \geq 2k$, if $n > q \cdot q_k$, then both $G_1(n,k)$ and $G_2(n,q,k)$ have maximum degree at most n/k.

Claim 5.2. *If $n > q \cdot q_k$, then $G_1(n,k)$ and $G_2(n,q,k)$ do not pack.*

Proof. Suppose that there exists a packing of $G_1(n,q)$ and $G_2(n,q,k)$, i.e., that there is an edge-disjoint placement f of the vertex set V_1 of G_1 onto V_2. Let $t_j = f(s_j)$ for $j = 1, \ldots, k$. By the definition of G_2, the neighbourhood of every of t_j contains some $H_{i(j)}$ (if $t_j \in H_i$, then it contains many H_i). Suppose that the set $T = \{t_1, \ldots, t_k\}$ contains exactly r vertices of H_1. Since any $k-r$ hyperplanes of W have a common r-dimensional subspace, the neighbourhoods in G_2 of the remaining $k-r$ elements of T have at least q_r vertices in common. Since $q_r > r$ and vertices of H_1 are adjacent to every vertex in W, there exists a common neighbour $t_0 \in W$ of all vertices in T. But then the vertex $s_0 = f^{-1}(t_0)$ cannot be adjacent in G_1 to any of $s_1, \ldots, s_k$. This contradicts the definition of G_1. □

These two claims prove Theorem 1.7.

Acknowledgement

We are grateful to the referees for their helpful comments.

References

[1] Aigner, M. and Brandt, S. (1993) Embedding arbitrary graphs of maximum degree two. *J. London Math. Soc.* **48** 39–51.
[2] Alon, N. and Fischer, E. (1996) 2-factors in dense graphs. *Discrete Math.* **152** 13–23.
[3] Bollobás, B. (1978) *Extremal Graph Theory*, Academic Press, London/New York.
[4] Bollobás, B. and Eldridge, S. E. (1976) Maximal matchings in graphs with given maximal and minimal degrees. *Congressus Numerantium* **XV** 165–168.
[5] Bollobás, B. and Eldridge, S. E. (1978) Packing of graphs and applications to computational complexity. *J. Combin. Theory Ser. B* **25** 105–124.
[6] Brandt, S. (1995) An extremal result for subgraphs with few edges. *J. Combin. Theory Ser. B* **64** 288–299.
[7] Caro, Y. (1979) New results on the independence number. Technical Report, University of Tel Aviv, Israel.
[8] Caro, Y. and Tuza, Z. (1991) Improved lower bounds on k-independence. *J. Graph Theory* **15** 99–107.
[9] Catlin, P. A. (1974) Subgraphs of graphs I. *Discrete Math.* **10** 225–233.
[10] Catlin, P. A. (1976) Embedding subgraphs and coloring graphs under extremal degree conditions. PhD Thesis, Ohio State University, Columbus, OH.
[11] Csaba, B., Shokoufandeh, A. and Szemerédi, E. (2003) Proof of a conjecture of Bollobás and Eldridge for graphs of maximum degree three. *Combinatorica* **23** 35–72.
[12] Hajnal, A. and Szemerédi, E. (1970) Proof of a conjecture of Erdős. In *Combinatorial Theory and its Applications*, Vol. II (P. Erdős, A. Rényi and V. T. Sós, eds), North-Holland, pp. 601–603.
[13] Sauer, N. and Spencer, J. (1978) Edge disjoint placement of graphs. *J. Combin. Theory Ser. B* **25** 295–302.
[14] Wei, V. K. (1981) A lower bound on the stability number of a simple graph. Technical Memorandum TM 81-11217-9, Bell Laboratories.
[15] Wozniak, M. (1997) Packing of graphs. *Dissertationes Math.* **362** 1–78.
[16] Yap, H. P. (1988) Packing of graphs: A survey. *Discrete Math.* **72** 395–404.

Approximate Counting and Quantum Computation

M. BORDEWICH,[1†] M. FREEDMAN,[2] L. LOVÁSZ[2] and D. WELSH[1‡]

[1]Mathematical Institute, University of Oxford, 24-29 St. Giles', Oxford, OX1
(e-mail: magnusb@comp.leeds.ac.uk, dwelsh@maths.ox.ac.uk)

[2]Microsoft Research, One Microsoft Way, Redmond, WA 98052, USA
(e-mail: michaelf@microsoft.com, lovasz@microsoft.com)

Received 15 October 2003; revised 7 February 2005

For Béla Bollobás on his 60th birthday

Motivated by the result that an 'approximate' evaluation of the Jones polynomial of a braid at a 5th root of unity can be used to simulate the quantum part of any algorithm in the quantum complexity class BQP, and results relating BQP to the counting class GapP, we introduce a form of additive approximation which can be used to simulate a function in BQP. We show that all functions in the classes #P and GapP have such an approximation scheme under certain natural normalizations. However, we are unable to determine whether the particular functions we are motivated by, such as the above evaluation of the Jones polynomial, can be approximated in this way. We close with some open problems motivated by this work.

1. Introduction

The quantum complexity class BQP consists of those decision problems that can be computed with bounded error, using quantum resources, in polynomial time. Relative to the polynomial hierarchy of classical computation, it is known that

$$\text{BPP} \subseteq \text{BQP} \subseteq \text{PP} \subseteq \text{PSPACE},$$

and at the moment none of these inclusions is known to be proper [1]. Recent work by Freedman, Kitaev, Larson and Wang [5] has shown that the 'quantum part' of any quantum computation can be replaced by an approximate evaluation of the Jones polynomial of a related braid. A classical polynomial time algorithm can convert a quantum circuit for an instance of such a problem, into a braid, such that the probability that the output of the quantum computation is zero is a simple (polynomial time) function

---

[†] Research funded by the EPSRC and Vodafone, and supported in part by ESPRIT Project RAND-APX.
[‡] Research supported in part by ESPRIT Project RAND-APX.

of the Jones polynomial of the braid at a 5th root of unity. For an exact statement of this see Freedman, Kitaev, Larsen and Wang [5], or the more detailed papers by Freedman, Kitaev and Wang [6], and Freedman, Larsen and Wang [7, 8].

It therefore follows that if we take $A(L, x)$ to be an oracle that returns the evaluation of the Jones polynomial of a braid L at a point x, any BQP computation can be replicated by a classical polynomial time algorithm with one call to A, i.e., BQP $\subseteq$ P^A. Since computing the Jones polynomial is in general a #P-hard problem, this does not help. However, it is not an exact evaluation of the Jones polynomial that is required, but an approximate evaluation at a specific point for braids of a specific class. Hence we may look for a weaker oracle A' such that BQP $\subseteq$ P$^{A'}$.

In a different approach Fortnow and Rogers [4] link quantum complexity to the classical complexity class GapP. In particular they show that for any quantum Turing machine M running in time $t(n)$ there is a GapP function f such that for all inputs x

$$\mathbf{Pr}(M(x) \text{ accepts}) = \frac{f(x)}{5^{2t(|x|)}}.$$

Again evaluating a general GapP function exactly is #P-hard; however, one can simulate M using a polynomial algorithm with access to an oracle A'', where A'' is an oracle giving an approximation to the GapP function f.

With this motivation we examine the type of approximation needed in order to simulate a quantum computation, and then consider the complexity of such approximations. It turns out that an *additive approximation* is sufficiently powerful. We should emphasize that a polynomial time additive approximation scheme is weaker than the familiar and much studied *fully polynomial randomized approximation scheme* (FPRAS). However, it is well known that any function which counts objects for which the corresponding decision problem is NP-complete cannot have an FPRAS (unless NP = RP). We show below that *all* #P functions do have polynomial time additive approximation schemes under natural normalizations. We also show that in two senses this is the best sort of approximation we can hope to achieve in polynomial time (see Theorems 4.1, 4.3 and 4.4).

2. Quantum computing

A *link* L is a smooth submanifold of S^3, consisting of $c(L)$ disjoint simple closed curves. A *braid* on m strings is constructed as follows. Take m distinct points in a horizontal line $(p_1, p_2, \ldots, p_m)$ and link them to m distinct points $(q_1, q_2, \ldots, q_m)$ lying on a parallel line, by m disjoint simple arcs f_i in $\mathbb{R}^3$, so that f_i starts at p_i and ends at $q_{\pi(i)}$ where π is a permutation. A braid can be closed in numerous ways, by identifying the points p_i and q_j in some way, creating a link. Similarly any link can be represented as a braid. In particular, the *plat closure* of a braid on $2m$ strings is obtained by identifying the points p_{2i-1} and p_{2i}, and q_{2i-1} and q_{2i} for $1 \leqslant i \leqslant m$.

2.1. Topological computing and the Jones polynomial

One of the major difficulties in building a quantum computer has been the sensitivity of the system to outside interference. Freedman, Kitaev, Larson and Wang [5] introduced the notion of *topological quantum computing*, in an attempt to make the computations less sensitive to small disturbances. The basic idea is as follows. One can create pairs of special quasi-particles, called anyons, in a 2-dimensional plane sandwiched between two blocks of a superconductor. The anyons have a certain probability of annihilating each other (leaving a vacuum) when brought together. However, this probability changes, according to the laws of quantum mechanics when one anyon is moved around the other before they are brought together. Even if it is moved in a complete circle around the other, on reaching its original position the probability of annihilation is changed. Thus a system of a large number of these particles can be used as a quantum computer for decision problems; pairs of anyons are created, moved around relative to each other, and then a predefined pair of the anyons is brought together. If this pair annihilate each other leaving a vacuum, this is taken to be an output of 0 (or rejection); if they do not it is taken to be an output of 1 (or acceptance). The paths in the 2-dimensional surface, combined with a time dimension, give rise to a 3-dimensional representation of the 'computation' as a braid. There remain major difficulties in constructing such a quantum computer, and controlling the movement of anyons. However, one of the important results of Freedman, Kitaev, Larsen and Wang is that small changes in the paths of the anyons do not affect the outcome of the computation; indeed it is determined by the isotopy class of the braid, and therefore stable under perturbations of the paths that do not change the braiding itself.

The way in which the probability changes is sufficiently subtle that such a quantum computer is universal in the following sense. The Kitaev–Solovay theorem [12, 14] together with the density theorem of Freedman, Larsen and Wang [7, 8] yields an algorithm which given any quantum circuit on $m/2$ qubits and error parameter ϵ, outputs a braid on m strings using a polynomial number of crossings (polynomial in m and $\log \epsilon^{-1}$). The topological quantum computation using this braid efficiently simulates the quantum circuit, (the probability of acceptance is within ϵ of the correct value). Since an algorithm for a BQP problem can be used to generate a quantum circuit for a given instance, the above result gives an explicit method for finding an equivalent topological quantum computation, and so the class BQP is the same under either model.

Hence a quantum computation on $m/2$ qubits is approximately represented by a braid b. In showing that the topological quantum computation depends on the isotopy class of b alone [5], the following link L is considered. L is the plat closure of the composition of b^{-1}, b and a small loop γ inserted (between b^{-1} and b) around the leftmost two strings (see Figures 1 and 2). Both b and b^{-1} are needed as any quantum computation must be reversible; the loop γ effects a measurement of the qubit represented by the leftmost pair of strings. The conclusions of [5] may then be summarized as the following theorem: refer to [5] for full details.

Theorem 2.1. *Let π be a problem in BQP, with a polynomial time quantum algorithm $\mathscr{A}$, and let $\mathscr{I}$ be an instance of π. For any $\epsilon > 0$, a link L may be determined in time polynomial in $|\mathscr{I}|$ and $\log \epsilon$ such that*

$$\left| \mathbf{Pr}(\mathscr{A}(\mathscr{I}) = 0) - \frac{1}{1 + [2]_5^2} \left(1 + \frac{(-1)^{c(L)+w(L)}(-a)^{3w(L)} V_L(e^{2\pi i/5})}{[2]_5^{m(L)-2}} \right) \right| < \epsilon, \qquad (2.1)$$

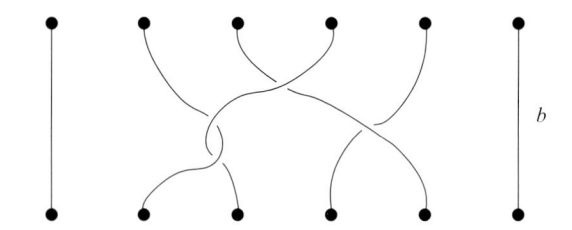

Figure 1. The braid b.

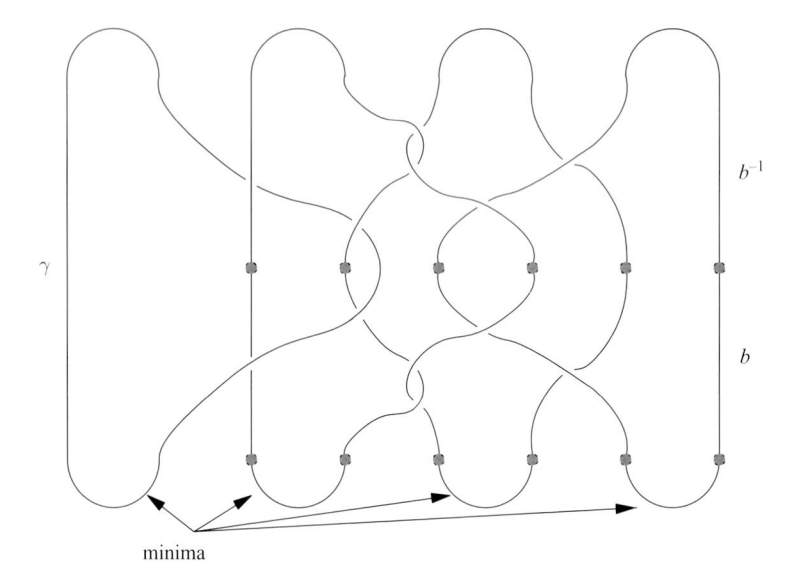

Figure 2. The link L.

where $a = e^{i\pi/10}$ and $[2]_5 = 2\cos \pi/5$ and $c(L), w(L)$ and $m(L)$ are the number of components, writhe and number of minima of the link L respectively.

The minima of [5] are the individual joins in the plat closure at the bottom of the braid, hence $m(L)$ is half the number of strings in b plus one. By construction, the number of strings in b is twice the number of (qu)bits in the input, hence $m(L) = |I| + 1$. The writhe is defined for an oriented link, and is the number of 'positively oriented' crossings minus the number of 'negatively oriented' crossings, with respect to the given orientation of L. It is easily computable. The Jones polynomial is also defined for oriented links; however, the formula above is independent of the orientation chosen for L. Since every crossing in b appears reversed in b^{-1}, these do not contribute to the writhe of L, hence the writhe is determined by the four crossings involving γ, and can only be $-4, 0,$ or 4 (depending on the orientation of the two strands passing through γ). If L^* denotes L with the orientation of one component reversed, then $V_{L^*}(t) = t^{-3\lambda/2} V_L(t)$, where λ is the contribution to the writhe of L from crossings of the reversed component over (or under) the rest of L. Hence only reversing γ or one of the leftmost two strings can affect the Jones evaluation, and it is easily checked that this is compensated by the change in the terms involving the writhe. To be consistent, we will retain the notation $[2]_5$ for $2\cos\pi/5$ from [5] throughout.

It is Theorem 2.1 that gives rise to our interest in approximating $V_L(t)$. Further explanation of the derivation of this equation is given in [18] where the special but sufficient case $w(L) = 0$ is considered (we could restrict attention to braids with writhe zero without affecting the main results). Although this formula involves an evaluation at $e^{2\pi i/5}$, similar results can be obtained for the nth root of unity for any $n \geqslant 5, n \neq 6$ but these involve multiple Ls.

The preceding paragraphs explain how an evaluation of a Jones polynomial can yield the answer to a (general) quantum computation. There is a weak converse to this. Suppose we have a quantum computer at our disposal with which to learn something about a Jones evaluation of a link L. We may assume without loss of generality (Freedman, Larsen and Wang [7]) that our quantum computer is of the topological kind and thus nicely adapted to braids. We can (easily) write L as the plat closure of a braid b by starting with the link diagram and pulling the overcrossings up and the undercrossings down. Let m be the number of strands of this braid b. If we wish to evaluate $V_L(\alpha)$, $\alpha = e^{2\pi i/r}$, we encounter an important constant $d = 2\cos\pi/r$. The norm $|V_L(\alpha)|$ is bounded from above by $d^{m/2}$ with $|V_L(\alpha)| = d^{m/2}$ achieved only when L is the unlink on $m/2$ components, a case which occurs when b is the identity braid. Our quantum computer will be able to provide an additive approximation (see below) of $|V_L(\alpha)|$ as a variable with range $[0, d^{m/2}]$.

Given m marked points in the horizontal plane and the number α, there is a finite dimensional Hilbert space H on which m-strand braids act through a Jones representation p. The $m/2$ maxima (in the plat closure) determine a vector c in this space and the $m/2$ minima (in the plat closure) determine a vector in the dual H^*, which when identified with H by the Hermitian inner product, is the same c.

We have

$$\frac{V_L(\alpha)}{d^{m/2}} = \langle c|p(b)|c\rangle.$$

Furthermore $\text{Prob}(|0\rangle) = |\langle c|p(b)|c\rangle|^2$, where $\text{Prob}(|0\rangle)$ refers to the physical probability that below the cups, after all the 'particles' have been fused in pairs, the vacuum $|0\rangle$ is observed, that is, no nontrivial particles result from these fusions. The last formula reflects the quantum mechanical rule that the probability of observing an outcome, in this case $|0\rangle$, is proportional to the square of the component of the state vector in the $|0\rangle$-direction.

Because the range of $|V_L(\alpha)|$ depends exponentially on the number of strings in the braid, $m(b)$, our quantum computer will give much better information (sooner) if we succeed in displaying L with, or nearly with, the minimal $m(b)$, called the *braid index* of L.

Turning to the computational question, it is a theorem of Thistlethwaite [15] that when L is an alternating link, with associated plane graph G, then

$$V_L(t) = \alpha T(G; -t, -t^{-1})$$

where α is an easily computable function, and T is the Tutte polynomial of the planar graph. It is known [17] that even for planar graphs, computing $T(G; x, y)$ is #P-hard, except when (x, y) is one of a few special points, or lies on a hyperbola satisfying $(x-1)(y-1) = q \in \{1, 2\}$.

Since $(-e^{2\pi i/n}, -e^{-2\pi i/n}), n \geq 5$, is not one of these 'easy' points, exact computation in polynomial time is not feasible (unless #P=P). It also seems unlikely that an FPRAS exists for these points. However, the notion of an FPRAS seems to be much stronger than the kind of approximation that is needed in the current context. For any BQP language L there is a quantum Turing machine M such that for all $x \in L$, M accepts with probability at least 3/4, and for all $x \notin L$, M accepts with probability at most 1/4. Therefore all that we require is to determine which quartile of its range $V_L(e^{2\pi i/5})$ lies in. We return to this topic in Section 5.

When considering algorithms on braids or links, the size of the input is taken to be the number of crossings. A quantum gate on $m/2$ qubits is converted into a braid on m strings of length polylog $(1/\epsilon)$, therefore the number of crossings in the braid associated with a BQP circuit is polynomially related to the number of gates in the BQP circuit. This in turn is bounded by a polynomial in the input size, hence an algorithm will either be polynomial with respect to both the number of crossings in the braid and the number of input qubits to the circuit, or neither.

2.2. GapP functions

The class of counting functions which constitute #P is the set of functions that count certificates of membership of a language belonging to NP, hence #P functions are constrained to evaluate to non-negative integers. The class of functions GapP can be regarded as the closure of #P under subtraction, that is to say a function $f : \mathscr{I} \mapsto \mathbb{Z}$ is in GapP if and only if there exist functions $g, h \in$ #P such that $f(I) = g(I) - h(I)$ for all $I \in \mathscr{I}$. The class AWPP can be defined as follows [3]. A language L is in AWPP if and only if there exist a polynomial p and a GapP function g such that for all $I \in \mathscr{I}$,

$$I \in L \Rightarrow \frac{3}{4} \leq \frac{g(I)}{2^{p(|I|)}} \leq 1,$$
$$I \notin L \Rightarrow 0 \leq \frac{g(I)}{2^{p(|I|)}} \leq \frac{1}{4}. \quad (2.2)$$

The increase in power of quantum computation over classical computation is that in a quantum computer there is an ability to cancel out computations paths. Fortnow and Rogers [4] show that this power is captured by the class GapP, in which a similar effect is seen. In particular they show that BQP ⊆ AWPP. It therefore follows that for a BQP language L, polynomial p and GapP function g satisfying (2.2), determining which quartile of the range $[0, 2^{p(|I|)}]$ contains $g(I)$ would be enough to determine membership of L.

To summarize, our foremost problems can be interpreted as finding a suitable approximation for the Jones polynomial of a link, $V_L(t)$, the Tutte polynomial of an associated planar graph, $T(G; x, y)$, at a particular point, or for the GapP functions arising from BQP languages.

3. Approximation

Given a function $\psi : \mathscr{I} \mapsto \mathbb{R}$ for which no efficient exact evaluation algorithm is known, one may be interested in an 'approximate' answer instead. A standard approach is to look for a fully polynomial randomized approximation scheme (FPRAS) for the problem. If ψ is such a function and $I \in \mathscr{I}$ is an input, then an FPRAS for ψ is a randomized algorithm that given any $I \in \mathscr{I}, \epsilon > 0$ will output $\hat{\psi}(I, \epsilon)$, such that

$$\mathbf{Pr}[|\hat{\psi}(I, \epsilon) - \psi(I)| > \epsilon \psi(I)] < 1/4,$$

and the running time is polynomial in $|I|$ and ϵ^{-1}.

Here one might be prompted to consider the following sort of approximation. Suppose we know a range in which the answer lies. Can we say where in that range the answer lies? Is it in the top or bottom half of the range, or in which quartile? We shall see that this approach is unlikely to be feasible, and in Section 3.1 we present an alternative. Clearly this type of approximation depends on the nature of the range. For the moment let us restrict our attention to the class of functions in #P. We will make the standard assumption that for a given NDTM M there exists a fixed polynomial p such that for any input x, all certificates have size $p(|x|)$ (so the total number of possible certificates of M is $2^{p(|x|)}$). We would like to answer the following problem, denoted by π_r: Given r, for which k is the number of accepting certificates for x between $\frac{(k-1)}{r} 2^{p(|x|)}$ and $\frac{k}{r} 2^{p(|x|)}$?

The problem π_2 is simply to determine which inputs have more than half of all certificates as accepting certificates. The set of languages in this class is exactly the set PP of probabilistic polynomial time languages. Furthermore, π_2 is clearly Turing reducible to π_{2s}, for any positive integer s, since if $\pi_{2s}(x) \leqslant s$ then $\pi_2(x) = 1$, otherwise $\pi_2(x) = 2$. Hence it is no surprise that this approach to approximation is NP-hard for #SAT, indeed the following lemma shows that any attempt to approximate #SAT in this way, or any problem with a parsimonious reduction to SAT, is unlikely to work. The proof is straightforward and we omit it; details may be found in [2].

Lemma 3.1. *For $k \in \mathbb{Z}$, deciding whether a CNF formula in n literals has more than 2^{n-k} solutions is NP-hard.*

When $k = 1$ the same decision for disjunctive normal form (DNF) formulae is equivalent to that for SAT, since the negation of a SAT formula is in DNF, and hence for an instance F, we have $\#\text{SAT}(F) = 2^n - \#\text{DNF}(\overline{F})$, where n is the number of literals. This observation leads to the following related lemma. Again, the proof is omitted and details may be found in [2].

Lemma 3.2. *For $k \in \mathbb{Z}$, deciding whether a DNF formula in n literals has at least 2^{n-k} solutions is NP-hard.*

This may seem more counterintuitive since not only is DNF in P, but also #DNF has an FPRAS [11]. On the other hand, the next lemma shows that the number of stable (independent) sets of vertices in a graph (#SS) can be approximated in this way, even though it is #P-complete and does not admit an FPRAS unless NP = RP. Essentially this is because the 'natural' upper bound on the number of stable sets, 2^n, is far too big unless the graph has very few edges. For details of the proof see [2].

Lemma 3.3. *Let G be a graph on n vertices. For $r \in \mathbb{Z}$, determining for which k, $\#SS(G) \in [\frac{(k-1)}{r}2^n, \frac{k}{r}2^n)$ is computable in time polynomial in n and r.*

Lemma 3.1 suggests that we cannot hope to fix a partition of the range and then determine in polynomial time in which section the answer lies; the difficulty associated with an NP-complete decision problem can be shifted to exactly the boundary between two parts of our partition. We therefore consider an alternative method of approximation which will meet our needs.

3.1. Additive approximation

Our approach to approximation consists of determining a small section of the range depending on the input, and in which we can say the answer lies with high probability. This gives rise to an additive approximation.

Definition 1 (Additive Approximation (AA)). Given any function $f : \mathscr{I} \mapsto \mathbb{C}$ and a normalization $u : \mathbb{Z}^+ \mapsto \mathbb{R}^+$, an additive approximation for (f, u) is a probabilistic algorithm which, given any $I \in \mathscr{I}, \epsilon > 0$, produces an output $\hat{f}(I)$ such that

$$\mathbf{Pr}[|f(I) - \hat{f}(I)| > \epsilon u(|I|)] < 1/4,$$

in time polynomial in $|I|$ and ϵ^{-1}.

Note that the 1/4 in the definition could be replaced by any $\delta \in (0, 1/2)$, since we could reduce this error probability in polynomial time by taking several runs of the algorithm. Note also that most of the time we shall be considering the case where f is real. In contrast to the set of functions admitting an FPRAS, which is closed under addition but not under subtraction (e.g., $\#DNF(f)$ has an FPRAS, but $\#SAT(f) = 2^n - \#DNF(\bar{f})$ does not), we have the following result, whose proof we leave to the reader.

Proposition 3.4. *Suppose (f, u) and (g, v) admit AA algorithms, then there exists AA algorithms for $(-f, u)$, $(f + g, u + v)$ and $(f - g, u + v)$. If, in addition, $|f(I)| \leqslant u(|I|)$ and $|g(I)| \leqslant v(|I|)$ for all I, then there is an AA algorithm for (fg, uv).*

The normalization is crucial. Since we are most interested in determining where in the range of possible values the answer lies, we shall usually be taking u to be an upper bound on $|f|$ depending only on input size. An additive approximation allows errors up to an absolute value of $\epsilon u(|I|)$, whereas an FPRAS allows only errors up to an absolute value of $\epsilon f(I)$. It is therefore a weaker notion of approximation, and it is easy to check that any function that admits an FPRAS also admits an AA algorithm under any upper bound.

Lemma 3.5. *Let $f : \mathscr{I} \mapsto \mathbb{R}$ be a function that admits an FPRAS, and let $u : \mathbb{Z}^+ \mapsto \mathbb{R}$ satisfy $|f(I)| \leqslant u(|I|)$ for all inputs $I \in \mathscr{I}$. Then (f, u) has an AA algorithm.*

Note also that a given function will have an AA with respect to some normalizations but not others. For example, we show later that for the number of proper 3-colourings of a connected graph G on n vertices, $P_G(3)$ where P_G is the chromatic polynomial of G, we have an AA for $(P_G(3), 2^n)$. However, for any constant $\delta > 0$, $(P_G(3), (2-\delta)^n)$ does not have an AA unless NP = RP (Theorem 4.4). In other words we can determine $P_G(3)$ to within an additive error $\epsilon 2^n$ in polynomial time, but we cannot approximate to within an additive error $\epsilon(2-\delta)^n$. Note that if $(f(I), u(|I|))$ has an AA, then for any fixed polynomial p, $(f(I), u(|I|)/p(|I|))$ also does, since we can absorb the polynomial factor in the normalization into ϵ at only a polynomial slowing of the algorithm.

It is the determination of the 'best' normalization for a given function that causes the greatest difficulties, particularly in relation to approximating $V_L(t)$. Nevertheless our first positive result shows that any function belonging to #P does have an AA algorithm under very natural normalizations.

4. Additive approximations for #P functions

The class of functions which constitute #P can be regarded as the set of functions that count certificates of membership of a language belonging to NP. For a given NP-language L there will be infinitely many NDTMs which check membership of L, and the certificates for a given input I will depend on the machine used in verification.

The main result of this section is that all such counting functions have additive approximation schemes under the 'natural normalization' associated with the corresponding NDTM. For example, if we take $f(G)$ to be the number of Hamiltonian circuits in a graph G, then two possible NDTMs for checking membership of L are M_1, which takes as certificates subsets of the edges, and checks that these form a cycle of length $|V|$, and M_2, which takes as certificates an ordering of the vertices $v_1, v_2, \ldots, v_n$ and checks that the edges between any two adjacent vertices in the ordering, and between v_n and v_1, do appear in the graph (to avoid double counting we must insist that relative to some fixed ordering of the vertices $\pi : V \mapsto 1, \ldots, n$, we have $\pi(v_1) = 1$ and $\pi(v_2) < \pi(v_n)$). In each case the number of good certificates for a given graph G is exactly the number of Hamiltonian circuits of G; however, M_1 has $2^{|E(G)|}$ possible certificates, while M_2 has $(|V|-1)!/2$ possible certificates. In either case the number of possible certificates is a natural upper bound on the number of Hamiltonian circuits. We show below that there is an additive approximation algorithm under the normalization associated with any such bound.

Theorem 4.1. *Let f be a function in the class #P, with an associated NDTM M, so that, for a given instance I, M has $f(I)$ accepting certificates, each of length $p(|I|)$. Then there exists an additive approximation algorithm for $(f, 2^{p(|I|)})$ that runs in time polynomial in $|I|$ and ϵ^{-1}.*

Proof. Given an instance I of f, we will select t computation paths, or certificates, uniformly at random from the $2^{p(|I|)}$ possible. We then run M using these inputs, and let $X_i, i = 1 \ldots t$ be indicator functions which take value 1 if and only if the ith computation path accepts I. The estimator for $f(I)$ is then $X = \frac{2^{p(|I|)}}{t} \sum_{i=1}^{t} X_i$.

Clearly $\mathbf{E}[X] = f(I)$. It remains to show that we can select t only polynomially large, such that the error bounds given in Definition 1 are satisfied. First note from Chebyshev's inequality that

$$\mathbf{Pr}\big[|X - f(I)| \geqslant \epsilon 2^{p(|I|)}\big] \leqslant \frac{\mathbf{Var}(X)}{\epsilon^2 2^{2p(|I|)}}$$

$$\leqslant \frac{1}{t} \frac{\frac{f(I)}{2^{p(|I|)}}\left(1 - \frac{f(I)}{2^{p(|I|)}}\right)}{\epsilon^2}$$

$$\leqslant \frac{1}{t\epsilon^2}.$$

Now if $t = 4\epsilon^{-2} + 1$, we have

$$\mathbf{Pr}\big[|X - f(I)| \geqslant \epsilon 2^{p(|I|)}\big] < 1/4. \qquad \square$$

Turning briefly to GapP functions, we get the following immediate corollary.

Theorem 4.2. *Let f be a GapP function such that $f = g - h$ where $g, h \in \#P$.*

(i) *Suppose that there are additive approximations for (g, u) and (h, v), then there is an additive approximation scheme for $(f, \max\{u, v\})$;*
(ii) *Suppose that $g(I)$ and $h(I)$ have certificates of length $p(|I|)$ for all I, then there is an additive approximation scheme for $(f, 2^p)$.*

Proof. From Proposition 3.4 we have that there is an additive approximation for $(f, u + v)$. From Definition 1 we can halve the permitted error for only a polynomial increase in running time, hence there is an AA for $(f, \frac{u+v}{2})$ and therefore also $(f, \max\{u, v\})$, which gives (i). When g and h have certificates of length p, by Theorem 4.1 there are AA schemes for $(g, 2^p)$ and $(h, 2^p)$, (ii) now follows from (i). $\qquad \square$

We have seen that all functions contained in #P have an AA algorithm relative to normalization by the size of the certificate space, and it is reasonable to ask if we could do better. However, we give two results that suggest this is already the best we can do in general. First we will see that sharpening our approximation to a logarithmic scale for the number of proper 5-colourings of a graph is NP-hard. Secondly, we show that the normalization by the number of possible certificates cannot be improved significantly in the case of the number of k-colourings of a graph.

Theorem 4.3. *Let $P_G(5)$ be the number of proper 5-colourings of a graph. Then for a general graph G on n vertices, there cannot be an additive approximation algorithm for $(\log(P_G(5) + 1), 3n)$ that runs in time polynomial in n and ϵ^{-1} unless $NP = RP$.*

Proof. Consider an NDTM for $P_G(5)$ that takes as certificates any 5-colouring of the graph, hence the certificates are of length $n\lceil \log 5 \rceil = 3n$, where n is the number of vertices in G. We show that an AA algorithm for $(\log(P_G(5) + 1), 3n)$ would be able to solve the NP-complete problem of determining whether a graph is 5-colourable.

Let G be a graph on n vertices and consider the following polynomial time transformation. We form G^+ by adding n isolated vertices to G. If G is not 5-colourable, then nor is G^+. However, if G is 5-colourable, each 5-colouring can be extended to 5^n 5-colourings of G^+. Therefore, if G is not 5-colourable $\log(P_{G^+}(5) + 1) = 0$, whereas if G is 5-colourable,

$$\log(P_{G^+}(5) + 1) \geqslant \log(5^n.5! + 1)$$
$$> 2n.$$

Hence an additive approximation algorithm for $(\log(P_{G^+}(5) + 1), 6n)$ could determine whether or not G is 5-colourable in random polynomial time. □

Theorem 4.1 shows that for connected graphs there exists an AA algorithm for $(P_G(k), (k-1)^n)$ as follows. We can take an arbitrary spanning tree on G, and take the set of certificates to define colourings relative to this spanning tree, giving $k(k-1)^{n-1}$ possible certificates. We can then adjust the normalization by a constant factor $(k-1)/k$. We show that this cannot be improved, in the sense that the normalization (and therefore the error) cannot be reduced by any exponential factor. We have already noted that the normalization can be improved by any fixed polynomial factor.

Theorem 4.4. *If $NP \neq RP$ then for any fixed $k \geqslant 3, \delta > 0$ there cannot be a polynomial time AA algorithm for $(P_G(k), \phi(n))$ for connected graphs G on n vertices, for any function $\phi(n)$ of order $O((k-1-\delta)^n)$.*

Proof. Let $\phi(n) \leqslant c(k-1-\delta)^n$ for sufficiently large n. Take r such that $(k-1-\delta) \leqslant (k-1)^{1-1/r}$. Given any graph G, we form a graph H by attaching a path of length $n(r-1)$ to a vertex of G. Now

$$P_G(k)(k-1)^{n(r-1)} = P_H(k), \tag{4.1}$$

$$P_G(k) = \frac{P_H(k)}{\left((k-1)^{1-1/r}\right)^{nr}}, \tag{4.2}$$

$$P_G(k) = \frac{cP_H(k)}{\phi(nr)} \frac{\phi(nr)}{c\left((k-1)^{1-1/r}\right)^{nr}}. \tag{4.3}$$

Now suppose that there is an AA algorithm for $(P_H(k), \phi(|H|))$, then we can get an approximation $\hat{P}_H(k)$ within an additive error of $\frac{1}{2c}\phi(nr)$. Using $\hat{P}_H(k)$ and equation (4.2), we obtain an approximation $\hat{P}_G(k)$. By equation (4.3), $\hat{P}_G(k)$ is within an additive error of $1/2$, since

$$\frac{\phi(nr)}{c\left((k-1)^{1-1/r}\right)^{nr}} \leqslant \frac{\phi(nr)}{c(k-1-\delta)^{nr}} \leqslant 1.$$

Since $P_G(k)$ is integral, we can therefore determine it exactly. □

5. Approximating $V_L(t)$ and related quantities

We have seen in Section 2.1 that our primary problem is to decide whether or not there exists an additive approximation scheme for $(V_L(e^{2\pi i/5}), [2]_5^{m/2})$ where L is the plat closure of a braid on m strings, indeed an additive approximation for the absolute value of the Jones polynomial suffices. We make this precise in the following theorem.

Theorem 5.1. *Let A be a oracle which takes as input a braid b on m strings and $\epsilon > 0$, and returns an additive approximation for $(|V_L(e^{2\pi i/5})|, [2]_5^{m/2})$, where L is the plat closure of b. Then* $\mathrm{BQP} = \mathrm{P}^A$.

Proof. Recall from Section 2.1 that given a braid on m strings, a topological quantum computer can be constructed such that the probability of output zero is $\frac{|V_L(e^{2\pi i/5})|^2}{[2]_5^m}$ and the computer runs in time polynomial in m. Given $\epsilon > 0$, using independent runs of this computer and a standard sampling approach, the probability of zero can be estimated to within an error of ϵ^2, where the number of runs is polynomial in ϵ^{-1}. Hence we may estimate $|V_L(e^{2\pi i/5})|$ to within an absolute error of $\epsilon [2]_5^{m/2}$ in polynomial time. Therefore $\mathrm{P}^A \subseteq \mathrm{BQP}$.

Secondly, suppose we have a BQP language and an input x. By Theorem 2.1 we can determine a link L, of size polynomial in $|x|$, such that L satisfies equation (2.1), and the number of minima of L is $|x|+1$. If x is in the language, $\mathbf{Pr}(0) < 1/4$, hence

$$0 \leqslant \frac{1}{1+[2]_5^2}\left(1 + \frac{(-1)^{c(L)+w(L)}(-a)^{3w(L)}V_L(e^{2\pi i/5})}{[2]_5^{|x|-1}}\right) < 1/4,$$

$$-[2]_5^{|x|+1}[2]_5^{-2} \leqslant (-1)^{c(L)+w(L)}(-a)^{3w(L)}V_L(e^{2\pi i/5}) < [2]_5^{|x|+1}[2]_5^{-2}\left(\frac{1+[2]_5^2}{4}-1\right),$$

$$\left|V_L(e^{2\pi i/5})\right| < [2]_5^{|x|+1} 0.39. \tag{5.1}$$

Whereas, if x is not in the language, $\mathbf{Pr}(0) > 3/4$, hence

$$\frac{1}{1+[2]_5^2}\left(1 + \frac{(-1)^{c(L)+w(L)}(-a)^{3w(L)}V_L(e^{2\pi i/5})}{[2]_5^{|x|-1}}\right) > 3/4,$$

$$(-1)^{c(L)+w(L)}(-a)^{3w(L)}V_L(e^{2\pi i/5}) > [2]_5^{|x|+1}[2]_5^{-2}\left(\frac{3(1+[2]_5^2)}{4}-1\right),$$

$$\left|V_L(e^{2\pi i/5})\right| > [2]_5^{|x|+1} 0.65. \tag{5.2}$$

Clearly use of an oracle giving an additive approximation for $(|V_L(e^{2\pi i/5})|, [2]_5^{|x|+1})$ will enable us to distinguish these two cases with probability at least $3/4$. Hence $\mathrm{BQP} \subseteq \mathrm{P}^A$. □

We saw in Section 2.1 the equivalence of the Jones polynomial and a specialization of the Tutte polynomial for alternating links, hence we would like an AA for a general planar graph G for

$$\left(T\left(G; -e^{2\pi i/5}, -e^{-2\pi i/5}\right), u\right),$$

where u is some reasonable upper bound.

Hyperbolae of the form $H_q := (x-1)(y-1) = q$ play a crucial role in the manipulation of the Tutte polynomial; loosely speaking, the process of performing a *tensor product* on an input graph G with some other fixed graph N enables us to 'move around' the Tutte plane. That is, the new graph $G \otimes N$ satisfies

$$T(G \otimes N; x, y) = f(N; x, y) T(G; X, Y), \tag{5.3}$$

where f and the arguments X, Y can be computed in time polynomial in $|x|, |y|$ and $|N|$. However, for any choice of N, the new points X, Y satisfy

$$(X-1)(Y-1) = (x-1)(y-1) = q. \tag{5.4}$$

Thus, such transformations restrict us to remain on the initial hyperbola H_q; see [10] for further details. The close relationship between points on the hyperbola enables us to use an additive approximation at one point to get an additive approximation at any other point (X, Y) on the same hyperbola for which there exists a suitable planar N which transforms (x, y) to (X, Y) by (5.3).

Proposition 5.2. *Let $x, y \in \mathbb{Q}$ and N a planar graph on k vertices be fixed. Suppose there is an AA scheme for $(T(G; x, y), u(n))$ for any planar G on n vertices and m edges. Then there is also an AA for*

$$(T(G; X, Y), u(n + m(k-2))),$$

where X and Y are the points determined by the transformation in (5.3) (depending only on x, y and N). □

Proof. Let X and Y be the points satisfying (5.3). Since $G \otimes N$ is planar (see the construction in [10]) we may use the AA scheme for

$$(T(G \otimes N; x, y), u(|V(G \otimes N)|))$$

to get an approximation to within an error $\epsilon u(|V(G \otimes N)|)$ with probability at least $3/4$ in polynomial time. Note that $|V(G \otimes N)| = n + m(k-2)$. Since the running time of the AA scheme is polynomial in ϵ^{-1}, and $f(N; x, y)$ is a constant, we can approximate to within an error of $\epsilon |f(N; x, y)| u(n + m(k-2))$ and still run in polynomial time. By (5.3) we have

$$T(G; X, Y) = \frac{T(G \otimes N; x, y)}{f(N; x, y)}.$$

Hence the AA for $T(G \otimes N; x, y)$ yields an AA for $T(G; X, Y)$ with error at most $\epsilon u(n + m(k-2))$ in polynomial time. □

Because of the important role of these hyperbolae, it is natural to look at the hyperbolae containing the roots of unity $(-e^{\frac{2\pi i}{n}}, -e^{-\frac{2\pi i}{n}})$. These are H_{q_n}, where $q_n = 2 + 2\cos(2\pi/n)$, which cut the x-axis at

$$x = -1 - 2\cos(2\pi/n), \tag{5.5}$$

corresponding to an evaluation of the chromatic polynomial at one of the well-known Beraha numbers $B_n = 2 + 2\cos(2\pi/n)$. Since for real x and y and any graph N, the related points X and Y will also be real, we cannot find an N such that we can directly relate $T(G \otimes N; 1 - B_5, 0)$ and $T(G; -e^{\frac{2\pi i}{5}}, -e^{-\frac{2\pi i}{5}})$. Whether or not we can find a point within absolute value ϵ of $(1 - B_5, 0)$ that can be directly related to $(-e^{\frac{2\pi i}{5}}, -e^{-\frac{2\pi i}{5}})$ is an interesting ongoing question. We present some positive results below, and return to these difficulties in Section 7.

First note that by Theorems 4.1 and 4.2 we know that $T(G; x, y)$, $x, y \in \mathbb{Z}$ will have an AA scheme with respect to an appropriate normalization, since evaluations at these points are GapP functions. However, the drawback is that often the naive normalization will be too large. This will not always be the case, for example the point $(1 - \lambda, 0), \lambda \in \mathbb{Z}$, gives the number of proper λ colourings, and by Theorems 4.3 and 4.4 here we have a best possible normalization.

When we consider the non-integer points, the situation is more complicated. A straightforward sampling approach gives the following result.

Proposition 5.3. *For rational (x, y) and a connected graph G, there exists a AA algorithm for the following:*

(i) $(T(G; x, y), y^{|E|}(y - 1)^{-|V|+1})$ when $\{x = 1, y > 1\}$,
(ii) $(T(G; x, y), x^{|E|}(x - 1)^{|V|-|E|-1})$ when $\{x > 1, y = 1\}$,
(iii) $(T(G; x, y), y^{|E|}(x - 1)^{|V|-1})$ when $\{x > 1, y > 1\}$.

In other regions, in particular where there are negative terms in the expansion of T, cancellation between terms means that there is no longer a natural upper bound by which to normalize. We return to this problem in Section 7.

6. An alternative approach

Returning to our original motivation, at the moment we are unable to determine whether there is an AA algorithm for $(V_L(e^{2\pi i/5}), [2]_5^{m/2})$ where L is the plat closure of a braid on m strings. However, we now show that in order to simulate a quantum computation it would be sufficient to determine the sign of the real part of $V_L(e^{2\pi i/5})$. Particularly in the case that the writhe of L is zero, and hence $V_L(e^{2\pi i/5})$ is real, this seems an easier problem.

Theorem 6.1. *Let $A(L)$ be an oracle that returns the sign of the real part of the Jones polynomial of the link L evaluated at $e^{2\pi i/5}$. Then $\text{BQP} \subseteq \text{P}^A$.*

Proof. This proof follows that of Theorem 5.1. Suppose we have a BQP language and an input x. By Theorem 2.1 we can determine a link L, of size polynomial in $|x|$, such that L satisfies equation (2.1), and the number of minima of L is $|x| + 1$. We now assume that $w(L) = 0$; the proof in the cases $w(L) = 4$ and $w(L) = -4$ follow by a similar argument. Simplifying equation (2.1) in the case $w(L) = 0$, we have

$$\mathbf{Pr}(0) = \frac{1}{1 + [2]_5^2} \left(1 + \frac{(-1)^{c(L)} V_L(e^{2\pi i/5})}{[2]_5^{m(L)-2}} \right). \tag{6.1}$$

If x is in the language, $\mathbf{Pr}(0) < 1/4$, hence

$$\frac{1}{1+[2]_5^2}\left(1 + \frac{(-1)^{c(L)}V_L\left(e^{2\pi i/5}\right)}{[2]_5^{|x|-1}}\right) < 1/4,$$

$$(-1)^{c(L)}V_L\left(e^{2\pi i/5}\right) < [2]_5^{|x|-1}\left(\frac{1+[2]_5^2}{4} - 1\right),$$

$$(-1)^{c(L)}V_L\left(e^{2\pi i/5}\right) < -[2]_5^{|x|-1}0.09 < 0. \tag{6.2}$$

Whereas, if x is not in the language, $\mathbf{Pr}(0) > 3/4$, hence

$$\frac{1}{1+[2]_5^2}\left(1 + \frac{(-1)^{c(L)}V_L\left(e^{2\pi i/5}\right)}{[2]_5^{|x|-1}}\right) > 3/4,$$

$$(-1)^{c(L)}V_L\left(e^{2\pi i/5}\right) > [2]_5^{|x|-1}\left(\frac{3(1+[2]_5^2)}{4} - 1\right),$$

$$(-1)^{c(L)}V_L\left(e^{2\pi i/5}\right) > [2]_5^{|x|-1}1.71 > 0. \tag{6.3}$$

Clearly use of an oracle giving an additive approximation for the sign of $V_L(e^{2\pi i/5})$ will enable us to distinguish these two cases with probability at least $3/4$. Hence BQP $\subseteq$ P^A. □

In the previous section we outlined the importance of the hyperbolae $H_q := (x-1)(y-1) = q$ to the Tutte polynomial. For x, y, X, Y and N related as in equation (5.3), we can determine the sign of $T(G; X, Y)$ if we can determine the sign of $T(G \otimes N; x, y)$.

This gives rise to the natural question of the complexity of determining whether a function is greater than or less than zero, in particular the Tutte polynomial, of which the Jones is a specialization. It is immediate from the definitions that the Tutte is non-negative in the region $x, y \geqslant 0$. At all other integer points on the axes the Tutte polynomial counts either colourings or flows, up to easy multiplicative factors. Since these factors may be positive or negative, we can always select one of either '$T(G; x, y)$ is non-negative' or '$T(G; x, y)$ is non-positive' that is true, in polynomial time. In the above situation we are not concerned with cases in which the value is exactly zero, hence this would suffice. We consider the situation at other points in the next section.

7. Some combinatorial and complexity questions

We close with the following questions which have been prompted by this work.

In Section 5 we noted that we are unable to find a suitable normalization for approximating the Tutte polynomial when the expansion included negative terms. We return to this here and examine the chromatic polynomial to highlight the difficulties. We have seen that for a connected graph G, we have an additive approximation for $(P_G(\lambda), (\lambda-1)^n))$ for all $\lambda \in \mathbb{Z}^+$. However, we are most interested in an additive approximation at the non-integral Beraha numbers. One might hope to achieve the above approximation for all $\lambda \in \mathbb{R}^{>1}$; however, this seems unlikely as $(\lambda-1)^n$ is not even close to being an upper bound for $P_G(\lambda)$. Indeed, consider the complete graphs: for small δ,

$$P_{K_n}(1+\delta) = (1+\delta)(\delta)(-1+\delta)\cdots(-n+\delta+1) \tag{7.1}$$

$$\approx (-1)^{n-2}\delta(n-2)!. \tag{7.2}$$

This prompts the first open question.

Question 1. *What is the best upper bound depending on λ, n and m for $|P_G(\lambda)|$ for all (planar) graphs G on n vertices and m edges?*

As far as we are aware the best upper bound known [19] is

$$|P_G(\lambda)| \leqslant |\lambda|^{n-m}(|\lambda|+1)^m \quad \lambda \in \mathbb{C}.$$

For general graphs we can make the following small improvement.

Proposition 7.1. *Let G be a graph and let $\lambda \in \mathbb{C}$, then*

$$|P_G(\lambda)| \leqslant \left(\frac{m}{n}-1\right)^{n-m}\left(\frac{m}{n}\right)^m \quad \text{for } \frac{m}{n} \geqslant |\lambda|+1,$$

$$|P_G(\lambda)| \leqslant (|\lambda|)^{n-m}(|\lambda|+1)^m \quad \text{for } \frac{m}{n} < |\lambda|+1.$$

If G is a connected graph then

$$|P_G(\lambda)| \leqslant \left(\frac{m}{n-1}-1\right)^{n-m-1}\left(\frac{m}{n-1}\right)^m|\lambda| \quad \text{for } \frac{m}{n-1} \geqslant |\lambda-1|,$$

$$|P_G(\lambda)| \leqslant (|\lambda-1|-1)^{n-m-1}(|\lambda-1|)^m|\lambda| \quad \text{for } \frac{m}{n-1} < |\lambda-1|.$$

These bounds hold for all $\lambda \in \mathbb{C}$; however, the Beraha numbers have special characteristics. The evaluations of the chromatic polynomial at these points have some beautiful, but not totally understood, properties [16]. The values begin $4, 0, 1, 2, 1+\tau, 3, \ldots$ and converge towards 4, where τ is the golden ratio $\frac{1+\sqrt{5}}{2}$. The integers in this series are clearly central to the theory of chromatic polynomials. Writing $B_5 = 1+\tau$, then for any plane triangulation T on n vertices:

$$|P_T(B_5)| \leqslant \tau^{5-n}. \tag{7.3}$$

For a connected graph G with average degree at least 3.24 (note that a planar triangulation has average degree $6 - 12/n$), the above proposition gives

$$|P_G(B_5)| \leqslant \left(\frac{m}{n-1}-1\right)^{n-m-1}\left(\frac{m}{n-1}\right)^m B_5.$$

Hence we ask the following.

Question 2. *Is there a better bound for $|P_G(B_n)|$ than there is for an evaluation at a general point?*

Following the results of Section 6 we are also prompted to examine the complexity of determining whether the Tutte polynomial is greater than or equal to, or less than zero at a given point. Recall that this decision problem is trivial for $x, y \geq 0$, and for integer points on the axes. Again considering the specialization to the chromatic polynomial we ask the following.

Question 3. *For fixed $\lambda \in \mathbb{Q}$, is it NP-hard to decide whether $P_G(\lambda)$ is greater than or equal to, or less than zero?*

Note that this is trivial for $\lambda \in \mathbb{Z}$. It is also P-time decidable for $\lambda < 32/27$ by the following theorem of Woodall [19] and Jackson [9].

Theorem 7.2. *Let G be a graph without loops on n vertices, κ components and b blocks.*
 (i) *If $\lambda < 0$, then $P_G(\lambda)$ is nonzero with the sign of $(-1)^n$.*
 (ii) *If $0 < \lambda < 1$, then $P_G(\lambda)$ is nonzero with the sign of $(-1)^{n-\kappa}$.*
 (iii) *If $1 < \lambda < \frac{32}{27}$, then $P_G(\lambda)$ is nonzero with the sign of $(-1)^{n-\kappa-b}$.*

Note that $P_G(\lambda) \neq 0$ for $\lambda \in \mathbb{Q} \backslash \mathbb{Z}$, since the chromatic polynomial has integer coefficients. It is easy to show the following.
- Let $\lambda \in \mathbb{Q} \backslash \mathbb{Z}$. If deciding whether $P_G(\lambda) > 0$ is NP-hard, then it is also NP-hard to decide whether $P_G(\lambda + 1) > 0$ for a general graph G.

However, since it is easy to decide for $\lambda < 32/27$, the converse cannot be true for all $\lambda \in \mathbb{Q} \backslash \mathbb{Z}$ unless these questions are all in P. It would be interesting to know the answer to the following questions.

Question 4. *Does there exist a critical $\alpha > 0$ such that deciding whether $P_G(\lambda)$ is greater than or less than zero is NP-hard for all rational $\lambda > \alpha$, $\lambda \notin \mathbb{Z}$?*

Question 5. *Is this critical α equal to $32/27$?*

As before we are more interested in evaluating the chromatic polynomial at the Beraha points than at general non-integers, and the graphs we are most interested in are planar. Hence we ask the following specific question.

Question 6. *For planar graphs, is the problem of deciding whether $P_G(B_n)$ is greater or less than zero NP-hard?*

For any graph G, not necessarily planar, it is known that $P_G(B_n) \neq 0$ for $n \geq 5, n \neq 6, 10$, [13]. Also Tutte [16] has shown that for any planar triangulation the following equation holds, writing $B_{10} = \tau\sqrt{5}$,

$$P_T(B_{10}) = \sqrt{5}\tau^{3(n-3)}(P_T(B_5))^2. \tag{7.4}$$

So $P_T(B_{10}) > 0$ for all plane triangulations T, indeed a simple reverse induction shows that for any planar graph G, $P_G(B_{10}) > 0$ holds. Further, for any outerplanar graph G, we can form the planar graph G^+ by adding a new vertex adjacent to all original vertices. Since

$$P_{G^+}(\lambda + 1) = (\lambda + 1)P_G(\lambda)$$

holds for all positive integers, it holds for all $\lambda \in \mathbb{R}$. Noting that $B_{10} = B_5 + 1$, we conclude that $P_G(B_5) > 0$ for all outerplanar graphs G.

Acknowledgements

The authors would like to thank the referee for many helpful comments and suggestions, in particular for suggesting a simpler proof of Theorem 4.3. We also thank Graham Brightwell and Colin McDiarmid for comments on an earlier version of this work [2].

References

[1] Adleman, L., DeMarris, J. and Huang, M. (1997) Quantum computability. *SIAM J. Computing* **26** 1524–1540.
[2] Bordewich, M. (2003) The complexity of counting and randomised approximation. PhD thesis, New College, Oxford University.
[3] Fenner, S. (2003) PP-lowness and a simple definition of AWPP. *Theory Comput. Syst.* **36** 199–212.
[4] Fortnow, L. and Rogers, J. (1999) Complexity limitations on quantum computation. *J. Comput. Syst. Sci.* **59** 240–252.
[5] Freedman, M., Kitaev, A., Larsen, M. and Wang, Z. (2003) Topological quantum computation. *Bull. Amer. Math. Soc.* **40** 31–38.
[6] Freedman, M., Kitaev, A. and Wang, Z. (2002) Simulation of topological field theories by quantum computers. *Commun. Math. Phys.* **227** 587–603.
[7] Freedman, M., Larsen, M. and Wang, Z. (2002) Density representations of braid groups and distribution of values of Jones invariants. *Commun. Math. Phys.* **228** 177–199.
[8] Freedman, M., Larsen, M. and Wang, Z. (2002) A modular functor which is universal for quantum computation. *Commun. Math. Phys.* **227** 605–622.
[9] Jackson, B. (1993) A zero-free interval for chromatic polynomials of graphs. *Combin. Probab. Comput.* **2** 325–336.
[10] Jaeger, F., Vertigan, D. and Welsh, D. J. A. (1990) On the computational complexity of the Jones and Tutte polynomials. *Math. Proc. Camb. Phil. Soc.* **108** 35–53.
[11] Karp, R., Luby, M. and Madras, N. (1989) Monte Carlo approximation algorithms for enumeration problems. *J. Algorithms* **10** 429–448.
[12] Kitaev, A. (2002) Quantum computations: algorithms and error correction. *Russian Math. Survey* **52:61** 1191–1249.
[13] Salas, J. and Sokal, A. (2000) Transfer matrices and partition-function zeros for antiferromagnetic Potts models. *J. Statist. Phys.* **98** 551–588.
[14] Solovay, R. Private communications.
[15] Thistlethwaite, M. B. (1987) A spanning tree expansion of the Jones polynomial. *Topology* **26** 297–309.
[16] Tutte, W. T. (1970) On chromatic polynomials and the golden ratio. *J. Combin. Theory Ser. B* **9** 289–296.
[17] Vertigan, D. and Welsh, D. J. A. (1992) The computational complexity of the Tutte plane: the bipartite case. *Combin. Probab. Comput.* **1** 181–187.
[18] Wang, Z. (2003) Addendum: Derivation of the formula in [FKLW]. www.tqc.iu.edu.
[19] Woodall, D. R. (1978) Zeros of chromatic polynomials. In *Combinatorial Surveys: Proceedings of the Sixth British Combinatorial Conference* (P. J. Cameron, ed.), Academic Press, London, pp. 199–223.

Absence of Zeros for the Chromatic Polynomial on Bounded Degree Graphs

CHRISTIAN BORGS

[1]Microsoft Research, One Microsoft Way, Redmond WA 98052, USA
(e-mail: borgs@microsoft.com)

Received 7 February 2005; revised 22 February 2005

For Béla Bollobás on his 60th birthday

In this paper, I give a short proof of a recent result by Sokal, showing that all zeros of the chromatic polynomial $P_G(q)$ of a finite graph G of maximal degree D lie in the disk $|q| < KD$, where K is a constant that is strictly smaller than 8.

1. Introduction

This paper is dedicated to Béla Bollobás on the occasion of his 60th birthday. Béla is a dear friend and wonderful collaborator. The work presented here concerns a new proof of absence of zeros of the chromatic polynomial on a finite graph of bounded degree, which grew out of one of ten CBMS-lectures I gave in Memphis in the early summer of 2002. Béla was the main organizer, and, as always, a great host. I also would like to thank him for encouraging me to write up this work.

For a finite graph G, the chromatic polynomial P_G is the unique polynomial $P_G(q)$ in q that is equal to the number of proper colourings of G with q colours when q is a positive integer. It was introduced by Birkhoff [6] in 1912, and can be used to express many properties of the graph G. For example, one of the most well-known uses of the chromatic polynomial (and Birkhoff's original motivation) is associated with an unsuccessful attempt to prove the 4-colour theorem by showing that in a region of the complex plane containing the point $q = 4$, the chromatic polynomial of a planar graph has no zeros. Of course, the zeros of the polynomial are interesting in their own right, and have been intensely studied in the combinatorics community. Most of the earlier mathematical results on the chromatic polynomial concern real zeros (see, e.g., [6, 7, 41, 43, 44, 25, 40, 45, 17]), but recently the study of complex zeros has also become quite popular [24, 3, 4, 1, 2, 19, 33, 42, 10, 11, 12, 13, 38, 35, 26, 39].

In this note I present a short proof of a recent result by Sokal [38], who showed that the chromatic polynomial on any finite graph of maximal degree D is free of zeros if q lies outside the disk $\{q \in \mathbb{C} : |q| \leqslant \tilde{K}D\}$, where $\tilde{K}$ is positive constant strictly smaller than 8. Sokal's work was motivated by the work of Biggs, Damarell and Sands, who conjectured [4] such a result for D-regular graphs, as well as a question of Brenti, Royle and Wagner [10], who asked whether it is true that a result of this form holds for arbitrary graphs.

2. Results and proof strategy

2.1. Sokal's theorem

Let $G = (V, E)$ be a finite graph. The *chromatic polynomial* P_G of the graph G is the polynomial

$$P_G(q) = \sum_{E' \subset E} q^{C(E')}(-1)^{|E'|}, \tag{2.1}$$

where $C(E')$ is the number of connected components of the graph $G' = (V, E')$. Note that $P_G(q)$ is the unique polynomial which is equal to the number of proper colourings of G for integer q, an fact which is easy to establish and was already known to Birkhoff.

In this note, I give a new short proof of Sokal's theorem.

Theorem 2.1. (Sokal) *Let G be a finite graph of maximal degree D, and let*

$$K(a) = \frac{a + e^a}{\log(1 + ae^{-a})}. \tag{2.2}$$

Then all zeros of P_G lie inside the disk $\{q \in \mathbb{C} : |q| < DK\}$, where $K = \min_{a \geqslant 0} K(a)$.

Sokal proved the theorem in terms of an *a priori* different constant $\tilde{K} = \min_{a \geqslant 0} \tilde{K}(a)$, where

$$\tilde{K}(a) = \inf\left\{K' : \sum_{n=1}^{\infty} \frac{1}{n!}\left(\frac{ne^a}{K'}\right)^{n-1} \leqslant 1 + ae^{-a}\right\}. \tag{2.3}$$

It turns out, however, that the two formulations are equivalent; in fact, I will show that $\tilde{K}(a) = K(a)$ for all $a > 0$. Note that our formulation allows us to get easy upper bounds on the constant K. Choosing, e.g., $a = 2/5$ we get that $K \leqslant K(2/5) = 7.964\cdots < 8$. By contrast, the representation (2.3) requires quite a bit of rigorous numerical mathematics to establish good upper bounds on $K = \tilde{K}$.

Remark. Having shown that the chromatic polynomial is nonzero, one can define the entropy per vertex as the quantity $s_G(q) = \frac{1}{|V|} \log P_G(q)$. This raises the question under which circumstances the entropy per vertex has a limit as the number of vertices tends to infinity, and whether this limit is analytic in $1/q$. For graphs which are induced subgraphs of an amenable quasi-transitive graph, these questions have been analysed in [32].

2.2. Basic proof strategy

The main idea of Sokal was to map the chromatic polynomial into the generating function for independent sets on the intersection graph for the subsets of V, i.e., the graph $\mathcal{G} = (\mathcal{V}, \mathcal{E})$ whose vertex set consists of all finite subsets $\gamma \subset V$, with an edge between γ

and γ' whenever $\gamma \cap \gamma' \neq \emptyset$. More precisely, he showed that it is possible to define complex weights $z(\gamma)$ for the vertices of $\mathscr{V}$ such that the chromatic polynomial can be rewritten as

$$P_G(q) = q^{|V|} \sideset{}{'}\sum_{I \subset \mathscr{V}} \prod_{\gamma \in I} z(\gamma), \quad (2.4)$$

where the sum runs over the independent sets I in $\mathscr{G}$. Having obtained such a representation, Sokal then referred to a powerful theorem by Dobrushin [16]. Dobrushin's theorem states that a sum of the above form is free of zeros, provided there is a set of constants $c(\gamma) \in [0, \infty)$, $\gamma \in \mathscr{V}$, such that $z(\cdot)$ lies in the polydisk defined by the condition

$$|z(\gamma)| \leqslant (1 - e^{-c(\gamma)}) \prod_{\substack{\gamma' \in \mathscr{V} \\ \gamma\gamma' \in \mathscr{E}}} e^{-c(\gamma')} \quad \forall \gamma \in \mathscr{V}. \quad (2.5)$$

The main technical task was then to show that one can choose the function $c(\cdot)$ in such a way that for $|q| \geqslant DK$, the weights $z(\cdot)$ obey Dobrushin's condition (2.5). Sokal based his proof on a lemma which goes back to Rota [34]; applied to the weights $z(\cdot)$, it implies that $|z(\gamma)|$ can be bounded by $|q|^{1-|\gamma|}$ times the number of spanning trees of the induced graph $G[\gamma]$. Using detailed estimates on the number of subtrees $T \subset G$ which have size s and contain a fixed vertex $x \in V$, he then proved that $z(\cdot)$ obeys Dobrushin's condition if $|q| \geqslant D\tilde{K}(a)$ for some $a > 0$.

The proof of Sokal's theorem in this note also uses Dobrushin's theorem. But it follows a different strategy to verify condition (2.5). Instead of using Rota's lemma, it uses an inductive approach which is based on a reduction formula that expresses the weight $z(\gamma)$ in terms of the weights of the subsets $\gamma' \subset \gamma$.

This approach leads to an *a priori* different condition, the condition that $|q| \geqslant DK(a)$, with $K(a)$ given by (2.2). When presenting this proof in my lecture in Memphis, I knew that the resulting constant K agreed with Sokal's constant to the accuracy which Sokal had calculated. While I could not verify at the time that K and $\tilde{K}$ were actually equal, I have since found a proof of this fact. Appropriately for this volume, it uses a function that is well known in random graph theory, where it describes the size of the giant component above the threshold. The equality of K and $\tilde{K}$ therefore provides another example where formulas from random graph theory lead to identities between *a priori* different functions which might be hard to prove without this formula.

3. Preliminaries

3.1. Dobrushin's theorem

Dobrushin's theorem states that under condition (2.5), a sum of the form (2.4) is nonzero. Dobrushin formulated this theorem in the language of abstract polymer systems. The term *abstract polymer system* was coined in [37], but the mathematical theory of these systems goes back much further [31, 21, 23, 28], and the main ideas are even older [29]. Later developments [15, 30, 20, 14, 22, 27] led to weaker and weaker conditions for the applicability of the theory, and culminated in Dobrushin's work in 1996 [16]. The applications of this theory in mathematical physics are probably as diverse as the applications of the Lovász Local Lemma [18] are in graph theory and computer science.

Interestingly, the connection is not purely sociological: as shown in [36], the *statement* of Dobrushin's theorem and that of the Lovász Local Lemma are equivalent!

In the language of graph theory, an *abstract polymer system* is just a weighted countable graph with complex vertex weights. More explicitly, a pair $(\mathcal{G}, z)$ is called an abstract polymer system if $\mathcal{G} = (\mathcal{V}, \mathcal{E})$ is a countable graph, and $z : \gamma \mapsto z(\gamma)$ is a complex-valued function on $\mathcal{V}$. The vertices of $\mathcal{G}$ are usually called *polymers*, and the complex number $z(\gamma)$ is called the *activity* or *weight* of γ. For a finite subset $\mathcal{U} \subset \mathcal{V}$, one then defines the *partition function*

$$\mathcal{Z}(\mathcal{U}) = {\sum_{I \subset \mathcal{U}}}' \prod_{\gamma \in I} z(\gamma), \tag{3.1}$$

where the sum goes over independent sets in $\mathcal{G}$.

Dobrushin's theorem gives both the statement that $\mathcal{Z}(\mathcal{U}) \neq 0$, and a bound on the logarithm of $\mathcal{Z}(\mathcal{U})$. In the literature, this statement is usual formulated as a statement about the principal branch of the logarithm, i.e., the version of the logarithm with imaginary part between $-\pi$ and π. Here we follow a slightly different route, and define the logarithm by analytic continuation as follows: assume that $\mathcal{Z}(\mathcal{U}) \neq 0$ inside a polydisk $\mathbb{D}$ of the form

$$|z(\gamma)| \leq R(\gamma) \quad \text{for all } \gamma \in \mathcal{U}, \tag{3.2}$$

where $R(\gamma) \geq 0$ for all $\gamma \in \mathcal{U}$. Since $\mathbb{D}$ is a compact set, we have that $|\mathcal{Z}(\mathcal{U})|$ is bounded from below by a strictly positive constant, implying that we can find a slightly larger open disk $\tilde{\mathbb{D}}$ such that $\mathcal{Z}(\mathcal{U}) \neq 0$ in $\tilde{\mathbb{D}}$. In $\tilde{\mathbb{D}}$, we then define $\log \mathcal{Z}(\mathcal{U})$ by analytic continuation from the intersection of $\tilde{\mathbb{D}}$ with the set $\{\gamma(\cdot) : z(\gamma) > 0 \text{ for all } \gamma \in \mathcal{U}\}$. Note that for this version of the logarithms, we have that

$$\log[\mathcal{Z}(\mathcal{U})/\mathcal{Z}(\tilde{\mathcal{U}})] = \log \mathcal{Z}(\mathcal{U}) - \log \mathcal{Z}(\tilde{\mathcal{U}}) \tag{3.3}$$

whenever $\tilde{\mathcal{U}} \subset \mathcal{U}$, a fact that will be used in our proof of Dobrushin's theorem below.

Theorem 3.1. (Dobrushin) *Let $(\mathcal{G}, z)$ be an abstract polymer system such that the weights $z(\cdot)$ obey condition (2.5) for some function $c : \gamma \mapsto c(\gamma)$ from the vertex set $\mathcal{V}$ of $\mathcal{G}$ into $[0, \infty)$. Let $\mathcal{U}$ be a finite set. Then $\mathcal{Z}(\mathcal{U}) \neq 0$, and*

$$|\log(\mathcal{Z}(\mathcal{U})/\mathcal{Z}(\tilde{\mathcal{U}}))| \leq \sum_{\gamma \in \mathcal{U} \setminus \tilde{\mathcal{U}}} c(\gamma) \tag{3.4}$$

for all $\tilde{\mathcal{U}} \subset \mathcal{U}$.

Proof. We prove the theorem by induction on the size of $\mathcal{U}$. If $|\mathcal{U}| = 0$, i.e., $\mathcal{U} = \emptyset$, we have $\mathcal{Z}(\mathcal{U}) = 1$ and the statements of the theorem are obvious.

Let $n \in \mathbb{N}$ and assume that the statements of the theorem hold for all $\mathcal{U} \subset \mathcal{V}$ with $|\mathcal{U}| \leq n$. Consider a set $\mathcal{U}$ with $|\mathcal{U}| = n + 1$, and let $\tilde{\mathcal{U}}$ be a strict subset of $\mathcal{U}$. Choose $\gamma_0 \in \mathcal{U}$ in such a way that $\gamma_0 \notin \tilde{\mathcal{U}}$. Decomposing the sum representing $\mathcal{Z}(\mathcal{U})$ into a sum over independent sets I not containing the polymer γ_0 and a sum over independent sets containing γ_0, we now rewrite $\mathscr{L}(\mathscr{U})$ as

$$\mathscr{L}(\mathscr{U}) = \mathscr{L}(\mathscr{U}') + z(\gamma_0)\mathscr{L}(\mathscr{U}_0) = \mathscr{L}(\mathscr{U}')\left(1 + z(\gamma_0)\frac{\mathscr{L}(\mathscr{U}_0)}{\mathscr{L}(\mathscr{U}')}\right) \quad (3.5)$$

where $\mathscr{U}' = \mathscr{U} \setminus \{\gamma_0\}$ and $\mathscr{U}_0 = \{\gamma \in \mathscr{U}' : \gamma\gamma_0 \notin \mathscr{E}\}$. By the bound (2.5), the inductive assumption (3.4) and the observation that $\mathscr{U}' \setminus \mathscr{U}_0 = \{\gamma \in \mathscr{U}' : \gamma\gamma_0 \in \mathscr{E}\} \subset \{\gamma \in \mathscr{V} : \gamma\gamma_0 \in \mathscr{E}\}$, we have

$$\left|z(\gamma_0)\frac{\mathscr{L}(\mathscr{U}_0)}{\mathscr{L}(\mathscr{U}')}\right| \leqslant (1 - e^{-c(\gamma_0)}) \prod_{\substack{\gamma \in \mathscr{V}: \\ \gamma\gamma_0 \in \mathscr{E}}} e^{-c(\gamma)} \exp\left(\sum_{\substack{\gamma \in \mathscr{U}': \\ \gamma\gamma_0 \in \mathscr{E}}} c(\gamma)\right) \leqslant (1 - e^{-c(\gamma_0)}) < 1. \quad (3.6)$$

Combined with (3.5) and the fact that $\mathscr{L}(\mathscr{U}') \neq 0$ by the inductive assumption, this clearly gives $\mathscr{L}(\mathscr{U}) \neq 0$. To prove the estimate (3.4), we rewrite $\log(\mathscr{L}(\mathscr{U})/\mathscr{L}(\tilde{\mathscr{U}}))$ as $\log(\mathscr{L}(\mathscr{U})/\mathscr{L}(\mathscr{U}')) + \log(\mathscr{L}(\mathscr{U}')/\mathscr{L}(\tilde{\mathscr{U}}))$. To bound the first term, we use (3.6) and (3.5) once more. Together with the observation that $|\log(1 + y)| \leqslant -\log(1 - |y|)$ whenever $|y| < 1$, this gives

$$|\log(\mathscr{L}(\mathscr{U})/\mathscr{L}(\mathscr{U}'))| \leqslant -\log(1 - (1 - e^{-c(\gamma_0)})) = c(\gamma_0). \quad (3.7)$$

Bounding $|\log(\mathscr{L}(\mathscr{U}')/\mathscr{L}(\tilde{\mathscr{U}}))|$ with the help of the inductive assumption (3.4), this gives the desired bound on $|\log(\mathscr{L}(\mathscr{U})/\mathscr{L}(\tilde{\mathscr{U}}))|$. □

The statement of Dobrushin's theorem clearly implies that $\log \mathscr{L}(\mathscr{U})$ is analytic in the interior of the disk defined by (2.5), allowing one to expand $\log \mathscr{L}(\mathscr{U})$ into an absolutely convergent Taylor series about $z(\cdot) \equiv 0$. For applications in statistical physics, one usually needs explicit expressions for the coefficients of this expansion; most treatments of abstract polymer systems therefore include a calculation of these coefficients; see, e.g., [37, 14, 22]. But for the application at hand, we only need the fact that under condition (2.5), the partition function $\mathscr{L}(\mathscr{U})$ is free of zeros.

3.2. A graph-theoretic lemma

As we will see in the next section, the chromatic polynomial of a graph $G = (V, E)$ can be rewritten in terms of an abstract polymer system with polymers consisting of subsets $\gamma \subset V$ with two or more elements, and weights involving a certain graph function $\phi_c(\cdot)$. In this subsection, we derive an inductive expression for this function; see Lemma 3.2 below.

Given a non-empty, finite graph $G = (V, E)$, let

$$\phi_c(G) = {\sum_{E' \subset E}}'(-1)^{|E'|}, \quad (3.8)$$

where the sum $\sum'$ goes over sets $E' \subset E$ such that the graph (V, E') is connected. If G is the empty graph we define $\phi_c(G) = 0$. Note that $\phi_c(G) = 1$ if G is a graph with a single vertex, since in this case the subgraph $(V, \emptyset)$ is a connected spanning graph. On the other hand $\phi_c(G) = 0$ if G is not connected, in accordance with the usual convention that an empty sum is considered to be zero.

To my knowledge the following lemma first appeared in [9]. It is somewhat reminiscent of Rota's Möbius lemmas [34] and certain lemmas of mathematical physics [14] relating 'connected' and 'disconnected' diagrams, but is also quite different from these earlier lemmas in that it expresses ϕ_c as a sum of terms involving ϕ_c alone.

Lemma 3.2. *Let $G = (V, E)$ be a non-empty finite graph, let $v \in V$, and let $V_0 = V \setminus \{v\}$. Then*

$$\phi_c(G) = \sum_{\pi \text{ of } V_0} \prod_{Y \in \pi} (-\phi_c(G[Y])I(v \sim Y)), \tag{3.9}$$

where the sum goes over all partitions π of V_0 into non-empty subsets and $I(v \sim Y)$ is the indicator function of the event that there exists at least one vertex $w \in Y$ such that $vw \in E$.

Proof. Let $E' \subset E$ be a set of edges contributing to the right-hand side of (3.8). Let E'_0 be the set of those edges in E' which do not contain the vertex v as an endpoint, and let $G_0 = (V_0, E'_0)$. The connected components of G_0 then induce a partition π of V_0. Summing over all $E' \subset E$ that induce a given partition π, we get a contribution that can be decomposed as the product

$$\prod_{Y \in \pi} \left[\phi_c(G[Y]) \sum_{\emptyset \neq E''_Y \subset E_Y} (-1)^{|E''_Y|} \right], \tag{3.10}$$

where E_Y denotes the set of edges in E that join Y to the vertex v. Observing that

$$\sum_{\emptyset \neq E''_Y \subset E_Y} (-1)^{|E''_Y|} = (1-1)^{|E_Y|} - 1 = -1 \tag{3.11}$$

if $E_Y \neq \emptyset$, while $\sum_{\emptyset \neq E''_Y \subset E_Y} (-1)^{|E''_Y|} = 0$ if $E_Y = \emptyset$, we obtain (3.9). □

4. Proof of Theorem 2.1

In this section, we prove Theorem 2.1. Following Sokal, we first show that the chromatic polynomial can be rewritten in terms of an abstract contour system. In a second step, we then verify condition (2.5). It is here that our approach is different from that of Sokal.

4.1. Mapping to a polymer system

We start from the representation (2.1), which we repeat here for the convenience of the reader:

$$P_G(q) = \sum_{E' \subset E} q^{C(E')}(-1)^{|E'|}. \tag{4.1}$$

Consider a set of edges E' contributing to the right-hand side. The connected components of the graph $G' = (V, E')$ then induce a partition of the vertex set V into $C(E')$ disjoint sets $Y_1, \ldots, Y_{C(E')}$. Consider the sum over all spanning subgraphs $G' = (V, E')$ that lead to the same partition π. Observing that the factor $q^{C(E')}$ can be rewritten as $q^{|\pi|}$, where $|\pi|$ is the number of elements of the partition π, this allows us to rewrite $P_G(q)$ as

$$P_G(q) = \sum_{\pi \text{ of } V} \prod_{\gamma \in \pi} (q\phi_c(G[\gamma])), \tag{4.2}$$

where the sum goes over partitions of V into connected subsets, and $G[\gamma]$ is the induced graph on γ. Extracting a factor $q^{|V|}$ and observing that $\phi_c(G[\gamma]) = 1$ if $|\gamma| = 1$, we get

$$P_G(q) = q^{|V|} \sum_{\pi \text{ of } V} \prod_{\substack{\gamma \in \pi: \\ |\gamma| \geq 2}} (q^{1-|\gamma|} \phi_c(G[\gamma])). \tag{4.3}$$

This is the desired representation in terms of an abstract polymer system. Indeed, let $\mathscr{V}$ be the set of all connected sets $\gamma \subset V$ with $|\gamma| \geq 2$, with an edge between $\gamma \subset V$ and $\gamma' \subset V$ whenever $\gamma \cap \gamma' \neq \emptyset$. Let $\mathscr{E}$ be the set of such edges, and $\mathscr{G} = (\mathscr{V}, \mathscr{E})$. The sum in (4.3) can then be rewritten as a sum over independent sets. Defining the activity of a set $\gamma \in \mathscr{V}$ as

$$z(\gamma) = q^{1-|\gamma|} \phi_c(G[\gamma]), \tag{4.4}$$

this gives the representation (2.4) for the chromatic polynomial.

4.2. Verification of Dobrushin's condition

Given the representation (2.4), Dobrushin's theorem implies that $P_G(q)$ is free of zeros whenever condition (2.5) is satisfied. To verify this condition, it is convenient to rewrite it in a slightly different form.

Let $a : \gamma \mapsto a(\gamma)$ be a function from the set of polymers $\mathscr{V}$ into $[0, \infty)$. Setting $c(\gamma) = \log(1 + |z(\gamma)|e^{a(\gamma)})$, one easily checks that condition (2.5) is equivalent to the condition that

$$\sum_{\substack{\gamma \in \mathscr{V}: \\ \gamma\gamma_0 \in \mathscr{E} \\ \text{or } \gamma = \gamma_0}} \log(1 + |z(\gamma)|e^{a(\gamma)}) \leq a(\gamma_0) \tag{4.5}$$

for all $\gamma_0 \in \mathscr{V}$ with $z(\gamma_0) \neq 0$. Setting $a(\gamma) = a|\gamma|$, where a is a positive real number and $|\gamma|$ denotes the number of elements in γ, it is clearly enough to show that

$$\sum_{\substack{\gamma \in \mathscr{V}: \\ \gamma \ni x}} \log(1 + |z(\gamma)|e^{a|\gamma|}) \leq a \tag{4.6}$$

for all $x \in V$.

The proof of the next lemma is the main technical step in the proof of Sokal's theorem.

Lemma 4.1. *Let G be a finite graph of maximal degree D, let $(\mathscr{G}, z)$ be the polymer system defined in Section 4.1, and let $a(\gamma) = a|\gamma|$. If $|q| \geq K(a)D$, where $K(a)$ is defined in (2.2), then*

$$\sum_{\substack{\gamma \in \mathscr{V}: \\ \gamma \ni x}} |z(\gamma)|e^{a|\gamma|} \leq a \tag{4.7}$$

for all $x \in V$, implying in particular that (4.6) and hence the Dobrushin condition (2.5) are satisfied.

With the help of Dobrushin's theorem, Theorem 2.1 clearly follows from Lemma 4.1. The proof of this lemma, and thus the technical meat of our proof of Theorem 2.1, is based on Lemma 3.2. By contrast, Sokal's proof is based on an inequality that goes back to Rota, stating that for an arbitrary graph $\tilde{G}$, the absolute value of $\phi_c(\tilde{G})$ is bounded by the number of spanning subtrees of $\tilde{G}$.

Proof of Lemma 4.1. Setting $\epsilon = e^a |q|^{-1}$ and recalling the definition (4.4) of $z(\gamma)$, we rewrite the terms on the left-hand side of (4.7) as $|z(\gamma)| e^{a|\gamma|} = e^a \epsilon^{|\gamma|-1} |\phi_c(G[\gamma])|$. Multiplying both sides of (4.7) by e^{-a} and adding 1 in the form $1 = |\phi_c(G[\{x\}])| \epsilon^{1-1}$ to both sides, we see that condition (4.7) is equivalent to the condition

$$\sum_{\substack{\gamma \subset \Lambda: \\ \gamma \ni x}} \epsilon^{|\gamma|-1} |\phi_c(G[\gamma])| \leqslant 1 + ae^{-a}, \tag{4.8}$$

where the sum runs over all connected subsets of V, including the set $\gamma = \{x\}$ containing only one point.

Given an arbitrary subset $\Lambda \subset V$ and a vertex $x_1 \in V$, let us define

$$F_\Lambda(x_1) = \sum_{\substack{\gamma \subset \Lambda: \\ \gamma \ni x_1}} \epsilon^{|\gamma|-1} |\phi_c(G[\gamma])|. \tag{4.9}$$

For the proof of the lemma, it is then enough to show

$$F_\Lambda(x_1) \leqslant 1 + ae^{-a} \tag{4.10}$$

for all $\Lambda \subset V$ and all $x_1 \in \Lambda$.

We prove this bound by induction on the size of Λ. As a first step, we establish the bound

$$F_\Lambda(x_1) \leqslant \exp\left(\epsilon \sum_{\substack{x_2 \in \Lambda \setminus \{x_1\}: \\ x_1 x_2 \in E}} F_{\Lambda \setminus \{x_1\}}(x_2) \right). \tag{4.11}$$

To this end, we rewrite the sum in (4.9) as a sum over sequences of pairwise distinct vertices. Since x_1 is fixed, each set γ of order n corresponds to $(n-1)!$ different sequences, leading to the representation

$$F_\Lambda(x_1) = \sum_{n=1}^{\infty} \frac{\epsilon^{n-1}}{(n-1)!} \sideset{}{'}\sum_{x_2,\ldots,x_n \in \Lambda_1} |\phi_c(G(x_1, x_2, \ldots, x_n))|, \tag{4.12}$$

where the sum $\sum'$ goes over sequences of pairwise distinct vertices $x_2, \ldots, x_n$ in $\Lambda_1 = \Lambda \setminus \{x_1\}$, and $G(x_1, x_2, \ldots, x_n)$ is the graph $G(x_1, x_2, \ldots, x_n) = ([n], E(x_1, x_2, \ldots, x_n))$, with $[n] = \{1, \ldots, n\}$ and $E(x_1, \ldots, x_n) = \{ij : x_i x_j \in E\}$. Note that we have replaced the induced subgraph $G[\{x_1, x_2, \ldots, x_n\}] \subset G$ by its 'image' $G(x_1, x_2, \ldots, x_n)$ on the vertex set $\{1, \ldots, n\}$. This will be convenient when summing over the vertices $x_2, \ldots, x_n$, since it decouples the graph structure of the graph $G[\{x_1, x_2, \ldots, x_n\}] \subset G$ from the locations of the vertices $x_1, \ldots, x_n \in V$.

Using Lemma 3.2, the right-hand side becomes

$$F_\Lambda(x_1) = \sum_{n \geq 1} \frac{1}{(n-1)!} {\sum_{x_2,\ldots,x_n \in \Lambda_1}}' \sum_\pi \prod_{Y \in \pi} (\epsilon^{|Y|} |\phi_c(G(x_Y))| I(x_1 \sim x_Y)), \qquad (4.13)$$

where the third sum goes over all partitions π of $\{2,\ldots,n\}$, x_Y denotes the subsequence of $x_2,\ldots,x_n$ with indices in Y, and $I(x_1 \sim x_Y)$ is the indicator function of the event that at least one of the vertices in $\{x_i\}_{i \in Y}$ is adjacent to x_1. Relaxing the condition of pairwise distinctness to include only pairwise distinctness within each group $\{x_i\}_{i \in Y}$ and exchanging the sum over partitions with the sum over the vertices $x_2,\ldots,x_n$ we then get the bound

$$F_\Lambda(x_1) \leq \sum_{n \geq 1} \sum_\pi \frac{1}{(n-1)!} \prod_{Y \in \pi} \left(\epsilon^{|Y|} {\sum_{\substack{x_Y \in \Lambda_1^Y: \\ x_1 \sim x_Y}}}' |\phi_c(G(x_Y))| \right). \qquad (4.14)$$

To continue, we note that the last sum on the right-hand side only depends on the size of Y, not the particular set $Y \subset \{2,\ldots,n\}$. Indeed, relabelling the vertices in the sum over x_Y, and using that a sum over sequences of $|Y|$ distinct vertices in Λ_1 is equal to a sum over subsets of size $|Y|$ times $|Y|!$, we see that the above bound can be rewritten as

$$F_\Lambda(x_1) \leq \sum_{n \geq 1} \sum_\pi \frac{1}{(n-1)!} \prod_{Y \in \pi} (|Y|! W_{|Y|}), \qquad (4.15)$$

where

$$W_\ell = \sum_{\substack{\gamma \subset \Lambda_1: \\ |\gamma| = \ell, \\ \gamma \sim x_1}} \epsilon^{|\gamma|} |\phi_c(G([\gamma]))|. \qquad (4.16)$$

Next we rewrite the sum over partitions $\pi = \{Y_1,\ldots,Y_k\}$ of order k as $\frac{1}{k!}$ times the sum over ordered partitions $(Y_1,\ldots,Y_k)$. Using the fact that the number of ordered partitions $\pi = (Y_1,\ldots,Y_k)$ of $\{2,\ldots,n\}$ with fixed sizes $|Y_1| = n_1,\ldots,|Y_k| = n_k$ is equal to

$$\frac{(n-1)!}{n_1! \ldots n_k!}, \qquad (4.17)$$

it is then not hard to see that the sums in (4.14) can be carried out explicitly, leading to the identity

$$\sum_{n \geq 1} \sum_\pi \frac{1}{(n-1)!} \prod_{Y \in \pi} (|Y|! W_{|Y|}) = 1 + \sum_{n \geq 1} \sum_{k \geq 1} \sum_{\substack{n_1,\ldots,n_k \geq 1: \\ \sum_i n_i = n-1}} \frac{1}{k!} \frac{1}{n_1! \ldots n_k!} \prod_{i=1}^k n_i! W_{n_i}$$

$$= 1 + \sum_{k \geq 1} \sum_{n_1,\ldots,n_k \geq 1} \frac{1}{k!} \prod_{i=1}^k W_{n_i} \qquad (4.18)$$

$$= \exp\left(\sum_{n=1}^\infty W_n \right) = \exp\left(\sum_{\substack{\gamma \subset \Lambda_1: \\ \gamma \sim x_1}} \epsilon^{|\gamma|} |\phi_c(G[\gamma])| \right).$$

This gives the estimate

$$F_\Lambda(x_1) \leq \exp\left(\sum_{\substack{\gamma \subset \Lambda_1: \\ \gamma \sim x_1}} \epsilon^{|\gamma|} |\phi_c(G[\gamma])|\right) \leq \exp\left(\epsilon \sum_{\substack{x_2 \in \Lambda_1: \\ x_2 \sim x_1}} \sum_{\substack{\gamma \subset \Lambda_1: \\ \gamma \ni x_2}} \epsilon^{|\gamma|-1} |\phi_c(G[\gamma])|\right), \quad (4.19)$$

and hence (4.11).

With the bound (4.11) in hand, the proof of (4.10) is an easy induction argument. Indeed, we clearly have $F_{\{x\}}(x) = 1 \leq 1 + ae^{-a}$. Thus consider a finite subset $\Lambda \subset V$ and some vertex $x_1 \in \Lambda$. Assume by induction that $F_{\Lambda \setminus \{x_1\}}(x_2) \leq 1 + ae^{-a}$ for all $x_2 \in \Lambda_1 = \Lambda \setminus \{x_1\}$. The bound (4.11) then implies

$$F_\Lambda(x_1) \leq \exp\left(\epsilon \sum_{\substack{x_2 \in \Lambda \setminus \{x_1\}: \\ x_2 \sim x_1}} (1 + ae^{-a})\right) \leq \exp(\epsilon D(1 + ae^{-a})). \quad (4.20)$$

Inserting the value of ϵ and using the assumption $|q| \geq K(a)D$, we have

$$\epsilon D(1 + ae^{-a}) = (a + e^a)\frac{D}{|q|} \leq \log(1 + ae^{-a}), \quad (4.21)$$

which proves $F_\Lambda(x_1) \leq 1 + ae^{-a}$, as desired. □

We close this section by proving that our bound and that of Sokal are equivalent. To this end, we consider the function

$$f(x) = \sum_{n=1}^{\infty} \frac{n^{n-1}}{n!} x^n. \quad (4.22)$$

The radius of convergence for $f(x)$ is clearly $1/e$, and for $x \in (0, 1/e]$, we may express x as ce^{-c} for a uniquely defined $c \in (0, 1]$. Using these two facts, we rewrite Sokal's constant $\tilde{K}(a)$ as

$$\tilde{K}(a) = \frac{e^{a+c(a)}}{c(a)}, \quad (4.23)$$

where

$$c(a) = \sup\left\{c \in (0, 1] : \frac{e^c}{c} f(ce^{-c}) \leq 1 + ae^{-a}\right\}. \quad (4.24)$$

To complete the proof, we need a simple fact that is well known in the random graph community (see, e.g., [8], p. 103, equation (5.6)):

$$f(ce^{-c}) = c \quad \text{for all } c \in (0, 1]. \quad (4.25)$$

As a consequence, $c(a)$ can be calculated explicitly, giving $c(a) = \log(1 + ae^{-a})$ and hence the desired equality of $\tilde{K}(a)$ and $K(a)$.

References

[1] Beraha, S. and Kahane, J. (1979) Is the four-color conjecture almost false? *J. Combin. Theory Ser. B* **27** 1–12.

[2] Beraha, S., Kahane, J. and Weiss, N. J. (1980) Limits of chromatic zeros of some families of maps. *J. Combin. Theory Ser. B* **28** 52–65.

[3] Berman, G. and Tutte, W. T. (1969) The golden root of a chromatic polynomial. *J. Combin. Theory* **6** 301–302.

[4] Biggs, N. L., Damerell, R. M. and Sands, D. A. (1972) Recursive families of graphs. *J. Combin. Theory Ser. B* **12** 123–131.

[5] Birkhoff, G. (1948) *Lattice Theory*, AMS, New York.

[6] Birkhoff, C. D. (1912) A determinantal formula for the number of ways of coloring a map. *Ann. Math.* **14** 42–46.

[7] Birkhoff, C. D. and Lewis, D. C. (1946) Chromatic polynomials. *Trans. Amer. Math. Soc.* **60** 335–451.

[8] Bollobás, B. (2001) *Random Graphs*, second edition, Vol. 73 of *Cambridge Studies in Advanced Mathematics*, Cambridge University Press, Cambridge.

[9] Borgs, C (2000) Polymer systems, Pirogov–Sinai theory and conductance bounds. Lecture notes, Microsoft Research, October 2000, unpublished.

[10] Brenti, F., Royle, G. F. and Wagner, D. G. (1994) Location of zeros of chromatic and related polynomials of graphs. *Canad. J. Math.* **46** 55–80.

[11] Brown, J. I. (1998) On the roots of chromatic polynomials. *J. Combin. Theory Ser. B* **72** 251–256.

[12] Brown, J. I. (1998) Chromatic polynomials and order ideals of monomials. *Discrete Math.* **189** 43–68.

[13] Brown, J., Hickman, C., Sokal, A. and Wagner, D. (2001) On the chromatic roots of generalized theta graphs. *J. Combin. Theory Ser. B* **83** 272–297.

[14] Brydges, D. (1986) A short course on cluster expansions. In *Critical Phenomena, Random Systems, Gauge Theories: Les Houches Summer School, Ser. XLIII, 1984* (K. Osterwalder and R. Stora, eds), North-Holland, Amsterdam, pp. 129–183.

[15] Cammarota, C. (1982) Decay of correlations for infinite range interactions in unbounded spin systems. *Comm. Math. Phys.* **85** 517–528.

[16] Dobrushin, R. L. (1996) Estimates of semi-invariants for the Ising model at low temperatures. In *Topics in Statistical and Theoretical Physics, Amer. Math. Soc. Trans., Ser. 2* **177** 59–81.

[17] Edwards, H., Hierons, R. and Jackson, B. (1998) The zero-free intervals for characteristic polynomials of matroids. *Combin. Probab. Comput.* **7** 153–165.

[18] Erdős, P. and Lovász, L. (1975) Problems and results on 3-chromatic hypergraphs and some related questions. In *Infinite and Finite Sets* (A. Hajnal et al., eds), North-Holland, Amsterdam, pp. 609–628.

[19] Farrell, E. J. (1980) Chromatic roots: some observations and conjectures. *Discrete Math.* **29** 161–167.

[20] Fredenhagen, K. and Marcu, M. (1983) Charged states in $\mathbb{Z}_2$ gauge theories. *Comm. Math. Phys.* **92** 81–119.

[21] Gallavotti, G. and Miracle-Solé, S. (1968) Correlation functions of a lattice system. *Comm. Math. Phys.* **7** 274–88.

[22] Glimm, J. and Jaffe, A. (1985) Expansions in statistical physics. *Comm. Pure Appl. Math.* **XXXVIII** 613–630.

[23] Gruber, C. and Kunz, M. (1971) General properties of polymer systems. *Comm. Math. Phys.* **22** 133–161.

[24] Hall, D. W., Siry, J. W. and Vanderslice, B. R. (1965) The chromatic polynomial of the truncated icosahedron. *Proc. Amer. Math. Soc.* **16** 620–628.

[25] Jackson, B. (1993) A zero-free interval for chromatic polynomials of graphs. *Combin. Probab. Comput.* **2** 325–336.

[26] Jacobsen, J., Salas, J. and Sokal, A. (2003) Transfer matrices and partition-function zeros for antiferromagnetic Potts models III: Triangular-lattice chromatic polynomial. *J. Statist. Phys.* **112** 921–1017.

[27] Kotecký, R. and Preiss, D. (1986) Cluster expansion for abstract polymer models. *Comm. Math. Phys.* **103** 491–498.

[28] Malyshev, V. A. (1979) Uniform cluster estimates for lattice models. *Comm. Math. Phys.* **64** 131–157.
[29] Mayer, J. E. (1947) Integral equations between distribution functions of molecules. *J. Chem. Phys.* **15** 187–201.
[30] Navrátil, J. (1982) Contour models and unicity of random fields (in Czech). Diploma thesis, Charles University, Prague.
[31] Penrose, O. (1967) Convergence of fugacity expansions for classical systems. In *Statistical Mechanics: Foundations and Applications* (T. A. Bak, ed.), Benjamin, New York/Amsterdam, pp. 101–109.
[32] Procacci, A, Scoppola, B. and Gerasimov, V. (2003) Potts model on infinite graphs and the limit of chromatic polynomials. *Comm. Math. Phys.* **235** 215–231.
[33] Read, R. C. and Royle, G. F. (1991) Chromatic roots of families of graphs. In *Graph Theory, Combinatorics, and Applications: Proc. Sixth Quadrennial International Conference on the Theory and Applications of Graphs, Western Michigan University, 1988*, Vol. 2 (Y. Alavi et al., eds), Wiley, New York, pp. 1009–1029.
[34] Rota, C. G. (1964) On the foundations of combinatorial theory I: Theory of Möbius functions. *Z. Wahrscheinlichkeitstheorie verw. Geb.* **2** 340–368.
[35] Salas, J. and Sokal, A. (2001) Transfer matrices and partition-function zeros for antiferromagnetic Potts models I: General theory and square-lattice chromatic polynomial. *J. Statist. Phys.* **104** 609–699.
[36] Scott, A. and Sokal, A. (2005) The repulsive lattice gas, the independent-set polynomial, and the Lovász local lemma. *J. Statist. Phys.* **118** 1151–1261.
[37] Seiler, E. (1982) *Gauge Theories as a Problem of Constructive Field Theory and Statistical Mechanics*, Vol. 159 of *Lecture Notes in Physics*, Springer, Berlin/Heidelberg/New York.
[38] Sokal, A. (2001) Bounds on the complex zeros of (di)chromatic polynomials and Potts-model partition functions. *Combin. Probab. Comput.* **10** 41–77.
[39] Sokal, A. (2004) Chromatic roots are dense in the whole complex plane. *Combin. Probab. Comput.* **13** 221–261.
[40] Thomassen, C. (1997) The zero-free intervals for chromatic polynomials of graphs. *Combin. Probab. Comput.* **6** 497–506.
[41] Tutte, W. T. (1970) On chromatic polynomials and the golden ratio. *J. Combin. Theory* **9** 289–296.
[42] Wakelin, C. D. and Woodall, D. R. (1992) Chromatic polynomials, polygon trees, and outerplanar graphs. *J. Graph Theory* **16** 459–466.
[43] Woodall, D. R. (1977) Zeros of chromatic polynomials. In *Combinatorial Surveys: Proc. Sixth British Combinatorial Conference* (P. J. Cameron, ed.), Academic Press, London, pp. 199–223.
[44] Woodall, D. R. (1992) A zero-free interval for chromatic polynomials. *Discrete Math.* **101** 333–341.
[45] Woodall, D. R. (1997) The largest real zero of the chromatic polynomial. *Discrete Math.* **172** 141–153.

Duality in Infinite Graphs

HENNING BRUHN and REINHARD DIESTEL

Mathematisches Seminar, Universität Hamburg, Bundesstraße 55, 20146 Hamburg, Germany
(e-mail: hbruhn@gmx.net, diestel@math.uni-hamburg.de)

Received 25 February 2004; revised 3 November 2004

For Béla Bollobás on his 60th birthday

The adaption of combinatorial duality to infinite graphs has been hampered by the fact that while cuts (or cocycles) can be infinite, cycles are finite. We show that these obstructions fall away when duality is reinterpreted on the basis of a 'singular' approach to graph homology, whose cycles are defined topologically in a space formed by the graph together with its ends and can be infinite. Our approach enables us to complete Thomassen's results about 'finitary' duality for infinite graphs to full duality, including his extensions of Whitney's theorem.

1. Introduction

The *cycle space* over $\mathbb{Z}_2$ of a finite graph G is the set of all symmetric differences of its *circuits*, the edge sets of the cycles in G. If G^* is another graph and there exists a bijection $E(G) \to E(G^*)$ that maps the circuits of G precisely to the minimal non-empty cuts (or *bonds*) of G^*, then G^* is called a *dual* of G. The classical result in this context is Whitney's theorem.

Theorem 1.1. (Whitney [14]) *A finite graph G has a dual if and only if it is planar.*

For infinite graphs, however, there is an obvious asymmetry between circuits and cuts that gets in the way of duality: while cuts can be infinite, cycles are finite. Indeed, let G be the half-grid shown in solid lines in Figure 1. Geometrically, the dotted graph G^* should be its dual. But then various infinite sets of edges in G, such as the edge sets of its horizontal 2-way infinite paths, should be circuits, because they correspond to bonds of G^*.

This problem ties in with similar recently observed difficulties about extending other homology aspects of finite graphs to infinite graphs [6]. In all those cases, the problems could be resolved in a topologically motivated extension of the cycle space, which takes as its basic circuits not only the edge sets of finite cycles but more generally the edge sets of

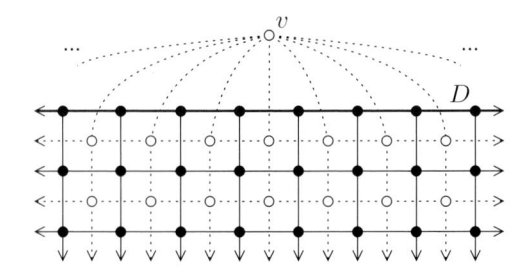

Figure 1. A dual with infinite bonds requires infinite circuits

all *circles* – the homeomorphic images of S^1 in the space $|G|$ formed by the graph together with its ends. (When G is locally finite, $|G|$ is known as its *Freudenthal compactification*. In topological terms, the key step is to use a singular-type homology in $|G|$ rather than the simplicial homology of the graph G.) In our example, every 2-way infinite horizontal path D in G forms such a circle in $|G|$ together with the unique end of G, since both its tails converge to this end. The edge set of D would thus be a circuit, as required by its duality to a bond of G^*.

We show that the obstructions to duality in infinite graphs can indeed be overcome in this way. As in those other cases, we build the cycle space not just on the edge sets of the finite cycles of G but on the edge sets of all its circles. This time, however, we take these circles not in $|G|$ itself but in a natural quotient space of $|G|$: whenever an infinite path R is *dominated* by a vertex v (that is, if G contains infinitely many v–R paths that are disjoint except in v) we identify v with the end of R, so that R converges to v rather than to a newly added point at infinity. In Figure 1, every infinite path in the dotted graph converges to the vertex v, so every maximal horizontal path and every maximal vertical path in G^* forms a circle through v.

Our results extend the duality of finite graphs in what appears to be a complete and best-possible way. However, all our results build on previous work of Thomassen [12, 13], who extended finite duality to infinite graphs as far as it will go without considering infinite circuits. Thomassen's approach was to overcome the disparity between finite cycles and infinite cuts by disregarding the latter. In his terms, G^* qualifies as a dual of G as soon as the edge bijection between the two graphs maps all the (finite) circuits of G to the finite bonds of G^*.

This weaker notion of duality already permits quite a satisfactory extension of Whitney's theorem to 2-connected graphs. Our stronger notion strengthens this (in one direction), and in addition re-establishes two aspects of finite duality that cannot be achieved for infinite graphs when infinite cuts (and cycles) are disregarded: the uniqueness of duals for 3-connected graphs, and symmetry, the fact that a graph is always a dual of its duals.

Concerning symmetry, note that taking duals may force us out of the class of locally finite graphs: while the graph G in Figure 1 is locally finite, its dual is not. But graphs with vertices of infinite degree do not, in general, have duals. Indeed, Thomassen showed that any infinite graph with a dual (even in his weaker sense) must satisfy the following condition:

> *No two vertices are joined by infinitely many edge-disjoint paths.* (∗)

To achieve symmetry, we thus need that the class of graphs satisfying (∗) is closed under taking duals. While this is not the case for Thomassen's notion, it will be true for ours.

Our paper is organized as follows. After providing the required terminology in Section 2 we continue the above discussion in more precise terms in Section 3, which leads up to the statement of our basic duality theorem, Theorem 3.4. We prove this theorem in Section 4. In Section 5 we characterize the graphs that have locally finite duals. In Section 6 we treat duality in terms of spanning trees. In Section 7 we apply our results to colouring-flow duality.

2. Definitions

The basic terminology we use can be found in [7]. However, *graphs* in this paper are the 'multigraphs' of [7]: they may have loops and multiple edges. When we contract an edge $e = uv$ then this may create loops (from edges parallel to e) and new parallel edges (if u and v had a common neighbour). Observe that contracting a loop is the same as deleting it. Following Thomassen [12, 13], we require 2-connected graphs to be loopless, and 3-connected graphs to have no parallel edges.[1] A 1-way infinite path is called a *ray*, a 2-way infinite path is a *double ray*, and the subrays of a ray or double ray are their *tails*.

A graph is said to be *planar* if it can be drawn in the plane in such a way that if the images of two edges e and f have a point in common then this point corresponds to a vertex incident with both e and f, and such that all the images of vertices are disjoint. By Kuratowski's theorem and compactness, a countable graph is planar if and only if it contains neither K_5 nor $K_{3,3}$ as a minor [10].

Let $G = (V, E)$ be a graph, fixed throughout this section. Two rays in G are *equivalent* if no finite set of vertices separates them; the corresponding equivalence classes of rays are the *ends* of G. We denote the set of these ends by $\Omega = \Omega(G)$. An end ω is said to be *dominated* by a vertex v if for some (equivalently: for every) ray $R \in \omega$ there are infinitely many v–R paths that meet only in v. The ends of G that are not dominated are precisely its topological ends as defined by Freudenthal [11]; see [8] for details.

We will now define two topological spaces. The first of these, denoted as $|G|$, has $V \cup \Omega \cup \bigcup E$ as its point set; we shall call its topology VTop. The other, denoted as $\tilde{G}$, will be a quotient space of $|G|$. Its point set can be viewed as $V \cup \Omega' \cup \bigcup E$, where Ω' is the set of undominated ends, and we call its topology ITop. When G is locally finite, the two spaces will coincide.

Let us start with $|G|$. We begin by viewing G itself (without ends) as the point set of a 1-complex. Then every edge is a copy of the real interval $[0, 1]$, and we give it the corresponding metric and topology. For every vertex v we take as a basis of open neighbourhoods the open stars of radius $1/n$ around v. (That is to say, for every integer

---

[1] This is motivated by matroid theory. Disallowing loops is also necessary for uniqueness: a graph with loops never has exactly one dual (unless it is itself a loop).

$n \geqslant 1$ we declare as open the set of all points on edges at v that have distance less than $1/n$ from v, in the metric of that edge.)[2] In order to extend this topology to Ω, we take as a basis of open neighbourhoods of a given end $\omega \in \Omega$ the sets of the form

$$C(S, \omega) \cup \Omega(S, \omega) \cup \mathring{E}(S, \omega),$$

where $S \subseteq V$ is a finite set of vertices, $C(S, \omega)$ is the unique component of $G - S$ in which every ray from ω has a tail, $\Omega(S, \omega)$ is the set of all ends $\omega' \in \Omega$ whose rays have a tail in $C(S, \omega)$, and $\mathring{E}(S, \omega)$ is the set of all inner points of edges between S and $C(S, \omega)$.[3] We shall freely view G and its subgraphs either as abstract graphs or as subspaces of $|G|$. Note that in $|G|$ every ray converges to the end of which it is an element.

Now let $\tilde{G}$ be the quotient space obtained from $|G|$ by identifying each vertex v with all the ends it dominates. Since G (*i.e.*, all the graphs we shall consider) will satisfy (∗), the equivalence class containing v contains no other vertex. We also denote this class by v, and think of it as the old vertex v to which now the rays dominated by v converge. This quotient space $\tilde{G}$ is easily seen to be Hausdorff (unlike $|G|$, where we cannot find disjoint open sets for an end and a vertex that dominates it), and if G is 2-connected then $\tilde{G}$ is compact [5].

For the definitions that follow we shall formally work in $\tilde{G}$, but bear in mind that they apply also to $|G|$ when G is locally finite (in which case no identification takes place and $\tilde{G} = |G|$).

A subset of $\tilde{G}$ is a *circle* (respectively, an *arc*) if it is homeomorphic to the unit circle S^1 in the Euclidean plane (respectively, to the real interval $[0, 1]$). For example, every horizontal double ray in the graph G of Figure 1 forms, together with the unique end of G, a circle in $\tilde{G} = |G|$, because its tails converge to this end. Similarly, every vertical ray in the dotted graph G^* that starts at v forms a circle in $\tilde{G}^*$ (but not in $|G^*|$), because in ITop its end – the unique end of G^* – is identified with its starting vertex v.

Note that a circle C includes every edge of which it contains an inner point, and thus it has a well-defined edge set, called its *circuit*. Conversely, it is not hard to show [9] that $C \cap G$ is dense in C, so every circle is the closure in $\tilde{G}$ of the union of the edges in its circuit, and hence defined uniquely by its circuit. Note that every finite circuit in G is also a circuit in this sense.

Call a family $(D_i)_{i \in I}$ of subsets of E *thin* if no edge appears in infinitely members of the family. Let the *sum* $\sum_{i \in I} D_i$ of this family be the set of all edges that lie in D_i for an odd number of indices i. Then the *cycle space* $\mathscr{C}(\tilde{G})$ of $\tilde{G}$ is the set of all sums of (thin families of) edge sets of circuits, finite or infinite. Symmetric difference as addition makes $\mathscr{C}(\tilde{G})$ into a $\mathbb{Z}_2$ vector space, which coincides with the usual cycle space of G over $\mathbb{Z}_2$ when G is finite. We remark that $\mathscr{C}(\tilde{G})$ is closed also under taking infinite thin sums [9], which is

---

[2] If G is locally finite, this is the usual identification topology of the 1-complex. Vertices of infinite degree, however, have a countable neighbourhood basis in VTop, which they do not have in the 1-complex.

[3] In the early papers on this topic, such as [9], some more basic open sets were allowed: in the place of $\mathring{E}(S, \omega)$ we could take an arbitrary union of open half-edges from C towards S, one from every S–C edge. When G is locally finite, this yields the same topology. When G has vertices of infinite degree, our topology is slightly sparser but still yields the same cycle space; see the end of this section for more discussion.

not obvious from the definitions. When G is finite or locally finite, we usually write $\mathscr{C}(G)$ instead of $\mathscr{C}(\tilde{G})$.

A set $F \subseteq E$ is a *cut* of G if there is a partition (A, B) of V such that F is the set of all the edges of G with one vertex in A and the other in B. We shall also denote this set by $E(A, B)$. A cut is called a *bond* if it is minimal among the non-empty cuts.

We shall need the following two results as tools in our proofs.

Theorem 2.1. ([9]) *Let G be a graph satisfying (∗). Then every element of $\mathscr{C}(\tilde{G})$ is a disjoint union of circuits.*

Theorem 2.2. ([9]) *Let G be a graph satisfying (∗). Then a set $Z \subseteq E(G)$ is an element of $\mathscr{C}(\tilde{G})$ if and only if Z meets every finite cut in an even number of edges.*

For the conscientious reader we remark that, although the topology for $|G|$ considered in [9] is slightly larger than ours (see the earlier footnote), the above two theorems are nevertheless applicable in our context. This is because the circuits in $\tilde{G}$ coincide for these topologies: as one readily checks, the identity on $\tilde{G}$ between the two spaces is bicontinuous when restricted to a circle in either space.

3. Duality in infinite graphs

As discussed in the introduction, Thomassen pursued an approach to duality in infinite graphs that is based solely on finite circuits and cuts. While being very successful in some respects, such as conditions for the existence of duals and extensions of Whitney's theorem, this approach leads to unavoidable problems in others, such as symmetry and the uniqueness of duals. Our aim in this section is to discuss these problems, to indicate why considering infinite circuits and working in ITop is both, in essence, necessary and sufficient to cure them, and to state our main result, Theorem 3.4.

Consider graphs G and G^*, possibly infinite, and assume that there is a bijection $^* : E(G) \to E(G^*)$. Given a set $F \subseteq E(G)$, put $F^* := \{e^* \mid e \in F\}$, and *vice versa*. (That is, given a subset of $E(G^*)$ denoted by F^*, we write F for the subset $\{e \mid e^* \in F^*\}$ of $E(G)$.) Call G^* a *finitary dual* of G if, for every finite set $F \subseteq E(G)$, the set F is a circuit in G if and only if F^* is a bond in G^*.

Expressed in these terms, Thomassen obtained the following extension of Whitney's theorem.

Theorem 3.1. (Thomassen [13]) *A 2-connected[4] graph G has a finitary dual if and only if G is planar and satisfies (∗).*

---

[4] We expect that Thomassen's theorem extends to graphs of smaller connectivity. There is no mention of this in [13], however, and we note that the canonical proof for the forward implication fails: when G^* is a finitary dual of G, then the duality map * need not map the blocks of G to blocks of G^*, so it is not obvious that the blocks of G have finitary duals too. Compare Lemma 4.8 below.

Going back to Figure 1, we see that the dotted graph G^* is a finitary dual of the half-grid G. However, splitting the vertex v into two vertices u and w, and making each of these adjacent to infinitely many neighbours of v in such a way that every neighbour of v is adjacent to exactly one of u and w, we obtain another finitary dual H of G. This violates the intended uniqueness of duals for 3-connected graphs such as G. (Recall that duals of 3-connected finite graphs are unique.)

Moreover, admitting H as a dual violates symmetry, since G is not a finitary dual of H. In fact, H has no finitary dual at all, and it might not even be planar, depending on how we join u and w to the neighbours of v.

Thomassen realized these problems, as is witnessed by the following two theorems.

Theorem 3.2. (Thomassen [12]) *Let G be a 2-connected graph having a finitary dual. Then G has a finitary dual G^* satisfying* (∗), *and every such finitary dual G^* has G as its finitary dual.*

We say that a graph H is a *finitary predual* of G if G is a finitary dual of H.

Theorem 3.3. (Thomassen [13]) *If G has a 3-connected finitary predual then this is its only predual, up to isomorphism.*

By considering infinite as well as finite circuits, however, we can restore uniqueness. In our example, consider the edge set F of the double ray D in G. In G^*, its dual set F^* (the set of edges incident with v) is a bond. But F^* is not a bond in H, because it contains the edges incident with u (say) as a proper subset. Thus, if F counts as a circuit, then G^* will be a dual of G but H will not, as should be our aim. Taking the circuits of G in $|G| = \tilde{G}$ achieves this.

To restore symmetry, we have to allow vertices of infinite degree. (Note that G^* has one, and we want G to be its dual.) We thus have to decide now whether to work in $|G|$ or in $\tilde{G}$. That is to say, should we take the circles that define our infinite circuits in the topology ITOP specifically designed for graphs satisfying (∗), or in the simpler VTOP?

To answer this question, let us consider the graph G shown in unbroken lines in Figure 2, and let G^* be a hypothetical dual of G (by the definition we are seeking). We want G to be a dual of G^*, and in particular a finitary dual. Thus, G^* will be a finitary predual of G. Now G is certainly a finitary dual of the dotted graph H shown in Figure 2, so H is also a finitary predual of G. Since H is 3-connected, Theorem 3.3 implies that $H = G^*$.

Since the dotted edges at v form a bond of G^*, we thus have to make the edge set F of the double ray D a circuit of G. Now in $|G|$ the set F is not a circuit, because G has two ends and D has a tail in each. In $\tilde{G}$, however, both ends of G are identified with the vertex v, so the double ray D and the vertex v together do form a circle, making F into a circuit as desired.

We therefore propose the following stronger notion of duality for infinite graphs, in which the duality condition is required of all sets of edges, finite or infinite, and circuits are defined as in $\tilde{G}$ under ITOP.

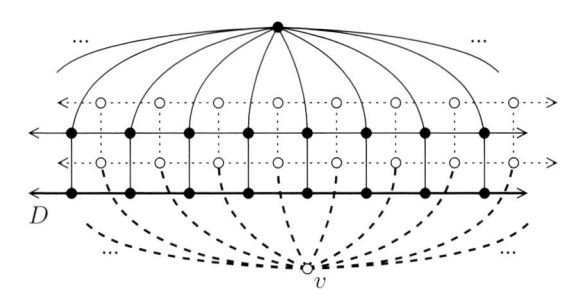

Figure 2. The self-dual graph G

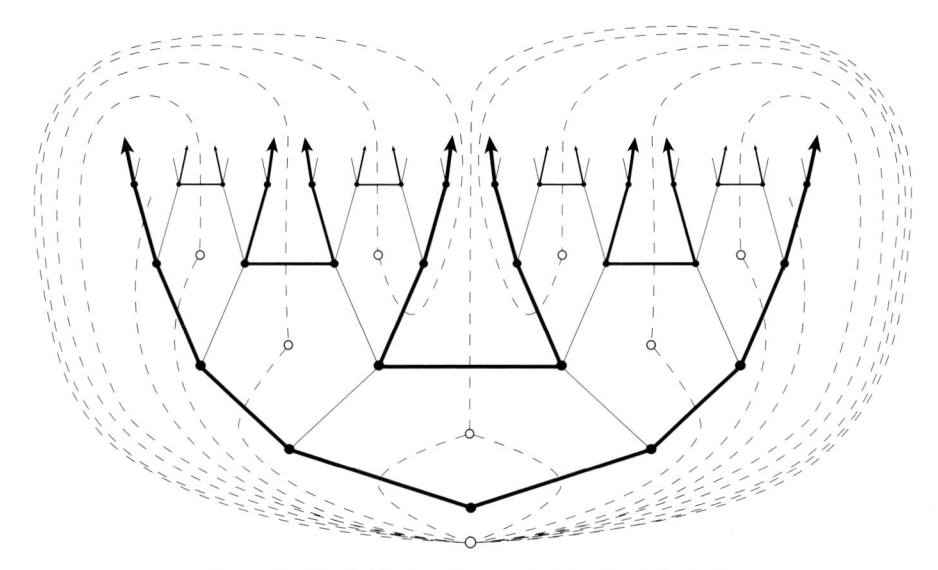

Figure 3. The bold edges form an infinite circuit in G, the broken edges indicate the corresponding cut in G^*

Definition. Let G be a graph satisfying (∗). Let G^* be another graph, with a bijection $^*: E(G) \to E(G^*)$. Call G^* a *dual* of G if the following holds for every set $F \subseteq E(G)$, finite or infinite: F is a circuit in $\tilde{G}$ if and only if F^* is a bond in G^*.

Note that every dual in this sense is also a finitary dual, but not conversely.

Figure 3 shows that infinite circuits can get pretty wild, even in locally finite graphs. The following theorem, which is our main result, can thus deviate more from the corresponding finite situation than it might at first appear.

Theorem 3.4. *Let G be a countable[5] graph satisfying (∗).*

---

[5] By Lemma 4.4 below, the countability assumption is redundant for 2-connected graphs, and therefore inessential. If we agree to call a graph 'planar' as soon as it has neither a K_5 nor a $K_{3,3}$ minor, then Theorem 3.4 becomes true also for uncountable graphs.

(i) G has a dual if and only if G is planar.
(ii) If G^* is a dual of G, then G^* satisfies (∗), G is a dual of G^*, and this is witnessed by the inverse bijection of *.
(iii) If G^* is a dual of G and $F \subseteq E(G)$, then $F \in \mathscr{C}(\tilde{G})$ if and only if F^* is a cut in G^*.

We shall prove Theorem 3.4 in the next section.

Since all finitary duals of a 3-connected graph are again 3-connected [13], Theorems 3.3 and 3.4(ii) together imply at once that the dual of a 3-connected graph is unique.

Corollary 3.5. *A 3-connected graph has at most one dual, up to isomorphism.*

4. Proof of the duality theorem

Recall that, by Theorem 3.1, any graph G with a finitary dual G^* satisfies (∗). Our first aim is to show that if G^* is a dual of G, then G^* too satisfies (∗).

We need two lemmas. The first we quote from [2].

Lemma 4.1. ([2]) *Let G be a 2-connected graph satisfying* (∗), *and let U be a finite set of vertices in G. Then we can contract edges of G so that no two vertices from U are identified, the graph H obtained has only finitely many edges and vertices, and every cut in H is also a cut of G.*

Lemma 4.2. *Let G be a 2-connected graph satisfying* (∗), *and let C be an infinite circuit in $\tilde{G}$. Let X be a finite set of edges meeting C in exactly one edge e. Then there is a finite circuit in G that meets X precisely in e.*

Proof. Apply Lemma 4.1 to G, taking as U the set of endvertices of the edges in X. Consider the finite graph H returned by the lemma. Applying Theorem 2.2 twice, we deduce from $C \in \mathscr{C}(\tilde{G})$ and the separation property of H that $C \cap E(H) \in \mathscr{C}(H)$. Let $C' \subseteq C \cap E(H)$ be a circuit containing e. As C meets X only in e, so does C'. Since no two vertices from U were identified when G was contracted to H, the branch sets of the contraction induce no edge from X in G. We can therefore expand C' to a finite circuit in G that still meets X only in e. □

We need the following strong version of Menger's theorem for countable graphs.

Theorem 4.3. (Aharoni [1]) *For any countable graph G and two sets A, B of vertices in G there exist a set $\mathscr{P}$ of disjoint A–B paths and an A–B separator X in G such that X consists of a choice of one vertex from each of the paths in $\mathscr{P}$.*

As every uncountable connected graph has a vertex of uncountable degree, it is easy to show that an uncountable 2-connected graph contains two vertices joined by uncountably many independent paths. (See, *e.g.*, [8, Lemma 2.1], or Thomassen [13].)

Lemma 4.4. *A 2-connected graph satisfying* (∗) *is countable.* □

We can now show that, unlike with finitary duals, the class of graphs satisfying (∗) is closed under taking duals.

Lemma 4.5. *Any dual G^* of a graph G satisfies* (∗).

Proof. Suppose G^* violates (∗). Then there are two vertices in G^*, x and y say, that cannot be separated by finitely many edges. We may assume that x and y lie in the same block B^* of G^*. Let B be the subgraph of G consisting of the edges e for which $e^* \in E(B^*)$ together with their incident vertices.

Let e be an edge of B, and consider an edge f that lies in the same block of G as e. It is not hard to see that there is a finite circuit containing both e and f. Thus, by duality, e^* and f^* lie in a common bond F^* of G^*. The edges in F^* that lie in B^* suffice to separate the endvertices of e^* in G^*. By the minimality of F^*, these are all its edges, including f^*, and thus $f \in E(B)$. This shows that B is the union of blocks of G.

Since an edge set is a circuit (resp. bond) of a graph if and only if it is a circuit (resp. bond) in one of its blocks, we see that B^* is a dual of the graph B. We may therefore assume that G^* is 2-connected, and thus, by Lemma 4.4, countable.

By Theorem 4.3 applied to the line graph of G^* (which is countable because G^* is), we can find in G^* an infinite set $\mathcal{P}$ of edge-disjoint x–y paths and a set C^* of edges separating x from y, such that C^* consists of a choice of one edge from each path in $\mathcal{P}$. Then C^* is a bond in G^*, and $C = \{e \mid e^* \in C^*\}$ is an infinite circuit in $\tilde{G}$.

Pick an edge $e \in C$. We claim the following:

There is an infinite sequence of distinct finite circuits $C_1, C_2, \ldots$ in G, each containing e, and such that $C_i \setminus C$ and $C_j \setminus C$ are non-empty and disjoint for all $i \neq j$. (4.1)

To prove (4.1), assume inductively that $C_1, \ldots, C_{i-1}$ have been constructed, and put

$$X := \{e\} \cup \bigcup_{j<i} C_j \setminus C.$$

Since C meets X only in e, Lemma 4.2 gives us a finite circuit C_i that contains e and does not meet $C_1 \cup \ldots \cup C_{i-1}$ outside C. As both C and C_i are circuits in $\tilde{G}$, neither contains the other properly, so $C_i \setminus C \neq \emptyset$. This proves (4.1).

Let u, v be the endvertices of e^* in G^*. Each of the sets C_i^* is a cut in G^* that contains e^*, and hence separates u from v in G^*. Denote by P the path in $\mathcal{P}$ that contains e^*. Since $E(P)$ meets C^* only in e^*, no edge of P other than e^* lies in more than one of the sets C_i^* (by (4.1)). Therefore only finitely many of the sets C_i^* meet $E(P - e^*)$; let C_n^* be one that does not. Since C_n^* is finite, there is a path $Q \in \mathcal{P}$ that has no edge in C_n^*. But then $(P - e^*) \cup Q$ is a connected subgraph of G^* that avoids C_n^* but contains both u and v, a contradiction. □

As pointed out before, every dual of a graph is also its finitary dual, and we have just seen that it satisfies (∗). Our next aim is to show that, conversely, every finitary dual satisfying (∗) is also a dual. We need the following lemma.

Lemma 4.6. *A set $F \subseteq E(G)$ in a graph G is a cut if and only if it meets every finite circuit in an even number of edges.*

Proof. Clearly, a cut meets every finite circuit in an even number of edges, so let us prove the other direction.

Let G' be the graph obtained from G by contracting every edge not in F. Then G' is bipartite (in particular, loopless), since any odd circuit would give rise to a finite circuit in G meeting F in an odd number of edges. The bipartition of G' induces a partition (A, B) of the vertex set of G such that every edge in F has one vertex in A and the other in B and such that no edge outside F has that property. Thus, $F = E(A, B)$ is a cut. □

Lemma 4.7. *Let G be a 2-connected graph, and let G^* be a finitary dual of G that satisfies (∗). Then the following assertions hold.*

(i) *G^* is a dual of G (witnessed by the same map *).*
(ii) *A set $F \subseteq E(G)$ lies in $\mathscr{C}(\tilde{G})$ if and only if F^* is a cut in G^*.*

Proof. We first prove (ii). Let $F \in \mathscr{C}(\tilde{G})$ be given. To show that F^* is a cut in G^*, it suffices by Lemma 4.6 to show that F^* meets every finite circuit Z^* in G^* in an even number of edges. Since G is a finitary dual of G^* (Theorem 3.2), Z is a finite cut in G. Hence by Theorem 2.2, $|F \cap Z| = |F^* \cap Z^*|$ is even.

Similarly, consider a cut F^* in G^*. To show that $F \in \mathscr{C}(\tilde{G})$, it suffices by Theorem 2.2 to show that F meets every finite cut D of G evenly. But $D^* \in \mathscr{C}(\tilde{G}^*)$ since G is a finitary dual of G^* (Theorem 3.2). So $|F \cap D| = |F^* \cap D^*|$ is even by the trivial direction of Lemma 4.6 and the fact that D^* is a disjoint union of circuits.

To prove (i), let now C be a circuit in $\tilde{G}$; we have to show that C^* is a bond in G^*. We have already shown that C^* is a cut in G^*, and that any cut $F^* \subseteq C^*$ corresponds to a set $F \in \mathscr{C}(G)$. Since F cannot be a proper subset of C unless it is empty, we deduce that C^* is a bond.

Conversely, let F^* be a bond in G^*. Then F is a minimal non-empty element of $\mathscr{C}(G)$. By Theorem 2.1, F must be a circuit. □

For a proof of Theorem 3.4, it remains to combine Lemmas 4.5 and 4.7 with Thomassen's results on finitary duals, and to extend the result from 2-connected to arbitrary graphs.

The latter is standard for finite graphs, but we have to be more careful here. (Indeed, Lemma 4.8 below fails for finitary duals.) If G has a finitary dual G^* and B is a block of G, let B^* denote the subgraph of G^* formed by the edges e^* with $e \in B$ and their incident vertices.

Lemma 4.8. *Let a graph G have a finitary dual G^* that satisfies (∗). If B is a block of G, then B^* is a block of G^* and a finitary dual of B. If G^* is also a dual of G then B^* is a dual of B.*

Proof. Two edges $e, f \in G$ lie in a common block of G if and only if they lie in a common finite circuit of $\tilde{G}$. For a proof that B^* is a block of G^*, it therefore suffices to show that the edges e^* and f^* lie in a common block of G^* if and only if they lie in a common finite bond of G^*.

If e^* and f^* lie in a common bond F^* of G^*, we proceed in a similar way as in the proof of Lemma 4.5.

Now suppose that e^* and f^* lie in a common block B^* of G^*. Then B^* has a finite circuit containing both e^* and f^*. Deleting e^* and f^* from this circuit, we obtain the edge sets of two paths, P and Q. Suppose that every $X \subseteq E(G^*)$ separating P and Q is infinite. Then there are also two vertices, one in P and the other in Q, that cannot be separated in G^* by finitely many edges. Consequently, we find infinitely many edge-disjoint paths connecting these two vertices, a contradiction to (∗). Therefore, there is a finite set $F^* \subseteq E(G^*)$ separating P from Q in G^*, which clearly contains e^* and f^*. If we choose F^* to be minimal then it is a bond.

It remains to show that B^* is a finitary dual (resp. dual) of B. But since B^* is a block of G^*, a set $F^* \subseteq E(B^*)$ is a bond of B^* if and only if it is a bond of G^*. Similarly, a set $F \subseteq B$ is a circuit in $\tilde{B}$ if and only if it is a circuit in $\tilde{G}$. The assertion therefore follows from the assumption that G^* is a finitary dual (resp. dual) of G. $\square$

Lemma 4.9. *If G and G^* are two graphs and $* : E(G) \to E(G^*)$ maps the blocks B of G to the blocks of G^* so that B^* is a dual of B, then G^* is a dual of G.*

Proof. It is easily checked that a subset of $E(G)$ is a circuit in $\tilde{G}$ if and only if it is a circuit in $\tilde{B}$ for some block B of G. Similarly, a subset of $E(G^*)$ is a bond in G^* if and only if it is a bond in some block of G^*. $\square$

Proof of Theorem 3.4. (i) By Lemmas 4.5, 4.8 and 4.9, G has a dual if and only if its blocks do. (To obtain a dual of G from duals of its blocks, take their disjoint union.) Similarly, a countable graph is planar if and only if its blocks are [10]. We may therefore assume that G is 2-connected.

If G is planar then, by Theorems 3.1 and 3.2, G has a finitary dual G^* that satisfies (∗). By Lemma 4.7, G^* is also a dual of G. Conversely, if G has a dual G^*, then G is planar by Theorem 3.1.

(ii) Suppose that G^* is a dual of G. By Lemma 4.5, G^* satisfies (∗). By Lemma 4.8, the subgraphs B^* of G^*, where B ranges over the blocks of G, are the blocks of G^*, and each B^* is a dual of B. We show that, conversely, B is a dual of B^*. Then, by Lemma 4.9, G is a dual of G^*.

By Lemma 4.5, every B^* satisfies (∗). (Hence so does G^*.) By Theorem 3.2, B is a finitary dual of B^*. Now B satisfies (∗), because G does so by assumption. Hence by Lemma 4.7, B is a dual of B^*.

(iii) For 2-connected graphs, this is Lemma 4.7(ii). The general case reduces easily to this with the help of Lemma 4.6 and Theorem 2.2. □

5. Locally finite duals

We started out by observing that a dual of a locally finite graph may have vertices of infinite degree. This raises the question under what circumstances the dual is locally finite. For 3-connected graphs, Thomassen gave the following characterization in terms of *peripheral* circuits, circuits C whose incident vertices do not separate the graph and do not span any edges not in C.

Theorem 5.1. (Thomassen [12]) *Let G be a locally finite 3-connected graph. Then G has a locally finite finitary dual if and only if G is planar and every edge lies in exactly two finite peripheral circuits.*

Since locally finite graphs trivially satisfy condition (∗), Lemma 4.7 implies that Theorem 5.1 still holds if the word 'finitary' is dropped.

To obtain another characterization, we need the following extension of Tutte's planarity criterion to locally finite graphs.

Theorem 5.2. ([3]) *Let G be a locally finite 3-connected graph. If G is planar then every edge appears in exactly two peripheral circuits. Conversely, if every edge appears in at most two peripheral circuits then G is planar.*

Theorem 5.3. *A locally finite 3-connected graph has a locally finite dual if and only if it is planar and all its peripheral circuits are finite.*

Proof. Let G be a locally finite 3-connected graph. If G has a locally finite dual then, by Theorem 5.1, G is planar and every edge lies in exactly two finite peripheral circuits. By Theorem 5.2, its edges cannot lie in any other peripheral circuits, so all peripheral circuits are finite.

Conversely, if G is planar and all its peripheral circuits are finite then, by Theorems 5.2 and 5.1, G has a locally finite finitary dual. By Lemma 4.7, this is in fact a dual. □

6. Duality in terms of spanning trees

In this section we show that our notion of duality permits the extension of another well-known duality theorem for finite graphs: that the complement of the edge set of any spanning tree of G defines a spanning tree in any dual of G, and conversely that any two graphs whose edge sets are in bijective correspondence so that their spanning trees complement each other as above form a pair of duals.

It is not difficult to see that the verbatim analogue of this fails for infinite graphs. Indeed, the edge set of an ordinary spanning tree of G might contain an infinite circuit C (such as the edges of the double ray D in Figure 1), in which case C^* would be a cut in G^*, and $G^* - C^*$ could not contain a spanning tree of G^*. However, the following adjustment to the notion of a spanning tree makes an extension possible.

Let us call a spanning tree T of G *acirclic* (under ITOP) if its closure in $\tilde{G}$ contains no circle – or equivalently, if its edges contain no circuit of $\tilde{G}$. (We remark that if G is locally finite then its acirclic spanning trees are precisely its end-faithful spanning trees [9].)[6]

Theorem 6.1. *Let $G = (V, E)$ and $G^* = (V^*, E^*)$ be connected[7] graphs satisfying* (∗), *and let* $^* : E \to E^*$ *be a bijection. Then the following two assertions are equivalent.*

(i) *G and G^* are duals of each other, and this is witnessed by the map * and its inverse.*
(ii) *Given a set $F \subseteq E$, the graph (V, F) is an acirclic spanning tree of G if and only if $(V^*, E^* \setminus F^*)$ is an acirclic spanning tree of G^* (both in ITOP).*

Before we prove Theorem 6.1, let us show that those acirclic spanning trees always exist. We need the following easy fact, whose proof is the same as for finite graphs [7, Lemma 1.9.4].

Lemma 6.2. *Every cut in a graph is a disjoint union of bonds.*

Theorem 6.3 below settles Problem 7.9 of [9].

Theorem 6.3. *Every connected graph G satisfying* (∗) *has a spanning tree whose closure in $\tilde{G}$ contains no circle.*

Proof. We may assume that $G = (V, E)$ is 2-connected, since the union T of acirclic spanning trees of the blocks of G is always an acirclic spanning tree of G. (Indeed, any circle C in the closure of T must contain two edges e, e' from different blocks; if x is a cutvertex separating these blocks in G, then $\mathring{e}$ and $\mathring{e}'$ are separated topologically in the space $\tilde{G} - x$, which contradicts the connectedness of the open arc $C - x$.)

By Lemma 4.4, E has an enumeration $e_1, e_2, \ldots$. Put $S_0 = T_0 = E$, and inductively for $n = 1, 2, \ldots$ define $S_n, T_n \subseteq E$ as follows. Given n, denote by i_n the least index i such that $\tilde{G}$ has a circuit $C \subseteq S_{n-1}$ that contains both e_n and e_i; if there is no such circuit, let $i_n = \infty$. Analogously, choose j_n minimum so that some bond $B \subseteq T_{n-1}$ contains both e_n and e_{j_n}; if there is no such bond, let $j_n = \infty$. If $i_n < j_n$ put $S_n := S_{n-1} - e_n$ and $T_n := T_{n-1}$; if $i_n > j_n$ put $S_n := S_{n-1}$ and $T_n := T_{n-1} - e_n$; if $i_n = j_n$, choose arbitrarily whether to delete e_n from

---

[6] In [9], we considered the more general concept of 'topological spanning trees'. These are acirclic and path-connected subspaces of $|G|$ or $\tilde{G}$, which however need not induce a connected subgraph of G. Our acirclic spanning trees defined above *are* meant to be connected as graphs: they are just graph-theoretical spanning trees of G with the additional property that their closure in $\tilde{G}$ contains no circle (and is therefore a topological spanning tree of $\tilde{G}$).
[7] This assumption is for convenience only. For disconnected graphs, one has to replace 'acirclic spanning tree' with 'subgraph inducing an acirclic spanning tree in every component'.

S_{n-1} or from T_{n-1}. (The ambiguity here is deliberate, to keep the definition symmetrical in S and T. This symmetry will be used later.) Then $S := \bigcap_{n=1}^{\infty} S_n$ and $T := \bigcap_{n=1}^{\infty} T_n$ partition E. More precisely:

For all $n \leq m$, the edge e_n lies in exactly one of the two sets S_m, T_m. (6.1)

We shall prove that S contains no circuit, and that T contains no bond. Then (V, S) is an acirclic spanning tree, completing the proof. (Indeed, if (V, S) is not connected there is a bond B in G such that $B \subseteq E \setminus S = T$, a contradiction.)

So, assume that S contains a circuit or that T contains a bond, and choose i minimum so that there is a set $C \subseteq E$ with $e_i \in C$, and such that $C \subseteq S$ is a circuit in $\tilde{G}$ or such that $C \subseteq T$ is a bond of G. We first assume that $C \subseteq S$, i.e., that C is a circuit.

If C is finite, consider the edge $e_k \in C$ with k maximum. As $e_k \in C \subseteq S \subseteq S_k$, we have $e_k \notin T_k$ by (6.1). Then $j_k \leq i < \infty$, so there is a bond $D \subseteq T_{k-1}$ containing e_k. As $e_k \in C \cap D$, Lemma 4.6 implies that $C \cap D$ contains another edge e_j, with $j \leq k-1$ by the choice of k. As $e_j \in C \cap D \subseteq S_{k-1} \cap T_{k-1}$, this contradicts (6.1).

Therefore C is infinite. Let $e_i, e_{k_1}, e_{k_2}, \ldots$ be distinct edges in C, and note that $i < k_l$ for each l, by the choice of i. Since S_{k_l-1} contains the circuit $C \ni e_{k_l}$ but $e_{k_l} \in S_{k_l}$, there is a bond $D_l \subseteq T_{k_l-1}$ containing e_{k_l} and an edge e_{m_l} with $m_l \leq i$; otherwise we would have deleted e_{k_l} from S_{k_l-1} to obtain S_{k_l}. Since $e_i \in S_{k_l-1}$ implies $e_i \notin T_{k_l-1}$, by (6.1), we cannot have $m_l = i$, so in fact $m_l < i$. Choose $m < i$ so that $m = m_l$ for infinitely many l, and let L_0 denote the set of these l. Thus, $e_m \in D_l$ for every $l \in L_0$.

Put $E^n := \{e_1, \ldots, e_n\}$. Inductively for $n = 1, 2, \ldots$, choose infinite index sets $L_0 \supseteq L_1 \supseteq \ldots$ so that, for each n, the sets $D^n := D_l \cap E^n$ coincide for all $l \in L_n$. We claim that $D = \bigcup_{n=1}^{\infty} D^n$ is a cut of G contained in T. The edge $e_m \in D$ will then lie in some bond $B \subseteq T$ (Lemma 6.2), which contradicts the minimal choice of i as $m < i$.

To show that D is a cut, it suffices to check that D meets every finite circuit C' in an even number of edges (Lemma 4.6). Choose n large enough that $C' \subseteq E^n$. Then

$$C' \cap D = C' \cap E^n \cap D = C' \cap D_l$$

for every $l \in L_n$. But $|C' \cap D_l|$ is even since D_l is a cut, again by Lemma 4.6.

To show that $D \subseteq T$, consider any edge $e_n \in D$. Then $e_n \in D_l$ for every $l \in L_n$. By definition, D_l is a subset of T_{k_l-1}. Now as L_n is infinite, we may assume that $k_l > n$. But then $e_n \in T_{k_l-1}$ implies that $e_n \in T$, as desired.

Finally, if $C \subseteq T$ is a bond rather than a circuit, the proof is analogous to the above, with the roles of S and T and of circuits and bonds interchanged. Instead of Lemmas 4.6 and 6.2 we use Theorems 2.2 and 2.1. □

Proof of Theorem 6.1. (i) ⇒ (ii) Let $T = (V, F)$ be an acirclic spanning tree of G in ITOP. Then $E^* \setminus F^*$ contains no circuit C^* of $\tilde{G}^*$, since C would then be a cut of G missed by T. Similarly $(V^*, E^* \setminus F^*)$ must be connected: if not, then F^* contains a bond of G^*, and F contains the corresponding circuit of $\tilde{G}$. The converse implication follows by symmetry, since G is a dual of G^*.

(ii) ⇒ (i) We show that the map * makes G^* a finitary dual of G. Then Lemma 4.8 implies that for each block B of G the block B^* of G^* is a finitary dual. Then B^* is also a dual of B (with the same map *, by Lemma 4.7), and by Theorem 3.4 the inverse of * makes B a dual of B^*. With Lemma 4.9 we obtain (i).

So consider a finite circuit C of G. We first show that C^* contains a cut of G^*. Using Theorem 6.3, choose an acirclic spanning tree S^* of G^*. If C^* contains no cut, we can join up the components of $S^* - C^*$ by finitely many edges from $E^* \setminus C^*$ to form another spanning tree T^* of G^*. Then T^*, too, is acirclic: any circle in its closure contains an arc A^* that contains infinitely many edges but avoids the (finitely many) new edges and hence lies in the closure of S^*, so the union of A^* with a suitable path from S^* contains a circle in the closure of S^*. Now use (ii) to find an acirclic spanning tree T of G corresponding to T^*. Since T^* contains no edge from C^*, the edges of T include C, a contradiction.

To show that C^* is also a minimal cut in G^*, we show that for every $e \in C$ and $A := C \setminus \{e\}$ the graph $G^* - A^*$ is connected. To do so, it suffices to find a spanning tree T^* of G^* with no edge in A^*, and hence by (ii) to find an acirclic spanning tree of G whose edges include A. Let S be any acirclic spanning tree of G. Since A is finite but contains no circuit, we can obtain another spanning tree T from S by adding all the edges from A and deleting some (finitely many) edges not in A. As before, T is acirclic in $\tilde{G}$ because S was, and hence is as desired.

It remains to show that if B^* is a finite bond in G^* then B is a circuit in G. As before, we first show that B contains a circuit. If not, we can modify an acirclic spanning tree of G into one whose edges include B, which by (ii) corresponds to a spanning tree of G^* that has no edge in B^* (contradiction). On the other hand, given any proper subset D^* of B^*, we can modify an acirclic spanning tree of G^* into one missing D^*, because D^* contains no cut of G^*. Then this tree corresponds by (ii) to a spanning tree of G whose edges include D, so D is not a circuit in G. □

7. Colouring-flow duality and circuit covers

As an application of Theorem 3.4 and our results from Section 4, we now show that the edge set of every bridgeless locally finite planar graph can be covered by two elements of its cycle space. For finite graphs, this is a well-known reformulation of the four colour theorem. For infinite graphs, of course, it must fail as long as the cycle space contains only finite sets of edges.

In our setting, however, Theorem 3.4 enables us to imitate the finite result (and its proof from the four colour theorem), because 4-colourability extends by compactness [4]. Rather than assuming that G is locally finite, we work slightly more generally in $\tilde{G}$.

Theorem 7.1. *Let G be a bridgeless planar graph satisfying (*). Then there are $Z_1, Z_2 \in \mathscr{C}(\tilde{G})$ such that $E(G) = Z_1 \cup Z_2$.*

Proof. Assume that we find for every block B of G elements Z_1^B, Z_2^B of the cycle space of B such that $E(B) = Z_1^B \cup Z_2^B$. From Theorem 2.2 follows that for $i = 1, 2$, $Z_i^B \in \mathscr{C}(G)$ and then also $Z_i := \sum_B Z_i^B \in \mathscr{C}(G)$, where the sum ranges over the blocks of G. Clearly, we get $E(G) = Z_1 \cup Z_2$. As G is bridgeless, we may therefore assume that G is 2-connected.

By Theorem 3.4, G has a dual G^* and is itself a dual of G^*, which therefore is planar too. By the four colour theorem and compactness [4], G^* has chromatic number at most 4. Choose a 4-colouring $c : V(G^*) \to \mathbb{Z}_2 \times \mathbb{Z}_2$ of G^*. For $i = 1, 2$, let $c_i : V(G^*) \to \mathbb{Z}_2$ be c followed by the projection to the ith coordinate, define $f_i : E(G) \to \mathbb{Z}_2$ by $f_i(e) := c_i(v) + c_i(w)$ where v and w are the endvertices of e^*, and put $Z_i := f_i^{-1}(1)$.

Let us show that every edge e of G lies in Z_1 or Z_2. If not, then $f_1(e) = f_2(e) = 0$, and hence $c(v) = c(w)$ for $e^* =: vw$. But this contradicts our assumption that c is a proper colouring of G^*.

Next we show that $Z_i \in \mathscr{C}(\tilde{G})$, for both $i = 1, 2$. By Theorem 2.2, it suffices to show that Z_i meets every finite cut F of G in an even number of edges, i.e., that

$$f_i(F) := \sum_{e \in F} f_i(e) = 0.$$

As every cut is a disjoint union of bonds (Lemma 6.2), we may assume that F is a bond. Then F^* is a circuit in G^*. Hence,

$$f_i(F) = \sum_{e^* = vw \in F^*} (c_i(v) + c_i(w)) = 2 \sum_{u \in U} c_i(u) = 0,$$

where U is the vertex set of the cycle in G^* whose edge set is F^*. $\square$

References

[1] Aharoni, R. (1987) Menger's theorem for countable graphs. *J. Combin. Theory Ser. B* **43** 303–313.

[2] Bruhn, H., Diestel, R. and Stein, M. (2005) Cycle-cocycle partitions and faithful cycle covers for locally finite graphs. *J. Graph Theory* **50** 150–161.

[3] Bruhn, H. and Stein, M. MacLane's planarity criterion for locally finite graphs. To appear in *J. Graph Theory*.

[4] de Bruijn, N. G. and Erdős, P. (1951) A colour problem for infinite graphs and a problem in the theory of relations. *Indag. Math.* **13** 371–373.

[5] Diestel, R. End spaces and spanning trees. To appear in *J. Combin. Theory Ser. B*.

[6] Diestel, R. (2005) The cycle space of an infinite graph. *Combin. Probab. Comput.* **14** 59–79.

[7] Diestel, R. (2005) *Graph Theory*, 3rd edition, Springer. Electronic edition available at: http://www.math.uni-hamburg.de/home/diestel/books/graph.theory/.

[8] Diestel, R. and Kühn, D. (2003) Graph-theoretical versus topological ends of graphs. *J. Combin. Theory Ser. B* **87** 197–206.

[9] Diestel, R. and Kühn, D. (2004) Topological paths, cycles and spanning trees in infinite graphs. *Europ. J. Combin.* **25** 835–862.

[10] Dirac, G. A. and Schuster, S. (1954) A theorem of Kuratowski. *Nederl. Akad. Wetensch. Proc. Ser. A* **57** 343–348.

[11] Freudenthal, H. (1942) Neuaufbau der Endentheorie. *Ann. of Math.* **43** 261–279.

[12] Thomassen, C. (1980) Planarity and duality of finite and infinite graphs. *J. Combin. Theory Ser. B* **29** 244–271.

[13] Thomassen, C. (1982) Duality of infinite graphs. *J. Combin. Theory Ser. B* **33** 137–160.

[14] Whitney, H. (1932) Non-separable and planar graphs. *Trans. Amer. Math. Soc.* **34** 339–362.

Homomorphism-Homogeneous Relational Structures

PETER J. CAMERON[1†] and JAROSLAV NEŠETŘIL[2‡]

[1]School of Mathematical Sciences, Queen Mary, University of London,
Mile End Road, London E1 4NS, UK
(e-mail: `p.j.cameron@qmul.ac.uk`)

[2]Department of Applied Mathematics and Institute of Theoretical Computer Sciences,
Charles University, Malostranské Nám. 25, 11800 Praha, Czech Republic
(e-mail: `nesetril@kam.mff.cuni.cz`)

Received 23 August 2004; revised 24 February 2005

For Béla Bollobás on his 60th birthday

We study relational structures (especially graphs and posets) which satisfy the analogue of homogeneity but for homomorphisms rather than isomorphisms. The picture is rather different. Our main results are partial characterizations of countable graphs and posets with this property; an analogue of Fraïssé's theorem; and representations of monoids as endomorphism monoids of such structures.

1. Introduction

A graph G (or more general relational structure) is *homogeneous* if any isomorphism between finite induced subgraphs of G can be extended to an automorphism of G. The homogeneous graphs can be recognized by the fact that their collections of finite subgraphs have the amalgamation property (Fraïssé's theorem). The finite homogeneous graphs were determined by Gardiner [8] and the countably infinite ones by Lachlan and Woodrow [13]. Other determinations of homogeneous structures in various classes include posets (Schmerl [17]), tournaments (Lachlan [12]), permutations (Cameron [3]), and digraphs (Cherlin [4]). These structures are important in many parts of mathematics: see Hubička and Nešetřil [15, 11] for the connection with Ramsey theory, for example.

In this paper we consider what happens if we replace 'isomorphism' in the definition of homogeneity by 'homomorphism'. A homomorphism of a graph, for example, is a

---

[†] This research was carried out while the first author was visiting KAM/ITI in Prague. He is grateful to the Department and Institute for their hospitality.
[‡] Supported by Czech grants LN00A056, 1M0021620808 and ICREA, Barcelona.

function which maps vertices to vertices and preserves the edges. A monomorphism is one-to-one homomorphism (see, *e.g.*, [10]).

There are several different conditions. We say that a graph G belongs to the class

HH if every homomorphism from a finite induced subgraph of G into G extends to a homomorphism from G to G;

MH if every monomorphism from a finite induced subgraph of G into G extends to a homomorphism from G to G;

MM if every monomorphism from a finite induced subgraph of G into G extends to a monomorphism from G to G.

Clearly both HH and MM are included in the class MH. So we begin with some structural results for MH graphs. Later we show that it is the class MM in which an analogue of Fraïssé's theory can be developed for arbitrary relational structures.

Proposition 1.1.

(a) *Any disjoint union of complete graphs all of the same size is HH, and hence MH. If a disjoint union of complete graphs is MH, then the complete graphs all have the same size.*
(b) *If an MH graph is disconnected, or if it is finite, then it is a disjoint union of complete graphs of the same size.*
(c) *If an MH graph is connected but not finite, then it has diameter at most 2 and contains no finite maximal clique (so, in particular, every edge is contained in a triangle).*

Proof. The first part of (a) is clear. Moreover, if an MH graph has two components A and B which are complete, then for $a \in A$ and $b \in B$, the map $a \mapsto b$ extends to a homomorphism which maps A injectively to B, so $|A| \leqslant |B|$. Similarly $|B| \leqslant |A|$. So $|A| = |B|$ by the Cantor–Schröder–Bernstein theorem.

Suppose that G is MH and has a component which is not complete: equivalently, it has an induced path x, y, z. For any two distinct vertices a, b, the map $x \mapsto a$, $z \mapsto b$ extends to a homomorphism, which maps y to a common neighbour of a and b. This shows that G is connected and has diameter 2 and every edge is in a triangle. This proves (b) for disconnected graphs, and also the first part of (c).

Let C be a finite clique in an MH graph and suppose there is a vertex v with degree at least $|C|$. Any injective map from $|C|$ neighbours of v to C extends to a homomorphism, which must map v to a common neighbour of C, so C is not maximal.

Suppose that C is a finite maximal clique. If C is not a component of G, then it contains a vertex v with a neighbour outside C; then the degree of v is at least $|C|$, contrary to the preceding paragraph. This finishes the proof of (c). □

The classification of finite MM graphs is simpler.

Proposition 1.2. *The only finite MM graphs are the complete and null graphs.*

Proof. Suppose that G is a finite MM graph which is neither complete nor null. Let the vertices a and b be adjacent, and the vertices c and d be non-adjacent. Then the map

$c \mapsto a$, $d \mapsto b$ extends to a monomorphism of G, which strictly increases the number of edges, which is clearly impossible. □

2. Graphs spanned by R

Let R be the countable random graph (the 'Rado graph' [16]). Recall that R is characterized as a countable graph with the property that, if U and V are finite disjoint sets of vertices, there is a vertex z joined to all vertices in U and to none in V. See [2] for more information about R.

Proposition 2.1.

(a) *A countable graph contains R as a spanning subgraph if and only if it has the property that any finite set of vertices has a common neighbour.*
(b) *Any graph containing R as a spanning subgraph is HH and MM, and hence MH.*
(c) *If G is an MH-graph which does not contain R as a spanning subgraph, then there is a bound on the size of claws (induced stars $K_{1,n}$) in G.*

Proof. (a) The property holds in R, and hence certainly in any graph obtained by adding extra edges.

Conversely, let G be a countable graph satisfying the property. Construct a bijection between R and G by the back-and-forth method, except that in going from R to G we do not insist that non-edges are preserved. In more detail: we define a map $f : R \to G$ recursively. At odd-numbered steps, take the first vertex of R on which f is not yet defined, and map it to a common neighbour of the range of f. At even-numbered stages, take the first vertex v not in the range of f, choose $v' \in R$ such that, for all u in the domain of f, $u \sim v'$ if and only if $f(u) \sim v$.

(b) If the property of (a) holds, then certainly homomorphisms extend: if we have defined f on $v_0, \ldots, v_{n-1}$, then choose $f(v_n)$ to be any vertex adjacent to all of $f(v_0), \ldots, f(v_{n-1})$. Moreover, if f is one-to-one, then so is the extension.

(c) Suppose that G is a countable MH graph which contains claws of unbounded size. Let U be a finite set of vertices, with $|U| = n$. Find a claw $K_{1,n}$ in G and map its independent vertices bijectively to U. The remaining vertex is mapped to a neighbour of U. So G satisfies the condition of (a). □

Corollary 2.2.

(a) *There is a countable graph which is homomorphism-homogeneous but is automorphism-rigid.*
(b) *There is a countable graph which is homomorphism-homogeneous but its complement is homomorphism-rigid.*

Proof. The graph shown in Figure 1 is automorphism-rigid and for every finite subset U there is a vertex joined to no vertex in U. So the complement is also automorphism-rigid and contains R as a spanning subgraph. This proves (a).

Figure 1.

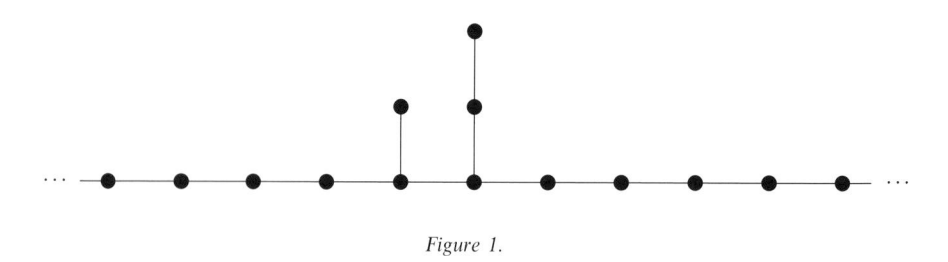

To prove (b) it suffices to note that there exists a countable homomorphism-rigid graph with all its degrees ≤ 3: see [9]. The complement then contains R as a spanning subgraph. □

Here are two questions which we have not been able to resolve.

Problem 1. Is there a graph which is MH but not HH?

We remark that for more general structures than graphs we prove below that the classes MH and HH are different.

Problem 2. Is there a countable graph which is HH but not a disjoint union of complete graphs and does not contain R as a spanning subgraph?

A positive answer to this problem would yield a graph which contains a finite set of vertices with no common neighbours, and there is a bound on the size of its claws. One famous class of graphs with bounded claw size consists of *line graphs* $L(G)$ of graphs; these contain no 3-claw $K_{1,3}$. We show that at least in this class we obtain no new examples.

Proposition 2.3. *Let G be a finite or countable graph with the property that $L(G)$ is MH. Then G is a disjoint union of stars of the same size (and hence $L(G)$ is a disjoint union of complete graphs of the same size).*

Proof. By Proposition 1.1, we can assume that G is infinite and connected with bounded diameter. So G contains a vertex v of infinite degree.

First we show that G is triangle-free. Suppose that $\{a,b,c\}$ is a triangle in G, and let p,q,r,s be neighbours of v. The map $vp \mapsto ab$, $vq \mapsto bc$, $vr \mapsto ca$ (defined on an induced subgraph of $L(G)$) extends to a homomorphism of $L(G)$, under which vs must map to an edge meeting all three edges of the triangle, which is impossible.

Now we show that any neighbour of v has degree 1. For suppose that p is a neighbour of v which is also adjacent to a vertex x (necessarily not adjacent to v), and let q and r be two further neighbours of v. The map $vq \mapsto vq$, $vr \mapsto vp$, $px \mapsto px$ extends to a homomorphism, which must map vp to an edge containing v and meeting px; this edge cannot be vp, and if it is vx then the graph contains a triangle. So no such vertex can exist.

Thus, the connected component containing v is an infinite star, and we are done. $\square$

More generally, let $\mathscr{K}(k,l)$ be the class of finite graphs defined as follows, where k and l are integers with $1 < l < k$: the vertex set is an arbitrary set $\mathscr{M}$ of k-sets; two vertices are adjacent if and only if they intersect in at least l points. We call this the $(\geqslant l)$-*intersection graph* of $\mathscr{M}$. (So line graphs form the class $\mathscr{K}(2,1)$. More generally, in the case $l = 1$, we refer to such a graph as the *intersection graph* of the collection of k-sets.)

Now if the k-sets $(X, Y_1, \ldots, Y_r)$ form a claw in the $(\geqslant l)$-intersection graph, then the intersections $X \cap Y_i$ for $1 \leqslant i \leqslant r$ all have size at least l and are all distinct; so certainly there is no 2^k-claw in such a graph. We would like to prove that no countable graph of this type can be homogeneous unless it is a disjoint union of complete graphs. Here is a first step.

Proposition 2.4. *For $k \geqslant 3$, let $\binom{X}{k}$ denote the set of all k-subsets of a countable set. Then the $(\geqslant l)$-intersection graph of $\binom{X}{k}$ is not MH.*

Proof. There always exists a $(\geqslant l)$-intersecting family of k-subsets with the property that no further k-set intersects every set in the family. For example, the set $\binom{Y}{k}$ of all k-subsets of a $(2k-l)$-set Y obviously has this property, since a set containing fewer than k points of Y has intersection smaller than l with some set in $\binom{Y}{k}$.

Now there is also an infinite star (a set of k-sets containing l common points). Now the map taking $\binom{2k-l}{k}$ sets of the star bijectively to $\binom{Y}{k}$ cannot be extended to a further set of the star. $\square$

A final observation about MM graphs:

Proposition 2.5. *Any infinite non-null MM graph contains an infinite complete subgraph.*

Proof. Let G be such a graph.

Suppose first that G is disconnected, and so a union of complete graphs. If the components are finite (of size $n > 1$, say), take two points a, b in different components and map them to two points in the same component; it is clear that no monomorphism can extend this map. So the components are infinite, as required.

The case where G is connected follows immediately from Proposition 1.1(c). $\square$

3. Posets

A homomorphism of posets is a map f such that, if $x < y$, then $f(x) < f(y)$. Now the definitions of the classes HH, MH and MM of posets are exactly as for graphs. Here is the start of an attempt to classify the MH posets.

First, the analogue of the fact that graphs containing R as spanning subgraph have the MM and HH properties also holds here. Let U denote the generic (universal and homogeneous) countable poset. An *extension* of U is a poset P on the same set such that, if $x \leqslant_U y$, then $x \leqslant_P y$.

Proposition 3.1.

(a) *A countable poset P is an extension of U if and only if it has the following property:*

   (†) *for any two finite sets A and B with $A < B$, there is a point z such that $A < z < B$.*

(b) *Any extension of U has the HH and MM properties.*

Proof. (a) The argument is similar to that for graphs. Recall that U is characterized by the property that, for any finite disjoint sets A, B, C with $A < B$ and, for all $a \in A$, $b \in B$, $c \in C$, we have $c \not< a$ and $b \not< c$, there is a point z with $A < z < B$ and z and c incomparable for all $c \in C$.

Now construct a bijection from U to P which preserves comparability as follows. Enumerate P and U. Suppose that f has been defined on $u_1, \ldots, u_n \in U$. If n is even, choose the first unused point x of U; let A and B be the subsets of $\{u_1, \ldots, u_n\}$ consisting of points less than and greater than x respectively. Then $A < B$, so $f(A) < f(B)$; by (†), there is a point z such that $A < z < B$, and we can map x to z. If n is odd, choose the first unused point y of P, and choose a point $z \in P$ incomparable with all of $u_1, \ldots, u_n$; then map z to y.

(b) Let P be an extension of U, and f a homomorphism between finite subsets of P. Now we can extend f to all of P as follows. Suppose that f has been defined on $p_1, \ldots, p_n$. Let A and B be the sets of elements of $\{p_1, \ldots, p_n\}$ which are respectively less than and greater than p_{n+1}. As before, choose z with $f(A) < z < f(B)$, and map p_{n+1} to z. The extension can be taken to be one-to-one everywhere except possibly on the points where f was initially defined. □

Now we gather a few facts about MH posets. Clearly any antichain is HH and MM.

Proposition 3.2. *Let P be a countable poset which is MH but not an antichain.*

(a) *Any maximal chain in P is dense and without endpoints.*

(b) *If P is disconnected then it is a disjoint union of incomparable chains each isomorphic to the rationals.*

(c) *If there is a 2-element antichain in P which has an upper bound, then any finite antichain has an upper bound.*

Proof. (a) By assumption, there exist a and b with $a < b$. Extending the map $a \mapsto x$ for any x, we see that there is an element above x. (Dually there is an element below x.) In particular, there is a 3-element chain $a < b < c$. Now, if $x < y$, extending the map $a \mapsto x$, $c \mapsto y$, we see that the image of b is an element z with $x < z < y$.

(b) If P has a component which is not a chain, then it contains a 2-element antichain with either an upper or a lower bound. By the MH property, every 2-element antichain has a bound; so there is only one connected component.

(c) As in (b), if some 2-element antichain has an upper bound, then so does every 2-element antichain. Now, inductively, let $a_1, \ldots, a_n$ be an antichain. Let b_i be an upper bound of a_i and a_{i+1} for $i = 1, \ldots, n-1$. Now the set of maximal elements among $b_1, \ldots, b_{n-1}$ is an

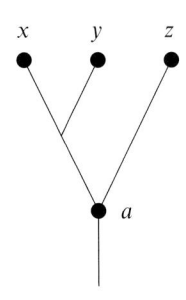

Figure 2.

antichain of size smaller than n; by induction it has an upper bound c, which is also an upper bound for $a_1, \ldots, a_n$. □

Consequently all homogeneous posets have the MM property. A bit surprisingly there is another example (which is also not an extension of the generic poset U, but is more analogous to a claw-free graph). Recall that a *tree order* is a poset in which the elements below any given element x form a chain. Tree orders may have large symmetry groups (see Droste [5]) while not being homogeneous (in the usual sense). We prove that they may have the MM property. Let T be a countable tree order satisfying the following additional properties.

(a) T is dense.
(b) T has neither maximal nor minimal element.
(c) If an element a of T is an infimum of a finite subset A of T then $a \in A$ (in other words: infima and minimal elements of finite subsets coincide).
(d) T is infinitely branching (*i.e.*, for every $x < y$ there exists an antichain $z_i; i = 1, 2, \ldots$, with $x < z_i$ and $y \not\leq z_i$ for every $i = 1, 2, \ldots$).

An example of T can be obtained as follows. First we split the set of rational numbers in two dense sets, say D, D'. We form an infinite tree of copies of D by adding infinitely many branching copies of D at any element of D' and continuing recursively in this way.

We have the following.

Proposition 3.3.

(a) *T is not homogeneous (in the usual sense).*
(b) *T has the MM property.*

Proof. The reason why T is not homogeneous in the ordinary sense can be seen from Figure 2.

(Of course a is just a reference point which does not belong to the tree T.) It is clear that no automorphism can interchange x with z and fix y. (But there is a monomorphism that will do this: simply map the points below x, y, z to suitable points below a.)

To prove (b), it suffices to prove that for any $x \notin A$ any monomorphism $f : A \longrightarrow B$ may be extended to a monomorphism $f' : A' \longrightarrow B'$ where $A' = A \cup \{x\}$. This may be

```
         mono
    B₁ ──────→ C
    ↑          ↑
emb │          │ emb
    │   mono   │
    A ──────→ B₂
```

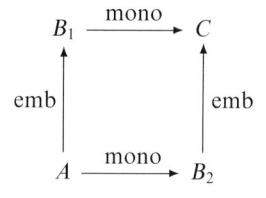

Figure 3.

seen as an extension of the above example. Given $x \notin A$ denote by A_x the set of all $z \in A$ such that $x \leqslant z$. Let $B_x = \{f(y); y \in A_x\}$. Consider the infimum a of the set $A_x \cup B_x$ (it may not belong to T). If $a \geqslant x$ then we put $f'(x) = x$. If $a < x$ then necessarily $B_x \setminus A_x$ is non-empty. In this situation we distinguish two cases.

- If there is an element z of A satisfying $a < z < x$, we choose maximal such z, and we let $f'(x)$ be the element above $f(z)$ close enough to be distinct from all of A and B.
- If there is no element of A satisfying $a < z < x$ we let $f'(x)$ be an element below a which is such that $f'(x) > z$ for all $z < a, z \in A \cup B$. □

4. Homogeneity and amalgamation

The definitions of the classes HH, MH and MM work in the same way for arbitrary relational structures as for graphs or posets. We are going to develop a characterization of MM structures in general. As usual, the *age* of a structure is the class of finite structures embeddable in it; and the *joint embedding property* or JEP is as usual: $\mathscr{C}$ has the JEP if any two members of $\mathscr{C}$ can be embedded in some member of $\mathscr{C}$.

The *mono-amalgamation property* (for short, MAP) of a class $\mathscr{C}$ of finite relational structures is the following assertion.

For any $A, B_1, B_2 \in \mathscr{C}$, and any maps $f_i : A \to B_i$ (for $i = 1, 2$) such that f_1 is an embedding (an isomorphism to an induced substructure) and f_2 a monomorphism, there exists $C \in \mathscr{C}$ and monomorphisms $g_i : B_i \to C$ for $i = 1, 2$ such that $g_1 \circ f_1 = g_2 \circ f_2$ and g_2 is an embedding (the diagram in Figure 3 commutes).

Note the asymmetry between B_1 and B_2!

The *mono-extension property* of a structure M with age $\mathscr{C}$ is the following property.

If $B \in \mathscr{C}$ and A is an induced substructure of B, then every monomorphism $A \to M$ extends to a monomorphism $B \to M$.

Now our analogue of Fraïssé's theorem is the following result.

Proposition 4.1.

(a) *A countable structure is MM if and only if it has the mono-extension property.*
(b) *The age of any MM-structure has the mono-amalgamation property.*
(c) *If a class $\mathscr{C}$ of finite relational structures is isomorphism-closed, closed under induced substructures, has only a countable number of isomorphism classes, and has the JEP and the MAP, then there is a countable MM structure M whose age is equal to $\mathscr{C}$.*

Proof. **(a)** If M has the mono-extension property, then clearly any monomorphism from a finite substructure of M can be extended (one point at a time) to a monomorphism of M.

Conversely, suppose that M is an MM structure. Let B be a structure in the age of M and $A \subseteq B$: without loss of generality, $B \subseteq M$. Suppose that f is any monomorphism from A into M. Then f extends to a monomorphism g of M, whose restriction to B is the required monomorphism $B \to M$.

(b) Suppose that M is an MM structure and let A, B_1, B_2, f_1, f_2 be as in the hypothesis of the mono-amalgamation property, with A, B_1, B_2 in the age of M. As in (a), we can assume that $B_1, B_2 \subseteq M$ and f_1 is the identity on A. Now extend f_2 to a monomorphism g of M; let $C = B_1 g$, g_1 the restriction of g to B_1, and g_2 the identity on B_2.

(c) We build the structure in stages; even and odd numbered stages achieve different parts of the construction. Suppose that the finite structure $M_i \in \mathscr{C}$ has been constructed at stage i.

If i is even, we can use the JEP to find a structure M_{i+1} containing M_i and any given structure $A \in \mathscr{C}$.

If i is odd, we select a pair (A, B) of structures in $\mathscr{C}$ with $A \subseteq B$. Now a given monomorphism $A \to M'$ can be extended to a monomorphism $B \to M''$ for some $M'' \supseteq M'$, by the MAP. Applying this successively for each monomorphism $A \to M_i$, we obtain the structure M_{i+1} so that every monomorphism $A \to M_i$ extends to a monomorphism $B \to M_{i+1}$.

Arranging the stages so that every structure in $\mathscr{C}$ occurs at some even stage and every pair (A, B) at infinitely many odd stages, we finally build a countable structure M. Every finite substructure of M is contained in one of the finite structures M_i and thus belongs to $\mathscr{C}$ by the induced substructure property. It follows also that $\mathscr{C}$ is the age of M with the mono-extension property. Hence M is an MM-structure. □

We say that a class $\mathscr{C}$ having the properties of part (c) of the theorem is a *mono-Fraïssé class*, and that an MM structure with age $\mathscr{C}$ is a *mono-limit* of $\mathscr{C}$. Unlike the usual form of Fraïssé's theorem, it is not the case that a class satisfying the mono-amalgamation property has a unique mono-limit (up to isomorphism). Indeed, there are many examples of graphs containing R as spanning subgraph whose age is the class of all finite graphs.

However, any two such structures M and M' must bear a certain resemblance to each other. For example, there is a monomorphism from M to M' and *vice versa*. In fact, more is true; we can characterize this equivalence as follows.

Let M and M' be two structures. We say that M and M' are *mono-equivalent* if

- $\mathrm{Age}(M) = \mathrm{Age}(M')$;
- every embedding of a finite substructure A of M into M' extends to a monomorphism from M to M', and *vice versa* (with M and M' reversed).

This turns out to be the relation which replaces isomorphism in our version of Fraïssé's theorem.

Proposition 4.2.

(a) *Suppose that M and M' are mono-equivalent structures. If M is an MM structure, then M' is an MM structure too.*

(b) *Conversely, if M and M' are MM structures with $\mathrm{Age}(M) = \mathrm{Age}(M')$, then they are mono-equivalent.*

Proof. (a) Suppose that M and M' are equivalent and that M has the MM property. Take $A, B \in \mathrm{Age}(M')$ with $A \subseteq B$, and let $f : A \to M'$ be a monomorphism. By assumption, we may assume that $B \subseteq M$.

Let A' be the image of f. Since $\mathrm{Age}(M) = \mathrm{Age}(M')$, we can find a copy A'' of A' within M; in other words, there is a monomorphism $\phi : A \to A''$ and an isomorphism $g : A'' \to A'$ such that $g \circ \phi = f$. Since M has the MM property, the monomorphism ϕ extends to a monomorphism $\phi^* : M \to M$. Let $B'' = \phi^*(B)$. Also, by assumption, the isomorphism g extends to a monomorphism $g^* : M \to M'$. Now the restriction of $g^* \circ \phi^*$ to B is a monomorphism $B \to M'$ extending the given monomorphism f. So M' has the mono-extension property, and hence it is an MM structure.

(b) Suppose that M and M' are MM structures with the same age. Let A be a finite substructure of M and $f : A \to M'$ an embedding. For any $B \supseteq A$, the mono-extension property in M' allows us to extend f to a monomorphism $B \to M'$. So there is a monomorphism $M \to M'$ extending f. Thus M and M' are equivalent. □

The above proof that the MM property for M implies that of M' uses only half of the definition of equivalence. Let us say that $M \leq M'$ holds if

- $\mathrm{Age}(M) \supseteq \mathrm{Age}(M')$, and
- any embedding of a finite substructure of A into M' extends to a monomorphism from M to M'.

This relation between structures defines a partial order on the set of equivalence classes. (The reverse ordering of ages looks strange, but consider graphs containing R as spanning subgraph: intuitively, the more extra edges we add, the smaller the age becomes.) Then the proof of Proposition 4.2(a) actually shows that, if M is an MM structure and $M \leq M'$, then M' is also an MM structure.

We mention one fact about this order for graphs.

Proposition 4.3. *Let R be the Rado graph. Then a countable graph G satisfies $R \leq G$ if and only if R is a spanning subgraph of G.*

Proof. Since R is universal, the condition $\mathrm{Age}(R) \supseteq \mathrm{Age}(G)$ is trivial.

Suppose that R is a spanning subgraph of G, and let $f : A \to G$ be an embedding of a finite subgraph A of R into G. Since there is a common neighbour of $f(A)$ in G by Proposition 2.1, it is always possible to extend f to a monomorphism on one extra point.

Conversely, suppose that $R \leq G$, and let U be a finite set of vertices in G. Let A be a subgraph of R isomorphic to the subgraph U (by means of the isomorphism f), and z a common neighbour of A in R. The map f extends to a monomorphism from R to G

(by assumption), and the image of z is a common neighbour of U. So G contains R as a spanning subgraph, again by Proposition 2.1. □

On the other hand, it is not true that if M and M' are MM structures and $\mathrm{Age}(M) \supseteq \mathrm{Age}(M')$, then $M \leq M'$. For example, let M be the Rado graph R, and M' the disjoint union of two infinite complete graphs. The map taking a non-edge in M to a non-edge in M' clearly cannot be extended.

5. Algebraic classes

We say that a class $\mathscr{K}$ of finite structures has the ME-property if the classes of monomorphisms and embeddings coincide for $\mathscr{K}$.

Proposition 5.1. *Let $\mathscr{K}$ be a mono-Fraïssé class of finite structures with the ME-property. Then any two mono-limits of $\mathscr{K}$ are isomorphic.*

Examples:
- all algebraic classes (containing only function and constant symbols)
- tournaments (see below).

Commentary. Clearly any monomorphism between finite algebras is an embedding. For if ρ is an r-ary operation in the algebra and f a monomorphism, then clearly $\rho(f(x_1), \ldots, f(x_r)) = f(\rho(x_1, \ldots, x_r))$.

Now, given any loopless directed graph D, it is possible to define an algebra on the vertex set of D with a single binary operation $\cdot$ by the rules

$$x \cdot x = x,$$

and if $x \neq y$, then $x \cdot y = \begin{cases} x & \text{if there is an arc } x \to y, \\ y & \text{otherwise.} \end{cases}$

Any monomorphism of this algebra is a digraph embedding. Note that not all digraph monomorphisms are algebra monomorphisms! If the digraph is a tournament then the two types of monomorphisms coincide: see [14].

This example shows that, even in the case of algebras of a given type, we may have $2^{\aleph_0}$ non-isomorphic MM-structures (in this case, the algebras of the homogeneous digraphs determined by Cherlin [4]).

A more familiar example of this phenomenon occurs with abelian groups. For any set Π of prime numbers, the direct product of the countable abelian group of exponent p for all $p \in \Pi$ is homogeneous.

The above example of tournament algebras (sometimes called *quasitrivial algebras*) also yield an example of structures showing MH $\neq$ HH: Consider the algebra A_T corresponding to the universal homogeneous tournament T. This algebra is obviously an MM structure. However it fails to be an HH structure as every finite tournament may be extended to a finite tournament which is *simple* (i.e., does not have any nontrivial congruence); see [6, 14]. (This may be seen as follows: if T_0 is a non-simple finite tournament, $f : T_0 \to T_0$ a homomorphism which is not one-to-one, $T_0 \subset T_1$ an inclusion with T_1 a simple tournament, then no homomorphism $f : T_0 \to T_0$ which is not one-to-one can be extended to a homomorphism $T \to T$.)

6. Representing closed monoids

There is a natural topology on X^X, for a countable set X, namely the product topology induced from the discrete topology on X. Thus the basic open sets are of the form

$$\{f \in X^X : f(x_i) = y_i \text{ for } i = 1, \ldots, n\},$$

where $x_1, \ldots, x_n, y_1, \ldots, y_n \in X$ and $x_1, \ldots, x_n$ are distinct. It is known that, in the induced topology on the symmetric group $\text{Sym}(X)$, a permutation group G is closed if and only if it is the automorphism group of a homogeneous relational structure on X (see [1]). A similar observation holds here.

Proposition 6.1.

(a) *A submonoid S of X^X is closed in the product topology on X^X if and only if S is the monoid $\text{End}(M)$ of endomorphisms of an HH relational structure M on X.*

(b) *A submonoid S of the monoid of one-to-one maps $X \to X$ is closed in the product topology if and only if S is the monoid of monomorphisms of an MM relational structure M on X.*

Proof. (a) First note that $\text{End}(M)$ is always closed in X^X. In the reverse direction assume that S is a fixed submonoid of X^X. For each n, and each $\bar{x} \in X^n$, we take an n-ary relation $R_{\bar{x}}$ defined by

$$R_{\bar{x}}(\bar{y}) \Leftrightarrow (\exists s \in S)(\bar{y} = s(\bar{x})).$$

Let M be the relational structure with relations $R_{\bar{x}}$ for all n-tuples $\bar{x}$ (and all n). We claim that S acts as endomorphisms of M, that M is HH, and $\text{End}(M) = S$.

For the first point, take $s \in S$ and $\bar{y} \in X^n$ such that $R_{\bar{x}}(\bar{y})$ holds; we must show that $R_{\bar{x}}(s(\bar{y}))$ holds. But $\bar{y} = s'(\bar{x})$ for some $s' \in S$; then $s(\bar{y}) = ss'(\bar{x})$, so the assertion is true.

Next, let f be a homomorphism between finite subsets of X, say $f(x_i) = y_i$ for $i = 1, \ldots, n$. Let $\bar{x} = (x_1, \ldots, x_n)$ and $\bar{y} = (y_1, \ldots, y_n)$. Now S is a monoid and so contains the identity mapping. Thus, by definition, $R_{\bar{x}}(\bar{x})$ holds. Since f is a homomorphism, $R_{\bar{x}}(\bar{y})$ holds. So by definition, there exists $s \in S$ such that $s(\bar{x}) = \bar{y}$. Now s is an endomorphism of M extending f. So M is HH.

Finally, to show that $\text{End}(M) = S$, we know already that $S \subseteq \text{End}(M)$ and have to prove the reverse inclusion. We must take $h \in \text{End}(M)$ and show that every basic neighbourhood of h contains an element of S, so that h is a limit point of S. Since S is assumed closed, we conclude that $h \in S$. Now each n-tuple $\bar{x}$ defines a basic neighbourhood of h, consisting of all functions g such that $g(\bar{x}) = h(\bar{x})$. Now $R_{\bar{x}}(\bar{x})$ holds; since h is a homomorphism, $R_{\bar{x}}(h(\bar{x}))$ also holds, and by definition of $R_{\bar{x}}$ this means that there exists $s \in S$ with $h(\bar{x}) = s(\bar{x})$, as required.

(b) The proof of this is entirely analogous, replacing homomorphisms by monomorphisms. $\square$

The relational structures constructed in the proof have infinitely many relations of each arity. It would be interesting to recognize the monoids which are the endomorphism monoids (or monomorphism monoids) of homogeneous structures with only finitely many relations of each arity (these would be the analogue of the closed *oligomorphic* permutation groups, [1]), or even those with only finitely many relations altogether.

Acknowledgement

We thank the anonymous referee for remarks which improved presentation of this paper.

References

[1] Cameron, P. J. (1990) *Oligomorphic Permutation Groups*, Vol. 152 of *London Math. Soc. Lecture Notes*, Cambridge University Press, Cambridge.
[2] Cameron, P. J. (1996) The random graph. In *The Mathematics of Paul Erdős* (J. Nešetřil and R. L. Graham, eds), Springer, Berlin, pp. 331–351
[3] Cameron, P. J. (2002) Homogeneous permutations. *Electron. J. Combin.* **9** #R2 (9pp.)
[4] Cherlin, G. L. (1998) The classification of countable homogeneous directed graphs and countable homogeneous n-tournaments. *Mem. Amer. Math. Soc.* **621**, AMS, Providence, RI.
[5] Droste, M. (1985) Structure of partially ordered sets with transitive automorphism groups. *Mem. Amer. Math. Soc.* **57**, AMS, Providence, RI.
[6] Erdős, P., Hajnal, A. and Milner, E. C. (1972) Simple one-point extensions of tournaments. *Mathematika* **19** 57–62.
[7] Fraïssé, R. (1953) Sur certains relations qui généralisent l'ordre des nombres rationnels. *CR Acad. Sci. Paris* **237** 540–542.
[8] Gardiner, A. D. (1976) Homogeneous graphs, *J. Combin. Theory Ser. B* **20** 94–102.
[9] Hell, P. and Nešetřil, J. (1973) Groups and monoids of regular graphs (and of graphs with bounded degrees). *Canad. J. Math.* **25** 239–251.
[10] Hell, P. and Nešetřil, J. (2004) *Graphs and Homomorphisms*, Oxford University Press.
[11] Hubička, J. and Nešetřil, J. Finite presentation of homogeneous graphs, posets and Ramsey classes. KAM-DIMATIA Series, pp. 204–675. To appear in *Israel J. Math.*
[12] Lachlan, A. H. (1984) Countable homogeneous tournaments. *Trans. Amer. Math. Soc.* **284** 431–461.
[13] Lachlan, A. H. and Woodrow, R. E. (1980) Countable ultrahomogeneous undirected graphs. *Trans. Amer. Math. Soc.* **262** 51–94.
[14] Müller, V., Nešetřil, J. and Pelant, J. (1975) Either tournaments or algebras? *Discrete Math.* **11** 37–66.
[15] Nešetřil, J. (2005) Ramsey classes and homogeneous structures. *Combin. Probab. Comput.* **14** 171–189.
[16] Rado, R. (1964) Universal graphs and universal functions. *Acta Arith.* **9** 331–340.
[17] Schmerl, J. H. (1979) Countable homogeneous partially ordered sets. *Algebra Universalis* **9** 317–321.

A Spectral Turán Theorem

FAN CHUNG[†]

Department of Mathematics, University of California, San Diego, CA 92093-0112, USA
(e-mail: fan@ucsd.edu)

Received 15 December 2003; revised 30 May 2004

For Béla Bollobás on his 60th birthday

If all nonzero eigenvalues of the (normalized) Laplacian of a graph G are close to 1, then G is t-Turán in the sense that any subgraph of G containing no K_{t+1} contains at most $(1 - 1/t + o(1))e(G)$ edges where $e(G)$ denotes the number of edges in G.

1. Introduction

One of the classical theorems in graph theory is Turán's theorem, which states that a graph on n vertices containing no K_{t+1} can have at most $(1 - 1/t + o(1))\binom{n}{2}$ edges. Sudakov, Szabó and Vu [6] consider a generalization of Turán's theorem. A graph G is said to be t-Turán if any subgraph of G containing no K_{t+1} has at most $(1 - 1/t + o(1))e(G)$ edges, where $e(G)$ denotes the number of edges in G. In [6], it is shown that a regular graph on n vertices with degree d is t-Turán if the second-largest eigenvalue of its adjacency matrix λ is sufficiently small.

In this paper, we consider Turán numbers for general graphs as introduced in [6]. For two given graphs G and H, the Turán number $t(G, H)$ is defined to be

$$t(G, H) = \max\{e(G') : G' \text{ is a subgraph of } G \text{ containing no } H\}.$$

The classical Turán number is the special case that G is a complete graph K_n. Turán's theorem implies

$$t(K_n, K_{t+1}) = \left(\frac{t-1}{t} + o(1)\right)\binom{n}{2}.$$

In this paper, we will show that

$$t(G, K_{t+1}) = \left(\frac{t-1}{t} + o(1)\right)e(G) \tag{1.1}$$

as long as certain spectral bounds of G are satisfied (to be specified in Section 4).

[†] Research supported in part by NSF grant DMS 0100472 and ITR 0205061.

Since any t-partite subgraph of G contains no K_{t+1}, the inequality $t(G, K_{t+1}) \geq (1 - \frac{1}{t} + o(1))e(G)$ always holds. Thus, equation (1.1) implies that a maximum t-partite subgraph of G is an extremal graph having the maximum number of edges among all subgraphs of G containing no K_{t+1}. In Section 4, we will show that our main theorem implies (the asymptotic version of) the classical Turán theorem as a special case. Another consequence of our main theorem is the result in [6] for d-regular graphs. Namely, if the second-largest eigenvalue μ of the adjacency matrix of a d-regular graph on n vertices satisfies $\mu \ll d^t/n^{t-1}$, then $t(G, K_{t+1}) = (1 - 1/t + o(1))dn/2$. This will also be proved in Section 4.

In order to derive the relationship between the spectral bounds and the Turán property, we will first consider eigenvalues of the (normalized) Laplacian. Detailed definitions will be given in the next section.

The connection between eigenvalues of the Laplacian and Turán numbers depends on a notion of generalized volumes: for a subset X of vertices in a graph G, the k-volume of X is defined by

$$\operatorname{vol}_k(X) = \sum_{v \in X} d_v^k,$$

where d_v denotes the degree of v in G. We will first describe several key properties of graphs which are consequences of spectral gaps. In particular, we will give several general isoperimetric inequalities in Section 3. These inequalities provide good estimates for the 'discrepancies' of a graph. We will use these inequalities to establish the relationship between eigenvalues and the Turán property. We will show that if the nonzero eigenvalues of the (normalized) Laplacian are bounded (depending on t and the volumes of G), then the graph is t-Turán. The proofs are given in Section 4.

2. Preliminaries on eigenvalues

For a graph G, there are several ways to evaluate eigenvalues by associating various matrices with G. A typical matrix is the adjacency matrix $A = A_G$ which has entries $A(u, v) = 1$ if u and v are adjacent, and 0 otherwise. Another matrix is the combinatorial Laplacian L which is defined as $L = D - A$ where D is the diagonal matrix with diagonal entries $D(v, v) = d_v$ where d_v is the degree of the vertex v. The well-known matrix-tree theorem of Kirchhoff [4] states that the number of spanning trees in a graph G is the product of all (except for the smallest) eigenvalues of L divided by the number of vertices of G. The eigenvalues of the adjacency matrix are useful in enumerating walks in a graph. For example, the largest eigenvalue of A, denoted by $\|A\|$, is the limit of the kth root of the number of k-walks in G, as k approaches infinity. In this paper, we will mainly focus on the (normalized) Laplacian $\mathscr{L}$, which is defined as follows:

$$\mathscr{L}(u, v) = \begin{cases} 1 - \frac{A(v,v)}{d_v} & \text{if } u = v \text{ and } d_v \neq 0, \\ -\frac{1}{\sqrt{d_u d_v}} & \text{if } u \text{ and } v \text{ are adjacent}, \\ 0 & \text{otherwise}. \end{cases}$$

We can write

$$\mathscr{L} = D^{-1/2} L D^{-1/2}$$

with the convention $D^{-1}(v,v) = 0$ for $d_v = 0$.

For a regular graph with degree d, we have

$$\mathscr{L} = I - \frac{1}{d}A.$$

Let g denote an arbitrary function which assigns to each vertex v of G a real value $g(v)$. We can view g as a column vector. Then

$$\frac{\langle g, \mathscr{L} g \rangle}{\langle g, g \rangle} = \frac{\langle g, D^{-1/2} L D^{-1/2} g \rangle}{\langle g, g \rangle}$$
$$= \frac{\langle f, Lf \rangle}{\langle D^{1/2}f, D^{1/2}f \rangle}$$
$$= \frac{\sum_{u \sim v}(f(u) - f(v))^2}{\sum_v f(v)^2 d_v}, \quad (2.2)$$

where $g = D^{1/2}f$ and $\sum_{u \sim v}$ denotes the sum over all unordered pairs $\{u,v\}$ for which u and v are adjacent. Here $\langle f, g \rangle = \sum_x f(x)g(x)$ denotes the standard inner product in $\mathbb{R}^n$. (We note that we can also use the inner product $\langle f, g \rangle = \sum \overline{f(x)}g(x)$ for complex-valued functions.) From equation (2.2), we see that all eigenvalues are nonnegative and 0 is an eigenvalue of $\mathscr{L}$. We denote the eigenvalues of $\mathscr{L}$ by $0 = \lambda_0 \leqslant \lambda_1 \leqslant \cdots \leqslant \lambda_{n-1}$. Let $\mathbf{1}$ denote the constant function which assumes the value 1 on each vertex. Then $D^{1/2}\mathbf{1}$ is an eigenfunction of $\mathscr{L}$ with eigenvalue 0.

Quite a few basic facts can be derived from the above definition (see [2]). All λ_i are between 0 and 2. The number of eigenvalues of $\mathscr{L}$ having value 0 is the same as the number of connected components in G. The maximum eigenvalue of $\mathscr{L}$ is 2 if and only if the graph is bipartite. In the next few sections, we will focus on the family $\mathscr{F}_\delta$ of graphs with Laplacian eigenvalues satisfying

$$\bar{\lambda} = \max_{i \neq 0}|1 - \lambda_i| < \delta \quad (2.3)$$

for $i \neq 0$. We note that for d-regular graphs, the eigenvalues of the adjacency matrix are just $d(1 - \lambda_i)$ so the so-called (n, d, λ)-graphs in [1, 5] are in $\mathscr{F}_\delta$ for $\delta = \bar{\lambda}/d$.

3. Eigenvalues and discrepancies

A main tool for investigating various graph invariants for $\mathscr{F}_\delta$ concerns the notion of *discrepancy* and the related discrepancy inequalities. A typical definition for discrepancy is the difference between the *actual quantity* and the *expected value*. The goal is to upper-bound the discrepancy in terms of eigenvalues. For example, in a given graph G, a quantity of concern is the number $e(X, Y)$ of edges between two subsets X and Y. In many situations (such as G is regular), the expected value of $e(X, Y)$ is taken to be the edge density multiplied by the cardinality of X and Y. The condition of the graph being regular is quite restrictive. In particular, such an inequality cannot be applied to (nonregular) subgraphs of a regular graph. Here we extend such a discrepancy inequality to general graphs by using the eigenvalues of the Laplacian.

Lemma 3.1. *Suppose a graph G on n vertices has eigenvalues λ_i of the Laplacian satisfying $\bar{\lambda} = \max_{i \neq 0} |1 - \lambda_i| < \delta$. For any two subsets X and Y of vertices, $e(X, Y)$ denotes the number of ordered pairs (x, y) so that $\{x, y\}$ is an edge and $x \in X$ and $y \in Y$. Then $e(X, Y)$ satisfies*

$$\left| e(X, Y) - \frac{\operatorname{vol}(X)\operatorname{vol}(Y)}{\operatorname{vol}(G)} \right| \leq \delta \sqrt{\operatorname{vol}(X)\operatorname{vol}(Y)}$$

where $\operatorname{vol}(X) = \sum_{x \in X} d_x$ and $\operatorname{vol}(G) = \sum_v d_v$.

The above lemma is a special case of the following.

Lemma 3.2. *Suppose k is a given real value (possibly negative) and a graph G on n vertices has Laplacian eigenvalues λ_i satisfying $\bar{\lambda} = \max_{i \neq 0} |1 - \lambda_i| < \delta$. Then, for any two subsets X and Y of vertices, the k-weight of X and Y, denoted by*

$$e_k(X, Y) = \sum_{u \in X} \sum_{v \in Y, v \sim u} d_u^k d_v^k$$

satisfies

$$\left| e_k(X, Y) - \frac{\operatorname{vol}_{k+1}(X)\operatorname{vol}_{k+1}(Y)}{\operatorname{vol}(G)} \right| \leq \delta \sqrt{\operatorname{vol}_{2k+1}(X)\operatorname{vol}_{2k+1}(Y)},$$

where $\operatorname{vol}_i(X) = \sum_{x \in X} d_x^i$ and $\operatorname{vol}_i(G) = \sum_v d_v^i$.

Proof. We define

$$\psi_X(u) = \begin{cases} d_u^k & \text{if } u \in X, \\ 0 & \text{otherwise.} \end{cases}$$

Then

$$\left| e_k(X, Y) - \frac{\operatorname{vol}_{k+1}(X)\operatorname{vol}_{k+1}(Y)}{\operatorname{vol}(G)} \right|$$

$$= \left| \langle \psi_X, A\psi_Y \rangle - \frac{\operatorname{vol}_{k+1}(X)\operatorname{vol}_{k+1}(Y)}{\operatorname{vol}(G)} \right|$$

$$= \left| \langle \psi_X, D^{1/2}(I - \mathcal{L})D^{1/2}\psi_Y \rangle - \frac{\operatorname{vol}_{k+1}(X)\operatorname{vol}_{k+1}(Y)}{\operatorname{vol}(G)} \right|$$

$$= \left| \langle \psi_X, D^{1/2}(I - \mathcal{L} - \phi_0^*\phi_0)D^{1/2}\psi_Y \rangle \right|$$

since the eigenvector ϕ_0 associated with eigenvalue 0 has coordinates $\sqrt{d_v/\operatorname{vol}(G)}$, where f^* denotes the transpose of f. Since G is in $\mathscr{F}_\delta$, we have

$$\|I - \mathcal{L} - \phi_0^*\phi_0\| \leq \delta.$$

Therefore
$$\left| e_k(X, Y) - \frac{\mathrm{vol}_{k+1}(X)\mathrm{vol}_{k+1}(Y)}{\mathrm{vol}(G)} \right| \leq \|D^{1/2}\psi_X\| \, \|I - \mathscr{L} - \phi_0^*\phi_0\| \, \|D^{1/2}\psi_Y\|$$
$$= \sqrt{\mathrm{vol}_{2k+1}(X)} \, \delta \, \sqrt{\mathrm{vol}_{2k+1}(Y)}$$

as desired. □

For a vertex v in a graph G, the neighbourhood $\Gamma(v)$ of v is defined by
$$\Gamma(v) = \{u \; : \; u \sim v\} = \{u \; : \; \{u, v\} \text{ is an edge}\}.$$
In general, the neighbourhood $\Gamma_X(v)$ of v in X is denoted by
$$\Gamma_X(v) = \{u \in X \; : \; u \sim v\}.$$
In addition to Lemma 3.2, we also need the following estimate.

Lemma 3.3. *Suppose a graph G on n vertices has Laplacian eigenvalues λ_i satisfying $\bar{\lambda} = \max_{i \neq 0} |1 - \lambda_i| < \delta$. For any real value k and any subset X of vertices of G, we have*
$$\sum_{v \in X} \frac{1}{d_v} \left(\mathrm{vol}_k(\Gamma_X(v)) - d_v \frac{\mathrm{vol}_{k+1}(X)}{\mathrm{vol}(G)} \right)^2 \leq \delta^2 \mathrm{vol}_{2k+1}(X),$$
where $\Gamma_X(v) = \{u \in X : u \sim v\}$.

Proof. We consider ψ_X as defined in the proof of Lemma 3.2. The difference of $(I - \mathscr{L})D^{1/2}\psi_X$ and the projection of $D^{1/2}\psi_X$ on $D^{1/2}\mathbf{1}$ can be written as
$$\left\| D^{-1/2} A \psi_X - \langle D^{1/2}\psi_X, D^{1/2}\mathbf{1} \rangle \frac{D^{1/2}\mathbf{1}}{\mathrm{vol}(G)} \right\| \leq \delta \|D^{1/2}\psi_X\|,$$
which implies
$$\sum_{v \in X} \frac{1}{d_v} \left(\mathrm{vol}_k(\Gamma_X(v)) - d_v \frac{\mathrm{vol}_{k+1}(X)}{\mathrm{vol}(G)} \right)^2 \leq \delta^2 \mathrm{vol}_{2k+1}(X)$$
as desired. □

Lemma 3.4. *Suppose a graph G on n vertices has Laplacian eigenvalues λ_i satisfying $\bar{\lambda} = \max_{i \neq 0} |1 - \lambda_i| < \delta$. Suppose X is a subset of vertices of G and v is a vertex in X. Let $\Gamma_X(v)$ denote the neighbourhood of v in X and let $R(v)$ denote a subset of $\Gamma(v)$. We have*
$$\sum_{v \in X} \frac{|\Gamma_X(v)|^2}{d_v} \geq \frac{\mathrm{vol}^3(X)}{\mathrm{vol}^2(G)} + O\left(\delta \frac{\mathrm{vol}^2(X)}{\mathrm{vol}(G)} \right) + O(\delta^2 \mathrm{vol}(X)). \qquad (3.1)$$
and
$$\sum_{v \in X} \frac{|\Gamma_X(v)| \, |R(v)|}{d_v} \leq \sum_{v \in X} \frac{|R(v)| \mathrm{vol}(X)}{\mathrm{vol}(G)} + O\left(\delta \frac{\mathrm{vol}^2(X)}{\mathrm{vol}(G)} \right) + O(\delta^2 \mathrm{vol}(X)). \qquad (3.2)$$

Proof. Using Lemma 3.3, we have

$$\sum_{v \in X} \frac{|\Gamma_X(v)|^2}{d_v} = \sum_{v \in X} \frac{d_v \operatorname{vol}^2(X)}{\operatorname{vol}^2(G)} + \sum_{v \in X} \frac{|\Gamma_X(v)|^2 - d_v^2 \operatorname{vol}^2(X)/\operatorname{vol}^2(G)}{d_v}$$

$$\geq \frac{\operatorname{vol}^3(X)}{\operatorname{vol}^2(G)} + O\left(\delta \frac{\operatorname{vol}^2(X)}{\operatorname{vol}(G)}\right) + O(\delta^2 \operatorname{vol}(X)),$$

$$\sum_{v \in X} \frac{|\Gamma_X(v)| \; |R(v)|}{d_v} = \sum_{v \in X} \frac{|R(v)| \operatorname{vol}(X)}{\operatorname{vol}(G)} + \sum_{v \in X} \frac{|R(v)|(|\Gamma_X(v)| - d_v \operatorname{vol}(X)/\operatorname{vol}(G))}{d_v}$$

$$\leq \sum_{v \in X} \frac{|R(v)| \operatorname{vol}(X)}{\operatorname{vol}(G)} + \delta \sqrt{\operatorname{vol}(X) \left(\sum_{v \in X} \frac{|R(v)|^2}{d_v}\right)}$$

$$\leq \sum_{v \in X} \frac{|R(v)| \operatorname{vol}(X)}{\operatorname{vol}(G)} + O\left(\delta \frac{\operatorname{vol}(X)^2}{\operatorname{vol}(G)}\right) + O(\delta^2 \operatorname{vol}(X)). \qquad \square$$

A useful generalization of Lemma 3.4 is the following.

Lemma 3.5. *Suppose a graph G on n vertices has Laplacian eigenvalues λ_i satisfying $\bar\lambda = \max_{i \neq 0} |1 - \lambda_i| < \delta$. Suppose X is a subset of vertices of G and i is a nonnegative value. We have*

$$\sum_{v \in X} \frac{1}{d_v^{i+1}} (\operatorname{vol}_{-i}(\Gamma_X(v)))^2 = \frac{\operatorname{vol}_{-i+1}^3(X)}{\operatorname{vol}^2(G)} + O\left(\delta \frac{\operatorname{vol}_{-i+1}(X) \operatorname{vol}_{-2i+1}(X)}{\operatorname{vol}(G)}\right)$$

and

$$\sum_{v \in X} \frac{1}{d_v^{i+1}} \operatorname{vol}_{-i}(\Gamma_X(v)) \operatorname{vol}_{-i}(R(v)) = \frac{\operatorname{vol}_{-i+1}(X)}{\operatorname{vol}(G)} \sum_{\{u,v\} \text{red}} \frac{1}{d_v^i d_u^i} + O\left(\bar\lambda \frac{\operatorname{vol}_{-i+1}(X) \operatorname{vol}_{-2i+1}(X)}{\operatorname{vol}(G)}\right).$$

Proof.

$$\sum_{v \in X} \frac{1}{d_v^{i+1}} (\operatorname{vol}_{-i}(\Gamma_X(v)))^2 = \frac{\operatorname{vol}_{-i+1}^3(X)}{\operatorname{vol}^2(G)} + \sum_{v \in X} \frac{\operatorname{vol}_{-i}^2(\Gamma_X(v)) - (d_v \operatorname{vol}_{-i+1}(X)/\operatorname{vol}(G))^2}{d_v^{i+1}}$$

$$= \frac{\operatorname{vol}_{-i+1}^3(X)}{\operatorname{vol}^2(G)} + O\left(\bar\lambda \frac{\operatorname{vol}_{-i+1}(X) \operatorname{vol}_{-2i+1}(X)}{\operatorname{vol}(G)}\right),$$

$$\sum_{v \in X} \frac{1}{d_v^{i+1}} \operatorname{vol}_{-i}(R(v)) \operatorname{vol}_{-i}(\Gamma_X(v)) = \sum_{v \in X} \frac{1}{d_v^i} \operatorname{vol}_{-i}(R(v)) \frac{\operatorname{vol}_{-i+1}(X)}{\operatorname{vol}(G)}$$

$$+ \sum_{v \in X} \frac{1}{d_v^{i+1}} \operatorname{vol}_{-i}(R(v)) \left(\operatorname{vol}_{-i}(\Gamma_X(v)) - \frac{d_v \operatorname{vol}_{-i+1}(X)}{\operatorname{vol}(G)}\right)$$

$$= \frac{\operatorname{vol}_{-i+1}(X)}{\operatorname{vol}(G)} \sum_{\{u,v\} \text{red}} \frac{1}{d_v^i d_u^i} + O\left(\bar\lambda \frac{\operatorname{vol}_{-i+1}(X) \operatorname{vol}_{-2i+1}(X)}{\operatorname{vol}(G)}\right).$$

$\square$

Lemma 3.6. *Suppose that X is a subset of vertices in a graph G and $\alpha \leqslant \beta$ are nonnegative values. Then*

$$\text{vol}_{-\alpha}(X)\text{vol}_{-\beta}(X) \leqslant \text{vol}_{-\alpha+1}(X)\text{vol}_{-\beta-1}(X). \tag{3.3}$$

Proof. The inequality (3.3) follows from the following general version of the Cauchy–Schwarz inequality for positive a_js and $0 \leqslant \alpha \leqslant \beta$:

$$\left(\sum_{j=1}^{k} a_j^{\alpha}\right)\left(\sum_{j=1}^{k} a_j^{\beta}\right) \leqslant \left(\sum_{j=1}^{k} a_j^{\alpha-1}\right)\left(\sum_{j=1}^{k} a_j^{\beta+1}\right).$$

By choosing a_js to be the reciprocal of the degrees, (3.3) is an immediate consequence. □

4. A generalization of Turán's theorem

We will now prove the main theorem.

Theorem 4.1. *Suppose a graph G on n vertices has eigenvalues $0 = \lambda_0 \leqslant \lambda_1 \leqslant \cdots \leqslant \lambda_{n-1}$ with $\bar{\lambda} = \max_{i \neq 0}|1 - \lambda_i|$ satisfying*

$$\bar{\lambda} = o\left(\frac{1}{\text{vol}_{-2t+3}(G)\text{vol}(G)^{t-2}}\right). \tag{4.1}$$

Then, G is t-Turán for $t \geqslant 2$; i.e., any subgraph of G containing no K_{t+1} has at most $(1 - 1/t + o(1))e(G)$ edges, where $e(G)$ is the number of edges in G.

There are expressions in terms of $o(\cdot)$s in the statement of Theorem 4.1. To be precise, the result in Theorem 4.1 can be restated as follows. *For any $\epsilon > 0$, there is a δ such that if the eigenvalues of the Laplacian of the graph G satisfies*

$$\bar{\lambda} = \max_{i \neq 0}|1 - \lambda_i| < \frac{\delta}{\text{vol}_{-2t+3}(G)\text{vol}(G)^{t-2}},$$

then any subgraph of G containing no K_{t+1} has at most $(1 - 1/t + \epsilon)e(G)$ edges where $e(G)$ is the number of edges in G.

We remark that the condition in (4.1) has the following implication for the minimum degree η of G:

$$\bar{\lambda} \ll \frac{1}{\text{vol}_{-2t+3}(G)\text{vol}(G)^{t-2}}$$

$$< \frac{\eta^{2t-3}}{\text{vol}(G)^{t-2}n} \leqslant \frac{\eta^{2t-3}}{n^{t-1}\eta^{t-2}} = \frac{\eta^{t-1}}{n^{t-1}}.$$

Since the inequality

$$\bar{\lambda} \geqslant 1/\sqrt{\eta n}$$

always holds, condition (4.1) implies that

$$\eta \gg n^{(2t-3)/(2t-1)}.$$

Thus, condition (4.1) may hold only if the minimum degree of the graph G is sufficiently large.

Theorem 4.1 implies the following two facts: the classical Turán theorem and the case for regular graphs.

Corollary 4.2. *A graph on n vertices containing no K_{t+1} has at most $(1 - 1/t + o(1))n^2/2$ edges.*

Proof. The complete graph on n vertices has Laplacian eigenvalues 0 and $n/(n-1)$ (with multiplicity $n-1$). Thus, for $G = K_n$, it is always true that

$$\bar{\lambda} = \frac{1}{n-1} = o(1).$$

Therefore, Theorem 4.1 implies that any graph on n vertices containing no K_{t+1} has at most $(1 - 1/t + o(1))n^2/2$ edges. □

Corollary 4.3. *If a graph is regular with degree d and has n vertices, then the condition in (4.1) is just*

$$\bar{\lambda} = o\left(\left(\frac{d}{n}\right)^{t-1}\right).$$

Then any subgraph of G containing no K_{t+1} has at most $(1 - 1/t + o(1))dn/2$ edges.

The proof is by direct substitution and will be omitted.

Suppose that R is a subset of edges so that every K_{t+1} in G contains at least one edge in R. In order to prove Theorem 4.1, we wish to show that $|R| \geq (1 + o(1))|E(G)|/t$.

To do so, we will prove the following stronger result.

Theorem 4.4. *Suppose a graph G on n vertices has eigenvalues satisfying (4.1). Then we have:*

(∗) *For any subset X of vertices in G and for nonnegative integer $k \leq t$, if R contains an edge from every complete subgraph on $k + 1$ vertices, then we have, for all i, $0 \leq i \leq k$,*

$$\sum_{v \in X, u \in X, \{u,v\} \in R} \frac{1}{d_u^i d_v^i} \geq \frac{\text{vol}_{-i+1}^2(X)}{k \text{vol}(G)} + O\left(\bar{\lambda} \sum_{j=0}^{k-1} \text{vol}_{-2i-2j+1}(X) \text{vol}^j(G)\right) \quad (4.2)$$

To derive Theorem 4.1 from Theorem 4.4, suppose that R contains an edge from every complete subgraph on $k + 1$ vertices. We apply (∗) with $i = 0$ and $k = t$. We then use (4.1) and have

$$|R| \geq \frac{\text{vol}(G)}{t} + O\left(\bar{\lambda} \sum_{j=0}^{t-1} \text{vol}_{-2j+1}(G) \text{vol}^j(G)\right)$$

$$\geq \frac{\text{vol}(G)}{t}(1 + O(\bar{\lambda} \text{vol}_{-2t+3}(G) \text{vol}^{t-2}(G)))$$

$$\geq (1 + o(1))\frac{\text{vol}(G)}{t}.$$

Therefore G is t-Turán, as claimed. □

5. Proof of the main theorem

In this section, we shall prove Theorem 4.4. The inequality in (∗) is somewhat complicated. We shall first deal with simpler cases (such as $i = 0$) which contain the main ideas.

Proof of Theorem 4.4. First we want to show that (∗) holds for $k = 1$. In this case we have $R = E(G)$. Lemma 3.2 implies that (∗) holds for $k = 1$.

Suppose that $k \geq 2$ and (∗) holds for $k' < k$. We wish to prove (∗) for k. Suppose R contains an edge from every complete subgraph on $k + 1$ vertices. We want to show that the inequality (4.2) holds for all i, $0 \leq i \leq k$.

We shall first prove the case that $i = 0$. Suppose that edges in R are coloured *red* and the rest of the edges of G are *blue*. We focus on edges inside the given set X. Recall that $\Gamma_X(v) = \Gamma(v) \cap X$. For each vertex v, let $R(v)$ and $B(v)$ denote the set of neighbours u of v in X with $\{u,v\}$ red and blue, respectively. For a vertex v, we consider the induced subgraph on $B(v)$ which does not contain a complete graph on k vertices. By induction we have

$$\sum_{u,w \in B(v), \{u,w\} \text{red}} \frac{1}{d_u d_w} \geq \frac{1}{(k-1)} \frac{|B_v|^2}{\text{vol}(G)} + O\left(\bar{\lambda} \sum_{j=0}^{k-2} \text{vol}_{-2j-1}(B(v)) \text{vol}^j(G)\right). \tag{5.1}$$

We consider the set T_j of all triangles in X containing exactly j red edges, for $j = 1, 2, 3$:

$$W_1 = \sum_{\{u,v,w\} \in T_1} \frac{2}{d_v d_u d_w}$$

$$= \sum_{v \in X} \frac{1}{d_v} \sum_{u,w \in B(v), \{u,w\} \text{red}} \frac{1}{d_u d_w}$$

$$\leq \sum_{v \in X} \frac{1}{d_v} \sum_{u \in R(v)} \sum_{w \in \Gamma_X(v) \cap \Gamma(u)} \frac{1}{d_u d_w} - \sum_{\{v,u,w\} \in T_2} \frac{2}{d_v d_u d_w} - \sum_{\{v,u,w\} \in T_3} \frac{3}{d_v d_u d_w}$$

$$\leq \sum_{v \in X} \frac{1}{d_v} \sum_{u \in R(v)} \sum_{w \in \Gamma_X(v) \cap \Gamma(u)} \frac{1}{d_u d_w} - 2W_2 - 3W_3.$$

We note that

$$W_2 + 3W_3 = \sum_{v \in X} \frac{1}{d_v} \sum_{u,w \in R(v), u \sim w} \frac{1}{d_u d_w}.$$

Thus,

$$W_1 \leq \sum_{v \in X} \frac{1}{d_v} \sum_{u \in R(v)} \sum_{w \in \Gamma_X(v), w \sim u} \frac{1}{d_u d_w} - W_2 - 3W_3$$

$$\leq \sum_{v \in X} \frac{1}{d_v} \sum_{u \in R(v)} \sum_{w \in \Gamma_X(v), w \sim u} \frac{1}{d_u d_w} - \sum_{v \in X} \frac{1}{d_v} \sum_{u,w \in R(v), u \sim w} \frac{1}{d_u d_w}.$$

By Lemma 3.2, we have

$$\sum_{u \in R(v)} \sum_{w \in \Gamma_X(v), w \sim u} \frac{1}{d_u d_w} = \frac{|R(v)| \, |\Gamma_X(v)|}{\text{vol}(G)} + O(\bar\lambda \sqrt{\text{vol}_{-1} R(v) \text{vol}_{-1}(\Gamma_X(v))})$$

$$\sum_{u,w \in R(v), u \sim w} \frac{1}{d_u d_w} = \frac{|R(v)|^2}{\text{vol}(G)} + O(\bar\lambda \text{vol}_{-1}(R(v))).$$

Therefore,

$$W_1 \leq \sum_{v \in X} \frac{1}{d_v} \frac{|\Gamma_X(v)| \, |R(v)|}{\text{vol}(G)} - \sum_{v \in X} \frac{1}{d_v} \frac{|R(v)|^2}{\text{vol}(G)} + O\left(\bar\lambda \sum_{v \in X} \frac{\text{vol}_{-1}(\Gamma_X(v))}{d_v}\right). \quad (5.2)$$

On the other hand, from (5.2) we have

$$W_1 = \sum_{\{u,v,w\} \in T_1} \frac{2}{d_v d_u d_w}$$

$$\geq \frac{1}{(k-1)} \sum_{v \in X} \frac{1}{d_v} \frac{|B(v)|^2}{\text{vol}(G)} - O\left(\bar\lambda \sum_{j=0}^{k-2} \frac{\text{vol}_{-2j-1}(B(v))\text{vol}^j(G)}{d_v}\right)$$

$$\geq \frac{1}{(k-1)} \sum_{v \in X} \frac{1}{d_v} \frac{|B(v)|^2}{\text{vol}(G)} - O\left(\bar\lambda \sum_{j=0}^{k-2} \frac{\text{vol}_{-2j-1}(\Gamma_X(v))\text{vol}^j(G)}{d_v}\right)$$

$$\geq \frac{1}{(k-1)} \sum_{v \in X} \frac{1}{d_v} \frac{(|\Gamma_X(v)| - |R(v)|)^2}{\text{vol}(G)} - O\left(\bar\lambda \sum_{j=0}^{k-2} \frac{|X| \text{vol}_{-2j}(X)\text{vol}^j(G)}{\text{vol}(G)}\right)$$

$$+ O\left(\bar\lambda^2 \sum_{j=0}^{k-2} \sqrt{\text{vol}_{-1}(X)\text{vol}_{-4j-1}(X)}\text{vol}^j(G)\right).$$

We note that the terms involving $\bar\lambda^2$ are of lower order by using the assumption on $\bar\lambda$. Combining the preceding upper and lower bounds for W_1, we have

$$(k+1) \sum_{v \in X} \frac{|\Gamma_X(v)| \, |R(v)|}{d_v}$$

$$\geq \sum_{v \in X} \frac{|\Gamma_X(v)|^2}{d_v} + k \sum_{v \in X} \frac{|R(v)|^2}{d_v} - O\left(\bar\lambda |X| \sum_{j=0}^{k-2} \text{vol}_{-2j}(X)\text{vol}^j(G)\right). \quad (5.3)$$

By Lemma 3.4 and inequality (3.2), we have

$$\sum_{v \in X} \frac{|\Gamma_X(v)| \, |R(v)|}{d_v} \leq \sum_{v \in X} \frac{|R(v)| \text{vol}(X)}{\text{vol}(G)} + O\left(\bar\lambda \frac{\text{vol}(X)^2}{\text{vol}(G)}\right) + O(\bar\lambda^2 \text{vol}(X)).$$

Also, by Lemma 3.4 and inequality (3.1), we have

$$\sum_{v \in X} \frac{|\Gamma_X(v)|^2}{d_v} \geq \frac{\text{vol}^3(X)}{\text{vol}^2(G)} + O\left(\bar\lambda \frac{\text{vol}(X)^2}{\text{vol}(G)}\right) + O(\bar\lambda^2 \text{vol}(X)).$$

Substituting into (5.3), we have

$$(k+1)\sum_v |R(v)|\frac{\text{vol}(X)}{\text{vol}(G)}$$
$$\geq \frac{\text{vol}^3(X)}{\text{vol}^2(G)} + k\sum_{v\in X}\frac{|R(v)|^2}{d_v} + O\left(\bar{\lambda}\left(\frac{\text{vol}(X)^2}{\text{vol}(G)} + \sum_{j=0}^{k-2}|X|\text{vol}_{-2j}(X)\text{vol}^j(G)\right)\right)$$
$$\geq \frac{\text{vol}^3(X)}{\text{vol}^2(G)} + k\frac{(\sum_{v\in X}|R(v)|)^2}{\sum_{v\in X}d_v} + O\left(\bar{\lambda}\left(\frac{\text{vol}(X)^2}{\text{vol}(G)} + \sum_{j=0}^{k-2}|X|\text{vol}_{-2j}(X)\text{vol}^j(G)\right)\right).$$

This implies

$$\left(\frac{\text{vol}^2(X)}{\text{vol}(G)} - \sum_v |R(v)|\right)\left(\frac{\text{vol}^2(X)}{\text{vol}(G)} - k\sum_v |R(v)|\right)$$
$$+ O\left(\bar{\lambda}\left(\frac{\text{vol}(X)^3}{\text{vol}(G)} + \sum_{j=0}^{k-2}|X|\text{vol}_{-2j}(X)\text{vol}^j(G)\text{vol}(X)\right)\right) < 0.$$

Thus we have

$$|R| = \frac{1}{2}\sum_v |R(v)|$$
$$\geq \frac{1}{2k}\frac{\text{vol}^2(X)}{\text{vol}(G)} + O\left(\bar{\lambda}\left(\text{vol}(X) + \sum_{j=0}^{k-2}\frac{|X|\text{vol}_{-2j}(X)\text{vol}^{j+1}(G)}{\text{vol}(X)}\right)\right)$$
$$\geq \frac{1}{2k}\frac{\text{vol}^2(X)}{\text{vol}(G)} + O\left(\bar{\lambda}\left(\text{vol}(X) + \sum_{j=0}^{k-2}\text{vol}_{-2j-1}(X)\text{vol}^{j+1}(G)\right)\right)$$
$$\geq \frac{1}{k}|E(X)| + O\left(\bar{\lambda}\left(\text{vol}(X) + \sum_{j=1}^{k-1}\text{vol}_{-2j+1}(X)\text{vol}^j(G)\right)\right)$$
$$\geq \frac{1}{k}|E(X)| + O\left(\bar{\lambda}\sum_{j=0}^{k-1}\text{vol}_{-2j+1}(X)\text{vol}^j(G)\right).$$

We have completed the proof for the case $i=0$.

Suppose $i \geq 1$. For $j = 1, 2, 3$, we consider

$$W_j^{(i)} = \sum_{\{u,v,w\}\in T_j}\frac{2}{d_v^i d_u^i d_w^i}.$$

As before, we have

$$W_1^{(i+1)} \leq \sum_{v\in X}\frac{1}{d_v^{i+1}}\sum_{u\in R(v)}\sum_{w\in \Gamma_X(v)\cap \Gamma(u)}\frac{1}{d_u^{i+1}d_w^{i+1}} - W_2^{(i+1)} - 3W_3^{(i+1)}$$
$$\leq \sum_{v\in X}\frac{1}{d_v^{i+1}}\sum_{u\in R(v)}\sum_{w\in \Gamma_X(v)\cap \Gamma(u)}\frac{1}{d_u^{i+1}d_w^{i+1}} - \sum_{v\in X}\frac{1}{d_v^{i+1}}\sum_{u,w\in R(v), u\sim w}\frac{1}{d_u^{i+1}d_w^{i+1}}$$

$$\leq \sum_{v \in X} \frac{1}{d_v^{i+1}} \frac{\mathrm{vol}_{-i}(R(v))\mathrm{vol}_{-i}(\Gamma_X(v))}{\mathrm{vol}(G)} - \sum_{v \in X} \frac{1}{d_v^{i+1}} \frac{\mathrm{vol}_{-i}^2(R(v))}{\mathrm{vol}(G)}$$

$$+ O\left(\bar{\lambda} \sum_{v \in X} \frac{\mathrm{vol}_{-2i-1}(\Gamma_X(v))}{d_v^{i+1}}\right)$$

$$\leq \sum_{v \in X} \frac{1}{d_v^{i+1}} \frac{\mathrm{vol}_{-i}(R(v))\mathrm{vol}_{-i}(\Gamma_X(v))}{\mathrm{vol}(G)} - \sum_{v \in X} \frac{1}{d_v^{i+1}} \frac{\mathrm{vol}_{-i}^2(R(v))}{\mathrm{vol}(G)}$$

$$+ O\left(\bar{\lambda} \frac{\mathrm{vol}_{-i}(X)\mathrm{vol}_{-2i}(X)}{\mathrm{vol}(G)} + \bar{\lambda}^2 (\mathrm{vol}_{-2i-1}(X)\mathrm{vol}_{-4i-1}(X))^{1/2}\right).$$

On the other hand, by induction we have

$$W_1^{(i+1)} = \sum_{\{u,v,w\} \in T_1} \frac{2}{d_v^{(i+1)} d_u^{(i+1)} d_w^{(i+1)}}$$

$$\geq \frac{1}{(k-1)} \sum_{v \in X} \frac{1}{d_v^{(i+1)}} \left(\frac{\mathrm{vol}_{-i}^2(B(v))}{\mathrm{vol}(G)} + O\left(\bar{\lambda} \sum_{j=0}^{k-2} \mathrm{vol}_{-2i-2j+1}(B(v))\mathrm{vol}^j(G)\right)\right)$$

$$\geq \frac{1}{(k-1)} \sum_{v \in X} \frac{1}{d_v^{i+1}} \frac{\mathrm{vol}_{-i}^2(\Gamma_X(v)) - \mathrm{vol}_{-i}(R(v))}{\mathrm{vol}(G)}$$

$$+ O\left(\bar{\lambda} \sum_{j=0}^{k-2} \frac{\mathrm{vol}_{-i}(X)\mathrm{vol}_{-2i-2j}(X)\mathrm{vol}^j(G)}{\mathrm{vol}(G)}\right.$$

$$+ \bar{\lambda}^2 \sum_{j=0}^{k-2} \sqrt{\mathrm{vol}_{-2i-1}(X)\mathrm{vol}_{-4i-4j-1}(X)}\mathrm{vol}^j(G)\right).$$

By combining the upper and lower bounds of $W_1^{(i+1)}$ (and multiplying by $\mathrm{vol}(G)$), we have

$$(k+1)A + B \leq kC + O(\bar{\lambda} D \mathrm{vol}(G)), \tag{5.4}$$

where A, B, C, D are as follows:

$$A = \sum_{v \in X} \frac{1}{d_v^{i+1}} (\mathrm{vol}_{-i}(\Gamma_X(v)))^2$$

$$\geq \frac{\mathrm{vol}_{-i+1}^3(X)}{\mathrm{vol}^2(G)} + O\left(\bar{\lambda} \frac{\mathrm{vol}_{-i+1}(X)\mathrm{vol}_{-2i+1}(\Gamma(v))}{\mathrm{vol}(G)}\right) \quad \text{by using Lemma 3.5;}$$

$$B = \sum_{v \in X} \frac{1}{d_v^{i+1}} \mathrm{vol}_{-i}^2(R(v))$$

$$\geq \left(\sum_{v \in X} \frac{\mathrm{vol}_{-i}(R(v))}{d_v^i}\right)^2 / \mathrm{vol}_{-i+1}(G) \quad \text{by the Cauchy–Schwarz inequality;}$$

$$C = \sum_{v \in X} \frac{1}{d_v^{i+1}} \mathrm{vol}_{-i}(\Gamma_X(v))\mathrm{vol}_{-i}(R(v))$$
$$\leq \frac{\mathrm{vol}_{-i+1}(X)}{\mathrm{vol}(G)} \sum_{\{u,v\}\mathrm{red}} \frac{1}{d_v^i d_u^i} + O\left(\bar{\lambda} \frac{\mathrm{vol}_{-i+1}(X)\mathrm{vol}_{-2i+1}(X)}{\mathrm{vol}(G)}\right);$$
$$D = \sum_{j=0}^{k-2} \mathrm{vol}_{-i}(X)\mathrm{vol}_{-2i-2j}(X)\mathrm{vol}^j(G) + \mathrm{vol}_{-i+1}(X)\mathrm{vol}_{-2i+1}(X).$$

By substituting A, B and C into (5.4), we have the following:

$$k\frac{\mathrm{vol}^3_{-i+1}(X)}{\mathrm{vol}^2(G)} + \frac{(\sum_{\{u,v\}\mathrm{red}} \frac{1}{d_v^i d_u^i})^2}{\mathrm{vol}_{-i+1}(X)} \leq (k+1)\frac{\mathrm{vol}_{-i+1}(X)}{\mathrm{vol}(G)} \sum_{\{u,v\}\mathrm{red}} \frac{1}{d_v^i d_u^i} + O(\bar{\lambda} D \mathrm{vol}(G)).$$

This implies

$$\sum_{\{u,v\}\mathrm{red}} \frac{1}{d_v^i d_u^i} \geq \frac{\mathrm{vol}^2_{-i+1}(X)}{k\mathrm{vol}(G)}$$
$$- O\left(\bar{\lambda}\left(\mathrm{vol}_{-2i+1}(X) + \sum_{j=0}^{k-2} \frac{\mathrm{vol}_{-i}(X)\mathrm{vol}_{-2i-2j}(X)\mathrm{vol}^{j+1}(G)}{\mathrm{vol}_{-i+1}(X)}\right)\right).$$

Now, by using Lemma 3.6 and inequality (3.3), we have

$$\sum_{\{u,v\}\mathrm{red}} \frac{1}{d_v^i d_u^i} \geq \frac{\mathrm{vol}^2_{-i+1}(X)}{k\mathrm{vol}(G)} + O\left(\bar{\lambda}\mathrm{vol}_{-2i+1}(X) + \sum_{j=0}^{k-2} \mathrm{vol}_{-2i-2j-1}(X)\mathrm{vol}^{j+1}(G)\right)$$
$$\geq \frac{\mathrm{vol}^2_{-i+1}(X)}{k\mathrm{vol}(G)} + O\left(\bar{\lambda}\sum_{j=0}^{k-1} \mathrm{vol}_{-2i-2j+1}(X)\mathrm{vol}^j(G)\right).$$

This completes the proof of Theorem 4.4. □

References

[1] Alon, N. (1986) Eigenvalues and expanders. *Combinatorica* **6** 86–96.
[2] Chung, F. (1997) *Spectral Graph Theory*, AMS Publications.
[3] Chung, F., Lu, L. and Vu, V. (2003) Spectra of random graphs with given expected degrees. *Proc. National Acad. Sci.* **100** 6313–6318.
[4] Kirchhoff, F. (1847) Über die Auflösung der Gleichungen, auf welche man bei der Untersuchung der linearen Verteilung galvanischer Ströme geführt wird. *Ann. Phys. Chem.* **72** 497–508.
[5] Krivelevich, M. and Sudakov, B. Pseudo-random graphs. Preprint.
[6] Sudakov, B., Szabo, T. and Vu, V. (2005) A generalization of Turán's theorem. *J. Graph Theory* **49** 187–195.

Automorphism Groups of Metacirculant Graphs of Order a Product of Two Distinct Primes

EDWARD DOBSON

Department of Mathematics and Statistics, PO Drawer MA,
Mississippi State University, Mississippi State, MS 39762, USA
(e-mail: dobson@math.msstate.edu)

Received 2 December 2003; revised 29 April 2005

For Béla Bollobás on his 60th birthday

Let p and q be distinct primes. We characterize transitive groups G that admit a complete block system of q blocks of size p such that the subgroup of G which fixes each block set-wise has a Sylow p-subgroup of order p. Using this result, we prove that the full automorphism group of a metacirculant graph Γ of order pq such that Aut(Γ) is imprimitive, is contained in one of several families of transitive groups. As the automorphism groups of vertex-transitive graphs of order pq that are primitive have been determined by several authors, this result implies that automorphism groups of vertex-transitive graphs of order pq are known. We also determine all nonnormal Cayley graphs of order pq, and all 1/2-transitive graphs of order pq.

In 1970, Elspas and Turner [13] posed the problem of giving a procedure to calculate the automorphism groups of circulant graphs of order n. We will consider the more general problem of determining the full automorphism groups of metacirculant graphs of order n in the special case where n is a product of two distinct primes. We remark that Klin and Pöschel [18], using the method of Schur (see [34]), have already solved Elspas and Turner's original problem for the case under consideration in this paper. Our method is quite general and classifies all imprimitive groups of degree pq under certain circumstances. As applications of our main results, we determine all nonnormal Cayley graphs of order pq, and all 1/2-transitive graphs of order pq, answering affirmatively a conjecture of Marušič [22, Conjecture 3.2].

1. Preliminaries

For definitions and properties of permutation groups the reader is referred to [10], and for graph-theoretic notation to [4]. We denote the symmetric group on n letters by S^n.

Definition 1. A graph Γ is *vertex-transitive* if Aut(Γ), the *automorphism group of* Γ, is a transitive permutation group on $V(\Gamma)$.

Definition 2. Let G be a group and $S \subset G$ such that $1 \notin S$ and $S^{-1} = S$. Define a graph $\Gamma(G,S)$ to be the graph with $V(\Gamma(G,S)) = G$, and edge set $E(\Gamma(G,S)) = \{(g,gs) : g \in G \text{ and } s \in S\}$. Let $G_L = \{g_L : G \to G : g_L(x) = gx, g \in G\}$. Clearly $G_L \leqslant \text{Aut}(\Gamma(G,S))$ and G_L is transitive. Hence $\Gamma(G,S)$ is a vertex-transitive graph. We say that $\Gamma(G,S)$ is a *Cayley graph* of G, and usually write $\Gamma = \Gamma(G,S)$.

We remark that Cayley digraphs are defined analogously, except that the condition that $S^{-1} = S$ is discarded.

Definition 3. A *circulant graph of order* n is a Cayley graph of $\mathbb{Z}_n$, the cyclic group of order n.

If has been shown that every vertex-transitive graph of prime order p is isomorphic to a circulant graph of order p [13]. Furthermore, the full automorphism groups of circulant graphs of order p can be deduced from the following classical result of Burnside [5].

Theorem 1.1. *Let G be a transitive permutation group acting on a set Ω of order p, p a prime. Then we may relabel the elements of Ω with elements of $\mathbb{Z}_p$, such that G is either doubly transitive or $G < \{ax + b : a \in \mathbb{Z}_p^*, b \in \mathbb{Z}_p\} = \text{AGL}(1,p)$.*

Of course, if Γ is a graph and Aut(Γ) is doubly transitive, then Γ is a complete graph or the complement of a complete graph. Hence if Γ is a circulant graph of order p, Aut(Γ) $= S^p$ or Aut(Γ) $<$ AGL$(1,p)$. We remark that this was first observed and exploited in some detail by Alspach [1].

There is one other form of Theorem 1.1 that we will exploit in this paper. Another classical result of Burnside (see, for example, [10, Theorem 4.1B]) states that the minimal normal subgroup of a finite doubly transitive group is either a regular elementary abelian p-group, or a nonregular non-abelian simple group. Thus Theorem 1.1 can be restated as follows.

Theorem 1.2. *Let G be a transitive permutation group acting on a set Ω of order p, p a prime. Then either G is solvable or G is doubly transitive and the minimal normal subgroup of G is a nonregular non-abelian simple group.*

Throughout the remainder of this paper, p and q are distinct primes, and Γ_p and Γ_q are circulant graphs of order p and q respectively. Let $\mathbb{Z}_n^*$ be the units of $\mathbb{Z}_n$. Let m, n be positive integers and set $\mu = \lfloor m/2 \rfloor$. Let $V = V(\Gamma) = \{v_j^i : i \in \mathbb{Z}_m, j \in \mathbb{Z}_n\}$, and $\alpha \in \mathbb{Z}_n^*$. Let $S_0, S_1, \ldots, S_\mu$ be subsets of $\mathbb{Z}_n$ satisfying the following conditions:

(i) $0 \notin S_0 = -S_0$,
(ii) $\alpha^m S_r = S_r$ for $0 \leqslant r \leqslant \mu$,
(iii) if m is even, then $\alpha^\mu S_\mu = -S_\mu$.

Let $E = \{(v_j^i, v_h^{i+r}) : 0 \leq r \leq \mu \text{ and } h - j \in \alpha^i S_r\}$.

Definition 4. We define the *metacirculant graph* $\Gamma = \Gamma(m, n, \alpha, S_0, \ldots, S_\mu)$ to be the graph with vertex set V and edge set E. We will also refer to Γ as an (m, n)-*metacirculant*.

Define two permutations ρ, τ on V by

$$\rho(v_j^i) = v_{j+1}^i$$

and

$$\tau(v_j^i) = v_{\alpha j}^{i+1}.$$

The following characterization of metacirculant graphs was proved by Alspach and Parsons in [2, Theorem 1].

Theorem 1.3. *The metacirculant graph* $\Gamma = \Gamma(m, n, \alpha, S_0, \ldots, S_\mu)$ *is vertex-transitive with* $\langle \rho, \tau \rangle \leq \mathrm{Aut}(\Gamma)$. *Conversely, any graph* Γ' *with vertex set* V *and* $\langle \rho, \tau \rangle \leq \mathrm{Aut}(\Gamma')$ *is an* (m, n)-*metacirculant.*

Throughout the paper, we will have recourse to the element τ as defined above. As a standard convention, we will always assume that $\tau(v_j^i) = v_{\alpha j}^{i+1}$. We remark that the particular value of α may vary in the paper, but will be used consistently within a given result.

In [2, Theorem 9], Alspach and Parsons gave the following sufficient condition for an (m, n)-metacirculant graph to be a Cayley graph. Note that this theorem also gives the relationship between metacirculant graphs and circulant graphs.

Theorem 1.4. *Let* Γ *be an* (m, n)-*metacirculant graph with* $a = |\alpha|$, *and let* $c = a/\gcd(a, m)$. *If* $\gcd(c, m) = 1$, *then* Γ *is a Cayley graph for the group* $\langle \rho, \tau^c \rangle$. *Furthermore, this group is abelian if* $\gcd(a, m) = 1$ *and it is cyclic if* $\gcd(a, m) = 1 = \gcd(m, n)$.

The following characterization of (q, p)-metacirculant graphs can be deduced [21, Theorem 3.4].

Theorem 1.5 (Marušič). *A vertex-transitive graph* Γ *of order* qp *is isomorphic to a metacirculant graph if and only if* $\mathrm{Aut}(\Gamma)$ *contains subgroups* H *and* $K \neq 1$ *such that* H *is transitive,* $K \triangleleft H$ *and* K *is not transitive.*

Definition 5. Let G be a transitive permutation group of degree mk that admits a complete block system $\mathscr{B}$ of m blocks of size k. If $g \in G$, then g permutes the m blocks of $\mathscr{B}$ and hence induces a permutation in S^m, which we denote by $g/\mathscr{B}$. We define $G/\mathscr{B} = \{g/\mathscr{B} : g \in G\}$. Let $\mathrm{fix}_\mathscr{B}(G) = \{g \in G : g(B) = B \text{ for every } B \in \mathscr{B}\}$. If $G \leq \mathrm{Aut}(\Gamma)$, for some vertex-transitive graph Γ, define a graph $\Gamma/\mathscr{B}$ with vertex set $V(\Gamma/\mathscr{B}) = \mathscr{B}$ and edge set

$$E(\Gamma/\mathscr{B}) = \{(B, B') : \text{some vertex of } B \text{ is adjacent to some vertex of } B', B \neq B'\}.$$

We observe that $G/\mathscr{B} < \mathrm{Aut}(\Gamma/\mathscr{B})$.

Definition 6. Let G be a transitive group acting on X and $K \triangleleft G$, $K \neq 1$, such that K is not transitive. Then G admits a complete block system $\mathscr{B}$ of m blocks each of size k, where the blocks are formed by the orbits of K. Note that $\mathrm{fix}_H(\mathscr{B})$ acts on each $B \in \mathscr{B}$ in a canonical fashion. If this action is faithful, then $\mathrm{fix}_G(\mathscr{B}) \cong J \leqslant S^k$. If this action is faithful and $\mathrm{Stab}_{\mathrm{fix}_G(\mathscr{B})}(x) \neq 1$, we define an equivalence relation $\equiv$ on X by $x \equiv x'$ if and only if $\mathrm{Stab}_{\mathrm{fix}_G(\mathscr{B})}(x) = \mathrm{Stab}_{\mathrm{fix}_G(\mathscr{B})}(x')$. One can easily show that the equivalence classes of $\equiv$ form a complete block system of G.

Definition 7. Let G be a finite group. The *socle* of G, denoted $\mathrm{soc}(G)$, is the product of all minimal normal subgroups of G. If G acts transitively on Ω and G is primitive but not doubly transitive, we say G is *simply primitive*. Let G be a transitive permutation group on a set Ω and let G act on $\Omega \times \Omega$ by $g(\alpha, \beta) = (g(\alpha), g(\beta))$. The orbits of G in $\Omega \times \Omega$ are called the *orbitals* of G. The orbit $\{(\alpha, \alpha) : \alpha \in \Omega\}$ is called the *trivial orbital*. Let Δ be a nontrivial orbit of G in $\Omega \times \Omega$. Define the *orbital digraph* Δ to be the graph with vertex set Ω and edge set Δ. Each orbital of G has a *paired orbital* $\Delta' = \{(\beta, \alpha) : (\alpha, \beta) \in \Delta\}$. Define the *orbital graph* Δ to be the graph with vertex set Ω and edge set $\Delta \cup \Delta'$.

Definition 8. Let Γ_1 and Γ_2 be vertex-transitive graphs. Let

$$E = \{((x, x'), (y, y')) : xy \in E(\Gamma_1), x', y' \in V(\Gamma_2) \text{ or } x = y \text{ and } x'y' \in E(\Gamma_2)\}.$$

Define the *wreath (or lexicographic) product* of Γ_1 and Γ_2, denoted $\Gamma_1 \wr \Gamma_2$, to be the graph such that $V(\Gamma_1 \wr \Gamma_2) = V(\Gamma_1) \times V(\Gamma_2)$ and $E(\Gamma_1 \wr \Gamma_2) = E$. We remark that the wreath product of a circulant graph of order m and a circulant graph of order n is circulant.

2. Group-theoretic results

The proof of the following useful, but technical fact can be found in [27].

Lemma 2.1. *Let $m = (q^d - 1)/(q - 1)$, with $d \geqslant 3$. Suppose $\mathrm{PSL}(d, q) \leqslant G \leqslant \mathrm{P\Gamma L}(d, q)$ with $d \geqslant 3$, and let H be a group that contains G as a normal subgroup. Suppose H acts imprimitively on a set Ω, with a complete block system $\mathscr{B} = \{B_1, B_2\}$ consisting of 2 blocks of cardinality m. Assume $G = \{h \in H \mid h(B_1) = B_1, h(B_2) = B_2\} = \mathrm{fix}_H(\mathscr{B})$. Assume G acts doubly transitively on B_i for each $i = 1, 2$, and that the action of G on B_1 is not equivalent to the action of G on B_2. Then H does not contain a transitive, cyclic subgroup.*

Lemma 2.2. *Let $n = mk$ and $G \leqslant S^n$ such that G is transitive and admits complete block systems $\mathscr{B}$ and $\mathscr{C}$ of m blocks of size k and k blocks of size m, respectively, such that if $B \in \mathscr{B}$ and $C \in \mathscr{C}$ then $|B \cap C| = 1$. Then G is equivalent to a subgroup of $S^k \times S^m$ in its natural action on $\mathbb{Z}_k \times \mathbb{Z}_m$.*

Proof. Note that G has a natural action on $\mathscr{B} \times \mathscr{C}$ given by $g(B,C) = (g(B), g(C))$, and that in this action each $g \in G$ induces a permutation contained in $S^{\mathscr{B}} \times S^{\mathscr{C}}$, namely, $(g/\mathscr{B}, g/\mathscr{C})$. Any element of G in the kernel of this representation of G must fix every block of $\mathscr{B}$ and every block of $\mathscr{C}$. As $|B \cap C| = 1$ for every $B \in \mathscr{B}$ and $C \in \mathscr{C}$, and there are exactly $mk = n$ such intersections, the kernel of this representation is the identity and the representation is faithful. Let $B \in \mathscr{B}$ and $C \in \mathscr{C}$. If $g \in G$ stabilizes the point (B,C) in this representation, then $g(B) = B$ and $g(C) = C$. Let $B \cap C = \{x\}$. Then $g(x) = x$. Conversely, if $g(x) = x$, then there exists $B \in \mathscr{B}$ and $C \in \mathscr{C}$ such that $x \in B$ and $x \in C$. Then $g(B,C) = (B,C)$ so $\mathrm{Stab}_G(x) = \mathrm{Stab}_G((B,C))$. It then follows by [10, Lemma 1.6B] that these two actions of G are equivalent. □

We remark that if $G \leqslant S^{pq}$ is transitive and admits a complete block system $\mathscr{B}$ of q blocks of size p and $\mathscr{C}$ of p blocks of size q, then, as the intersection of two blocks of G is again a block of G and the order of a block divides the degree of the group, we have that $|B \cap C| = 1$ for every $B \in \mathscr{B}$ and $C \in \mathscr{C}$. Hence in this case G is equivalent to a subgroup of $S^q \times S^p$ by Lemma 2.2.

Let $V^i = \{v_j^i : j \in \mathbb{Z}_p\}$, and $V_j = \{v_j^i : i \in \mathbb{Z}_q\}$. If Γ is a (q,p)-metacirculant and $\mathrm{Aut}(\Gamma)$ admits a complete block system of q blocks of size p, then the blocks are the sets V^i, $i \in \mathbb{Z}_q$, as these sets are the orbits of the unique Sylow p-subgroup of $\langle \rho, \tau \rangle$. If Γ is circulant and $\mathrm{Aut}(\Gamma)$ admits a complete block system of p blocks of size q, the blocks are the sets V_j, $j \in \mathbb{Z}_p$, as then $\langle \rho, \tau \rangle$ (for the choice $\alpha = 1$) has a unique Sylow q-subgroup. For example, if $\alpha = 1$, the blocks of $\langle \rho, \tau \rangle$ are formed by the orbits of the normal subgroups $\langle \rho \rangle$ and $\langle \tau \rangle$. Furthermore, the pq-cycle $\rho\tau$ would correspond to the left (and right) regular representation of $\mathbb{Z}_{pq}$ in $\mathrm{Aut}(\Gamma)$, so that translation by 1 in $\mathbb{Z}_{pq}$ corresponds to the function $v_j^i \to v_{j+1}^{i+1}$. For an integer m, let $N(m) = \{f : \mathbb{Z}_m \to \mathbb{Z}_m \text{ where } f(x) = ax + b, a \in \mathbb{Z}_m^*, b \in \mathbb{Z}_m\}$. If m is prime, this group is usually denoted $\mathrm{AGL}(1,m)$.

Lemma 2.3. *Let Γ be a vertex-transitive graph of order pq with $G \leqslant \mathrm{Aut}(\Gamma)$ a transitive subgroup that admits a complete block system $\mathscr{B}$ of q blocks of size p. Assume $\mathrm{fix}_G(\mathscr{B}) \neq 1$. If the Sylow p-subgroups of $\mathrm{fix}_G(\mathscr{B})$ have order at least p^2 or $\mathrm{fix}_G(\mathscr{B})$ does not act faithfully on each $B \in \mathscr{B}$, then $\Gamma \cong \Gamma_q \wr \Gamma_p$.*

Proof. It follows by Theorem 1.5 that Γ is isomorphic to a metacirculant graph, and it was shown in [2, Lemma 10] that if $p^2 \mid |\mathrm{fix}_G(\mathscr{B})|$ then $\Gamma \cong \Gamma_q \wr \Gamma_p$ as required. Define $\pi : \mathrm{fix}_G(\mathscr{B}) \to S^B$ by $\pi(g) = g|_B$, for some fixed $B \in \mathscr{B}$. As $\mathrm{fix}_G(\mathscr{B})$ does not act faithfully on each $B \in \mathscr{B}$, $\mathrm{Ker}(\pi) \neq 1$. Let $B' \in \mathscr{B}$ such that $\mathrm{Ker}(\pi)|_{B'} \neq 1$. As $|B| = p$ is prime, it follows by [34, Theorem 8.3] that $\mathrm{fix}_G(\mathscr{B})|_{B'}$ is primitive. As $\mathrm{Ker}(\pi)|_{B'} \triangleleft \mathrm{fix}_G(\mathscr{B})|_{B'}$, it follows by [34, Theorem 8.8] that $\mathrm{Ker}(\pi)|_{B'}$ is transitive. As the size of an orbit divides the order of a group, we have that $p \mid |\mathrm{Ker}(\pi)|$ so that $p^2 \mid |\mathrm{fix}_G(\mathscr{B})|$. The result then follows by previous arguments. □

Theorem 2.4. *Let G be a transitive permutation group of degree pq that admits a complete block system $\mathcal{B}$ of q blocks of size p. Assume $p \mid |\text{fix}_G(\mathcal{B})|$ but p^2 does not divide $|\text{fix}_G(\mathcal{B})|$. Then one of the following is true:*

(1) *G is equivalent to a subgroup of $S^q \times S^p$, or*
(2) *$q = 2$ and $G = \mathbb{Z}_2 \ltimes \text{PSL}(2, 11)$ or $\mathbb{Z}_2 \ltimes \text{PSL}(n, k) \leqslant G \leqslant \mathbb{Z}_2 \ltimes \text{P}\Gamma\text{L}(n, k)$, where n is prime, $k = r^m$, r a prime, $(n, k-1) = 1$, and m is a power of n, or*
(3) *$p < q$, and if $H = \langle g \in G : |g/\mathcal{B}| = q$ and $|g| = q\rangle$, then $H/\mathcal{B}$ is a doubly transitive non-abelian simple group, H is a transitive faithful representation of $H/\mathcal{B}$ of degree pq, and $\langle H, \rho\rangle \cong H \times \mathbb{Z}_p$. Thus G contains a pq-cycle. Finally, if Γ is a vertex-transitive graph of order pq with $G \leqslant \text{Aut}(\Gamma)$, then $\Gamma \cong \Gamma_q \wr \Gamma_p$, where $\Gamma_q = K_q$ or it's complement $\bar{K}_q$.*

Furthermore, if G does not contain a pq-cycle then $G/\mathcal{B} \cong \mathbb{Z}_q$.

Proof. Let $\rho' \in \text{fix}_G(\mathcal{B})$ be of order p, and $\tau' \in G$ such that $\tau'/\mathcal{B}$ is a q-cycle and $|\tau'|$ is a power of q. As $\langle\rho'\rangle$ is a Sylow p-subgroup of $\text{fix}_G(\mathcal{B})$, there exists $\delta \in \text{fix}_G(\mathcal{B})$ such that $\delta^{-1}(\tau')^{-1}\rho'\tau'\delta \in \langle\rho'\rangle$. We thus assume without loss of generality that $(\tau')^{-1}\rho'\tau' = (\rho')^a$ for some $a \in \mathbb{Z}_p$. If $a = 1$, then $\langle\rho'\tau'\rangle$ is cyclic. If $a \neq 1$, then the commutator subgroup of $\langle\rho', \tau'\rangle$ is $\langle\rho'\rangle$ and $\langle\rho', \tau'\rangle/\mathcal{B}$ is cyclic of order q. It follows by a result in [9, p. 10] that every such group G contains a subgroup isomorphic to $\langle\rho, \tau\rangle$ as in the preliminaries, for some choice of α where $|\alpha|$ is a power of q. Hence we may assume without loss of generality that $\langle\rho, \tau\rangle \leqslant G$ (so that G permutes V) for some choice of α where $|\alpha|$ is a power of q.

We first show that if G does not contain a pq-cycle, then $G/\mathcal{B} \cong \mathbb{Z}_q$. Suppose $G/\mathcal{B}$ is not contained in $\text{AGL}(1, q)$. It follows from Theorem 1.1 that $G/\mathcal{B}$ is doubly transitive and contains at least two Sylow q-subgroups. Let $\gamma \in G$ such that $\langle\gamma/\mathcal{B}\rangle$ is a Sylow q-subgroup of $G/\mathcal{B}$, but $\gamma/\mathcal{B} \notin \langle\tau/\mathcal{B}\rangle$ and $|\gamma|$ is a power of q. As $\langle\rho\rangle$ and $\langle\gamma^{-1}\rho\gamma\rangle$ are Sylow p-subgroups of $\text{fix}_G(\mathcal{B})$, there exists $\delta \in \text{fix}_G(\mathcal{B})$ such that $\delta^{-1}\langle\gamma^{-1}\rho\gamma\rangle\delta = \langle\rho\rangle$. Hence by replacing γ with $\gamma\delta$, we may assume that $\langle\rho\rangle \triangleleft \langle\gamma, \tau, \rho\rangle$. Thus $\gamma(v_j^i) = v_{\beta j+b_i}^{\sigma(i)}$, $\sigma \in S^q$, $\beta \in \mathbb{Z}_p^*$ and $|\beta|$ a power of q, $b_i \in \mathbb{Z}_p$. Then for some $r \in \mathbb{Z}_q^*$, $\tau^r\gamma/\mathcal{B}$ has a fixed point and

$$\tau^r\gamma(v_j^i) = v_{\alpha^r\beta j+\alpha^r b_i}^{\sigma(i)+r}.$$

Let $m = |\tau\gamma/\mathcal{B}|$. As $\gamma/\mathcal{B} \notin \langle\tau/\mathcal{B}\rangle$, $m > 0$. Then $(\tau^r\gamma)^m(v_j^i) = v_{(\alpha^r\beta)^m j+c_i}^i$, $c_i \in \mathbb{Z}_p$. Now, $|\alpha| = q^k$, $k \geqslant 1$, and $\mathbb{Z}_p^*$ is cyclic. Hence if $|(\alpha^r\beta)^m| = q^{k+c}$, $c \geqslant 0$, then there exists $s \in \mathbb{Z}$ such that $(\alpha^r\beta)^{ms} = \alpha^{-1}$, in which case $(\tau^r\gamma)^{ms}\tau(v_j^i) = v_{j+d_i}^{i+1}$, $d_i \in \mathbb{Z}_p$. Let $\gamma' = (\tau^r\gamma)^{ms}\tau$. Then $|\gamma'| = q$ or $|\gamma'| = pq$. If $|\gamma'| = pq$, then clearly G contains a pq-cycle. If $|\gamma'| = q$, then $\langle\gamma'\rangle \triangleleft \langle\rho, \gamma'\rangle$, $\langle\rho\rangle \triangleleft \langle\rho, \gamma'\rangle$, and $\langle\gamma'\rangle \cap \langle\rho\rangle = 1$. Thus $\langle\gamma', \rho\rangle \cong \mathbb{Z}_{pq}$ and is transitive, and so G contains a pq-cycle as required. We may thus assume that $|(\alpha^r\beta)^m| = q^n$, $n < k$, so that $(\alpha^r\beta)^m = \alpha^t$, for some t, where $q|t$, and $\beta = \alpha^s$, for some s. Then

$$(\alpha^r\beta)^m = \alpha^{rm}\beta^m = \alpha^{rm}\alpha^{sm} = \alpha^t,$$

and so $rm + sm \equiv t \equiv 0 \pmod{q}$. As $\gcd(m, q) = 1$, $r \equiv -s \pmod{q}$. Now, as $\gamma/\mathcal{B} \notin \langle\tau/\mathcal{B}\rangle$, there exists $r' \in \mathbb{Z}_q^*$, $r' \neq r$ such that $\tau^{r'}\gamma/\mathcal{B}$ also has a fixed point and $0 \neq m' = |\tau^{r'}\gamma/\mathcal{B}|$. By the argument above, $r' \equiv -s \pmod{q}$, and so $r \equiv r' \pmod{q}$, a contradiction. Thus $G/\mathcal{B} \leqslant \text{AGL}(1, q)$.

Suppose there exists $\gamma \in G$ such that $\gamma/\mathcal{B} \neq 1$ and does not have order q. By replacing γ with $\gamma\delta$, for a suitable δ as above, we may assume that γ normalizes $\langle \rho \rangle$. Then $\gamma(v_j^i) = v_{\beta j + b_i}^{\delta i + c}$, $\delta \in \mathbb{Z}_q^*$, $\delta \neq 1$, $c \in \mathbb{Z}_q$, $\beta \in \mathbb{Z}_p^*$, $b_i \in \mathbb{Z}_p$. Note then that $\delta^{-1}(v_j^i) = v_{\beta^{-1}(j-b_{\delta^{-1}(i-c)})}^{\delta^{-1}(i-c)}$. Elementary calculations will show that if $\gamma' = \gamma^{-1}\tau^{-1}\gamma\tau$, then

$$\gamma'(v_j^i) = v_{j+\beta^{-1}(\alpha^{-1}b_{i+1}-b_{i+1-\delta^{-1}})}^{i+1-\delta^{-1}}.$$

As $\delta \neq 1$, $\delta^{-1} \neq 1$ and $|\gamma'| = q$ or $|\gamma'| = pq$. By arguments above, G contains a pq-cycle, a contradiction. Thus, if G does not contain a pq-cycle then $G/\mathcal{B} \cong \mathbb{Z}_q$.

If $\text{fix}_G(\mathcal{B})$ acts faithfully on each $B \in \mathcal{B}$, then either $\text{fix}_G(\mathcal{B}) \cong \mathbb{Z}_p$ or $\text{Stab}_{\text{fix}_G(\mathcal{B})}(v_0^0) \neq 1$. If $\text{Stab}_{\text{fix}_G(\mathcal{B})}(v_0^0) \neq 1$, then the equivalence relation $\equiv$ is defined (see Definition 6). Hence $\text{fix}_G(\mathcal{B}) \cong J \cong 1_{S^q} \times J$, where $J \leqslant S^p$. If the equivalence classes of $\equiv$ have order 1, then $\text{fix}_G(\mathcal{B})|_B \cong \text{fix}_G(\mathcal{B})|_{B'}$ for every $B, B' \in \mathcal{B}$ and the actions of $\text{fix}_G(\mathcal{B})|_B$ and $\text{fix}_G(\mathcal{B})|_{B'}$ are faithful on B and B' respectfully. By [10, Lemma 1.6B], the actions of $\text{fix}_G(\mathcal{B})|_B$ and $\text{fix}_G(\mathcal{B})|_{B'}$ are inequivalent actions. If $\text{fix}_G(\mathcal{B})|_B \leqslant \text{AGL}(1,p)$, then let $|\text{fix}_G(\mathcal{B})|_B| = pr$, $r|(p-1)$. As $\text{AGL}(1,p)$ is solvable and $|\text{AGL}(1,p)| = (p-1)p$, $\text{fix}_G(\mathcal{B})|_B$ is solvable so that any two subgroups of $\text{fix}_G(\mathcal{B})|_B$ of order $r|(p-1)$ are conjugate in $\text{fix}_G(\mathcal{B})|_B$. We conclude that the actions of $\text{fix}_G(\mathcal{B})|_B$ and $\text{fix}_G(\mathcal{B})|_{B'}$ are equivalent for every $B, B' \in \mathcal{B}$ so by Theorem 1.2 $\text{fix}_G(\mathcal{B})|_B$ is doubly transitive and nonsolvable for every $B \in \mathcal{B}$.

If $\text{fix}_G(\mathcal{B})|_B$ is doubly transitive and nonsolvable, then $\text{soc}(\text{fix}_G(\mathcal{B})|_B)$ is one of the groups listed in [24, Proposition 2.4] and also has more than one representation on B. Perusing the list of doubly transitive groups in [6, Theorem 5.3], we see that either $\text{soc}(\text{fix}_G(\mathcal{B})|_B) = \text{PSL}(2,11)$ and $p = 11$ or $\text{soc}(\text{fix}_G(\mathcal{B})|_B) = \text{PSL}(n,k)$, where k is a prime power, $p = (k^n - 1)/(k - 1)$, and each of these latter groups has exactly two representations unless $n = 2$, in which case there is exactly one representation. Thus $n \neq 2$. As p is prime, by [10, Exercise 3.5.11], we have that n is prime and $\gcd(n, k-1) = 1$. As $\text{fix}_G(\mathcal{B})|_B$ and $\text{fix}_G(\mathcal{B})|_{B'}$ are inequivalent actions for every $B, B' \in \mathcal{B}$, $B \neq B'$, we conclude $q = 2$.

If $\text{soc}(\text{fix}_G(\mathcal{B})|_B) = \text{PSL}(2,11)$, then by [10, Table B.2], $\text{fix}_G(\mathcal{B})|_B = \text{PSL}(2,11) = \text{soc}(\text{fix}_G(\mathcal{B})|_B)$. As $|\text{PSL}(2,11)| = 660$, a straightforward computation will then show that the normalizer of a Sylow 11-subgroup of $\text{PSL}(2,11)$ has order $5 \cdot 11$. Thus, by raising τ to an appropriate power, we may assume that $|\tau| = 2$. Then $\langle \tau \rangle \cap \text{fix}_G(\mathcal{B}) = 1$, $\text{fix}_G(\mathcal{B}) \triangleleft G$, and $G = \langle \tau \rangle \cdot \text{fix}_G(\mathcal{B})$. It then follows by [10, Exercise 2.5.3], that $G \cong \mathbb{Z}_2 \ltimes \text{PSL}(2,11)$ as required.

If $\text{soc}(\text{fix}_G(\mathcal{B}))|_B = \text{PSL}(n,k)$, then $\text{PSL}(n,k) \leqslant \text{fix}_G(\mathcal{B})|_B \leqslant \text{P}\Gamma\text{L}(n,k)$ (see [10, p. 245]). Choose $\alpha \in \mathbb{Z}_p^*$ such that $\tau \in G$ and $|\alpha|$ is minimum. Hence $|\tau|$ is a power of 2. If $\alpha = 1$, then G contains a pq-cycle, contradicting Lemma 2.1. Hence $\alpha \neq 1$ and $|\alpha| \neq 1$. By [17, Satz 7.3], $|N_{\text{PSL}(n,k)}(\Pi)| = np$, where $\Pi \leqslant \text{PSL}(n,k)$ is cyclic of order $p = (k^n - 1)/(k - 1)$. By choice of α, no power that is not a power of 2 of the function $\bar{\alpha} : V \to V$ by $\bar{\alpha}(v_j^i) = v_{\alpha j}^i$ is contained in $\text{fix}_{\langle \tau, \text{soc}(\text{fix}_G(\mathcal{B}))\rangle}(\mathcal{B})$. As $n \neq 2$, we conclude that no power of τ, other than one yielding the identity, is contained in $\text{soc}(\text{fix}_G(\mathcal{B}))$. Let $k = r^m$ be a power of a prime r. By [10, Exercise 3.5.11], we then have that m is a power of n. As $\text{P}\Gamma\text{L}(n,k)$ is generated by $\text{PSL}(n,k)$ together with the field automorphisms of the field $\mathbb{F}_k$, we have that $|\text{P}\Gamma\text{L}(n,k)/\text{PSL}(n,k)| = m$. As $n \neq 2$, we have that m is odd. Hence no power of τ, other than one yielding the identity, is contained in $\text{fix}_G(\mathcal{B})$. As $\tau^2 \in \text{fix}_G(\mathcal{B})$, we have that $|\tau| = 2$. It then follows by arguments in the immediately preceding paragraph that $G \cong \mathbb{Z}_2 \ltimes \text{fix}_G(\mathcal{B})$ so that $\mathbb{Z}_2 \ltimes \text{PSL}(n,k) \leqslant G \leqslant \mathbb{Z}_2 \ltimes \text{P}\Gamma\text{L}(n,k)$ as required. We thus assume that there are fewer than pq equivalence classes of $\equiv$.

If there is more than 1 equivalence class of $\equiv$, then there are p equivalence classes of $\equiv$ and each has cardinality q, i.e., if E is an equivalence class of $\equiv$ then E contains exactly one element from each orbit of $\text{fix}_G(\mathcal{B})$. Denote the equivalence classes of $\equiv$ by $E_0, E_1, \ldots, E_{p-1}$. Note that the equivalence classes $E_0, E_1, \ldots, E_{p-1}$ are blocks of G of size q and the result follows by Lemma 2.2. Hence we may assume that $\text{fix}_G(\mathcal{B}) \cong \mathbb{Z}_p$. This implies that $|\tau| = q$.

If $G/\mathcal{B} \leqslant \text{AGL}(1,q)$, then $G/\mathcal{B}$ contains a normal subgroup of order q, and hence G contains a normal subgroup of order pq [16, Corollary 5.12]. If G does not contain a pq-cycle, then by previous arguments $G/\mathcal{B} = \langle \tau \rangle/\mathcal{B}$ and so $G = \langle \rho, \tau \rangle$. Hence $G \leqslant S^q \times S^p$ as required. Otherwise, we may take $\alpha = 1$ (so that $\langle \rho, \tau \rangle = \langle \rho\tau \rangle$ is cyclic of order pq) in which case $\langle \rho, \tau \rangle \triangleleft G$ so that $G \leqslant N(pq) = \text{AGL}(1,p) \times \text{AGL}(1,q) \leqslant S^q \times S^p$.

If $G/\mathcal{B} \not\leqslant \text{AGL}(1,q)$, then $G/\mathcal{B}$ is doubly transitive and there are at least two Sylow q-subgroups of $G/\mathcal{B}$. Furthermore, by previous arguments, G contains a pq-cycle so we may again take $\alpha = 1$. Choose $\beta \in G$ such that $\beta/\mathcal{B}$ is a q-cycle, but $\langle \beta/\mathcal{B} \rangle \notin \langle \tau/\mathcal{B} \rangle$. Without loss of generality, we assume that $|\beta| = q$. As $\text{fix}_G(\mathcal{B}) = \langle \rho \rangle$, $\beta(v_j^i) = v_{\iota j + b_i}^{\sigma(i)}$, $\sigma \in S^q$, $\iota \in \mathbb{Z}_p^*$, and $b_i \in \mathbb{Z}_p$. We first show that $\beta(v_j^i) = v_{j+b_i}^{\sigma(i)}$, $\sigma \in S^q$, $b_i \in \mathbb{Z}_p$.

Suppose $\beta(v_j^i) = v_{\iota j + b_i}^{\sigma(i)}$, $\sigma \in S^q$, $\iota \in \mathbb{Z}_p^*$, $c_i \in \mathbb{Z}_p$. As $|\beta| = q$, we must have $|\iota| = q$, or $|\iota| = 1$. If $q > p$, then $q \nmid |\mathbb{Z}_p^*| = p - 1$. Hence $\alpha = 1$. If $q < p$, suppose $|\iota| = q$. Then for some $r \in \mathbb{Z}_q$, $\tau^r \beta/\mathcal{B}$ has a fixed point and hence does not have order q. But $\tau^r \beta(v_j^i) = v_{\iota j + c_i}^{\sigma(i)+r}$ so $q \mid |\tau^r \beta|$. Hence $q \mid |\text{fix}_G(\mathcal{B})|$, a contradiction. Thus $|\iota| = 1$.

Let $X = \{\beta : \beta(v_j^i) = v_{j+b_i}^{\sigma(i)}$ and $\Sigma_{i=0}^{q-1} b_i \equiv 0 \pmod{p}\}$, and $H = \langle X \rangle$. Let $\delta \in G$. Then $\delta(v_j^i) = v_{\iota j + c_i}^{\omega(i)}$. As for every $\beta \in X$, $|\beta| = q$, we have that $\delta^{-1}\beta\delta(v_j^i) = v_{j+c_i}^{\omega^{-1}\sigma\omega(i)}$, where $|\omega^{-1}\sigma\omega| = q$ and $\Sigma_{i=0}^{q-1} c_i \equiv 0 \pmod{p}$. Hence $H \triangleleft G$. Furthermore, if $h \in H$ and $h(v_j^i) = v_{j+d_i}^{\psi(i)}$, then as $h = \beta_1^{a_1} \beta_2^{a_2} \ldots \beta_r^{a_r}$, $\beta_i \in X$ and $a_i \in \mathbb{Z}_q$, we have that $\Sigma_{i=1}^{q-1} d_i \equiv 0 \pmod{p}$ as well. Hence $\rho \notin H$ so that $H \cap \langle \rho \rangle = 1$. Thus $\langle H, \rho \rangle \cong H \times \langle \rho \rangle \cong H \times \mathbb{Z}_p$ and $\langle H, \rho \rangle$ contains a pq-cycle. If H is intransitive, then the orbits of H form a complete block system of p blocks of size q. Thus, by Lemma 2.2, $G \leqslant S^q \times S^p$ and (1) follows. If H is transitive, then $H/\mathcal{B}$ is a faithful representation of H, and is thus a subgroup of S^q. As H is transitive, $p \mid |H|$ so that $p \mid q!$. We conclude that $p < q$.

It now only remains to be shown that if H is transitive and Γ is a vertex-transitive graph of order pq with $G \leqslant \text{Aut}(\Gamma)$, then $\Gamma = \Gamma_q \wr \Gamma_p$, where $\Gamma_q = K_q$ or $\bar{K}_q$. As H is transitive, the orbits of H have length pq, so that $\text{Stab}_H(B)|_B$ is transitive for every $B \in \mathcal{B}$. Hence there exists $h \in H$ such that $h(v_j^i) = v_{j+b_i}^{\sigma(i)}$, and for some i, $\sigma(i) = i$ and $b_i \neq 0$. This then implies that p divides $|h/\mathcal{B}|$. By raising h to an appropriate power s relatively prime to p, we may assume without loss of generality that h has order a power of p. Note then that $h/\mathcal{B} \neq 1$, as otherwise $1 \neq h^s \in \text{fix}_G(\mathcal{B}) = \langle \rho \rangle$ and it was shown above that $\rho \notin H$. We may also assume that $h/\mathcal{B}$ is of minimal order while preserving the property that $\sigma(i) = i$ and $b_i \neq 0$ for some $i \in \mathbb{Z}_q$. Let $\mathcal{O}$ be a non-singleton orbit of σ. We now show that $\sum_{i \in \mathcal{O}} b_i \equiv 0 \pmod{p}$.

If $\mathcal{O}$ is an orbit of σ of maximal length and $|\mathcal{O}| = p^k$, then $h^{p^k} \in \mathrm{fix}_H(\mathcal{B}) = 1$. We conclude that for such orbits, that $\sum_{i \in \mathcal{O}} b_i \equiv 0 \pmod{p}$. If $\mathcal{O}$ is an orbit of σ of length p^ℓ, $0 < \ell < k$, then $h^{p^\ell}/\mathcal{B} \neq 1$. Let $K = \cup_{i \in \mathcal{O}} V^i$. If $\sum_{i \in \mathcal{O}} b_i \equiv a \not\equiv 0 \pmod{p}$, then $h^{p^\ell}|_K = \rho^a|_K$. As $\mathcal{O}$ is not an orbit of σ of maximal length, $h^{p^\ell}/\mathcal{B} \neq 1$ and fixes every $i \in \mathcal{O}$. However, if $h^{p^\ell}(v_j^i) = v_{j+c_i}^{\sigma^{p^\ell}(i)}$, then $c_i \neq 0$ for every $i \in \mathcal{O}$, contradicting our choice of h. Thus $\sum_{i \in \mathcal{O}} b_i \equiv 0 \pmod{p}$ for every non-singleton orbit of h.

Note that as $|h|$ is a power of p and $\gcd(p, q) = 1$, $h/\mathcal{B}$ must have at least one singleton orbit and the number of singleton orbits of $h/\mathcal{B}$ cannot be a multiple of q. Furthermore, $b_i \neq 0$ for at least one singleton orbit $\{i\}$ of $h/\mathcal{B}$. As $\sum_{i \in \mathbb{Z}_q} b_i \equiv 0 \pmod{p}$, and for every non-singleton orbit $\mathcal{O}$ of $h/\mathcal{B}$ we have that $\sum_{i \in \mathcal{O}} b_i \equiv 0 \pmod{p}$, it must follow that h has at least two singleton orbits, and if $L = \{i \in \mathbb{Z}_q : \{i\}$ is a singleton orbit of $\sigma\}$, then $\sum_{i \in L} b_i \equiv 0 \pmod{p}$. We conclude that there exists $k, \ell \in L$ such that $b_k \neq b_\ell$. As $h \in \mathrm{Aut}(\Gamma)$, if there is an edge between some vertex of V^k and some vertex of V^ℓ in Γ, then every vertex of V^k is adjacent to every vertex of V^ℓ. As q is prime and $G/\mathcal{B}$ is nonsolvable, we have that $H/\mathcal{B}$ is nonsolvable. Then $H/\mathcal{B}$ is doubly transitive, and so if there is some edge between a vertex of block V^k and some vertex of block V^ℓ of $\mathcal{B}$, $k \neq \ell$, then every vertex of V^k is adjacent to every vertex of V^ℓ for every $V^k, V^\ell \in \mathcal{B}$, $k \neq \ell$. We conclude that $\Gamma = K_q \wr \Gamma_p$ or $\Gamma = \bar{K}_q \wr \Gamma_p$ and (3) follows. □

Let $\alpha \in \mathbb{Z}_p^*$ with $|\alpha| = q$. For the remainder of this paper, define $\Psi_\alpha : V \to V$ by $\Psi_\alpha(v_j^0) = v_j^0$ and $\Psi_\alpha(v_j^i) = v_{j - \sum_{k=0}^{i-1} \alpha^k}^i$ for $i \neq 0$. We now prove a few results which establish a variety of facts about $\langle \rho, \tau \rangle$, for various choices of α, that will be needed later in the paper. Note that we do not consider any choices of α such that $q \nmid |\alpha|$, as in all such cases any vertex-transitive graph Γ with $\langle \rho, \tau \rangle \leqslant \mathrm{Aut}(\Gamma)$ is by Theorem 1.4 necessarily isomorphic to a circulant graph. We first compute the normalizers of $\langle \rho, \tau \rangle$ in S^V.

Lemma 2.5. *Let $\alpha \in \mathbb{Z}_p^*$ such that $q \mid |\alpha|$, and β be a generator of $\mathbb{Z}_p^*$. Define $\omega : V \to V$ by $\omega(v_j^i) = v_{\beta j}^i$.*
(1) *If $|\alpha| \neq q$, then $N_{S^V}(\langle \rho, \tau \rangle) = \langle \rho, \tau, \omega \rangle$.*
(2) *If $|\alpha| = q$, then $N_{S^V}(\langle \rho, \tau \rangle) = \langle \rho, \tau, \omega, \Psi_\alpha \rangle$.*

Proof. As $\langle \rho \rangle$ is the unique Sylow p-subgroup of $\langle \rho, \tau \rangle$ (for any choice of α), $\langle \rho \rangle$ is fully invariant so that $\langle \rho \rangle \triangleleft N_{S^V}(\langle \rho, \tau \rangle)$. Thus $N_{S^V}(\langle \rho, \tau \rangle)$ admits a complete block system $\mathcal{B}$ formed by the orbits of $\langle \rho \rangle$. As $q \mid |\alpha|$, $\langle \rho, \tau \rangle$ does not contain a pq-cycle so by Theorem 2.4 $N_{S^V}(\langle \rho, \tau \rangle)/\mathcal{B} \cong \mathbb{Z}_q$. Thus, if $\gamma \in N_{S^V}(\langle \rho, \tau \rangle)$ then $\gamma(v_j^i) = v_{ij + a_i}^{i+r}$, $r \in \mathbb{Z}_q$, $\iota \in \mathbb{Z}_p^*$ and $a_i \in \mathbb{Z}_p$. As $\tau \in N_{S^V}(\langle \rho, \tau \rangle)$, we may assume without loss of generality that $r = 0$. It is straightforward to verify that $\omega \in N_{S^V}(\langle \rho, \tau \rangle)$ so that we may also assume that $\iota = 1$. Thus $\gamma(v_j^i) = v_{j + a_i}^i$, $a_i \in \mathbb{Z}_p$. As $\rho \in N_{S^V}(\langle \rho, \tau \rangle)$, we may furthermore assume that $a_0 = 0$. A straightforward computation will show that γ centralizes $\langle \rho \rangle$ so that $\gamma \in N_{S^V}(\langle \rho, \tau \rangle)$ if and only if $\gamma^{-1}\tau\gamma \in \langle \rho, \tau \rangle$. As $\gamma^{-1}\tau\gamma(v_j^i) = v_{\alpha j + \alpha a_i - a_{i+1}}^{i+1}$ and $\gamma^{-1}\tau\gamma = \tau\rho^c$ for some $c \in \mathbb{Z}_p$, we have that $\alpha a_i - a_{i+1} \equiv c \pmod{p}$ for every $i \in \mathbb{Z}_q$. As $a_0 = 0$, we have for $1 \leqslant i \leqslant q - 1$, that $a_i = -c \sum_{k=0}^{i-1} \alpha^k$ and $\alpha a_{q-1} - a_0 \equiv c \pmod{p}$. If $c = 0$, then it is easy to see that each $a_i = 0$ and $\gamma \in \langle \rho, \tau \rangle$. If $c \neq 0$, then c is invertible, and, as $a_0 = 0$ we have that

$-c\alpha(\sum_{k=0}^{q-2}\alpha^k) \equiv c \pmod{p}$. Thus $-\alpha\sum_{k=0}^{q-2}\alpha^k \equiv 1 \pmod{p}$. As $\sum_{k=0}^{q-2}\alpha^k = \frac{\alpha^{q-1}-1}{\alpha-1}$, we must have that $-\alpha(\alpha^{q-1} - 1) \equiv \alpha - 1 \pmod{p}$. Thus it is equivalent to $\alpha^q \equiv 1 \pmod{p}$, and this occurs if and only if $|\alpha| = q$. Hence if $|\alpha| \neq q$, then $N_{S^V}(\langle\rho,\tau\rangle) = \langle\rho,\tau,\omega\rangle$. If $|\alpha| = q$, then $\gamma = \Psi_\alpha^c$ and $N_{S^V}(\langle\rho,\tau\rangle) = \langle\rho,\tau,\omega,\Psi_\alpha\rangle$. □

We now compute all complete block systems of $\langle\rho,\tau\rangle$, as well as of some overgroups of $\langle\rho,\tau\rangle$ that are of interest here.

Lemma 2.6. *Let $\alpha \in \mathbb{Z}_p^*$ such that $q \mid |\alpha|$.*
(1) *If $|\alpha| \neq q$, then $\langle\rho,\tau\rangle$ admits exactly two complete block systems $\mathscr{B}$ and $\mathscr{C}$, where if $B \in \mathscr{B}$ and $v_0^0 \in B$, then B is the orbit of $\langle\rho\rangle$ that contains v_0^0, and if $C \in \mathscr{C}$ and $v_0^0 \in C$, then C is the orbit of $\langle\tau\rangle$ that contains v_0^0. Thus $\mathscr{B}$ consists of q blocks of size p and $\mathscr{C}$ consists of p blocks of size q.*
(2) *If $|\alpha| = q$, then $\langle\rho,\tau\rangle$ admits exactly $p+1$ complete block systems $\mathscr{B}$ and $\mathscr{C}_i$, $0 \leq i \leq p-1$, where if $B \in \mathscr{B}$ and $v_0^0 \in B$, then B is the orbit of $\langle\rho\rangle$ that contains v_0^0, and if $C_i \in \mathscr{C}_i$ and $v_0^0 \in C_i$, then C_i is the orbit of $\langle\tau\rho^i\rangle$ that contains v_0^0, $0 \leq i \leq p-1$. Thus $\mathscr{B}$ consists of q blocks of size p and each $\mathscr{C}_i$ consists of p blocks of size q, $0 \leq i \leq p-1$.*
(3) *If $|\alpha| = q$ and $H \leq N_{S^V}(\langle\rho,\tau\rangle)$ such that $\langle\rho,\tau\rangle \leq H$, $|H| = apq$, $a > 1$, and $\gcd(a, pq) = 1$, then H admits exactly two complete block systems, and there exists $b \in \mathbb{Z}_p$ such that $\Psi_\alpha^{-b} H \Psi_\alpha^b$ admits $\mathscr{B}$ and $\mathscr{C}_0$ as complete block systems, where $\mathscr{B}$ and $\mathscr{C}_0$ are as in part (2) of this lemma.*

Proof. (2) By [10, Theorem 1.5A], there exists a bijection f from the set T of all blocks of $\langle\rho,\tau\rangle$ that contain v_0^0 onto the set S of all subgroups of $\langle\rho,\tau\rangle$ that contain $\text{Stab}_{\langle\rho,\tau\rangle}(v_0^0)$ whose inverse mapping is given by $f^{-1}(H) = \{h(v_0^0) : h \in H\}$, $H \in S$. If $|\alpha| = q$, then $\langle\rho,\tau\rangle$ is regular (so that $\text{Stab}_{\langle\rho,\tau\rangle}(v_0^0) = 1$) and has exactly one proper subgroup of order p (namely $\langle\rho\rangle$) and exactly p subgroups of order q (namely $\langle\rho\tau^i\rangle$, $0 \leq i \leq p-1$). Thus, if $|\alpha| = q$, the block systems of $\langle\rho,\tau\rangle$ are as claimed in (2).

Much of the work for (1) and (3) can be handled simultaneously. With this in mind, if $|\alpha| \neq q$, then we set $H = \langle\rho,\tau\rangle$ and $a = |\alpha|/q$. Note that $\langle\rho\rangle \triangleleft H$. Then H admits a complete block system of q blocks of size p formed by the orbits of $\langle\rho\rangle$, $|\text{fix}_H(\mathscr{B})| = |H|/q = ap$, and, of course, $\text{Stab}_H(v_0^0) \leq \text{fix}_H(\mathscr{B})$. Let $K \leq H$ such that $\text{Stab}_H(v_0^0) \leq K$ and $\{k(v_0^0) : k \in K\}$ has order p. As $\{k(v_0^0) : k \in K\}$ is an orbit of K, p divides $|K|$. As H contains a unique Sylow p-subgroup which is $\langle\rho\rangle$, $\rho \in K$. As K is not transitive, $|K| < |H| = pq \cdot a$. Thus $|K| \leq p \cdot a$. As $\text{Stab}_H(v_0^0) \leq K$, $\langle\rho\rangle \leq K$, and $|\text{Stab}_H(v_0^0)| = a$, we have that $K = \langle\rho, \text{Stab}_H(v_0^0)\rangle$. Furthermore, $\langle\rho\rangle, \text{Stab}_H(v_0^0) \leq \text{fix}_H(\mathscr{B})$ so that $K = \text{fix}_H(\mathscr{B})$. Thus H admits a unique complete block system of q blocks of size p formed by the orbits of $\langle\rho\rangle$. We now consider (1) and (3) separately.

(1) If $|\alpha| \neq q$, then note that $\langle\tau^q\rangle \leq \text{Stab}_H(v_0^0)$ and $|\langle\tau^q\rangle| = a$, so that $\text{Stab}_H(v_0^0) = \langle\tau^q\rangle$. Let $K \leq H$ such that $\text{Stab}_H(v_0^0) \leq K$ and $\{k(v_0^0) : k \in K\}$ has order q. As $|\text{Stab}_H(v_0^0)| = a$, we have that $|K| = q \cdot a = |\alpha|$. As $|H| = qp \cdot a$, we have that K is a p-complement of $\langle\rho\rangle$. Note that $\langle\tau, \text{Stab}_H(v_0^0)\rangle = \langle\tau\rangle$ has order $q \cdot a$ so that $\rho^{-c} K \rho^c = \langle\tau\rangle$ for some $c \in \mathbb{Z}_p$ by the Schur–Zassenhaus theorem [15, Theorem 6.2.1]. Hence $K = \rho^c\langle\tau\rangle\rho^{-c} = \langle\tau\rho^m\rangle$ for

some $m \in \mathbb{Z}_p$. Let $b = \sum_{\ell=0}^{q-1} \alpha^\ell$. Then $(\tau \rho^m)^q(v_j^i) = v_{\alpha^q j + \alpha mb}^i$. As $|\alpha| \ne q$, it follows that $\alpha mb \not\equiv 0 \pmod{p}$ unless $m = 0$. We conclude that $(\tau \rho^m)^q \in \mathrm{Stab}_{\langle \rho, \tau \rangle}(v_0^0)$ if and only if $m = 0$. As $\mathrm{Stab}_{\langle \rho, \tau \rangle}(v_0^0) \leqslant K = \langle \tau \rho^m \rangle$, we have that $K = \langle \tau \rangle$. Thus the block systems of $\langle \rho, \tau \rangle$ are as claimed in (1).

(3) If $|\alpha| = q$, then let $\delta \in \mathrm{fix}_H(\mathcal{B})$ such that $|\delta| = a$. Such a δ exists, as by Lemma 2.5, $N_{S^V}(\langle \rho, \tau \rangle) = \langle \rho, \tau, \omega, \Psi_\alpha \rangle$ and $\mathrm{fix}_{N_{S^V}(\langle \rho, \tau \rangle)}(\mathcal{B}) = \langle \rho, \omega, \Psi_\alpha \rangle$. By conjugating δ by an appropriate power of ρ, we assume without loss of generality that $\delta \in \mathrm{Stab}_H(v_0^0)$. As $|H| = pqa$, we have that $\mathrm{Stab}_H(v_0^0) = \langle \delta \rangle$. Note that $\mathrm{Stab}_{N_{S^V}(\langle \rho, \tau \rangle)}(v_0^0) = \langle \omega, \Psi_\alpha \rangle$, $\langle \Psi_\alpha \rangle \triangleleft \langle \omega, \Psi_\alpha \rangle$ and $\langle \Psi_\alpha \rangle$ is the unique Sylow p-subgroup of $\langle \omega, \Psi_\alpha \rangle$. It follows by the Schur–Zassenhaus theorem that $\langle \Psi_\alpha \rangle$ has a p-complement in $\langle \omega, \Psi_\alpha \rangle$, and every p-complement of $\langle \Psi_\alpha \rangle$ in $\langle \omega, \Psi_\alpha \rangle$ is conjugate to $\langle \omega \rangle$ by an element of $\langle \Psi_\alpha \rangle$. As $\langle \omega \rangle$ is one such p-complement, there exists $b \in \mathbb{Z}_p$ such that $\Psi_\alpha^{-b} \delta \Psi_\alpha^b \in \langle \omega \rangle$. We now show that $\Psi_\alpha^{-b} H \Psi_\alpha^b = \langle \rho, \tau' \rangle$, where $\tau'(v_j^i) = v_{\gamma j}^{i+1}$, for some $\gamma \in \mathbb{Z}_p^*$ of order qa, and $\Psi_\alpha^{-b} H \Psi_\alpha^b$ has $\mathscr{C}_0$ as the unique complete block system of $\Psi_\alpha^{-b} H \Psi_\alpha^b$ of p blocks of size q.

As $\Psi_\alpha \in N_{S^V}(\langle \rho, \tau \rangle)$, $\langle \rho, \tau \rangle \leqslant \Psi_\alpha^{-b} H \Psi_\alpha^b$. Let $\tau' = \tau \Psi_\alpha^{-b} \delta \Psi_\alpha^b$. Then $\tau'(v_j^i) = v_{\alpha \omega^g j}^{i+1}$ for some integer g. As $\langle \rho \rangle \triangleleft \Psi_\alpha^{-b} H \Psi_\alpha^b$ and $|\langle \tau' \rangle| = qa$, we have that $\Psi_\alpha^{-b} H \Psi_\alpha^b = \langle \rho, \tau' \rangle$. Clearly $\mathrm{Stab}_{\langle \rho, \tau' \rangle}(v_0^0) \leqslant \langle \tau' \rangle$ and one can easily check that the corresponding complete block system is $\mathscr{C}_0$. By part (1) of the lemma, $\langle \rho, \tau' \rangle$ has exactly one complete block system with blocks of size q and the result follows. $\square$

Lemma 2.7. *Let $\alpha \in \mathbb{Z}_p^*$ such that $q \mid |\alpha|$, and β be a generator of $\mathbb{Z}_p^*$. Define $\omega : V \to V$ by $\omega(v_j^i) = v_{\beta j}^i$.*

(1) *If $|\alpha| \ne q$ and $H \leqslant N_{S^V}(\langle \rho, \tau \rangle)$ is transitive and does not contains a pq-cycle, then there exists $\gamma \in \mathbb{Z}_p^*$ such that $|\gamma| = |H|/p$ and $H = \langle \rho, \tau' \rangle$ where $\tau'(v_j^i) = v_{\gamma j}^{i+1}$.*

(2) *If $|\alpha| = q$ and $H \leqslant N_{S^V}(\langle \rho, \tau \rangle)$ such that $\langle \rho, \tau \rangle \leqslant H$, $|H| = apq$, $a > 1$, and $\gcd(a, pq) = 1$, then there exists $\gamma \in \mathbb{Z}_p^*$ such that $|\gamma| = qa$ and if $\tau' : V \to V$ by $\tau'(v_j^i) = v_{\gamma j}^{i+1}$, then $\Psi_\alpha^{-b} H \Psi_\alpha^b = \langle \rho, \tau' \rangle$ for some $b \in \mathbb{Z}_p$. Let $L = \langle \rho, \tau, \omega \rangle$. Then $L \cap H = \langle \rho, \tau \rangle$ or $b = 0$ and $L \cap H = \langle \rho, \tau' \rangle$.*

Proof. (1) If $|\alpha| \ne q$ and $H \leqslant N_{S^V}(\langle \rho, \tau \rangle)$ is transitive and does not contain a pq-cycle, then by previous arguments, as $N_{S^V}(\langle \rho, \tau \rangle) = \langle \rho, \tau, \omega \rangle$, we have that $|\alpha|$ does not divide $|\mathrm{fix}_H(\mathcal{B})|$. This follows as, if $|\tau| = |\alpha|$ divides $|\mathrm{fix}_H(\mathcal{B})|$, then as $\mathbb{Z}_p^*$ is cyclic, the function ϕ defined by $\phi(v_j^i) = v_{\alpha j}^i$ is contained in $\mathrm{fix}_H(\mathcal{B})$. Then $\langle \rho, \tau \phi^{-1} \rangle$ contains a pq-cycle, a contradiction. Again, as $N_{S^V}(\langle \rho, \tau \rangle) = \langle \rho, \tau, \omega \rangle$, there exists $\iota \in \mathbb{Z}_p^*$ such that $|\iota| = |\mathrm{fix}_H(\mathcal{B})|/(pq^c)$, where $q^c \mid |\mathrm{fix}_H(\mathcal{B})|$ but q^{c+1} does not divide $|\mathrm{fix}_H(\mathcal{B})|$. Then the function defined by $v_j^i \to v_{ij}^i$ is contained in $\mathrm{fix}_H(\mathcal{B})$. Let $\gamma = \alpha \iota$. Then $\tau' \in H$, and $|\tau'| = |\gamma| = |\iota| \cdot |\alpha| = |H|/p$.

(2) If $|\alpha| = q$, then it was shown in last two paragraphs of Lemma 2.6 that there exists $\gamma \in \mathbb{Z}_p^*$ such that $|\gamma| = qa$ and if $\tau' : V \to V$ by $\tau'(v_j^i) = v_{\gamma j}^{i+1}$, then $\Psi_\alpha^{-b} H \Psi_\alpha^b = \langle \rho, \tau' \rangle$ for some $b \in \mathbb{Z}_p$. If $b = 0$ then clearly $L \cap H = \langle \rho, \tau' \rangle$. If $b \ne 0$, then $\Psi_\alpha^b (\tau')^q \Psi_\alpha^{-b} \in H$ and $H = \langle \rho, \tau, \Psi_\alpha^b (\tau')^q \Psi_\alpha^{-b} \rangle$. It is then easy to check that $\Psi_\alpha^b (\tau')^q \Psi_\alpha^{-b} \notin L$, so that $H \cap L = \langle \rho, \tau \rangle$. $\square$

Part of the motivation for the above results is that there are at least two results in the literature (one by the author) that are slightly incorrect, and these errors are caused by neglecting to consider that if $|\alpha| = q$, then $\langle \rho, \tau \rangle$ has p complete block systems of size q. We will discuss (and correct) both of these errors in this paper, although we will put off examining one of these errors until a more appropriate time. The first error that we will discuss occurred in the author's paper [11]. In that paper, the following result was proved (see [11] for terms not defined in this paper).

Theorem 2.8 ([11], Theorem 9). *Let X be a Cayley object of G with $|G| = pq$, $p > q$ and q divides $p - 1$, such that $p^2 \nmid |\text{Aut}(X)|$. If $G = \mathbb{Z}_{pq}$ then X is a CI-object for G. If $G = \langle \rho, \tau \rangle$, for some $\alpha \in \mathbb{Z}_p^*$ such that $|\alpha| = q$, then either X is a CI-object for G or X is also a Cayley object of $\mathbb{Z}_{pq}$.*

This result is somewhat correct, but the proof only shows that if X is a Cayley object of G with $G = \langle \rho, \tau \rangle$ for some α, then either X is a CI-object for G or X is *isomorphic* to a Cayley object of $\mathbb{Z}_{pq}$. Specifically, either X is a CI-object for G or there exists $\tau' \in \text{Aut}(X)$ of order q such that $\tau'(v_j^i) = v_{j+b_i}^{i+1}$, $b_i \in \mathbb{Z}_p$ for all $i \in \mathbb{Z}_q$. This author then concluded that $b_i = b_j$ for all $i, j \in \mathbb{Z}_p$, based on the erroneous belief that $\langle \rho, \tau, \tau' \rangle$ admitted $\mathscr{C}_0$ as a unique complete block system of p blocks of size q. (If this were the case, then the argument in [11, Theorem 9] would be correct.) In fact, $\langle \rho, \tau, \tau' \rangle$ does admit a unique complete block system $\mathscr{C}$ of p blocks of size q, as $\langle \rho, \tau' \rangle$ is a regular abelian group, and $\mathscr{C}$ must also be a complete block system of $\langle \rho, \tau \rangle$. Thus $\mathscr{C} = \mathscr{C}_i$, $0 \leqslant i \leqslant p - 1$. Furthermore, τ' centralizes $\langle \rho \rangle$, $(\tau')^{-1}\tau\tau'\tau^{-1} \in \text{fix}_{\langle \rho, \tau, \tau' \rangle}(\mathscr{B})$ and has order p. As $p^2 \nmid |\text{Aut}(X)|$, $(\tau')^{-1}\tau\tau'\tau^{-1} \in \langle \rho \rangle$ so that $\tau' \in N_{S^V}(\langle \rho, \tau \rangle)$. It then follows by Lemma 2.6(3) that $\Psi_\alpha^b \langle \rho, \tau, \tau' \rangle \Psi_\alpha^{-b}$ admits only $\mathscr{B}$ and $\mathscr{C}_0$ as complete block systems. Thus, if we replace X with $\Psi_\alpha^b(X)$, then $\langle \rho, \tau, \tau' \rangle$ does admit $\mathscr{C}_0$ as a complete block system. Hence the argument given in [11] only shows that either X is a CI-object for G or $\Psi_\alpha^b(X)$ is a Cayley object of $\mathbb{Z}_{pq}$ for some $b \in \mathbb{Z}_p$. There are two other results in [11] that are based on [11, Theorem 9] that have similar defects, namely, Corollary 10 and Corollary 12. Below we give correct statements of all three results.

Theorem 2.9 ([11], Theorem 9 corrected). *Let X be a Cayley object of G with $|G| = pq$, $p > q$ and q divides $p - 1$, such that $p^2 \nmid |\text{Aut}(X)|$. If $G = \mathbb{Z}_{pq}$ then X is a CI-object for G. If $G = \langle \rho, \tau \rangle$, for some $\alpha \in \mathbb{Z}_p^*$ such that $|\alpha| = q$, then either X is a CI-object for G or $\Psi_\alpha^b(X)$ is also a Cayley object of $\mathbb{Z}_{pq}$ for some $b \in \mathbb{Z}_p$.*

Corollary 2.10 ([11], Corollary 10 corrected). *Let $X = X(q, p, \alpha)$ and $X' = X'(q, p, \alpha')$ be metacirculant combinatorial objects, $q < p$ such that X and X' are Cayley objects, $p^2 \nmid |\text{Aut}(X)|$, $|\alpha| = 1, q$, and $|\alpha'| = 1, q$. Then X is isomorphic to X' if and only if*

(1) *if X is isomorphic to a Cayley object of $\mathbb{Z}_{pq}$ then there exists $\delta \in \text{Aut}(\mathbb{Z}_{pq})$ and $b, c \in \mathbb{Z}_p$ such that $\Psi_{\alpha'}^c \delta \Psi_\alpha^b(X) = X'$;*
(2) *if X is not isomorphic to a Cayley object of $\mathbb{Z}_{pq}$ then there exists $\delta \in \text{Aut}(\langle \rho, \tau \rangle)$ and $\gamma : V \to V$, where $\gamma(v_j^i) = v_j^{ri}$, $r \in \mathbb{Z}_q^*$ and $\gamma\delta(X) = X'$.*

Further, if X and X' are isomorphic, then X is circulant if and only if $\Psi_{\alpha'}^c(X')$ is circulant for some $c \in \mathbb{Z}_p$.

Corollary 2.11 ([11], Corollary 12 corrected). *Let $\Gamma = \Gamma(q, p, \alpha, S_0, \ldots, S_\mu)$ and $\Gamma' = \Gamma'(q, p, \alpha', S_0', \ldots, S_\mu')$ be metacirculant graphs that are Cayley graphs, with $q < p$. Then Γ is isomorphic to Γ' if and only if*

(1) *if $p^2 \nmid |\mathrm{Aut}(\Gamma)|$, and $\Psi_\alpha^b(\Gamma)$ and $\Psi_{\alpha'}^c(\Gamma')$ are circulant for some $b, c \in \mathbb{Z}_p$, then there exists $\delta \in \mathrm{Aut}(\mathbb{Z}_{pq})$ such that $\Psi_{\alpha'}^c \delta \Psi_\alpha^b(\Gamma) = \Gamma'$;*
(2) *if Γ is not circulant and $\Gamma \not\cong \Gamma_1 \wr \Gamma_2$, Γ_1 an order q-circulant and Γ_2 an order p-circulant, then there exists $\delta \in \mathrm{Aut}(\langle \rho, \tau \rangle)$ and $\gamma : V \to V$ where $\gamma(v_j^i) = v_j^{ri}$, $r \in \mathbb{Z}_q^*$, and $\gamma\delta(\Gamma) = \Gamma'$;*
(3) *if Γ is not circulant and $\Gamma \cong \Gamma_1 \wr \Gamma_2$, then define $\gamma_1, \gamma_2 : V \to V$ by $\gamma_1(v_j^i) = v_{\alpha^i j}^i$ and $\Gamma_2(v_j^i) = v_{(\alpha')^i j}^i$. Then $\gamma_2 \delta \gamma_1(\Gamma) = \Gamma'$, for some $\delta \in \mathrm{Aut}(\mathbb{Z}_{pq})$.*

3. Automorphism groups of metacirculant graphs

Theorem 2.4 will allow us to determine, without too much effort, the automorphism groups of almost all vertex-transitive graphs of order pq, and, together with results already in the literature, the automorphism groups of all vertex-transitive graphs of order pq. We remark that the automorphism groups of two important families of vertex-transitive graphs of order $2p$ [8] and $3p$ [3, 7], namely 1/2-transitive graphs and symmetric graphs, are known, along with their full automorphism groups, while the automorphism groups of vertex-transitive graphs of order $2p$ were at least implicitly determined in [20]. A vertex-transitive graph Γ is *1/2-transitive* if $\mathrm{Aut}(\Gamma)$ is transitive on the set of unordered pairs of vertices, and *symmetric* if $\mathrm{Aut}(\Gamma)$ is transitive on the set of ordered pairs of vertices. These classes are important from the point of view of this paper for the following reason. Given a vertex-transitive graph Γ, we may let $\mathrm{Aut}(\Gamma)$ act on the edges of the graph in canonical fashion. This action may have several orbits, which we denote by $\mathcal{O}_1, \ldots, \mathcal{O}_r$. Define graphs Γ_i by $V(\Gamma_i) = V(\Gamma)$ and $E(\Gamma_i) = \mathcal{O}_i$, $1 \leq i \leq r$. Clearly $\mathrm{Aut}(\Gamma) \leq \mathrm{Aut}(\Gamma_i)$, $1 \leq i \leq r$, and so each Γ_i is an edge-transitive graph, which is necessarily symmetric or 1/2-transitive. Furthermore, it is easily seen that $\mathrm{Aut}(\Gamma) = \cap_{i=1}^r \mathrm{Aut}(\Gamma_i)$. Thus, if the full automorphism groups of all symmetric and 1/2-transitive graphs of a given order are known, then the automorphism groups of all vertex-transitive graphs of the given order are also known. Thus the previously cited results imply that the full automorphism group of every vertex-transitive graph of order $2p$ and $3p$ are known. Of course, as previously mentioned, the full automorphism groups of circulant graphs of order pq were determined by Klin and Pöschel [18]. As it will not be difficult, we begin with a new proof of Klin and Pöschel's result.

Theorem 3.1. *Let Γ be a circulant digraph of order pq. Then one of the following assertions is true:*

(i) $\mathrm{Aut}(\Gamma) = S^{pq}$,
(ii) $\mathrm{Aut}(\Gamma) \leq N(pq)$,

(iii) $\operatorname{Aut}(\Gamma) = \operatorname{Aut}(\Gamma_q) \wr \operatorname{Aut}(\Gamma_p)$,
(iv) $\operatorname{Aut}(\Gamma) = \operatorname{Aut}(\Gamma_q) \times \operatorname{Aut}(\Gamma_p)$,

where Γ_q and Γ_p are circulant digraphs of order q and p respectively.

Proof. Let ρ and τ be defined as in the preliminaries, with $\alpha = 1$. As Γ is circulant, then, as $\mathbb{Z}_{pq}$ is a Burnside group [34, Theorem 25.3], $\operatorname{Aut}(\Gamma)$ is either doubly transitive or imprimitive. If $\operatorname{Aut}(\Gamma)$ is doubly transitive, then clearly $\Gamma = K^{pq}$ or E^{pq} and $\operatorname{Aut}(\Gamma) = S^{qp}$ and (i) follows. If $\operatorname{Aut}(\Gamma)$ is imprimitive, $\operatorname{Aut}(\Gamma)$ admits a complete block system $\mathcal{B}$ of, say, q blocks each of size p, where the blocks are formed by the orbits of ρ. By previous comments, the blocks of $\operatorname{Aut}(\Gamma)$ of size p are the sets V^i, $i \in \mathbb{Z}_q$. As Γ is circulant, we may thus take H of Theorem 1.5 to be $\operatorname{Aut}(\Gamma)$. If $\operatorname{fix}_{\operatorname{Aut}(\Gamma)}(\mathcal{B})$ does not act faithfully on each $B \in \mathcal{B}$ or has Sylow p-subgroups of order at least p^2, then by Lemma 2.3 $\Gamma = \Gamma_q \wr \Gamma_p$ and by [31] $\operatorname{Aut}(\Gamma) = \operatorname{Aut}(\Gamma_q) \wr \operatorname{Aut}(\Gamma_p)$, where Γ_q is a circulant digraph of order q and Γ_p a circulant digraph of order p, and (ii) follows. Thus we may assume that $\operatorname{fix}_{\operatorname{Aut}(\Gamma)}(\mathcal{B})$ acts faithfully on each $B \in \mathcal{B}$ and has Sylow p-subgroups of order p. Hence by Theorem 2.4 $\operatorname{Aut}(\Gamma) \leqslant S^q \times S^p$ or $\operatorname{Aut}(\Gamma) = \operatorname{Aut}(\Gamma_q) \wr \operatorname{Aut}(\Gamma_p)$ in which case (iii) follows. (Note that Theorem 2.4(2) cannot occur as these groups do not contain a regular cyclic subgroup.) We thus need only consider when $\operatorname{Aut}(\Gamma) \leqslant S^q \times S^p$.

As $\operatorname{Aut}(\Gamma) \leqslant S^q \times S^p$, $\operatorname{Aut}(\Gamma)$ admits complete block systems $\mathcal{B}$ and $\mathcal{C}$ of q blocks of size p and p blocks of size q, respectively. As $\langle \rho, \tau \rangle = \langle \rho\tau \rangle \leqslant \operatorname{Aut}(\Gamma)$, $\mathcal{B}$ and $\mathcal{C}$ are unique and $\mathcal{B} = \{V^i : i \in \mathbb{Z}_q\}$, $\mathcal{C} = \{V_j : j \in \mathbb{Z}_p\}$. Let $K = \operatorname{fix}_{\operatorname{Aut}(\Gamma)}(\mathcal{B})$ and $K' = \operatorname{fix}_{\operatorname{Aut}(\Gamma)}(\mathcal{C})$. Then $K, K' \neq 1$ as $\rho \in K$ and $\tau \in K'$.

If $K|_{V^0}$ is doubly transitive then $\operatorname{Stab}_K(v_j^i)$ is transitive on $V^i - \{v_j^i\}$, and if $v_j^i \in V^i$ and V^ℓ is any other block of size p, then v_j^i is adjacent to $0, 1, p-1$, or p vertices of V^ℓ, where v_j^i is adjacent to one vertex of V^ℓ if and only if v_j^i is adjacent to only v_j^ℓ, and v_j^i is adjacent to $p-1$ vertices of V^ℓ if and only if v_j^i is adjacent to every vertex of V^ℓ except v_j^ℓ. Consider the function $\gamma : V \to V$ by $\gamma(v_j^i) = v_{\delta(j)}^{\sigma(i)}$, where $\sigma \in \operatorname{Aut}(\Gamma)/\mathcal{B}$ and $\delta \in S^p$. Note that, as $\sigma \in \operatorname{Aut}(\Gamma)/\mathcal{B}$, if V^r and V^s are blocks of size p and each vertex of V^r is adjacent to t vertices of V^s, then each vertex of $V^{\sigma(r)}$ is adjacent to t vertices of $V^{\sigma(s)}$. It follows from the comments above that $\gamma \in \operatorname{Aut}(\Gamma)$. Thus, if $K'|_{V_0}$ is not doubly transitive and $K|_{V^0}$ is, then $\operatorname{Aut}(\Gamma) = A \times S^p$, $A \leqslant \operatorname{AGL}(1, q)$. By analogous arguments, if $K'|_{V_0}$ is doubly transitive but $K|_{V^0}$ is not, then $\operatorname{Aut}(\Gamma) = S^q \times B$, $B \leqslant \operatorname{AGL}(1, p)$, and if both $K|_{V^0}$ and $K'|_{V_0}$ are doubly transitive, $\operatorname{Aut}(\Gamma) = S^q \times S^p$. Note that we must have $A < \operatorname{AGL}(1, p)$, $B < \operatorname{AGL}(1, p)$, as if r is a prime, $\operatorname{AGL}(1, r)$ is itself doubly transitive. It is then easy to show that $A = \operatorname{Aut}(\Gamma_q)$ and $B = \operatorname{Aut}(\Gamma_p)$ for some Γ_p and Γ_q.

Otherwise, both $K|_{V^0}$ and $K'|_{V_0}$ are transitive but not doubly transitive, and so by Theorem 1.1 $K|_{V^0}$ and $K'|_{V_0}$ are proper subgroups of $\operatorname{AGL}(1, p)$ and $\operatorname{AGL}(1, q)$, respectively. Let H and H' be minimal subgroups of $S^q \times S^p$ such that $\operatorname{Aut}(\Gamma) \leqslant H' \times H$. (Note that then $K|_{V^0} \leqslant H$ and $K'|_{V_0} \leqslant H'$.) Then $K|_{V^0} \triangleleft H$ and $K'|_{V_0} \triangleleft H'$. As $\operatorname{AGL}(1, p)$ and $\operatorname{AGL}(1, q)$ have a unique Sylow p-subgroup and a unique Sylow q-subgroup respectively, and the normalizer of the normalizer of a Sylow r-subgroup is itself, $H \leqslant \operatorname{AGL}(1, p)$ and $H' \leqslant \operatorname{AGL}(1, q)$. Thus $\operatorname{Aut}(\Gamma) \leqslant \operatorname{AGL}(1, q) \times \operatorname{AGL}(1, p) = N(pq)$. □

Using Theorem 3.1, it is straightforward to give a procedure for calculating the full automorphism group of a circulant digraph of order pq, solving the problem of Elspas and Turner in this case. We now investigate the imprimitive automorphism groups of vertex-transitive graphs Γ of order pq that are not circulant. In what follows, if Γ is a (q,p)-metacirculant graph that is not isomorphic to a circulant graph, we will say that $\alpha \in \mathbb{Z}_p^*$ is *maximal* in $\mathrm{Aut}(\Gamma)$ if $\tau(v_j^i) = v_{\alpha j}^{i+1}$ and whenever $\tau'(v_j^i) = v_{\beta j}^{i+1}$ for some $\beta \in \mathbb{Z}_p^*$ and $\tau' \in \mathrm{Aut}(\Gamma)$, then $\langle \rho, \tau' \rangle \leqslant \langle \rho, \tau \rangle$. The existence of a maximal α is guaranteed by Lemma 2.7.

Theorem 3.2. *Let Γ be a vertex-transitive graph of order pq, $p > q$, that is not isomorphic to a circulant graph such that $\mathrm{Aut}(\Gamma)$ is imprimitive with complete block system $\mathscr{B}$. Then*

(1) *if $\mathscr{B}$ consists of p blocks of size q for every complete block system $\mathscr{B}$ of $\mathrm{Aut}(\Gamma)$, then $p = 2^{2^s} + 1$ is a Fermat prime, and q divides $2^{2^s} - 1$. Further, the minimal transitive subgroup of $\mathrm{Aut}(\Gamma)$ that admits only a complete block system of p blocks of size q is isomorphic to $\mathrm{SL}(2, 2^s)$ and $\mathrm{Aut}(\Gamma)$ is isomorphic to a subgroup of $\mathrm{Aut}(\mathrm{SL}(2, 2^s))$, or*

(2) *if $\mathscr{B}$ consists of q blocks of size p, then Γ is isomorphic to a metacirculant graph, and if $V(\Gamma)$ is labelled with elements of V so that Γ is metacirculant with α maximal in $\mathrm{Aut}(\Gamma)$, then*

  (a) *if $\mathrm{fix}_{\mathrm{Aut}(\Gamma)}(\mathscr{B})|_B$ is doubly transitive for $B \in \mathscr{B}$, then*

   (i) $\mathrm{Aut}(\Gamma) = \mathbb{Z}_2 \ltimes \mathrm{PSL}(2,11)$, *or*

   (ii) $\mathrm{Aut}(\Gamma) = \mathbb{Z}_2 \ltimes \mathrm{P\Gamma L}(n,k)$, *where n is prime, $k = r^m$, r a prime, $(n, k-1) = 1$, and m is a power of n, or*

  (b) *if $\mathrm{fix}_{\mathrm{Aut}(\Gamma)}(\mathscr{B})|_B$ is not doubly transitive, then*

   (i) *if $|\alpha| \neq q$ then $\mathrm{Aut}(\Gamma) = \langle \rho, \tau \rangle$,*

   (ii) *if $|\alpha| = q$, then there exists $\gamma \in \mathbb{Z}_p^*$ such that $|\gamma| = qa$, where $|\mathrm{Aut}(\Gamma)| = aqp$, $\gcd(a,q) = 1$, and $b \in \mathbb{Z}_p$ such that if $\tau' : V \to V$ by $\tau'(v_j^i) = v_{\gamma j}^{i+1}$, then $\Psi_\alpha^{-b} \mathrm{Aut}(\Gamma) \Psi_\alpha^b = \langle \rho, \tau' \rangle$ for some $b \in \mathbb{Z}_p$.*

Proof. (1) If $\mathrm{Aut}(\Gamma)$ admits only a complete block system $\mathscr{B}$ of p blocks of size q and Γ is a (p,q)-metacirculant then, as $\gcd(p-1,p) = 1$, it follows by Theorem 1.4 that Γ is isomorphic to a circulant graph, a contradiction. Thus Γ is not isomorphic to a metacirculant graph and it follows by [24, Theorem] that $p = 2^{2^s} + 1$ is a Fermat prime, q divides $2^{2^s} - 1$, and by [24, Proposition 1.4] that the minimal transitive subgroup of $\mathrm{Aut}(\Gamma)$ that admits a complete block system of p blocks of size q is isomorphic to $\mathrm{SL}(2, 2^s)$. Further, by [21, Theorem 3.4], $\mathrm{fix}_{\mathrm{Aut}(\Gamma)}(\mathscr{B}) = 1$ and $\mathrm{Aut}(\Gamma)/\mathscr{B}$ is a nonsolvable doubly transitive group. It then follows by [24, Proposition 2.5] that $\mathrm{SL}(2, 2^s) \leqslant \mathrm{Aut}(\Gamma)/\mathscr{B} \leqslant \mathrm{Aut}(\mathrm{SL}(2, 2^s))$. As $\mathrm{Aut}(\Gamma) \cong \mathrm{Aut}(\Gamma)/\mathscr{B}$, the result follows.

(2) If $\mathrm{Aut}(\Gamma)$ admits a complete block system of q blocks of size p, then by [24, Theorem] Γ is isomorphic to a (q,p)-metacirculant graph and $\mathscr{B}$ is formed by the orbits of $\langle \rho \rangle$. If Γ is a metacirculant graph of order pq, $p > q$, that is not isomorphic to a circulant graph, then by Theorem 1.4 Γ is a (q,p)-metacirculant and $q \mid |\alpha|$, where α is chosen to be maximal in $\mathrm{Aut}(\Gamma)$. As $\rho \in \mathrm{fix}_{\mathrm{Aut}(\Gamma)}(\mathscr{B})$, $\mathrm{fix}_{\mathrm{Aut}(\Gamma)}(\mathscr{B}) \neq 1$. As Γ is not isomorphic to a circulant graph,

it follows by Lemma 2.3 that the Sylow p-subgroups of $\text{fix}_{\text{Aut}(\Gamma)}(\mathscr{B})$ have order p. As Γ is not isomorphic to a circulant graph, $\text{Aut}(\Gamma) \not\cong \text{Aut}(\Gamma_q) \wr \text{Aut}(\Gamma_p)$, and so by Theorem 2.4 $\text{Aut}(\Gamma) \leqslant S^q \times S^p$ or $q = 2$ and $\text{Aut}(\Gamma) = \mathbb{Z}_2 \ltimes \text{PSL}(2,11)$ or $\mathbb{Z}_2 \ltimes \text{PSL}(n,k) \leqslant \text{Aut}(\Gamma) \leqslant \mathbb{Z}_2 \ltimes \text{P}\Gamma\text{L}(n,k)$, where n is prime, $k = r^m$, r a prime, $(n, k-1) = 1$, m is a power of n, and $\text{Aut}(\Gamma)/\mathscr{B} \cong \mathbb{Z}_p$. If $\text{fix}_{\text{Aut}(\Gamma)}(\mathscr{B})|_{V^0}$ is doubly transitive, we claim that Γ is isomorphic to a circulant graph or that $\text{Aut}(\Gamma) = \mathbb{Z}_2 \ltimes \text{PSL}(2, 11)$ or $\mathbb{Z}_2 \ltimes \text{P}\Gamma\text{L}(n, k)$, where n is prime, $k = r^m$, r a prime, $(n, k-1) = 1$, and m is a power of n.

If $\text{Aut}(\Gamma) = \mathbb{Z}_2 \ltimes \text{PSL}(2, 11)$ and $p = 11$, then by [8] there exists a vertex-transitive graph with automorphism group $\mathbb{Z}_2 \ltimes \text{PSL}(2, 11)$. If $\mathbb{Z}_2 \ltimes \text{PSL}(n,k) \leqslant \text{Aut}(\Gamma) \leqslant \mathbb{Z}_2 \ltimes \text{P}\Gamma\text{L}(n,k)$, then the action of $\text{fix}_{\text{Aut}(\Gamma)}(\mathscr{B})|_B$ is not equivalent to the action of $\text{fix}_{\text{Aut}(\Gamma)}(\mathscr{B})|_{B'}$, where $\mathscr{B} = \{B, B'\}$. The two inequivalent doubly transitive actions of $\text{PSL}(n,k)$ of degree $(k^n - 1)/(k-1)$ are $\text{PSL}(n,k)$ acting on the one-dimensional subspaces of $\mathbb{F}_k^n$ and also on the hyperplanes of $\mathbb{F}_k^n$ (the $n-1$ dimensional subspaces). An elementary exercise in linear algebra will then show that the stabilizer of a one-dimensional subspace of $\mathbb{F}_k^n$ has two orbits in its natural action on the hyperplanes of $\mathbb{F}_k^n$. These two orbits consist of those hyperplanes that contain the fixed one-dimensional subspace of $\mathbb{F}_k^n$ and those that do not contain the fixed one-dimensional subspace. Thus $\text{Stab}_{\text{fix}_{\text{Aut}(\Gamma)}(\mathscr{B})}(x)|_{B'}$ has two orbits for every $x \in V$ and $B' \in \mathscr{B}$ with $x \notin B'$. As Γ is not isomorphic to a circulant graph, x cannot be adjacent to every vertex of B', so that x is adjacent to every vertex in one orbit of $\text{Stab}_{\text{fix}_{\text{Aut}(\Gamma)}(\mathscr{B})}(x)|_{B'}$. As $\Gamma[B]$ is complete or has no edges, we conclude that either Γ or its complement is an edge-transitive graph. It then follows by [8] that $\text{Aut}(\Gamma) = \mathbb{Z}_2 \ltimes \text{P}\Gamma\text{L}(n,k)$. We may thus assume that $\text{Aut}(\Gamma)$ is equivalent to a subgroup of $S^q \times S^p$. Hence $\text{Aut}(\Gamma)$ also admits a complete block system $\mathscr{C}'$ of p blocks of size q.

Observe that $\text{Stab}_{\text{fix}_{\text{Aut}(\Gamma)}(\mathscr{B})}(v_j^i)|_{V^i}$ is transitive on $V^i - \{v_j^i\}$. We conclude that if v_j^i is adjacent to some vertex of V^0 then v_j^i is either adjacent to only v_j^0, or v_j^i is adjacent to v_t^0 where $t \neq j$, and v_j^i is not adjacent to v_j^0, or v_j^i is adjacent to every vertex of V^0. It then follows by arguments analogous to those in Theorem 3.1 that $\text{fix}_{\text{Aut}(\Gamma)}(\mathscr{B}) = S^p$, in which case Γ is circulant as claimed. We thus assume $\text{fix}_{\text{Aut}(\Gamma)}(\mathscr{B})|_{V^0} < \text{AGL}(1, p)$.

As $\text{Aut}(\Gamma)$ is not isomorphic to a circulant graph, by Theorem 2.4 we have that $\text{Aut}(\Gamma)/\mathscr{B}$ is cyclic of order q. If $|\text{fix}_{\text{Aut}(\Gamma)}(\mathscr{B})| = p$, then $\text{Aut}(\Gamma) = \langle \rho, \tau \rangle$ (with $|\alpha| = q$) and the result follows. Otherwise, $\text{fix}_{\text{Aut}(\Gamma)}(\mathscr{B}) \neq \langle \rho \rangle$ so that $\text{Stab}_{\text{fix}_{\text{Aut}(\Gamma)}(\mathscr{B})}(v_0^0) \neq 1$ and the equivalence relation $\equiv$ (as in Definition 6) is defined. As $\text{fix}_{\text{Aut}(\Gamma)}(\mathscr{B})|_B$ is not doubly transitive, by Theorem 1.1, $\text{fix}_{\text{Aut}(\Gamma)}(\mathscr{B})|_B \leqslant \text{AGL}(1, p)$ for every $B \in \mathscr{B}$, and, as $\text{AGL}(1, p)$ is solvable, we have $|\text{fix}_{\text{Aut}(\Gamma)}(\mathscr{B})| = |\text{fix}_{\text{Aut}(\Gamma)}(\mathscr{B})|_B| = ap$ for every $B \in \mathscr{B}$, where $a > 1$. Clearly if $B, B' \in \mathscr{B}$, then $|\text{Stab}_{\text{fix}_{\text{Aut}(\Gamma)}(\mathscr{B})|_B}(b)| = |\text{Stab}_{\text{fix}_{\text{Aut}(\Gamma)}(\mathscr{B})|_{B'}}(b')| = a$ for $b \in B$ and $b' \in B'$. As $\text{AGL}(1, p)$ is solvable and $\gcd(a, p) = 1$, any two subgroups of order a of the unique subgroup W of $\text{AGL}(1, p)$ of order ap are conjugate in W. We conclude that the action of W on B and the action of W on B' are equivalent. Hence if $x \in B$ then there exists $x' \in B'$ such that $\text{Stab}_{\text{fix}_{\text{Aut}(\Gamma)}(\mathscr{B})|_B}(x) = \text{Stab}_{\text{fix}_{\text{Aut}(\Gamma)}(\mathscr{B})|_{B'}}(x')$. We conclude that the equivalence classes of $\equiv$ each have q elements, so we may assume without loss of generality that these equivalence classes form the complete block system $\mathscr{C}'$, and trivially, $\mathscr{C}'$ is a complete block system of $\langle \rho, \tau \rangle$.

If $|\alpha| \neq q$, then by Lemma 2.6 $\mathscr{C}' = \mathscr{C}$, and if $|\alpha| = q$, then there exists $b \in \mathbb{Z}_p$ such that the complete block system of $\Psi_\alpha^{-b} \text{Aut}(\Gamma) \Psi_\alpha^b$ corresponding to $\mathscr{C}'$ is also $\mathscr{C}$. If $|\alpha| \neq q$, let $H = \text{Aut}(\Gamma)$, and if $|\alpha| = q$, let $H = \Psi_\alpha^{-b} \text{Aut}(\Gamma) \Psi_\alpha^b$. If $\delta \in \text{Stab}_{\text{fix}_H(\mathscr{B})}(v_0^0)$, then $\delta(v_j^i) = v_{\beta j}^i$, where $\beta \in \mathbb{Z}_p^*$. A straightforward computation will then show that $\delta \in N_{S^V}(\langle\rho,\tau\rangle)$ so that $\langle\rho,\tau\rangle \triangleleft H$. As $|H| = pqa$, by Lemma 2.7, the result follows. □

Corollary 3.3. *Let Γ be a vertex-transitive graph of order pq, $q < p$ primes, and $G = \text{Aut}(\Gamma)$. If G is primitive, then G is one of the groups in Table 1. If $\text{Aut}(\Gamma) \cong G$ is imprimitive, then one of the following is true.*

(1) *If Γ is isomorphic to a circulant graph, then*
   (a) $G = S^q \times S^p$,
   (b) $G = A \times S^p$,
   (c) $G = S^q \times B$,
   (d) $G < N(pq)$,
   (e) $G = \text{Aut}(\Gamma_q) \wr \text{Aut}(\Gamma_p)$ or $\text{Aut}(\Gamma_p) \wr \text{Aut}(\Gamma_q)$,
   for some $A < \text{AGL}(1,q)$, $B < \text{AGL}(1,p)$.
(2) *If $\Gamma = \Gamma(q,p,\alpha,S_0,\ldots,S_\mu)$ is metacirculant but not isomorphic to a circulant graph with α maximal in G, then*
   (a) *if G is not solvable, then*
      (i) $G = \mathbb{Z}_2 \ltimes \text{PSL}(2,11)$, *or*
      (ii) $G = \mathbb{Z}_2 \ltimes \text{P}\Gamma\text{L}(n,k)$, *where n is prime, $k = r^m$, r a prime, $(n, k-1) = 1$, and m is a power of n;*
   (b) *if G is solvable, then*
      (i) *if $|\alpha| \neq q$ then $G = \langle\rho,\tau\rangle$,*
      (ii) *if $|\alpha| = q$, then there exists $\gamma \in \mathbb{Z}_p^*$ such that $|\gamma| = qa$, where $|G| = aqp$, $\gcd(a,q) = 1$, and $b \in \mathbb{Z}_p$ such that if $\tau' : V \to V$ by $\tau'(v_j^i) = v_{\gamma j}^{i+1}$, then $\Psi_\alpha^{-b} G \Psi_\alpha^b = \langle\rho,\tau'\rangle$.*
(3) $p = 2^{2^s} + 1$ *is a Fermat prime, and q divides $2^{2^s} - 1$. Further, the minimal transitive subgroup of G that admits only a complete block system of p blocks of size q is isomorphic to $\text{SL}(2,2^s)$ and G is isomorphic to a subgroup of $\text{Aut}(\text{SL}(2,2^s))$.*

Proof. If $\text{Aut}(\Gamma)$ is imprimitive, the result follows from Theorems 3.1, 3.2. If $\text{Aut}(\Gamma)$ is primitive, then the result follows from [24], [30], [33]. □

It is worthwhile pointing out that some of the graphs whose automorphism groups are primitive are metacirculant graphs, and some are Marušič–Scapellato graphs (as in part (3). Indeed, there is a relatively short list of vertex-transitive graphs of order pq whose automorphism group does not contain an imprimitive subgroup given in [26, Table 2]. Most of the vertex-transitive graphs of order pq whose automorphism group is primitive and contains an imprimitive subgroup are metacirculant, with the exceptions being the graphs corresponding to the action of A_6 on pairs and $\text{PSL}(2,9)$ (which is isomorphic to A_6), and the rank 3 action on $\text{PSp}(4,k)$ with $p = k^2 + k$ and $q = k + 1$ both being Fermat primes. This information can be found in the proof of [26, Theorem 2.1].

Table 1. Automorphism groups G of vertex-primitive graphs of order pq.

pq	G	soc(G)
pq	S^{pq}	A^{pq}
$p(p-1)/2$	S^p	A^p
$p(p+1)/2$	S^{p+1}	A^{p+1}
$(2^d \pm 1)(2^{d-1} \pm 1)$	$O^{\pm}(2d, 2)$	$\Omega^{\pm}(2d, 2)$
$(k+1)(k^2+1)$	$P\Gamma S_p(4, k)$	$PSp(4, k)$
$q(q^2+1)/2$	$P\Sigma L(2, k^2)$	$PSL(2, k^2)$
$p(p \pm 1)/2$	$PSL(2, p)$	$PSL(2, p)$
$3 \cdot 7$	$PGL(2, 7)$	$PGL(2, 7)$
$5 \cdot 7$	S^7	A^7
$5 \cdot 7$	$P\Gamma L(4, 2)$	$PSL(4, 2)$
$5 \cdot 11$	$PGL(2, 11)$	$PGL(2, 11)$
$7 \cdot 11$	$\text{Aut}(M_{22})$	M_{22}
$5 \cdot 31$	$P\Gamma L(5, 2)$	$PSL(5, 2)$
$7 \cdot 29$	$PSL(2, 29)$	$PSL(2, 29)$
$11 \cdot 23$	$PSL(2, 23)$	$PSL(2, 23)$
$29 \cdot 59$	$PSL(2, 59)$	$PSL(2, 59)$
$31 \cdot 61$	$PSL(2, 61)$	$PSL(2, 61)$
$3 \cdot 19$	$PSL(2, 19)$	$PSL(2, 19)$

Now, suppose that Γ is a vertex-transitive graph with $G \leqslant \text{Aut}(\Gamma)$ a transitive subgroup. Let P be a Sylow p-subgroup of G, and Q a Sylow q-subgroup of G, where $q < p$. If $N_G(P)$ or $N_G(Q)$ is transitive, then Γ is isomorphic to a metacirculant graph by Theorem 1.5. Otherwise, Γ either has a primitive automorphism group or is a Marušič–Scapellato graph, and the former occurs if G is primitive and the latter otherwise. Thus metacirculant graphs, Marušič–Scapellato graphs and graphs with primitive automorphism groups can be distinguished, as any intersections among these families of graphs are known by the preceding paragraph (provided that we know a transitive subgroup $G \leqslant \text{Aut}(\Gamma)$). Thus, if Γ is a Marušič–Scapellato graph or has a primitive automorphism group, then $\text{Aut}(\Gamma)$ can be determined.

If Γ is metacirculant, then in order to calculate $\text{Aut}(\Gamma)$ one needs to determine if the graph is isomorphic to a circulant graph or not. If p^2 divides $|\text{Aut}(\Gamma)|$, then of course it is isomorphic to a circulant graph. If not and q^2 divides $|\alpha|$, then [11, Theorem 5] tells how to determine if the graph is isomorphic to a circulant graph. If q^2 does not divide $|\alpha|$, then, by raising τ to an appropriate power, if necessary, we may assume that $|\alpha| = q$. If $\Gamma \cong \Gamma'$, where Γ' is circulant, then by Theorem 2.9, $\Psi_\alpha^b(\Gamma)$ is circulant for some $b \in \mathbb{Z}_p$. Define $\tau' : V \to V$ by $\tau'(v_j^i) = v_j^{i+1}$. Then $\langle \rho, \tau' \rangle \leqslant \text{Aut}(\Psi_\alpha^b(\Gamma))$, and, as Ψ_α^b normalizes $\langle \rho, \tau \rangle$, we also have $\langle \rho, \tau \rangle \leqslant \text{Aut}(\Psi_\alpha^b(\Gamma))$. Whence $\bar{\alpha} : V \to V$ by $\bar{\alpha}(v_{\alpha j}^i)$ is in $\text{Aut}(\Psi_\alpha^b(\Gamma))$ as $\bar{\alpha} = \tau(\tau')^{-1}$. The following result, which is interesting in its own right, will resolve how to determine if a metacirculant graph Γ with $|\alpha| = q$ is isomorphic to a circulant graph (and hence how to determine $\text{Aut}(\Gamma)$).

Theorem 3.4. *Let* $\Gamma = \Gamma(q, p, \alpha, S_0, \ldots, S_\mu)$ *be a metacirculant graph, where* $q | (p-1)$ *and* $|\alpha|$ *is a power of* q. *Then* Γ *is isomorphic to a Cayley graph of both groups of order* pq *if* *and only if one of the following occurs:*

(1) $p^2 \mid |\text{Aut}(\Gamma)|$, in which case Γ is isomorphic to the wreath product of a circulant graph of order q and a circulant graph of order p,
(2) $p^2 \nmid |\text{Aut}(\Gamma)|$, $|\alpha| \neq q$, and $\Gamma = \Gamma(q, p, 1, S_0, \ldots, S_\mu)$,
(3) $p^2 \nmid |\text{Aut}(\Gamma)|$, $|\alpha| = q$, and $\Psi_\alpha^b(\Gamma) = \Gamma(q, p, 1, S_0', \ldots, S_\mu')$, for some $S_0', \ldots S_\mu'$.

Proof. It follows by [2, Lemma 10] that if $p^2 \mid |\text{Aut}(\Gamma)|$, then Γ is isomorphic to the wreath product of a circulant graph of order q and a circulant graph of order p. Thus Γ is isomorphic to a circulant graph of order pq. Without loss of generality, we thus assume that $\alpha = 1$. Let $\rho = z_0 z_1 \cdots z_{q-1}$ where each z_i is a p-cycle that permutes v_0^i. Then each $z_i \in \text{Aut}(\Gamma)$. Hence $\rho' = \Pi_{i=0}^{q-1} z_i^{\alpha^i} \in \text{Aut}(\Gamma)$. Then $\tau^{-1} \rho' \tau(v_j^i) = v_{j+\alpha^{i+1}}^i$ so that $\tau^{-1} \rho' \tau = (\rho')^\alpha$. Then $\langle \rho', \tau \rangle$ is isomorphic to the non-abelian group of order pq. Clearly, if (3) holds then Γ is isomorphic to a Cayley graph of both groups of order qp. If (2) holds, then let $|\alpha| = q^b$. Then $\tau^{q^{b-1}}(v_j^i) = v_{\alpha^{q^{b-1}} j}^i$ and as $\Gamma = \Gamma(q, p, 1, S_0, \ldots, S_\mu)$, then function $\tau': V \to V$ by $\tau'(v_j^i) = v_j^{i+1}$ is in $\text{Aut}(\Gamma)$. Then $\langle \tau' \tau^{q^{b-1}}, \rho \rangle$ is a transitive, non-abelian group of order pq. Thus, if (1), (2), or (3) hold, then Γ is isomorphic to a Cayley graph of both groups of order qp.

Conversely, if Γ is isomorphic to both groups of order qp, then either $p^2 \mid |\text{Aut}(\Gamma)|$ or $p^2 \nmid |\text{Aut}(\Gamma)|$. If $p^2 \mid |\text{Aut}(\Gamma)|$, then (1) follows from [2, Lemma 10]. If $p^2 \nmid |\text{Aut}(\Gamma)|$ and $q^2 \mid |\alpha|$, then, by [11, Theorem 5], (2) follows. If $p^2 \nmid |\text{Aut}(\Gamma)|$ and $q^2 \nmid |\text{Aut}(\Gamma)|$, then the result follows by arguments in the paragraph immediately preceding this result. □

Remark. We remark that if $p^2 \nmid |\text{Aut}(\Gamma)|$ in the previous result, then q divides $\text{fix}_{N_{\text{Aut}(\Gamma)}(\langle \rho \rangle)}(\mathscr{B})$, where $\mathscr{B}$ is the complete block system of $N_{\text{Aut}(\Gamma)}(\langle \rho \rangle)$ formed by the orbits of $\langle \rho \rangle$. As $\mathbb{Z}_p^*$ is cyclic, this implies that there exists $\bar{\alpha}: V \to V$ by $\bar{\alpha}(v_j^i) = v_{\alpha j + b_i}^i$, $b_i \in \mathbb{Z}_p$, $b_0 = 0$ as an automorphism of Γ. This then implies that $\alpha S_0 = S_0$. A similar statement follows from [11, Theorem 5] if $|\alpha| \neq q$. Thus, if $|\alpha|$ is a power of q, then a (q, p, α)-metacirculant graph will not be isomorphic to a circulant graph if $\alpha S_0 \neq S_0$.

We now illustrate the use of these results with an example. Let q and $p \geq 7$ be primes such that $q \mid (p-1)$, and no vertex-transitive graph of order pq (other than the complete graph or its complement) has a primitive automorphism group. Such values exist as we may take $q = 5$ and $p = 31$, for example. Let $\alpha \in \mathbb{Z}_p^*$ such that $|\alpha| = q$. Let $S_0 = \{1, p-1\}$, $S_1 = \{0, 1, d : d \neq p-1, 2, \text{ or } 2^{-1}\}$, and for $2 \leq i \leq \mu$, let $S_i = \emptyset$. Note that as $p \geq 7$, a d as in the definition of S_1 does indeed exist. Let $\Gamma = \Gamma(q, p, \alpha, S_0, \ldots, S_\mu)$ so that Γ is a Cayley graph of the non-abelian group of order qp. Clearly Γ cannot be written as $\Gamma_q \wr \Gamma_p$, so if Γ is isomorphic to a circulant graph by Theorem 3.4 and the Remark following it we must have $\alpha S_0 = S_0$. Thus $\alpha = -1$. This then implies that $q = 2$. It is easy to see that $\iota': V \to V$ given by $\iota'(v_j^i) = v_{-j}^i$ is not an automorphism of Γ, and so Γ is not a circulant graph. Then by Theorem 3.4, we must have that $\Psi_{-1}^b(\Gamma)$ is a circulant graph or Γ is only a Cayley graph of $\langle \rho, \tau \rangle$. If $\Psi_{-1}^b(\Gamma)$ is a circulant graph, then as $|S_1| = 3$, $|S_1'| = 3$ and so S_1' must contain 0. Note that $\Psi_{-1}^b(v_0^1) = v_0^1$ implies that $\Psi_{-1}^b = 1$, which is not possible. Thus $\Psi_{-1}^b(v_1^1) = v_0^1$ or $\Psi_{-1}^b(v_d^1) = v_0^1$. If $\Psi_{-1}^b(v_1^1) = v_0^1$ then $\Psi_{-1}^b(v_j^1) = v_{j+p-1}^1$, in which case $S_1' = \{0, p-1, d+p-1\}$. Then $d+p-1 = 1$ so that $d = 2$, a contradiction.

If $\Psi^b_{-1}(v^1_d) = v^1_0$ then $\Psi^b_{-1}(v^1_j) = v^1_{j+p-d}$, in which case $S'_1 = \{0, p-d, p-d+1\}$. But then $p-d = -(p-d+1)$, which is impossible. We conclude that Γ is not isomorphic to a circulant graph.

Suppose that $\text{Aut}(\Gamma) \neq \langle \rho, \tau \rangle$. It is easy to see that $\text{fix}_{\text{Aut}(\Gamma)}(\mathcal{B})|_B$ is not doubly-transitive, so by Corollary 3.3 either α is not maximal in $\text{Aut}(\Gamma)$ and for β maximal in $\text{Aut}(\Gamma)$ we have $\text{Aut}(\Gamma) = \langle \rho, \tau_\beta \rangle$ ($\tau_\beta(v^i_j) = v^{i+1}_{\beta j}$) or there exists $\gamma \in \mathbb{Z}^*_p$ such that $|\gamma| = qa$, where $|\text{Aut}(\Gamma)| = aqp$, $\gcd(a, q) = 1$, and $b \in \mathbb{Z}_p$ such that if $\tau' : V \to V$ by $\tau'(v^i_j) = v^{i+1}_{\gamma j}$, then $\Psi^{-b}_\alpha \text{Aut}(\Gamma) \Psi^b_\alpha = \langle \rho, \tau' \rangle$. As $\text{Aut}(\Gamma[B])$ is dihedral of order $2p$ for every $B \in \mathcal{B}$, either $\beta = -\alpha$ or $a = 2$. It is easy to see that $\iota' : V \to V$ given by $\iota'(v^i_j) = v^i_{-j}$ is not an automorphism of Γ, so that α is maximal in $\text{Aut}(\Gamma)$ and $a = 2$. If $a = 2$, then $\iota : V \to V$ by $\iota(v^i_j) = v^i_{-j+b_i}$, $b_i \in \mathbb{Z}_p$, $b_0 = 0$, is an automorphism of Γ, $\iota = \Psi^b_\alpha \iota' \Psi^{-b}_\alpha$, and $\iota(S_1) = S_1$ (as $\iota(v^0_0) = v^0_0$). As $|S_1| = 3$, S_1 must consist of an orbit of ι of order 2 together with a fixed point of ι. If v^1_0 is the fixed point of ι in S_1, then $b_i = 0$ for all $i \in \mathbb{Z}_p$ (as this would imply that $b = 0$) and α is not maximal in $\text{Aut}(\Gamma)$, a contradiction. If v^1_1 is the fixed point of ι in S_1, then $\iota(v^1_1) = v^1_{-1+b_1} = v^1_1$ and $-1 + b_1 = 1$. Hence $b_1 = 2$. Furthermore, an orbit of ι is $\{v^1_0, v^1_d\}$. As $d \neq 2$, $\iota(v^1_0) \neq v^1_d$, a contradiction. Finally, if $\iota(v^1_d) = v^1_d$ then $\iota(v^1_0) = v^1_1$. Then $v^1_{-d+b_1} = v^1_d$ so that $b_1 = 2d$. But then $\iota(v^1_0) = v^1_{2d} = v^1_1$ so that $d = 2^{-1}$, a contradiction. Whence $\text{Aut}(\Gamma) = \langle \rho, \tau \rangle$ and is of order pq. A Cayley graph Γ of G such that $\text{Aut}(\Gamma) = G$ is a *graphical regular representation*. These have been studied extensively, and groups that have a graphical regular representation have been characterized [14].

We now discuss the second error in the literature caused by assuming $\langle \rho, \tau \rangle$ has a unique complete block system of p blocks of size q. This error produces a 'gap' in a proof rather than an incorrect result. In [19], it was shown that if Γ is a self-complementary vertex-transitive graph of order pq, then $p, q \equiv 1 \pmod 4$. The proof is essentially broken into two parts, one dealing with graphs whose automorphism group is primitive, and the other with graphs whose automorphism group is imprimitive. The imprimitive case is handled by letting Γ be a self-complementary vertex-transitive graph of order pq whose automorphism group is imprimitive, with $\mathcal{B}$ a complete block system of $\text{Aut}(\Gamma)$. The author then states that it follows that $\mathcal{B}$ is the unique complete block system of $\mathcal{B}$ with blocks of size $|B|$, $B \in \mathcal{B}$. In fact, this statement is true, as this author has shown in [12] that every self-complementary vertex-transitive graph of order pq is isomorphic to a circulant graph of order pq. However, no justification for this is given in [19]. In view of Lemma 2.6 and Corollary 3.3, if there exist (q, p)-metacirculant graphs of order pq whose automorphism group is $\langle \rho, \tau \rangle$ where $|\alpha| = q$, then there is a 'gap' in the proof given in [19], as the argument given there does not show that such graphs are not self-complementary. The example in the immediately preceding paragraph shows that such graphs exist. We remark that Muzychuk [28] has shown that a self-complementary vertex-transitive graph of order n exists if and only if, whenever p^k is the highest power of the prime p that divides n, then $p^k \equiv 1 \pmod 4$.

4. Nonnormal Cayley graphs

In [36], a Cayley graph Γ of a group G is *normal* if $G_L \triangleleft \text{Aut}(\Gamma)$. There has been some interest in normal Cayley graphs: see [36] for a survey of such results. In particular, in

[36, Problem 2] Ming-Yao Xu poses the problem of determining all nonnormal Cayley graphs of order pq. The edge-transitive nonnormal Cayley graphs of order pq are known [36, Theorem 2.12]. While the result just cited is correct, it is only correct in some sense. That is, there exists edge-transitive graphs Γ which are Cayley graphs of $\mathbb{Z}_{pq}$ and the non-abelian group F_{pq} of order pq such that $(\mathbb{Z}_{pq})_L \triangleleft \operatorname{Aut}(\Gamma)$ but $(F_{pq})_L$ is not normal in $\operatorname{Aut}(\Gamma)$. In other words, this definition of a normal Cayley graph is not invariant under isomorphism. In particular, it was shown in [33, Example 3.4], that if $S_1 = \langle c \rangle$ is the unique subgroup of $\mathbb{Z}_p^*$ of order r, $r \neq p-1$, then the graph $\Gamma = \Gamma(3, p, c, \emptyset, S_1) = \Gamma(3, p, 1, \emptyset, S_1)$ is symmetric (and so edge-transitive) with full automorphism group $\langle v_j^i \to v_{dj+e}^{i+b}$ or $v_{-j}^{-i} : b \in \mathbb{Z}_3, d \in \langle c \rangle, e \in \mathbb{Z}_p \rangle$ (the full automorphism group could also be obtained using Corollary 3.3). If $3|(p-1)$ and $r = 3$, then these graphs are Cayley graphs of both groups of order $3p$, with regular non-abelian subgroup $\langle \rho, \tau \rangle$. Let $\iota : V \to V$ by $\iota(v_j^i) = v_{-j}^{-i}$. A straightforward computation will then show that $\iota^{-1} \tau \iota \notin \langle \rho, \tau \rangle$. Other such examples can be found for suitable graphs $G(q \cdot p; r, s, u)$ defined in [29, 3.4] with $\gcd(q, r) = q$. With these examples in mind, we make the following definition.

Definition 9. Let Γ be a Cayley graph. We say Γ is a *normal* Cayley graph if there exists a group G such that Γ is a Cayley graph of G and $G_L \triangleleft \operatorname{Aut}(\Gamma)$.

We remark that using this definition of a normal Cayley graph, Theorem 2.12 of [36] is correct (and this definition of a normal Cayley graph is invariant under isomorphism). In the following result, we will slightly abuse notation by denoting, for a subset H of $\mathbb{Z}_n$, the units of $\mathbb{Z}_n$ contained in H as H^*.

Theorem 4.1. *Let $q < p$ be prime. Then a nonnormal Cayley graph is isomorphic to one of the following:*

(1) a vertex-primitive Cayley graph of order pq;
(2) a circulant graph Γ with connection set S such that, for some prime divisor r of pq, if H is the unique subgroup of $\mathbb{Z}_{pq}$ of order r and for every $a \in \mathbb{Z}_{pq}$ such that $a \not\equiv 0 \pmod{r}$, then

  (a) $r \neq 2, 3$, and $H \cap S = \emptyset$ or $H - 0$, and $(a + H) \cap S = \emptyset, a + H, (a + H)^*$, or $(a + H) - (a + H)^*$, or
  (b) $(a + H) \cap S = \emptyset$ or $a + H$;

(3) a graph whose automorphism group is isomorphic to $\mathbb{Z}_2 \ltimes \operatorname{PSL}(2, 11)$ or $\mathbb{Z}_2 \ltimes \operatorname{P\Gamma L}(n, k)$, with n prime, $k = t^m$, t a prime, $(n, k - 1) = 1$, and m a power of n (these graphs, or their complements, can be found in [8]).

Proof. It is easily seen that if $\operatorname{Aut}(\Gamma)$ is primitive and Γ is Cayley, then Γ is not normal and (1) follows. Furthermore, if $\operatorname{Aut}(\Gamma) = S^q \times S^p$, $q \neq 2$, $p \neq 3$, $\operatorname{Aut}(\Gamma) = A \times S^p$, $p \neq 3$ or $\operatorname{Aut}(\Gamma) = S^q \times B$, $q \neq 2, 3$, $A < \operatorname{AGL}(1, q)$, $B < \operatorname{AGL}(1, p)$, then Γ is not normal. If any of the three immediately preceding possibilities occur, then Γ is isomorphic to a circulant graph with connection set S and there exists a complete block system $\mathscr{B}$ of $\operatorname{Aut}(\Gamma)$ such that $\operatorname{fix}_{\operatorname{Aut}(\Gamma)}(\mathscr{B}) = S^r$ for some prime divisor r of pq, $r \neq 2, 3$. Whence if $H \leqslant \mathbb{Z}_{pq}$ is the unique subgroup of $\mathbb{Z}_{pq}$ of order r, then $\mathscr{B}$ is formed by the orbits of H_L and $\Gamma[H]$ is a complete graph or its complement, so that $S \cap H = \emptyset$ or $H - 0$. Furthermore, by arguments in Theorem 3.1 each vertex of a block $B \in \mathscr{B}$ is adjacent to $0, 1, r-1$ or r vertices of any other block. It is also not difficult to see that if $0 \in B_0 \in \mathscr{B}$ is adjacent to exactly one vertex of another block B of $\mathscr{B}$, then these vertices must be congruent modulo r. Thus, if $B = a + H$, $a \neq 0$, then $B \cap S = (a+H) - (a+H)^*$. If each vertex of a block of $B = a + H$ is adjacent to $r-1$ vertices of another block of $\mathscr{B}$, then the same argument in the complement of Γ will establish (2a).

If Γ can be written as a nontrivial wreath product, then by Theorem 3.3 $\text{Aut}(\Gamma) = \text{Aut}(\Gamma_s) \wr \text{Aut}(\Gamma_r)$, where Γ_r and Γ_s are prime order circulants with $rs = pq$. Note that $N(pq)$ admits a complete block system $\mathscr{B}$ of s blocks of size r formed by the unique subgroup H_L of $(\mathbb{Z}_{pq})_L$ of order r. Then $\text{fix}_{N(pq)}(\mathscr{B})$ acts faithfully on each $B \in \mathscr{B}$. However, $\text{fix}_{\text{Aut}(\Gamma)}(\mathscr{B})$ does not act faithfully on any $B \in \mathscr{B}$ as a Sylow r-subgroup of $\text{fix}_{\text{Aut}(\Gamma)}(\mathscr{B})$ has order r^s. Whence $(\mathbb{Z}_{pq})_L$ is not normal in $\text{Aut}(\Gamma)$. If $\text{Aut}(\Gamma)$ contains a regular non-abelian subgroup, then we may without loss of generality write $\Gamma = \Gamma(s, r, \alpha, S_0, \ldots, S_\mu)$ with $|\alpha| = s$. Then $\langle \rho, \tau \rangle$ is a regular non-abelian subgroup of $\text{Aut}(\Gamma)$. If $r = 2$, then $\text{Aut}(\Gamma)$ contains an element of order 2 that has $pq - 2$ fixed points while by Lemma 2.5 $N_{S^V}(\langle \rho, \tau \rangle)$ contains no such element. If $r \neq 2$, then $\text{Aut}(\Gamma_r)$ contains a non-semiregular element of order 2. Then a Sylow 2-subgroup of $\text{fix}_{\text{Aut}(\Gamma)}(\mathscr{B})$ has order at least 2^s, and it is easy to see that not every element of such a Sylow 2-subgroup normalizes the unique Sylow r-subgroup of H. Thus (2b) follows.

As (3) follows from [36, Theorem 2.12], it only remains to show that every Cayley graph Γ of order pq not satisfying (1)–(3) is a normal Cayley graph. Hence $\text{Aut}(\Gamma)$ is imprimitive, and by Corollary 3.3 the only such Cayley graphs have automorphism group $\text{Aut}(\Gamma) \leq N(pq)$ or $\text{Aut}(\Gamma) = \langle \rho, \tau \rangle$ (or a conjugate) for some choice of α. The result then follows. $\square$

5. 1/2-transitive graphs

In recent years, 1/2-transitive graphs have received a considerable amount of attention. The interested reader is referred to [22] for a relatively recent survey of activity regarding 1/2-transitive graphs. Briefly, this work has focused on providing examples of 1/2-transitive graphs, classifying 1/2-transitive graphs of valency at most 4, and classifying all 1/2-transitive graphs of certain orders. We shall be interested in work on all of these areas of study, and will characterize 1/2-transitive graphs of order pq.

First, it has previously been shown by Cheng and Oxley [8] that there are no 1/2-transitive graphs of order $2p$ and all 1/2-transitive graphs of order $3p$ have been found by Alspach and Xu [3], along with their full automorphism groups. Wang then considered the case of 1/2-transitive graphs of order pq with $q \geq 5$, but unfortunately there is a flaw in his argument, as pointed out in [22, Section 3], and in fact, the statement of his main result is incorrect. Also, Xu [35, Table 1], has determined all 1/2-transitive graphs of order pq whose automorphism group is primitive – there are 10 such graphs. In [22, Conjecture 3.2], Marušič then made a conjecture on the characterization of 1/2-transitive graphs of order pq. We will show using the following result that Marušič's conjecture is true.

Theorem 5.1. *Let $2 < q < p$ be prime and $\Gamma(q, p, \alpha, S_0, \ldots, S_\mu)$ a metacirculant graph such that the following conditions hold:*

(1) $|\alpha| = mq$, $m > 1$,
(2) $S_0 = \emptyset$,
(3) $S_k \neq \emptyset$ for exactly one of the integers $1 \leqslant k \leqslant \mu$, and S_k is an orbit of length m of $\langle \tau^q \rangle$.

Then $\Gamma(q, p, \alpha, S_0, \ldots, S_\mu)$ is an edge-transitive graph. Conversely, if Γ is a 1/2-transitive metacirculant graph such that $\mathrm{Aut}(\Gamma)$ is not primitive, then there exists $b \in \mathbb{Z}_p$ such that $\Psi_\alpha^b(\Gamma) = \Gamma(q, p, \alpha, S_0, \ldots, S_\mu)$.

Proof. Let $\Gamma = \Gamma(q, p, \alpha, S_0, \ldots, S_\mu)$ be such that the three conditions listed above are true. Let $e_1, e_2 \in E(\Gamma)$. As $\langle \rho, \tau \rangle \leqslant \mathrm{Aut}(\Gamma)$, we may assume, by applying an appropriate element of $\langle \rho, \tau \rangle$ to e_1 and an appropriate element of $\langle \rho, \tau \rangle$ to e_2, that $e_1 = (v_0^0, v_{j_1}^{i_1})$ and $e_2 = (v_0^0, v_{j_2}^{i_2})$. Then $i_1 = i_2 = k$ and $j_1, j_2 \in S_k$. Let $S_k = \{b, \alpha^q b, \ldots, \alpha^{q(m-1)} b\}$. Then $j_1 = \alpha^{c_1 q} b$ and $j_2 = \alpha^{c_2 q} b$. Then $(\tau^q)^{c_2 - c_1}(v_0^0) = v_0^0$ and

$$(\tau^q)^{c_2 - c_1}\left(v_{j_1}^{i_1}\right) = v_{\alpha^{(c_2-c_1)q} j_1}^{i_1} = v_{\alpha^{(c_2-c_1)q} \alpha^{c_1 q} b}^{i_2} = v_{\alpha^{c_2 q} b}^{i_2} = v_{j_2}^{i_2}.$$

Thus $(\tau^q)^{c_2 - c_1}(e_1) = e_2$ and Γ is edge-transitive.

It follows by [2, Lemma 10] that since $S_k \neq \emptyset$ and $S_k \neq \mathbb{Z}_p$ (as $0 \notin S_k$), the Sylow p-subgroups of $\mathrm{Aut}(\Gamma)$ have order p. Suppose that Γ is isomorphic to a circulant graph of order pq. By Theorem 3.1, $\mathrm{Aut}(\Gamma) \cong A \times S^p$, $S^q \times B$, $S^q \times S^p$ or $\mathrm{Aut}(\Gamma) \leqslant N(pq)$. In any case, $\mathrm{Aut}(\Gamma)$ admits a complete block system $\mathscr{B}$ of q blocks of size p formed by the orbits of $\langle \rho \rangle$. Then $\Gamma/\mathscr{B}$ is q-cycle, and, as the automorphism group of a q-cycle is a dihedral group, we have that $\mathrm{Aut}(\Gamma) \cong A \times S^p$, $A \leqslant \mathrm{AGL}(1, q)$, or $\mathrm{Aut}(\Gamma) \leqslant N(pq)$. As $1 < m < mq \leqslant p - 1$, we have that $\mathrm{fix}_{\mathrm{Aut}(\Gamma)}(\mathscr{B})$ does not act doubly transitively on any $B \in \mathscr{B}$, so that $\mathrm{Aut}(\Gamma) \leqslant N(pq)$. As Γ is isomorphic to a circulant graph, and $\mathrm{fix}_{\mathrm{Aut}(\Gamma)}(\mathscr{B})$ does not contain a pq-cycle, there exists $\tau' \in \mathrm{Aut}(\Gamma)$ such that $\langle \tau', \rho \rangle$ is cyclic of order pq so that τ' centralizes ρ. Hence $\tau'(v_j^i) = v_{j+b_i}^{i+r}$, and by replacing τ' with a power, we assume $r = 1$. Then $(\tau')^{-1}\tau \in \mathrm{fix}_{\mathrm{Aut}(\Gamma)}(\mathscr{B})$ and has orbits of length $|\alpha| = mq$ or 1. We conclude that v_0^0 is adjacent to at least mq elements of V^k, a contradiction. Thus Γ is not isomorphic to a circulant graph.

It now follows from Corollary 3.3 that $\mathrm{Aut}(\Gamma) = \langle \rho, \tau' \rangle$, where $\tau' : V \to V$ by $\tau'(v_j^i) = v_{\alpha' j}^{i+1}$, for some $\alpha' \in \mathbb{Z}_p^*$ with $\alpha \in \langle \alpha' \rangle$. As $\tau \in \mathrm{Aut}(\Gamma)$, $\tau^{-1}\tau' \in \mathrm{fix}_{\mathrm{Aut}(\Gamma)}(\mathscr{B})$, where $\mathscr{B}$ is the complete block system of $\mathrm{Aut}(\Gamma)$ formed by the orbits of $\langle \rho \rangle$. Furthermore, $\tau^{-1}\tau'(v_j^i) = v_{\alpha'\alpha^{-1}j}^i$. We conclude that as S_k is an orbit of $\langle \tau^q \rangle$, the orbits of $\tau^{-1}\tau'$ are contained in the orbits of τ^q, so that $\alpha'\alpha^{-1} \in \langle \alpha^q \rangle$. Whence $\tau' \in \langle \rho, \tau \rangle$ so that $\mathrm{Aut}(\Gamma) = \langle \rho, \tau \rangle$.

We need only show that Γ is not symmetric. Note that Γ is $2m$-regular and so has mpq edges. If Γ is symmetric, then $\mathrm{Aut}(\Gamma)$ would have an orbit of length $2mpq$ and so $2mpq$ must divide $|\mathrm{Aut}(\Gamma)|$. However, as $\mathrm{Aut}(\Gamma) = \langle \rho, \tau \rangle$, and $|\alpha| = mq$, we have that $|\mathrm{Aut}(\Gamma)| = mpq$, a contradiction. Thus Γ is not symmetric, and as Γ is edge-transitive, Γ is 1/2-transitive as required.

Conversely, let Γ be a 1/2-transitive metacirculant graph such that $\mathrm{Aut}(\Gamma)$ is not primitive. As an edge-transitive Cayley graph of an abelian group is symmetric, we have that Γ is not isomorphic to a circulant graph. It then follows by Corollary 3.3

that $\text{Aut}(\Psi_\alpha^b(\Gamma)) = \langle \rho, \tau \rangle$ (or a conjugate) for some choice of α and some $b \in \mathbb{Z}_p$, and, as Γ is not isomorphic to a circulant graph, q must divide $|\alpha|$. If $S_0 \neq \emptyset$, then, either Γ is disconnected (in which case Γ is isomorphic to a circulant graph) or there exists $1 \leq k \leq \mu$ such that $S_k \neq \emptyset$. Then $(v_0^0, v_j^0), (v_0^0, v_\ell^k) \in E(\Gamma)$ for some $j, k \in \mathbb{Z}_p$. However, as $\text{Aut}(\Gamma) = \langle \rho, \tau \rangle$, $\text{Aut}(\Gamma)$ admits a complete block system $\mathcal{B}$ of q blocks of size p formed by the orbits of $\langle \rho \rangle$. There thus can be no $\delta \in \text{Aut}(\Gamma)$ such that $\gamma(v_0^0, v_j^0) = (v_0^0, v_\ell^k)$. This contradicts the hypothesis that Γ is edge-transitive. Thus $S_0 = \emptyset$.

If $S_k \neq \emptyset$ and $S_\ell \neq \emptyset$ for $1 \leq k, \ell \leq \mu$ and $k \neq \ell$, then, again as $\text{Aut}(\Gamma) = \langle \rho, \tau \rangle$ (or a conjugate), we have that $\text{Aut}(\Gamma)/\mathcal{B}$ is cyclic. As Γ is edge-transitive, $\text{Aut}(\Gamma)/\mathcal{B}$ acts transitively on the edges of $\Gamma/\mathcal{B}$. As $\text{Aut}(\Gamma)/\mathcal{B}$ is cyclic, $\Gamma/\mathcal{B}$ is regular of degree 2. However, as $S_k \neq \emptyset$ and $S_\ell \neq \emptyset$, $\Gamma/\mathcal{B}$ is regular of degree at least 4, a contradiction. Thus $S_k \neq \emptyset$ for exactly one integer $1 \leq k \leq \mu$.

Finally, if $S_k \neq \emptyset$ and $m = 1$, then as Γ is edge-transitive, Γ contains pq edges so is regular of degree 2. However, such a graph is a cycle or a disjoint union of cycles. In either case, we have that Γ is isomorphic to a circulant graph, a contradiction. Thus $m > 1$ and S_k is a union of orbits of τ^q. If S_k is a union of at least two orbits of τ^q, then Γ has at least $(m+1)pq$ edges. However, $|\text{Aut}(\Gamma)| = |\langle \rho, \tau \rangle| = mqp$. Thus Γ is not edge-transitive, a contradiction. $\square$

Definition 10. Following [22], a (q, p)-metacirculant graph satisfying conditions (1)–(3) of Theorem 5.1 with $k = 1$ and S_1 the unique subgroup of $\mathbb{Z}_p^*$ of order m will be denoted by $X(\alpha, m, q, p)$.

The following result characterizes 1/2-transitive graphs of order pq, and the statement is the above mentioned conjecture of Marušič.

Corollary 5.2. Let $3 \leq q < p$ be primes and let Γ be a 1/2-transitive graph of order pq. Then either $\text{Aut}(\Gamma)$ is primitive and Γ is isomorphic to one of the ten graphs of [35, Table 1] or $\Gamma \cong X(s, m; q, p)$, where $p \equiv 1 \pmod{q}$, $m > 1$ is a divisor of $(p-1)/q$, and $(m, q, p) \neq (2, 3, 7)$ or $(3, 3, 19)$.

Proof. Let Γ be a 1/2-transitive graph of order pq. Clearly, if $\text{Aut}(\Gamma)$ is primitive then the result holds. Otherwise, Γ is isomorphic to a metacirculant graph or a Marušič–Scapellato graph. By [25], all suborbits of appropriate actions of $SL(2, 2^s)$ are self-paired, and an edge-transitive Marušič-Scapellato graph is symmetric. If Γ is isomorphic to a metacirculant graph, then by Theorem 5.1 we may assume that Γ satisfies the conclusion of Theorem 5.1. If $k \neq 1$, then it is easy to check that $\delta(\Gamma)$ satisfies the conclusion of Theorem 5.1 with $k = 1$, where $\delta(v_j^i) = v_j^{k^{-1}i}$, so we assume that $k = 1$. If S_1 is not the unique subgroup H of $\mathbb{Z}_p^*$ of order m, then $S_1 = \ell \cdot H$ for some $\ell \in \mathbb{Z}_p^*$. Hence $\gamma(\Gamma)$ satisfies the conclusion of Theorem 5.1 with $k = 1$ and $S_1 = H$, where $\gamma(v_j^i) = v_{\ell^{-1}j}^i$.

Conversely, we clearly need only show that $X(s, m; q, p)$ is 1/2-transitive provided that $(m, q, p) \neq (2, 3, 7)$ or $(3, 3, 19)$. By Theorem 5.1, $X(s, m; q, p)$ is 1/2-transitive provided that $\text{Aut}(X(s, m; q, p))$ is not primitive. If $\text{Aut}(X(s, m; q, p))$ is primitive, then $X(s, m; q, p)$ is symmetric, and is given in [30, Table 1], and is also metacirculant, so $\text{soc}(\text{Aut}(X(s, m; q, p)))$

cannot appear in [26, Table 2]. We must also have that $X(s,m;q,p)$ is regular of degree $2m$ and m is a divisor of $(p-1)/q$. It was shown in [23] (see also [22]) that every $X(s,2;q,p)$ is 1/2-transitive, so $2m > 4$. Checking the degrees given in [30, Table 1], we see that $\text{Aut}(X(s,m;q,p))$ can be primitive only if $(m,q,p) = (2,3,7)$ or $(3,3,19)$, and by [3] these graphs are symmetric. □

Acknowledgement

As usual, the author is indebted to the anonymous referee, particularly for excellent comments regarding Sections 3 and 5.

References

[1] Alspach, B. (1973) Point-symmetric graphs and digraphs of prime order and transitive permutation groups of prime degree. *J. Combin. Theory* **15** 12–17.

[2] Alspach, B. and Parsons, T. D. (1982) A construction for vertex-transitive graphs. *Canad. J. Math* **24** 307–318.

[3] Alspach, B. and Xu, M.Y. (1994) 1/2-transitive graphs of order $3p$. *J. Algebraic Combin.* **3** 347–355.

[4] Bollobás, B. (1979) *Graph Theory*, Springer.

[5] Burnside, W. (1901) On some properties of groups of odd order. *J. London Math. Soc.* **33** 162–185.

[6] Cameron, P. J. (1981) Finite permutation groups and finite simple groups. *Bull. London Math. Soc.* **13** 1–22.

[7] Chen, J. Z. (1999) The symmetric graphs of order $3p$ (Chinese). *Math. Theory Appl.* **19** 24–27.

[8] Cheng, Y. and Oxley, J. (1987) On weakly symmetric graphs of order twice a prime. *J. Combin. Theory Ser. B* **42** 196–211.

[9] Coxeter, H. S. M. and Moser, W. O. T. (1965) *Generators and Relations for Discrete Groups*, Springer.

[10] Dixon, J. D. and Mortimer, B. (1996) *Permutation Groups*, Vol. 163 of *Springer Graduate Texts in Mathematics*.

[11] Dobson, E. (1998) Isomorphism problem for metacirculant graphs of order a product of distinct primes. *Canad. J. Math.* **50** 1176–1188.

[12] Dobson, E. (2004) On self-complementary vertex-transitive graph of order a product of distinct primes. *Ars Combin.* **71** 249–256.

[13] Elspas, B. and Turner, J. (1970) Graphs with circulant adjacency matrices. *J. Combin. Theory Ser. B* **9** 297–307.

[14] Godsil, C. D. (1981) GRRs for nonsolvable groups. In *Algebraic Methods in Graph Theory, Vol. I, II* (Szeged, 1978), Vol. 25 of *Colloq. Math. Soc. János Bolyai*, North-Holland, pp. 221–239.

[15] Gorenstein, D. (1980) *Finite Groups*, Chelsea.

[16] Hungerford, T. (1974) *Algebra*, Holt, Rinehart and Winston.

[17] Huppert, B, (1967) *Endliche Gruppen I*, Springer.

[18] Klin, R. H. and Pöschel, R. (1981) The König problem, the isomorphism problem for cyclic graphs and the method of Schur. In *Algebraic Methods in Graph Theory, Vol. I, II* (Szeged, 1978), Vol. 25 of *Colloq. Math. Soc. János Bolyai*, North-Holland, pp. 405–434.

[19] Li, C. H. (1997) On self-complementary vertex-transitive graphs. *Comm. Algebra* **25** 3903–3908.

[20] Marušič, D. (1981) On vertex symmetric digraphs. *Disc. Math.* **36** 69–81.

[21] Marušič, D. (1988) On vertex-transitive graphs of order qp. *J. Combinat. Math. Combinat. Comp.* **4** 97–114.

[22] Marušič, D. (1998) Recent developments in half-transitive graphs. *Discrete Math.* **182** 219–231.

[23] Marušič, D. (1998) Half-transitive group actions on finite graphs of valency 4. *J. Combin. Theory Ser. B* **73** 41–76.
[24] Marušič, D. and Scapellato, R. (1992) Characterizing vertex transitive *pq*-graphs with imprimitive automorphism group. *J. Graph Theory* **16** 375–387.
[25] Marušič, D. and Scapellato, R. (1993) Imprimitive representations of SL$(2, 2^k)$. *J. Combin. Theory Ser. B* **58** 46–57.
[26] Marušič, D. and Scapellato, R. (1994) Classifying vertex-transitive graphs whose order is a product of two primes. *Combinatorica* **14** 187–201.
[27] Morris, J. (1999) Isomorphisms of Cayley Graphs, PhD Thesis, Simon Fraser University.
[28] Muzychuk, M. (1999) On Sylow subgraphs of vertex-transitive self-complementary graphs. *Bull. London Math. Soc.* **31** 531–533.
[29] Praeger, C. E., Wang, R. J. and Xu, M. Y. (1993) Symmetric graphs of order a product of two distinct primes. *J. Combin. Theory Ser. B* **58** 299–318.
[30] Praeger, C. E. and Xu, M. Y. (1993) Vertex-primitive graphs of order a product of two distinct primes. *J. Combin. Theory Ser. B* **59** 245–266.
[31] Sabidussi, G. (1959) The composition of graphs. *Duke Math J.* **26** 693–696.
[32] Wang, R. J. (1994) Half-transitive graphs of order a product of two distinct primes. *Comm. Algebra* **22** 915–927.
[33] Wang, R. J. and Xu, M. Y. (1993) A classification of symmetric graphs of order 3*p*. *J. Combin. Theory Ser. B* **58** 197–216.
[34] Wielandt, H. (trans. by R. Bercov) (1964) *Finite Permutation Groups*, Academic Press.
[35] Xu, M. Y. (1994) Some new results on 1/2-transitive graphs. *Adv. Math. China* **23** 505–516.
[36] Xu, M. Y. (1998) Automorphism groups and isomorphism of Cayley digraphs. *Discrete Math.* **182** 309–319.

On the Number of Hamiltonian Cycles in a Tournament

EHUD FRIEDGUT[1†] and JEFF KAHN[2‡]

[1]Institute of Mathematics, Hebrew University, Jerusalem, Israel
(e-mail: ehudf@math.huji.ac.il)

[2]Department of Mathematics and RUTCOR, Rutgers University,
New Brunswick NJ 08854, USA
(e-mail: jkahn@math.rutgers.edu)

Received 13 November 2003; revised 1 September 2004

For Béla Bollobás on his 60th birthday

Let $P(n)$ and $C(n)$ denote, respectively, the maximum possible numbers of Hamiltonian paths and Hamiltonian cycles in a tournament on n vertices. The study of $P(n)$ was suggested by Szele [14], who showed in an early application of the probabilistic method that $P(n) \geqslant n!2^{-n+1}$, and conjectured that

$$\lim (P(n)/n!)^{1/n} = 1/2.$$

This was proved by Alon [2], who observed that the conjecture follows from a suitable bound on $C(n)$, and showed $C(n) < O(n^{3/2}(n-1)!2^{-n})$. Here we improve this to

$$C(n) < O\big(n^{3/2-\xi}(n-1)!2^{-n}\big),$$

with $\xi = 0.2507\ldots$. Our approach is mainly based on entropy considerations.

1. Introduction

A tournament is a complete directed graph. For a tournament T we denote by $C(T)$ the number of (directed) Hamiltonian cycles in T and by $P(T)$ the number of Hamiltonian paths. Let $C(n) = \max C(T)$, the maximum taken over tournaments with n vertices, and define $P(n)$ similarly. The problem of estimating $P(n)$ seems to have been first suggested by Szele [14], whose proof that $P(n) \geqslant n!2^{-n+1}$ is considered the first combinatorial application of the probabilistic method (see, e.g., [4]). The same argument – namely that $P(n)$ is at least the expected number of Hamiltonian paths in a *random* T – shows

---

[†] Research supported in part by the Israel Science Foundation, grant no. 0329745. Part of this research was carried out while the author was visiting Microsoft Research and the Institute for Advanced Study in Princeton.

[‡] Research supported in part by NSF. Part of this research was carried out while the author was visiting Microsoft Research.

$C(n) \geq (n-1)! 2^{-n}$. Szele also showed that $\lim(P(n)/n!)^{1/n}$ exists and conjectured that its value is $1/2$. This was proved by Alon [2], who derived from the following theorem of Brégman (formerly the *Minc Conjecture*) the upper bound $C(n) < O(n^{1/2} n! 2^{-n})$ (we will not worry about the constants in the 'big Ohs'), and used this to show

$$P(n) < O(n^{3/2} n! 2^{-n}). \tag{1.1}$$

In what follows we set $\Psi(n) = n^{1/2} n! 2^{-n}$. This is the order of magnitude in Alon's bound on $C(n)$ and the benchmark against which we will measure our results.

Theorem 1.1. (Brégman's theorem [5]) *For an $n \times n$ $\{0,1\}$-matrix A with row sums r_i,*

$$\operatorname{per}(A) \leq \prod_{i=1}^{n} (r_i!)^{1/r_i}$$

(where per *means permanent).*

If one regards A in Theorem 1.1 as the adjacency matrix of a digraph T on $[n]$ ($A_{ij} = 1$ if and only if there is an arc from i to j), then the permanent of A is the number of *1-factors* of T – that is, spanning subdigraphs with all in- and outdegrees equal to 1 – so in particular is a bound on the number of Hamiltonian cycles of T. It is also not hard to see (again, see [2]) that, for given $\sum r_i$, the bound in Theorem 1.1 is maximized by taking the r_i to be as equal as possible ($|r_i - r_j| \leq 1 \; \forall i, j$). This gives Alon's bound on $C(n)$. Alon then shows that

$$4P(n) \leq C(n+1) \tag{1.2}$$

(his argument for this is repeated in the proof of Corollary 3.3 below), which, with the trivial

$$P(n) \geq nC(n), \tag{1.3}$$

says the two problems are not much different. Our main purpose here is to slightly improve the upper bounds.

Theorem 1.2. *For any $\xi < 2(1 - \exp[\sqrt{3/4} - 1]) = 0.2507\ldots,$*

$$C(n) < O(n^{1/2-\xi} n! 2^{-n}),$$

and (consequently)

$$P(n) < O(n^{3/2-\xi} n! 2^{-n}).$$

For convenience we will prove the theorem with

$$\xi = (1 + o(1)) 2 (1 - \exp[\sqrt{1 - (0.99)(0.24)} - 1]), \tag{1.4}$$

but it is easy to see that the same argument can be used to prove the theorem as stated. Our proof of this follows the beautiful entropy proof of Brégman's theorem given by J. Radhakrishnan in [11] (itself inspired by Schrijver's proof [13], especially as presented in [4]). In retrospect, at least, the entropy approach is quite natural in such settings, where one wants to bound the size of some subset of a product of sets (as is the case here, the set of 1-factors of a digraph being a subset of $[n]^{[n]}$). See, e.g., [12] for a discussion of other results of this type. It turns out that a convenient way to explain the present argument is to review what Radhakrishnan does and see where there might be room for improvement in our special situation. This is done in Section 2; in a few words the plan is as follows. We treat 1-factors as functions in the obvious way: $f(x) = y$ means the arc (x, y) is part of the 1-factor. In the entropy approach we consider **f** chosen uniformly at random from the set of 1-factors and try to bound the entropy $H(\mathbf{f})$, which is simply the log of the number of 1-factors. (For entropy see Section 2.) This is done by taking the vertices in some order $x_1, \ldots, x_n$ and considering for $i = 1, \ldots, n$ the conditional entropy of $\mathbf{f}(x_i)$ given $(\mathbf{f}(x_j) : j < i)$, the idea being that some *a priori* possibilities for $\mathbf{f}(x_i)$ are ruled out because they have been chosen earlier in the sequence. The mild improvement in the present situation then derives from the fact that when we allow only Hamiltonian cycles, we have the additional restriction that $\mathbf{f}(x_i)$ cannot close a cycle of length less than n. Details of the implementation of this plan are contained in Sections 3 and 4.

The actual growth rate of $C(n)$ (or $P(n)$) – and some related questions suggested by it – seem quite interesting; see Section 5. We tend to think Szele's lower bound is close(r) to the truth. A small but ingenious improvement – multiplication by $e - o(1)$ – in Szele's bound was given by Adler, Alon and Ross in [1]; and very recently Wormald [15], motivated by the question at the end of the present paper, improved this to $2.85584\ldots$, which he conjectures to be close to the truth. Both these improvements are based on analysis of suitable random *regular* tournaments.

Notation and conventions. We use $\mathcal{H}$ or $\mathcal{H}(T)$ for the set of Hamiltonian cycles of T (treated, as above, as functions) and often abbreviate 'Hamiltonian cycle' to 'cycle'. We use $\Gamma^+(x)$ for the set of out-neighbours of x, $\Gamma^-(x)$ for the set of in-neighbours and $d(x) = |\Gamma^+(x)|$. We write $[n]$ for $\{1, \ldots, n\}$, S_n for the symmetric group (here the set of orderings of the vertex set of our tournament), $\mathbb{E}$ for expectation, and log for $\log_2$. For simplicity we pretend throughout that all large numbers are integers.

2. Review and preview

Here we recall Radhakrishnan's proof of Brégman's theorem and indicate what needs to be added for the proof of Theorem 1.2. The *entropy* of a discrete random variable **X** (meaning, simply, a random variable taking values in some countable set) is

$$H(\mathbf{X}) = \sum_x p(x) \log(1/p(x)),$$

where we write $p(x)$ for $\Pr(\mathbf{X} = x)$. For an event A, $H(\mathbf{X}|A)$ is the entropy of X given A; that is,

$$H(\mathbf{X}|A) = \sum_x p(x|A) \log(1/p(x|A))$$

(where $p(x|A) = \Pr(\mathbf{X} = x|A)$). The *conditional entropy* of $\mathbf{X}$ given $\mathbf{Y}$ is (with the obvious meanings for $p(y), p(x|y)$)

$$H(\mathbf{X}|\mathbf{Y}) = \mathbb{E}H(\mathbf{X}|\mathbf{Y} = y) = \sum_y p(y) \sum_x p(x|y) \log \frac{1}{p(x|y)}.$$

The only entropy facts we will need here are, first, that for a random vector $\mathbf{X} = (\mathbf{X}_1, \ldots, \mathbf{X}_n)$ (note this is also a random variable), one has

$$H(\mathbf{X}) = H(\mathbf{X}_1) + H(\mathbf{X}_2|\mathbf{X}_1) + \cdots + H(\mathbf{X}_n|\mathbf{X}_1, \ldots, \mathbf{X}_{n-1}) \tag{2.1}$$

(this is the 'chain rule'), and, second, that for any $\mathbf{X}$,

$$H(\mathbf{X}) \leqslant \log |\text{range}(\mathbf{X})|. \tag{2.2}$$

(Note this is sharp if $\mathbf{X}$ is chosen uniformly from its range.) For more entropy background see [10] or [6].

We deal with Brégman's theorem in its digraph formulation: we are given a digraph T on $[n]$ and $\mathbf{f}$ drawn from an (*arbitrary*) probability measure μ on $\mathscr{F}$, the set of 1-factors in T, and want to bound the entropy $H(\mathbf{f})$. (For Brégman's theorem μ is uniform measure, which according to (2.2) is actually the worst case; but it will later be helpful to have this more general version.) Fix $\sigma \in S_n$. We think of sequentially exposing the values $\mathbf{f}(x)$ according to the order on vertices x given by σ. By (2.1) we have

$$H(\mathbf{f}) = \sum_{i=1}^n H[\mathbf{f}(\sigma^{-1}(i)) \mid \mathbf{f}(\sigma^{-1}(1)), \ldots, \mathbf{f}(\sigma^{-1}(i-1))]$$

$$= \sum_x H[\mathbf{f}(x) \mid (\mathbf{f}(y) : \sigma(y) < \sigma(x))].$$

Suppose now that

$\mathbf{f}$ is chosen uniformly from some $\mathscr{M} \subseteq \mathscr{F}$ with $|\mathscr{M}| = m$.

Then with f ranging over $\mathscr{M}$,

$$H(\mathbf{f}) = \frac{1}{m} \sum_f \sum_x H[\mathbf{f}(x) \mid \mathbf{f}(y) = f(y) \ \forall \sigma(y) < \sigma(x)]. \tag{2.3}$$

Set

$$k_f(x, \sigma) := |\Gamma^+(x) \setminus \{f(y) : \sigma(y) < \sigma(x)\}|. \tag{2.4}$$

This quantity, the number of possibilities for $\mathbf{f}(x)$ at the time it is exposed given agreement with f up to that point, will play a central role in this paper. By (2.2) the summand in (2.3) is at most

$$\log k_f(x, \sigma). \tag{2.5}$$

So if we now unfix σ, choosing it at random according to some probability measure on S_n (which in practice will just be uniform measure), then

$$H(\mathbf{f}) \leq \mathbb{E}_\sigma \frac{1}{m} \sum_f \sum_x \log k_f(x,\sigma)$$

$$= \frac{1}{m} \sum_f \sum_x \mathbb{E}_\sigma \log k_f(x,\sigma).$$

If σ is chosen uniformly from S_n then the summand does not depend on f, and is equal to

$$\frac{1}{d(x)} \sum_{k=1}^{d(x)} \log k = \frac{1}{d(x)} \log(d(x)!),$$

since – this is perhaps the main point – for fixed f,

$$k_f(x,\sigma) = |\{y \in f^{-1}(\Gamma^+(x)) : \sigma(y) \geq \sigma(x)\}|$$

is uniform from $[d(x)]$. This gives Theorem 1.1. □

To do better in the case of Hamiltonian cycles we intend to strengthen the bound (2.5) in some cases (with $\mathscr{M}$ now a subset of $\mathscr{H}(T)$). Given f, define for each x and σ (with $\sigma(x) < n$),

$$i(x,\sigma) = \min\{i \geq 1 : \sigma(f^{-i}(x)) > \sigma(x)\}$$

and

$$b(x,\sigma) = f^{-i(x,\sigma)+1}(x).$$

That is, if we follow f backwards from x, $b(x,\sigma)$ is the last vertex we see before we first encounter a vertex that follows x in σ. If $b(x,\sigma) \in \Gamma^+(x)$ then (given $\mathbf{f}(y) = f(y) \ \forall \sigma(y) < \sigma(x)$ as in (2.3)) we cannot have $\mathbf{f}(x) = b(x,\sigma)$, since this choice is incompatible with what is known at the time we choose $\mathbf{f}(x)$ (it would imply that f contains a cycle of length $i(x,\sigma) < n$). So, defining the event

$$A(x,\sigma) = \{b(x,\sigma) \in \Gamma^+(x)\},$$

we may replace the bound (2.5) on $H\big[\mathbf{f}(x)|\mathbf{f}(y) = f(y) \ \forall \sigma(y) < \sigma(x)\big]$ in the proof of Theorem 1.1 by $\log(k_f(x,\sigma) - \mathbf{1}_{A(x,\sigma)})$ (where $\mathbf{1}_Q$ is the indicator of event Q). Setting

$$p_k(x) = \Pr_\sigma[b(x,\sigma) \in \Gamma^+(x) \mid k_f(x,\sigma) = k],$$

this gives

$$H(\mathbf{f}) \leq \frac{1}{m} \sum_f \sum_x \frac{1}{d(x)} \sum_{k=1}^{d(x)} \left[\log k - p_k(x) \log \frac{k}{k-1}\right]$$

$$= \sum_x \frac{1}{d(x)} \log(d(x)!) - \frac{1}{m} \sum_f \sum_x \sum_k \frac{1}{d(x)} p_k(x) \log \frac{k}{k-1}. \quad (2.6)$$

So we would like to show that the subtracted expression is substantial. In practice this will go as follows. We first introduce the notion of a 'good' Hamiltonian cycle, taking $\mathscr{M}$

to be the set of such cycles (and $m = |\mathcal{M}|$). We show (Lemma 3.4) that the number of bad cycles is negligible:

$$|\mathcal{H} \setminus \mathcal{M}| \leq (1 - 10^{-4} + o(1))^{0.01n}\Psi(n) \qquad (2.7)$$

(Ψ was defined following (1.1)). This is essentially a consequence of Theorem 1.1, as is the observation that we may assume T is fairly regular (see (4.5)). The final and main point of the proof is showing that, given this regularity, the desired gain in (2.6) even holds 'locally':

$$\text{for each } f \in \mathcal{M}, \quad \sum_x \sum_k \frac{1}{d(x)} p_k(x) \log \frac{k}{k-1} > \xi \log n \qquad (2.8)$$

(ξ as in (1.4)). Combining these with (2.6) gives Theorem 1.2.

3. Few bad cycles

Here we say what is meant by a bad cycle and prove (2.7). This requires a few preliminaries. We first observe that Theorem 1.1 gives the following.

Corollary 3.1. *For a tournament T on n vertices with outdegrees $r_1, \ldots, r_n$, and r the geometric mean of the r_is, the number of 1-factors in T is less than $\exp[(1/4 + o(1))\ln^2 n](\frac{r}{e})^n$.*

Proof. This is a simple calculation using Theorem 1.1 and, for instance, the fact that the sum of any t of the r_is is at least $\binom{t}{2}$. (Briefly: setting $(k!)^{1/k} = h(k)k/e$, we should show $\prod h(r_i) < \exp[(1/4 + o(1))\ln^2 n]$. Stirling's formula gives $\prod h(r_i) \approx \exp[\sum (\ln r_i)/(2r_i)]$, and the aforementioned restriction on small r_is (together with $\sum r_i = \binom{n}{2}$) are easily seen to imply $\sum (\ln r_i)/(2r_i) < (1/4 + o(1))\ln^2 n$.) □

Say an arc (j, i) in a tournament on $[n]$ is *oriented backwards* if $i < j$.

Lemma 3.2. *Suppose T is a tournament on $[n]$ in which at most $(1/4 - \alpha)\binom{n}{2}$ arcs are oriented backwards. Then the number of 1-factors in T is less than $(1 - \alpha^2 + o(1))^n \Psi(n)$, provided $\alpha > (\ln n)/\sqrt{n}$.*

(In what follows α will just be a positive constant. Note that the cyclic tournament – that with arc set $\{(i, j) : 1 \leq j - i \leq (n-1)/2 \pmod{n}\}$, where we assume n is odd – has about 1/4 of its edges oriented backwards and seems to have *many* Hamiltonian cycles [15]; so Lemma 3.2 is probably about the best one can hope for here.)

Proof. Suppose for simplicity that n is even, and again write $r_1, \ldots, r_n$ for the outdegrees. The arithmetic mean of those r_is with $n/2 < i \leq n$ is less than $(1/2 - \alpha)n$ (it may be helpful to picture the adjacency matrix with at most $(1/4 - \alpha)\binom{n}{2}$ 1s below the main diagonal); so since the arithmetic mean of *all* the r_is is $(n-1)/2$, concavity of the logarithm implies that the *geometric* mean of (all) the r_is is less than $n\sqrt{(1/2 - \alpha)(1/2 + \alpha)} < (1 - 2\alpha^2)n/2$.

Inserting this in Corollary 3.1 we find that the number of 1-factors in T is less than

$$\exp[(1/4 + o(1))\ln^2 n]\left((1 - 2\alpha^2)\frac{n}{2e}\right)^n,$$

which gives the statement in the lemma. □

The reason for shifting our attention to $P(T)$ in the following corollary is that our proof of a bound on the number of 'bad' cycles (Lemma 3.4) will involve breaking such cycles into paths.

Corollary 3.3. *For T, α as in Lemma 3.2, $P(T) < (1 - \alpha^2 + o(1))^n \Psi(n)$.*

Proof. We follow [2]: let T' be the random tournament on $[n+1]$ obtained from T by choosing uniformly from the 2^n possible orientations of the edges joining $n + 1$ to vertices from $[n]$. Then each Hamiltonian path of T extends to a Hamiltonian cycle of T' with probability $1/4$, so that

$$P(T)/4 = \mathbb{E}C(T') < (1 - \alpha^2 + o(1))^n \Psi(n),$$

the inequality following from Lemma 3.2. This gives the stated bound. (Note that the $o(1)$ term can be used to absorb minor factors such as $\Psi(n+1)/\Psi(n) \approx n$.) □

We take the *length* of an interval I (of a cycle) to be the number of edges (arcs) in I, and the *length* of a chord to be the number of edges in the shorter interval connecting the ends of the chord. We then use *j-interval* (*j-chord*) to mean an interval (chord) of length j. Let ω be a function of n tending slowly to infinity ($\omega < n^{o(1)}$ will be slow enough). From now on k will always denote an integer satisfying

$$\omega < k < n/(2\omega). \tag{3.1}$$

For a cycle f we say a chord (x, y) goes backwards if $x = f^i(y)$ for some $i < n/2$. Given $f \in \mathcal{H}$, we say an interval of f is *bad* if less than 0.24 of its chords (say of length at least 2, though this makes no significant difference) go backwards (and of course, here and elsewhere, 'good' means 'not bad'). A cycle is *k-bad* if more than 0.01 of its k-intervals are bad, and *bad* if it is k-bad for some k in our range. This is the notion of 'bad' for which we prove (2.7).

Lemma 3.4. *In any tournament on $[n]$ the number of bad cycles is less than $(1 - 10^{-4} + o(1))^{0.01n} \Psi(n)$.*

Proof. It is enough to show that the number of k-bad cycles is less than

$$(1 - 10^{-4} + o(1))^{0.01n} \Psi(n). \tag{3.2}$$

Say a k-set K (of vertices) is bad if it has an ordering with respect to which the fraction of edges in K which are oriented backwards in T is less than 0.24. If a cycle H is k-bad, then it may be partitioned into n/k k-intervals at least 0.01 of which are bad, which in particular implies that their underlying k-sets are bad. So we may specify a bad cycle H as follows.

(i) Choose a cyclically ordered partition $I_1 \cup \cdots \cup I_{n/k}$ of $[n]$ into k-sets and some $0.01n/k$ of these which are bad (requiring that H traverse the I_js in the cyclic order).
(ii) For each j choose the Hamiltonian path which is the restriction of H to I_j.

The number of possibilities in (i) is at most

$$\frac{k}{n}\binom{n}{k,\ldots,k}\binom{n/k}{0.01n/k}, \tag{3.3}$$

while (1.1) and Corollary 3.3 bound the number of possibilities in (ii) by

$$(O(k\Psi(k)))^{n/k}(1 - 10^{-4} + o(1))^{0.01n}; \tag{3.4}$$

and the product of (3.3) and (3.4) is easily seen to be no more than (3.2). (We again bury minor factors – in this case mainly something like $k^{n/k}$ – in the $o(1)$.) $\square$

4. Not too many good cycles

Here we prove Theorem 1.2; that is, for a tournament T on $[n]$,

$$C(T) < O(n^{1/2-\xi}n!2^{-n}). \tag{4.1}$$

We need one preliminary observation. For a cycle f let $\text{back}_j(f)$ be the fraction of j-chords of f that go backwards. Notice that if f is k-good then

$$\sum_{j=2}^{k} \text{back}_j(f)(k+1-j) \geq \beta \sum_{j=2}^{k}(k+1-j), \tag{4.2}$$

where $\beta = (0.99)(0.24)$, since each j-chord belongs to exactly $k+1-j$ intervals of length k.

Lemma 4.1. *For any good cycle f, k satisfying (3.1) and $q = 1 - \frac{2k}{n}$,*

$$\sum_{j=2}^{\frac{n}{2k}} \text{back}_j(f)q^j \geq (\gamma - o(1))\frac{n}{2k},$$

where $\gamma = 1 - \exp[\sqrt{1-\beta} - 1] = 0.119\ldots$.

Proof. Set $k' = \frac{n}{2k}$. We have (4.2) for k', that is,

$$\sum_{j=2}^{k'} \text{back}_j(f)(k'+1-j) \geq \beta \sum_{j=2}^{k'}(k'+1-j), \tag{4.3}$$

and should minimize $\sum_{j=2}^{k'} \text{back}_j(f)q^j$. But since $\frac{q^j}{k'+1-j}$ is increasing in j for $j \leq k'$, the minimum is achieved by taking $\text{back}_j = 1$ if $j \leq m$ and 0 otherwise, where m is the smallest value for which this assignment achieves (4.3). An easy calculation then gives

$m = (1 - \sqrt{1-\beta})k'$, and

$$\sum_{j=2}^{\frac{n}{2k}} \text{back}_j(f) q^j = \sum_{j=2}^{m} q^j = \frac{q^2 - q^{m+1}}{1-q} = (1-o(1))\gamma \frac{n}{2k}. \qquad \square$$

Remark 1. The value of γ can presumably be improved by exploiting (4.2) in general (rather than just for k'), but we will not try to optimize here.

We now turn to (4.1). To begin, notice we may assume that r, the geometric mean of the outdegrees in T, satisfies

$$r > \left(1 - \frac{\ln^2 n}{n}\right) n/2,$$

since when this is not the case Corollary 3.1 immediately gives (4.1). This implies that T is fairly regular, as follows. Let $r_i = (1 + \delta_i)\frac{n-1}{2}$. Then

$$\prod (1+\delta_i)^{1/n} = \frac{2r}{n-1} > 1 - \frac{\ln^2 n}{7n}.$$

On the other hand, letting $\delta = \frac{1}{n}\sum |\delta_i|$, and using $\sum \delta_i = 0$, the Taylor expansion of $\ln(1+x)$ and Cauchy–Schwarz, we have

$$\frac{1}{n}\sum \ln(1+\delta_i) \leq \frac{1}{n}\sum\left(\delta_i - \frac{\delta_i^2}{2} + \frac{\delta_i^3}{3}\right) \leq \frac{1}{n}\sum\left(-\frac{\delta_i^2}{6}\right) \leq -\frac{\delta^2}{6}.$$

So

$$1 - \frac{\ln^2 n}{7n} < e^{-\delta^2/6} < 1 - \frac{\delta^2}{7},$$

implying $\delta < n^{-1/2} \ln n$. So if we set $\zeta = n^{-1/4}\sqrt{\ln n}$ (say) and call a vertex x *good* if

$$|d(x) - (n-1)/2| < \zeta n, \qquad (4.4)$$

then

$$|\{x : x \text{ bad}\}| < \zeta n = o(n). \qquad (4.5)$$

As in (2.8), write $\mathscr{M}$ for the set of good cycles in T, and set $m = |\mathscr{M}|$. In view of Lemma 3.4 (= (2.7)) the proof of (4.1) will be complete if we can show (2.8) for T satisfying (4.5). So we consider a fixed $f \in \mathscr{M}$ and write $k(x,\sigma)$ for $k_f(x,\sigma)$ (see (2.4) for k_f). The main remaining point is as follows.

Lemma 4.2. *If x is a good vertex, k satisfies (3.1) and $0 \leq j \leq n/(2k)$, then*

$$\Pr_\sigma[b(x,\sigma) = f^{-j}(x) \mid k(x,\sigma) = k] > (1-o(1))\frac{2k}{n}\left(1 - \frac{2k}{n}\right)^j.$$

This easily gives (2.8): writing $\text{back}_j(x)$ for the indicator of $\{f^{-j}(x) \in \Gamma^+(x)\}$, Lemma 4.2 says that for good x,

$$p_k(x) > (1-o(1))\frac{2k}{n}\sum_{j=2}^{\frac{n}{2k}}\left(1 - \frac{2k}{n}\right)^j \text{back}_j(x), \qquad (4.6)$$

whence the sum in (2.8) (even restricted to the good vertices, of which there are more than $(1-\zeta)n$, is at least

$$(1-o(1))\frac{2}{n}\sum_{k}\frac{2k}{n}\sum_{j=2}^{\frac{n}{2k}}\left(1-\frac{2k}{n}\right)^{j}(\mathrm{back}_{j}(f)-\zeta)n\log\frac{k}{k-1}$$
$$> (1-o(1))2\gamma\sum_{k}\log\frac{k}{k-1}=(2\gamma-o(1))\log n,$$

where k ranges over $(\omega, n/(2\omega))$, and we used Lemma 4.1 for the inequality and $\omega < n^{o(1)}$ for the $\log n$. □

Proof of Lemma 4.2. The intuition behind the proof is as follows. Notice that '$b(x,\sigma) = f^{-j}(x)$' means that in the sequence

$$(f^{-i}(x) : i = 1, \ldots), \qquad (4.7)$$

$f^{-j}(x)$ is the first vertex y for which

$$\sigma(f^{-1}(y)) > \sigma(x). \qquad (4.8)$$

We are given k, the number of $y \in \Gamma^+(x)$ for which (4.8) holds, and will show – as one would expect given (4.4) – that the number of such y in $\Gamma^-(x)$ is typically also close to k. Thus (typically) *each* y (other than $f(x)$) satisfies (4.8) with probability approximately $2k/n$, so we may guess that the number of y from (4.7) that need to be examined before finding one satisfying (4.8) is approximately geometric with mean $n/(2k)$, which is what the lemma says. □

Some additional notation will be helpful. Let

$$S = f^{-1}(\Gamma^+(x)),$$
$$T = [n] \setminus S$$

(the f-preimages of x and its in-neighbours), $s = |S|$ ($= d(x)$) and $t = |T| = n - s$. Note that since x is good (see (4.4)), we have

$$\left|\frac{s}{t} - 1\right| < o(1) \qquad (4.9)$$

(with $o(1) \approx 4\zeta$). Let $A_\sigma = \{y : \sigma(y) \geqslant \sigma(x)\}$. Then

$$k(x,\sigma)(= |\Gamma^+(x) \setminus \{f(y) : \sigma(y) < \sigma(x)\}|) = |S \cap A_\sigma(x)|,$$

and we use ℓ for the analogous quantity for in-neighbours:

$$\ell(x,\sigma) = |T \cap A_\sigma(x)|.$$

Our first task is to say something about the conditional distribution of $\ell(x,\sigma)$ given $k(x,\sigma)$. This is in fact an instance of 'Pólya's urn' (see, e.g., [9]): we may think of starting with an urn containing $b = k$ blue balls and $r = s + 1 - k$ red balls, and repeatedly drawing balls (uniformly) at random, following the rule that after each draw we replace the chosen ball together with an additional ball of the same colour. Then the number of blue balls chosen in t draws is distributed as $\ell(x,\sigma)$ (given $k(x,\sigma) = k$), and we have (e.g., [9], p. 240, Prob. 30)

$$\mathbb{E}[\ell(x,\sigma) \mid k(x,\sigma) = k] = \frac{kt}{s+1} \sim k, \tag{4.10}$$

$$\mathrm{Var}[\ell(x,\sigma) \mid k(x,\sigma) = k] = \frac{tk(s-k+1)(s+t+1)}{(s+1)^2(s+2)} \sim 2k \tag{4.11}$$

(using (4.9) and (3.1)). This gives sufficient concentration for our purposes.

Lemma 4.3. *For any good x, k satisfying (3.1), and $\lambda > 0$,*

$$\mathrm{Pr}_\sigma\left[|\ell(x,\sigma) - \frac{kt}{s+1}| > \lambda \mid k(x,\sigma) = k\right] < (1 + o(1))\lambda^{-2} 2k,$$

and in particular

$$\mathrm{Pr}_\sigma[|\ell(x,\sigma) - k| < o(k) \mid k(x,\sigma) = k] > 1 - o(1).$$

Proof. This follows from (4.10) and (4.11) via Chebyshev's inequality. □

Write $Q(k,\ell)$ for the event $\{k(x,\sigma) = k \text{ and } \ell(x,\sigma) = \ell\}$. By Lemma 4.3, Lemma 4.2 will follow if we show

$$\mathrm{Pr}_\sigma[b(x,\sigma) = f^{-j}(x) \mid Q(k,\ell)] > (1 - o(1))\frac{2k}{n}\left(1 - \frac{2k}{n}\right)^j \tag{4.12}$$

whenever

$$|\ell - k| < o(k). \tag{4.13}$$

Let

$$I = \{f^{-i}(x) : 1 \leq i \leq j\},$$

$a = |S \cap I|$ and $b = |T \cap I| = j - a$. Let $w = f^{-j-1}(x)$. We will assume that $w \in S$ since this is the only case that contributes to the sum in (4.6). If we condition on $Q(k,\ell)$ then

$$K := S \cap A_\sigma(x) \setminus \{x\} \text{ and } L := T \cap A_\sigma(x)$$

are chosen uniformly (and independently) from $\binom{S \setminus \{x\}}{k-1}$ and $\binom{T}{\ell}$ respectively, and we have $b(x,\sigma) = f^{-j}(x)$ if and only if

$$w \in K \text{ and } K \cap I = \emptyset = L \cap I.$$

Thus, with $(m)_u := m(m-1)\cdots(m-u+1)$,

$$\mathrm{Pr}_\sigma[b(x,\sigma) = f^{-j}(x) \mid Q(k,\ell)] = \frac{\binom{s-a-2}{k-2}\binom{t-b}{\ell}}{\binom{s-1}{k-1}\binom{t}{\ell}} = \frac{k-1}{s-1}\frac{(s-a-2)_{k-2}}{(s-2)_{k-2}}\frac{(t-b)_\ell}{(t)_\ell},$$

which for ℓ satisfying (4.13) (and x,k,j as in the lemma) is easily seen to be at least the right-hand side of (4.12). □

5. Remarks

Recall we began with an $O(n^{3/2})$ gap between Szele's lower bound, $C(n) \geq (n-1)! 2^{-n} := R(n)$ (the expected number of Hamiltonian cycles in a random tournament) and Alon's upper bound, $O(\Psi(n)) = O(n^{3/2} R(n))$. Theorem 1.2 replaces $n^{3/2}$ by $n^{3/2-\xi}$. The value of ξ given here is certainly not optimal (even without going beyond the present ideas), but of course the real question at this point is: What is the true behaviour? We think it likely that $C(n)$ is no more than $O(n^{1/2} R(n))$, and would not rule out the perhaps surprising possibility that $C(n) = O(R(n))$, i.e., that *the number of cycles in a tournament cannot exceed the expected number of cycles in a random tournament by more than a multiplicative constant.* (These natural guesses have also been suggested by Noga Alon [3].) A heuristic reason for the first guess – i.e., that one can reduce the bound one gets from Brégman's theorem by at least a factor like n – is the feeling that the Hamiltonian cycles ought to account for no more than about a $1/n$ proportion of the 1-factors. (It seems reasonable to expect this for more general digraphs, but we have no concrete conjecture here.) A more 'practical' reason is that one can see where the proof should be strengthened to give essentially this gain; namely, it should be the case that for typical f one has $\text{back}_j(f) \approx 1/2$ for most j (more precisely, the number of f violating this should be small compared to $R(n)$), which would allow us to replace the ξ in (2.8) by something close to 1 and our upper bound by about $n^{-1} \Psi(n)$. (This would also require somewhat expanding the range of j in Lemma 4.2, but such a change causes no difficulty.) As to a further reduction, we do feel that $n^{-1} \Psi(n)$ is still too large. A vague reason for this is that the matrices associated with tournaments are (in various senses) quite unlike the block diagonal matrices which are the tight cases for Brégman's theorem. Thus (another conjecture) the number of 1-factors in a tournament should be significantly less than $\Psi(n)$, say at most $O(n^{-\delta} \Psi(n))$ for some positive constant δ, suggesting, according to the preceding paragraph, that $C(n) < O(n^{-1-\delta} \Psi(n))$. But we have no particular reason (other than the appeal of a clean answer) to expect $\delta = 1/2$.

A related, also seemingly quite interesting question is: What can one say about *lower* bounds on $C(T)$ for *regular* T? There is of course no such lower bound for general T; but as far as we know, it could even be that one has $C(T) > \Omega(R(n))$ for all regular T. For a start, it seems plausible that using the Van der Waerden Conjecture [8], one may show that $R(n)$ gives the right asymptotics for the logarithm: $C(T) > n^{(1-o(1))n}$ for regular T.

Acknowledgement

We would like to thank Noga Alon for helpful conversations, and the referee for a careful reading.

References

[1] Adler, I., Alon, N. and Ross, S. M. (2001) On the maximum number of Hamiltonian paths in tournaments. *Random Struct. Alg.* **18** 291–296.

[2] Alon, N. (1990) The maximum number of Hamiltonian paths in tournaments. *Combinatorica* **10** 319–324.

[3] Alon, N. Personal communication.

[4] Alon, N. and Spencer, J. (2000) *The Probabilistic Method*, second edition, Wiley-Interscience, New York.
[5] Brégman, L. (1973) Some properties of nonnegative matrices and their permanents. *Math. Doklady* **14** 945–949.
[6] Csiszár, I. and Körner, J. (1981) *Information Theory: Coding Theorems for Discrete Memoryless Systems*, Akadémiai Kiadó, Budapest.
[7] Egorychev, G. P. (1981) The solution of Van der Waerden's problem for permanents. *Adv. Math.* **42** 299–305.
[8] Falikman, D. I. (1981) Proof of the Van de Waerden's conjecture on the permanent of a doubly stochastic matrix. *Mat. Zametki* **29** 931–938, 957 (in Russian).
[9] Feller, W. (1968) *An Introduction to Probability Theory and its Applications*, Vol. 1, Wiley, New York.
[10] McEliece, R. J. (1977) *The Theory of Information and Coding*, Addison-Wesley, London.
[11] Radhakrishnan, J. (1997) An entropy proof of Bregman's theorem. *J. Combin. Theory Ser. A* **77** 161–164.
[12] Radhakrishnan, J. Entropy and counting. To appear in the IIT Kharagpur Golden Jubilee Volume.
[13] Schrijver, A. (1978) A short proof of Minc's conjecture, *J. Combin. Theory Ser. A* **25** 80–83.
[14] Szele, T. (1943) Kombinatorikai vizsgalatok az iranyitott teljes graffal. *Kapcsolatban, Mt. Fiz. Lapok* **50** 223–256.
[15] Wormald, N. Tournaments with many Hamiltonian cycles. Preprint.

The Game of JumbleG

ALAN FRIEZE,[1][†] MICHAEL KRIVELEVICH,[2][‡]
OLEG PIKHURKO[1] and TIBOR SZABÓ[3]

[1]Department of Mathematical Sciences, Carnegie Mellon University, Pittsburgh PA 15213, USA
(e-mail: alan@random.math.cmu.edu, pikhurko@andrew.cmu.edu)

[2]Department of Mathematics, Raymond and Beverly Sackler Faculty of Exact Sciences,
Tel Aviv University, Tel Aviv 69978, Israel
(e-mail: krivelev@post.tau.ac.il)

[3]Institut für Theoretische Informatik, ETH Zentrum, IFW B48.1, CH-8092 Zürich, Switzerland
(e-mail: szabo@inf.ethz.ch)

Received 31 October 2003; revised 19 November 2004

For Béla Bollobás on his 60th birthday

JumbleG is a Maker–Breaker game. Maker and Breaker take turns in choosing edges from the complete graph K_n. Maker's aim is to choose what we call an ϵ-regular graph (that is, the minimum degree is at least $(\frac{1}{2} - \epsilon)n$ and, for every pair of disjoint subsets $S, T \subset V$ of cardinalities at least ϵn, the number of edges $e(S, T)$ between S and T satisfies $\left|\frac{e(S,T)}{|S||T|} - \frac{1}{2}\right| \leqslant \epsilon$.) In this paper we show that Maker can create an ϵ-regular graph, for $\epsilon \geqslant 2(\log n/n)^{1/3}$. We also consider a similar game, JumbleG2, where Maker's aim is to create a graph with minimum degree at least $(\frac{1}{2} - \epsilon)n$ and maximum co-degree at most $(\frac{1}{4} + \epsilon)n$, and show that Maker has a winning strategy for $\epsilon > 3(\log n/n)^{1/2}$. Thus, in both games Maker can create a pseudo-random graph of density $\frac{1}{2}$. This guarantees Maker's win in several other positional games, also discussed here.

1. Introduction

JumbleG is a Maker–Breaker game. Maker and Breaker take turns in choosing edges from the complete graph K_n on n vertices. Maker's aim is to choose a graph which is ϵ-regular (the definition follows).

[†] Supported in part by NSF grant CCR-0200945.
[‡] Research supported in part by USA–Israel BSF grant 2002-133 and by grant 64/01 from the Israel Science Foundation.

Let $G = (V, E)$ be a graph of order n. We usually assume that the vertex set is $[n] = \{1, \ldots, n\}$. We call a pair S, T of non-empty disjoint subsets of $[n]$ ϵ-unbiased if

$$\left| \frac{e_G(S,T)}{|S||T|} - \frac{1}{2} \right| \leqslant \epsilon, \tag{1.1}$$

where $e_G(S, T)$ is the number of $S - T$ edges in G. The graph G is ϵ-regular if

P1: $\delta(G) \geqslant (\frac{1}{2} - \epsilon)n$,

P2: any pair S, T of disjoint subsets of $[n]$ with $|S|, |T| \geqslant \epsilon n$ is ϵ-unbiased.

Theorem 1.1. *Maker has a winning strategy in JumbleG provided $\epsilon \geqslant 2(\log n/n)^{1/3}$ and n is sufficiently large.*

We consider also a similar game, which we denote by JumbleG2. In this game Maker's aim is to create a graph with properties P1 and P3, where

P3: maximum co-degree is at most $(\frac{1}{4} + \epsilon)n$.

(The co-degree of vertices $u, v \in V(G)$ is the number of common neighbours of u and v in G.)

Here, too, Maker can win provided ϵ is not too small.

Theorem 1.2. *Maker has a winning strategy in JumbleG2 for all $\epsilon \geqslant 3(\log n/n)^{1/2}$ if n is sufficiently large.*

Theorems 1.1 and 1.2 are proved in Section 2. As shown in Section 3, our restrictions on ϵ are best possible, up to a logarithmic factor.

Although the goals of the above two games appear to be quite different, they are in fact very similar to each other: in both, Maker tries to create a *pseudo-random graph* of density around $\frac{1}{2}$. Informally speaking, a pseudo-random graph $G = (V, E)$ is a graph on n vertices whose edge distribution resembles that of a truly random graph $G(n, p)$ of the same edge density $p = e(G)\binom{n}{2}^{-1}$. The reader can consult [12] for a recent survey on pseudo-random graphs. The fact that an ϵ-regular graph is pseudo-random with density $\frac{1}{2}$ is apparent from the definition. To see that degrees and co-degrees can guarantee pseudo-randomness, we need to recall some notions and results due to Thomason. He introduced the notion of *jumbled* graphs [17]. A graph G with vertex set $[n]$ is (α, β)-jumbled if, for every $S \subseteq [n]$, we have

$$\left| e_G(S) - \alpha \binom{|S|}{2} \right| \leqslant \beta |S|$$

where $e_G(S)$ is the number of edges of G contained in S.

Thomason showed that one can check for pseudo-randomness via jumbledness by checking degrees and co-degrees. Suppose that $G = (V, E)$ has minimum degree at least αn and no two vertices have more than $\alpha^2 n + \mu$ common neighbours. Then (see Theorem 1.1 of [17] and its proof), for every $s \leqslant n$, every set $S \subseteq V$ of size $|S| = s$ satisfies

$$\left| e(S) - \alpha \binom{s}{2} \right| \leqslant \frac{((s-1)\mu + \alpha n)^{1/2} + \alpha}{2} s, \tag{1.2}$$

and therefore G is (α, β)-jumbled with $\beta = ((\alpha n + (n-1)\mu)^{1/2} + \alpha)/2$.

Now suppose that for some $\epsilon = \Omega(1/n)$ a graph G on n vertices has minimum degree at least $\alpha n = \left(\frac{1}{2} - \epsilon\right)n$ and maximum co-degree at most $\left(\frac{1}{4} + \epsilon\right)n = \alpha^2 n + (2\epsilon - \epsilon^2)n$. Then a routine calculation, based on (1.2), shows that G is ϵ'-regular for $\epsilon' = \Omega(\epsilon^{1/4})$. Thus Theorem 1.2 can be used to show that Maker can create an ϵ-regular graph with $\epsilon = n^{-1/8+o(1)}$, a weaker result than that provided by the direct application of Theorem 1.1. Indeed, let $|S| = s$, $|T| = t \geq \epsilon'n$, $\mu = (2\epsilon - \epsilon^2)n$, and $\epsilon' \geq \Omega(\epsilon^{1/4}) \geq \Omega(n^{-1/4})$. Then

$$\left|\frac{e_G(S,T)}{st} - \frac{1}{2}\right| = \frac{1}{st}|e_G(S \cup T) - e_G(S) - e_G(T) - (\alpha + \epsilon)st|$$

$$\leq \frac{1}{st}\left(\left|e_G(S \cup T) - \alpha\binom{s+t}{2}\right| + \left|e_G(S) - \alpha\binom{s}{2}\right| + \left|e_G(T) - \alpha\binom{t}{2}\right|\right) + \epsilon$$

$$\leq \frac{((s+t)\mu + \alpha n)^{1/2} + \alpha}{2st}(s+t) + \frac{(s\mu + \alpha n)^{1/2} + \alpha}{2t} + \frac{(t\mu + \alpha n)^{1/2} + \alpha}{2s} + \epsilon$$

$$\leq \frac{(s+t)^{3/2}\mu^{1/2}}{2st} + \frac{(s\mu)^{1/2}}{2t} + \frac{(t\mu)^{1/2}}{2s} + \frac{4((\alpha n)^{1/2} + \alpha)}{2\min\{s,t\}} + \epsilon$$

$$\leq \frac{((1+\epsilon')n)^{1/2}(2\epsilon n)^{1/2}}{\epsilon'n} + \frac{(2n\epsilon n)^{1/2}}{2\epsilon'n} + \frac{(2n\epsilon n)^{1/2}}{2\epsilon'n} + \frac{2n^{1/2}}{\epsilon'n} + \epsilon$$

$$\leq c\frac{\epsilon^{1/2}}{\epsilon'}$$

$$\leq \epsilon'.$$

Pseudo-random graphs are known to have many nice properties. Hence, Maker's ability to create a pseudo-random graph guarantees his win in several other positional games. For example, using a result of [11], one can guarantee Maker's success in creating $\frac{n}{4} - O(n^{5/6}\log^{1/6} n)$ pairwise edge-disjoint Hamiltonian cycles. This is trivially best possible up to the error-term and confirms a conjecture of Lu [13] in a strong form. We will discuss this and other games in Section 4.

2. Playing JumbleG

In this section we prove Theorems 1.1 and 1.2. The proofs are quite similar and are based on the approach of Erdős and Selfridge [9] via potential functions.

Lemma 2.1. *If the edges of a hypergraph $\mathscr{F}$ satisfy $\sum_{X \in \mathscr{F}} 2^{-|X|} < 1/4$ then Maker can force a 2-colouring of $\mathscr{F}$.*

Proof. Let a round consist of a move of Maker followed by a move of Breaker. At the start of a round, let C_M, C_B denote the set of edges chosen so far by Maker and Breaker, let R denote the unchosen edges and for $X \in \mathscr{F}$ let $\delta_{X,M}, \delta_{X,B}$ be the indicators of $X \cap C_M \neq \emptyset$, $X \cap C_B \neq \emptyset$ respectively. Let $\delta_X = \delta_{X,M} + \delta_{X,B}$. We use the potential function

$$\Phi = \sum_{\substack{X \in \mathscr{F} \\ \delta_X \leq 1}} 2^{-|X \cap R| + 1 - \delta_X}.$$

This represents the expected number of monochromatic sets if the unchosen edges are coloured at random. Our assumption is that $\Phi < \frac{1}{2}$ at the start and we will see that it can be kept this way until the end of the last complete round. If n is odd, Maker with his last choice can at most double the value of Φ. In any case, at the end of the play $\Phi < 1$. Also, at the end $R = \emptyset$; thus $\delta_X \geqslant 2$ for all $X \in \mathcal{F}$, showing that Maker has achieved his objective.

It remains to show that Maker can ensure that the value of Φ never increases after one complete round is played. Suppose that in some round Maker chooses an edge a and Breaker chooses an edge b. Let Φ' be the new value of Φ. Then

$$\Phi' - \Phi = - \sum_{\substack{a,b \in X \\ \delta_X = 0}} 2^{1-|X \cap R|} - \sum_{\substack{a \in X \\ \delta_{X,B} = 1}} 2^{-|X \cap R|} - \sum_{\substack{b \in X \\ \delta_{X,M} = 1}} 2^{-|X \cap R|}$$

$$+ \sum_{\substack{a \in X, b \notin X \\ \delta_{X,M} = 1}} 2^{-|X \cap R|} + \sum_{\substack{a \notin X, b \in X \\ \delta_{X,B} = 1}} 2^{-|X \cap R|}$$

$$\leqslant - \left(\sum_{\substack{a \in X \\ \delta_{X,B} = 1}} 2^{-|X \cap R|} - \sum_{\substack{a \in X \\ \delta_{X,M} = 1}} 2^{-|X \cap R|} \right) + \left(\sum_{\substack{b \in X \\ \delta_{X,B} = 1}} 2^{-|X \cap R|} - \sum_{\substack{b \in X \\ \delta_{X,M} = 1}} 2^{-|X \cap R|} \right),$$

which is nonpositive if Maker chooses a to maximize

$$\sum_{\substack{a \in X \\ \delta_{X,B} = 1}} 2^{-|X \cap R|} - \sum_{\substack{a \in X \\ \delta_{X,M} = 1}} 2^{-|X \cap R|}. \qquad \square$$

Lemma 2.2. *Let $\epsilon = \epsilon(n)$ tend to zero with n. Let $\delta > 1$ be fixed. Let $t = \lceil \delta \epsilon^{-2} \log n \rceil$. Then, for all sufficiently large n, Maker can ensure that any pair of disjoint subsets of V, both of size at least t, is ϵ-unbiased.*

Proof. Assume that $t \leqslant n/2$, for otherwise there is nothing to prove. This means that $\epsilon > \left(\frac{2 \log n}{n} \right)^{1/2}$.

Let $k = \lceil (\frac{1}{2} + \epsilon) t^2 \rceil$. Let $\mathcal{T}$ consist of pairs (S, T) of disjoint subsets of V, both of size exactly t. Recall that $e_M(S, T)$ counts the number of Maker's edges connecting S to T. A simple averaging argument shows that it is enough to show that Maker can guarantee that

$$t^2 - k < e_M(S, T) < k, \quad \text{for all } (S, T) \in \mathcal{T}. \tag{2.1}$$

(Indeed, let S', T' have size at least t each. The expectation of $\frac{e_M(S,T)}{t^2}$, where S, T are random t-subsets of S', T', is $\frac{e_M(S', T')}{|S'||T'|}$. By (2.1) this cannot differ from $\frac{1}{2}$ by more than ϵ, as required.)

If Maker is able to ensure that all k-element subsets of the edge-set $S : T = \{\{x, y\} \mid x \in S, y \in T\}$ are properly 2-coloured (*i.e.*, not monochromatic) for every $(S, T) \in \mathcal{T}$, then he has achieved his goal. A direct application of Lemma 2.1 is not possible, however: there are simply too many of these k-sets and the criterion does not hold. We need to cut down on the number of sets.

Define $\ell = \lceil 2t^2\epsilon \rceil$ and $\lambda = \lceil 2^\ell n^{-2t} \rceil$. For $(S, T) \in \mathcal{T}$ we prove the existence of a collection $\mathcal{X}_{S,T}$, of ℓ-subsets of the edge-set $S : T = \{\{x, y\} \mid x \in S, y \in T\}$ such that (i) $|\mathcal{X}_{S,T}| = \lambda$ and (ii) each k-set $B \subseteq S : T$ contains at least one member of $\mathcal{X}_{S,T}$. Let us show that if the elements of $\mathcal{X}_{S,T}$ are chosen at random, independently with replacement, then this property is almost surely satisfied. In estimating this probability we will use the following auxiliary inequalities: $\ell = o(t^2)$ and

$$\frac{\binom{k}{\ell}}{\binom{t^2}{\ell}} = \prod_{i=0}^{\ell-1} \frac{k-i}{t^2-i} = \left(\frac{k}{t^2}\right)^\ell \prod_{i=0}^{\ell-1}\left(1 - \frac{i(t^2-k)}{t^2k - ki}\right)$$

$$\geq \left(\frac{1}{2} + \epsilon\right)^\ell \exp\left\{-\frac{\ell^2}{2t^2} + O(\epsilon^2\ell)\right\}.$$

The probability that there is a k-subset of $S : T$ which does not contain a member of $\mathcal{X}_{S,T}$ is at most

$$\binom{t^2}{k}\left(1 - \frac{\binom{k}{\ell}}{\binom{t^2}{\ell}}\right)^\lambda \leq 2^{t^2} \exp\left\{-\lambda\left(\frac{1}{2}+\epsilon\right)^\ell e^{-\frac{\ell^2}{2t^2}+O(\epsilon^2\ell)}\right\}$$

$$= 2^{t^2} \exp\{-n^{-2t} e^{2\epsilon\ell - \frac{\ell^2}{2t^2}+O(\epsilon^2\ell)}\}$$

$$= 2^{t^2} \exp\{-e^{-2t\log n + (2+o(1))\epsilon^2 t^2}\} = o(1),$$

so a family $\mathcal{X}_{S,T}$ with the required property does exist.

Let $\mathcal{F} = (\binom{[n]}{2}, \mathcal{E})$ be the hypergraph with hyper-edges $\mathcal{E} = \bigcup_{(S,T)\in\mathcal{T}} \mathcal{X}_{S,T}$. (We will use the term *hyper-edges* to distinguish them from the edges of K_n.) To complete the proof it is enough to show that Maker can ensure that the choices $E_M, E_B \subset \binom{[n]}{2}$ of Maker, Breaker respectively are a 2-colouring of $\mathcal{F}$. This follows from Lemma 2.1 in view of the inequality

$$|\mathcal{E}| 2^{-\ell} \leq \binom{n}{t}^2 \lambda 2^{-\ell} = o(1). \tag{2.2}$$

$\square$

Proof of Theorem 1.1. To ensure that all degrees of Maker's graph are appropriate we use a trick similar to the one in the proof of the previous lemma. Let $k = \lceil(1/2 + \epsilon)n\rceil$. Maker again would like to use Lemma 2.1 and ensure that all k-subsets of the edges incident with vertex i are properly 2-coloured. These are again too many; we define $\ell = \lceil 10\epsilon^{-1}\log n\rceil$, $M = \lceil 2^\ell/n^2\rceil$, and $\mu = nM$. We want to find a collection $A_1, A_2, \ldots, A_\mu$ of ℓ-sets such that, for $1 \leq i \leq n$, every k-subset of the edges incident with i contains at least one of $A_{(i-1)M+j}$, $1 \leq j \leq M$. As before, we construct the sets A_i randomly. The probability that there is a bad k-subset (containing no chosen ℓ-set) is at most

$$\binom{n-1}{k}\left(1 - \frac{\binom{k}{\ell}}{\binom{n-1}{\ell}}\right)^M \leq 2^n \exp\left\{-\frac{2^\ell}{n^2}\frac{k^\ell}{n^\ell}e^{-\ell^2/n}\right\} \leq 2^n \exp\{-((1+\epsilon)e^{-\ell/n})^\ell\} < n^{-2}$$

for large n, and so the desired sets exist.

For property P2 let $t = \lceil 6\epsilon^{-2}\log n\rceil$. By our assumption on ϵ, we have $t < \epsilon n$. Define $\mathcal{X}_{S,T}$ as in the proof of Lemma 2.2. Namely, let $\ell' = \lceil 2t^2\epsilon \rceil$ and $\lambda = \lceil 2^{\ell'} n^{-2t}\rceil$. For

$(S, T) \in \mathcal{T}$ (that is, S, T are disjoint t-sets) let $\mathscr{X}_{S,T}$ be a collection ℓ-subsets of $S : T$ such that (i) $|\mathscr{X}_{S,T}| = \lambda$ and (ii) every $\lceil(\frac{1}{2}+\epsilon)t^2\rceil$-set contains at least one member of $\mathscr{X}_{S,T}$.

Let $\mathscr{F}$ be the hypergraph with the edge set $\mathscr{E}_1 \cup \mathscr{E}_2 = \{A_1, A_2, \ldots, A_\mu\} \cup \bigcup_{(S,T) \in \mathcal{T}} \mathscr{X}_{S,T}$.

Lemma 2.2 (or rather its proof) implies that it suffices for Maker to force a 2-colouring of $\mathscr{F}$. Indeed, the definition of the sets A_i will imply property P1. To see that P2 will also hold, observe that for any $S, T \in \mathcal{T}$, we will have

$$\left|\frac{e_M(S,T)}{t^2} - \frac{1}{2}\right| \leq \epsilon,$$

while the claim for general $|S|, |T| \geq t$ follows by averaging.

It remains to check that $\mathscr{F}$ satisfies the conditions of Lemma 2.1 for large n. The initial value Φ of the potential function satisfies

$$\Phi \leq Mn2^{-\ell} + \Phi(\mathscr{E}_2) = o(1). \tag{2.3}$$

(Here we have used (2.2).) This completes the proof of Theorem 1.1. □

Proof of Theorem 1.2. This time for property P1 we define $\ell = \lfloor \epsilon n \rfloor$; as before, $M = \lceil 2^\ell/n^2 \rceil$, $\mu = nM$, $k = \lceil(1/2+\epsilon)n\rceil$. The family $A_1, A_2, \ldots, A_\mu$ should satisfy: For $1 \leq i \leq n$, every k-subset of the edges incident with i contains at least one of $A_{(i-1)M+j}$, $1 \leq j \leq M$. We construct the A_i randomly. Suppose that we randomly choose M ℓ-subsets of $[n-1]$ independently with replacement. The probability that there is a k-subset of $[n-1]$ which contains no chosen ℓ-set is at most

$$\binom{n-1}{k}\left(1 - \frac{\binom{k}{\ell}}{\binom{n-1}{\ell}}\right)^M$$

$$\leq 2^n \exp\left\{-\frac{\binom{k}{\ell}M}{\binom{n}{\ell}}\right\}$$

$$= 2^n \exp\left\{-M\frac{k\cdots(k-\lfloor \ell/2\rfloor+1)}{n\cdots(n-\lfloor \ell/2\rfloor+1)} \cdot \frac{(k-\lfloor \ell/2\rfloor)\cdots(k-\ell+1)}{(n-\lfloor \ell/2\rfloor)\cdots(n-\ell+1)}\right\}$$

$$\leq 2^n \exp\left\{-M\left(\frac{k-\ell/2}{n}\right)^{\lfloor \ell/2\rfloor} \cdot \left(\frac{k-\ell}{n}\right)^{\lceil \ell/2\rceil}\right\}$$

$$\leq \exp\left\{n\log 2 - \frac{2^\ell}{n^2}\left(\frac{1}{2}+\frac{\epsilon}{2}\right)^{\lfloor \ell/2\rfloor}\left(\frac{1}{2}\right)^{\lceil \ell/2\rceil}\right\}$$

$$= \exp\left\{n\log 2 - \frac{(1+\epsilon)^{\lfloor \ell/2\rfloor}}{n^2}\right\}$$

$$< n^{-2}$$

for $\epsilon \geq 3(\log n/n)^{1/2}$, so the required family exists.

For property P3 we take a collection $B_1, B_2, \ldots, B_\rho$ of ℓ-sets where $\rho = \binom{n}{2}N$ and $N = \lceil 4^\ell/n^3 \rceil$. For each pair $i, j \in [n]$ select N random ℓ-subsets of $[n] \setminus \{i,j\}$ so that each $\lceil(1/4+\epsilon)n\rceil$-set contains at least one of them. The hyper-edges are $\{(i,x) : x \in A\} \cup \{(j,x) : x \in A\}$ for each random $A \subseteq [n] \setminus \{i,j\}$. $B_1, B_2, \ldots, B_\rho$ are chosen randomly and now with $k = \lceil(1/4+\epsilon)n\rceil$ the probability that there is a k-subset of $[n-2]$ which contains no chosen ℓ-set is at most

$$\binom{n-2}{k}\left(1-\frac{\binom{k}{\ell}}{\binom{n-2}{\ell}}\right)^N \leq \exp\left\{n\log 2 - \frac{(1+2\epsilon)^{\lfloor \ell/2 \rfloor}}{n^3}\right\} < n^{-3}$$

for large n, and so the sets exist.

We will use Lemma 2.1 and so we need to check that the initial potential is less than $1/4$. Now the initial value of the potential function is at most

$$Mn2^{1-\ell} + Nn^2 2^{1-2\ell} = o(1)$$

and this completes the proof of Theorem 1.2. $\square$

3. Breaker's strategies

In this section we show that up to a small power of $\log n$, our restrictions on ϵ are sharp in both Theorems 1.1 and 1.2 or, even more strongly, with respect to each of properties P1–P3.

Property P1

Theorem 1.2 gives immediately that Maker can guarantee a graph with minimum degree at least $n/2 - 3\sqrt{n \log n}$. A similar result has been previously obtained by Székely [16], by applying a lemma of Beck [2, Lemma 3], which in turn is based on the Erdős–Selfridge method. This comes quite close to a result of Beck [3] who proved that Breaker can force the minimum degree of Maker's graph to be $n/2 - \Omega(\sqrt{n})$.

Property P2

Let $c > 0$ be any constant which is less than $6^{-1/3}$, n be large, and $\epsilon = cn^{-1/3}\log^{1/3} n$.

Here we prove that *no* graph of order n can satisfy property P2 for this ϵ, which shows that the restriction on ϵ in Theorem 1.1 is sharp up to a multiplicative constant. The proof is based on ideas of Erdős and Spencer [10].

Let G be an arbitrary graph of order n. Let $m = \lceil \epsilon n \rceil$. Let X be a random m-subset of $V(G)$ chosen uniformly. For $y \in V(G)$, let $\mathcal{E}_y$ be the event that $y \notin X$ and $||\Gamma(y) \cap X| - m/2| > \epsilon m$, where $\Gamma(y)$ denoted the set of neighbours of y in G.

Let us show that for every y,

$$\mathbf{Pr}(\mathcal{E}_y) \geq \frac{2m}{n}. \tag{3.1}$$

Let $d = d(y)$ be the degree of y. By symmetry, we can assume that $d \leq \frac{n-1}{2}$. For such d we bound from below the probability p that $y \notin X$ and $|\Gamma(y) \cap X| \leq m/2 - \epsilon m$, which equals

$$p = \sum_{i < m/2 - \epsilon m} \binom{d}{i}\binom{n-1-d}{m-i}\binom{n}{m}^{-1}.$$

The combinatorial meaning of p implies that it decreases with d, so it is enough to bound p for $d = \lfloor \frac{n-1}{2} \rfloor$ only. Let us consider the summands s_h corresponding to $i = m/2 - h$, with,

say, $\epsilon m < h \leq \epsilon m + n^{1/3}$. Let
$$f(x) = (1+x)^{\frac{1+x}{2}}(1-x)^{\frac{1-x}{2}}.$$

Its Taylor series at 0 is $1 + \frac{x^2}{2} + O(x^4)$. By Stirling's formula, we obtain that each summand

$$s_h = \Omega\left(\frac{n^{-1/3}(\log n)^{1/6}}{f^m(\frac{2h}{m}) f^{2d-m}(\frac{2h}{2d-m})}\right)$$
$$= \exp\left(-\frac{1}{3}\log n - \frac{2h^2}{m} - \frac{2h^2}{2d-m} + O(\log \log n)\right)$$
$$= n^{-1/3 - 2c^3 - o(1)}.$$

Thus
$$\sum_{h=\epsilon m}^{\epsilon m + n^{1/3}} s_h = n^{-2c^3 - o(1)} \geq \frac{2m}{n}.$$

It follows that there is a choice of an m-set X such that $|Y| \geq 2m$, where Y consists of the vertices for which R_x holds. By definition $Y \cap X = \emptyset$.

Assume without loss of generality that we have $d_X(y) < m - \epsilon m$ for at least half of the vertices of Y. Let $Z \subset Y$ consist of any m of these vertices. This pair (X, Z), both sets having at least ϵn elements, has the required bias.

Property P3

Here we show that Breaker can force Maker to create a co-degree of at least $\frac{n}{4} + c\sqrt{n}$. Our argument is based on a theorem of Beck [5], which states that Breaker can force Maker's graph to have maximum degree at least $n/2 + \sqrt{n}/20$. Then the following lemma shows that Breaker also succeeds in forcing a high co-degree in Maker's graph.

Lemma 3.1. *Assume that $c_1 > 0$ is constant. Then, for sufficiently large n, the following holds. Let $G = (V, E)$ be a graph on n vertices with $n(n-1)/4$ edges. If G has a vertex of degree at least $n/2 + c_1\sqrt{n}$, then G has a pair of vertices w_1, w_2 whose co-degree is at least $n/4 + c_1\sqrt{n}/10$.*

Proof. Let $c_2 = c_1/10$. Let v be a vertex of maximum degree in G. Denote $N_1 = N(v)$, $N_2 = V - N_1$. Then $|N_2| \leq n/2 - c_1\sqrt{n}$. If there is $u \in V$ such that $d(v, N_1) \geq n/4 + c_2\sqrt{n}$, we are done. Otherwise, for every u, $d(u, N_1) \leq n/4 + c_2\sqrt{n}$, implying:

$$A \stackrel{\text{def}}{=} \sum_{u \in V} d(u, N_2)$$
$$\geq \sum_{u \in V}(d(u) - d(u, N_1) - 1)$$
$$\geq 2|E| - n(n/4 + c_2\sqrt{n}) - n$$
$$= n^2/4 - c_2 n^{3/2} - 3n/2.$$

Therefore by convexity,

$$B \stackrel{\text{def}}{=} \sum_{u \in V} \binom{d(u, N_2)}{2} \geq n \binom{A/n}{2} \geq n^3/32 - c_2 n^{5/2} - O(n^2).$$

On the other hand,

$$B = \sum_{w_1 \neq w_2 \in N_2} \text{co-degree}(w_1, w_2),$$

and thus there is a pair $w_1, w_2 \in N_2$ such that:

$$\text{co-degree}(w_1, w_2) \geq |B| / \binom{|N_2|}{2}$$
$$\geq \frac{n^3/32 - c_2 n^{5/2} - O(n^2)}{\binom{n/2 - c_1 n^{1/2}}{2}}$$
$$\geq n/4 + c_2 \sqrt{n}. \qquad \square$$

4. Consequences

As we have already mentioned in the Introduction, Maker's ability to create a pseudo-random graph of density about $\frac{1}{2}$ allows him to win quite a few other combinatorial games. We will describe some of them below. All these games are played on the complete graph K_n unless stated otherwise; Maker and Breaker choose one edge alternately, Maker's aim being to create a graph that possesses a desired graph property.

Edge-disjoint Hamilton cycles. In this game Maker's aim is to create as many pairwise edge-disjoint Hamilton cycles as possible. Lu proved [13] that Maker can always produce at least $\frac{1}{16}n$ Hamilton cycles and conjectured that Maker should be able to make $(\frac{1}{4} - \epsilon)n$ for any fixed $\epsilon > 0$. This conjecture follows immediately from our Theorem 1.1 and Theorem 2 of [11]. In [11], Frieze and Krivelevich show that a 2ϵ-regular graph contains at least $(\frac{1}{2} - 6.5\epsilon)n$ edge-disjoint Hamilton cycles, for all $\epsilon > 10(\log n/n)^{1/6}$. Our argument applies equally to the bipartite version of the problem where the game is played on the complete bipartite graph $K_{n,n}$. Thus Maker can always produce at least $(\frac{1}{4} - \epsilon)n$ edge-disjoint Hamilton cycles, verifying another conjecture of Lu [14, 15]. Finally, there is an analogous game that can be played on the complete digraph D_n and here Maker can always produce at least $(\frac{1}{2} - \epsilon)n$ edge-disjoint Hamilton cycles.

Vertex-connectivity. Theorem 1.2 can be used to show that Maker can always force an $(n/2 - 3\sqrt{n \log n})$-vertex-connected graph. Indeed, let Maker's graph M have minimum degree at least $n/2 - 3\sqrt{n \log n}$ and maximum co-degree at most $n/4 + 3\sqrt{n \log n}$. Suppose that the removal of some set R disconnects M, say $V(M) \setminus R = A \cup B$ with $|A| \leq |B|$. If $|A| = 1$, then obviously all neighbours of $a \in A$ are in R, implying $|R| \geq \delta(M) \geq n/2 - 3\sqrt{n \log n}$. If $|A| \geq 2$, let a_1, a_2 be two distinct vertices in A. Then all neighbours of a_1, a_2 lie in $A \cup R$, and therefore

$$|A| + |R| \geq \deg_M(a_1) + \deg_M(a_2) - \text{co-deg}_M(a_1, a_2) \geq \frac{3n}{4} - 9\sqrt{n \log n}.$$

If $|A| \geq n/4 - 6\sqrt{n\log n}$, then $|B| \geq |A| \geq n/4 - 6\sqrt{n\log n}$ as well, and by the $o(1)$-regularity of M there is an edge between A and B, a contradiction. We conclude that $|A| \leq n/4 - 6\sqrt{n\log n}$, implying $|R| \geq n/2 - 3\sqrt{n\log n}$, as required.

The result of Beck [3] showing that Breaker can force a vertex which has degree at most $n/2 - \Omega(\sqrt{n})$ in Maker's graph indicates that the error term in our result about the connectivity game is tight up to a logarithmic factor.

$c \log n$-**universality.** A graph G is called r-*universal* if it contains an induced copy of every graph H on r vertices. We can show the following result.

Theorem 4.1. *Let $r = r(n)$ be an integer, which satisfies*

$$\frac{n-r+1}{r}\left(\frac{1}{2} - \epsilon\right)^{r-1} \geq \frac{2\log n}{\epsilon^2},$$

for some $\epsilon = \epsilon(n) \to 0$. Then for all sufficiently large n Maker can ensure that his graph M is r-universal.

Proof. Let $t = \lfloor \frac{2\log n}{\epsilon^2} \rfloor$. Let n be sufficiently large so that the conclusion of Lemma 2.2 is valid. Let M be an arbitrary graph satisfying this property, that is, any pair of disjoint subsets of $V(M)$, both of size at least t, is ϵ-unbiased. Let G be any graph on $[r]$. We will show that G is an induced subgraph of M.

Partition $V(M) = \cup_{i=1}^{r} V_i$ into r parts, each having at least $\frac{n-r+1}{r}$ vertices. Initially, let $A_i = V_i$, $i \in [r]$. We define $f : [r] \to V(M)$ with $f(i) \in A_i$ inductively.

Suppose we have already defined f on $[i-1]$. It will be the case that $|A_j| \geq \frac{n-r+1}{r}\eta^{i-1}$ for any $j \geq i$, where for brevity $\eta = \frac{1}{2} - \epsilon$. We will choose $f(i) = v \in A_i$ so that for any $j > i$ we have

$$|A_{ji}(v)| \geq \eta |A_j|, \tag{4.1}$$

where we define $A_{ji}(v) = A_j \cap \Gamma_M(v)$ if $\{i,j\} \in E(G)$ and $A_{ji}(v) = A_j \setminus \Gamma_M(v)$ otherwise. (Here $\Gamma_M(v)$ is the set of neighbours of v in M.)

Let B_{ji} be the set of vertices of A_i violating (4.1), i.e., $\{v \in A_i : |A_{ji}(v)| < \eta|A_j|\}$. Then $|B_{ji}| < t$ as the pair (B_{ji}, A_j) is not ϵ-unbiased. (Observe that $|A_j| \geq \frac{n-r+1}{r}\eta^{r-1} \geq t$.) Update A_i by deleting B_{ji} for all $j \in [i+1, r]$. Thus at least $\frac{n-r+1}{r}\eta^{i-1} - (r-i)t \geq t$ vertices still remain in A_i. This inequality is true for $i = r$ by our assumption and for any other i, because $\eta \leq \frac{1}{2}$. So a suitable $f(i)$ can always be found. Now, replace A_j with $A_{ji}(f(i))$ for $j > i$. This completes the induction step. At the end of the process $f([r])$ induces a copy of G in M. □

It follows from Theorem 4.1 that Maker can create an r-universal graph with $r = (1 + o(1))\log_2 n$. On the other hand, Maker cannot achieve $r = 2\log_2 n - 2\log_2 \log_2 n + C$ because, as was shown by Beck [4, Theorem 4], Breaker can prevent K_r in Maker's graph.

There is a remarkable parallel between random graphs and Maker–Breaker games: see e.g., Chvátal and Erdős [8], Beck [3, 4] and Bednarska and Łuczak [6]. As shown by Bollobás and Thomason [7], the largest r such that a random graph of order n is almost surely r-universal is around $2\log_2 n$. We conjecture that games have the same universality threshold (asymptotically).

Conjecture 4.2. *Maker can claim an r-universal graph with* $r = (2 + o(1)) \log_2 n$.

The following related result improves the unbiased case of Theorem 4 in Beck [3]. (His assumption $n \geqslant 100r^3 v 3^{r+1}$ is stronger than ours.)

Theorem 4.3. *Let integers r, v and a real $\epsilon > 0$ (all may depend on n) satisfy $\epsilon \to 0$ and*

$$\frac{n-r+1}{r}\left(\frac{1}{2} - \epsilon\right)^{r-1} \geqslant v + \frac{2\log n}{\epsilon^2}.$$

Then for sufficiently large n, Maker can ensure that any graph G of order at most v and maximum degree less than r is a subgraph (not necessarily induced) *of Maker's graph M.*

Outline of proof. Use the method of Theorem 4.1 with the following changes. Take a proper colouring $c : V(G) \to [r]$. The desired f will map $i \in V(G)$ into $A_{c(i)}$. The proof goes the same way except that when choosing $f(i)$ we have to worry only about those $j \geqslant i$ which are neighbours of i in G and make sure that there are at least v good choices for $f(i) \in A_{c(i)}$ (so that we can ensure that f is injective). The details are left to the reader. □

Acknowledgement

The authors wish to thank the anonymous referee for his/her helpful criticism.

References

[1] Alon, N. and Spencer, J. (2000) *The Probabilistic Method*, second edition, Wiley.
[2] Beck, J. (1981) Van der Waerden and Ramsey type games. *Combinatorica* **1** 103–116.
[3] Beck, J. (1994) Deterministic graph games and a probabilistic intuition. *Combin. Probab. Comput.* **3** 13–26.
[4] Beck, J. (2002) Positional games and the second moment method. *Combinatorica* **22** 169–216.
[5] Beck, J. (2004) Arithmetic progressions and tic-tac-toe like games. Manuscript.
[6] Bednarska, M. and Łuczak, T. (2000) Biased positional games for which random strategies are nearly optimal. *Combinatorica* **20** 477–488.
[7] Bollobás, B. and Thomason, A. (1981) Graphs which contain all small graphs. *European J. Combin.* **2** 13–15.
[8] Chvátal, V. and Erdős, P. (1978) Biased positional games. *Ann. Discrete Math.* **2** 221–229.
[9] Erdős, P. and Selfridge, J. L. (1973) On a combinatorial game. *J. Combin. Theory Ser. A* **14** 298–301.
[10] Erdős, P. and Spencer, J. H. (1971/2) Imbalances in k-colorations. *Networks* **1** 379–385.
[11] Frieze, A. M. and Krivelevich, M. (2005) On packing Hamilton cycles in ϵ-regular graphs. *J. Combin. Theory Ser. B* **94** 159–172.
[12] Krivelevich, M. and Sudakov, B. Pseudo-random graphs. In *Proc. Conference on Finite and Infinite Sets*, Bolyai Society Mathematical Studies, in press.
[13] Lu, X. (1992) Hamiltonian games. *J. Combin. Theory Ser. B* **55** 18–32.
[14] Lu, X. (1995) A Hamiltonian game on $K_{n,n}$. *Discrete Math.* **142** 185–191.
[15] Lu, X. (1995) Hamiltonian cycles in bipartite graphs. *Combinatorica* **15** 247–254.
[16] Székely, L. A. (1984) On two concepts of discrepancy in a class of combinatorial games, finite and infinite sets. Vol. 37 of *Colloq. Math. Soc. János. Bolyai*, North-Holland, pp. 679–683.
[17] Thomason, A. G. (1987) Pseudo-random graphs. *Ann. Discrete Math.* **33** 307–331.

2-Bases of Quadruples

ZOLTÁN FÜREDI[1†] and GYULA O. H. KATONA[2‡]

[1]Rényi Institute of Mathematics of the Hungarian Academy of Sciences,
Budapest, PO Box 127, Hungary-1364
and
Department of Mathematics, University of Illinois at Urbana-Champaign,
Urbana, IL61801, USA
(e-mail: furedi@renyi.hu, z-furedi@math.uiuc.edu)

[2]Rényi Institute of Mathematics of the Hungarian Academy of Sciences,
Budapest, PO Box 127, Hungary-1364
(e-mail: ohkatona@renyi.hu)

Received 18 October 2004; revised 6 July 2005

For Béla Bollobás on his 60th birthday

Let $\mathcal{B}(n, \leqslant 4)$ denote the subsets of $[n] := \{1, 2, \ldots, n\}$ of at most 4 elements. Suppose that $\mathcal{F}$ is a set system with the property that every member of $\mathcal{B}$ can be written as a union of (at most) two members of $\mathcal{F}$. (Such an $\mathcal{F}$ is called a 2-base of $\mathcal{B}$.) Here we answer a question of Erdős proving that

$$|\mathcal{F}| \geqslant 1 + n + \binom{n}{2} - \left\lfloor \frac{4}{3}n \right\rfloor,$$

and this bound is best possible for $n \geqslant 8$.

1. 2-bases

The n-element set $\{1, 2, \ldots, n\}$ is denoted by $[n]$. The family of all subsets of $[n]$ is called the Boolean lattice and is denoted by $\mathcal{B}(n)$. Its kth level is $\mathcal{B}(n, k) := \{B : B \subset [n] : |B| = k\}$, and $\mathcal{B}(n, \leqslant k) := \cup_{0 \leqslant i \leqslant k} \mathcal{B}(n, i)$. The set system $\mathcal{F}$ is called a 2-base of $\mathcal{A}$ if every member $A \in \mathcal{A}$ can be obtained as a union of two members of $\mathcal{F}$, in other words $A = F_1 \cup F_2$, $F_1, F_2 \in \mathcal{F}$. Note that we allow $F_1 = F_2$ and we do not insist that the 2-base is a subset of the set system.

[†] Research supported in part by Hungarian National Science Foundation grant OTKA T 032452, T 037846 and by National Science Foundation grant DMS 0140692.
[‡] Research supported by Hungarian National Science Foundation grants OTKA T 037846, T 038210, T 034702.

The interest is in how small a base one can find. Let $f(\mathcal{A}) := \min\{|\mathcal{F}| : \mathcal{F}$ is a 2-base of $\mathcal{A}\}$. This is known exactly in very few cases, even when the set system is a natural one. For example, it is not known even for the power-set itself (the discrete cube). In 1993 Erdős [2] proposed the problem of determining $f(\mathcal{B}(n))$ and also the problem of determining the minimum size of a 2-base of the small sets, $f(\mathcal{B}(n, \leqslant k))$. We also use $f_k(n)$ for $f(\mathcal{B}(n, \leqslant k))$. Erdős conjectured that

$$f(\mathcal{B}(n)) = 2^{\lfloor n/2 \rfloor} + 2^{\lceil n/2 \rceil} - 1,$$

and that the extremal family consists of all subsets of V_1 and V_2 where $V_1 \cup V_2 = [n]$ is a partition of $[n]$ into two almost equal parts. A lower bound $f(\mathcal{B}(n)) \geqslant (1 + o(1))2^{(n+1)/2}$ is obvious from the fact that

$$|\mathcal{A}| \leqslant \binom{|\mathcal{F}|}{2} + |\mathcal{F}|,$$

which holds for any 2-base $\mathcal{F}$ of $\mathcal{A}$.

The aim in this paper is to answer this question for the family $\mathcal{B}(n, \leqslant 4)$. The question of the smallest base for $\mathcal{B}(n, \leqslant k)$ is trivial for $k \leqslant 2$, and for $k = 3$ it turns out to be a question about graphs whose answer follows immediately from Turán's theorem. So the case $k = 4$ is the first nontrivial case. It boils down to an interesting question about 3-graphs (3-regular hypergraphs), and it might be somewhat surprising that it is possible to give an exact answer.

Let $f_4(n) := 1 + n + \binom{n}{2} - h(n)$. The main result of this paper can be summarized in the following table:

n	0	1	2	3	4	5	6	7	$n \geqslant 8$
$h(n)$	0	0	1	2	4	5	7	8	$\lfloor \frac{4}{3} n \rfloor$

Theorem 1.1. *For $n \geqslant 8$, $f_4(n) = 1 + n + \binom{n}{2} - \lfloor \frac{4}{3} n \rfloor$.*

Let $g_k(n) := f(\mathcal{B}(n, 4))$, the size of a minimum 2-base for the k-tuples. We will deduce from Theorem 1.1 that $g_4(n) + n + 1 = f_4(n)$ for $n \geqslant 5$.

Theorem 1.2. *We have $g_4(5) = 4$, $g_4(6) = 8$, $g_4(7) = 13$ and for $n \geqslant 8$, $g_4(n) = \binom{n}{2} - \lfloor \frac{4}{3} n \rfloor$.*

In the following section we discuss $f_k(n)$ in the (easy) case $k \leqslant 3$. Then give constructions for $f_4(n)$ separating the cases $n \leqslant 7$ and $n \geqslant 8$ and thus providing lower bounds for $h(n)$. In Section 2 the structure of minimal bases of $\mathcal{B}(n, \leqslant 4)$ is investigated, namely those with minimum deficiency with at least 2, and then (the upper bounds for) the values of $h(n)$ in the above table is proved in Section 3. In Section 4 the uniform case (the case of g_4) is considered, and in Section 5 we close with a few remarks on the case $k > 4$.

1.1. The case $\mathcal{B}(n, \leqslant 3)$

For $k \geqslant 1$ every 2-base of $\mathcal{B}(n, \leqslant k)$ must contain the $\emptyset$ and all singletons. This easily leads to

$$f_0(n) = 1, \quad f_1(n) = 1 + n, \quad f_2(n) = 1 + n.$$

Suppose that $\mathscr{F}$ is a 2-base of $\mathscr{B}(n, \leq k)$, $1 < k \leq n$, such that $|\mathscr{F}| = f_k(n)$ and $\sum_{F \in \mathscr{F}} |F|$ is minimal. Such bases are called *minimal*. Then

(i) $\emptyset \in \mathscr{F}$, $\mathscr{B}(n, 1) \subset \mathscr{F}$,
(ii) for every $F \in \mathscr{F}$ we have $|F| \leq k - 1$.

Indeed, one need only observe that for $F \in \mathscr{F}$, $|F| = k$, $x \in F$ one can replace F by $F' := F \setminus \{x\}$, i.e., $\mathscr{F} \setminus \{F\} \cup \{F'\}$ is also a 2-base.

Construction 1.3. *Consider a 2-partition $V_1 \cup V_2$ of $[n]$ with $\lfloor n/2 \rfloor \leq |V_1| \leq |V_2| \leq \lceil n/2 \rceil$ and let $\mathscr{F}$ be all the subsets of V_1 and V_2 of size at most 2. Every triple from $[n]$ meets a V_i in at least 2 elements so it also contains a 2-element member of $\mathscr{F}$. Hence $\mathscr{F}$ is a 2-base of $\mathscr{B}(n, \leq 3)$.*

Claim 1.4. $f_3(n) = 1 + n + \binom{\lfloor n/2 \rfloor}{2} + \binom{\lceil n/2 \rceil}{2}$.

Proof of Claim 1.4. Suppose that $\mathscr{F}$ is a minimal 2-base of $\mathscr{B}(n, \leq 3)$ satisfying (i) and (ii). Split it into subfamilies according to the sizes of its members, $\mathscr{F} = \mathscr{F}_0 \cup \mathscr{F}_1 \cup \mathscr{F}_2$ where $\mathscr{F}_i := \mathscr{F} \cap \mathscr{B}(n, i)$. Then $\mathscr{F}_2$ is a graph (i.e., a 2-graph) with the property that every triple contains an edge, so its complement $\mathscr{H}_2$ is triangle-free ($\mathscr{H}_2 := \mathscr{B}(n, 2) \setminus \mathscr{F}_2$). Then Turán's theorem [7] implies that $|\mathscr{H}_2| \leq \lfloor n^2/4 \rfloor$, hence

$$|\mathscr{F}| = |\mathscr{F}_0| + |\mathscr{F}_1| + |\mathscr{F}_2| \geq 1 + n + \binom{n}{2} - \left\lfloor \frac{n^2}{4} \right\rfloor.$$ □

1.2. Constructions for $\mathscr{B}(n, \leq 4)$ if $n \leq 7$

Let $\mathscr{F}$ be a minimal 2-base of $\mathscr{B}(n, \leq 4)$ satisfying (i) and (ii). Let $\mathscr{F}_i := \mathscr{F} \cap \mathscr{B}(n, i)$; then $\mathscr{F} = \mathscr{F}_0 \cup \mathscr{F}_1 \cup \mathscr{F}_2 \cup \mathscr{F}_3$ where $\mathscr{F}_0 = \{\emptyset\}$, $\mathscr{F}_1 = \mathscr{B}(n, 1)$. Use the notation $\mathscr{H}_2 := \mathscr{B}(n, 2) \setminus \mathscr{F}_2$. Then

$$|\mathscr{F}| = 1 + n + \binom{n}{2} - |\mathscr{H}_2| + |\mathscr{F}_3| := 1 + n + \binom{n}{2} - h(n).$$

Since $\mathscr{B}(n, \leq 2)$ is a 2-base of $\mathscr{B}(n, \leq 4)$ we have $h(n) \geq 0$.

Let us summarize the properties of $\mathscr{F}_2 \cup \mathscr{F}_3$:

for every triple $T \subset [n]$ either T contains a pair from $\mathscr{F}_2$ (1.1)
                               or $T \in \mathscr{F}_3$,   (1.2)
for every quadruple $Q \subset [n]$ either Q contains a triple from $\mathscr{F}_3$ (1.3)
                               or Q is a union of two edges from $\mathscr{F}_2$.   (1.4)

Construction 1.5. *For $n \geq 4$ let $\mathscr{H}_2$ be a Hamilton cycle, $|\mathscr{F}_3| = 0$.*

It is easy to show that this family $\mathscr{F}_2$ satisfies (1.1) and (1.4) so (together with $\mathscr{B}(n, \leq 1)$) it is a 2-base. This construction shows that $h(n) \geq n$ (for ≥ 4), and one can see that this is the best possible for $n = 4$ and $n = 5$.

Claim 1.6. $h(0) = h(1) = 0$, $h(2) = 1$, $h(3) = 2$, $h(4) = 4$ *and* $h(5) = 5$.

The proof of this (and the following two claims concerning $n = 6$ and 7) is a short, finite process. For completeness we sketch them in Section 3.

Construction 1.7. *For $n = 6$ let $\mathscr{F}_3$ be two disjoint triples F_1, F_2 and let $\mathscr{F}_2$ be the six pairs contained in either F_1 or F_2.*

Another construction of the same size can be obtained by considering a Hamilton cycle $\mathscr{F}_2 := \{12, 23, 34, 45, 56, 16\}$ with two triples $\mathscr{F}_3 := \{135, 246\}$.

Claim 1.8. $h(6) = 7$.

Construction 1.9. *For $n = 7$ label the seven elements by two coordinates, $V := \{v(1,1), v(1,2), v(1,3), v(2,1), v(2,2), v(3,1)\}$. Let $\mathscr{F}_2$ be the ten pairs $v(\alpha, \beta)v(\alpha', \beta')$ with $\alpha \neq \alpha'$ and $\beta \neq \beta'$, and let $\mathscr{F}_3$ be formed by the three triples having a constant coordinate, i.e., $\{v(1,1), v(1,2), v(1,3)\}$, $\{v(2,1), v(2,2), v(2,3)\}$ and $\{v(1,1), v(2,1), v(3,1)\}$. (This is a truncated version of Construction 1.13 for $n = 9$.)*

Claim 1.10. $h(7) = 8$.

Construction 1.11. *Let n_1, n_2 be nonnegative integers, $V^1 \cup V^2$ a partition of $[n]$ with $|V^i| = n_i$, $\mathscr{F}^i$ a minimal 2-base on V_i. Define $\mathscr{F}$ as $\mathscr{F}^1 \cup \mathscr{F}^2$ together with all pairs joining V^1 and V^2.*

It is easy to see that this construction satisfies (1.1)–(1.4): it is a 2-base. Indeed, it is sufficient to check a triple T and a quadruple Q meeting both V_1 and V_2. Then T contains a pair joining V^1 and V^2; thus it satisfies (1.1). If $|Q \cap V^1| = |Q \cap V^2| = 2$, then it is a union of two crossing pairs. Finally, if $Q = \{a, b, c, d\}$ and $Q \cap V^1 = \{a, b, c\}$, then since $\mathscr{F}^1$ is a 2-base, $Q \cap V^1$ satisfies either (1.1) or (1.2). In the first case $Q \cap V^1$ it contains a pair, say ab from $\mathscr{F}^1$; then $\{a,b\} \cup \{c,d\}$ is a partition of Q satisfying (1.4). In the second case $Q \cap V^1 \in \mathscr{F}^1$, so Q satisfies (1.3). We obtained the following.

Claim 1.12. *For n_1, n_2 nonnegative integers $h(n_1 + n_2) \geq h(n_1) + h(n_2)$.* □

1.3. Constructions for $n \geq 8$

Construction 1.13. *Suppose that $\mathscr{F}_3$ is a triple system on $[n]$ of girth at least 4, i.e., $|F' \cap F''| \leq 1$ for $F', F'' \in \mathscr{F}_3$, $F_1, F_2, F_3 \in \mathscr{F}_3$ and $F_1 \cap F_2 \neq \emptyset$, $F_1 \cap F_3 \neq \emptyset$, $F_2 \cap F_3 \neq \emptyset$ imply $F_1 \cap F_2 \cap F_3 \neq \emptyset$. Suppose further that every degree of $\mathscr{F}_3$ is at most two, i.e., every singleton is contained in at most two triples. Define $\mathscr{H}_2$ as the pairs covered by the members of $\mathscr{F}_3$.*

This construction (together with $\mathscr{B}(n, \leq 1)$) forms a 2-base. Indeed, if a triple $T \subset [n]$ contains no edge from $\mathscr{F}_2$, then it belongs to $\mathscr{F}_3$, so either (1.1) or (1.2) holds. Moreover, if $Q = \{a, b, c, d\} \subset [n]$ is a quadruple and contains no triple from $\mathscr{F}_3$, then the induced graph $\mathscr{H}_2|Q$ contains no triangle. So $\mathscr{F}_2|Q$ contains two disjoint edges (and thus fulfils (1.4)) unless $\mathscr{H}_2|Q$ has a vertex of degree 3, say, $ab, ac, ad \in \mathscr{H}_2$. Since the degree of $\mathscr{F}_3$ at the vertex a is at most two and the edges of $\mathscr{H}_2$ are obtained from the triples of $\mathscr{F}_3$ we get that there exists a triple $T \in \mathcal{F}_3$ with $a \in T \subset Q$. We have proved that Construction 1.13 indeed defines a 2-base.

For $n = 3k$, $k \geq 3$ we obtain $h(3k) \geq 4k$ as follows. Let $[n] = \{a_1, a_2, \ldots, a_k\} \cup \{b_1, b_2, \ldots, b_k\} \cup \{c_1, c_2, \ldots, c_k\}$. Define $\mathcal{F}_3$ as all triples of the form $a_i b_i c_i$ and $a_i b_{i+1} c_{i+2}$ (indices are taken modulo k). This satisfies the constraint of Construction 1.13. Since $|\mathcal{H}_2| = 3|\mathcal{F}_3|$, we get $h(n) \geq 2|\mathcal{F}_3| = 4k$.

If we leave out from the above construction the 2 triples of $\mathcal{F}_3$ and the 4 pairs of $\mathcal{H}_2$ containing the element $3k$ we obtain that $h(3k - 1) \geq 4k - 2$. Thus we already have the cases $n = 3k$ and $n = 3k - 1$ in the following claim.

Claim 1.14. $h(n) \geq \lfloor \frac{4}{3}n \rfloor$ for $n \geq 8$.

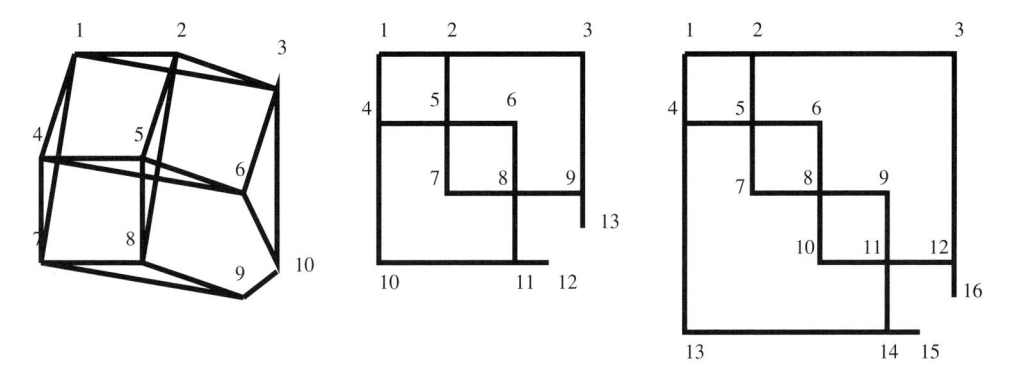

Figure 1.

Proof. We only need a construction for $n = 3k + 1$, $k \geq 3$ to show $h(3k + 1) \geq 4k + 1$. It is enough to show $h(10) \geq 13$, $h(13) \geq 17$ and $h(16) \geq 21$; then the general case follows from $h(9) \geq 12$ using Claim 1.12.

Define the six triples of $\mathcal{F}_3$ as $\{1, 2, 3\}$, $\{4, 5, 6\}$, $\{7, 8, 9\}$, $\{1, 4, 7\}$, $\{2, 5, 8\}$ and $\{3, 6, 10\}$ and $\mathcal{H}_2$ as the 18 pairs covered by these triples and $\{9, 10\}$. The graph $\mathcal{H}_2$ has only these 6 triangles, so (1.1)–(1.2) hold, and it is not difficult to check the four-tuples, too.

The other cases are similar: for $n = 13$ we can define $\mathcal{F}_3 := \{1, 2, 3\}$, $\{4, 5, 6\}$, $\{7, 8, 9\}$, $\{10, 11, 12\}$ and $\{1, 4, 10\}$, $\{2, 5, 7\}$, $\{6, 8, 11\}$, $\{3, 9, 13\}$ and $\mathcal{H}_2$ consists of these triangles and the pair $\{12, 13\}$.

Finally, for $n = 16$ we define $\mathcal{F}_3$ as $\{1, 2, 3\}$, $\{4, 5, 6\}$, $\{7, 8, 9\}$, $\{10, 11, 12\}$, $\{13, 14, 15\}$ and $\{1, 4, 13\}$, $\{2, 5, 7\}$, $\{6, 8, 10\}$, $\{9, 11, 14\}$, and $\{3, 12, 16\}$. Again $\mathcal{H}_2$ consists of the triangles obtained from $\mathcal{F}_3$ and the edge $\{15, 16\}$. □

2. Bases with deficiency at least 2

The aim of this paper is to prove Theorem 1.1, so suppose that $\mathcal{F}$ is a minimal 2-base of $\mathcal{B}(n, \leq 4)$ and that $\mathcal{F}_2 \cup \mathcal{F}_3$ satisfies (1.1)–(1.4).

Lemma 2.1. *If $abc \in \mathscr{F}_3$, then either $\{ab, bc, ca\} \subset \mathscr{F}_2$ or $\{ab, bc, ca\} \subset \mathscr{H}_2$.*

Proof. Suppose, on the contrary, that $ab \in \mathscr{F}_2$, $ac \notin \mathscr{F}_2$. Replace abc by ac in $\mathscr{F}$. Since $\sum_{F \in \mathscr{F}} |F|$ is minimal the family $\mathscr{F}' := \mathscr{F} \setminus \{abc\} \cup \{ac\}$ is not a 2-base. What can go wrong? Since we added a new pair, conditions (1.1) and (1.2) still hold. The only condition we can violate is (1.3)–(1.4). We removed abc, so there exists an $Q = abcd$ not a union of two members of $\mathscr{F}'$. So $abcd$ does not contain any triple from $\mathscr{F}'$ and also $bd, cd \notin \mathscr{F}'$. Consider bcd. We have $bcd \notin \mathscr{F}$ so (1.1) implies that $bc \in \mathscr{F}_2$. Consider acd. Since ac, cd, and $acd \notin \mathscr{F}$ again (1.1) implies that $ad \in \mathscr{F}_2$. However, then $Q = ad \cup bc$, a contradiction. □

Use the notation $\deg_2^-(x)$ for the degree of the vertex x in the graph $\mathscr{H}_2$ and $\deg_3(x)$ for the degree of x in $\mathscr{F}_3$. The difference $\deg_2^-(x) - \deg_3(x)$ is called the *deficiency* of the vertex $x \in V$. From now on in this section we suppose that

$$\deg_2^-(x) - \deg_3(x) \geq 2 \text{ for every } x \in [n]. \tag{2.1}$$

Let $N(x)$ denote the neighbourhood of x in $\mathscr{H}_2$, $N(x) := \{y : xy \in \mathscr{H}_2\}$, $\deg_2^-(x) = |N(x)|$. Let $\mathscr{T}(x)$ denote the set of triples T from $\mathscr{F}_3$ with $x \in T \subset N(x) \cup \{x\}$, and let $t(x) := |\mathscr{T}(x)|$. Suppose that $D = \max_{x \in [n]} \deg_2^-(x)$, and a has maximum degree in $\mathscr{H}_2$. Consider $A = \{a\} \cup N(a)$, $|A| = D + 1$: let $t := t(a)$. Then (2.1) implies $t, t(x) \leq D - 2$.

2.1. Eliminating the case $D \geq 5$

Claim 2.2. *(2.1) implies that $D \leq 4$.*

Proof. Consider the $\binom{D}{3}$ four-tuples of A containing x: let $\mathscr{B} := \{Q : a \in Q \subset A, |Q| = 4\}$. Note that none of these can satisfy (1.4), so each of them contains a member of $\mathscr{F}_3$. Classify them into two groups as follows:

$\mathscr{B}_1 := \{abcd : b, c, d \in A \text{ and there exists a } T \in \mathscr{F}_3 \text{ with } a \in T \subset \{a, b, c, d\}\}$,
$\mathscr{B}_2 := \{abcd : abcd \subset A, abc, abd, acd \notin \mathscr{F}_3\}$.

Each $Q \in \mathscr{B}_2$ contains a member of $\mathscr{F}_3 | N(a)$, hence

$$|\mathscr{B}_2| \leq |\mathscr{F}_3 | N(a)|.$$

Each member of $\mathscr{T}(a)$ is contained in $D - 2$ four-tuples from $\mathscr{B}_1$, hence

$$|\mathscr{B}_1| \leq t(D-2). \tag{2.2}$$

Here the sum of the left-hand sides is $\binom{D}{3}$. The sum of the right-hand sides can be estimated by the degrees of $\mathscr{F}_3$ on A. Using $\deg_3(x) \leq D - 2$ we obtain

$$\binom{D}{3} = |\mathscr{B}_1| + |\mathscr{B}_2| \leq t(D-2) + |\mathscr{F}_3|N(a)| = t(D-3) + |\mathscr{F}_3|A|$$

$$\leq t(D-3) + \frac{1}{3} \sum_{x \in A} \deg_3(x) \leq t(D-3) + \frac{1}{3}(t + D(D-2)). \tag{2.3}$$

Hence
$$\frac{1}{6}D(D-2)(D-3) \leqslant t\frac{3D-8}{3}. \quad (2.4)$$

Since $t \leqslant D-2$ we get $D \leqslant 6$. In the case of $t \leqslant D-3$ (2.4) implies $D \leqslant 4$. So two cases are left in the proof of the claim, namely $(D,t) = (6,4)$ and $(5,3)$.

In the case of $D = 6$, $t = 4$ the right-hand side of (2.2) can be improved by 2, since there are at least 2 coincidences when estimating the cardinality of $\mathcal{B}_1$. So $|\mathcal{B}_1| \leqslant 14$, and we can decrease the right-hand sides of (2.3) and (2.4) by 2, and that leads to the contradiction $12 \leqslant 4 \times \frac{10}{3} - 2$.

In the case of $D = 5$, $t = 3$ we use two things. The first one is implied by Lemma 2.1 and (1.1).

(C1) If $abc \in \mathcal{T}(a)$ then $bc \in \mathcal{H}_2$; if $abc \notin \mathcal{T}(a)$ and $b,c \in N(a)$ then $bc \in \mathcal{F}_2$. Thus $\mathcal{F}_2|N(a)$ has exactly $\binom{D}{2} - t$ edges.

(C2) If $\deg_3(x) \geqslant 3$, then $t(x) = 3$. Indeed, (2.1) implies $\deg_2^-(x) \geqslant \deg_3(x) + 2 \geqslant 5$.

Consequently $\deg_2^-(x) = 5 = D$, x has maximum degree, D, and then the previous considerations for a are valid for x, too, i.e., (2.4) implies that $t(x) = 3$ is the only possibility.

Now we are ready to show that, in fact, $(D,t) = (5,3)$ is impossible. Suppose, on the contrary, that there is such a construction and let $N(a) = \{b,c,d,e,f\}$. Consider the 3-edge graph $G := \{xy : axy \in \mathcal{F}_3\}$. There are 4 non-isomorphic possibilities for G:

(α) G is a triangle, $\{bc, cd, bd\}$,
(β) G is a path of length 3, $\{bc, cd, de\}$,
(γ) G is a star, $\{bc, bd, be\}$,
(δ) G has 2 components, $\{bc, cd, ef\}$.

In each case we will find one or more $x \in N(a)$ with $t(x) = 3$. Then the triples containing x cover no pair from $\mathcal{F}_2$ and this will lead to a contradiction.

For (α), by (1.3) we have $bef, cef, def \in \mathcal{F}_3$. Hence $\deg_3(f) \geqslant 3$. Then (C2) implies that $t(f) = 3$ and then Lemma 2.1 gives that $\{b,c,d,e\} \subset N(f)$, $ef \notin \mathcal{F}_2$. However, $ef \in \mathcal{F}_2$ by (C1), a contradiction.

The other cases can be handled in the same way. For (β) we have $bdf, bef, cef \in \mathcal{F}_3$, hence $\deg_3(f) \geqslant 3$. Then $t(f) = 3$ and $\{b,c,d,e\} \subset N(f)$, $ef \notin \mathcal{F}_2$. For (γ) we have $cdf, cef, def \in \mathcal{F}_3$, hence $\deg_3(f) \geqslant 3$. Then $t(f) = 3$ and $\{c,d,e\} \subset N(f)$, $ef \notin \mathcal{F}_2$. For (δ) we have $bde, bdf \in \mathcal{F}_3$, hence $\deg_3(b) \geqslant 3$. Then $t(b) = 3$ and $\{c,d,e,f\} \subset N(b)$, $bf \notin \mathcal{F}_2$. This final contradiction completes the proof of the case $(D,t) = (5,3)$ and Claim 2.2. □

2.2. The case $D \leqslant 4$

From now on in this section we suppose that $D \leqslant 4$.

Claim 2.3. (2.1) and $\deg_2^-(a) = 4$ imply that $t(a) = \deg_3(a) = 2$ and the two triples containing the element a meet only in a, e.g., $N(a) = bcde$ and $\mathcal{T}(a) = \{abc, ade\}$.

Proof. Suppose first that $t(a) = 0$. Then all the four triples of the form xyz, $x,y,z \in N(a)$ belong to $\mathcal{F}_3$. Hence $\deg_3(b) \geqslant 3$, contradicting $D \geqslant \deg_2(x) \geqslant 2 + \deg_3(x)$. If $t(a) = 1$, say $abc \in \mathcal{T}(a)$, then $bde, cde \in \mathcal{F}_3$ is implied by (1.3). Hence $\deg_3(e) \geqslant 2$, so $\deg_2^-(e) = 4$. Since

(1.1) implies that $be, ce, de \in \mathcal{F}_2$ we get that $N(e) \cap \{b,c,d\} = \emptyset$, so $t(e) = 0$. However, we have seen that $\deg_2^-(e) = D = 4$ implies $t(e) > 0$.

So we get $t(a) \geq 2$, i.e., by $t(x) \leq D - 2$ we have $t(a) = 2$. The only case left to exclude is when the triples in $\mathcal{T}(a)$ meet in two elements, say $\mathcal{T}(a) = \{abc, acd\}$. Then $bde \in \mathcal{F}_3$, so $\deg_3(b) \geq 2$. Hence we get $\deg_2^-(b) = 4$, this implies $t(b) = 2$ and $\{c,d,e\} \subset N(b)$. We get $ab, ae, be \in \mathcal{H}_2$, $abe \notin \mathcal{F}_3$, contradicting (1.1). $\square$

Claim 2.4. (2.1) and $\deg_2^-(x) = 3$ imply that $\deg_3(x) = 1$.

Proof. Suppose, on the contrary, that $\deg_3(x) = 0$. Consider $N(x) = abc$, we have $ab, bc, ca \in \mathcal{F}_2$ by (1.1) and $abc \in \mathcal{F}_3$ by (1.3). Then $ab \in \mathcal{F}_2$ implies that $abc \notin \mathcal{T}(a)$. Therefore $t(a)$ cannot be $D - 2 = 2$. So Claim 2.3 gives that $\deg_2^-(a) \neq 4$. Since $\deg_3(a) \geq 1$ we get that $\deg_2^-(a) = 3$. Consider $N(a) = xyz$. Note that $y, z \notin \{x, a, b, c\}$. Then $xyz \in \mathcal{F}_3$ by (1.3). This contradicts $\deg_3(x) = 0$, so we have $\deg_3(x) \geq 1$. On the other hand, (2.1) implies $\deg_3(x) \leq 1$. $\square$

Claim 2.5. (2.1) implies that $h(\mathcal{F}) \leq \frac{4}{3}n$.

Proof. For $x \in [n]$ define $\varphi(x) := \frac{1}{2}\deg_2^-(x) - \frac{1}{3}\deg_3(x)$. We are going to prove that $\varphi(x) \leq 4/3$ for every x. This implies the claim as follows:

$$h(\mathcal{F}) = |\mathcal{H}_2| - |\mathcal{F}_3| = \sum_{x \in [n]} \varphi(x) \leq \frac{4}{3}n. \tag{2.5}$$

Using the previous three claims one can split $[n]$ into three parts, $[n] = P \cup Q \cup R$, where $P := \{x : \deg_2^-(x) = 4, \deg_3(x) = 2\}$, $Q := \{x : \deg_2^-(x) = 3, \deg_3(x) = 1\}$, and $R := \{x : \deg_2^-(x) = 2, \deg_3(x) = 0\}$. For each case we have $\varphi \leq 4/3$. $\square$

Note that $h(\mathcal{F}) = \frac{4}{3}n$ in Claim 2.5 is only possible for Construction 1.13, especially

$$P = [n] \text{ and } Q = R = \emptyset. \tag{2.6}$$

3. Proof of the main result

Let $\mathcal{F}$ be a minimal 2-base for $\mathcal{B}(n, \leq 4)$. Then

$$1 + n + \binom{n}{2} - h(n) = |\mathcal{F}| = |\mathcal{F}|([n] \setminus \{x\})| + 1 + (n - 1 - \deg_2^-(x)) + \deg_3(x)$$

$$\geq 1 + n + \binom{n}{2} - h(n-1) - (\deg_2^-(x) - \deg_3(x)) \tag{3.1}$$

gives that the deficiency of every vertex is at least $h(n) - h(n-1)$.

Proof of Theorem 1.1. We use induction on n to show that $h(n) \leq \frac{4}{3}n$. This is certainly true for $n \leq 2$. Suppose that $h(n-1) \leq \frac{4}{3}(n-1)$ and consider $h(n)$. If $h(n) \leq h(n-1) + 1$, then we are done. If $h(n) \geq h(n-1) + 2$, then, as we have seen in (3.1), there exists a minimal 2-base $\mathscr{F}$ on $[n]$ with deficiency at least 2. Then Claim 2.5 gives $h(n) = h(\mathscr{F}) \leqslant \frac{4}{3}n$. □

Proofs of Claims 1.6, 1.8 and 1.10. The case $n \leqslant 4$ is trivial. Suppose that $5 \leqslant n \leqslant 7$ and let $\mathscr{F}$ be a minimal 2-base on n vertices.

The case $n = 5$ is easy. $h(\mathscr{F}) \geqslant 6$ implies $|\mathscr{F}_2| + |\mathscr{F}_3| \leqslant 4$. If $|\mathscr{F}_2| = 4$, then there is a unique way to satisfy (1.1) (namely, $\mathscr{F}_2$ is a union of an edge and a triangle) and then (1.4) is violated. If $|\mathscr{F}_2| = 3$, then there are at least 2 triples not containing any member of $\mathscr{F}_2$, so (1.2) gives $|\mathscr{F}_3| \geqslant 2$. If $|\mathscr{F}_2| \leqslant 2$, then they satisfy (1.1) with at most $3|\mathscr{F}_2|$ triples. Hence, (1.2) gives $|\mathscr{F}_3| \geqslant 10 - 3|\mathscr{F}_2|$. Then $|\mathscr{F}_2| + |\mathscr{F}_3|$ exceeds 4, a final contradiction.

If the minimum deficiency of $\mathscr{F}$ is (at most) 1 then (3.1) gives $h(n) \leqslant h(n-1) + 1$, and we are done. From now on suppose that the deficiency of $\mathscr{F}$ is at least 2, i.e., (2.1) holds.

For $n = 6$ Claim 2.5 gives that $h(\mathscr{F}) \leqslant \frac{4}{3} \times 6 = 8$. By (2.6) $h(\mathscr{F}) = 8$ is only possible if $P = [n]$, i.e., $\mathscr{H}_2$ is a 4-regular graph, and $\mathscr{F}_3$ consists of four triples. Then $\mathscr{F}_2$ is a matching, say, $\mathscr{F}_2 = \{a_1a_2, b_1b_2, c_1c_2\}$. Then (1.2) implies that all the eight triples of the form $a_ib_jc_k$ should belong to $\mathscr{F}_3$, a contradiction. We have obtained $h(\mathscr{F}) = h(6) \leqslant 7$.

For $n = 7$ Theorem 1.1 implies $h(\mathscr{F}) \leqslant \lfloor 7 \times \frac{4}{3} \rfloor = 9$. We claim that $h(7) = 8$. Suppose, on the contrary, that $h(\mathscr{F}) = 9$. Consider the partition of $[n] = P \cup Q \cup R$ defined in the proof of Claim 2.5. For $R \neq \emptyset$ (2.5) gives $|R| = 1$, $|P| = 6$, $Q = \emptyset$. Then $\mathscr{H}_2|P$ is a 4-regular graph, not joined to R, so $\deg_2^-(R) = 2$ is impossible. Finally, if $R = \emptyset$, $|Q| = 2$ and $|P| = 5$ then we get $|\mathscr{F}_3| = 4$. The four members of $\mathscr{F}_3$ can pairwise meet in at most 1 vertex (by Claims 2.3 and 2.4) and have girth 4. But such an $\mathscr{F}_3$ does not exist on 7 vertices.

So we have obtained the exact value of $h(n)$ for every n. □

4. 2-bases for quadruples

Here we prove Theorem 1.2. Suppose that $\mathscr{F}$ is an extremal 2-base for $\mathscr{B}(n,4)$, i.e., $|\mathscr{F}| = g_4(n)$, such that $|\mathscr{F}_1| + |\mathscr{F}_4|$ is minimal. The case $n = 5$ is a short finite process, the unique 2-base with 4 members $\{12, 34, 135, 245\}$.

In the case $n = 6$ the 6 pairs of a hexagon and the 2 disjoint triples of the second example in Construction 1.7 shows $g_4(6) \leqslant 8$. Consider a minimal 2-base $\mathscr{F}$. If $\deg_{\mathscr{F}}(x) \geqslant 3$, then

$$|\mathscr{F}| = \deg_{\mathscr{F}}(x) + |\mathscr{F}|([n] \setminus \{x\})| \geqslant \deg_{\mathscr{F}}(x) + g_4(n-1) \quad (4.1)$$

implies $|\mathscr{F}| \geqslant 3 + 4$. The impossibility of this case with $|\mathscr{F}| = 7$ follows easily from the uniqueness of the 2-base on 5 elements. Moreover, it is easy to check that a hypergraph of 7 edges on 6 elements with maximum degree 2 cannot be a 2-base, so $g_4(6) \geqslant 8$. From now on we may suppose that $n \geqslant 7$.

The upper bounds for $g_4(n)$ follows by leaving out the singletons and the empty set from Constructions 1.9 and 1.13 in Section 1. To prove a lower bound we proceed as in Section 2. The main idea of the proof is that we first investigate the minimal 2-bases with a maximum degree condition

$$\deg_{\mathscr{F}}(x) \leqslant n - 3 \quad (4.2)$$

for all $x \in [n]$.

We claim that (4.2) implies that $\mathscr{F}_4 = \emptyset$. Indeed, suppose, on the contrary, that $Q \in \mathscr{F}_4$. If Q contains any proper subset $F \in \mathscr{F}$, $x \in F \subset Q$, $Q \neq F$, then one can replace Q by $Q \setminus \{x\}$ to obtain another 2-base with smaller $|\mathscr{F}_1| + |\mathscr{F}_4|$. So we may suppose that such a proper subset does not exist. Consider $Q \setminus \{x\} \cup \{y\}$ for some $x \in Q$, $y \in [n] \setminus Q$. This is a union of (at most) two sets $A, B \in \mathscr{F}$. Both of them contain y. We obtain that the sets $\{F : y \in F \subset Q \cup \{y\}, |F| > 1\}$ cover Q, and some vertex of Q is covered at least twice. Hence there exists an $x \in Q$ covered by these sets more than $n - 4$ times while y runs through $[n] \setminus Q$. Taking Q itself, we get that $\deg_{\mathscr{F}}(x) > n - 3$, contradicting (4.2).

Use the notation of the previous section, e.g., $D := \max \deg_2^-(x)$ and $\deg_2^-(a) = D$. We claim that (4.2) implies that

$$D \leqslant 4.$$

In the proof of this one cannot use Lemma 2.1, either (1.1) or (1.2); however, (2.2)–(2.4) still hold, implying $D \leqslant 6$. Furthermore, $ab, ac, ad \notin \mathscr{F}_2$, and $abc, abd, acd \notin \mathscr{F}_3$ imply not only $bcd \in \mathscr{F}_3$ but $a \in \mathscr{F}_1$. Thus, in the case $\mathscr{B}_2 \neq \emptyset$ (e.g., for $D > 4$), one gets $a \in \mathscr{F}_1$. Then (4.2) gives $t(a) \leqslant \deg_2^-(a) - 3 = D - 3$. So (2.4) gives $D \leqslant 4$.

Using the same idea one can see that Claim 2.3 remains true. The following analogue of Claim 2.4 is obviously true: $\deg_2^-(x) = 3$ implies $\deg_1(x) + \deg_3(x) = 1$.

As in Claim 2.5 we show that (4.2) implies

$$|\mathscr{F}| \geqslant \binom{n}{2} - \frac{4}{3}n. \tag{4.3}$$

Indeed, for $x \in [n]$ define $\varphi(x) := \frac{1}{2} \deg_2^-(x) - \frac{1}{3} \deg_3(x) - \deg_1(x)$. As before we have that (4.2) implies that $\varphi(x) \leqslant 4/3$ for every x, completing the proof of (4.3) for this case.

Finally, for hypergraphs with maximum degree at least $n - 2$ one can use induction on n. Inequality (4.1) implies that (4.3) always holds.

The case $n = 7$ can be finished as in the proof of Claim 2.5, by considering a partition of $[n]$ into three parts, $[n] = P \cup Q \cup R$, where now $Q := \{x : \deg_2^-(x) = 3, \deg_1(x) + \deg_3(x) = 1\}$. The details are omitted. □

5. More hypergraphs

Let $T(n, k, r)$ denote the minimum size of a hypergraph $\mathscr{F} \subseteq \mathscr{B}(n, r)$ such that every k-subset of $[n]$ contains a member of $\mathscr{F}$. The determination of $T(n, k, r)$ is proposed by Turán [8], who solved the case $r = 2$ (the case of graphs – see [7]) and has a longstanding conjecture $T(n, 4, 3) = \left(\frac{4}{9} + o(1)\right)\binom{n}{3}$. For a survey on this see Sidorenko [6].

One can prove for every odd integer k that our $f_k(n)$ equals $(1 + o(1))T(n, k, (k + 1)/2)$, but the even case is more involved and apparently leads to a new Turán-type problem. The authors intend to return to this topic in a future work.

References

[1] Bollobás, B. (1978) *Extremal Graph Theory*, Academic Press, London.
[2] Erdős, P. personal communication.

[3] Füredi, Z. (1991) Turán type problems. In *Surveys in Combinatorics, 1991* (A. D. Keedwell, ed.), Cambridge University Press, pp. 253–300.

[4] Katona, G., Nemetz, T. and Simonovits, M. (1964) A new proof of a theorem of P. Turán and some remarks on a generalization of it (in Hungarian). *Mat. Lapok* **15** 228–238.

[5] Mubayi, D. and Rödl, V. (2002) On the Turán number of triple systems. *J. Combin. Theory Ser. B* **100** 135–152.

[6] Sidorenko, A. F. (1995) What we know and what we do not know about Turán numbers. *Graphs Combin.* **11** 179–199.

[7] Turán, P. (1941) On an extremal problem in graph theory (in Hungarian). *Math. Fiz. Lapok* **48** 436–452.

[8] Turán, P. (1961) Research problems. *MTA Mat. Kutató Int. Közl.* **6** 417–423.

On Triple Systems with Independent Neighbourhoods

ZOLTÁN FÜREDI,[1,†] OLEG PIKHURKO[2]
and MIKLÓS SIMONOVITS[3,‡]

[1]Rényi Institute of Mathematics of the Hungarian Academy of Sciences
Budapest, PO Box 127, Hungary-1364
and
Department of Mathematics, University of Illinois at Urbana-Champaign
Urbana, IL61801, USA
(e-mail: furedi@renyi.hu, z-furedi@math.uiuc.edu)

[2]Department of Mathematical Sciences, Carnegie Mellon University
Pittsburgh, PA 15213-3890, USA
(URL address: http://www.math.cmu.edu/~pikhurko)

[3]Rényi Institute of Mathematics of the Hungarian Academy of Sciences
Budapest, PO Box 127, Hungary-1364
(e-mail: miki@renyi.hu)

Received 18 February 2004; revised 12 March 2004

For Béla Bollobás on his 60th birthday

Let $\mathcal{H}$ be a 3-uniform hypergraph on an n-element vertex set V. The neighbourhood of $a, b \in V$ is $N(ab) := \{x : abx \in E(\mathcal{H})\}$. Such a 3-graph has independent neighbourhoods if no $N(ab)$ contains an edge of $\mathcal{H}$. This is equivalent to $\mathcal{H}$ not containing a copy of $\mathbb{F}_{3,2} := \{abx, aby, abz, xyz\}$.

In this paper we prove an analogue of the Andrásfai–Erdős–Sós theorem for triangle-free graphs with minimum degree exceeding $2n/5$. It is shown that any $\mathbb{F}_{3,2}$-free 3-graph with minimum degree exceeding $(\frac{4}{9} - \frac{1}{125})\binom{n}{2}$ is bipartite, (for $n > n_0$), i.e., the vertices of $\mathcal{H}$ can be split into two parts so that every triple meets both parts.

This is, in fact, a Turán-type result. It solves a problem of Erdős and T. Sós, and answers a question of Mubayi and Rödl that

$$\mathrm{ex}(n, \mathbb{F}_{3,2}) = \max_\alpha (n - \alpha) \binom{\alpha}{2}.$$

[†] Research supported in part by the Hungarian National Science Foundation grants OTKA T 032452, T 037846 and by a National Science Foundation grant DMS 0140692.
[‡] Research supported in part by the Hungarian National Science Foundation grants OTKA T 026069, T 038210, and T 0234702.

Here the right-hand side is $\frac{4}{9}\binom{n}{3} + O(n^2)$. Moreover $e(\mathcal{H}) = \operatorname{ex}(n, \mathbb{F}_{3,2})$ is possible only if $V(\mathcal{H})$ can be partitioned into two sets A and B so that each triple of $\mathcal{H}$ intersects A in exactly two vertices and B in one.

1. Independent neighbourhoods

Consider a 3-uniform hypergraph $\mathcal{H}$, and let a and b be two distinct vertices. The *neighbourhood*, $N(ab)$, of the pair ab consists of the vertices z for which $\{a, b, z\} \in E(\mathcal{H})$. Its size, $\mu(ab) := |N(ab)|$, is called the *codegree* of ab. We say that the neighbourhoods of $\mathcal{H}$ are *independent* if $N(ab)$ contains no triple from $E(\mathcal{H})$ for any pair ab. The *link graph* G_a of a in $\mathcal{H}$ is defined as an ordinary graph on $V(\mathcal{H})$ consisting of the pairs bc for which $\{a, b, c\} \in E(\mathcal{H})$. The *degree* of a, $\deg_{\mathcal{H}}(a) = |E(G_a)|$. We use $d_{\min}(\mathcal{H})$ for the minimum degree and $\operatorname{maxcodeg}(\mathcal{H})$ or μ^* for the max $\mu(ab)$.

When it is possible, we shall use simplified notation, discarding parentheses and commas, e.g., we shall often abbreviate a triple $\{a, b, c\}$ to abc. As usual, $G[A]$ denotes the subgraph (subhypergraph) of G induced by the vertices of A. For a graph G, $G[A, B]$ denotes the bipartite subgraph defined by the edges joining A to B.

Construction 1.1. ((2,1)-colourings and $\mathcal{H}_{2,1}(A,B)$) *A hypergraph $\mathcal{H}$ has a (2,1)-colouring if there exists a partition $V = A \cup B$ such that each triple in $E(\mathcal{H})$ meets A in exactly two vertices, and meets B in one vertex. Then the neighbourhoods of $\mathcal{H}$ are independent. Denote by $\mathcal{H}_{2,1}(A, B)$ the hypergraph consisting of all (2,1)-coloured triples.*

Let $\mathbb{H}_n$ denote the class of n-vertex (2,1)-colourable hypergraphs with maximum number of edges. Then $||A| - \frac{2}{3}n| < 1$, $e(\mathcal{H}) = \frac{4}{9}\binom{n}{3} + O(n^2)$ and every degree is $\frac{4}{9}\binom{n}{2} + O(n)$. For $n = 3k + 2$ the choices $|A| = 2k + 1$ and $2k + 2$ give the same edge-number, $\mathbb{H}_n$ has 2 members. $\mathbb{H}_n$ consists of a single hypergraph for $n = 3k, 3k + 1$.

Construction 1.2. (A hypergraph without (2,1)-colourings) *Let $V = A \cup B \cup D \cup \{x\}$, $|V| = n$, $|A| + |B| + |D| = n - 1$, $|A| = \lfloor \frac{2}{3}n \rfloor$, $|B| > 0$ and $|D| \geq 2$. Define $E(\mathcal{H})$ by taking the edges of $\mathcal{H}_{2,1}(A, B \cup D)$ and having the link graph of x as follows: $E(G_x) := \{ab : a \in$*

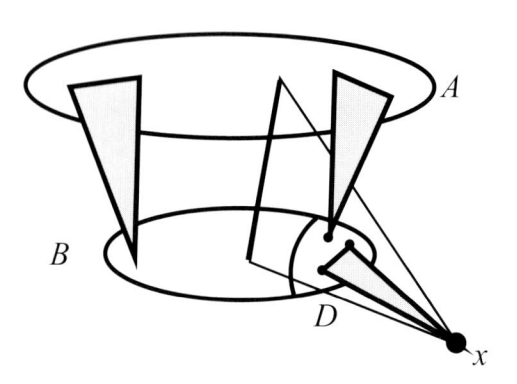

Figure 1. A hypergraph without a (2,1)-colouring

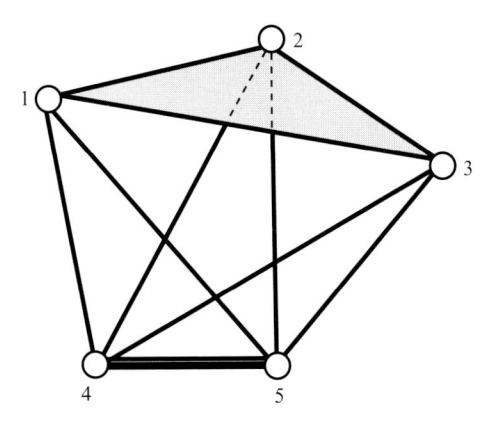

Figure 2. The excluded hypergraph $\mathbb{F}_{3,2}$

$A, b \in B\} \cup \{b_1b_2 : b_1, b_2 \in D\}$. Then the neighbourhoods of $\mathcal{H}$ are independent but $\mathcal{H}$ has no (2,1)-colouring, although (for $|D| = O(1)$) each vertex has degree at least $\frac{4}{9}\binom{n}{2} - O(n)$.

The vertex x can be replaced by a set X to get a hypergraph containing all triples of types AAB, AAD, ABX and DDX.

The notion of 'independent neighbourhoods' can be considered as one of the hypergraph extensions of triangle-free graphs. Andrásfai, Erdős and T. Sós [1] showed that if an n-vertex graph is triangle-free and its minimum degree exceeds $\frac{2}{5}n$, then it is bipartite. The main result of this paper is the following analogue of this important theorem. A hypergraph is *bipartite* if there exists a partition $A \cup B$ of the vertices for which every hyperedge meets both parts.

Theorem 1.3. *Let $\gamma \leqslant 1/125$ be fixed and $n > n_0$. Let $\mathcal{H}$ be an n-vertex 3-uniform hypergraph. Suppose that the neighbourhoods of $\mathcal{H}$ are independent and*

$$d_{\min}(\mathcal{H}) > \left(\frac{4}{9} - \gamma\right)\binom{n}{2}. \tag{1.1}$$

Then $\mathcal{H}$ is bipartite.

We use $n_0 = 25000$ throughout this paper, but the statements (probably) hold for much smaller values of n, too. Construction 1.2 shows that the above min-degree condition does not imply that $\mathcal{H}$ must have a (2,1)-colouring. The next example shows that $\gamma \geqslant 5/72 = 0.069\ldots$ cannot be chosen in Theorem 1.3.

Construction 1.4. (A non-bipartite triple system) *Let V be an n-element set, $V = A \cup B$, with $|A| = (3/4)n + O(1)$, $|B| = n/4 + O(1)$, and let $E_0 = \{z_1, z_2, z_3\} \subset B$. Split the pairs of A into 3 almost equal parts: $\cup_{1 \leqslant i \leqslant 3} \mathcal{E}_i = \{ab : a, b \in A\}$, $|\mathcal{E}_i| \sim \frac{1}{3}\binom{|A|}{2}$. Let $E(\mathcal{H})$ consist of E_0, the hyperedges of $\mathcal{H}_{2,1}(A, B \setminus E_0)$, and all triples of the form $\{abz_i : a, b \in A, z_i \in E_0, \text{ and } ab \in \mathcal{E}_j, 1 \leqslant j \neq i \leqslant 3\}$. Then $\mathcal{H}$ is not bipartite, its neighbourhoods are independent, and $d_{\min}(\mathcal{H}) = \frac{3}{8}\binom{n}{2} + O(n)$.*

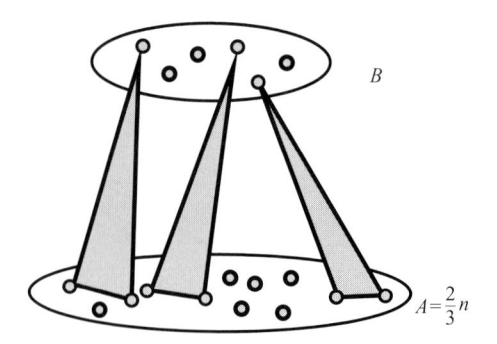

Figure 3. The extremal hypergraph

Conjecture 1.5. *Theorem 1.3 holds for every $\gamma < 5/72$.*

2. Turán's problem

Let $\mathbb{F}_{3,2}$ be the hypergraph on the vertices $1, 2, 3, 4, 5$ having 4 triples $\{1,2,3\}$, $\{1,4,5\}$, $\{2,4,5\}$ and $\{3,4,5\}$. In an ordinary graph the neighbourhoods are independent if and only if it is triangle-free. Similarly, in a 3-uniform hypergraph the neighbourhoods of pairs are independent if and only if it is $\mathbb{F}_{3,2}$-free.

More generally, given a 3-uniform hypergraph $\mathscr{F}$, let $\text{ex}(n, \mathscr{F})$ denote the maximum possible size of a 3-uniform hypergraph of order n that does not contain any subhypergraph isomorphic to $\mathscr{F}$. An averaging argument shows that the ratio $\text{ex}(n, \mathscr{F})/\binom{n}{3}$ is a non-increasing sequence [15], therefore $\pi(\mathscr{F}) := \lim_{n\to\infty} \text{ex}(n, \mathscr{F})/\binom{n}{3}$ always exists.

Erdős and T. Sós, in connection with Ramsey–Turán problems for hypergraphs, investigated $\text{ex}(n, \mathbb{F}_{3,2})$. (In [7] $\mathbb{F}_{3,2}$ is denoted by $G^{(3)}(5,4)$, and in [8, Theorem 2] a more general class of hypergraphs is considered.) Mubayi and Rödl [20] showed that $4/9 \leqslant \pi(\mathbb{F}_{3,2}) \leqslant 1/2$ and conjectured that

$$\text{ex}(n, \mathbb{F}_{3,2}) = \frac{4}{9}\binom{n}{3} + o(n^3). \tag{2.1}$$

They also conjectured that $\mathscr{H}_{2,1}(A, B)$ is the extremal hypergraph. Equation (2.1) was verified by the present authors.

Theorem 2.1. (Turán density [14]) $\pi(\{abc, ade, bde, cde\}) = 4/9$.

Here we will give a new proof. The new method also leads to structure theorems for $\mathbb{F}_{3,2}$-free hypergraphs: to the min-degree stability theorem (Theorem 1.3) stated in the previous section, and to the following further refinements and exact solutions.

Theorem 2.2. (The finer structure) *Let $\mathscr{H}$ be an n-vertex 3-uniform hypergraph not containing $\mathbb{F}_{3,2}$ and suppose that it satisfies the min-degree condition (1.1) with $\gamma \leqslant 1/125$. Suppose that n is sufficiently large, e.g., $n > \max\{n_0, 1/\gamma\}$ (where $n_0 := 25000$). Let A be a maximum independent set of vertices; denote its complement by B. Then we have the following.*

(i) B is also independent; (A, B) is a 2-colouring.
(ii) $||A| - \frac{2}{3}n| < \sqrt{\gamma}n$; $||B| - \frac{1}{3}n| < \sqrt{\gamma}n$.
(iii) The structure of $\mathcal{H}$ is very close to the extremal one; all but at most $\sqrt{\gamma}n^3$ hyperedges have type AAB, the rest being of type ABB.
(iv) Consider any two-colouring (A_1, B_1) of $\mathcal{H}$ with $|A_1| \geq |B_1|$, the existence of which was stated in Theorem 1.3. If $|A| \geq 0.65n$ then $A_1 \subset A$ and $|A \setminus A_1| \leq \sqrt{\gamma}n$.

Theorem 2.3. (Extremal) If $\mathcal{H}$ is an n-vertex 3-uniform hypergraph not containing $\mathbb{F}_{3,2}$ and n is sufficiently large, $n > n_0$, then $e(\mathcal{H}) \leq \max_\alpha (n-\alpha)\binom{\alpha}{2}$. In case of equality $\mathcal{H} \in \mathbb{H}_n$.

We show that if $e(\mathcal{H})$ is sufficiently large, then the structure of $\mathcal{H}$ is close to that of $\mathcal{H}_{2,1}(A, B)$.

Theorem 2.4. (Global stability) Suppose that $\mathcal{H}$ is an n-vertex 3-uniform hypergraph not containing $\mathbb{F}_{3,2}$ and $e(\mathcal{H}) > (\frac{4}{9} - c)\binom{n}{3}$, where $c < 10^{-4}$. Then one can delete $O(c^{1/3})n^3$ triples from $\mathcal{H}$ so that the remaining hypergraph has a $(2,1)$-colouring (A, B) with $||A| - \frac{2}{3}n| < O(c^{1/3})n$ and $||B| - \frac{1}{3}n| < O(c^{1/3})n$.

Taking $|A| = (\frac{2}{3} - \frac{1}{2}c)n$, $|B| = (\frac{1}{3} - \frac{1}{2}c)n$, $|D| = |X| = \frac{1}{2}cn$ in Construction 1.2, one can obtain a hypergraph satisfying the constraint of Theorem 2.4; however, one needs to remove at least $\Omega(c^3)n^3$ triples from it to make it $(2, 1)$-colourable.

Theorem 2.5. (Codegree stability) For every $\varepsilon > 0$ there exists a $c > 0$ such that the following holds. If $\mathcal{H}$ is an n-vertex 3-uniform hypergraph not containing an $\mathbb{F}_{3,2}$ and satisfying

$$\left| \mathrm{maxcodeg}(\mathcal{H}) - \frac{2}{3}n \right| > \varepsilon n,$$

then

$$e(\mathcal{H}) \leq \left(\frac{4}{9} - c\right)\binom{n}{3}.$$

Theorem 1.3 is not the first Turán-type stability result concerning hypergraphs. Based on earlier works of Füredi and Kündgen [12], de Caen and Füredi proved [5] that $\pi(\mathcal{L}_7) = \frac{3}{4}$, where $\mathcal{L}_7$ is the hypergraph formed by the 7 lines of a Fano plane. This result was sharpened in [13] to a min-degree result of the same type as Theorem 1.3. A slightly weaker form of this was also proved independently by Keevash and Sudakov [17]. Further, Keevash and Mubayi obtained in [16] a min-degree version of a Turán-type result of Bollobás [2]. Keevash and Sudakov [18] also obtained a stability result, improving an extremal result of Frankl [9] on the hypergraph-triangle problem. In general, stability may not hold (see the constructions of W. G. Brown [4], and Kostochka [19] for $K_4^{(3)}$). As far as we know, at present these are the only hypergraph results of *this* type.

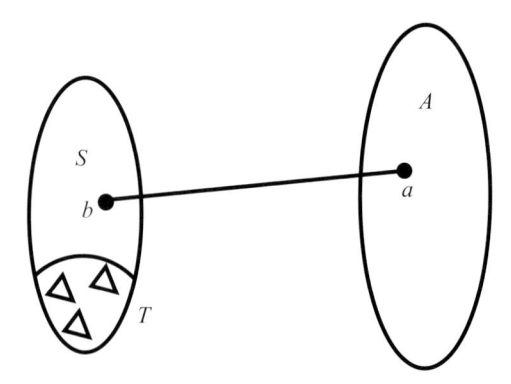

Figure 4. The partition. A and S are independent. The line (a,b) indicates that we shall sum multiplicities $\mu(a,b)$.

3. Proofs

3.1. The partition

Let $\mathcal{H}$ be an $\mathbb{F}_{3,2}$-free triple system on the vertex set V, $|V| = n$, satisfying (1.1), the assumption of Theorem 1.3. Let α denote the maximum size of an independent (i.e., hyperedge-free) subset of V. Let A be a maximal independent set with $|A| = \alpha$ and denote its complement by B, $B = V(\mathcal{H}) \setminus A$. Take as many independent (i.e., pairwise disjoint) triples in B as possible, denote their number by ν. That is, $\mathcal{M} := \{E^1, E^2, \ldots, E^\nu\} \subseteq E(\mathcal{H})$, $E^i \cap E^j = \emptyset$, $E^i \cap A = \emptyset$. Their vertices form T, the remaining part is S. So V is partitioned into $A \cup S \cup T$, where A and S are independent, and $|T| = 3\nu$.

Since the neighbourhoods are independent, we have $\alpha \geq \mu^\star$. Clearly,

$$n \times d_{\min}(\mathcal{H}) \leq \sum_{x \in V} \deg_{\mathcal{H}}(x) = 3e(\mathcal{H}) = \sum_{a,b \in V} \mu(ab) \leq \binom{n}{2} \mu^\star.$$

Then (1.1) implies that

$$\alpha \geq \mu^\star \geq \frac{2d_{\min}}{n-1} > \left(\frac{4}{9} - \gamma\right) n > \frac{2}{5} n.$$

Define the rectangular domain

$$D_1 := \{(x,y) : 2/5 \leq x \leq 1, \; 0 \leq y \leq 1/5\}.$$

Then $\alpha \geq \frac{2}{5}n$ gives $\nu \leq \frac{1}{3}(n-\alpha) \leq \frac{1}{5}n$. Hence, for all possible values of the pairs (α, ν),

$$\frac{1}{n}(\alpha, \nu) \in D_1. \tag{3.1}$$

3.2. Sketch of the proof of Theorem 1.3

We shall classify the triples according to their positions relative to A, S and T. The number of hyperedges of type XYZ is denoted by Δ_{XYZ}. Since A and S are independent sets,

$$e(\mathcal{H}) = \Delta_{AAS} + \Delta_{AAT} + \Delta_{ASS} + \Delta_{AST} + \Delta_{ATT} + \Delta_{SST} + \Delta_{STT} + \Delta_{TTT}.$$

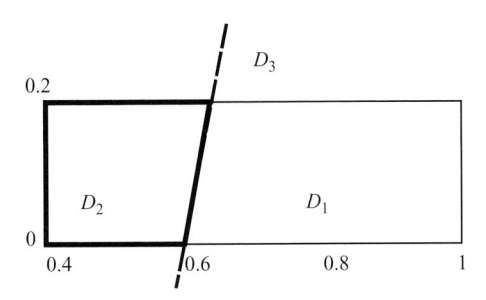

Figure 5.

In Section 3.3 we collect some inequalities for degree-3 polynomials in two variables we will use later. In Sections 3.4–3.9 we are going to give several upper bounds for different combinations of Δs which hold for every $\mathbb{F}_{3,2}$-free hypergraph. In Section 3.10 we show that the min-degree condition (1.1), used for the vertices of $A \cup B$, implies bipartiteness. This will complete the proof of Theorem 1.3. Meanwhile, in Section 3.6 we make a short digression to prove $\pi(\mathbb{F}_{3,2}) = 4/9$.

3.3. Three inequalities on polynomials

Define the polynomial

$$f(x,y) = \frac{3}{2}\left((1-x)x^2 - \frac{4}{27}\right) + y\left(\frac{2}{3} - 2x^2 + (1-x-y)(1-x-3y)\right).$$

Lemma 3.1. $f(x,y) \leq 0$ for every point (x,y) in the rectangle D_1.

Proof (standard optimization). The determinant of the Hessian matrix of $f(x,y)$ is negative for every $(x,y) \in D_1$ (see Section A.1). This implies that f does not have local minima or maxima inside D_1: the extrema must be on the boundary ∂D_1. Then the maximum can be found by checking the behaviour of $f(x,y)$ on the four boundary segments of ∂D_1. On each line $f(x,y)$ reduces to a third-degree polynomial of one variable, whose extrema can be identified by the roots of its derivative, a second-degree polynomial. The details of the calculations are postponed to the Appendix (Section A.1). □

The very same optimization method establishes the next two lemmas, too. Details are postponed to the Appendix. Define the domain $D_2 \subset D_1$ (a trapezium) by the lines $y = 0$, $y = 1/5$, $x = 2/5$, and $y = 5x - 3$ (the vertices are $(2/5, 0)$, $(2/5, 1/5)$, $(3/5, 0)$ and $(64/100, 2/5)$). Define the polynomial

$$F(x,y) := f(x,y) + \frac{1}{2}y(1-3y).$$

Lemma 3.2. $F(x,y) < -1/4000$ for every $(x,y) \in D_2$.

Let D_3 be the open half-plane $\{(x,y) : y < 5x - 3\}$. Note that $D_2 \cap D_3 = \emptyset$. Define

$$g(x,y) := -7x^2 - xy - \frac{3}{2}y^2 + 7x + y - \frac{31}{18} + 5\gamma. \tag{3.2}$$

Lemma 3.3. $g(x,y) < -1/2500$ *for every* $(x,y) \in D_3$.

3.4. Estimating $e(\mathcal{H})$

(1) Let $X \cup Y \cup Z = V(\mathcal{H})$ be a partition of the vertices. Add up the codegrees of the pairs (x,y) with $x \in X, y \in Y$. For every x,y we have $\mu(xy) \leqslant \mu^\star \leqslant \alpha$, so

$$\sum_{x \in X, y \in Y} \mu(xy) \leqslant |X||Y|\mu^\star \leqslant |X||Y|\alpha. \tag{3.3}$$

In the left-hand side of this sum we count the hyperedges of type XXY and XYY exactly twice, the types XYZ occur only once. Therefore

$$2\Delta_{XXY} + 2\Delta_{XYY} + \Delta_{XYZ} \leqslant |X||Y|\alpha.$$

Apply this to $(X,Y,Z) = (A,S,T)$:

$$2\Delta_{AAS} + 2\Delta_{ASS} + \Delta_{AST} \leqslant (n - \alpha - 3v)\alpha^2. \tag{3.4}$$

For $X = A, Y = B$ (and $Z = \emptyset$) we obtain

$$\Delta_{AAB} + \Delta_{ABB} \leqslant \frac{1}{2}(n-\alpha)\alpha^2. \tag{3.5}$$

The main goal in the rest of the proof (Sections 3.4–3.10) is to show that B is independent, because (3.5) and $\Delta_{BBB} = 0$ imply the upper bound $e(\mathcal{H}) \leqslant (2/27)n^3$. Next we show (in Sections 3.11–3.12) that $\mu^\star$ is strictly less than $|A|$. Then $\mu^\star < \alpha$ and (3.3) will imply the exact upper bound for $\text{ex}(n, \mathbb{F}_{3,2})$.

(2) For each pair $a, b \in V$ and edge $E \in E(\mathcal{H})$ we have $|N(ab) \cap E| \leqslant 2$, otherwise we would have an $\mathbb{F}_{3,2} \subseteq \mathcal{H}$. This implies that $|\{z \in T : abz \in E(\mathcal{H})\}| \leqslant 2v$. Hence

$$\Delta_{AAT} = \sum_{\{a,b\} \subset A} |\{z \in T : abz \in E(\mathcal{H})\}| \leqslant 2v \binom{\alpha}{2} \leqslant v\alpha^2. \tag{3.6}$$

(3) We claim that

$$\Delta_{SST} \leqslant \frac{1}{2}v(n - \alpha - 3v)^2 + 2v. \tag{3.7}$$

Consider again the maximum independent family $\mathcal{M} = \{E^1, \ldots, E^v\}$. Define the multigraph G^i with vertex set S by $E(G^i) := \{E \setminus E^i : E \in E(\mathcal{H}), |E \cap E^i| = 1, |E \cap S| = 2\}$, where the multiplicity of ab is the number of Es with $E \setminus E^i = \{a,b\}$. Clearly, $\Delta_{SST} = \sum_{1 \leqslant i \leqslant v} e(G^i)$. We claim that

$$e(G^i) \leqslant \frac{1}{2}|S|^2 + 2,$$

which implies (3.7). Indeed, every edge of G^i has multiplicity at most 2. If all the edges are of multiplicity 0 or 1, then $e(G^i) \leqslant \binom{|S|}{2}$ and we are done. If, on the other hand, there exists a pair s_1s_2 with multiplicity 2, say $x_1s_1s_2, x_2s_1s_2 \in E(\mathcal{H})$, $x_1x_2x_3 = E^i$, then every other edge of G^i must meet s_1s_2. If not, say $xs_3s_4 \in E(\mathcal{H})$ with $x \in E^i$, then either $x \neq x_1$ and we can replace E^i in $\mathcal{M}$ by the triples $x_1s_1s_2$ and xs_3s_4, obtaining $v + 1$ triples in $S \cup T$, which contradicts the maximality of $\mathcal{M}$ or $x = x_1$ and we can replace E^i in $\mathcal{M}$ by the triples $x_2s_1s_2$ and $x_1s_3s_4$. Hence $|E(G^i)| \leq \deg(s_1) + \deg(s_2) - 2 \leq 2 \times 2(|S| - 1) - 2 \leq |S|^2/2 + 2$. □

(4) We show that
$$\Delta_{STT} \leq 2v^2(n - \alpha - 3v) + v(2n - 4). \tag{3.8}$$

Indeed, pick two hyperedges of the maximum matching $E^i, E^j \in \mathcal{M}$. Consider the 6-vertex, bipartite link graph $F_z := G_z[E^i, E^j]$ for every $z \in S$. Let $\mathcal{F}^{i,j}$ be the multigraph formed by their union, the multiplicity of the edge ab (where $a \in E^i, b \in E^j$) is $|\{z : z \in S, abz \in E(\mathcal{H})\}| \leq |S|$. Let $\mathcal{F}_3^{i,j}$ be the multigraph defined by the edges of multiplicities at least 3. Counting the edges with multiplicities, we have

$$e(\mathcal{F}^{i,j}) = e(\mathcal{F}_3^{i,j}) + \text{(the number of edges of multiplicities} \leq 2) \leq e(\mathcal{F}_3^{i,j}) + 18.$$

The graph $\mathcal{F}_3^{i,j}$ contains no 3 disjoint edges, otherwise they could be extended by vertices from S to 3 disjoint triples. Then we could replace E^i, E^j in $\mathcal{M}$ by these 3 triples, obtaining a larger matching: this would contradict the maximality of v. The König–Hall theorem (applied to this 6-vertex graph) implies that the edges of the bipartite graph $\mathcal{F}_3^{i,j}$ can be covered by two vertices, say, a and b; every edge of $\mathcal{F}_3^{i,j}$ is adjacent to either a or b. In every F_z every degree is at most 2 (because $\mathbb{F}_{3,2} \not\subset \mathcal{H}$). Hence F_z has at most 2×2 edges in $\mathcal{F}_3$ (namely, those adjacent to either a or b). Thus $e(\mathcal{F}_3) = \sum_{z \in S} e(\mathcal{F}_3 \cap F_z) \leq 4|S|$.

$$\Delta_{STT} = \sum_{i \neq j} |\{abz \in E(\mathcal{H}) : a \in E^i, b \in E^j, z \in S\}|$$
$$+ \sum_i |\{abz \in E(\mathcal{H}) : a, b \in E^i, z \in S\}|$$
$$\leq \binom{v}{2}(4|S| + 18) + 3v|S| \leq 2v^2(n - \alpha - 3v) + v(2n - 4).$$

In the last step we used $\alpha \geq n/3$ (see (3.1)) and wrote the upper bound in a form convenient to use later.

3.5. Estimating the degrees in $A \cup S$

Add up the degrees in A and S:

$$\sum_{x \in A} \deg_{\mathcal{H}}(x) = 2\Delta_{AAB} + \Delta_{ABB},$$

$$\sum_{a \in S} \deg_{\mathcal{H}}(a) = 2\Delta_{ASS} + 2\Delta_{SST} + \Delta_{AAS} + \Delta_{AST} + \Delta_{STT}.$$

Adding them up and using $\Delta_{AAB} = \Delta_{AAS} + \Delta_{AAT}$, we obtain

$$|A \cup S| \times d_{\min}(\mathcal{H}) \leq (\Delta_{AAB} + \Delta_{ABB}) + (2\Delta_{AAS} + 2\Delta_{ASS} + \Delta_{AST})$$
$$+ \Delta_{AAT} + \Delta_{STT} + 2\Delta_{SST}.$$

Here the right-hand side can be estimated by (3.5), (3.4), (3.6), (3.8), and (3.7):

$$(n - 3v) \times d_{\min}(\mathcal{H}) \leq \frac{1}{2}(n - \alpha)\alpha^2 + (n - \alpha - 3v)\alpha^2$$
$$+ v\alpha^2 + 2v^2(n - \alpha - 3v) + 2vn + v(n - \alpha - 3v)^2.$$

Rearranging, we get

$$(n - 3v) \times \left(d_{\min}(\mathcal{H}) - \frac{2}{9}n^2 \right) \leq \frac{3}{2}\left((n - \alpha)\alpha^2 - \frac{4}{27}n^3 \right)$$
$$+ v\left(\frac{2}{3}n^2 - 2\alpha^2 + (n - \alpha - v)(n - \alpha - 3v) \right) + 2vn$$
$$= n^3 f\left(\frac{\alpha}{n}, \frac{v}{n} \right) + 2vn. \quad (3.9)$$

Using that $12v \leq 7n$ (see (3.1)) and rearranging again,

$$(n - 3v) \times \left(d_{\min}(\mathcal{H}) - \left(\frac{4}{9} - \gamma \right)\binom{n}{2} \right) \leq \frac{3}{2}\left((n - \alpha)\alpha^2 - \frac{4}{27}n^3 \right)$$
$$+ v\left(\frac{2}{3}n^2 - 2\alpha^2 + (n - \alpha - v)(n - \alpha - 3v) \right) + \frac{1}{2}\gamma n^2(n - 3v) + n^2$$
$$= n^3 F\left(\frac{\alpha}{n}, \frac{v}{n} \right) + n^2. \quad (3.10)$$

3.6. Detour: The asymptotic density
Here we prove Theorem 2.1. We have to prove only an upper bound.

Lemma 3.4. *Let $\mathcal{H}$ be an arbitrary $\mathbb{F}_{3,2}$-free hypergraph. Then*

$$d_{\min}(\mathcal{H}) \leq \frac{2}{9}n^2 + n.$$

Proof. If the min-degree condition (1.1) does not hold then there is nothing to prove. Otherwise $\frac{1}{n}(\alpha, v) \in D_1$ by (3.1). Then Lemma 3.1 gives that $f(\alpha/n, v/n) \leq 0$. Since $(n - 3v) \geq \frac{2}{5}n$, from (3.9) we get that

$$\frac{2}{5}n \times \left(d_{\min}(\mathcal{H}) - \frac{2}{9}n^2 \right) \leq 2vn \leq \frac{2}{5}n^2. \quad \square$$

Proof of Theorem 2.1. Since the above lemma gives that

$$\text{ex}(n, \mathbb{F}_{3,2}) \leq \text{ex}(n - 1, \mathbb{F}_{3,2}) + \frac{2}{9}n^2 + n,$$

it follows that $\text{ex}(n, \mathbb{F}_{3,2}) \leq \sum_{i \leq n} (\frac{2}{9}i^2 + i) = \frac{2}{27}n^3 + O(n^2)$. $\square$

Recall the following lemma from [10]. If $\mathcal{F}$ is a k-uniform hypergraph such that every pair of its vertices is contained in some edge, then for every n

$$\pi(\mathcal{F})\binom{n}{k} \leqslant \mathrm{ex}(n,\mathcal{F}) \leqslant \pi(\mathcal{F})\frac{n^k}{k!}.$$

This can be applied to $\mathbb{F}_{3,2}$. So $\pi(\mathbb{F}_{3,2}) = \frac{4}{9}$ gives the following.

Corollary 3.5. $\mathrm{ex}(n, \mathbb{F}_{3,2}) \leqslant \frac{2}{27}n^3$ holds for every n. □

3.7. Estimating the degrees in T

Corollary 3.5 gives

$$\Delta_{TTT} \leqslant \mathrm{ex}(3v, \mathbb{F}_{3,2}) \leqslant 2v^3. \tag{3.11}$$

Consider the link graphs $G_z[B]$ restricted to B. A linear combination of (3.7), (3.8) and (3.11) with a little calculation give that

$$\sum_{z \in T} e(G_z[B]) = \Delta_{SST} + 2\Delta_{STT} + 3\Delta_{TTT}$$

$$\leqslant \frac{1}{2}v(n - \alpha - 3v)^2 + 4v^2(n - \alpha - 3v) + 4vn + 6v^3.$$

Since T is the union of v edges, for $v > 0$ we can divide the above inequality by v. There exists an edge $E^\star \in \mathcal{M}$ meeting at most that many triples of type BBB. We obtain

$$\sum_{z \in E^\star} e(G_z[B]) \leqslant \frac{1}{2}(n - \alpha - v)^2 + 2v(n - \alpha - v) + 4n. \tag{3.12}$$

3.8. Common links in A

Consider an arbitrary $z \in B$. Since A is maximal, there exists an $abz \in E(\mathcal{H})$, $a, b \in A$. Since $\mathcal{H}$ is $\mathbb{F}_{3,2}$-free, no pair uv with $u \in A$, $v \in B$ belongs to all the three bipartite graphs $G_a[A,B]$, $G_b[A,B]$ and $G_z[A,B]$. We obtain that

$$e(G_a[A,B]) + e(G_b[A,B]) + e(G_z[A,B]) \leqslant 2|A||B| = 2\alpha(n - \alpha). \tag{3.13}$$

Note that the link graph G_a has no edges in A and its maximum degree is at most $\mu^\star$. Therefore

$$e(G_a[B]) + e(G_a) = \sum_{y \in B} \deg_{G_a}(y) \leqslant |B|\mu^\star. \tag{3.14}$$

Similarly for b,

$$e(G_b[B]) + e(G_b) \leqslant |B|\mu^\star. \tag{3.15}$$

Adding up (3.13), (3.14) and (3.15) and using $|B|\mu^\star = (n-\alpha)\mu^\star \leqslant (n-\alpha)\alpha$, we get

$$e(G_a[A,B]) + e(G_b[A,B]) + e(G_z[A,B]) + e(G_a[B]) + e(G_a) + e(G_b[B]) + e(G_b)$$
$$\leqslant 2\alpha(n-\alpha) + 2|B|\mu^\star \leqslant 4\alpha(n-\alpha). \tag{3.16}$$

On the left-hand side we can use

$$\deg(a) = e(G_a) = e(G_a[B]) + e(G_a[A,B]),$$

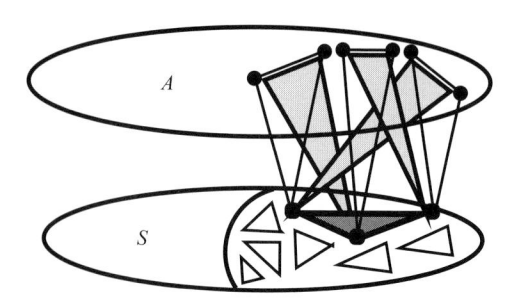

Figure 6. The special configuration H_9

since $G_a[A]$ has no edge. Similarly, $\deg(b) = e(G_b[B]) + e(G_b[A, B])$. So the left-hand side of (3.16) is $e(G_z[A, B]) + 2\deg(a) + 2\deg(b)$. This gives

$$e(G_z[A, B]) \leqslant 4\alpha(n - \alpha) - 4d_{\min}(\mathcal{H}) \quad \text{for every} \quad z \in B. \tag{3.17}$$

Lemma 3.6. *Suppose that $\mathcal{H}$ satisfies the min-degree condition (1.1) and $E^\star \subset B$ satisfies (3.12). Suppose further that $\frac{1}{n}(\alpha, v) \in D_3$, $n > n_0$. Then, for every pair $z_1, z_2 \in E^\star$, there are at least 4α common edges of $G_{z_1}[A]$ and $G_{z_2}[A]$.*

Proof. Indeed, assuming the contrary, we would also have that $e(G_{z_1}[A]) + e(G_{z_2}[A]) < \binom{\alpha}{2} + 4\alpha$. Using this, (3.17) and (3.12) we get

$$2d_{\min} \leqslant \deg_{\mathcal{H}}(z_1) + \deg_{\mathcal{H}}(z_2) = e(G_{z_1}[A]) + e(G_{z_2}[A])$$
$$+ e(G_{z_1}[A, B]) + e(G_{z_2}[A, B]) + e(G_{z_1}[B]) + e(G_{z_2}[B])$$
$$\leqslant \binom{\alpha}{2} + 4\alpha + 8\alpha(n - \alpha) - 8d_{\min} + \frac{1}{2}(n - \alpha - v)^2 + 2v(n - \alpha - v) + 4n.$$

Rearranging, we get

$$0 \leqslant 5\gamma n^2 - \frac{31}{18}n^2 + 7\alpha(n - \alpha) + v(n - \alpha) - \frac{3}{2}v^2 + 10n = n^2 g\left(\frac{\alpha}{n}, \frac{v}{n}\right) + 10n.$$

By Lemma 3.3, $g \leqslant -1/2500$ on D_3. This is a contradiction for $n > 25000$. □

3.9. Small substructures

Suppose that $\mathcal{H}$ is an arbitrary $\mathbb{F}_{3,2}$-free hypergraph, A is a maximum independent set, $\alpha := |A|$, $B := V \setminus A$. Define a *well-positioned* $\mathbf{H}_9 \subseteq \mathcal{H}$ as follows. There are 6 vertices $x_1, \ldots, x_6 \in A$ and 3 vertices $\{z_1, z_2, z_3\} \subseteq B$ and the following 7 edges belong to $E(\mathcal{H})$: (z_1, z_2, z_3), $(z_i, x_{2i-1}x_{2i})$ and $(z_{i+1}, x_{2i-1}x_{2i})$ for $i = 1, 2, 3$ (where $z_4 = z_1$).

Lemma 3.7. *Suppose that $\alpha > \frac{3}{5}n$, $E^\star \subset B$ belongs to a well-positioned $\mathbf{H}_9$ and it satisfies (3.12). Then $d_{\min}(\mathcal{H}) < 0.434\binom{n}{2}$ holds for $n > n_0$.*

Proof. We claim that

$$\frac{3}{2}\sum_{k=1,2,3} e(G_{z_k}[A,B]) + \sum_{1\leqslant j\leqslant 6} e(G_{x_j}[A,B]) \leqslant 6\alpha(n-\alpha). \qquad (3.18)$$

To see this, let us define the weight $w(ab)$ for $a \in A$, $b \in B$ as $w(ab) := \frac{3}{2}m_E(ab) + m_X(ab)$ where $m_E(ab)$ is the number of triples of the form abz_k and $m_X(ab)$ denotes the number of triples of the form abx_j ($1 \leqslant k \leqslant 3$, $1 \leqslant j \leqslant 6$). The left-hand side of (3.18) is equal to $\sum_{a\in A, b\in B} w(ab)$. We will show that $w(ab) \leqslant 6$. This is certainly true if $m_E(ab) = 0$ since, obviously, $m_X(ab) \leqslant 6$. For the case $m_E(ab) = 2$, i.e., if $N(ab)$ meets E^* in two vertices, observe that $\mathbb{F}_{3,2} \not\subset \mathcal{H}$ implies that $N(ab)$ contains at most one from each pair x_{2i-1}, x_{2i}, hence $m_X(ab) \leqslant 3$. Finally, it is easy to check that $m_E(ab) = 1$ implies $m_X \leqslant 4$, completing the proof of (3.18).

Consider the identity

$$\frac{3}{2}\sum_{k=1,2,3} \deg_\mathcal{H}(z_k) + 2\sum_{1\leqslant j\leqslant 6} \deg_\mathcal{H}(x_j)$$

$$= \frac{3}{2}\sum_{k=1,2,3} e(G_{z_k}[A]) + \frac{3}{2}\sum_{k=1,2,3} e(G_{z_k}[A,B]) + \sum_{1\leqslant j\leqslant 6} e(G_{x_j}[A,B])$$

$$+ \sum_{1\leqslant j\leqslant 6} (e(G_{x_j}) + e(G_{x_j}[B])) + \frac{3}{2}\sum_{k=1,2,3} e(G_{z_k}[B]).$$

Estimating the right-hand side, for the first sum we can use that no pair of vertices belongs to all the three $E(G_{z_i})$s. In the second and third sum we use (3.18), in the fourth sum each term is at most $\alpha(n-\alpha)$ by (3.14), and for the last sum we use (3.12). We obtain

$$\frac{33}{2}d_{\min} \leqslant \frac{3}{2} \times 2\binom{\alpha}{2} + 12\alpha(n-\alpha) + \frac{3}{2}\left(\frac{1}{2}(n-\alpha-v)^2 + 2v(n-\alpha-v) + 4n\right).$$

Using that $\frac{3}{2}(\frac{1}{2}(x-v)^2 + 2v(x-v)) \leqslant x^2$ holds for all x, v, we get

$$\frac{33}{2}d_{\min} \leqslant \frac{3}{2}\alpha^2 + 12\alpha(n-\alpha) + (n-\alpha)^2 + 6n = n^2 + 10\alpha n - \frac{19}{2}\alpha^2 + 6n.$$

Here $10\alpha n - (19/2)\alpha^2$ is monotone decreasing for $\alpha > \frac{10}{19}n$. Substituting $\alpha = \frac{3}{5}n$ we get the upper bound $d_{\min}(\mathcal{H}) \leqslant (179/825)n^2 + (12/33)n$, implying our claim. □

3.10. $\mathcal{H}$ is bipartite

Here we prove Theorem 1.3. The above lemmas hold, except Lemma 3.7, for every $\mathbb{F}_{3,2}$-free hypergraph satisfying $\alpha \geqslant \frac{2}{5}n$. From now on we will use that $\mathcal{H}$ satisfies the min-degree condition (1.1). Consider the partition $V = A \cup B = A \cup S \cup T$ defined in Section 3.1. We will prove that B is independent. (3.1) asserts $\frac{1}{n}(\alpha, v) \in D_1$. We will show that $v = 0$.

Our estimates could not handle all cases of (α, v) simultaneously, so we divide the domain D_1 into two parts by the line $y = 5x - 3$ and use different estimates for each of them. We chose this line to (somewhat) maximize the value of γ achievable with the method presented, and, simultaneously, to keep the required calculations minimal. Taking $y = 5x - 3.0053$ one can push γ up to 0.0084.

Consider first the case when $\frac{1}{n}(\alpha,v) \in D_2$. Then Lemma 3.2 gives that $F(\alpha/n, v/n) \leq -1/4000$. Since $(n-3v) \geq \frac{2}{5}n$, we get from (3.10) that

$$\frac{2}{5}n \times \left(d_{\min}(\mathcal{H}) - \left(\frac{4}{9}-\gamma\right)\binom{n}{2}\right) \leq -\frac{1}{4000}n^3 + n^2.$$

Here the left-hand side is nonnegative, but the right-hand side is negative for $n > 4000$. This contradiction implies that $\frac{1}{n}(\alpha,v) \in D_3 \cap D_1$, in particular $\alpha > \frac{3}{5}n$.

Suppose, on the contrary, that $v > 0$. Then there exists an edge $E^* = (z_1, z_2, z_3)$, $E^* \subset B$ satisfying (3.12). Lemma 3.6 implies that E^* can be extended into a well-positioned $\mathbf{H}_9$. Therefore we can apply Lemma 3.7. This leads to the contradiction $d_{\min}(\mathcal{H}) < 0.434\binom{n}{2}$. Thus only $v = 0$ is possible and $\mathcal{H}$ is bipartite. □

3.11. The finer structure of $\mathcal{H}$

In this part we prove Theorem 2.2(i)–(iii). Let A be a maximum independent set. The proof of Theorem 1.3 implies that $\mathcal{H}$ is bipartite, and B is independent (for $n > n_0$). Then all triples have types AAB or ABB. So the min-degree condition (1.1) and (3.5) give

$$\frac{1}{3}n\left(\frac{4}{9}-\gamma\right)\binom{n}{2} \leq e(\mathcal{H}) \leq \frac{1}{2}(n-\alpha)\alpha^2.$$

Multiplying by 6 and rearranging, we obtain

$$\left(\frac{2}{3}n - \alpha\right)^2 (n + 3\alpha) \leq \left(\gamma + \left(\frac{4}{9}-\gamma\right)\frac{1}{n}\right)n^3,$$

implying Theorem 2.2(ii) for $n > \max\{n_0, 1/\gamma\}$:

$$\left|\alpha - \frac{2}{3}n\right| < \sqrt{\gamma} \cdot n. \tag{3.19}$$

Proof of Theorem 2.2(iii). Use that $\mathcal{H}$ is bipartite, Corollary 3.5, (1.1) and (3.19):

$$\Delta_{BBA} = 2(\Delta_{BBA} + \Delta_{BAA}) - (\Delta_{BBA} + 2\Delta_{BAA})$$
$$= 2e(\mathcal{H}) - \sum_{x \in A} \deg_{\mathcal{H}}(x) \leq 2 \times \frac{2}{27}n^3 - \alpha\left(\frac{4}{9}-\gamma\right)\binom{n}{2}$$
$$= \frac{2}{9}\left(\frac{2}{3}n - \alpha\right)n^2 + \gamma\alpha\binom{n}{2} + \frac{2}{9}n\alpha < \sqrt{\gamma}n^3. \quad \square$$

3.12. The case $|A| \geq 0.65n$

In this section we suppose that $\mathcal{H}$ is an $\mathbb{F}_{3,2}$-free hypergraph satisfying the conditions of Theorem 2.2 and also suppose that $\alpha \geq 0.65n$.

Claim 3.8. *If C is an independent set of vertices with $|C \cap A| \geq \frac{1}{4}n$ then $C \subseteq A$.*

Proof. Assuming the contrary, we may fix a $z \in C \cap B$. Consider G_z. It has no edges in B (by Theorem 1.3), or in $C \cap A$, and only very few edges joining A and B (by (3.17)).

We get

$$d_{\min} \leqslant e(G_z[A,B]) + e(G_z[A]) \leqslant 4\alpha(n-\alpha) - 4d_{\min} + \binom{\alpha}{2} - \binom{|A \cap C|}{2}.$$

This is a contradiction for $d_{\min} > (\frac{4}{9} - \gamma)\binom{n}{2}$, $\alpha \geqslant 0.65n$ and $|A \cap C| \geqslant 0.25n$. □

Claim 3.9. $\mu^\star < \alpha$.

Proof. Consider a pair $x, y \in V$ with $\mu(xy) = \mu^\star$, and let $C := N(xy)$. If $|C| < 0.65n \leqslant |A|$, then there is nothing to prove. Otherwise $|C| > |B| + \frac{1}{4}n$, so it has at least $n/4$ common vertices with A. Since C is independent, too, Claim 3.8 implies that $C \subseteq A$. Thus $\mu^\star = \alpha$ is only possible if $C = A$ and $x, y \in B$.

Consider $E(G_x[A])$ and $E(G_y[A])$. If they have a common triangle, say abc, then the hyperedges xab, xac, xay and bcy form an $\mathbb{F}_{3,2}$, a contradiction. So we can apply the Turán–Mantel theorem for $E(G_x[A]) \cap E(G_y[A])$. We get

$$e(G_x[A]) + e(G_y[A]) \leqslant \binom{|A|}{2} + |E(G_x[A]) \cap E(G_y[A])| \leqslant \binom{|A|}{2} + \left\lfloor \frac{1}{4}\alpha^2 \right\rfloor.$$

Since B is independent (by Theorem 2.2(i)) we get $e(G_x) = e(G_x[A]) + e(G_x[A,B])$. Use this for x and y and apply (3.17):

$$2d_{\min} \leqslant e(G_x[A]) + e(G_y[A]) + e(G_x[A,B]) + e(G_y[A,B]) \leqslant \frac{3}{4}\alpha^2 + 8\alpha(n-\alpha) - 8d_{\min}.$$

For $\alpha \geqslant 0.65n$ this gives $d_{\min} \leqslant 0.43\binom{n}{2}$. This contradiction shows that $\alpha = \mu^\star$ is not possible. □

Proof of Theorem 2.2(iv). Consider an arbitrary 2-colouring (A_1, B_1) of $V(\mathcal{H})$ with A_1 and B_1 being independent. Then one of the colour classes meets A in at least $\frac{1}{4}n$ vertices, say $|A_1 \cap A| \geqslant n/4$. Claim 3.8 states that $A_1 \subseteq A$. We show that

$$|A \setminus A_1| \leqslant \sqrt{\gamma}n.$$

Indeed, every triple meeting $A \setminus A_1$ must meet both A_1 and B. Thus for $x \in A \setminus A_1$ we have $d_{\min} \leqslant \deg_{\mathcal{H}}(x) \leqslant |B||A_1|$. This is equivalent to $|B|(|A| - |A_1|) \leqslant |B||A| - d_{\min} = \alpha(n-\alpha) - d_{\min}$. Rearranging, we get

$$|B|(|A| - |A_1|) \leqslant \left(\frac{2}{3}n - \alpha\right)\left(-\frac{1}{3}n + \alpha\right) + \left(\frac{4}{9}\binom{n}{2} - d_{\min}\right) + \frac{2}{9}n.$$

Then (3.19) implies $|A| - |A_1| \leqslant \sqrt{\gamma}n$. □

3.13. The extremal hypergraph

Here we prove Theorem 2.3. Suppose that $\mathcal{H}$ is an n-vertex $\mathbb{F}_{3,2}$-free triple system of maximum cardinality,

$$e(\mathcal{H}) \geqslant \max_a \frac{1}{2}(n-a)a(a-1) := e(n).$$

The degrees of any two vertices of $\mathcal{H}$ differ by at most $n-2$. Otherwise one can delete the vertex of smaller degree and duplicate the other, thus increasing the size of $\mathcal{H}$. Thus we may suppose that (for $n > n_0$) $d_{\min}(\mathcal{H}) > (\frac{4}{9} - 10^{-4})\binom{n}{2}$. Apply Theorems 1.3 and 2.2(i). We obtain that $\mathcal{H}$ has a 2-colouring (A, B) where $|A|$ is the maximal independent set, $|A| = \alpha$. Then Theorem 2.2(ii) implies that $\alpha > 0.65n$. Then Claim 3.9 gives $\mu^\star \leqslant \alpha - 1$. We obtain the desired upper bound:

$$2e(\mathcal{H}) = \sum_{\substack{a \in A \\ b \in B}} \mu(ab) \leqslant \alpha(n-\alpha)\mu^\star \leqslant 2(n-\alpha)\binom{\alpha}{2} \leqslant 2e(n).$$

Moreover, equality can hold only if $\mu(ab) = \alpha - 1$ for every crossing pair $a \in A$, $b \in B$. Then $|N(ab)|$ is large, so by Claim 3.8 it is contained in A. Thus all hyperedges must be of type AAB. □

3.14. Reduction to minimum degree

Here we prove Theorems 2.4 and 2.5. Both deal with hypergraphs satisfying

$$e(\mathcal{H}) > \left(\frac{4}{9} - c\right)\binom{n}{3}. \tag{3.20}$$

Lemma 3.10. *Let $\gamma = c^{2/3}$. Then for $n > n_0(c)$ one can find a subset $V_1 \subseteq V$, $|V_1| = n_1 > (1 - c^{1/3})n$, such that, for $\mathcal{H}_1 := \mathcal{H}[V_1]$,*

$$\deg_{\mathcal{H}_1}(x) > \left(\frac{4}{9} - \gamma\right)\binom{n_1}{2} \tag{3.21}$$

holds for every $x \in V_1$.

Proof. Delete a vertex from V if its degree is at most $(\frac{4}{9} - \gamma)\binom{|V|}{2}$. Repeat this if we can find another vertex of small degree. This way the average degree goes up slightly, but it cannot go too high. A routine counting shows that the process stops within $c^{1/3}n$ steps. We omit the details. □

To prove Theorem 2.4 consider the hypergraph $\mathcal{H}$. Using Lemma 3.10, deleting at most $c^{1/3}n$ vertices (and $O(c^{1/3}n^3)$ edges) we get the hypergraph $\mathcal{H}_1$ satisfying (3.21). Apply Theorem 1.3 to $\mathcal{H}_1$ to obtain a bipartition (A, B). Apply Theorem 2.2(iii) to $\mathcal{H}_1$ to obtain a $(2, 1)$-colourable hypergraph after deleting another $O(\sqrt{\gamma}n^3) = O(c^{1/3}n^3)$ edges. □

To prove Theorem 2.5 consider a triple system $\mathcal{H}$ satisfying (3.20). Apply Lemma 3.10 to get $\mathcal{H}_1$. Then Theorem 2.2(ii), more exactly (3.19), implies that $\alpha(\mathcal{H}_1)$ is about $\frac{2}{3}|V_1|$. Since $|V \setminus V_1|$ is small this gives an upper bound for the maximum codegree,

$$\mathrm{maxcodeg}(\mathcal{H}_1) \leqslant \mathrm{maxcodeg}(\mathcal{H}) \leqslant \alpha(\mathcal{H}) \leqslant \alpha(\mathcal{H}_1) + |V \setminus V_1| \leqslant \frac{2}{3}n + O(\sqrt{\gamma})n.$$

Finally, since B_1 is independent,
$$e(\mathcal{H}_1) = \frac{1}{2} \sum_{a \in A_1, b \in B_1} \mu(ab) \leq \frac{1}{2}\alpha(\mathcal{H}_1)(|V_1| - \alpha(\mathcal{H}_1))\mu^\star(\mathcal{H}_1).$$

Then (3.20), and $|V_1| \sim n$, $\alpha(\mathcal{H}_1) \sim \frac{2}{3}n$ give the lower bound for $\mathsf{maxcodeg}(\mathcal{H}_1)$. □

Acknowledgements

The authors are grateful to one of the referees for helpful comments.

Appendix: Proof of the lemmas on polynomials

A.1. Proof of Lemma 3.1

First, we show that the Hessian of f is indefinite on the open halfplane $D_4 := \{(x,y) : x > 1/3\}$, so the extrema of f on $D_1 \subset D_4$ must be on the four boundary line-segments of ∂D_1. Let f_{xx}, f_{xy}, f_{yy} denote the partial derivatives, and let $\mathbf{J}(x,y)$ be the determinant of the Hessian. Then $f_{xx} = -9x - 2y + 3$, $f_{yy} = 8x + 18y - 8$, $f_{xy} = -2x + 8y - 2$ and

$$\mathbf{J}(x,y) = f_{xx}f_{yy} - f_{xy}^2 = 88x - 76x^2 - 146xy - 28 + 102y - 100y^2.$$

We have that $\mathbf{J}(x,y) \neq 0$ for every $x > 1/3$, because solving $\mathbf{J}(x,y) = 0$ for y, the discriminant $-2271x^2 + 1354x - 199$ is negative for $x > 1/3$. Since there is at least one point where $\mathbf{J}$ is negative, *e.g.*, $\mathbf{J}(1,0) = -16 < 0$, continuity implies that $\mathbf{J}$ is negative for every point of D_4.

Let us check the behaviour of $f(x,y)$ on the boundary ∂D_1. On each segment it reduces to a third-degree polynomial of one variable, whose extrema can be identified by the roots of its derivative, a second-degree polynomial.

(1) First check the lower horizontal boundary, $I_1 := \{(x,0) : 2/5 \leq x \leq 1\}$. Consider $f(x,0) = \frac{3}{2}((1-x)x^2 - \frac{4}{27}) = -\frac{3}{2}(\frac{2}{3} - x)^2(\frac{1}{3} + x)$. So $f \leq 0$ on I_1.

(2) Next check the left vertical boundary line $x = 2/5$. Let $\varphi_2(y) := f(2/5, y) = (225y^3 - 180y^2 + 53y)/75 - 88/1125$ and $I_2 := \{(2/5, y) : 0 \leq y \leq 1/5\}$. The derivative φ_2' has no real roots. So φ_2 is strictly increasing and takes its maximum on $[0, 1/5]$ at $y = 1/5$. Then $\varphi_2(1/5) = -2/225$ implies $f < 0$ on I_2.

(3) The third part to be checked is the boundary segment on $x = 1$. Let $\varphi_3(y) := f(1,y) = (27y^3 - 12y - 2)/9$ and $I_3 := \{(1,y) : 0 \leq y \leq 1/5\}$. We have $\varphi_3'(y) = (27y^2 - 4)/3$; its roots are $\pm \frac{2}{9}\sqrt{3} \sim \pm 0.3849$. So φ_3' is negative on $[0, 1/5]$ and φ_3 is decreasing and takes its maximum at $y = 0$. Then $\varphi_3(0) = -2/9$ implies $f < 0$ on I_3.

(4) Finally, check the segment on $y = 1/5$. Let $\varphi_4(x) := f(x, 1/5) = (-75x^3 + 65x^2 - 12x)/50 + 28/1125$ and $I_4 := \{(x, 1/5) : 2/5 \leq x \leq 1\}$. The derivative of φ_4 is $(-225x^2 + 130x - 12)/50$ and it has one root, $x_2 := (13 + \sqrt{61})/45 \sim 0.462$ in I_4. Hence φ_4' is positive on $[2/5, x_2)$ and negative on $(x_2, 1]$, and φ_4 has its maximum at x_2. We have $\varphi_4(x_2) = (61\sqrt{61} - 665)/30375 \sim -0.0062$ so $f < 0$ on I_4. □

A.2. Proof of Lemma 3.2
It is enough to prove the next two lemmas for $\gamma = 1/125$.

Since $F(x, y)$ and $f(x, y)$ differ by a linear term, their Hessians coincide. So **J** is negative on $D_2 \subset D_1$, too. So the extrema of F must be on some of the four boundary segments of ∂D_2.

(5) Consider the lower horizontal boundary, $I_5 := \{(x, 0) : 2/5 \leqslant x \leqslant 3/5\}$. Define $\varphi_5(x) := F(x, 0) = \frac{3}{2}((1-x)x^2 - \frac{4}{27}) + \frac{1}{2}\gamma$. Then $\varphi_5' = \frac{3}{2}(2x - 3x^2)$ and it is positive on I_5. So φ_5 is increasing and takes its maximum at $x = \frac{3}{5}$. Then $\varphi_5(3/5) = \frac{1}{2}\gamma - 7/1125 = -1/450$, implying $F < -1/4000$ on I_5.

(6) Check F on the left vertical boundary line $x = 2/5$. Let $\varphi_6(y) := F(2/5, y) = (6750y^3 - 5400y^2 + 1563y - 167)/2250$ and $I_6 := \{(2/5, y) : 0 \leqslant y \leqslant 1/5\} = I_2$. Then φ_6' has no real roots; it is positive on $[0, 1/5]$. So φ_6 takes its maximum at $y = 1/5$. We have $\varphi_6(1/5) = -41/5625 \sim -0.007289$, so $F < -1/4000$ on I_6.

(7) Consider the boundary segment on the tilted line $y = 5x - 3$. Let $\varphi_7(x) := F(x, 5x - 3) = (210825x^3 - 405225x^2 + 258873x - 54982)/450$ and $I_7 := \{(x, 5x - 3) : 3/5 \leqslant x \leqslant 0.64\}$. The derivative of φ_7 is $(210825x^2 - 270150x + 86291)/150$, and it has one root $x_1 = (9005 - \sqrt{235358})/14055 \sim 0.6061$ in I_7. Then φ_7' is positive on $[3/5, x_1]$ and negative on $(x_1, 0.64]$. So φ_7 takes its maximum at x_1. We have $\varphi_7(x_1) \sim -2.58 \times 10^{-4}$, so $F < -1/4000$ on I_7.

(8) Finally, check the upper boundary on $y = 1/5$. Let $\varphi_8(x) := F(x, 1/5) = \varphi_4(x) + 1/625$ and $I_8 := \{(x, 1/5) : 2/5 \leqslant x \leqslant 0.64\} \subset I_4$. Since $\varphi_4 < -0.0062$, on I_4 we get $\varphi_4(x) + 0.0016 = \varphi_8(x) < -0.0046$. So $F < -1/4000$ on I_8. □

A.3. Proof of Lemma 3.3
Consider the closed half-plane $\overline{D_3} := \{(x, y) : y \leqslant 5x - 3\}$. We claim that

$$g(x, y) = -7x^2 - xy - \frac{3}{2}y^2 + 7x + y - \frac{31}{18} + 5\gamma,$$

defined in (3.2), takes its maximum on $\overline{D_3}$ at the point $(x_0, y_0) := (20/33, 1/33)$ on the boundary line and $g(x_0, y_0) = -4/99 + 5\gamma = -1/2475 < -1/2500$. Indeed, $g(x, y) = 0$ is an ellipse lying outside D_3, its centre being $(20/41, 7/41)$. Further, g has an absolute maximum at this centre; it is a concave function so its maximum on $\overline{D_3}$ should be on the boundary. Finally, $g(x, 5x - 3) = -99x^2/2 + 60x - 4091/225$ and its maximum can easily be calculated. □

References

[1] Andrásfai, B., Erdős, P. and T. Sós, V. (1974) On the connection between chromatic number, maximal clique and minimal degree of a graph. *Discrete Math.* **8** 205–218.

[2] Bollobás, B. (1974) Three-graphs without two triples whose symmetric difference is contained in a third. *Discrete Math.* **8** 21–24.

[3] Bollobás, B. (1978) *Extremal Graph Theory*, Academic Press, London.
[4] Brown, W. G. (1983) On an open problem of Paul Turán concerning 3-graphs. In *Studies in Pure Mathematics*, Birkhäuser, Basel, pp. 91–93.
[5] de Caen, D. and Füredi, Z. (2000) The maximum size of 3-uniform hypergraphs not containing a Fano plane. *J. Combin. Theory Ser. B* **78** 274–276.
[6] Erdős, P. and Simonovits, M. (1973) On a valence problem in extremal graph theory. *Discrete Math.* **5** 323–334.
[7] Erdős, P. and T. Sós, V. (1980) Problems and results on Ramsey–Turán type theorems (preliminary report). In *Proc. West Coast Conference on Combinatorics, Graph Theory and Computing: Humboldt State Univ., Arcata, CA, 1979. Congress. Numer.* **XXVI** 17–23.
[8] Erdős, P. and T. Sós, V. (1982) On Ramsey–Turán type theorems for hypergraphs. *Combinatorica* **2** 289–295.
[9] Frankl, P. (1990) Asymptotic solution of a Turán-type problem. *Graphs Combin.* **6** 223–227.
[10] Frankl, P. and Füredi, Z. (1988) Extremal problems and the Lagrange function for hypergraphs. *Bull. Inst. Math. Acad. Sinica* **16** 305–313.
[11] Füredi, Z. (1991) Turán type problems. In *Surveys in Combinatorics, 1991* (A. D. Keedwell, ed.), Cambridge University Press, pp. 253–300.
[12] Füredi, Z. and Kündgen, A. (2002) Turán problems for integer-weighted graphs. *J. Graph Theory* **40** 195–225.
[13] Füredi, Z. and Simonovits, M. (2005) Triple systems not containing a Fano configuration. *Combin. Probab. Comput.* **14**, to appear.
[14] Füredi, Z., Pikhurko, O. and Simonovits, M. (2003) The Turán density of the hypergraph $\{abc, ade, bde, cde\}$. *Electronic J. Combin.* **10** # R18 (electronic). http://www.combinatorics.org/
[15] Katona, Gy., Nemetz, T. and M. Simonovits, M. (1964) A new proof of a theorem of P. Turán and some remarks on a generalization of it (in Hungarian). *Mat. Lapok* **15** 228–238.
[16] Keevash, P. and Mubayi, D. (2004) Stability theorems for cancellative hypergraphs. *J. Combin. Theory Sci. B* **1** 163–175.
[17] Keevash, P. and Sudakov, B. The exact Turán number of the Fano plane. *Combinatorica*, to appear.
[18] Keevash, P. and Sudakov, B., On a hypergraph Turán problem of Frankl. *Combinatorica*, to appear.
[19] Kostochka, A. V. (1982) A class of constructions for Turán's (3, 4)-problem. *Combinatorica* **2** 187–192.
[20] Mubayi, D. and Rödl, V. (2002) On the Turán number of triple systems. *J. Combin. Theory Ser. A* **100** 135–152.
[21] Simonovits, M. (1997) Paul Erdős' influence on extremal graph theory. In *The Mathematics of Paul Erdős II* (R. L. Graham and J. Nešetřil, eds), Vol. 14 of *Algorithms and Combinatorics*, Springer, pp. 148–192.
[22] Turán, P. (1961) Research problems. *MTA Mat. Kutató Int. Közl.* **6** 417–423.

Quasirandomness, Counting and Regularity for 3-Uniform Hypergraphs

W. T. GOWERS

Department of Pure Mathematics and Mathematical Statistics,
Wilberforce Road, Cambridge CB3 0WB, UK
(e-mail: W.T.Gowers@dpmms.cam.ac.uk)

Received 17 February 2005; revised 4 August 2005

For Béla Bollobás on his 60th birthday

The main results of this paper are regularity and counting lemmas for 3-uniform hypergraphs. A combination of these two results gives a new proof of a theorem of Frankl and Rödl, of which Szemerédi's theorem for arithmetic progressions of length 4 is a notable consequence. Frankl and Rödl also prove regularity and counting lemmas, but the proofs here, and even the statements, are significantly different. Also included in this paper is a proof of Szemerédi's regularity lemma, some basic facts about quasirandomness for graphs and hypergraphs, and detailed explanations of the motivation for the definitions used.

1. Introduction

One of the most important tools in extremal graph theory is Szemerédi's regularity lemma. Given a graph G with vertex set V, the regularity lemma provides a partition of V into sets $V_1, \ldots, V_K$, such that for almost every pair (i, j) the induced bipartite graph $G(V_i, V_j)$ (that is, the restriction of G to the set of edges xy such that $x \in V_i$ and $y \in V_j$) behaves like a typical random graph of the same density. This result has been such an important tool in graph theory and has so many applications that it is very natural to try to find a generalization that will help with extremal problems for hypergraphs. Indeed, there is an application that, even on its own, provides sufficient motivation for a project of this kind. In 1976, Ruzsa and Szemerédi discovered a simple way to deduce Roth's theorem (Szemerédi's theorem for progressions of length 3) from the regularity lemma [13], and for many years, Vojta Rödl, with several collaborators, has had a very promising programme for obtaining a new proof of the full Szemerédi theorem by generalizing the Ruzsa–Szemerédi argument. In order to do this, one needs not just a regularity lemma but also a so-called 'counting lemma', which says, very roughly, that a sufficiently quasirandom hypergraph will contain small subhypergraphs with approximately the correct frequency. (This way of putting it, though it will do for now, is in fact so rough as to be positively misleading – see Section 5 for an explanation of why.) In 2002, Frankl and Rödl carried out this programme for 3-uniform hypergraphs, thereby giving a new proof of Szemerédi's theorem in the case of progressions of length four [5]. Ten years earlier, they had proved a regularity lemma for k-uniform hypergraphs [4], but obtaining an appropriate counting lemma to go with it was much harder.

The main purpose of this paper is expository: we shall explain the ideas that lie behind a paper of the author [10], the main advance of which is to establish a counting lemma for k-uniform hypergraphs, but also a regularity lemma that is not quite the same as that of Frankl and Rödl. These two results imply not just Szemerédi's theorem in full generality, but also, as has already been observed by Solymosi [15], its multidimensional version. This result had previously been obtained only by Furstenberg's ergodic theory approach [6]. To explain how these results are proved, we shall concentrate on the case of 3-uniform hypergraphs, since the generalization to k-uniform hypergraphs, though notationally more complicated, is not different in an essential way. The main results of this paper are therefore not new, but the point is that the definitions and proofs are different from those of Frankl and Rödl and more readily generalized. Thus, the principal novelty of this paper occurs at the technical level. However, since this is an area where it is easy to become overwhelmed by technical difficulties, technical simplifications are of more than merely technical interest. That said, Nagle, Rödl and Schacht have also recently generalized the Frankl–Rödl approach to k-uniform hypergraphs [11], in work independent of the work discussed here. In the final section of this paper, we explain what the main difference is between the Frankl–Rödl approach and the one given here.

The methods of this paper have their roots in part of the analytic proof of Szemerédi's theorem given by the author in [8, 9]. Another purpose of the paper is to explain this connection. We shall occasionally assume a nodding acquaintance with the ideas of [8] (which are not used here in any formal way, but which shed light on some of our arguments): otherwise this paper is self-contained.

The reader is urged not to be put off by the length of the paper. The main results – counting and regularity lemmas for 3-uniform hypergraphs – are dealt with in Sections 6 and 8 respectively. The rest of the paper consists of discussion, motivating examples and well-known background results in graph theory. Even the sections with the main results contain quite a bit of discussion, rather than being written as densely as possible. If you are looking for a short proof of Szemerédi's theorem for progressions of length four, then the 'true' length of this paper is shorter than that of other papers that establish the same result. Almost all the proofs are straightforward applications of the second-moment method, otherwise known as the Cauchy–Schwarz inequality.

To begin with, here is a brief sketch of a variant of the argument of Ruzsa and Szemerédi that was the starting point for Rödl's programme. Their first step was to prove the following simple-looking statement about graphs.

Theorem 1.1. *For every constant $c > 0$ there exists a constant $a > 0$ with the following property. If G is any graph with n vertices that contains at most an^3 triangles, then it is possible to remove at most cn^2 edges from G to make it triangle-free.*

Sketch proof. For anybody with experience of the regularity lemma, this is an easy and standard argument (but it was of course a serious achievement of Ruzsa and Szemerédi to notice that this kind of result was both easy and significant). First, apply the regularity lemma to obtain a $c/4$-regular partition of G into vertex sets $V_1, \ldots, V_K$ of almost equal size. Remove all edges that belong to pairs (V_i, V_j) that fail to be $c/4$-regular or that have density at most $c/2$. The result is a subgraph G' of G such that every edge belongs to a pair (V_i, V_j) that is $c/4$ regular and has density at least $c/2$, and the number of edges we have removed from G to achieve this is less than cn^2.

It remains to show that G' is triangle-free. But if xyz is a triangle in G' then we must have $x \in V_i$, $y \in V_j$ and $z \in V_k$ with the pairs (V_i, V_j), (V_j, V_k) and (V_i, V_k) all $c/4$ regular and of density at least $c/2$. It is not hard to deduce from this that the number of triangles in G is at least $c^3 N^3 / 256 K^3$. Since K depends on c only, the result is proved. □

Remark. In the above argument we are allowing i, j and k to coincide, so strictly speaking the triangles we obtain should have labellings on their vertices. But this affects the bound by a factor of at most 6. It is more common to insist that K is large, so that there are very few edges joining vertices in the same V_i and one can afford to remove them. In order to allow vertex sets to coincide, we must use a slightly nonstandard version of the regularity lemma – a precise statement can be found in Theorem 7.6 below.

The bound one gets for the dependence of a on c in Theorem 1.1 is extremely weak, and it is a fascinating problem to find a proof that does not use the regularity lemma and therefore, one hopes, gives a better bound. The reason this would be more than a minor curiosity is that Theorem 1.1 implies Roth's theorem – that is, Szemerédi's theorem in the case of progressions of length three. There are several ways to demonstrate this – the way we give here is a small modification of an argument of Solymosi [14], which is a precursor to the argument mentioned earlier. He was the first to observe that one could obtain the following two-dimensional statement as well.

Corollary 1.2. *For every $\delta > 0$ there exists N such that every subset $A \subset [N]^2$ of size at least δN^2 contains a triple of the form (x, y), $(x + d, y)$, $(x, y + d)$ with $d > 0$.*

Proof. First, note that an easy argument allows us to replace A by a set B that is symmetric about some point. Briefly, if the point (x, y) is chosen at random then the intersection of A with $(x, y) - A$ has expected size $c\delta^2 N^2$ for some absolute constant $c > 0$, lives inside the grid $[-N, N]^2$, and has the property that $B = (x, y) - B$. So B is still reasonably dense, and if it contains a subset K then it also contains a translate of $-K$. So we shall not worry about the condition $d > 0$. (I am grateful to Ben Green for bringing this trick to my attention.)

Without loss of generality, the original set A is symmetric in this sense. Let X be the set of all vertical lines through $[N]^2$, that is, subsets of the form $\{(x, y) : x = u\}$ for some $u \in [N]$. Similarly, let Y be the set of all horizontal lines. Define a third set, Z, of diagonal lines, that is, lines of constant $x + y$. These sets form the vertex sets of a tripartite graph, where a line in one set is joined to a line in another if and only if their intersection belongs to A. For example, the line $x = u$ is joined to the line $y = v$ if and only if $(u,v) \in A$ and the line $x = u$ is joined to the line $x + y = w$ if and only if $(u, w - u) \in A$.

Suppose that the resulting graph G contains a triangle of lines $x = u$, $y = v$, $x + y = w$. Then the points (u,v), $(u, w - u)$ and $(w - v, v)$ all lie in A. Setting $d = w - u - v$, we can rewrite them as (u,v), $(u, v + d)$, $(u + d, v)$, which shows that we are done unless $d = 0$. When $d = 0$, we have $u + v = w$, which corresponds to the degenerate case when the vertices of the triangle in G are three lines that intersect in a single point. Clearly, this can happen in at most $|A| = o(N^3)$ ways.

Therefore, if A contains no configuration of the desired kind, then the hypothesis of Theorem 1.1 holds, and we can remove $o(N^2)$ edges from G to make it triangle-free. But this is a contradiction, because there are at least δN^2 degenerate triangles and they are edge-disjoint. $\square$

It is straightforward to deduce Roth's theorem from the above result. Note that for this deduction we do not mind if the d obtained in Corollary 1.2 is negative, but it is interesting to know that we can ask for the right angle to be in the bottom left corner of the relevant triangle.

Corollary 1.3. *For every $\delta > 0$ there exists N such that every subset A of $\{1, 2, \ldots, N\}$ of size at least δN contains an arithmetic progression of length 3.*

Proof. Define $B \subset [N]^2$ to be the set of all (x, y) such that $x - y \in A$. It is straightforward to show that B has density at least $\eta > 0$ for some η that depends on δ only. Applying Corollary 1.2 to B we obtain inside it three points (x, y), $(x + d, y)$ and $(x, y + d)$. Then the three numbers $x - y - d$, $x - y$, $x + d - y$ belong to A and form an arithmetic progression. $\square$

Rödl's programme, outlined in [12], was to generalize Theorem 1.1 to hypergraphs, using generalized regularity and counting lemmas to prove it. The various ways of deducing Roth's theorem from Theorem 1.1 can then be straightforwardly modified to give deductions of the full Szemerédi theorem.

To be more precise about this, let H be a 3-uniform hypergraph. By a *simplex* in H we mean a collection of four edges of the form $\{xyz, xyw, xzw, yzw\}$, that is, a complete subhypergraph on four vertices. If one thinks of the edges of H as (two-dimensional) triangles, then the 'edges' of a simplex can be thought of as the faces of a tetrahedron. However one looks at it, this is the natural generalization of the notion of a triangle in a graph. The result of Frankl and Rödl mentioned earlier is the following.

Theorem 1.4. *For every constant $c > 0$ there exists a constant $a > 0$ with the following property. If H is any 3-uniform hypergraph with n vertices that contains at most an^4 simplices, then it is possible to remove at most cn^3 edges from H to make it simplex-free.*

As Solymosi demonstrated, it is easy to adapt the proof of Corollary 1.2 and show that Theorem 1.4 has the following consequence.

Theorem 1.5. *For every $\delta > 0$ there exists N such that every subset $A \subset [N]^3$ of size at least δN^3 contains a quadruple of points of the form*

$$\{(x, y, z), (x + d, y, z), (x, y + d, z), (x, y, z + d)\}$$

with $d > 0$.

Similarly, Szemerédi's theorem for progressions of length four is an easy consequence of Theorem 1.5.

Thus, once one has appropriate generalizations of the regularity and counting lemmas to hypergraphs, the rest of the argument goes through quite easily. However, as will become clear over the next three sections, even to come up with the right statements is harder than one might at first think. To begin with, however, we shall provide some essential background by discussing several notions of quasirandomness and the relationships between them.

2. Quasirandom graphs, hypergraphs and subsets of $\mathbb{Z}_N$: some definitions

Every known proof of Szemerédi's theorem involves somewhere a notion of quasirandomness and a two-case argument of the following kind: if a certain structure is quasirandom then it contains several configurations of the kind one is looking for (just as one expects from a random structure), and if it is not then one can exploit the non-quasirandomness and pass to the next stage of an iteration. In this section we review certain notions of quasirandomness and point out connections between them (some of which are well known).

The first is the definition of a quasirandom graph, which was introduced by Chung, Graham and Wilson [3]. (A similar notion was discovered independently by Thomason [16].) There are in fact several different definitions, and the main purpose of their paper was to show that they are all equivalent. Here we restrict attention to bipartite graphs and focus on just two of the definitions. (Chung, Graham and Wilson state their results in the case $p = 1/2$ only but the generalization to arbitrary p is not hard.) A proof of a slightly modified¡ result will be given in the next section.

Theorem 2.1. *Let G be a bipartite graph with vertex sets X and Y. Let $|X| = M$ and $|Y| = N$ and suppose that G has pMN edges. Then the following properties of G are equivalent.*

(i) *The number of labelled 4-cycles in G that start in X (that is, quadruples $(x_1, x_2, y_1, y_2) \in X^2 \times Y^2$ such that x_1y_1, x_1y_2, x_2y_1 and x_2y_2 are all edges of G) is at most $p^4M^2N^2 + c_1M^2N^2$.*

(ii) *If X' and Y' are any two subsets of X and Y respectively, then the number of edges from X' to Y' differs from $p|X||Y|$ by at most c_2MN.*

The meaning of the equivalence is that for any $c_2 > 0$ there exists $c_1 > 0$ such that if (i) holds for c_1 then (ii) holds for c_2 – and the other way round as well. A graph that satisfies property (i) with $c_1 = \alpha$ is often called α-quasirandom. For the time being, we shall adopt this definition, though in the next section we shall choose a different, equivalent definition that is more convenient.

Chung and Graham [2] went on to define a notion of quasirandomness for subsets of $\mathbb{Z}_N$. Again, they gave the definition in the form of a theorem that asserted the equivalence of various randomness properties, and once again we shall focus on just two of these and consider the case of an arbitrary 'probability' p rather than just $p = 1/2$. By a *mod-N progression* we mean a set of the form $(a, a+d, \ldots, a+(m-1)d)$, where addition is in the group $\mathbb{Z}_N$.

Theorem 2.2. *Let A be a subset of $\mathbb{Z}_N$ of size pN. Then the following properties of A are equivalent.*

(i) *The number of quadruples $(a,b,c,d) \in A^4$ such that $a+b = c+d$ is at most $p^4 N^3 + c_1 N^3$.*
(ii) *If X is any mod-N progression then $|A \cap X| = p|X| + c_2 N$.*

Chung and Graham also pointed out that there is a close connection between quasirandom subsets of $\mathbb{Z}_N$ and quasirandom graphs. To see this, let A be a subset of $\mathbb{Z}_N$ and define a bipartite graph G with vertex sets $X = Y = \mathbb{Z}_N$ by letting $(x,y) \in X \times Y$ be an edge if and only if $x + y \in A$. Then, given a 4-cycle (x_1, x_2, y_1, y_2) of the kind counted in property (i) of Theorem 2.1, we know that $x_1 + y_1$, $x_1 + y_2$, $x_2 + y_1$ and $x_2 + y_2$ all belong to A, and moreover that

$$(x_1 + y_1) + (x_2 + y_2) = (x_1 + y_2) + (x_2 + y_1),$$

from which it is easy to see that there is an N-to-one correspondence between the 4-cycles from Theorem 2.1(i) and the quadruples from Theorem 2.2(i). Thus, the set A is quasirandom if and only if the corresponding graph G is quasirandom.

Why are quasirandom sets useful for Szemerédi's theorem? Briefly, the reason is that a quasirandom set A of density p must contain approximately the same number of arithmetic progressions of length 3 as a typical random set of that density, while if it fails to be quasirandom, then property (ii) of Theorem 2.2 tells us that there are well-structured subsets of $\mathbb{Z}_N$ inside which A has density substantially different from p (and hence, by an averaging argument, sometimes substantially larger). This makes possible an iteration argument of the kind mentioned at the beginning of this section.

In all known proofs of Szemerédi's theorem, the difficulty increases sharply when the length of the progression increases from 3 to 4. In the analytic approach of [8, 9], the reason for this difficulty is that the quasirandomness of Theorem 2.2 is not a sensitive enough property to detect that a set has too many or too few arithmetic progressions of length 4 (compared with a random set of the same density). The idea of [8] is to define a stronger property, known there as *quadratic uniformity*, which *is* sensitive to progressions of length 4. However, because it is a stronger property, it is much harder to say anything about sets that fail to be quadratically uniform – a point to which we shall return.

The main purpose of this section, however, is to show how the notion of quadratic uniformity leads naturally to a definition, also due to Chung and Graham [1] (but arrived at in a different way), of quasirandomness for 3-uniform hypergraphs. First, then, here is the definition of a quadratically uniform subset of $\mathbb{Z}_N$.

Definition 2.3. Let $\alpha > 0$. A subset $A \subset \mathbb{Z}_N$ of size pN is α-*quadratically uniform* if A^8 contains at most $(p^4 + \alpha)N^4$ octuples of the form

$$(x, x+a, x+b, x+c, x+a+b, x+a+c, x+b+c, x+a+b+c).$$

Since quadruples (a, b, c, d) with $a + b = c + d$ are in one-to-one correspondence with quadruples of the form $(x, x+a, x+b, x+a+b)$, this definition is a natural generalization of property (i) of Theorem 2.2.

In order to define quasirandomness for 3-uniform hypergraphs, we shall 'complete the square', by finding a property that stands in relation to quadratic uniformity as quasirandomness for graphs does to quasirandomness for subsets of $\mathbb{Z}_N$. To establish that connection we started with a set A and defined the bipartite graph G to consist of all pairs (x, y) such that $x + y \in A$. Now we would like to find an associated 3-uniform hypergraph, and it seems sensible to try the tripartite hypergraph H with vertex sets $X = Y = Z = \mathbb{Z}_N$, with the triple (x, y, z) forming an edge of H if and only if $x + y + z \in A$.

In order to generalize Definition 2.3, we now want to find a structure in H that corresponds to octuples of the given kind. It is natural to think of these octuples as labelling vertices of a cube (as they would if a, b and c denoted three orthogonal vectors of the same length). It is also natural to think of the edges of a 3-uniform hypergraph as triangles. Is there a configuration of triangles naturally associated with a cube?

Indeed there is: the dual of a cube is an octahedron and an octahedron is made of triangles. This suggests that we should regard a quadruple $(x, x+a, x+b, x+a+b)$ in $\mathbb{Z}_N$ as a kind of square, and a 4-cycle (x_1, x_2, y_1, y_2) in a bipartite graph as its dual, which happens, confusingly, to look like a square as well. (This dual square gives rise to the square $(x_1 + y_1, x_1 + y_2, x_2 + y_1, x_2 + y_2)$ in $\mathbb{Z}_N$.) The points x_i and y_i of the 4-cycle are vertices, so the vertices of the square in $\mathbb{Z}_N$ correspond to (1-dimensional) faces, and the value attached to a face is the sum of the values at the vertices. This gives us a construction that generalizes easily to all dimensions.

The above remarks provide the justification for the following construction. Given a subset $A \subset \mathbb{Z}_N$, define a tripartite 3-uniform hypergraph H with vertex sets $X = Y = Z = \mathbb{Z}_N$ to be the set of all triples (x, y, z) such that $x + y + z \in A$. Define an *octahedron* in this (or any other) hypergraph to be a set of eight 3-edges of the form $\{(x_i, y_j, z_k) : i, j, k \in \{1, 2\}\}$, where $x_1, x_2 \in X$, $y_1, y_2 \in Y$, $z_1, z_2 \in Z$. Equivalently, an octahedron is a complete tripartite subhypergraph with two vertices from each vertex set of H.

It is now reasonable to guess that a good definition of quasirandomness for 3-uniform hypergraphs is the following.

Definition 2.4. Let H be a tripartite 3-uniform hypergraph with vertex sets X, Y and Z of size L, M and N respectively and suppose that H has $pLMN$ edges. Then H is α-*quasirandom* if it contains at most $(p^8 + \alpha)L^2M^2N^2$ octahedra.

This suggestion can be motivated further (to do so one attempts to generalize arguments about graphs and sees what one needs) but for now it is perhaps enough simply to say that the guess turns out to be correct, and to point out that the number of octahedra must be at least $p^8 L^2 M^2 N^2$. (This last observation follows easily from the Cauchy–Schwarz inequality, as will become clear later in this paper.)

A first sign that the definition is a good one is that the eight numbers $x_i + y_j + z_k$ do form a cube, in the sense discussed earlier. It follows that a subset A of $\mathbb{Z}_N$ gives rise to an α-quasirandom 3-graph H if and only if A is α-quadratically uniform.

As it happens, there is a different definition (Definition 4.3 below) that is even better. However, it is equivalent apart from the precise value of α, and the advantage is a technical one, so for the purposes of the discussion in this section we shall stick with the more obvious Definition 2.4.

Theorem 2.1 stated that two quasirandomness properties of graphs are equivalent. We have just generalized the first of these properties. What about the second? It is here that hypergraph quasirandomness springs a surprise. The property that most obviously generalizes property (ii) of Theorem 2.1 is the following, which we shall call vertex-uniformity.

Definition 2.5. Let H be a 3-uniform hypergraph with vertex sets X, Y and Z of sizes L, M and N respectively and suppose that H has $pLMN$ edges. Then H is β-*vertex-uniform* if, for any choice of subsets $X' \subset X$, $Y' \subset Y$ and $Z' \subset Z$, the number of triples $(x, y, z) \in X' \times Y' \times Z'$ that belong to H differs from $p|X'||Y'||Z'|$ by at most βLMN.

In the next section we shall see that an α-quasirandom 3-graph is β-vertex-uniform for some β depending on α only, but the reverse is not true. In Section 4 we shall give two examples that demonstrate the failure of this implication. We end this section with one last definition, of a property that generalizes property (ii) of Theorem 2.1 in a less naive way. This property *will* turn out to be equivalent to quasirandomness – another result to be proved in the next section.

Definition 2.6. Let H be a 3-uniform hypergraph with vertex sets X, Y and Z of sizes L, M and N respectively and suppose that H has $pLMN$ edges. Then H is γ-*edge-uniform* if, for every $t \in [0, 1]$ and every tripartite graph G with vertex sets X, Y and Z and $tLMN$ triangles, the number of triangles in G that belong to H differs from $ptLMN$ by at most γLMN.

Here, of course, a triangle in G is said to belong to H if its vertices form a triple that belongs to H.

Why is this a generalization of property (ii)? Well, that property says of a bipartite graph that it doesn't significantly correlate with graphs that are induced by sets of vertices (that is, complete bipartite graphs on subsets of the vertex sets). Edge-uniformity says of a 3-uniform hypergraph that it doesn't correlate with 3-uniform hypergraphs that are induced by sets of edges (as opposed to hyperedges).

Although in this paper we are concentrating on 3-uniform hypergraphs, it should be clear how the definitions of this section are generalized. A k-partite k-uniform hypergraph H of density p with vertex sets X_i of size N_i is quasirandom if it contains at most $(p^{2^k} + c)(N_1 \ldots N_k)^2$ k-dimensional octahedra and c is small; this turns out to be equivalent to the assertion that H is $((k-1)$-edge)-uniform, in the sense that H does not significantly correlate with any k-uniform hypergraph induced from a $(k-1)$-uniform hypergraph. Using this kind of language, one could say that a quasirandom graph is a 2-uniform hypergraph that is (1-edge)-uniform. The right generalization of the pair $(2, 1)$ is not $(k, 1)$ as one might at first suppose, but $(k, k-1)$. Thus, Definition 2.6 is more useful to us than Definition 2.5.

3. Quasirandom functions and a counting lemma

Let G be a tripartite graph with vertex sets X, Y and Z, of sizes L, M and N respectively, and write $G(X, Y)$, $G(Y, Z)$ and $G(X, Z)$ for the three bipartite parts of G. Suppose that these parts are quasirandom with probabilities p, q and r respectively. How many triangles does G contain?

If we use genuinely random graphs as our guide, we should expect the answer to be about $pqrLMN$, and indeed it is easy to use property (ii) of Theorem 2.1 to prove that this is approximately right: a typical vertex in X has about pM neighbours in Y and rN neighbours in Z; these two neighbourhoods are linked by about $q(pM)(rN)$ edges; summing over all $x \in X$ we obtain the desired estimate.

The argument just sketched can be generalized quite easily from triangles to copies of any small graph, and this generalization is what we shall refer to as the *counting lemma* for graphs. The first aim of this section will be to give it a different and less transparent proof. Why should we wish to do something so apparently perverse? Because the alternative proof has a number of advantages: first, it is closely modelled on the analytic arguments of [8] and related arguments from Furstenberg's ergodic-theory proof of Szemerédi's theorem (see for example [7]); more importantly, this analytic approach is much easier to generalize, an advantage that is very noticeable even for 'the first nontrivial case', that of 3-uniform hypergraphs. Thus, the reader who is prepared to make the small effort needed to understand the proof of Theorem 3.5 below will understand the basic structure of the longer arguments of Sections 4 and 6, and also that of [10]: the main lemmas and arguments there all have their prototypes here.

An important technicality in the analytic approach to these arguments is to think of sets as $\{0, 1\}$-valued functions and to generalize set-theoretic arguments to functions taking values in more general sets such as $[0, 1]$, $[-1, 1]$, $\mathbb{R}_+$, $\mathbb{R}$, $\{z \in \mathbb{C} : |z| \leqslant 1\}$ or $\mathbb{C}$. We shall do that now, proving results about $[-1, 1]$-valued functions and only occasionally pausing to deduce results about graphs and hypergraphs. Our first result is a version of Theorem 2.1 for functions. It is not quite a direct generalization of Theorem 2.1 because the conclusion is generalized to functions as well as the hypothesis. However, as we shall see, if G is regular then it is easy to deduce Theorem 2.1 from it. We shall not bother with the irregular case, because we shall base our later arguments on Theorem 3.1 and not use Theorem 2.1.

Theorem 3.1. Let X and Y be sets of sizes M and N respectively and let $f : X \times Y \to [-1, 1]$. Then the following statements are equivalent.

(i) $\sum_{x,x' \in X} \sum_{y,y' \in Y} f(x, y) f(x', y) f(x, y') f(x', y') \leqslant c_1 M^2 N^2$.

(ii) For any pair of functions $u : X \to [-1, 1]$ and $v : Y \to [-1, 1]$ we have the inequality
$|\sum_{x,y} f(x, y) u(x) v(y)| \leqslant c_2 MN$.

(iii) For any pair of sets $X' \subset X$ and $Y' \subset Y$ we have the inequality

$$\left| \sum_{x \in X'} \sum_{y \in Y'} f(x, y) \right| \leqslant c_3 MN.$$

Moreover, if $\sum_{x,y} f(x, y) = 0$ then they are also equivalent to the following further statement.

(iv) For any pair of sets $X' \subset X$ and $Y' \subset Y$ we have the inequality

$$\sum_{x \in X'} \sum_{y \in Y'} f(x, y) \leqslant c_4 MN.$$

Proof. We shall begin with a very simple argument that shows that (ii) implies (i). Let us assume that (i) is false, or in other words that

$$\sum_{x,x' \in X} \sum_{y,y' \in Y} f(x, y) f(x', y) f(x, y') f(x', y') > c_1 M^2 N^2.$$

If we choose $x' \in X$ and $y' \in Y$ randomly and independently, then the average of the sum $\sum_{x,y} f(x, y) f(x', y) f(x, y') f(x', y')$ is greater than $c_1 MN$. Therefore, we may fix a choice of x' and y' such that the sum itself is greater than $c_1 MN$. But then (ii) is false if we set $c_2 = c_1$, $u(x) = f(x, y')$ and $v(y) = f(x', y) f(x', y')$. (What is important here is that fixing x' and y' turns all the terms except for $f(x, y)$ into constants or functions of one variable.)

The reverse implication uses the Cauchy–Schwarz inequality several times. It is this technique that, when suitably generalized, lies at the heart of the proof of the counting lemma for hypergraphs. (To be more precise, what we keep using is not the Cauchy–Schwarz inequality directly, but the inequality $\left(\sum_{i=1}^{N} a_i \right)^2 \leqslant N \sum_{i=1}^{N} a_i^2$, which follows from it.) In the expressions that follow, sums involving x and x' are over X and sums involving y and y' are over Y:

$$\left| \sum_{x,y} f(x, y) u(x) v(y) \right|^4 = \left(\left(\sum_x \sum_y f(x, y) u(x) v(y) \right)^2 \right)^2$$

$$\leqslant \left(M \sum_x \left(\sum_y f(x, y) u(x) v(y) \right)^2 \right)^2$$

$$\leqslant \left(M \sum_x \left(\sum_y f(x, y) v(y) \right)^2 \right)^2$$

$$= M^2 \left(\sum_x \sum_{y,y'} f(x, y) f(x, y') v(y) v(y') \right)^2$$

$$\leqslant M^2 N^2 \sum_{y,y'} \left(\sum_x f(x,y) f(x,y') v(y) v(y') \right)^2$$

$$\leqslant M^2 N^2 \sum_{y,y'} \left(\sum_x f(x,y) f(x,y') \right)^2$$

$$= M^2 N^2 \sum_{x,x'} \sum_{y,y'} f(x,y) f(x,y') f(x',y) f(x',y').$$

These calculations show that if (i) is true then (ii) is true for $c_2 = c_1^{1/4}$.

It is obvious that (ii) implies both (iii) and (iv), since if either were false then one could take u and v to be the characteristic functions of X' and Y' respectively. It therefore remains to prove that (iii) implies (ii), and that (iv) implies (ii) if the $f(x,y)$ add up to 0.

Suppose, then, that (ii) is false and we have functions $u : X \to [-1,1]$ and $v : Y \to [-1,1]$ such that $|\sum_{x,y} f(x,y) u(x) v(y)| > c_2 MN$. We can write $u = u_+ - u_-$ with u_+ and u_- disjointly supported and taking values in $[0,1]$, and similarly we can write $v = v_+ - v_-$. It follows that there are $[0,1]$-valued functions s and t such that $|\sum_{x,y} f(x,y) s(x) t(y)| > c_2 MN/4$. Now let X_1 and Y_1 be random subsets of X and Y respectively, with their elements chosen independently with probabilities given by the functions s and t. Then the expectation of $\sum_{x \in X_1} \sum_{y \in Y_1} f(x,y)$ is $\sum_{x,y} f(x,y) s(x) t(y)$, so there must exist a choice of X_1 and Y_1 such that $|\sum_{x \in X_1} \sum_{y \in Y_1} f(x,y)|$ is greater than $c_2 MN/4$. This contradicts (iii) if $c_3 \leqslant c_2/4$.

Now let us continue with the argument under the extra assumption that $\sum_{x,y} f(x,y) = 0$. If the sum we have obtained is positive, then we are done. Otherwise, let $X_2 = X \setminus X_1$ and $Y_2 = Y \setminus Y_1$, and for $i,j \in \{1,2\}$ let $S_{ij} = \sum_{x \in X_i} \sum_{y \in Y_j} f(x,y)$. Then $S_{11} + S_{12} + S_{21} + S_{22} = 0$ from which it follows that there exists a pair $(i,j) \neq (1,1)$ such that $\sum_{x \in X_i} \sum_{y \in Y_j} f(x,y) > c_2 MN/12$. This contradicts (iv) if $c_4 \leqslant c_2/12$. $\square$

Theorem 3.1 motivates the following definition.

Definition 3.2. Let X and Y be sets of size M and N. A function f is α-*quasirandom* if $\sum_{x,x' \in X} \sum_{y,y' \in Y} f(x,y) f(x',y) f(x,y') f(x',y') \leqslant \alpha M^2 N^2$.

If G is a bipartite graph with vertex sets X and Y, let us write $G(x,y)$ for the function that is 1 if xy is an edge of G and 0 otherwise. Suppose that $|X| = M$, $|Y| = N$ and every vertex in X has degree pN, and set $f(x,y) = G(x,y) - p$. Then it is easy to verify that

$$\sum_{x,x',y,y'} G(x,y) G(x,y') G(x',y) G(x',y')$$

$$= \sum_{x,x',y,y'} f(x,y) f(x,y') f(x',y) f(x',y') + p^4 M^2 N^2.$$

It follows that G is α-quasirandom, in the sense of Section 2, if and only if f is α-quasirandom. From now on we shall adopt this as our definition of quasirandomness even in the nonregular case.

Definition 3.3. Let G be a bipartite graph with vertex sets X and Y of size M and N and suppose that G has pMN edges. Then G is α-*quasirandom* if the function $f(x, y) = G(x, y) - p$ is α-quasirandom.

Notice also that, with this definition of f,

$$\sum_{x \in X'} \sum_{y \in Y'} G(x, y) = \sum_{x \in X'} \sum_{y \in Y'} f(x, y) + p|X'||Y'|,$$

so property (iii) of Theorem 3.1 and property (ii) of Theorem 2.1 are trivially equivalent. We have therefore proved Theorem 2.1 in the case of regular graphs.

Now let us state and prove a counting lemma for graphs. We begin with the special case of triangles in order to demonstrate the argument without getting tied up with notation. It is also the case of most immediate interest.

Lemma 3.4. *Let G be a tripartite graph with vertex sets X, Y and Z, of sizes L, M and N respectively. Suppose that the bipartite graphs $G(X, Y)$, $G(Y, Z)$ and $G(X, Z)$ are α-quasirandom with densities p, q and r respectively. Then the number of triangles in G differs from $pqrLMN$ by at most $4\alpha^{1/4}LMN$.*

Proof. Let the variables x, y and z always stand for elements of X, Y and Z respectively, so that we do not keep needing to specify this. Define a function $f : X \times Y \to [-1, 1]$ by $f(x, y) = G(x, y) - p$, and similarly let $g(y, z) = G(y, z) - q$ and $h(x, z) = G(x, z) - r$. In terms of this notation, the number of triangles in G is given by the sum

$$\sum_{x,y,z} (p + f(x, y))(q + g(y, z))(r + h(x, z)).$$

This sum splits naturally into eight parts, and the idea of the proof is that if α is small then only the main term $pqrLMN$ makes a significant contribution to it. To see this, let us consider any one of the four terms that involves $f(x, y)$ rather than p. It will have the form $\sum_{x,y,z} f(x, y)u(y, z)v(x, z)$, where u is either q or g and v is either r or h.

If we now fix z, we obtain an expression of the form $\sum_{x,y} f(x, y)u(y)v(x)$. The deduction of property (ii) from property (i) in Theorem 3.1 tells us that this is at most $\alpha^{1/4}LM$, which gives us that $\sum_{x,y,z} f(x, y)u(y, z)v(x, z)$ is at most $\alpha^{1/4}LMN$. There are seven terms other than the main one, of this kind, three of which are easily seen to be zero, so the result follows. □

The general counting lemma is proved in essentially the same way. We shall state it in an equivalent form, as a 'probability lemma'. Notice that the probability in the conclusion of the lemma is what one would expect in the case of random graphs. The conclusion of the lemma therefore says that there are about as many copies of H in G as one would expect.

Theorem 3.5. *Let G be an m-partite graph with vertex sets $X_1, \ldots, X_m$ and write N_i for the size of X_i. Suppose that for each pair (i, j) the induced bipartite graph $G(X_i, X_j)$ is*

α-quasirandom with density p_{ij}. Let H be any graph with vertex set $\{1, 2, \ldots, m\}$ and let $(x_1, \ldots, x_m)$ be a random element of $X_1 \times \cdots \times X_m$. Then the probability that the function $i \mapsto x_i$ is an isomorphic embedding of H into G differs from $\prod_{ij \in E(H)} p_{ij} \prod_{ij \notin E(H)} (1 - p_{ij})$ by at most $2^{\binom{m}{2}} \alpha^{1/4}$. The probability that $x_i x_j$ is an edge of G whenever ij is an edge of H (but not necessarily conversely) differs from $\prod_{ij \in E(H)} p_{ij}$ by at most $2^{|E(H)|} \alpha^{1/4}$.

Proof. This time, let x_i always stand for an element of X_i. Let $f_{ij}(x, y) = G(x, y) - p_{ij}$ for each pair (i, j). Then the probability in question is

$$(N_1 \ldots N_m)^{-1} \sum_{x_1, \ldots, x_m} \prod_{ij \in E(H)} (p_{ij} + f_{ij}(x_i, x_j)) \prod_{ij \notin E(H)} (1 - p_{ij} - f_{ij}(x_i, x_j)).$$

Once again, we have a sum that splits up into several terms. The main term is

$$\prod_{ij \in E(H)} p_{ij} \prod_{ij \notin E(H)} (1 - p_{ij})$$

and it remains to show that all other terms are small. But any other term must choose $f_{ij}(x_i, x_j)$ from at least one bracket, and since only one bracket involves both x_i and x_j, if we fix all the other x_k we obtain an expression of the form $\sum_{x_i, x_j} f_{ij}(x_i, x_j) u(x_i) v(x_j)$, with u and v taking values in the interval $[-1, 1]$. It follows from the α-quasirandomness of f_{ij} and Theorem 3.1 that this is at most $\alpha^{1/4} N_i N_j$. Summing over the other $m - 2$ variables and multiplying by $(N_1 \ldots N_m)^{-1}$ we find that each term apart from the main one has size at most $\alpha^{1/4}$. Since there are $2^{\binom{m}{2}}$ terms, the result follows (and could be improved slightly since some of the terms are zero).

The proof of the second assertion is similar, but slightly simpler. $\square$

4. A counting lemma for quasirandom 3-graphs

The virtue of the arguments in Section 3 is that they generalize easily. Our first demonstration of this takes the form of very similar proofs of corresponding results for quasirandom 3-uniform hypergraphs. As in Section 3, we begin with a result about functions that serves as a definition of quasirandomness. The statement is very similar to that of Theorem 3.1, and the proofs of the implications are also very similar to the corresponding proofs in Theorem 3.1. Properties (i) and (iii) below are functional versions of Definitions 2.4 and 2.6 respectively.

During part of the proof, we shall make use of a nonstandard but very convenient 'product convention'. If g is any function of k variables $x_1, \ldots, x_k$, then $g_{x, x'}(x_2, \ldots, x_k)$ will be shorthand for $g(x, x_2, \ldots, x_k) g(x', x_2, \ldots, x_k)$. What's more, we shall iterate this, writing $g_{x, x', y, y'}$ for $(g_{x, x'})_{y, y'}$ and so on. For instance, if g is a function of three variables, then

$$g_{x, x', y, y'}(z) = g(x, y, z) g(x', y, z) g(x, y', z) g(x', y', z).$$

If we iterate k times, then the resulting function is a function of no variables, that is, a constant. To be precise, $g_{x_1, x'_1, \ldots, x_k, x'_k}$ is the number $\prod_{\epsilon \in \{0,1\}^k} g(u_1(\epsilon), \ldots, u_k(\epsilon))$, where $u_i(\epsilon) = x_i$ if $\epsilon_i = 0$ and $u_i(\epsilon) = x'_i$ if $\epsilon_i = 1$.

Theorem 4.1. Let X, Y and Z be sets of sizes L, M and N and let $f : X \times Y \times Z \to [-1, 1]$. Then the following statements are equivalent.

(i) $\sum_{x_0,x_1 \in X} \sum_{y_0,y_1 \in Y} \sum_{z_0,z_1 \in Z} \prod_{(i,j,k) \in \{0,1\}^3} f(x_i, y_j, z_k) \leqslant c_1 L^2 M^2 N^2$.
(ii) For any three functions $u : X \times Y \to [-1, 1]$, $v : Y \times Z \to [-1, 1]$ and $w : X \times Z \to [-1, 1]$ we have the inequality $|\sum_{x,y,z} f(x, y, z) u(x, y) v(y, z) w(x, z)| \leqslant c_2 LMN$.
(iii) For any tripartite graph G with vertex sets X, Y and Z, the sum of $f(x, y, z)$ over all triangles xyz of G is at most $c_3 LMN$ in magnitude.

Moreover, if $\sum_{x,y,z} f(x, y, z) = 0$ then the following additional statement is equivalent to the above three.

(iv) For any tripartite graph G with vertex sets X, Y and Z, the sum of $f(x, y, z)$ over all triangles xyz of G is at most $c_4 LMN$.

Proof. Assume that (i) is false, so that

$$\sum_{x_0,x_1 \in X} \sum_{y_0,y_1 \in Y} \sum_{z_0,z_1 \in Z} \prod_{(i,j,k) \in \{0,1\}^3} f(x_i, y_j, z_k) > c_1 L^2 M^2 N^2.$$

Choose x_1, y_1 and z_1 randomly and independently. Then the expectation of the sum

$$\sum_{x_0,y_0,z_0} \prod_{(i,j,k) \in \{0,1\}^3} f(x_i, y_j, z_k)$$

is greater than $c_1 LMN$, so we may fix x_1, y_1 and z_1 in such a way that the sum itself is greater than $c_1 LMN$. But then (ii) is false if we set $c_2 = c_1$, $u(x, y) = f(x, y, z_1)$, $v(y, z) = f(x_1, y, z) f(x_1, y, z_1)$ and $w(x, z) = f(x, y_1, z) f(x, y_1, z_1) f(x_1, y_1, z_1)$.

Again, the details do not matter here: the point is that if we write x, y and z for x_0, y_0 and z_0, then $f(x, y, z)$ is the only term in the product that depends on all of x, y and z, so fixing x_1, y_1 and z_1 results in an expression of the form $\sum_{x,y,z} f(x, y, z) u(x, y) v(y, z) w(x, z)$.

Now let us prove the reverse inequality by making repeated use of the Cauchy–Schwarz inequality. This is where we shall use the notation introduced before the statement of the theorem:

$$\left(\sum_{x,y,z} f(x, y, z) u(x, y) v(y, z) w(x, z) \right)^8$$

$$\leqslant \left(MN \sum_{y,z} \left(\sum_x f(x, y, z) u(x, y) v(y, z) w(x, z) \right)^2 \right)^4$$

$$\leqslant M^4 N^4 \left(\sum_{y,z} \left(\sum_x f(x, y, z) u(x, y) w(x, z) \right)^2 \right)^4$$

$$= M^4 N^4 \left(\sum_{x,x'} \sum_{y,z} f(x, y, z) f(x', y, z) u(x, y) u(x', y) w(x, z) w(x', z) \right)^4$$

$$= M^4N^4 \left(\sum_{x,x'} \sum_{y,z} f_{x,x'}(y,z) u_{x,x'}(y) w_{x,x'}(z) \right)^4$$

$$\leqslant M^4N^4L^6 \sum_{x,x'} \left(\sum_{y,z} f_{x,x'}(y,z) u_{x,x'}(y) w_{x,x'}(z) \right)^4.$$

We now perform on the inner sum the steps from the corresponding part of the proof of Theorem 3.1. The last expression is equal to

$$M^4N^4L^6 \sum_{x,x'} \left(\left(\sum_z \sum_y f_{x,x'}(y,z) u_{x,x'}(y) w_{x,x'}(z) \right)^2 \right)^2$$

$$\leqslant M^4N^4L^6 \sum_{x,x'} \left(N \sum_z \left(\sum_y f_{x,x'}(y,z) u_{x,x'}(y) \right)^2 \right)^2$$

$$= M^4N^6L^6 \sum_{x,x'} \left(\sum_{y,y'} \sum_z f_{x,x'}(y,z) f_{x,x'}(y',z) u_{x,x'}(y) u_{x,x'}(y') \right)^2$$

$$= M^4N^6L^6 \sum_{x,x'} \left(\sum_{y,y'} \sum_z f_{x,x',y,y'}(z) u_{x,x',y,y'} \right)^2$$

$$\leqslant M^4N^6L^6 \sum_{x,x'} M^2 \sum_{y,y'} \left(\sum_z f_{x,x',y,y'}(z) u_{x,x',y,y'} \right)^2$$

$$\leqslant M^6N^6L^6 \sum_{x,x'} \sum_{y,y'} \left(\sum_z f_{x,x',y,y'}(z) \right)^2$$

$$= M^6N^6L^6 \sum_{x,x'} \sum_{y,y'} \sum_{z,z'} f_{x,x',y,y',z,z'}.$$

This is another way of writing

$$M^6N^6L^6 \sum_{x_0,x_1 \in X} \sum_{y_0,y_1 \in Y} \sum_{z_0,z_1 \in Z} \prod_{(i,j,k) \in \{0,1\}^3} f(x_i, y_j, z_k).$$

It follows that if (i) is true then (ii) is true with $c_2 = c_1^{1/8}$.

It is obvious that (ii) implies (iii), since one can take u, v and w to be the characteristic functions of the bipartite graphs $G(X,Y)$, $G(Y,Z)$ and $G(X,Z)$. It therefore remains to prove that (iii) implies (ii).

Suppose, then, that (ii) is false and we have functions $u : X \times Y \to [-1,1]$, $v : Y \times Z \to [-1,1]$ and $w : X \times Z \to [-1,1]$ such that

$$\left| \sum_{x,y,z} f(x,y,z) u(x,y) v(y,z) w(x,z) \right| > c_2 LMN.$$

One can write u as $u_+ - u_-$ with u_+ and u_- disjointly supported and taking values in $[0,1]$, and one can do the same for v and w. It follows that there are functions a, b and c

taking values in [0, 1] such that

$$\left|\sum_{x,y,z} f(x,y,z)a(x,y)b(y,z)c(x,z)\right| > c_2 LMN/8.$$

Now let A_1, B_1 and C_1 be random subsets of $X \times Y$, $Y \times Z$ and $X \times Z$, their elements chosen randomly and independently with probabilities given by the functions a, b and c. Writing A_1, B_1 and C_1 for the characteristic functions of the sets as well, we have that the expectation of

$$\left|\sum_{x,y,z} f(x,y,z)A_1(x,y)B_1(y,z)C_1(x,z)\right|$$

is greater than $c_2 LMN/8$. Choose A_1, B_1 and C_1 such that the absolute value of the sum is at least this big. Then we have disproved (iii) for any $c_3 \leqslant c_2/8$, since we may think of A_1, B_1 and C_1 as the edge sets of a tripartite graph.

Now suppose that $f(x, y, z)$ sums to 0 and let the complements of A_1, B_1 and C_1 be A_2, B_2 and C_2. Write S_{ijk} for the sum $\sum_{x,y,z} f(x,y,z)A_i(x,y)B_j(y,z)C_k(x,z)$. Then the sums S_{ijk} add up to zero. Since S_{111} has absolute value at least $c_2 LMN/8$, it follows that at least one S_{ijk} exceeds $c_2 LMN/56$. If we let G be the tripartite graph with edge sets A_i, B_j and C_k, then we have disproved (iv) for any $c_3 \leqslant c_2/56$. Since (iii) obviously implies (iv), the proof is complete. □

It is now very natural to make the following pair of definitions.

Definition 4.2. Let X, Y and Z be sets of sizes L, M and N. A function $f : X \times Y \times Z$ is α-*quasirandom* if

$$\sum_{x_0,x_1 \in X} \sum_{y_0,y_1 \in Y} \sum_{z_0,z_1 \in Z} \prod_{(i,j,k) \in \{0,1\}^3} f(x_i, y_j, z_k) \leqslant \alpha L^2 M^2 N^2.$$

Definition 4.3. Let H be a tripartite 3-uniform hypergraph with vertex sets X, Y and Z of sizes L, M and N. Let the number of edges of H be $pLMN$ and let $f(x, y, z) = H(x, y, z) - p$. Then H is α-*quasirandom* if f is α-quasirandom.

As we commented earlier, this definition is not identical to Definition 2.4, but it is equivalent (give or take the precise value of α) and it is the one we shall use. The advantage it has is that it is easier to use when proving a counting lemma. Note that if H fails to be quasirandom in this sense, then we have the easy deduction that f fails property (ii) of Theorem 4.1, which in turn shows easily that H is not α-edge uniform (in the sense of Definition 2.6).

We come now to a counting lemma for quasirandom 3-uniform hypergraphs. For simplicity, we prove it only in one special case, that of simplices (these were defined just before the statement of Theorem 1.4), but it is an easy exercise to generalize this case to a full counting lemma, just as we generalized Lemma 3.4 to Theorem 3.5.

Lemma 4.4. *Let H be a quadripartite 3-uniform hypergraph with vertex sets X, Y, Z and W, of sizes L, M and N and P respectively. Suppose that the induced subhypergraphs $H(X,Y,Z)$, $H(X,Y,W)$, $H(X,Z,W)$ and $H(Y,Z,W)$ are α-quasirandom with densities p, q, r and s respectively. Then the number of simplices in H differs from pqrsLMNP by at most $15\alpha^{1/8}LMNP$.*

Proof. Let the variables x, y, z and w stand for elements of X, Y, Z and W. Let the letter H stand for the characteristic function of the hypergraph H as well as the hypergraph itself. Define functions f, g, h and k (with obvious domains) by $f(x,y,z) = H(x,y,z) - p$, $g(x,y,w) = H(x,y,w) - q$, $h(x,z,w) = H(x,z,w) - r$ and $k(y,z,w) = H(y,z,w) - s$. Then the number of simplices in H is

$$\sum_{x,y,z,w} (p + f(x,y,z))(q + g(x,y,w))(r + h(x,z,w))(s + k(y,z,w)).$$

The main term in this sum is $pqrsLMNP$. We shall now show that all other terms are significantly smaller. Consider, for example, any term that chooses $f(x,y,z)$ rather than p from the first bracket. For each fixed w this results in a sum of the form

$$\sum_{x,y,z} f(x,y,z)t(x,y)u(y,z)v(z,x),$$

with t, u and v taking values in the interval $[-1, 1]$. Therefore, by Theorem 4.1 (and the bound obtained in the proof), the entire sum comes to at most $\alpha^{1/8}LMNP$. The same argument works for g, h and k, and that is enough to show that all terms apart from the main term have modulus at most $\alpha^{1/8}LMNP$, which proves the result. □

5. Why quasirandom 3-graphs are not enough

To prove Roth's theorem (that is, Szemerédi's theorem for progressions of length 3), one uses a combination of Szemerédi's regularity lemma and Lemma 3.4. The reader who has followed the paper so far may be disappointed to learn that Lemma 4.4 is not very useful when it comes to generalizing that argument. However, any effort spent on understanding it will pay dividends later, since the result that *is* useful is a further generalization, proved by a similar technique, and the steps of that result will make much more sense if they are compared with the steps in the proofs of Theorem 4.1 and Lemma 4.4.

What, then, is inadequate about quasirandomness of 3-uniform hypergraphs? The answer is not that the property is too weak – as Lemma 4.4 demonstrates – but rather that it is too strong. In other words, we are delighted if we are lucky enough to be presented with a quasirandom hypergraph, but in general it is too much to hope for. If we wish to generalize the proof for graphs, triangles and progressions of length 3, then we shall need two components: a regularity lemma and an associated counting lemma. The regularity lemma will tell us that we can divide any 3-uniform hypergraph H into random-like pieces (whatever this turns out to mean, it should somehow be analogous to the statement of the usual regularity lemma for graphs) and the counting lemma will allow us to use this information to approximate the number of simplices in H. One might think that 'random-like pieces' should simply be quasirandom sub-hypergraphs, but any sensible statement along these lines turns out to be false.

Here is a simple example of a tripartite 3-uniform hypergraph that has no large quasirandom subhypergraph, and which therefore cannot be decomposed into a small number of them. Let X, Y and Z be three sets of size N and let G be a random tripartite graph with vertex sets X, Y and Z. Let H be the hypergraph consisting of all triangles in G, that is, all triples (x, y, z) such that xy, yz and xz are edges of G. Then the density of H is $1/8$, but the number of octahedra in H is about $2^{-12}N^6$ (because an octahedron, considered as a graph, has 12 edges) rather than $8^{-8}N^6$ as it should have if H is quasirandom.

Now, given any large subsets $X' \subset X$, $Y' \subset Y$ and $Z' \subset Z$, the graphs $G(X', Y')$, $G(Y', Z')$ and $G(X', Z')$ are (with high probability) quasirandom, and therefore the same reasoning shows that the induced subhypergraph $H(X', Y', Z')$ still fails to be quasirandom.

Indeed, the situation is even worse, as it is not just induced subhypergraphs that fail to be quasirandom. Let H' be *any* subhypergraph of H and let the density of H' be p. Since H' is a subhypergraph of H,

$$\sum_{x,y,z} H'(x, y, z)G(x, y)G(y, z)G(x, z) = pN^3.$$

If we set $f(x, y, z) = H'(x, y, z) - p$, then we can deduce that

$$\sum_{x,y,z} f(x, y, z)G(x, y)G(y, z)G(x, z) = pN^3 - pN^3/8,$$

which is a clear violation of property (iii) of Theorem 4.1. This argument remains valid even if one starts by restricting to large subsets X', Y' and Z' of X, Y and Z.

This looks like bad news, and in a way it is, because it makes life more complicated, but it is not as bad as all that. To see why not, just look back at the discussion of the example above. Although the hypergraph H was not quasirandom, we had absolutely no difficulty calculating roughly how many octahedra it should have, and the reason was that we were able to use the quasirandomness of the graphs $G(X, Y)$, $G(Y, Z)$ and $G(Z, X)$. More generally, suppose we construct a hypergraph H as follows. First, we take random graphs $G(X, Y)$, $G(Y, Z)$ and $G(Z, X)$ with densities p, q and r and let G be the tripartite graph formed by their union. Next, we define H_0 to be the hypergraph consisting of all triangles of G. Finally, we let H be a random subhypergraph of H_0, choosing each edge of H_0 with probability s and making all choices independently.

How many octahedra do we expect H to contain? Well, if we choose $x_0, x_1 \in X$, $y_0, y_1 \in Y$ and $z_0, z_1 \in Z$ at random, then the probability that a pair $x_i y_j$ belongs to G is p, and these probabilities are more or less independent, so the probability that all four pairs belong to G is almost exactly p^4. Similar statements hold for the $y_i z_j$ and the $x_i z_j$, so the probability that all the 2-edges of the octahedron belong to G is almost exactly $(pqr)^4$. If this happens, then there are eight 3-edges, or faces, each of which has a probability s of lying in H. We therefore expect the number of octahedra in H to be about $(pqr)^4 s^8 (|X||Y||Z|)^2$.

It is clear from that calculation that quasirandom hypergraphs are not the only ones for which it ought to be possible to prove a counting lemma. That is, one ought to be able to relax the assumptions of Lemma 4.4 so that the induced subhypergraphs are not necessarily quasirandom, but are built rather like the hypergraph considered in the last paragraph. It is a lemma of this kind that we shall state and prove in the next section.

This will deal with the difficulty that we have just discussed. If we have proved a counting lemma for a wider class of hypergraphs than just the quasirandom ones, then it is enough, when proving a regularity lemma, to show that a hypergraph can be decomposed into subgraphs from this wider class. And this assertion is weak enough to be true, which is of course a huge advantage.

The situation we have just encountered occurs in other parts of mathematics – indeed, something like it seems to happen for almost any class of mathematical objects that do not have too rigid a structure but are well-endowed with subobjects. In such a situation, it is very useful to find, for any object X in the class, a subobject $Y \subset X$ that is in some way 'stable', in the sense that any further subobject $Z \subset Y$ does not differ interestingly from Y. To do this one must first identify the stable objects and then prove that every object contains a stable subobject. Here the structure we have been talking about is an approximate one (it is not hard to define a notion of *approximate isomorphism* to make it precise).

More exact instances are usually called canonical Ramsey theorems, of which the most famous example concerns arbitrary colourings of the edges of the complete graph on $\mathbb{N}$. Here, one cannot expect to find a monochromatic infinite clique, but one can find an infinite set X such that the restriction of the colouring to the clique $X^{(2)}$ has one of four simple forms. Write all edges as xy with $x < y$ and write $xy \sim zw$ if xy and zw have the same colour. Then one of the following four statements is true for all pairs xy, zw of edges with $x, y, z, w \in X$:

(i) $xy \sim zw$ if and only if $x = z$ and $y = w$,
(ii) $xy \sim zw$ if and only if $x = z$,
(iii) $xy \sim zw$ if and only if $y = w$,
(iv) $xy \sim zw$.

It is easy to check that if one of these statements holds for X then it holds for all subsets $Y \subset X$, so X (together with its colouring) is stable.

A second class of examples arises in Banach space theory. There are several theorems in the subject that allow one to pass from a Banach space X, perhaps with some extra properties, to a subspace Y that is in some way easier to handle. And in many cases, the property that Y has is a stability property, in the sense that all its subspaces are in some important way similar to the space itself.

Before we embark on the main results of this paper, here is a second hypergraph example to consider. It is not completely obvious that the first one matters, since it is not derived from a subset of $\mathbb{Z}_N$ in the way shown after Definition 2.3. Perhaps hypergraphs that come from sets have some extra property that makes them decomposable into quasirandom pieces.

It turns out that they don't. Rather than show precisely this, we shall briefly discuss a similar result for functions, because it is much easier technically. It is derived in a simple way from an example that plays a similar role in the analytic proof of Szemerédi's theorem [8].

Let N be an odd positive integer and let $\omega = e^{2\pi i/N}$. Define a function $f : \mathbb{Z}_N^3 \to \mathbb{C}$ by $f(x,y,z) = \omega^{(x+y+z)^2}$. This can be decomposed as $\omega^{(x+y)^2}\omega^{(y+z)^2}\omega^{(x+z)^2}\omega^{-x^2}\omega^{-y^2}\omega^{-z^2}$, or alternatively as $g(x,y)g(y,z)g(x,z)$, where $g(x,y) = \omega^{2^{-1}(x^2+y^2)+2xy}$. It is a straightforward exercise to deduce from the fact that the function ω^{x^2} has very small Fourier coefficients that g is a quasirandom function. So once again, we have a function of three variables that is a product of three quasirandom functions of two variables and therefore not quasirandom itself, even after restriction to any large set.

6. A counting lemma for two-dimensional quasirandom simplicial complexes

What the previous section shows is that we should consider objects that are slightly more complicated than 3-uniform hypergraphs. We need to look instead at 3-uniform hypergraphs that are obtained as subhypergraphs H of the set of all triangles in some tripartite graph G, paying attention to both H and G. Let us write $\Delta(G)$ for the set of triangles of G. Then one of these objects can be defined more formally as an ordered pair (G, H) such that $H \subset \Delta(G)$. It should be considered as quasirandom if the three bipartite parts of the graph G are quasirandom and H in some way 'sits quasirandomly' inside $\Delta(G)$. Our first task is to make this idea precise. Once we have done that, we shall prove another sequence of results, again following the scheme of Sections 3 and 4.

A slightly better way to think of our objects (G, H) is as two-dimensional simplicial complexes: that is, as collections Σ of sets of size at most 3 with the property that if $A \in \Sigma$ and $B \subset A$ then $B \in \Sigma$. Of course, to do this we need to take not just a graph and a hypergraph, but also a set of vertices (and, to be strictly correct, the empty set). What makes this a better way to think about it is partly that it is more natural when one comes to generalize to k-uniform hypergraphs, and partly that the regularity lemma we shall eventually prove involves restricting vertex sets. Despite all this, it will be simpler to stick to pairs (G, H) for now and bear in mind that our results will later be applied to pairs with restricted vertex sets. Let us make a formal definition.

Definition 6.1. An r-partite *chain* is a pair (G, H), where G is an r-partite graph, H is an r-partite hypergraph with the same vertex sets as G, and $H \subset \Delta(G)$.

Suppose, then, that we have a chain (G, H). We know what it means for the bipartite parts of G to be quasirandom, but must now say what it means for H to sit quasirandomly inside $\Delta(G)$. As before, we shall define this in terms of functions. For convenience, let us make another definition.

Definition 6.2. Let X, Y and Z be three sets and let $f : X \times Y \times Z \to \mathbb{R}$ be a function. Then $\operatorname{oct}(f)$ is defined to be $f_{x,x',y,y',z,z'}$, which equals the sum

$$\sum f(x,y,z)f(x,y,z')f(x,y',z)f(x,y',z')f(x',y,z)f(x',y,z')f(x',y',z)f(x',y',z')$$

taken over all $x, x' \in X$, $y, y' \in Y$ and $z, z' \in Z$.

Recall that the notation $f_{x,x',y,y',z,z'}$ was introduced before the statement of Theorem 4.1. We shall use it again when proving Lemma 6.7 below.

If f is the characteristic function of a hypergraph H then $\mathrm{oct}(f)$ is the number of octahedra in H. We shall write $\mathrm{oct}(H)$ for this quantity. With this notation it is easy to express precisely what it means for H to sit quasirandomly in G.

Definition 6.3. Let G be a tripartite graph with vertex sets X, Y and Z of sizes L, M and N, and let $f : X \times Y \times Z \to [-1, 1]$ be a function such that f is supported in $\Delta(G)$. Let the densities of $G(X, Y)$, $G(Y, Z)$ and $G(X, Z)$ be p, q and r respectively. Then f is α-*quasirandom relative to* G if $\mathrm{oct}(f) \leqslant \alpha(pqr)^4(LMN)^2$. Now let H be a tripartite 3-uniform hypergraph with vertex sets X, Y and Z and suppose that $H \subset \Delta(G)$ and $|H| = \gamma|\Delta(G)|$. Let $f(x, y, z) = H(x, y, z) - \gamma$ for $(x, y, z) \in \Delta(G)$ and 0 otherwise. Then H is α-*quasirandom relative to* G if f is α-quasirandom relative to G.

It might be more natural to say that f is α-quasirandom relative to G if $\mathrm{oct}(f) \leqslant \alpha \,\mathrm{oct}(G)$. If G is quasirandom, then this is roughly what the definition does say, and we shall apply the definition only to quasirandom graphs, where the formulation we have given turns out to be slightly more convenient for technical reasons.

These definitions are very similar to those of Section 4, but now everything takes place inside $\Delta(G)$. What we shall show is that if the bipartite parts of G are sufficiently quasirandom, then relative quasirandomness has consequences that are also similar to those of Section 4, though for reasons that will be explained more fully later, some of them are a bit more complicated.

The main result of this section is a counting lemma for simplices in quadripartite chains. The reader who follows the proof will see that it can be generalized easily to a counting lemma for arbitrary subchains. However, the case of simplices is easier to present, and is enough for the application to arithmetic progressions of length four.

Before we prove this counting lemma, we need to prepare for it with a technical lemma (Lemma 6.6 below), which itself needs a small amount of preparation.

Definition 6.4. Let G and H be k-partite graphs with vertex sets $X = X_1 \cup \cdots \cup X_k$ and $A = A_1 \cup \cdots \cup A_k$ respectively. A *homomorphism* from H to G is a map $\phi : A \to X$ such that $\phi(A_i) \subset X_i$ for each i and such that $\phi(v)\phi(w)$ is an edge of G whenever vw is an edge of H.

Note that in the above definition we say nothing about what happens if vw is not an edge of H. Nor do we insist that ϕ is an injection.

Let G be a quadripartite graph with vertex sets X, Y, Z and W of sizes L, M, N and P respectively. Let H be another quadripartite graph, with vertex sets A, B, C and D of sizes q, r, s and t, and let $a, a' \in A$, $b, b' \in B$ and $c, c' \in C$ be six vertices and suppose that the set $\{a, a', b, b', c, c'\}$ is independent. For any $x, x' \in X$, $y, y' \in Y$ and $z, z' \in Z$ let $h(x, x', y, y', z, z')$ be the number of homomorphisms ϕ from H to G such that $\phi(a) = x$, $\phi(a') = x'$, $\phi(b) = y$, $\phi(b') = y'$, $\phi(c) = z$ and $\phi(c') = z'$. Lemma 6.6 will tell us that if G is α-quasirandom for a sufficiently small α, then the function h is approximately constant.

This is a simple application of the counting lemma for graphs (Theorem 3.5) and the second-moment method.

Before we embark on the lemma, let us think about the constant we expect to obtain. For each pair i, j let the density of the graph $G(X_i, X_j)$ be δ_{ij}. Given any edge e of H, let us set $\delta(e)$ to be the δ_{ij} for which A_i and A_j are the vertex sets containing the two vertices joined by e. The number of ways of choosing a function ϕ that respects the partitions of G and H is $L^q M^r N^s P^t$. If we fix the images of a, a', b, b', c, c' then the number of possible extensions is $L^{q-2} M^{r-2} N^{s-2} P^t$. Given an edge e of H, it joins A_i to A_j for some $1 \leqslant i < j \leqslant 4$. The probability that $\phi(e)$ is an edge of G is the density δ_{ij} of the bipartite subgraph $G(X_i, X_j)$. If G behaves like a random graph, then the probability that ϕ is a homomorphism will be roughly the product, δ, of all these individual edge-probabilities. We shall call this the *expected H-density* of G. In other words, the expected H-density of G is $\prod_{e \in H} \delta(e)$. (The assumption that $\{a, a', b, b', c, c'\}$ is an independent set means that we do not have to worry about whether there are edges joining vertices in the set $\{x, x', y, y', z, z'\}$.) We expect the approximately constant value of h to be about $\delta L^{q-2} M^{r-2} N^{s-2} P^t$.

We shall be using second moments, so for convenience here first is an easy technical lemma that encapsulates what we need for the main lemma.

Lemma 6.5. *Let $\alpha, \delta \in [0, 1]$, let R be a real number and let $a_1, \ldots, a_n$ be real numbers such that $\sum_{i=1}^{n} a_i \geqslant (\delta - \alpha)Rn$ and $\sum_{i=1}^{n} a_i^2 \leqslant (\delta^2 + \alpha)R^2 n$. Then $|a_i - \delta R| \leqslant R\alpha^{1/4}$ for all but at most $3n\sqrt{\alpha}$ values of i.*

Proof. Using our hypotheses, we find that

$$\sum_{i=1}^{n}(a_i - \delta R)^2 = \sum_{i=1}^{n} a_i^2 - 2R \sum_{i=1}^{n} a_i \delta + n\delta^2 R^2$$
$$\leqslant R^2 n\left(\delta^2 + \alpha - 2\delta(\delta - \alpha) + \delta^2\right)$$
$$= R^2 n\alpha(1 + 2\delta) \leqslant 3R^2 n\alpha.$$

The result follows immediately. □

The statement that follows will look somewhat peculiar: we are giving the particular case that happens to arise later of a more general statement which, though more natural, is perhaps harder to digest. Some readers may wish to jump to Lemma 6.7 and then come back to this point of the paper when the motivation for it has become clear.

Lemma 6.6. *Let G, H and δ be as in the remarks preceding Lemma 6.5. Let $0 < \epsilon \leqslant 1$, let $\alpha > 0$ be such that $2^m \alpha^{1/16} \leqslant \epsilon\delta/3$ and suppose that G is α-quasirandom. Let m be the number of edges of H and let δ be the expected H-density of G. Then the number of sextuples (x, x', y, y', z, z') for which $h(x, x', y, y', z, z')$ differs from $\delta L^{q-2} M^{r-2} N^{s-2} P^t$ by more than $\epsilon \delta L^{q-2} M^{r-2} N^{s-2} P^t$ is at most $\epsilon \delta L^2 M^2 N^2$.*

Proof. First, we estimate $\sum_{x,x',y,y',z,z'} h(x, x', y, y', z, z')$. This is the number of homomorphisms ϕ from H to G, which, by Theorem 3.5, is at least $(\delta - 2^m \alpha^{1/4}) L^q M^r N^s P^t$.

Next, we estimate $\sum_{x,x',y,y',z,z'} h(x,x',y,y',z,z')^2$. For this we can use the counting lemma again, but first we must define an auxiliary graph J. For each vertex v of H apart from a, a', b, b', c and c', let v_1 be a copy of v. Let γ be a function defined on the vertex set V of H that takes the vertices a, a', b, b', c and c' to themselves and takes any other vertex v to its copy v_1. The vertex set of J is $V \cup \gamma(V)$ and the edges of J are the edges of H together with all pairs $\gamma(u)\gamma(v)$ such that uv is an edge of H. It is easy to see that $\sum_{x,x',y,y',z,z'} h(x,x',y,y',z,z')^2$ is the number of homomorphisms from J to G, which is of course why we defined J. Since every edge of H has been doubled up in J, the product of the edge-probabilities of J is δ^2, so Theorem 3.5 tells us that the number of homomorphisms from J to G is at most $(\delta^2 + 2^{2m}\alpha^{1/4})L^{2q-2}M^{2r-2}N^{2s-2}P^{2t}$.

Let us now apply the previous lemma, with $n = L^2M^2N^2$, $R = L^{q-2}M^{r-2}N^{s-2}P^t$, δ as it is and α replaced by $2^{2m}\alpha^{1/4}$. Then the hypotheses of the lemma are satisfied and it tells us that the number of sextuples (x,x',y,y',z,z') for which $h(x,x',y,y',z,z')$ differs from $\delta L^{q-2}M^{r-2}N^{s-2}P^t$ by more than $2^{m/2}\alpha^{1/16}L^{q-2}M^{r-2}N^{s-2}P^t$ is at most $3.2^m\alpha^{1/8}L^2M^2N^2$. Our upper bound on α then implies the result. $\square$

A few words of explanation are needed before the next lemma, to draw attention to how it differs from the implication of (ii) from (i) in Theorem 4.1. As in that proof we have a sum and we wish to show that it is small, subject to a quasirandomness assumption about the function f. As in that proof we shall use the Cauchy–Schwarz inequality several times, and the manipulations will be very similar. However, this time the functions we look at are supported in the set of triangles of a quasirandom quadripartite graph G, and the bound we obtain is stronger because it depends on the densities of the six bipartite parts of G. The significance of this is that the theorem says something even when the quasirandomness parameter η below is much larger than any of these densities. This extra strength is very significant when G is sparse, as it will be if it is one of the graphs given to us by the hypergraph regularity lemma proved later. To obtain the extra strength, we shall be very careful to use the full strength of the Cauchy–Schwarz inequality whenever we apply it: if the number of i such that $a_i \neq 0$ is m, then we shall bound $\left(\sum_{i=1}^n a_i\right)$ above by $m \sum_{i=1}^n a_i^2$ rather than by $n \sum_{i=1}^n a_i^2$.

We shall use the following notation. The graph G will have vertex sets X, Y, Z and W, but we shall also think of them as X_1, X_2, X_3 and X_4 respectively. (Sometimes one notation is easier to handle, sometimes the other.) The density of G_{ij} will again be denoted δ_{ij}. We shall write h_i for $|X_i|$, h_{ij} for $|G(X_i, X_j)|$ and h_{ijk} for the number of triangles in $G(X_i, X_j) \cup G(X_j, X_k) \cup G(X_i, X_k)$. If $x, x' \in X_1$ and $i > 1$ then we shall also write $h_i(x, x')$ for the number of vertices in X_i that are joined to both x and x'. Similarly, if $1 < i < j$ then we shall write $h_{ij}(x, x')$ for the set of all edges yz in $G(X_i, X_j)$ such that y and z are both joined to both of x and x'. Similarly, if $x, x' \in X_1$, $y, y' \in X_2$ and $i > 2$, then we shall write $h_i(x, x', y, y')$ for the number of $z \in X_i$ that are joined to all of x, x', y and y'. It is numbers such as these that will appear when we make our more efficient uses of the Cauchy–Schwarz inequality. Expressions such as $G_{x,x'}G_{y,y'}$ will also appear in the calculations below: these are just pointwise products, so for example $G_{x,x'}G_{y,y'}(w) = G(x,w)G(x',w)G(y,w)G(y',w)$.

Lemma 6.7. *Let G be as just described, and let $f : X \times Y \times Z \to [-1,1]$, $g : X \times Y \times W \to [-1,1]$, $h : X \times Z \times W \to [-1,1]$ and $k : Y \times Z \times W \to [-1,1]$ be functions that are nonzero only at triples that form triangles in G. Suppose that G is α-quasirandom and that f is η-quasirandom relative to the tripartite graph $G(X,Y,Z)$. Suppose also that $2^{36}\alpha^{1/16} \leqslant \eta(\delta_{12}\delta_{23}\delta_{13}\delta_{14}\delta_{24}\delta_{34})^8/6$. Then*

$$\left| \sum_{x,y,z,w} f(x,y,z)g(x,y,w)h(x,z,w)k(y,z,w) \right| \leqslant (2\eta)^{1/8}\delta_{12}\delta_{23}\delta_{13}\delta_{14}\delta_{24}\delta_{34}LMNP.$$

Proof. We begin with several applications of the Cauchy–Schwarz inequality, of a similar kind to ones that we have seen already:

$$\left(\sum_{x,y,z,w} f(x,y,z)g(x,y,w)h(x,z,w)k(y,z,w) \right)^8$$

$$\leqslant \left(h_{234} \sum_{y,z,w} \left(\sum_x f(x,y,z)g(x,y,w)h(x,z,w)k(y,z,w) \right)^2 \right)^4$$

$$\leqslant \left(h_{234} \sum_{y,z,w} \left(\sum_x f(x,y,z)g(x,y,w)h(x,z,w) \right)^2 \right)^4$$

$$= h_{234}^4 \left(\sum_{x,x'} \sum_{y,z,w} f_{x,x'}(y,z)g_{x,x'}(y,w)h_{x,x'}(z,w) \right)^4$$

$$\leqslant h_{234}^4 h_1^6 \sum_{x,x'} \left(\sum_{y,z,w} f_{x,x'}(y,z)g_{x,x'}(y,w)h_{x,x'}(z,w) \right)^4$$

$$\leqslant h_{234}^4 h_1^6 \sum_{x,x'} \left(h_{34}(x,x') \sum_{z,w} \left(\sum_y f_{x,x'}(y,z)g_{x,x'}(y,w)h_{x,x'}(z,w) \right)^2 \right)^2$$

$$\leqslant h_{234}^4 h_1^6 \sum_{x,x'} \left(h_{34}(x,x') \sum_{z,w} G(z,w) \left(\sum_y f_{x,x'}(y,z)g_{x,x'}(y,w) \right)^2 \right)^2$$

$$= h_{234}^4 h_1^6 \sum_{x,x'} h_{34}(x,x')^2 \left(\sum_{y,y'} \sum_{z,w} f_{x,x',y,y'}(z)g_{x,x',y,y'}(w)G(z,w) \right)^2$$

$$\leqslant h_{234}^4 h_1^6 \sum_{x,x'} h_{34}(x,x')^2 h_2(x,x')^2 \sum_{y,y'} \left(\sum_{z,w} f_{x,x',y,y'}(z)g_{x,x',y,y'}(w)G(z,w) \right)^2$$

$$\leqslant h_{234}^4 h_1^6 \sum_{x,x'} h_{34}(x,x')^2 h_2(x,x')^2 \sum_{y,y'} h_4(x,x',y,y')$$

$$\sum_w \left(\sum_z f_{x,x',y,y'}(z)g_{x,x',y,y'}(w)G(z,w) \right)^2$$

$$\leqslant h_{234}^4 h_1^6 \sum_{x,x'} h_{34}(x,x')^2 h_2(x,x')^2 \sum_{y,y'} h_4(x,x',y,y')$$
$$\sum_w \left(\sum_z f_{x,x',y,y'}(z) G_{x,x'} G_{y,y'}(w) G(z,w) \right)^2$$
$$= h_{234}^4 h_1^6 \sum_{x,x'} h_{34}(x,x')^2 h_2(x,x')^2 \sum_{y,y'} h_4(x,x',y,y')$$
$$\sum_{z,z'} f_{x,x',y,y',z,z'} \sum_w G_{x,x'} G_{y,y'} G_{z,z'}(w)$$
$$= h_{234}^4 h_1^6 \sum_{x,x'} h_{34}(x,x')^2 h_2(x,x')^2 \sum_{y,y'} h_4(x,x',y,y') \sum_{z,z'} f_{x,x',y,y',z,z'} h_4(x,x',y,y',z,z').$$

The main idea of the proof has now been given. The rest of the argument consists in showing that the h-terms are all approximately constant and calculating what the result would be if they *were* constant. For this we use the lemma about graphs proved earlier.

The final line above can be written as
$$\sum_{x,x',y,y',z,z'} h(x,x',y,y',z,z') f_{x,x',y,y',z,z'} = \langle h, F \rangle,$$
where $F(x,x',y,y',z,z') = f_{x,x',y,y',z,z'}$ and
$$h(x,x',y,y',z,z') = h_{234}^4 h_1^6 h_{34}(x,x')^2 h_2(x,x')^2 h_4(x,x',y,y') h_4(x,x',y,y',z,z').$$

This last quantity is the number of homomorphisms from a certain quadripartite graph H to G, given that a particular six of its vertices (none of which are joined to each other) map to x, x', y, y', z and z'. To see this, note first that it is true of each individual term in the product. For example, to understand the term $h_{34}(x,x')^2$ in this way, take the graph J with vertex set $\{a, a', b, b', c, c', d, e\}$ and edges $ad, ae, a'd, a'e$ and de. The number of homomorphisms ϕ from J to G such that $\phi(a) = x$, $\phi(a') = x'$, $\phi(b) = y$, $\phi(b') = y'$, $\phi(c) = z$, $\phi(c') = z'$, $\phi(d) \in Z$ and $\phi(e) \in W$ is the number of pairs $(z, w) \in Z \times W$ such that z is joined to w and both z and w are joined to both x and x', which is the definition of $h_{34}(x,x')$. To obtain the product, one takes disjoint copies of all the graphs J constructed in this way and identifies the vertices a, a', b, b', c and c' from each one. (That is, the a in one graph is the same as the a in another, and so on.)

Let the vertex set of H be $A \cup B \cup C \cup D$, and let us look at the sizes of A, B, C and D. The set A contains a and a', and receives an additional six (isolated) vertices from the term h_1^6. B contains b and b', and receives in addition four vertices from h_{234}^4 and two from $h_2(x,x')^2$. C contains c and c' and receives four vertices from h_{234}^4 and two from $h_{34}(x,x')^2$. Finally, D receives four vertices from h_{234}^4, two from $h_{34}(x,x')^2$ and one each from $h_4(x,x',y,y')$ and $h_4(x,x',y,y',z,z')$.

In a similar way one can work out how many edges there are between each pair from A, B, C and D. For example, the number of edges between B and D is $4 + 2 + 2 = 8$, since h_{234}^4 contributes four, $h_4(x,x',y,y')$ contributes two and $h_4(x,x',y,y',z,z')$ contributes two. As another example, the number of edges between A and C is 4, all coming from $h_{34}(x,x')^2$. It turns out that there are eight edges between any pair of sets that includes D and four between any other pair.

We wish to apply Lemma 6.6. It follows from the simple calculations we have just made that the expected H-density of G is
$$\delta = (\delta_{12}\delta_{23}\delta_{13})^4(\delta_{14}\delta_{24}\delta_{34})^8.$$
We also have $q = r = s = t = 8$ and $m = 36$. Let $\epsilon = \eta(\delta_{12}\delta_{23}\delta_{13})^4/2$ and call a sextuple (x, x', y, y', z, z') *bad* if $|h(x, x', y, y', z, z') - \delta L^6 M^6 N^6 P^8| > \epsilon \delta L^6 M^6 N^6 P^8$, and *good* otherwise.

By Lemma 6.6, the number of bad sextuples is at most $\epsilon \delta L^2 M^2 N^2$, and a trivial upper bound for each $h(x, x', y, y', z, z')$ is $L^6 M^6 N^6 P^8$. Therefore, if we let h' be a new function that equals h for every good sextuple and takes the value $\delta L^6 M^6 N^6 P^8$ otherwise, then $\|h - h'\|_1 \leqslant \epsilon \delta (LMNP)^8$. Writing d for the constant function $\delta L^6 M^6 N^6 P^8$, we also have that $\|h' - d\|_\infty \leqslant \epsilon \delta L^6 M^6 N^6 P^8$. As for F, we know that $\|F\|_1$ is trivially at most $(LMN)^2$, since $\|F\|_\infty \leqslant 1$. Putting all these facts together, we find that

$$\begin{aligned}|\langle h, F\rangle - \langle d, F\rangle| &\leqslant |\langle h - h', F\rangle| + |\langle h' - d, F\rangle| \\ &\leqslant \|h - h'\|_1 \|F\|_\infty + \|h' - d\|_\infty \|F\|_1 \\ &\leqslant \epsilon \delta (LMNP)^8 + \epsilon \delta (LMNP)^8 \\ &= \eta(\delta_{12}\delta_{23}\delta_{13}\delta_{14}\delta_{24}\delta_{34}LMNP)^8.\end{aligned}$$

But the the relative quasirandomness assumption on f tells us that

$$\begin{aligned}\langle d, F\rangle &= \delta L^6 M^6 N^6 P^8 \operatorname{oct}(f) \\ &\leqslant \eta \delta (\delta_{12}\delta_{23}\delta_{13})^4 (LMNP)^8 \\ &= \eta(\delta_{12}\delta_{23}\delta_{13}\delta_{14}\delta_{24}\delta_{34}LMNP)^8.\end{aligned}$$

The result follows. $\square$

We are now ready to prove a generalization of Lemma 4.4 from quasirandom hypergraphs to quasirandom chains. Notice the dependence of parameters in the statement. The graph G is α-quasirandom and the hypergraph H is relatively η-quasirandom. Both α and η need to be small for the conclusion to hold and be useful, but whereas the condition on α depends on η, the density of G and the relative density of H, the smallness of η depends only on the last of these. In particular, as we have already mentioned, η can be much larger than the density of G. This is critically important, since it is all that can be guaranteed by the regularity lemma later.

Theorem 6.8. *Let X, Y, Z and W be sets of size L, M, N and P respectively. Let G be a quadripartite graph with vertex sets X, Y, Z and W and suppose that the six bipartite parts of G are α-quasirandom. Write δ_{12} for the density of the graph $G(X, Y)$, and similarly for the other parts. Let H_{123} be a tripartite hypergraph with vertex sets X, Y and Z that is η-quasirandom relative to $\Delta\bigl(G(X, Y, Z)\bigr)$ and similarly for H_{124}, H_{134} and H_{234}. For each triple ijk let the relative density of H_{ijk} be δ_{ijk}. Let H be the union of the hypergraphs H_{ijk}. Suppose that α satisfies the condition $2^{36}\alpha^{1/16} \leqslant \eta(\delta_{12}\delta_{23}\delta_{13}\delta_{14}\delta_{24}\delta_{34})^8/6$. Then the number of simplices in H differs from $\delta_{12}\delta_{23}\delta_{13}\delta_{14}\delta_{24}\delta_{34}\delta_{123}\delta_{124}\delta_{134}\delta_{234}LMNP$ by at most $8\eta^{1/8}\delta_{12}\delta_{23}\delta_{13}\delta_{14}\delta_{24}\delta_{34}LMNP$.*

Proof. We wish to estimate the sum

$$\sum_{x,y,z,w} H(x,y,z)H(x,y,w)H(x,z,w)H(y,z,w).$$

For each triple $1 \leq i < j < k \leq 4$ let $d_{ijk}(x,y,z) = \delta_{ijk}G(x,y)G(y,z)G(x,z)$. Then

$$\sum_{x,y,z,w} d_{123}(x,y,z)d_{124}(x,y,w)d_{134}(x,z,w)d_{234}(y,z,w)$$

is $\delta_{123}\delta_{124}\delta_{134}\delta_{234}$ times the number of simplices in G. Since G is α-quasirandom, Theorem 3.5 tells us that the number of simplices in G is $\delta_{12}\delta_{23}\delta_{13}\delta_{14}\delta_{24}\delta_{34}LMNP$, to within an error of at most $64\alpha^{1/4}LMNP$, which is certainly at most $\delta_{12}\delta_{23}\delta_{13}\delta_{14}\delta_{24}\delta_{34}\eta^{1/8}LMNP$.

If we let $f(x,y,z) = H(x,y,z) - d_{123}(x,y,z)$, then our hypothesis implies that f is η-quasirandom relative to G. If we take the sum we wish to estimate and change $H(x,y,z)$ into $d_{123}(x,y,z)$, then the difference we make to the sum is

$$\left| \sum_{x,y,z,w} f(x,y,z)H(x,y,w)H(x,z,w)H(y,z,w) \right|.$$

By Lemma 6.7, this is at most $(2\eta)^{1/8}\delta_{12}\delta_{23}\delta_{13}\delta_{14}\delta_{24}\delta_{34}LMNP$. By a similar argument we can replace $H(x,y,w)$ by $d_{124}(x,y,w)$, again making a difference of at most

$$(2\eta)^{1/8}\delta_{12}\delta_{23}\delta_{13}\delta_{14}\delta_{24}\delta_{34}LMNP.$$

Repeating this process twice more, we find that

$$\left| \sum_{x,y,z,w} H(x,y,z)H(x,y,w)H(x,z,w)H(y,z,w) \right.$$

$$\left. - \sum_{x,y,z,w} d_{123}(x,y,z)d_{124}(x,y,w)d_{134}(x,z,w)d_{234}(y,z,w) \right|$$

is at most $4(2\eta)^{1/8}\delta_{12}\delta_{23}\delta_{13}\delta_{14}\delta_{24}\delta_{34}LMNP$. Combining this with the estimate of the previous paragraph, we obtain the desired result. $\square$

We have now finished the hardest part of the proof, by identifying a class of stable hypergraphs and proving a counting lemma for them. It remains to prove a regularity lemma, which says, roughly speaking, that every dense 3-uniform hypergraph can be decomposed into stable subhypergraphs.

7. A proof of Szemerédi's regularity lemma

The version of Szemerédi's regularity lemma that we used to prove Theorem 1.1 was, as commented there, not quite standard, but it can be proved more cleanly and is better suited for generalizing to hypergraphs. To demonstrate the first of these assertions, to keep this paper self-contained and to illuminate the proof of hypergraph regularity, we shall give in this statement a precise statement and a complete proof of the result.

Let G be a bipartite graph with vertex sets X and Y of sizes M and N respectively. Let $X_1 \cup \cdots \cup X_m$ and $Y_1 \cup \cdots \cup Y_n$ be partitions of X and Y, with $|X_i| = \alpha_i M$ and $|Y_j| = \beta_j N$.

Write $d(X_i, Y_j)$ for the *density* of the induced subgraph $G(X_i, Y_j)$, that is, $|X_i|^{-1}|Y_j|^{-1}$ times the number of edges from X_i to Y_j. The *mean-square density* of G with respect to the partitions is defined to be $\sum_{i,j} \alpha_i \beta_j d(X_i, Y_j)^2$. Sometimes, when G is clear from the context, we shall call this the mean-square density of the partitions. This concept can also be viewed probabilistically. Choose a random $x \in X$ and $y \in Y$. Then x belongs to some X_i and y to some Y_j and the mean-square density is the expectation of the square of the density $d(X_i, Y_j)$.

Our first steps are very simple – all they say is that certain projections on certain Hilbert spaces have norm at most 1. However, let us quickly establish them in our particular context.

Lemma 7.1. *Let U be a finite set and let $f : U \to \mathbb{R}$ be a function with mean d. Let $U = U_1 \cup \cdots \cup U_r$ with $|U_i| = \gamma_i |U|$, and let d_i be the mean of f restricted to U_i. Then $d^2 \leqslant \sum_{i=1}^{r} \gamma_i d_i^2$.*

Proof. By the Cauchy–Schwarz inequality,

$$\left(\sum_{i=1}^{r} \gamma_i d_i \right)^2 \leqslant \left(\sum_{i=1}^{r} \gamma_i \right) \left(\sum_{i=1}^{r} \gamma_i d_i^2 \right).$$

The left-hand side is d^2 and the first bracket on the right-hand side is 1, so the result is proved. $\square$

Lemma 7.2. *Let U, f and $U_1, \ldots, U_r$ be as in Lemma 7.1. Suppose that each U_i is partitioned further into sets U_{ij}, let $|U_{ij}| = \gamma_{ij}|U|$ and let d_{ij} be the mean of f restricted to U_{ij}. Then $\sum_i \gamma_i d_i^2 \leqslant \sum_{ij} \gamma_{ij} d_{ij}^2$.*

Proof. By Lemma 7.1, we have for each i the inequality

$$d_i^2 \leqslant \sum_j \frac{\gamma_{ij}}{\gamma_i} d_{ij}^2.$$

Multiplying both sides by γ_i and summing over i gives the result. $\square$

Corollary 7.3. *Let G be a bipartite graph of density d with vertex sets X and Y of sizes M and N respectively. Let $X_1 \cup \cdots \cup X_m$ and $Y_1 \cup \cdots \cup Y_n$ be partitions of X and Y. Let each X_i be partitioned further into sets X_{ik} and each Y_j into sets Y_{jl}. Then the mean-square density of G with respect to the partitions $\{X_{ik}\}$ and $\{Y_{jl}\}$ is at least the mean-square density of G with respect to the partitions $\{X_i\}$ and $\{Y_j\}$.*

Proof. Let U be the set $X \times Y$ and let f be the characteristic function of G. Then the result follows from Lemma 7.2 if we take as our cruder partition of U all sets of the form $X_i \times Y_j$ and as our finer one all sets of the form $X_{ik} \times Y_{jl}$, since the quantities compared in that lemma are the mean-square densities of G with respect to the two sets of partitions. $\square$

Lemma 7.4. *Let G be a bipartite graph of density d with vertex sets X and Y of sizes M and N, and suppose that G fails to be ϵ-quasirandom. Then there are partitions $X = X_1 \cup X_2$ and $Y = Y_1 \cup Y_2$ of the vertex sets that have mean-square density at least $d^2 + \epsilon^2/16$.*

Proof. Let $f(x, y) = G(x, y) - d$. The proof of Theorem 3.1 provides us with subsets $X_1 \subset X$ and $Y_1 \subset Y$ such that $|\sum_{x \in X_1} \sum_{y \in Y_1} f(x, y)| \geqslant \epsilon MN/4$. Let us write $\phi(X_i, Y_j)$ for the 'density' of f when restricted to $X_i \times Y_j$, that is, for $|X_i|^{-1}|Y_j|^{-1} \sum_{x \in X_i} \sum_{y \in Y_j} f(x, y)$. Then the mean-square density of the partitions is

$$\sum_{i,j=1}^{2} \alpha_i \beta_j (d + \phi(X_i, Y_j))^2 = \sum_{i,j=1}^{2} \alpha_i \beta_j (d^2 + 2d\phi(X_i, Y_j) + \phi(X_i, Y_j)^2).$$

The first term adds up to d^2. The second adds up to zero, since the average of f is zero. The third adds up to at least $\alpha_1 \beta_1 (\alpha_1 M)^{-2} (\beta_1 N)^{-2} (\epsilon MN/4)^2$, which is at least $\epsilon^2/16$. □

Lemma 7.5. *Let G be a bipartite graph with vertex sets X and Y of sizes M and N respectively. Let $X_1 \cup \cdots \cup X_m$ and $Y_1 \cup \cdots \cup Y_n$ be partitions of X and Y, with $|X_i| = \alpha_i M$ and $|Y_j| = \beta_j N$. Suppose that the mean-square density of these partitions is d^2. Let B be the set of all pairs (i, j) such that the subgraph $G(X_i, Y_j)$ fails to be ϵ-quasirandom, and suppose that $\sum_{(i,j) \in B} \alpha_i \beta_j > \epsilon$. Then one can find partitions $X_i = X_{i1} \cup \cdots \cup X_{is}$ and $Y_j = Y_{j1} \cup \cdots \cup Y_{jt}$ such that the refined partitions $\{X_{ik}\}$ and $\{Y_{jl}\}$ have mean-square density at least $d^2 + \epsilon^3/16$. Moreover, s and t are uniformly bounded above by 2^n and 2^m respectively.*

Proof. For every pair (X_i, Y_j) that fails to be ϵ-quasirandom Lemma 7.4 provides us with partitions $X_i = X_{i1}^{(j)} \cup X_{i2}^{(j)}$ and $Y_j = Y_{j1}^{(i)} \cup Y_{j2}^{(i)}$ of mean-square density at least $d(X_i, Y_j)^2 + \epsilon^2/16$. For each i let $X_i = X_{i1} \cup \cdots \cup X_{is}$ be a partition with $s \leqslant 2^n$ that simultaneously refines all the partitions $X_{i1}^{(j)} \cup X_{i2}^{(j)}$, and for each j let $Y_j = Y_{j1} \cup \cdots \cup Y_{jt}$ be a partition with $t \leqslant 2^m$ that simultaneously refines all the partitions $Y_j = Y_{j1}^{(i)} \cup Y_{j2}^{(i)}$.

For each $(i, j) \in B$, Corollary 7.3 implies that the mean-square density of G with respect to the partitions $X_{i1} \cup \cdots \cup X_{is}$ and $Y_{j1} \cup \cdots \cup Y_{jt}$ is at least $d(X_i, Y_j)^2 + \epsilon^2/16$. For every other (i, j), Corollary 7.3 implies that it is at least $d(X_i, Y_j)^2$. Multiplying by $\alpha_i \beta_j$ and summing over all i, j tells us that the mean-square density of the partitions $\{X_{ik}\}$ and $\{Y_{jl}\}$ is at least

$$\sum_{(i,j) \notin B} \alpha_i \beta_j d(X_i, Y_j)^2 + \sum_{(i,j) \in B} \alpha_i \beta_j (d(X_i, X_j)^2 + \epsilon^2/16),$$

which is at least $d^2 + \epsilon^3/16$, by our hypothesis on the size of B. □

The next result is our nonstandard statement of Szemerédi's regularity lemma.

Theorem 7.6. *Let $\epsilon > 0$ and let G be any bipartite graph with vertex sets X and Y. Then there are partitions $X = X_1 \cup \cdots \cup X_m$ and $Y = Y_1 \cup \cdots \cup Y_n$ with m and n bounded above by functions of ϵ, with the following property. For each i and j let $|X_i| = \alpha_i |X|$ and $|Y_j| = \beta_j |Y|$, and let B be the set of all pairs (i, j) such that the subgraph $G(X_i, Y_j)$ fails to be*

ϵ-quasirandom. Then $\sum_{(i,j)\in B} \alpha_i \beta_j \leq \epsilon$. Equivalently, the probability that a random pair $(x, y) \in X \times Y$ belongs to an $X_i \times Y_j$ for which $G(X_i, Y_j)$ fails to be ϵ-quasirandom is at most ϵ.

Proof. If G itself is ϵ-quasirandom then we are done. Otherwise, let d be the density of G. Then Lemma 7.4 gives us partitions $X = X_1 \cup X_2$ and $Y = Y_1 \cup Y_2$ of mean-square density at least $d^2 + \epsilon^3/16$. In general, given any pair of partitions $X = X_1 \cup \cdots \cup X_m$ and $Y = Y_1 \cup \cdots \cup Y_n$, either $\sum_{(i,j)\in B} \alpha_i \beta_j \leq \epsilon$ and we are done (where B is defined as in the statement of the theorem) or we can find refinements for which the mean-square density is greater by at least $\epsilon^3/16$. This allows us to construct a sequence of partitions of ever-increasing mean-square density, each refining the one before. Since mean-square density is bounded above by 1, this sequence must terminate in at most $16\epsilon^{-3}$ steps, and it terminates at a pair of partitions that satisfy the conclusion of the theorem. By the bound in Lemma 7.5, the number of sets in these partitions is bounded above by a function of ϵ only. (This function is given by a tower of 2s of height proportional to ϵ^{-3}.) □

To end this section, we prove a (known and easy) generalization of Szemerédi's regularity lemma, which we shall need later.

Lemma 7.7. *Let X and Y be sets and let $G_1, \ldots, G_r$ be bipartite graphs that form a partition of the edges of the complete bipartite graph $K(X, Y)$. Let $X_1, \ldots, X_m$ and $Y_1, \ldots, Y_n$ be partitions of X and Y respectively and suppose that the sum of the mean-square densities of all the graphs G_u with respect to these partitions is D. Choose an element (x, y) of the set $X \times Y$ uniformly at random, and suppose that with probability at least ϵ it lies in some $X_i \times Y_j$ for which not all the graphs $G_u(X_i \times Y_j)$ are ϵ-quasirandom. Then one can find partitions $X_i = X_{i1} \cup \cdots \cup X_{is_i}$ and $Y_j = Y_{j1} \cup \cdots \cup Y_{jt_j}$ such that the sum of the mean-square densities of the G_u with respect to the refined partitions $\{X_{ik}\}$ and $\{Y_{jl}\}$ is at least $D + \epsilon^3/16$. Moreover, all the s_i are bounded above by 2^n and all the t_j are bounded above by 2^m.*

Proof. This is very similar to the proof of Lemma 7.5 so we shall be brisk. Let (i, j) be a pair such that some graph $G_u(X_i \times Y_j)$, of density d, say, fails to be ϵ-quasirandom. Then by Lemma 7.4 we can find partitions of X_i and Y_j into two sets each in such a way that the mean-square density of $G_u(X_i \times Y_j)$ with respect to these partition is at least $d^2 + \epsilon^2/16$, and that will be true of any refinements of them, by Corollary 7.3.

Now let us find such partitions for every pair (i, j) for which a suitable u exists. Then each set X_i has been partitioned into two in at most n ways and each Y_j has been partitioned into two in at most m ways. Let $X_i = X_{i1} \cup \cdots \cup X_{is_i}$ and $Y_j = Y_{j1} \cup \cdots \cup Y_{jt_j}$ be common refinements of these partitions, into at most 2^n sets and 2^m sets respectively.

For each pair (i, j) for which there was a non-ϵ-quasirandom graph $G_u(X_i, X_j)$ of density d the mean-square density of $G_u(X_i, X_j)$ with respect to the partitions $X_i = X_{i1} \cup \cdots \cup X_{is_i}$ and $Y_j = Y_{j1} \cup \cdots \cup Y_{jt_j}$ is at least $d^2 + \epsilon^2/16$. Since a random pair (x, y) has a probability of at least ϵ of belonging to $X_i \times X_j$ for such a pair (i, j), a calculation similar to that of Lemma 7.5 shows that the sum of the mean-square densities of all the graphs G_u with respect to the partitions $\{X_{ik}\}$ and $\{Y_{jl}\}$ is at least $D + \epsilon^3/16$, as claimed. □

Theorem 7.8. Let $\epsilon > 0$, let $X_1, \ldots, X_k$ be finite sets. For each i let $X'_{i1}, \ldots, X'_{im_i}$ be a partition of X_i, and for each i, j let $G_{ij}(1), \ldots, G_{ij}(r_{ij})$ be bipartite graphs that form a partition of the complete bipartite graph $K(X_i, X_j)$. Then for each i one can find a partition $X_{i1}, \ldots, X_{in_i}$ of X_i that refines the partition $X'_{i1}, \ldots, X'_{im_i}$, and this can be done in such a way that, for every i and j, if a random pair (x, y) is chosen from $X_i \times X_j$, then with probability at least $1 - \epsilon$ it lies in some set $X_{is} \times X_{jt}$ for which all the r_{ij} induced subgraphs $G_{ij}(u)(X_{is}, X_{jt})$ are ϵ-quasirandom. Moreover, all the n_i are bounded above by a function that depends on ϵ, the m_i and the r_{ij} only.

Proof. Suppose that we have partitions $X_{i1}, \ldots, X_{in_i}$ of each X_i, and suppose that for these partitions the conclusion of the theorem is false. Then there exist i and j such that at least $\epsilon |X_i||X_j|$ of the pairs $(x, y) \in X_i \times X_j$ lie in sets $X_{is} \times X_{jt}$ for which at least one of the graphs $G_{ij}(u)(X_{is}, X_{jt})$ is not ϵ-quasirandom. By Lemma 7.7 we can refine the partitions $X_{i1}, \ldots, X_{in_i}$ and $X_{j1}, \ldots, X_{jn_j}$ in such a way that the sum of the mean-square densities of the graphs $G_{ij}(u)$ with respect to the refined partitions is greater by at least $\epsilon^3/16$ than it was for the original ones. Moreover, the numbers of sets in the new partitions are bounded above by an exponential function of the numbers in the old ones.

Since the sum of the mean-square densities of the graphs $G_{ij}(u)$ (over all i, j and u) cannot exceed $\sum_{i<j} r_{ij}$, this procedure must terminate after at most $16\epsilon^{-3}\sum_{i<j} r_{ij}$ steps. At that point we have a partition with the desired properties. If we start the iteration with the partitions given in the first place, then we end up proving the theorem. $\square$

8. Regularity for 3-uniform hypergraphs

As ever, the picture for hypergraphs is more complicated. One of the reasons for this we have already met – we shall split our hypergraphs into 'stable' subhypergraphs (see the discussion at the end of Section 5) rather than quasirandom ones, and this forces us to discuss chains (G, H) as well as hypergraphs. We shall find ourselves partitioning not just the vertex sets of H (and G) but also the edge sets of the graphs G, so the statements we prove are rather more elaborate.

A more technical complication, but nevertheless a fundamental one, arises out of the fact that we must consider chains (G, H) for which G is very sparse. This makes it hard to generalize Lemma 7.4 adequately. To see why, let G be an α-quasirandom tripartite graph of density p with vertex sets X, Y and Z, and suppose that p is very small. Let $H \subset \Delta(G)$ be a hypergraph that fails to be η-quasirandom relative to $\Delta(G)$. If α is small enough, then an averaging argument similar to that of Theorem 4.1 allows us to find sets $X' \subset X$, $Y' \subset Y$ and $Z' \subset Z$ such that the restriction of H to $X' \times Y' \times Z'$ is of significantly greater density than H itself. (This statement is true both relative to G and, since G is quasirandom, in absolute terms.) Unfortunately, the sets X', Y' and Z' that this argument gives are contained in certain neighbourhoods of vertices of G, and therefore their sizes depend not just on η but on p as well. Therefore, any increase in mean-square density that we can hope to get from the dense hypergraph $H(X', Y', Z')$ will also depend on p.

This matters a lot, because as the iteration proceeds in the hypergraph regularity lemma, we are forced to consider a sequence of graphs G_i with rapidly decreasing densities p_i, so if the increase in mean-square density at stage i depends on p_i, there is no guarantee that the iteration will come to an end.

The solution to this problem is to squeeze a bit more out of the proof of Theorem 4.1. Instead of choosing just one triple (X', Y', Z'), we shall choose several, and prove that they are sufficiently spread out to provide us with an increase in mean-square density that is strong enough to use.

The statement of the hypergraph regularity lemma is somewhat complicated, so we shall postpone it until after we have made the above remarks precise in Lemma 8.4 below. At that point, the formulation of the regularity lemma will be better motivated and the rest of the proof quite easy.

To begin with, here is a simple and general criterion that we can use when we are trying to establish that a partition gives us an increase in mean-square density.

Lemma 8.1. *Let U be a set of size n and let f and g be functions from U to the interval $[-1, 1]$. Let $B_1, \ldots, B_r$ be a partition of U and suppose that g is constant on each B_j. Then the mean-square density of f with respect to the partition $B_1, \ldots, B_r$ is at least $\langle f, g \rangle^2 / n \|g\|_2^2$.*

Proof. For each j let a_j be the value taken by g on the set B_j. Then, by the Cauchy–Schwarz inequality,

$$\langle f, g \rangle = \sum_j a_j \sum_{x \in B_j} f(x)$$

$$\leqslant \left(\sum_j |B_j| a_j^2 \right)^{1/2} \left(\sum_j |B_j|^{-1} \left(\sum_{x \in B_j} f(x) \right)^2 \right)^{1/2}$$

$$= \|g\|_2 \left(\sum_j |B_j| \left(|B_j|^{-1} \sum_{x \in B_j} f(x) \right)^2 \right)^{1/2}.$$

But $\sum_j |B_j| \left(|B_j|^{-1} \sum_{x \in B_j} f(x) \right)^2$ is n times the mean-square density of f with respect to the partition $B_1, \ldots, B_r$ (by definition), so the lemma follows. □

So that it does not clutter up the proof of Lemma 8.4, here is a second very simple technical lemma.

Lemma 8.2. *Let $0 < \delta < 1$ and let r be an integer greater than or equal to δ^{-1}. Let $v_1, \ldots, v_n$ be vectors in ℓ_2^n such that $\|v_i\|^2 \leqslant n$ for each i and such that $\|\sum v_i\|_2^2 \leqslant \delta n^3$. Let r vectors $w_1, \ldots, w_r$ be chosen uniformly and independently from the v_i. (To be precise, for each w_j an index i is chosen randomly between 1 and n and w_j is set equal to v_i.) Then the expectation of $\|\sum w_j\|_2^2$ is at most $2\delta r^2 n$.*

Proof. The expectation of $\|\sum w_j\|_2^2$ is the expectation of $\sum_{i,j} \langle w_i, w_j \rangle$. If $i \neq j$ then the expectation of $\langle w_i, w_j \rangle$ is $n^{-2} \|\sum v_i\|_2^2$ which, by hypothesis, is at most δn. If $i = j$, then $\langle w_i, w_j \rangle$ is at most n, again by hypothesis. Therefore, the expectation we are trying to bound is at most $(\delta r(r-1) + r)n$. Since $\delta r \geqslant 1$, this is at most $2\delta r^2 n$, as claimed. □

Quasirandomness, Counting and Regularity for 3-Uniform Hypergraphs   361

Definition 8.3. Let G be a tripartite graph with vertex sets X, Y and Z and let the bipartite graphs $G(X,Y)$, $G(Y,Z)$ and $G(X,Z)$ be partitioned into subgraphs $G_i(X,Y)$, $G_j(Y,Z)$ and $G_k(X,Z)$ respectively. For each triangle $(x,y,z) \in \Delta(G)$, define its *index* to be the triple (i,j,k) such that $xy \in G_i(X,Y)$, $yz \in G_j(Y,Z)$ and $xz \in G_k(X,Z)$. The *induced partition* of $\Delta(G)$ is the partition of the triples of $\Delta(G)$ according to their index. If $f : X \times Y \times \to [-1,1]$, then the *mean-square density of f relative to* the partitions $(G_i(X,Y))$, $(G_j(Y,Z))$ and $(G_k(X,Z))$ is defined to be the mean-square density of the restriction of f to $\Delta(G)$, relative to the induced partition of $\Delta(G)$.

Note that a typical cell of the induced partition is of the form $\Delta\big(G_i(X,Y) \cup G_j(Y,Z) \cup G_k(X,Z)\big)$.

Lemma 8.4. *Let G be a tripartite graph with vertex sets X, Y and Z of sizes L, M and N respectively, let the densities $G(X,Y)$, $G(Y,Z)$ and $G(X,Z)$ be δ_{12}, δ_{23} and δ_{13}, let $\delta = \delta_{12}\delta_{23}\delta_{13}$ and suppose that these three graphs are α-quasirandom. Let H be a tripartite 3-uniform hypergraph with the same vertex sets as G, let the relative density $|H \cap \Delta(G)|/|\Delta(G)|$ of H in G be d, and suppose that H is not η-quasirandom relative to $\Delta(G)$. Suppose also that $2^{21}\alpha^{1/4} \leqslant \delta^7$. Then there are partitions $G_1(X,Y) \cup \cdots \cup G_l(X,Y) = G(X,Y)$, $G_1(Y,Z) \cup \cdots \cup G_m(Y,Z) = G(Y,Z)$ and $G_1(X,Z) \cup \cdots \cup G_n(X,Z) = G(X,Z)$, relative to which the mean-square density of H is at least $d^2 + 2^{-10}\eta^2$. Moreover, if $r \geqslant \delta^{-4}$ is a positive integer then l, m and n can all be taken to be at most 3^r.*

Proof. Let $f(x,y,z) = H(x,y,z) - d$ whenever $(x,y,z) \in \Delta(G)$ and let it be zero otherwise. Then the hypothesis of the lemma is that

$$\sum_{x,x',y,y',z,z'} f_{x,x',y,y',z,z'} \geqslant \eta\delta^4 L^2 M^2 N^2.$$

Let $U = \Delta(G)$ and for each triple $(x,y,z) \in U$ define $G_{xyz}(x',y',z')$ to be

$$f_{x,x',y,y',z,z'}/f(x',y',z'),$$

and let $F(x',y',z') = \sum_{x,y,z} G_{xyz}(x',y',z')$. With this notation, the hypothesis can be rewritten $\langle f, F \rangle \geqslant \eta\delta^4 (LMN)^2$.

Let us now build some new functions E_{xyz}. These will have similar properties to the F_{xyz} but will take values 0, 1 and -1 only. A vital property of each function F_{xyz} is that it can be written in the form $u(x',y')v(y',z')w(x',z')$, since of the eight terms in the product $f_{x,x',y,y',z,z'}$ the only one that depends on all three of x', y' and z' is $f(x',y',z')$, which is absent from $F_{xyz}(x',y',z')$. Moreover, we can do this with u, v and w taking values in the interval $[-1,1]$.

Fix a triple (x,y,z) as above, and for each pair (x',y') define $u'(x',y')$ randomly according to the following simple rule. If $u(x',y')$ is positive then $u'(x',y')$ is 1 with probability $u(x',y')$ and 0 otherwise. If $u(x',y')$ is negative then $u'(x',y')$ is -1 with probability $|u(x',y')|$ and 0 otherwise. If $u(x',y') = 0$ then $u'(x',y') = 0$ as well. Construct functions v' and w' similarly, and let all the random choices that have been made be independent. Finally, let

$E_{xyz}(x', y', z') = u'(x', y')v'(y', z')w'(x', z')$, and note that the expectation of $E_{xyz}(x', y', z')$ is $F_{xyz}(x', y', z')$.

Let E be the sum of all the functions E_{xyz}. The expectations of $\langle f, F_{xyz}\rangle$ and $\langle f, E_{xyz}\rangle$ are the same, so we can make the random choices in such a way that $\langle f, E\rangle \geq \eta\delta^4(LMN)^2$. The other property we shall use is that $E_{xyz}(x', y', z')$ is nonzero only if x, x', y, y', z and z' are the vertices of an octahedron in the graph G. (This is true despite the fact that we have divided by $f(x', y', z')$: for each pair there is still a triple containing it for which f is required to be nonzero.) So that our notation will be reasonably concise, let us set $G_{x,x',y,y',z,z'}$ to be 1 if x, x', y, y', z and z' are the vertices of an octahedron and 0 otherwise. (Strictly speaking, it would be more accurate to write $\Delta(G)_{x,x',y,y',z,z'}$.)

Our next aim is to obtain an upper bound for $\|E\|_2^2$. To start with, we have

$$\|E\|_2^2 = \sum_{x',y',z'} \left(\sum_{x,y,z} E_{xyz}(x', y', z')\right)^2$$

$$\leq \sum_{x',y',z'} \left(\sum_{x,y,z} G_{x,x',y,y',z,z'}\right)^2$$

$$= \sum_{x',y',z'} \sum_{x_1,y_1,z_1,x_2,y_2,z_2} G_{x_1,x',y_1,y',z_1,z'} G_{x_2,x',y_2,y',z_2,z'}.$$

This counts the number of 9-tuples $(x', y', z', x_1, y_1, z_1, x_2, y_2, z_2)$ such that the sextuples $(x', x_1, y', y_1, z', z_1)$ and $(x', x_2, y', y_2, z', z_2)$ both form octahedra in G. That is, it counts the number of copies in G of a certain graph with nine vertices and seven edges between each pair of vertex sets (four for each octahedron but intersecting in one). By Theorem 3.5 and our upper bound for α, there are at most $2\delta^7(LMN)^3$ of these.

We are not yet in a position to apply Lemma 8.1: E is the sum of $|U|$ functions E_{xyz} and we are unlikely to find a partition into just a few sets on which it is constant. The next stage of the argument is to make a random selection of a small number of E_{xyz} in such a way that their sum preserves the good properties of E. Let $r \geq \delta^{-4}$, choose $E_1, \ldots, E_r$ randomly from the E_{xyz} and let $D = \sum_{i=1}^r E_i$. Since the expectation of $\langle f, E_i\rangle$ is $|U|^{-1}\langle f, E\rangle$ and we know that $\langle f, E\rangle \geq \eta\delta^4(LMN)^2$, the expectation of $\langle f, D\rangle$ is at least $\eta\delta^4 r|U|^{-1}(LMN)^2$. By Theorem 3.5 and our upper bound for α, we know that $|U|$ lies between $\delta LMN/2$ and $2\delta LMN$, so this is at least $\eta r\delta^2|U|/4$.

We shall now apply Lemma 8.2, with $n = |U|$, to the functions E_{xyz}. We have shown that $\|E\|_2^2 \leq 2\delta^7(LMN)^3$, which is at most $16\delta^4|U|^3$. Moreover, each E_{xyz} is supported in U and takes values in $[-1, 1]$, so $\|E_{xyz}\|_2^2$ is at most $|U|$. Therefore, Lemma 8.2 implies that the expectation of $\|D\|_2^2$ is at most $32\delta^4 r^2|U|$. Therefore, the expectation of $256\delta^2 r\langle f, D\rangle - \eta\|D\|_2^2$ is at least $64\eta\delta^4 r^2|U|$. From this it follows that we can choose D in such a way that $\langle f, D\rangle \geq \eta\delta^2 r|U|/4$ and $\|D\|_2^2 \leq 256\delta^2 r\langle f, D\rangle\eta^{-1}$. For such a D, we have

$$\frac{\langle f, D\rangle}{\|D\|_2^2} \geq \frac{\langle f, D\rangle}{256\delta^2 r\eta^{-1}} \geq \frac{\eta\delta^2 r|U|}{2^{10}\delta^2 r\eta^{-1}} = \frac{\eta^2|U|}{2^{10}}.$$

We finish the proof by applying Lemma 8.1 to a suitable partition of $U = \Delta(G)$. Each of the r functions E_{xyz} that we added up to make D is of the form $u'(x', y')v'(y', z')w'(x', z')$, where u', v' and w' take values in the set $\{-1, 0, 1\}$. We can partition the bipartite graph

$G(X, Y)$ into at most 3^r subgraphs $G_i(X, Y)$ such that every u' is constant on each $G_i(X, Y)$. In a similar way we can partition $G(Y, Z)$ into subgraphs $G_j(Y, Z)$ and $G(X, Z)$ into subgraphs $G_k(X, Z)$. For each triangle $(x, y, z) \in \Delta(G)$, define its *index* to be the triple (i, j, k) such that $xy \in G_i(X, Y)$, $yz \in G_j(Y, Z)$ and $xz \in G_k(X, Z)$. Then partition $\Delta(G)$ into triples according to their index. This partition has at most 3^{3r} cells, each of the form $\Delta(G_i(X, Y) \cup G_j(Y, Z) \cup G_k(X, Z))$.

The function D is constant on each cell, so Lemma 8.1 and our estimate for $\langle f, D \rangle / \|D\|_2^2$ imply that the mean-square density of f with respect to this partition is at least $2^{-10}\eta^2$. It follows that the mean-square density of H is at least $d^2 + 2^{-10}\eta^2$, which proves the lemma. □

Now let us prepare for the statement of our regularity lemma for 3-uniform hypergraphs. As with the counting lemma, we will keep things simple by restricting to the case of quadripartite hypergraphs, but the result can easily be generalized. The idea will be to take a quadripartite 3-uniform hypergraph H and 'decompose it into quasirandom chains'. Before we say exactly what this means, let us try to motivate the slightly technical definition we shall give in a moment of a quasirandom chain.

What we ultimately want from a chain is that it should satisfy the conditions for Theorem 6.8, the counting lemma for chains that we proved earlier. There we had a quadripartite graph G with vertex sets of sizes L, M, N and P and a quadripartite 3-uniform hypergraph $H \subset \Delta(G)$. Writing δ for the product of the densities of the six bipartite parts of G, the assumptions of the theorem were that each tripartite part of H was η-quasirandom relative to G, each bipartite part of G was α-quasirandom and $2^{36}\alpha^{1/16} \leqslant \eta\delta^8/6$. Writing γ for the product of the relative densities of the four tripartite parts of H, the conclusion was that the number of simplices in H differed from $\gamma\delta LMNP$ by at most $8\eta^{1/8}\delta LMNP$. We therefore consider this to be a small error if $\eta^{1/8}$ is small compared with γ.

Definition 8.5. Let $\psi(\eta, \theta)$ be a polynomial in η and θ that vanishes when either η or θ is zero. A tripartite chain (G', H') is (η, θ, ψ)-*quasirandom* if the three parts of G' each have density at least θ and are $\psi(\eta, \theta)$-quasirandom, and H' is η-quasirandom relative to G'. A quadripartite chain (G, H) is (η, θ, ψ)-*quasirandom* each of its four constituent tripartite chains is (η, θ, ψ)-quasirandom.

Later we shall use this definition in the case where $\psi(\eta, \theta) = (2^{-40}\eta\theta^{48})^{16}$. Then, if $\alpha = \psi(\eta, \theta)$, the condition $2^{36}\alpha^{1/16} \leqslant \eta\delta^8/6$, discussed above, will be satisfied, since δ is at least θ^6.

In Szemerédi's regularity lemma one decomposes a graph into quasirandom pieces using partitions of its vertex sets. As we have already mentioned, the decompositions we shall consider of hypergraphs are more complicated, so let us say precisely what they are. We shall use the following notation for this purpose and throughout the rest of the paper. If G is a quadripartite graph, X and Y are two of its vertex sets and $X' \subset X$ and $Y' \subset Y$, then $G(X', Y')$ stands for the induced bipartite subgraph of G with vertex sets X' and Y'. Similarly, if H is a quadripartite 3-uniform hypergraph and X', Y' and Z' are subsets of three of its vertex sets then $H(X', Y', Z')$ stands for the induced tripartite subhypergraph with vertex sets X', Y' and Z'. The complete k-partite graph with vertex sets $X_1, \ldots, X_k$ will be denoted $K(X_1, \ldots, X_k)$.

Definition 8.6. Let X_1, X_2, X_3 and X_4 be four sets. A *decomposition* of the complete quadripartite graph $K(X_1, X_2, X_3, X_4)$ consists of the following:

(a) for each vertex set X_i a partition into subsets $X_{i1}, \ldots, X_{in_i}$;
(b) for each bipartite graph $K(X_i, X_j)$ a partition into subgraphs $G_{ij}(1), \ldots, G_{ij}(m_{ij})$.

Such a decomposition provides us with a collection of tripartite graphs. A typical one of these graphs has vertex sets of the form X_{ir}, X_{js} and X_{kt}, and an edge set of the form

$$G_{ij}(u)(X_{ir}, X_{js}) \cup G_{jk}(v)(X_{js}, X_{kt}) \cup G_{ik}(w)(X_{ir}, X_{kt}).$$

If H is a quadripartite 3-uniform hypergraph with vertex sets X_1, X_2, X_3 and X_4, then the decomposition also provides us with a collection of tripartite hypergraphs. A typical one of these has vertex sets as above and edge set $H \cap \Delta(G)$, where G is one of the tripartite graphs of the form the collection just defined. The pair (G, H) will then be a chain.

Definition 8.7. Let H be a quadripartite 3-uniform hypergraph with vertex sets $X_1, \ldots, X_4$ and suppose that we have a decomposition of $K(X_1, X_2, X_3, X_4)$. Then the associated *chain decomposition* of H is the set of all chains formed in the way explained above.

Every triple $(x, y, z) \in (X_i, X_j, X_k)$ belongs to exactly one of the tripartite graphs defined above, and hence we can associate with it exactly one of the chains. Hence, to each (x, y, z, w) we can associate four chains, one for each triple.

Definition 8.8. Let H be as above, suppose that we have a decomposition of $K(X_1, X_2, X_3, X_4)$ and let $1 \leqslant i < j < k \leqslant 4$. Then the associated chain decomposition of $H(X_i, X_j, X_k)$ is (ϵ, η, ψ)-*quasirandom* if for all but $\epsilon |X_i||X_j||X_k|$ of its edges (x, y, z) the associated tripartite chain is (η, θ, ψ)-quasirandom for some $\theta > 0$ (which one can take to be the minimum of the densities of the three parts of the tripartite graph in the chain). The associated chain decomposition of H itself is (ϵ, η, ψ)-*quasirandom* if the decompositions of all its four parts are.

Just before we prove the regularity lemma, here is a generalization of Corollary 7.3 that we will need in the proof.

Lemma 8.9. *Let G be a tripartite graph with vertex sets X, Y and Z, let H be a tripartite 3-uniform hypergraph with the same vertex sets, and let d be the relative density $|H \cap \Delta(G)|/|\Delta(G)|$ of H in $\Delta(G)$. Let $E_1 \cup \cdots \cup E_l$, $F_1 \cup \cdots \cup F_m$ and $G_1 \cup \cdots \cup G_n$ be partitions of $G(X, Y)$, $G(Y, Z)$ and $G(X, Z)$ respectively. Let each E_i, F_j and G_k be partitioned further into sets E_{ir}, F_{js} and G_{kt}. Then the mean-square density of H relative to the partitions $\{E_{ir}\}$, $\{F_{js}\}$ and $\{G_{kt}\}$ is at least as big as the mean-square density of H relative to $\{E_i\}$, $\{F_j\}$ and $\{G_k\}$.*

Proof. Like Corollary 7.3, this result is an immediate consequence of Lemma 7.2. This time, let $U = \Delta(G)$ and let f be the characteristic function of H. The partition of U induced by the partitions $\{E_{ir}\}$, $\{F_{js}\}$ and $\{G_{kt}\}$ is a refinement of the partition induced by the partitions $\{E_i\}$, $\{F_j\}$ and $\{G_k\}$, and again the quantities compared in Lemma 7.2 are the mean-square densities we wish to compare here. $\square$

We are now ready for the main result of this section.

Theorem 8.10. *Let H be a quadripartite hypergraph with vertex sets X, Y, Z and W, let $\epsilon, \eta > 0$ and let $\psi(\eta, \theta)$ be a function of the form $c\eta^r \theta^s$ with $r, s \in \mathbb{N}$ and $0 < c \leqslant 1$. Let the densities of $H(X, Y, Z)$, $H(X, Y, W)$, $H(X, Z, W)$ and $H(Y, Z, W)$ be δ_{123}, δ_{124}, δ_{134} and δ_{234} respectively. Then there is a decomposition of the complete quadripartite graph $K(X, Y, Z, W)$ such that the associated chain decomposition of H is (ϵ, η, ψ)-quasirandom. Moreover, the number of bipartite graphs in the decomposition is bounded above by a number that depends only on ϵ and η, while the number of sets in the partitions of X, Y, Z and W is bounded above by a function of ϵ, η, ψ and the densities δ_{123}, δ_{124}, δ_{134} and δ_{234}.*

Proof. Suppose that we have a decomposition of the vertices and edges of $K(X, Y, Z, W)$, with each vertex set partitioned into at most n sets and each complete bipartite graph formed from two of the vertex sets partitioned into at most m bipartite subgraphs.

Let $\alpha = \psi(\eta, \epsilon/48m)$. By Theorem 7.8 we can refine the partitions of X, Y, Z and W so that if a random quadruple (x, y, z, w) is chosen from $X \times Y \times Z \times W$, then, with probability at least $1 - \alpha$, xy lies in a product $X_i \times Y_j$ of cells such that all the induced subgraphs $G(X_i, Y_j)$, where G is a graph from the decomposition of $K(X, Y)$, are α-quasirandom, and similarly for the other five pairs from (x, y, z, w). Suppose that we have passed to such a refinement. Then each vertex set is now partitioned into at most N sets, where N depends on ϵ, m, n, η and ψ only.

We do not want to consider chains that are too sparse, so let us show that these do not occur very often. To any quadruple $(x, y, z, w) \in X \times Y \times Z \times W$ there is an associated quadripartite chain. Let its vertex sets be A_1, A_2, A_3 and A_4 and let its edge-sets be G_{ij} for $1 \leqslant i < j \leqslant 4$. Here, A_4 is the cell that contains w from the partition of W, G_{23} is the bipartite graph from the decomposition that contains the edge yz, and so on. The graphs G_{ij} form the edges of the quadripartite chain. The hyperedges come from the union of the four tripartite hypergraphs

$$H \cap \Delta\bigl(G_{ij}(A_i, A_j) \cup G_{jk}(A_j, A_k) \cup G_{ik}(A_i, A_k)\bigr),$$

where $1 \leqslant i < j < k \leqslant 4$.

Suppose we choose (x, y, z, w) at random. Then there are at most m possibilities for each G_{ij}. It follows that if we condition on the set $A_1 \times A_2 \times A_3 \times A_4$ in which (x, y, z, w) lies, then for each ij the probability that G_{ij} has density less than $\epsilon/48m$ in $A_i \times A_j$ is at most $\epsilon/48$. Therefore, with probability at least $1 - \epsilon/8$, each G_{ij} has density at least $\epsilon/48m$ in $A_i \times A_j$.

Suppose that the chain decomposition we now have of H is not (ϵ, η, ψ)-quasirandom. Then, without loss of generality, for at least $\epsilon |X||Y||Z|/4$ edges of $H(X, Y, Z)$ the associated chain is not (η, θ, ψ)-quasirandom for any $\theta > 0$. It follows that there are at least $\epsilon|X||Y||Z|/8$ edges (x,y,z) of $H(X,Y,Z)$ such that the density conditions above are satisfied for the associated chains, so that we may take $\theta = \epsilon/48m$, and such that either $H(X,Y,Z)$ is not η-quasirandom relative to the associated tripartite graph or the three parts of the tripartite graph are not all α-quasirandom.

However, we have arranged for the second possibility to apply to at most $\alpha|X||Y||Z|$ triples, so there are at least $\epsilon|X||Y||Z|/16$ edges such that $H(X,Y,Z)$ is not η-quasirandom relative to the associated tripartite graph.

Whenever this happens we can use Lemma 8.4 to partition the three bipartite parts of the graph in such a way that the mean-square density of H relative to these partitions is greater by at least $2^{-10}\eta$ than the square of the density of H relative to the tripartite graph. Moreover, by Lemma 8.9 this property is maintained if we refine these partitions. Hence, we can choose a common refinement of all the partitions of all the bipartite parts of all the graphs, preserving all the mean-square density increases. Since at least $\epsilon|X||Y||Z|/16$ triples belong to tripartite graphs where such increases have taken place, we find that the mean-square density of H relative to the new decomposition is at least $2^{-14}\epsilon\eta$ greater than it was relative to the old one. The number of bipartite graphs in the new partitions is bounded above by a function of ϵ, η and m only.

This procedure can be iterated. When we refine the partitions of the vertex sets, the mean-square density of H relative to the decomposition does not decrease, so the iteration must eventually come to an end and the theorem is proved. □

Remark. An examination of the above proof shows that the bound for the numbers of cells in the eventual partition is of 'wowzer' type. The wowzer function W is defined in two steps as follows. First let the tower function T be defined by $T(1) = 2$, $T(n) = 2^{T(n-1)}$. Next, let $W(1) = 2$ and $W(n) = T(W(n-1))$. In general, the hypergraph regularity lemma for k-uniform hypergraphs iterates the bound for $(k-1)$-uniform hypergraphs, and thus advances one level in the Ackermann hierarchy. This means that the bounds that it gives for the theorems of van der Waerden and Szemerédi are of Ackermann type. Similar bounds have recently been achieved for Szemerédi's theorem by Tao, who has produced a discretization of Furstenberg's ergodic-theory proof. This answered a question that many people had asked, and the insights gained played an important role in his spectacular result with Green that the primes contain arbitrarily long arithmetic progressions.

We finish the section with a corollary that is designed to make the regularity lemma easy and convenient to use. As with our earlier results, the result we give is not the most general possible, but more general versions can be proved with only small adaptations.

Let H be a quadripartite 3-uniform hypergraph with vertex sets X_1, X_2, X_3 and X_4, and for each i let the size of X_i be N_i. Suppose that for each i we have a subset $A_i \subset X_i$ of size $\delta_i|X_i|$ and that G is a quadripartite graph with vertex sets A_1, A_2, A_3 and A_4. For each pair $i < j$ let δ_{ij} be the density $|G(A_i, A_j)|/|A_i||A_j|$ and for each $i < j < k$ let δ_{ijk} be the relative density of H inside $\Delta\big(G(A_i, A_j) \cup G(A_j, A_k) \cup G(A_i, A_k)\big)$. Let us call the product of the densities δ_i, δ_{ij} and δ_{ijk} the *expected simplex density* of H in G.

Corollary 8.11. Let $\epsilon > 0$. Then there exist positive integers m and n, depending on ϵ only, such that the following result holds. Let H be any quadripartite graph with vertex sets X_1, X_2, X_3 and X_4, and for each i let the size of X_i be N_i. Then for each triple $1 \leqslant i < j < k \leqslant 4$ one can remove at most $\epsilon |H(X_i, X_j, X_k)|$ edges of $H(X_i, X_j, X_k)$, and one can find a decomposition of the complete quadripartite graph $K = K(X_1, X_2, X_3, X_4)$, with the following property. Let each X_i be partitioned into at most n parts and let the number of bipartite graphs in any of the six parts of K be at most m. Let H' equal H after the edges have been removed and let G be any quadripartite graph arising from the decomposition with vertex sets A_i. Let σ be the expected simplex density of H' in G. Then either $\sigma = 0$ or $\sigma \geqslant (\epsilon/48m)^6 (\epsilon/8)^4$, each A_i has size at least $\epsilon N_i / 24n$ and the number of simplices in the hypergraph $H' \cap \Delta(G)$ differs from $\sigma |A_1||A_2||A_3||A_4|$ by at most $\epsilon\sigma |A_1||A_2||A_3||A_4|$.

Proof. Let γ be $(\epsilon/8)^3$, let $\eta = (\epsilon\gamma/16)^8$ and for any $\theta > 0$ let $\psi(\eta, \delta) = (2^{-40}\eta\theta^{48})^{16}$. Using Theorem 8.10, let us take a decomposition of the graph $K = K(X_1, X_2, X_3, X_4)$ such that the associated chain decomposition of H is $(\epsilon/2, \eta, \psi)$-quasirandom, and let m be the maximum number of bipartite graphs in any of the six parts of K. For each $i < j < k$ we shall remove at most $\epsilon |X_i||X_j||X_k|$ edges of H, using the conclusion of Theorem 8.10 and simple averaging arguments. The result will be a subhypergraph H' such that each edge (x_i, x_j, x_k) has several good properties. To describe these properties, let us take an arbitrary such edge, let $A_i \subset X_i$, $A_j \subset X_j$ and $A_k \subset X_k$ be the vertex sets from the decomposition that contain x_i, x_j and x_k respectively, and let G_{ij}, G_{jk} and G_{ik} be the bipartite graphs containing $x_i x_j$, $x_j x_k$ and $x_i x_k$. The properties are then as follows.

 (i) The densities of A_i, A_j and A_k inside X_i, X_j and X_k are all at least $\epsilon/24n$.
 (ii) The densities of G_{ij}, G_{jk} and G_{ik} inside $K(A_i, A_j)$, $K(A_j, A_k)$ and $K(A_i, A_k)$ are all at least $\epsilon/48m$.
(iii) The relative density of H' inside the tripartite graph

$$G = G_{ij}(A_i, A_j) \cup G_{jk}(A_j, A_k) \cup G_{ik}(A_i, A_k)$$

is at least $\epsilon/8$.
(iv) The chain $(G, H' \cap \Delta(G))$ is $(\epsilon/2, \eta, \psi)$-quasirandom.

Now let us see how we can get these properties.

(i) Since X_i is partitioned into at most n sets A_i, the number of vertices that belong to an A_i of density less than $e/24n$ is at most $\epsilon |X_i|/24$. Therefore, the number of triples (x_i, x_j, x_k) in $X_i \times X_j \times X_k$ such that at least one of A_i, A_j and A_k has density less than $\epsilon/24n$ is at most $\epsilon |X_i||X_j||X_k|/8$.

(ii) This is obtained by an argument similar to that for (i). In fact, we gave the argument as part of the proof of Theorem 8.10, which implies that the number of triples (x_i, x_j, x_k) in $X_i \times X_j \times X_k$ for which this density condition fails is again at most $\epsilon |X_i||X_j||X_k|/8$.

(iii) Since $K(X_i, X_j, X_k)$ is partitioned into sets $\Delta(G)$, where G is a tripartite graph of the given form, at most $\epsilon |X_i||X_j||X_k|/8$ edges of $H(X_i, X_j, X_k)$ can live in a $\Delta(G)$ where the relative density of H is less than $\epsilon/8$.

(iv) The conclusion of Theorem 8.10 (with ϵ replaced by $\epsilon/2$) and Definition 8.8 together tell us that this is true for all but at most $\epsilon|X_i||X_j||X_k|/2$ edges of $H(X_i, X_j, X_k)$.

It follows that we may remove at most $\epsilon|X_i||X_j||X_k|$ edges from each part $H(X_i, X_j, X_k)$ of H and obtain a hypergraph H' such that properties (i), (ii), (iii) and (iv) hold for every single edge of H'.

We now apply Theorem 6.8. Let G be any quadripartite graph arising from the decomposition, let its vertex sets be A_1, A_2, A_3 and A_4 and let δ be the product of the densities of its edge sets. If all four hypergraphs of the form $\Delta(G(A_i, A_j, A_k))$ contain edges of H' then the densities of all the $G(A_i, A_j)$ inside $K(A_i, A_j)$ are at least $\epsilon/48m$, the relative density of H' inside each $G(A_i, A_j) \cup G(A_j, A_k) \cup G(A_i, A_k)$ is at least $\epsilon/8$ and H' is $(\epsilon/2, \eta, \psi)$-quasirandom there. By our choices of η, ψ and the lower bounds for the densities, this means that the conditions for Theorem 6.8 are satisfied and we may conclude that the number of simplices in $H \cap \Delta(G)$ differs from $\sigma|A_1||A_2||A_3||A_4|$ by at most $8\eta^{1/8}\delta|A_1||A_2||A_3||A_4|$. By our choice of η, this is at most $\epsilon\sigma|A_1||A_2||A_3||A_4|$. $\square$

9. Szemerédi's theorem for progressions of length 4

It is now straightforward to generalize the proof of Theorem 1.1 to give a proof of Theorem 1.4. Let us modify the statement a little bit so that it fits better with the statements of the last section.

Theorem 9.1. *For every $a > 0$ there exists $c > 0$ with the following property. Let H be any quadripartite hypergraph H with vertex sets X_1, X_2, X_3 and X_4 of sizes N_1, N_2, N_3 and N_4 respectively, and suppose that the number of simplices in H is at most $cN_1N_2N_3N_4$. Then it is possible to remove at most $aN_iN_jN_k$ triples from each of the four induced subhypergraphs $H(X_i, X_j, X_k)$ in such a way that the resulting subhypergraph of H is simplex-free.*

Proof. Apply Corollary 8.11 with $\epsilon = a$ and suppose that the resulting hypergraph H' contains a simplex with vertices x_1, x_2, x_3 and x_4. Let G be the associated quadripartite graph coming from the decomposition. Then each tripartite part of G contains an edge of H', so the expected simplex density σ of H in G is nonzero. Corollary 8.11 tells us that it is therefore at least $(1-a)(a/8n)^4(a/48m)^6(a/8)^4$. If c is less than this, then we have obtained a contradiction, which shows that H' could not after all have contained a simplex. Since the numbers m and n depend on ϵ only, the theorem is proved. $\square$

To conclude, let us deduce Szemerédi's theorem for progressions of length four. First we prove a tiny weakening of Theorem 1.5. (The difference is that we prove that $d \neq 0$ rather than that $d > 0$. Ben Green's trick mentioned in the proof of Corollary 1.2 works here as well, but Szemerédi's theorem does not need the positivity of d.)

Theorem 9.2. *For every $\delta > 0$ there exists N such that every subset $A \subset [N]^3$ of size at least δN^3 contains a quadruple of points of the form*

$$\{(x, y, z), (x+d, y, z), (x, y+d, z), (x, y, z+d)\}$$

with $d \neq 0$.

Proof. Define a quadripartite hypergraph with vertex sets $X = Y = Z = [N]$ and $W = [3N]$ as follows. A triple $(x, y, z) \in X \times Y \times Z$ belongs to H if and only if $(x, y, z) \in A$. A triple $(x, y, w) \in X \times Y \times W$ belongs to H if and only if $(x, y, w - x - y) \in A$, and similarly for triples (x, z, w) and (y, z, w).

Suppose now that H contains a simplex (x, y, z, w) and let $d = w - x - y - z$. Then the points (x, y, z), $(x, y, z + d)$, $(x, y + d, z)$ and $(x + d, y, z)$ all belong to A. This proves the theorem unless d is always 0. But in that case, there are at most N^3 simplices in H: in fact, there are exactly $|A|$ of them. Now we apply Theorem 9.1 with $a = \delta/20$. It gives us a $c > 0$ such that, if H contains at most cN^4 simplices, then we can remove at most aN^3 edges from each part of H and remove all of them. Since our hypergraph contains at most N^3 simplices, if $N^{-1} < c$ we may apply the theorem. However, the number of simplices with $d = 0$ is δN^3, and no two of these share a face, since a simplex with $d = 0$ is uniquely determined by any one of its four faces. Therefore, if we remove at most a proportion a of the faces from each of the four parts of H, we end up removing at most $10aN^3 = \delta N^3/2$ simplices, which is not all of them. (The number 10 comes from the fact that $|W| = 3N$.) That is a contradiction, and the theorem is proved. $\square$

Corollary 9.3. *For every $\delta > 0$ there exists N such that every subset $A \subset [N]$ of size at least δN contains an arithmetic progression of length four.*

Proof. Define a subset $B \subset [N]^3$ to consist of all triples (x, y, z) such that $x + 2y + 3z \in A$. It is a straightforward exercise to show that B has density bounded below by a function of δ, so Theorem 9.2 yields the arithmetic progression

$$x + 2y + 3z, \quad (x + d) + 2y + 3z, \quad x + 2(y + d) + 3z, \quad x + 2y + 3(z + d).$$

$\square$

10. Concluding remarks

The technical details in the papers of Rödl and his co-authors are very different from those here and it is possible and instructive to pinpoint where the difference arises. The answer is that the two approaches use different notions of quasirandomness that are equivalent for graphs and dense hypergraphs but *not* equivalent for the sparse hypergraphs we are forced to consider here.

Briefly, whereas in this paper we have focused on 'octahedral quasirandomness', they concentrate their attention on edge-uniformity (see Definition 2.6). These two definitions are not equivalent for chains (G, H), because, as we noted at the beginning of Section 8, if G is sparse, then the failure of H to be relatively quasirandom does not imply that there is a significant increase in mean-square density.

This presented us with a technical problem (dealt with in Lemma 8.4) that does not arise if one uses edge-uniformity instead: that property is weaker, so its denial has stronger consequences. However, one pays the price for this weakness when proving a counting lemma. (In fact, this is an oversimplification: edge-uniformity does not seem to be enough on its own, but a more complicated variant of it can be used instead, more complicated in a way that is rather similar to the way that Lemma 8.4 is more complicated than Lemma 7.4.) Octahedral quasirandomness appears to be exactly the right concept for proving a counting lemma in an analytic style modelled on arguments that have already been used for proving Szemerédi's theorem; edge-uniformity is the concept one naturally comes up with if one wishes to model one's arguments on the usual proof of Szemerédi's regularity lemma. It is just a pity that one cannot use both!

Thus, even though the two approaches share many features, in that they both prove regularity and counting lemmas for chains, they are also genuinely different: for Rödl and his co-authors the counting lemma is harder than the regularity lemma, whereas for us it is the other way round.

References

[1] Chung, F. R. K. and Graham, R. L. (1990) Quasi-random hypergraphs. *Random Struct. Alg.* **1** 105–124.

[2] Chung, F. R. K. and Graham, R. L. (1992) Quasi-random subsets of $\mathbb{Z}_n$. *J. Combin. Theory Ser. A* **61** 64–86.

[3] Chung, F. R. K., Graham, R. L. and Wilson, R. M. (1989) Quasi-random graphs. *Combinatorica* **9** 345–362.

[4] Frankl, P. and Rödl, V. (1992) The uniformity lemma for hypergraphs. *Graphs Combin.* **8** 309–312.

[5] Frankl, P. and Rödl, V. (2002) Extremal problems on set systems. *Random Struct. Alg.* **20** 131–164.

[6] Furstenberg, H. and Katznelson, Y. (1978) An ergodic Szemerédi theorem for commuting transformations. *J. Analyse Math.* **34** 275–291.

[7] Furstenberg, H., Katznelson, Y. and Ornstein, D. (1982) The ergodic theoretical proof of Szemerédi's theorem *Bull. Amer. Math. Soc.* **7** 527–552.

[8] Gowers, W. T. (1998) A new proof of Szemerédi's theorem for arithmetic progressions of length four. *Geom. Funct. Anal.* **8** 529–551.

[9] Gowers, W. T. (2001) A new proof of Szemerédi's theorem. *Geom. Funct. Anal.* **11** 465–588.

[10] Gowers, W. T. Hypergraph regularity and Szemerédi's theorem. Submitted.

[11] Nagle, B., Rödl, V. and Schacht, M. The counting lemma for regular k-uniform hypergraphs. *Random Struct. Alg.*, to appear.

[12] Rödl, V. (1991) Some developments in Ramsey theory. In *Proc. International Congress of Mathematicians*, Vol. I, II (Kyoto, 1990), Math. Soc. Japan, Tokyo, pp. 1455–1466.

[13] Ruzsa, I. Z. and Szemerédi, E. (1978) Triple systems with no six points carrying three triangles. In *Combinatorics* (Proc. Fifth Hungarian Colloq., Keszthely, 1976), Vol. II, pp. 939–945.

[14] Solymosi, J. (2003) Note on a generalization of Roth's theorem. In *Discrete and Computational Geometry*, Vol. 25 of *Algorithms and Combinatorics*, Springer, Berlin, pp. 825–827.

[15] Solymosi, J. (2004) A note on a question of Erdős and Graham. *Combin. Probab. Comput.* **13** 263–267.

[16] Thomason, A. G. (1987) Pseudo-random graphs. In *Proc. Random Graphs*, Poznán 1985 (M. Karonski, ed.), *Ann. Discrete Math.* **33** 307–331.

Triangle-Free Hypergraphs

ERVIN GYŐRI[†]

Alfréd Rényi Institute of Mathematics, Budapest, Hungary
(e-mail: gyori@renyi.hu)

Received 6 September 2004; revised 15 May 2005

For Béla Bollobás on his 60th birthday

In this paper, we give sharp upper bounds on the maximum number of edges in very unbalanced bipartite graphs not containing any cycle of length 6. To prove this, we estimate roughly the sum of the sizes of the hyperedges in triangle-free multi-hypergraphs.

Introduction

In [2], Erdős, Sárközy and Sós studied some number-theoretic problems on product representations of squares. To prove a number-theoretic conjecture, they needed the following conjecture.

Let $G(X, Y)$ be a bipartite graph with colour classes X, Y where $|X| = m \leqslant |Y| = n$. If the graph G does not contain any cycle of length 6 then $e(G) \leqslant 2n + c(mn)^{2/3}$ for some constant c.

In [4], the present author proved that $e(G) \leqslant 2n + m^2/2$. (We note that this result was proved for any m and n but it is weak if $n = o(m^2)$. De Caen and Székely [1] and Faudree and Simonovits [3] proved $e(G) \leqslant c(mn)^{2/3}$ for some constant c if $m^2 \geqslant n$.) The result above was sufficient to prove the number-theoretic conjecture but not sharp. It is clear that the coefficient 2 of n cannot be improved, but the term involving m can be improved. It was conjectured in [4] that $e(G) \leqslant 2n + m - 2$ holds and this is the best possible. However, we show that this conjecture is false. In this paper, we prove the estimate

$$e(G) \leqslant 2n + m^2/8$$

which is sharp if $n \geqslant m^2/16$.

Note that bipartite graphs can be considered as incidence bipartite graphs of multi-hypergraphs. ('Multi' means that we may have several copies of the same hyperedge.) The incidence bipartite graph is C_6-free if and only if the multi-hypergraph is triangle-free.

[†] Research partially supported by OTKA Research Fund 34702.

We use the weakest possible definition of triangle: three hyperedges h_1, h_2, h_3 constitute a *triangle* if there are vertices v_1, v_2, v_3 such that $v_1, v_2 \in h_1$, $v_2, v_3 \in h_2$ and $v_3, v_1 \in h_3$.

Let the weight $w(e)$ of a hyperedge e be $|e| - 2$ and the total weight $w(H)$ of the hypergraph H be the sum of the weights of the hyperedges. Then we prove the following.

Theorem 1. *Let $H = (V, E)$ be a triangle-free multi-hypergraph. Then*
$$w(H) = \sum_{e \in E} (|e| - 2) \leqslant n^2/8$$
if n is large enough.

Remarks. (1) We prove the statement for $n \geqslant 100$, but for the sake of brevity we do not make special efforts to get the best estimate for n. On the other hand, we skip the proof of the upper estimate $(n^2 + c)/8$ with the best constant $c = 16$ for $n \leqslant 84$.

(2) There are some counterexamples if n is small.

Example 1. For $n = 9$, let the elements be arranged in a 3×3 grid and take two copies of the triplets in each row and each column. The resulting 3-uniform multi-hypergraph has 12 edges, and its total weight is 12, which is more than the required upper bound 10.

Example 2. Let $n = 12$, let the elements be arranged in a 3×4 grid and take two copies of each row and each column. The resulting multi-hypergraph has 14 edges, and its total weight is 20, which is more than the required upper bound 18.

Remark. Theorem 1 is sharp, as the following example shows.

Example 3. For n divisible by 4, take the hypergraph H with
$$V(H) = \{x_1, x_1', x_2, x_2', \ldots, x_{n/4}, x_{n/4}', y_1, y_1', y_2, y_2', \ldots, y_{n/4}, y_{n/4}'\},$$
$$E(H) = \{\{x_i, x_i', y_j, y_j'\} : 1 \leqslant i \leqslant n/4, 1 \leqslant j \leqslant n/4\}.$$

If 4 does not divide n then take the hypergraph above with $4\lceil n/4 \rceil$ vertices and truncate H by deleting the first $4\lceil n/4 \rceil - n$ vertices of x_1, x_1', y_1, y_1' from all hyperedges of H. Further examples can be obtained by taking some vertex pairs x_i, x_i' and replacing all 4-edges $\{x_i, x_i', y_j, y_j'\}$ by the 3-edges $\{x_i, y_j, y_j'\}, \{x_i', y_j, y_j'\}$.

Taking the incidence bipartite graph of the appropriate hypergraph, we obtain the following result.

Corollary. *Let $G(X, Y)$ be a C_6-free bipartite graph with colour classes X, Y, such that $|X| = m > 100$, $|Y| = n > 100$. Then*
$$e(G) \leqslant 2m + n^2/8.$$

This estimate is sharp too if $m \geqslant n^2/16$. Take the incidence bipartite graph of the hypergraph in Example 3 if $m = n^2/16$. If $m > n^2/16$ then first add $m - n^2/16$ copies of the 2-edge $\{x_1, x_1'\}$ and again take the incidence bipartite graph of the resulting hypergraph.

Before the proof of Theorem 1, we prove a special case, in which the idea of the proof is very clear and neat.

Theorem 2. *Let $H = (V, E)$ be a triangle-free 3-uniform hypergraph without multiple edges. Then*

$$|E| = \sum_{e \in E} (|e| - 2) \leqslant n^2/8.$$

Remark added in proof. We learned that P. Frankl, A. Füredi and G. Simonyi also proved Theorem 2 later, but independently.

Proof of Theorem 2. Note that for an arbitrary hyperedge $e \in E$, if $f \in E$ and $g \in E$ are hyperedges such that $|e \cap f| = |e \cap g| = 2$ then $e \cap f = e \cap g$, otherwise e, f, and g constitute a triangle on the vertices of e, a contradiction.

Now, if $e = xyz \in E$ is an arbitrary hyperedge, then we may assume that, say, $\{x, y\}, \{x, z\} \not\subseteq f$ for any hyperedge $f \neq e$ in H. Then replace the hyperedge e by the 2-edges xy, xz, and similarly every hyperedge is replaced by two 2-edges. Note that we do not take the same edge to replace two different hyperedges, since then this edge is the intersection of these two hyperedges and they must have been replaced by other 2-edge pairs.

Claim 1. *The resulting graph G is triangle-free.*

Proof of Claim 1. Suppose that G contains a triangle xyz. If, say, xy and yz are obtained from the same hyperedge e and xz is obtained from a hyperedge $\{x, z, u\}$, then $e \cap f = \{x, z\}$, and then by definition f should have been replaced by the edges ux, uz, a contradiction. If xy, yz, and xz are obtained from three different hyperedges, then these hyperedges constitute a triangle on the vertices x, y, z, a contradiction.

Thus, Turán's [5] theorem implies that G has at most $n^2/4$ edges and so H has at most $n^2/8$ hyperedges. □

Proof of Theorem 1

Obviously, we may assume that each hyperedge contains at least three vertices, and we cannot have 3 or more copies any of the hyperedges because they would constitute a triangle on three vertices of the hyperedge.

We will proceed by induction on n. Note that we cannot start the induction with some small values of n since, as we have seen, there are counterexamples for small n. Instead, when $n \geqslant 84$, we use an upper bound $((n-k)^2 + c)/8$ for $n - k$, with some small constant c and use this to prove the upper bound $(n^2 + c - k)/8$ for n (we call this 'strong' induction). After a small number of inductive steps, we will have $w(H) \leqslant n^2/8$. The existence of such a constant c is trivial by finiteness, and we skip the proof that $w(H) \leqslant (n^2 + c)/8$ with $c = 16$ for $n \leqslant 84$ because it is lengthy (although $c \leqslant 43$ falls out of the proof).

First, we prove a simple, but very useful lemma.

Lemma 1. *Let x be a vertex of a triangle-free multi-hypergraph H and let $h_1, h_2, \ldots, h_m$ be edges such that $|h_i| \geq 3$ and $x \in h_i$ for $i = 1, \ldots, m$. Then $|\cup_{i=1}^m h_i| > m$.*

Proof. Take two hyperedges $h_1, h_2 \in H$ and suppose that $h_1 \cap h_2 \neq \{x\}$, say $y \in h_1 \cap h_2 - \{x\}$. Then for any hyperedge h_3 we have $h_1 \cap h_3, h_2 \cap h_3 \subseteq \{x, y\}$, since otherwise h_1, h_2, h_3 constitute a triangle, a contradiction. It implies that if $|h_1 \cap h_2| \geq 3$ then for any hyperedge $h_3 \in H$, we have $h_1 \cap h_3 = h_2 \cap h_3 = \{x\}$. So, the components of the hypergraph H' with edge set $E(H') = \{h - \{x\} : h \in E(H)\}$ either have at most two hyperedges or the hyperedges of the component constitute a 'star', i.e., every two hyperedges meet in the same vertex. Thus, the number of hyperedges in H' is at most $|\cup_{h' \in E(H')} h'|$ and equality holds if and only if each component of H' is a double copy of a 2-edge. □

Now, we prove Theorem 1 by distinguishing a few cases concerning the hyperedge sizes.

Case 1. *For $k > 4$, there is a k-edge $e = x_1 \cdots x_k$ in H.*

Let H_i denote the subhypergraph of the hyperedges containing x_i and let $n_i + 1 = |X_i|$, where $X_i = \cup_{h \in H_i} h$ for $i = 1, \ldots, k$. Note that if $f \in H_i$ and $g \in H_j$ are distinct hyperedges (two identical copies are OK!) for some $i \neq j$ then $f \cap g \subseteq \{x_i, x_j\}$, since otherwise e, f and g constitute a triangle.

If x_i is not contained in any hyperedge $h \in H_j, j \neq i$, then X_i does not meet any other X_j and Lemma 1 implies that if we delete x_i from all hyperedges $h \neq e$ of H then the weight decreases by at most $|X_i| - 1$.

Let us consider the at least two element intersections of e with the hyperedges of H, i.e., the sets $\{S_f = e \cap f : f \in H, |e \cap f| > 1\}$. (If $S_f = S_g$ then we take two, or in general, several copies of the same set!)

If $|S_f| > 2$ then S_f is not met by another set S_g, because then e, f and g constitute a triangle in H, a contradiction. Then by applying Lemma 1 for the pairwise vertex-disjoint hypergraphs $H_i - f$, $x_i \in S_f$ we obtain that $n_i \leq |X_i| - |f|$, and if we delete all the vertices $x_i \in S_f$ from all hyperedges $h \neq e$ of H then the weight decreases by at most $|\cup_{x_i \in S_f} X_i| - 2$.

If $|S_f| = 2$ and S_f is met by another set S_g then $S_f = S_g$ (actually $f \cap g = S_f$ since otherwise e, f and g constitute a triangle in H). Suppose that exactly the sets $f_1 = f, f_2, \ldots, f_s$ have $e \cap f_i = S_f$. Then, by applying Lemma 1 for the vertex-disjoint hypergraphs $H_i - \{f_1, \ldots, f_s\}$, $x_i \in S_f$, we obtain that if we delete all the vertices $x_i \in S_f$ from all hyperedges $h \neq e$ of H then the weight decreases again by at most $|\cup_{x_i \in S_f} X_i| - 2$.

So if we delete all vertices of e from all hyperedges of $H - e$ then the weight decreases by at most $n - 2$ and equality can hold only if there is a hyperedge f containing e. Applying the induction hypothesis estimate $((n - k)^2 + c)/8$ for the resulting hypergraph, we obtain that $w(H) \leq ((n - k)^2 + c)/8 + n - 2 + k - 2$, which is at most $(n^2 + c - k)/8$ if $n \geq 19$. (The difference increases as k increases and it is easy to get the bound for $k = 5$.)

So we may assume that every hyperedge in H has three or four vertices.

Case 2. *There are 4-edges $e = x_1x_2x_3x_4$ and $f = x_1x_2x_3x_5$ of H sharing exactly three vertices, or there is a 4-edge $e = x_1x_2x_3x_4$ containing a 3-edge $f = x_1x_2x_3$.*

Then it is easy to see that replacing f by one more copy of e we would get a triangle-free hypergraph of not smaller weight.

So we may assume that H contains only 3- and 4-edges, and apart from the double copies, all intersections contain at most two vertices. In this case we prove a more general lemma than the statement of Theorem 1.

Lemma 2. *Let H be a triangle-free hypergraph of 2-, 3- and 4-edges on n vertices such that there are no parallel 2-edges; moreover, no 2-edge is contained in any other hyperedge and the intersection of any two hyperedges contains at most two vertices unless they are two copies of the same 3- or 4-edge. Let us define the weight f of the a hyperedge e by $2(|e| - 2)$ if e is a 3- or 4-edge and by 1 if it is a 2-edge. Then the total weight $f(H)$ of the edges is at most $n^2/4$ if n is large enough.*

Note that Lemma 2 implies the required statement. Theorem 1 is just the special case with no 2-edge. Again we have some counterexamples for small n (see the examples above). We try prove the lemma again by 'strong' induction, i.e., if we have the induction hypothesis upper bound $((n-m)^2 + c)/4$ for $n-m$ with some integer c then we prove the upper bound $(n^2 + c - m)/4$ for n. We find a direct proof when it does not work.

Proof of Lemma 2. The proof relies on the idea of the proof of Theorem 2. Take all the single (= no parallel copy) edges e met by another edge f in two vertices and fix such an f. Take the sets $X = e \cap f$ and $Y = e - X$. Replace the hyperedge e by the 2-edges joining X and Y. Note that e is replaced by $|X||Y| \geq 2(|e| - 2)$ 2-edges, and these 2-edge sets are pairwise disjoint we do not create parallel 2-edges. (If not, then there is a hyperedge g meeting both X and Y, and then e, f, g constitute a triangle on some vertices $x \in X \cap g$, $y \in Y \cap g$ and $z \in X - \{x\}$, a contradiction.) If e is a single 3- or 4-edge not met by any other hyperedge in two vertices, then take an arbitrary set $X \subset e$ of two vertices and replace e again by the 2-edges joining X and $Y = e - X$. These 2-edge sets are still pairwise disjoint and each hyperedge is replaced by 2-edges except the double 3- or 4-edges. What we have obtained is special hypergraph H_0: a set of single 2-edges and some double 3- and 4-edges. (If there is no such double 3- or 4-edge then we are done, by Turán's theorem and the forthcoming Claim 2.)

Claim 2. *The hypergraph H_0 is triangle-free.*

Proof of Claim 2. If H_0 contains a triangle of the hyperedges e, f, g on the vertices x, y, z then the appropriate hyperedges clearly constitute a triangle on the same vertex set in H unless two 2-edges, say $e = xy$ and $f = xz$, are obtained from the very same hyperedge h. By the construction, hyperedge g is obtained from another hyperedge. If g is a 3- or 4-edge containing vertices y, z, w, then h, g and the other copy g' of g constitute a triangle on $\{y, z, w\}$, a contradiction. If yz is a 2-edge obtained from a hyperedge e_0, then h meets

e_0 in at least two elements, and we have replaced the hyperedge e_0 by 2-edges because it was met by another hyperedge in two vertices. So there was a hyperedge f_0 containing, say, y but not containing z, which defined the edges replacing e_0. Then there is a vertex $w \in e_0 \cap f_0 - \{y\}$ and e_0, f_0, h constitute a triangle on y, z, w, a contradiction. □

The obtained hypergraph H_0 has a nice structure: it has double 3- and 4-edges and single 2-edges such that the intersection of any two hyperedges has at most one vertex unless these hyperedges are parallel, i.e., have the same vertex set.

Now we show that we may assume that H_0 does not contain any 4-edge because the 'strong' induction step can be done in this case. (As usual, the proof works only if n is large.)

Suppose that e and f are 4-edges of H_0 on the same vertex set $\{x_1, x_2, x_3, x_4\}$. Let H_i denote the subhypergraph of $H_0 - \{e, f\}$ consisting of the hyperedges containing x_i and let $X_i = \cup_{h \in H_i} h$ for $i = 1, 2, 3, 4$. Note that if $g \in H_i$ and $h \in H_j$ are distinct hyperedges (two identical copies are OK!) for some $i \neq j$ then $g \cap h = \emptyset$, since otherwise e, g and h constitute a triangle. So the sets X_i are pairwise disjoint and, say, X_1 has at most $n/4$ vertices by the pigeonhole principle. Suppose that H_1 contains m_1 2-edges, $2m_2$ 3-edges (m_2 parallel pairs) and $2m_3$ 4-edges (again, m_3 parallel pairs). Then we have $m_1 + 2m_2 + 3m_3 \leq n/4 - 1$ by Lemma 1, the fact that no 2-edge is contained in any other edge and $|X_1| \leq n/4$. Let us delete x_1 from all hyperedges of H_0 and keep just one copy of the arising 2-edges if we get two parallel 2-edges. Then the resulting hypergraph H_0' is triangle-free and has only double 3- and 4-edges and single 2-edges such that the intersection of any two hyperedges has at most one vertex unless these hyperedges are parallel, i.e., have the same vertex set. So we can apply the induction hypothesis that $f(H_0') \leq ((n-1)^2 + c)/4$. Deleting x_1, the total weight of the hyperedges decreases by $4 + m_1 + 3m_2 + 4m_3 \leq 4 + 3(m_1 + 2m_2 + 3m_3)/2 \leq 3n/8 + 5/2$. So $f(H_0) \leq ((n-1)^2 + c)/4 + 3n/8 + 5/2$, which is at most $(n^2 + c - 1)/4$ if $n \geq 24$.

What remains to be proved is as follows.

Claim 3. *If H_0 is a triangle-free multi-hypergraph of n vertices, n_3 double 3-edges (altogether $2n_3$ 3-edges) and n_2 single 2-edges then $f(H_0) = 4n_3 + n_2 \leq n^2/4$ (if n is large).*

Remark. There are some counterexamples for small n. Example 1 is just one of them, for which $4n_3 + n_2 = 24 > n^2/4 = 81/4$.

Proof of Claim 3. If the hypergraph is a graph (i.e., there are no 3-edges) then we are done, by Turán's theorem. Suppose that H_0 has a double hyperedge $e = x_1 x_2 x_3$. Then we try to proceed by 'strong' induction.

Let $e = f = x_1 x_2 x_3$ be a double 3-edge of H_0. The other hyperedges in H_0 meet e in at most one vertex. Let $2k$ be the number of 3-edges meeting e in exactly one vertex. If g and h are any two hyperedges meeting e in one vertex, but not two copies of the same triple, then clearly $(g - e) \cap (h - e) = \emptyset$ by the triangle-freedom of H_0. If we delete the vertices x_1, x_2, x_3 from the hyperedges of H_0 then we get 2-edges from the triples meeting e in one vertex, so the sum $4n_3 + n_2$ decreases by at most $4 + 3k + (n - 2k - 3)$ and by the induction hypothesis, the total weight is at most $((n-3)^2 + c)/4$ in the resulting multi-hypergraph. So $f(H_0) \leqslant (n^2 + c - 3)/4$ if $4 + 3k + (n - 2k - 3) \leqslant (6n - 12)/4$, i.e., if $2k \leqslant n - 8$.

So, we may assume that $2k \geqslant n - 7$ for any choice of $x_1 x_2 x_3$ and at least $n - 4$ vertices are contained in 3-edges.

Again, let H_1, H_2 and H_3 denote the hypergraphs of the hyperedges $h \neq e, f$ of H_0 containing x_1, x_2 and x_3, respectively, and let X_1, X_2 and X_3 be the (pairwise disjoint) vertex sets of H_1, H_2 and H_3, resp., with $|X_1| = n_1, |X_2| = n_2, |X_3| = n_3$. Suppose that $2k_1, 2k_2$ and $2k_3$ are the numbers of 3-edges in H_1, H_2 and H_3, respectively. Deleting the vertices x_1, x_2 from the hyperedges of H_0 and applying the induction hypothesis estimate $((n-2)^2 + c)/4$, we get $f(H_0) \leqslant 4 + 3k_1 + 3k_2 + (n_1 - 2k_1 - 1) + (n_2 - 2k_2 - 1) + ((n-2)^2 + c)/4 \leqslant 2 + 3n_1/2 + 3n_2/2 + ((n-2)^2 + c)/4$, which is at most $(n^2 + c - 2)/4$ if $3n_1 + 3n_2 \leqslant 2n - 7$. So, we may assume that $n_1 + n_2 \geqslant 2n/3 - 2$ and so $n_3 \leqslant n/3 + 2$. Similarly, we may assume that $n_1, n_2 \leqslant n/3 + 2$, and in general, if x is a vertex contained in a 3-edge, then the union of the hyperedges containing x has at most $n/3 + 4$ vertices.

Finally, deleting the vertex x_1 from the hyperedges of H_0 and applying the induction hypothesis estimate $((n-1)^2 + c)/4$, we get $f(H_0) \leqslant 3 + 3k_1 + (n_1 - 2k_1 - 1) + ((n-1)^2 + c)/4 \leqslant 3/2 + 3n_1/2 + ((n-1)^2 + c)/4$, which is at most $(n^2 + c - 1)/4$ if $n_1 \leqslant n/3 - 4/3$. So, we may assume that $n_1 \geqslant n/3 - 1$, and similarly we may assume that $n_2, n_3 \geqslant n/3 - 1$, and in general, if x is a vertex contained in a 3-edge, then the union of the hyperedges containing x has at least $n/3 - 1$ vertices.

Let y be a vertex not contained in any 3-edge. (There are at most four such vertices, as we have seen.) The vertex y is contained in at least four 2-edges, otherwise the induction goes trivially. So there is a 2-edge yz such that z is contained in a double 3-edge. So the hypergraph of the hyperedges containing z has at least $n/3 + 1$ vertices. Actually, it can be improved to $n/3 + 4/3$, since $n_1 - 2k_1 \geqslant 1$ in this case. From among these at least $n/3 + 4/3$ vertices just z is joined to y by a 2-edge by the triangle-freedom, so the number of 2-edges containing y is at most $2n/3 - 4/3$.

Now, we add up the weights of the hyperedges vertexwise: $4/3$ for the double 3-edges, $1/2$ for 2-edges containing the vertex. Since this sum is smaller for vertices contained in 3-edges, we may assume that there are four vertices not contained in a 3-edge. We obtain that $f(H_0) \leqslant 4(n-4)(n/6 + 3/2)/3 + 4(n/3 - 2/3)$, which is less than $n^2/4$ if $n \geqslant 84$. (It also yields that $f(H_0) \leqslant n^2/4 + 43$ for every n but the constant 43 as well the constant 84 can be improved easily, but it is very tiresome to get the best constants.) □

References

[1] de Caen, D. and Székely, L. A. (1992) The maximum size of 4- and 6-cycle free bipartite graphs on m, n vertices. In *Sets, Graphs and Numbers* (Proc. Coll. dedicated to the 60th birthday of A. Hajnal and V. T. Sós, Budapest, 1991), North-Holland, Amsterdam, pp. 135–142.

[2] Erdős, P., Sárközy, A. and Sós, V. T. (1995) On product representation of powers I. *European J. Combin.* **16** 323–331.

[3] Faudree, R. and Simonovits, M. On a class of degenerate extremal graph problems II. In preparation

[4] Győri, E. (1997) C_6-free bipartite graphs and product representation of squares. *Discr. Math.* **165/166** 371–375.

[5] Turán, P. (1941) On an extremal problem in graph theory (in Hungarian). *Mat. Fiz. Lapok* **48** 436–452.

Odd Independent Transversals are Odd

PENNY HAXELL[1†] and TIBOR SZABÓ[2]

[1] Department of Combinatorics and Optimization,
University of Waterloo, Waterloo Ont. Canada N2L 3G1
(e-mail: pehaxell@math.uwaterloo.ca)

[2] Department of Computer Science, ETH Zürich, 8092 Switzerland
(e-mail: szabo@inf.ethz.ch)

Received 26 September 2003; revised 27 April 2005

For Béla Bollobás on his 60th birthday

We put the final piece into a puzzle first introduced by Bollobás, Erdős and Szemerédi in 1975. For arbitrary positive integers n and r we determine the largest integer $\Delta = \Delta(r, n)$, for which any r-partite graph with partite sets of size n and of maximum degree less than Δ has an independent transversal. This value was known for all even r. Here we determine the value for odd r and find that $\Delta(r, n) = \Delta(r - 1, n)$. Informally this means that the addition of an oddth partite set does not make it any harder to guarantee an independent transversal.

In the proof we establish structural theorems which could be of independent interest. They work for *all* $r \geqslant 7$, and specify the structure of slightly sub-optimal graphs for *even* $r \geqslant 8$.

1. Introduction

Let G be a graph, and suppose the vertex set of G is partitioned into r parts $V(G) = V_1 \cup \cdots \cup V_r$. An *independent transversal* of G is an independent set in G containing exactly one vertex from each V_i. Let $\Delta = \Delta(r, n)$ be the largest integer such that any such G has an independent transversal whenever $|V_i| = n$ for each i, and the maximum degree $\Delta(G)$ satisfies $\Delta(G) < \Delta$. Define $\Delta_r = \lim_{n\to\infty} \Delta(r, n)/n$, where the limit is easily seen to exist. Clearly any edges of G that lie inside the classes V_i are irrelevant as far as the functions $\Delta(r, n)$ and Δ_r are concerned, so for simplicity we will consider only r-partite graphs.

The problem of determining the functions $\Delta(r, n)$ and Δ_r was raised and first studied by Bollobás, Erdős and Szemerédi [7] in 1975. This question is a very basic one, and it has come up in the study of various other combinatorial parameters such as linear arboricity

[†] Research supported in part by NSERC, and in part by ETH where much of this work was done.

and strong chromatic number. Throughout the years, continuing work on these problems has been done by several researchers [2, 4, 5, 6, 8, 9, 11, 14, 15] and steady progress has been made.

Trivially $\Delta(2, n) = n$, thus $\Delta_2 = 1$. Graver (cf. [7]) showed $\Delta_3 = 1$. In their original paper, Bollobás, Erdős and Szemerédi [7] proved that

$$\frac{2}{r} \leqslant \Delta_r \leqslant \frac{1}{2} + \frac{1}{r-2},$$

thus establishing $\mu = \lim_{r \to \infty} \Delta_r \leqslant 1/2$. They conjectured $\mu = 1/2$. Alon [4] was the first to separate μ from 0 by showing $\Delta_r \geqslant 1/(2e)$ for every r using the Local Lemma. This was improved to $\Delta_r \geqslant 1/2$ in [9], which settled the conjecture of [7] and established $\mu = 1/2$. Despite the significant progress on the asymptotic behaviour of Δ_r, knowledge about the exact values of Δ_r, even for very small values of r, was very sparse. Until very recently, the value of Δ_r was known only for $r = 2, 3, 4, 5$ [11]. The argument of Jin for 4- and 5-partite graphs, showing $\Delta_4 = \Delta_5 = 2/3$, is intricate and seems difficult to generalize.

Besides proving lower bounds for $r = 4$ and 5, Jin [11] also gave promising examples (with low maximum degree and no independent transversal), when r is a power of 2. (Later Yuster [15] also found the same construction.) Jin in fact conjectured that these examples provide the extremum for every r, i.e., $\Delta_{2^j} = \Delta_{2^j+i}$ for any $i \leqslant 2^j - 1$. Recently Alon [6] observed that the method of [9], which gives $\mu = 1/2$, actually implies the slightly stronger bound $\Delta_r \geqslant \frac{r}{2(r-1)}$. This implies Jin's construction [11] is optimal for powers of 2 and then one has $\Delta_r = \frac{r}{2(r-1)}$. For other integers r, Alon gave improvements on the constructions of Jin, thus disproving his conjecture in general.

Very recently, a construction matching the $\frac{r}{2(r-1)}$ lower bound was found [14] – but only for an even number of parts. After this discovery, taking into account that $\Delta_2 = \Delta_3$ and $\Delta_4 = \Delta_5$, it was natural to conjecture that $\Delta_{2t} = \Delta_{2t+1}$ for every t. Here we confirm this intuition by determining not only Δ_r, but all the values $\Delta(r, n)$ for every n when r is odd.

Theorem 1.1. *For every integer $n \geqslant 1$ and $r \geqslant 2$ odd,*

$$\Delta(r, n) = \Delta(r-1, n) = \left\lceil \frac{(r-1)n}{2(r-2)} \right\rceil.$$

In particular, for every r odd we have

$$\Delta_r = \frac{r-1}{2(r-2)}.$$

The construction of [14] determined $\Delta_6 = 3/5$, so we will concentrate on the case in which $r \geqslant 7$. Our argument consists of two parts. First we establish a structural theorem about minimal counterexamples. It was known that every r-partite graph with parts of size n and no independent transversal must have maximum degree at least $\frac{r}{2(r-1)}n$. What we prove here is that (for $r \geqslant 7$) any graph without an independent transversal, even if its maximum degree is a bit more than the threshold $\frac{r}{2(r-1)}n$ (but not more than $\frac{r-1}{2(r-2)}n$) is the vertex-disjoint union of $r - 1$ complete bipartite graphs together with some extra edges. Moreover, if the graph is minimal with respect to not having an independent transversal then it *is* the disjoint union of $r-1$ complete bipartite graphs. This theorem, Theorem 3.7, is valid for any number of parts ($\geqslant 7$) and for an even number parts it proves that any near-extremal example has this structure. We remark that this is in accordance with the known independent transversal-free examples: The graphs of [11], [15], and [14] are all $r-1$ copies of $K_{r/2,r/2}$. It is worthwhile to note though that the extremal examples are not unique, at least not for powers of 2. Although the graphs of Jin [11] and Yuster [15] are the same as the ones in [14], the partitions are very different.

In the second part of our proof we show that if r is odd, G is the union of $r-1$ complete bipartite graphs and $\Delta(G) < \frac{r-1}{2(r-2)}n$, then G has an independent transversal.

The organization of the paper is as follows. In Section 2 *induced matching configurations* are introduced, which are the basic structural tool of our proof. The technical Theorem 2.2 is applied in three different contexts throughout our paper, not always for the original graph with its vertex partition. In Section 3 we still deal with r-partite graphs where r is not necessarily odd and prove our structural theorems for independent transversal-free graphs. In Section 4 we prove Theorem 1.1. Here we finally make use of the fact that r is odd in the sense that an integer is odd if and only if every tree on r vertices could be considered a rooted tree in which every subtree not containing the root has strictly fewer than half of the vertices.

Throughout the paper, the neighbourhood of a vertex v is denoted by $N(v)$. For a set $T \subseteq V(G)$ we write $N_T(v) = N(v) \cap T$ and $G[T]$ for the subgraph of G induced by T. We denote the degree of a vertex v by $\deg(v) = |N(v)|$ and write $\deg_T(v) = |N_T(v)|$. We say that a vertex v is *dominated* by a set T if $N_T(v) \neq \emptyset$. If $N_T(v) = \emptyset$, then we say that v is *independent* of T.

2. Induced matching configurations

As in the Introduction, we consider r-partite graphs G, and to avoid trivialities we assume each part is nonempty. The notion of independent transversal naturally presupposes that the vertex partition $V(G) = V_1 \cup \cdots \cup V_r$ is fixed. However, for simplicity we will not refer explicitly to the vertex partition when the term independent transversal is used, unless there is a danger of confusion. By a *partial* independent transversal of G we mean an independent set U in G of size less than r, such that $|V_i \cap U| \leqslant 1$ for each i. Sometimes for emphasis we will refer to an independent transversal as a *complete* independent transversal. Often we will use the abbreviation IT. The aim of this section is to introduce and prove the existence of *induced matching configurations*, the basic structure employed in our proof.

Let G be an r-partite graph with vertex partition $V(G) = V_1 \cup \cdots \cup V_r$. A class V_i is called *active* for a subset $I \subseteq V(G)$ of the vertices if $V_i \cap I \neq \emptyset$, and $S(I)$ denotes the set of active classes of I. The *class-graph* $\mathscr{G}_I$ of a subset $I \subseteq V(G)$ of the vertices is obtained from $G[I]$ by contracting all the vertices of $V_i \cap I$ into one vertex, which, with slight abuse of notation, we also call V_i. Thus the vertex set of $\mathscr{G}_I$ is $S(I)$.

A set of vertices I is called an *induced matching configuration* (*IMC*), if $G[I]$ is a perfect matching and the graph $\mathscr{G}_I$ is a tree on r vertices. In particular every class is active, $|I| = 2(r-1)$, and $G[I]$ has at most one edge joining each pair of classes (since a

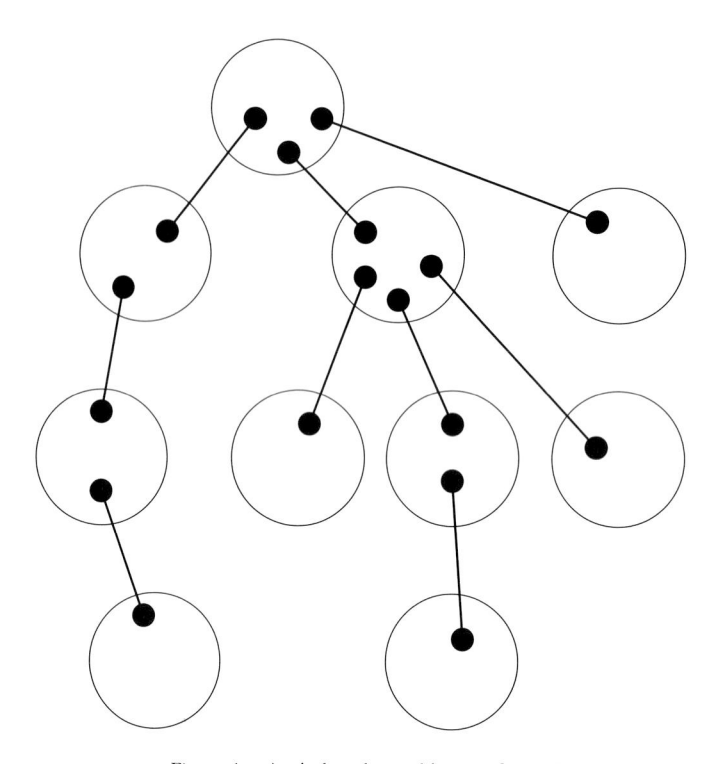

Figure 1. An induced matching configuration.

tree has no multiple edges). The following lemma is a simple but important observation. It describes circumstances under which a complete independent transversal could be obtained from an IMC.

Lemma 2.1. *Let G be an r-partite graph with vertex partition $V(G) = V_1 \cup \cdots \cup V_r$, and let I be an IMC in G. For any index $i \in [r]$, there is a partial independent transversal $T_i^I \subseteq I$ of G, such that $T_i^I \cap V_j = \emptyset$ if and only if $j = i$.*

Moreover, for any vertex v not dominated by I, there exists an independent transversal T_v^I containing v.

Proof. Consider the rooted tree produced from $\mathcal{G}_I$ by selecting V_i as the root. For every $j \neq i$ include in T_i^I the (unique) element of $I \cap V_j$ whose neighbour in I is in the parent class of V_j in $\mathcal{G}_I$.

For the second part, let $v \in V_i$ be a vertex not dominated by I. Then $T_v^I := T_i^I \cup \{v\}$ is an independent transversal. □

The main theorem of this section, Theorem 2.2, gives certain technical information about vertex-partitioned graphs that do not have independent transversals. In particular, we show that they *do* have induced matching configurations if their maximum degree is

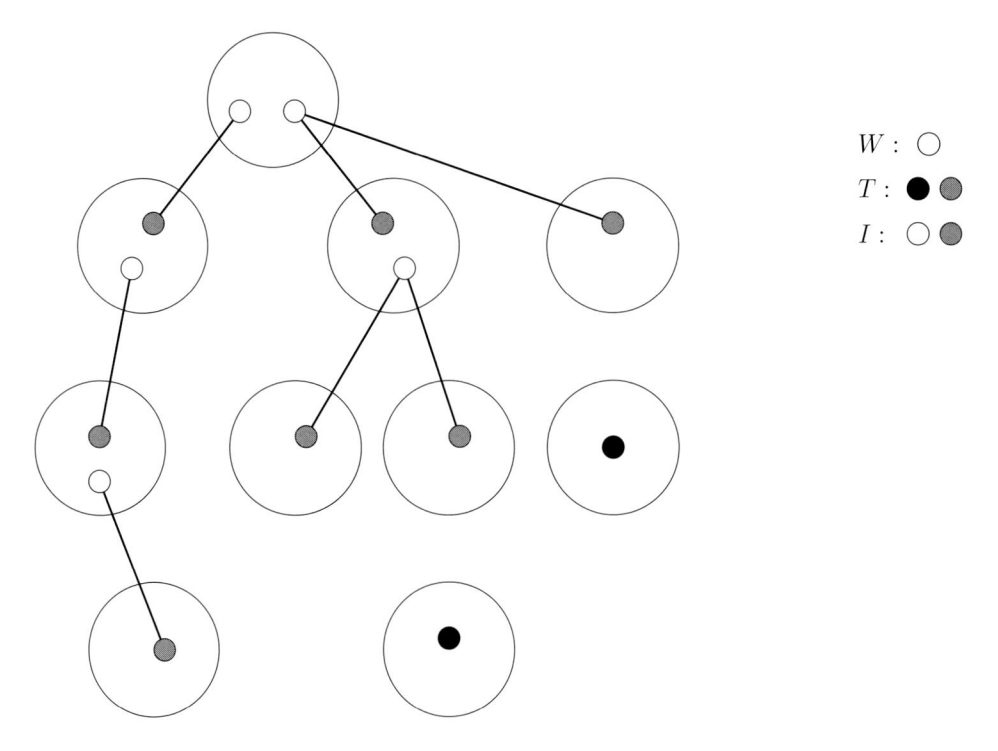

Figure 2. A feasible pair.

not too large. The proof of Theorem 2.2 we give here is based on the proof given in [10] that $\Delta_r \geq 1/2$.

Before we state Theorem 2.2 we need to establish some definitions and notation. Let G be an r-partite graph with vertex partition $V(G) = V_1 \cup \cdots \cup V_r$ that does not have an independent transversal. For a subset $A \subseteq \{V_1, \ldots, V_r\}$ let $\mathcal{T}_A$ be the set of partial independent transversals T which satisfy $|T \cap V_i| = 1$ if and only if $V_i \in A$. For a partial independent transversal T and a vertex $v \notin T$, we denote by $C(v, T)$ the vertex set of the component of $G[\{v\} \cup T]$ that contains v (so $G[C(v, T)]$ is always a star with centre v).

To prove the existence of an IMC we need to deal with the more general definition of a feasible pair (which is slightly different and stronger than the one in [10]). Despite looking awkwardly complicated, the following definition captures a relatively simple concept (see Figure 2). Below, by a *nontrivial* star we mean a star with at least 2 vertices. We call the pair (I, T) *feasible* if

(a) $I \subseteq V(G)$ and T is a partial independent transversal of maximum size,
(b) $S(I \cap T) = S(I) \cap S(T)$,
(c) $G[I]$ is a forest, whose components are the $|W|$ vertex-disjoint nontrivial stars $G[C(v, T)]$, with $v \in W$, where $W = I \setminus T$,
(d) (tree property) the graph $\mathcal{G}_I$ is a tree on the vertex set $S(I)$,
(e) (minimality property) there is no $v_0 \in W$ and $T' \in \mathcal{T}_{S(T)}$ with $T' \cap W = \emptyset$ such that $|C(v_0, T')| < |C(v_0, T)|$, but $C(v, T') = C(v, T)$ for $v \in W - \{v_0\}$.

Feasible pairs always exist, as $(\emptyset, T)$ is feasible if T is any partial transversal of maximum size (we consider the empty graph to be a tree).

The following theorem not only establishes the existence of an IMC, but it is used to derive our structural theorem, when it is applied for an auxiliary vertex-partitioned graph different from G.

Theorem 2.2. *Let G be an r-partite graph with vertex partition $V(G) = V_1 \cup \cdots \cup V_r$, and suppose G does not have an independent transversal of these classes. Let (I_0, T_0) be a feasible pair for G. Then there exists a feasible pair (I, T) in G such that*

  (i) $I_0 \subseteq I$, $|S(I)| \geq 2$, and $T \cap V_i = T_0 \cap V_i$ for every $V_i \in S(I_0)$,
  (ii) *I dominates all vertices in the active classes $\bigcup\{V_i : V_i \in S(I)\}$. In particular all vertices in $\bigcup\{V_i : V_i \in S(I)\}$ are dominated by $|I| \leq 2|S(I)| - 2$ vertices, and each class $V_i \in S(I)$ contains at least one of these dominating vertices from I.*

If in addition $r \geq 3$, $|V_i| = n$ for $i = 1, \ldots, r$, and the maximum degree Δ of G satisfies $\Delta < \frac{r-1}{2(r-2)} n$, then

  (iii) *I is an IMC in G, that dominates every vertex of G.*

Proof. To prove the theorem we apply the following algorithm.

ALGORITHM.

Input: The feasible pair (I_0, T_0).

Maintain: A feasible pair (I, T).
The set $W = I \setminus T$.
The set $S = S(I)$ of active classes for I.
The set $\mathcal{T} \subseteq \mathcal{T}_{S(T_0)}$ of transversals $T' \in \mathcal{T}_{S(T_0)}$, for which $T' \cap W = \emptyset$ and $C(v, T) = C(v, T')$ for every $v \in W$.

Initialization: $I := I_0$, $T := T_0$.

Idea: Iteratively grow I and change T accordingly in the non-active classes of I.

Iteration: If $I = \emptyset$, select a vertex $w \in \bigcup_{V_i \notin S(T)} V_i$ and transversal $T' \in \mathcal{T}$, such that $\deg_{T'}(w)$ is minimal. Update I by adding $C(w, T')$, update $T := T'$ and **iterate**.
If I dominates all vertices $w \in \bigcup_{V_i \in S} V_i$ in its active classes, then **stop** and **return** (I, T).
Otherwise select a vertex $w \in \bigcup_{V_i \in S} V_i$ and transversal $T' \in \mathcal{T}$, such that w is not dominated by I and $\deg_{T'}(w)$ is minimal. Update I by adding $C(w, T')$, update $T := T'$ and **iterate**.

First we prove that the pair (I, T) maintained by the algorithm is feasible throughout. Suppose that (I, T) is feasible, and w and T' are as defined in the Iteration. (Note that $\mathcal{T} \neq \emptyset$ since $T \in \mathcal{T}$.) We claim that with $I' = I \cup C(w, T')$, (I', T') is feasible as well.

Suppose first that $I = \emptyset$. The conditions (a), (b), (d), (e) are satisfied, by definition of w and T'. For condition (c) we must check that the star $G[C(w, T')]$ is nontrivial. This in fact is the case, because otherwise $T' \cup \{w\}$ would be an independent transversal of size larger than $|T|$.

Suppose now that $I \neq \emptyset$. Condition (a) holds since $|T'| = |T|$. For condition (b) note that $I' \cap T' = (I \cap T) \cup N_{T'}(w)$ and $I' = I \cup C(w, T')$. Then

$$S(I' \cap T') = S(I \cap T) \cup S(N_{T'}(w)) = (S(I) \cap S(T)) \cup S(N_{T'}(w))$$
$$= (S(I) \cup S(N_{T'}(w))) \cap (S(T) \cup S(N_{T'}(w)))$$
$$= (S(I) \cup S(C(w, T'))) \cap S(T') = S(I') \cap S(T').$$

For condition (c), consider the sets $C(v, T') = C(v, T)$ for $v \in W$; these are pairwise disjoint and of order at least 2. The last set $C(w, T')$ is disjoint from any set $C(v, T')$ ($v \in W$), since w is independent of I and T' agrees with T on the active classes of I.

Now assume for contradiction that $|C(w, T')| = 1$, i.e., $\deg_{T'}(w) = 0$. The class of w must contain a vertex of T', otherwise T' is not a maximum independent transversal, because w could be appended to it. Let u be the element of T' in the class of w. Since this class is active (we chose w from an active class) and T and T' agree on active classes, $u \in T$ as well. Then by (b) $u \in I$ and the degree of u in $G[I]$ is exactly one by (c). Let $v_0 \in W$ be its neighbour. Then $T'' = T' - \{u\} \cup \{w\}$ is a partial independent transversal in $\mathcal{T}_{S(T_0)}$ contradicting condition (e) of the feasibility of (I, T). Indeed, $C(v_0, T'') = C(v_0, T') - \{u\} = C(v_0, T) - \{u\}$, while $C(v, T'') = C(v, T') = C(v, T)$ for $v \in W - \{v_0\}$. Note also that $w \notin W$ because every $v \in W$ is dominated by a vertex of I (in $G[I]$ there is no isolated vertex). So $T'' \cap W = \emptyset$. Thus T'' provides the contradiction sought after, and this proves property (c) for the feasibility of (I', T').

For condition (d), it is enough to observe that w is in a class active for I, while all its T'-neighbours are in non-active classes. That is we obtain $\mathcal{G}_{I'}$ from $\mathcal{G}_I$ by appending $\deg_{T'}(w)$ leaves.

Now we check that condition (e) holds for (I', T'). Choose $v_0 \in W \cup \{w\}$, and suppose on the contrary that there exists $T'' \in \mathcal{T}_{S(T_0)}$ with $T'' \cap (W \cup \{w\}) = \emptyset$ such that $|C(v_0, T'')| < |C(v_0, T')|$ and $C(v, T'') = C(v, T')$ for every $v \in W \cup \{w\} \setminus \{v_0\}$. If $v_0 \in W$ then $|C(v_0, T'')| < |C(v_0, T')| = |C(v_0, T)|$ and $C(v, T'') = C(v, T') = C(v, T)$ for all $v \in W \setminus \{v_0\}$, contradicting condition (e) in the fact that (I, T) is feasible. If $v_0 = w$ then by definition $T'' \in \mathcal{T}$, since then $C(v, T'') = C(v, T') = C(v, T)$ for every $v \in W$. But $|C(w, T'')| < |C(w, T')|$ contradicts our choice of T'. Therefore no such v_0 can exist and we have verified that (I', T') is a feasible pair.

Thus the algorithm maintains and upon termination returns a feasible pair. Moreover the algorithm always terminates because in each step I increases in size.

To prove part (i) for the feasible pair (I, T) output by the algorithm, we note that the algorithm constructs (I, T) from (I_0, T_0) just by adding new vertices to I_0 and changing the transversal only outside $S(I_0)$. Since I_0 is the union of nontrivial stars, $|S(I)| \geqslant |S(I_0)| \geqslant 2$ unless $I_0 = \emptyset$. If I_0 is empty, the first step of the algorithm adds a nontrivial star to it, thus $|S(I)| \geqslant 2$ in this case as well.

Part (ii) is immediately implied by the stopping rule for the algorithm. The statement about the domination of the active classes follows from the facts that $\mathcal{G}_I$ is a tree on $S(I)$ and that $G[I]$ contains no isolated vertices, so $|I| \leqslant 2|S(I)| - 2$.

For part (iii), assume each V_i has size n and $\Delta < \frac{r-1}{2(r-2)}n$. We claim that the algorithm can terminate only when every class is active, i.e., if $|S| = r$. Since I dominates every vertex in $\cup_{V_i \in S} V_i$, we know $|\cup_{V_i \in S} V_i| = |S|n \leqslant |I|\Delta \leqslant (2|S| - 2)\Delta$. Therefore $\Delta \geqslant \frac{|S|}{2|S|-2}n$. But if $|S| \leqslant r - 1$ then $\frac{|S|}{2|S|-2}n \geqslant \frac{r-1}{2r-4}n$, contradicting our assumption. Hence $|S| = r$.

To complete the proof it remains to show that I induces a matching in G, in other words each $C(v, T)$ with $v \in W$ has size exactly 2. If this is not the case then $|W| \leqslant |I \cap T| - 1$ and thus $|I| = |W| + |I \cap T| \leqslant 2|I \cap T| - 1 \leqslant 2|T| - 1 \leqslant 2r - 3$. Then as above $rn \leqslant |I|\Delta \leqslant (2r-3)\Delta$, since I now dominates the whole graph by part (ii). This implies $\Delta \geqslant \frac{r}{2r-3}n \geqslant \frac{r-1}{2r-4}$ for $r \geqslant 3$, a contradiction. Therefore I is an IMC. □

3. Structural results

Our aim in this section is to show that if G is an r-partite graph with no independent transversal and Δ is not too large then G is the union of vertex-disjoint complete bipartite graphs, together with a few extra edges that can join two vertices in the same partite set, or cross between partite sets. Moreover if G also does not contain any *unnecessary* edges, where unnecessary means 'not preventing an independent transversal', then G is precisely a union of vertex-disjoint complete bipartite graphs.

Let G be an r-partite graph with vertex partition $V_1 \cup \cdots \cup V_r$, and let I be an IMC in G. We define the sets A_v of vertices, which are uniquely dominated by I. More formally, for $v \in I$ let $A_v(I) := \{y \in V(G) : N(y) \cap I = \{v\}\}$. If there is no possibility of confusion we omit from the notation the reference to the IMC and write simply A_v.

Because the first few lemmas of this section will all have the same assumptions, to avoid repetition we define the following. (Here by $G[A, B]$ we mean the bipartite subgraph of G consisting of all edges of G joining A and B.)

Setup. Let G be an r-partite graph with vertex partition $V_1 \cup \cdots \cup V_r$, where $|V_i| = n$ for each i, that does not have an IT. Denote the maximum degree of G by Δ. Let $I = \{v_i, w_i : 1 \leqslant i \leqslant r - 1\}$ be an IMC in G, where v_i and w_i are adjacent. Let $A_v = A_v(I)$ for each $v \in I$.

Lemma 3.1. *Let G be as in the Setup. Then*

(i) *for each i we have $w_i \in A_{v_i}$ and $v_i \in A_{w_i}$, and $G[A_{v_i}, A_{w_i}]$ is a complete bipartite graph,*
(ii) *for any $a \in A_{v_i}$, $b \in A_{w_i}$ the set $I' = I \setminus \{v_i, w_i\} \cup \{a, b\}$ is an IMC,*
(iii) *the number $|V(G) \setminus \cup_{v \in I} A_v|$ of vertices that are dominated more than once by I satisfies $|V(G) \setminus \cup_{v \in I} A_v| \leqslant 2(r-1)\Delta - rn$,*
(iv) *for any subset $\mathscr{S}$ of $\{A_v : v \in I\}$ we have $|\bigcup_{C \in \mathscr{S}} C| \geqslant (|\mathscr{S}| - 4r + 4)\Delta + 2rn$.*

(To give an idea of the size of these quantities, when we use this lemma to prove our main structural result Theorem 3.7, we have $\Delta < \frac{r-1}{2r-4}n$. Then the upper bound in (iii) becomes $\frac{n}{r-2}$ and the lower bound in (iv) is $|\mathscr{S}|\Delta - \frac{2n}{r-2}$.)

Proof. The first assertion of part (i) holds since v_i and w_i are adjacent by definition, and I is an *induced* matching.

Now we show that $G[A_{v_i}, A_{w_i}]$ is a complete bipartite graph. We fix i, and for convenience write $v = v_i$, $w = w_i$. The deletion of the edge vw disconnects the tree $\mathcal{G}_I$ into two components $\mathcal{G}_v$ and $\mathcal{G}_w$, where $\mathcal{G}_u$ is the tree containing the class of u for $u = v, w$. Then $I - \{v, w\}$ induces two IMCs I_v and I_w on the two sets of classes corresponding to the vertices of $\mathcal{G}_v$ and $\mathcal{G}_w$, respectively. Note that it is possible that, say, $\mathcal{G}_w$ consists of only one class, in which case I_w is empty.

Let $a \in A_v$, so its only neighbour in I is v. We claim that if the class of a were in $\mathcal{G}_v$, then G would have an independent transversal. To see this, note that a is not dominated by I_v and w is not dominated by I_w, so we can apply Lemma 2.1. Then the union of transversals $T_a^{I_v}$ and $T_w^{I_w}$ is an independent transversal in G, because a and w are not adjacent. This contradiction establishes our claim. Therefore the class of a is in $\mathcal{G}_w$. Similarly, if $b \in A_w$ then the class of b is in $\mathcal{G}_v$.

Now for any $a \in A_v$ and $b \in A_w$, if a and b were not adjacent, then the union of the transversals $T_b^{I_v}$ and $T_a^{I_w}$ would be an independent transversal in G. Thus a and b are adjacent and $G[A_v, A_w]$ is a complete bipartite graph.

For (ii), to prove that I' is an IMC we first note that by definition the only neighbour of a in I' is b and vice versa, so I' is an induced matching. Also, by the previous discussion, the class of a is in $\mathcal{G}_w$ and the class of b is in $\mathcal{G}_v$ (otherwise there is an IT), so the induced graph $\mathcal{G}_{I'}$ of I' is a reconnection of the two subtrees $\mathcal{G}_v$ and $\mathcal{G}_w$, thus a tree itself.

To establish (iii), we note that each element of I is adjacent to at most Δ vertices, but by Lemma 2.1 together they dominate $V(G)$. Hence at most $|I|\Delta - rn = 2(r-1)\Delta - rn$ vertices can be joined to more than one vertex from I.

For (iv), since each A_v has size at most Δ, we see that $\mathscr{S}$ must contain in its union at least $rn - (2(r-1)\Delta - rn) - (2(r-1) - |\mathscr{S}|)\Delta = (|\mathscr{S}| - 4r + 4)\Delta + 2rn$ vertices. $\square$

Next we prove a couple of technical lemmas. Let H be the spanning subgraph of G obtained by erasing all edges joining A_{v_i} to A_{w_i} for every i, and also all edges inside each A_v, $v \in I$. For a subset $\mathscr{S}$ of $\{A_v : v \in I\}$, we denote by $H_\mathscr{S}$ the induced subgraph of H on the vertex set $\cup_{C \in \mathscr{S}} C$. We will consider $H_\mathscr{S}$ as a vertex-partitioned graph, where the partition classes are the *partite sets* $C \in \mathscr{S}$ (and *not* related to the usual V_i).

Lemma 3.2. *Let G be as in the Setup, let $\mathscr{S}$ be a subset of $\{A_v : v \in I\}$, and let H be as above. Suppose we have a set Q of $2|\mathscr{S}| - 2$ vertices in $H_\mathscr{S}$ such that each member of $\mathscr{S}$ contains at least one of them. Then the number of vertices dominated by Q in $H_\mathscr{S}$ is at most* $(|\mathscr{S}| - 1)((4r - 4)\Delta - 2rn)$.

(When $\Delta < \frac{r-1}{2r-4}n$ this upper bound becomes $(|\mathscr{S}| - 1)\frac{2n}{r-2}$.)

Proof. Fix an $|\mathscr{S}|$-element subset Q' of Q containing exactly one vertex from each member of $\mathscr{S}$. The partite sets opposite these $|\mathscr{S}|$ vertices are distinct and total at least $(|\mathscr{S}| - 4r + 4)\Delta + 2rn$ vertices by Lemma 3.1(iv). These represent neighbours of Q' in G, which are lost in H, thus the number of vertices dominated by Q' in $H_\mathscr{S}$ is at most $|\mathscr{S}|\Delta - ((|\mathscr{S}| - 4r + 4)\Delta + 2rn) = (4r - 4)\Delta - 2rn$. The remaining $|\mathscr{S}| - 2$ vertices lie in partite sets opposite partite sets whose sizes are at least $(5 - 4r)\Delta + 2rn$ (since this is a lower bound on the size of any partite set by Lemma 3.1(iv)). Hence each of these dominates at most $\Delta - ((5-4r)\Delta + 2rn) = (4r-4)\Delta - 2rn$, for a total of at most $(|\mathcal{S}|-1)((4r-4)\Delta - 2rn)$. □

Lemma 3.3. *Let G be as in the Setup, let $S \subseteq [r-1]$ and let $\mathcal{S} = \{A_{v_i}, A_{w_i} : i \in S\}$. Suppose we have vertices $a_i \in A_{v_i}$ and $b_i \in A_{w_i}$ for each $i \in S$, with the property that $\{a_i, b_i : i \in S\}$ forms an IT in $H_\mathcal{S}$. Then $\{v_i, w_i : i \notin S\} \cup \{a_i, b_i : i \in S\}$ is an IMC in G.*

Proof. Suppose $\emptyset \subseteq R \subset S$ and we know that $\{v_i, w_i : i \notin R\} \cup \{a_i, b_i : i \in R\}$ is the vertex set of an IMC I_R in G. Let j be such that $j \in S \setminus R$. Then by definition of $a_j \in A_{v_j}$, $b_j \in A_{w_j}$, the only neighbour of a_j in $\{v_i, w_i : i \notin R\}$ is v_j, and the only neighbour of b_j in $\{v_i, w_i : i \notin R\}$ is w_j. Moreover, neither a_j nor b_j has a neighbour in $\{a_i, b_i : i \in R\}$ because of the IT condition on $H_\mathcal{S}$. Therefore $a_j \in A_{v_j}(I_R)$ and $b_j \in A_{w_j}(I_R)$, so by Lemma 3.1(ii) applied to I_R we have that $\{v_i, w_i : i \notin R \cup \{j\}\} \cup \{a_i, b_i : i \in R \cup \{j\}\}$ is an IMC in G. Repeating this argument we obtain the statement of the lemma. □

We will prove the first structural result stated at the beginning of this section in two steps. We are now ready to make the first step, in which we show that for our fixed IMC I, each vertex that is not in any A_v, $v \in I$ is completely joined to some A_v, $v \in I$.

Lemma 3.4. *Let G be as in the Setup, and suppose $r \geq 7$. Suppose the maximum degree Δ of G satisfies $\Delta < \frac{r-1}{2r-4}n$. Let x be a vertex lying outside $\bigcup_{v \in I} A_v$. Then x is joined to every vertex in A_v for some $v \in I$.*

Proof. For convenience we set $\mathscr{C} = \{A_v : v \in I\}$. For $v \in I$ let $A'_v \subseteq A_v$ be those vertices in A_v not adjacent to x. Suppose on the contrary that x is not joined completely to any A_v, i.e., each A'_v is nonempty. We claim that there exist $a_i \in A'_{v_i}$ and $b_i \in A'_{w_i}$ forming an IT in the subgraph $H'_\mathscr{C}$ of $H_\mathscr{C}$ induced by $V(H'_\mathscr{C}) = \bigcup_{v \in I} A'_v$, with vertex classes $\{A'_v : v \in I\}$. This would then be an IT in $H_\mathscr{C}$, and would by Lemma 3.3 be an IMC in G, which, by definition, would not dominate x. Lemma 2.1 then implies that there is an independent transversal in G, a contradiction.

Suppose there is no IT in $H'_\mathscr{C}$. Let $I_0 = \emptyset$ and T_0 be a partial transversal of $H'_\mathscr{C}$ of maximum size. Then by Theorem 2.2(ii) applied to $H'_\mathscr{C}$ with (I_0, T_0) and vertex classes $\{A'_v : v \in I\}$, we know that there exists a subset $\mathcal{S}$ of $\mathscr{C}$ and $2|\mathcal{S}|-2$ vertices in $\bigcup_{C \in \mathcal{S}} C'$ (each member of $\mathcal{S}$ containing at least one of them), that dominate all vertices in $\bigcup_{C \in \mathcal{S}} C'$ in $H_\mathcal{S}$. Therefore these vertices together with x dominate all of $H_\mathcal{S}$. But x dominates at most Δ, and the rest dominate at most $(|\mathcal{S}|-1)((4r-4)\Delta - 2rn)$ by Lemma 3.2. Hence $H_\mathcal{S}$, which contains at least $(|\mathcal{S}|-4r+4)\Delta + 2rn$ vertices by Lemma 3.1(iv), has size at most $(|\mathcal{S}|-1)((4r-4)\Delta - 2rn) + \Delta$. This implies $(|\mathcal{S}|-4r+4)\Delta + 2rn \leq (|\mathcal{S}|-1)((4r-4)\Delta - 2rn) + \Delta$, from which we conclude $|\mathcal{S}|2rn \leq \Delta((4r-5)|\mathcal{S}|+1)$. But then $\frac{|\mathcal{S}|2rn}{(4r-5)|\mathcal{S}|+1} \leq \Delta < \frac{r-1}{2r-4}n$ giving $(r-5)(|\mathcal{S}|-1) < 4$.

Note that by Theorem 2.2(i) we have $|\mathcal{S}| \geq 2$. For $|\mathcal{S}| \geq 3$ we have a contradiction because $r \geq 7$. Suppose that $|\mathcal{S}| = 2$. Observe that $\mathcal{S} \neq \{A_{v_i}, A_{w_i}\}$ for any index i, since otherwise $H_\mathcal{S}$ would have no edges, so there would be no dominating set either. Assume without loss of generality that $\mathscr{S} = \{A_{v_i}, A_{v_j}\}$ for some $i \neq j$. Let $Q = \{q_i, q_j\}$ be a set dominating A'_{v_i} and A'_{v_j} in $H_\mathscr{S}$. By Theorem 2.2(ii) they must be in different partite sets, say $q_i \in A'_{v_i}$ and $q_j \in A'_{v_j}$. Then q_i, q_j and x dominate the four partite sets $A_{v_i}, A_{v_j}, A_{w_i}, A_{w_j}$. Thus the size of these four partite sets, which is at least $2rn - (4r - 8)\Delta$ by Lemma 3.1(iv), must be at most 3Δ. We conclude $2rn \leqslant (4r - 5)\Delta$, a contradiction for $r \geqslant 5$.

Our contradiction implies the existence of the IT in $H_\mathscr{C}$, which is an IMC in G, not dominating x, a contradiction. □

Finally we can complete the proof by showing that extra vertices such as x in the previous lemma that are completely joined to A_{v_i} are also adjacent to all extra vertices y joined completely to A_{w_i}. For convenience we state the lemma below for $i = 1$, but the same argument gives the result for each i.

Lemma 3.5. *Let G be as in the Setup. Suppose $r \geqslant 7$ and $\Delta < \frac{r-1}{2r-4}n$. Suppose a is adjacent to all of A_{w_1}, and b to all of A_{v_1}. Then a is adjacent to b.*

Proof. Suppose on the contrary that a and b are not joined. Now for $v \in I \setminus \{v_1, w_1\}$ let $A'_v \subseteq A_v$ be those vertices not adjacent to a nor to b. Certainly these sets are all nonempty because otherwise a and b together would dominate three members of $\{A_v : v \in I\}$, which by Lemma 3.1(iv) have total size at least $(7 - 4r)\Delta + 2rn$, which is impossible since $\Delta < \frac{r-1}{2r-4}n$ implies $(7 - 4r)\Delta + 2rn < 2\Delta$.

We claim that there exist $a_i \in A'_{v_i}$ and $b_i \in A'_{w_i}$ forming an IT in $H_{\tilde{\mathscr{C}}}$, where $\tilde{\mathscr{C}} = \{A_v : v \in I \setminus \{v_1, w_1\}\}$. If not then by Theorem 2.2(ii) we know there exists a nonempty subset $\mathscr{S}$ of $\tilde{\mathscr{C}}$ and $2|\mathscr{S}| - 2$ vertices in $\bigcup_{C \in \mathscr{S}} C'$ that dominate $\bigcup_{C \in \mathscr{S}} C'$, and each member of $\mathscr{S}$ contains at least one of them. But then these vertices together with a and b dominate all of $\bigcup_{C \in \mathscr{S}} C \cup A_{v_1} \cup A_{w_1}$. Here a and b can each dominate at most Δ (and they dominate A_{v_1} and A_{w_1}), and the rest dominate at most $(|\mathscr{S}| - 1)((4r - 4)\Delta - 2rn)$ in $\bigcup_{C \in \mathscr{S}} C$ by Lemma 3.2. But $|\bigcup_{C \in \mathscr{S}} C \cup A_{v_1} \cup A_{w_1}| \geqslant (|\mathscr{S}| - 4r + 6)\Delta + 2rn$ by Lemma 3.1(iv). Thus we conclude $(|\mathscr{S}| - 4r + 6)\Delta + 2rn \leqslant (|\mathscr{S}| - 1)((4r - 4)\Delta - 2rn) + 2\Delta$ which implies $|\mathscr{S}|2rn \leqslant (4r - 5)|\mathscr{S}|\Delta$. But this contradicts our assumption on Δ for $r \geqslant 5$. Therefore the a_i and b_i exist as claimed.

Now by Lemma 3.3 applied with the subset $\tilde{\mathscr{C}}$, we find that $I' = \{v_1, w_1\} \cup \{a_i, b_i : 2 \leqslant i \leqslant r - 1\}$ is an IMC of G, with the property that all neighbours of a and b in I' lie in $\{v_1, w_1\}$. First we note that neither a nor b is completely joined to both $A_{v_1}(I')$ and $A_{w_1}(I')$. Indeed, if this were true then its degree would be at least $2rn - (4r - 6)\Delta$ by Lemma 3.1(iv). But then this would imply $\frac{2r}{4r-5}n \leqslant \Delta$, contradicting our assumption on Δ.

Now if we can find $v'_1 \in A_{v_1}(I')$ and $w'_1 \in A_{w_1}(I')$, such that v'_1 is not adjacent to a and w'_1 is not adjacent to b, then by Lemma 3.1(ii) $I'' = \{v'_1, w'_1\} \cup \{a_i, b_i : 2 \leqslant i \leqslant r - 1\}$ is an IMC in which a is joined only to w'_1 and b is joined only to v'_1. Thus $a \in A_{v_1}(I'')$ and $b \in A_{w_1}(I'')$, so by Lemma 3.1(i) they are adjacent.

Therefore, to complete the proof, we just need to show that such v'_1 and w'_1 exist. If not, then (without loss of generality) each of a and b is completely joined to $A_{v_1}(I')$. Recall that A_{w_1} is defined to be the set of vertices joined only to w_1 in I. But $v_1 \in I$ and each vertex of $A_{v_1}(I')$ is joined to v_1 by definition, so we conclude $A_{w_1} \cap A_{v_1}(I') = \emptyset$. Then, since by assumption a is joined to all of $A_{w_1} \cup A_{v_1}(I')$ we find $\deg(a) \geq |A_{w_1}| + |A_{v_1}(I')|$.

To estimate $|A_{v_1}(I')|$ we observe that each vertex of $A_{v_1} \setminus A_{v_1}(I')$ is joined to v_1, and hence by definition also joined to another vertex of I'. Therefore by Lemma 3.1(iii) we know $|A_{v_1} \setminus A_{v_1}(I')| \leq 2(r-1)\Delta - rn$, and so $|A_{v_1}(I')| \geq |A_{v_1}| - |A_{v_1} \setminus A_{v_1}(I')| \geq |A_{v_1}| - 2(r-1)\Delta + rn$. Therefore $\deg(a) \geq |A_{w_1}| + |A_{v_1}| - 2(r-1)\Delta + rn \geq (6-4r)\Delta + 2rn - 2(r-1)\Delta + rn$ by Lemma 3.1(iv), which tells us that $\Delta \geq (8-6r)\Delta + 3rn$. But then $\Delta \geq \frac{3r}{6r-7}n$, which contradicts our assumption on Δ for $r \geq 7$. Therefore v_1' and w_1' exist as required. $\square$

Now we are ready to prove that every minimal counterexample to Theorem 1.1 has to be the vertex-disjoint union of complete bipartite graphs. The following lemma is an easy consequence of Theorem 2.2.

Lemma 3.6. *Let G be an r-partite graph with vertex partition $V_1 \cup \cdots \cup V_r$ that does not have an IT, and suppose $|V_i| = n$ for each i and $\Delta < \frac{r-1}{2r-4}n$. Let e be an edge of G and suppose e prevents an IT (i.e., $G - e$ has an IT). Then e lies in an IMC.*

Proof. Let $U = \{v_1, \ldots, v_r\}$, $v_i \in V_i$, be an almost independent transversal inducing the lone edge $e = v_1 v_2$. Then $T_0 = \{v_2, \ldots, v_r\}$ is a maximum size partial independent transversal. Let $I_0 := \{v_1, v_2\}$. The pair (I_0, T_0) is easily seen to be feasible, because the failure of condition (e) would immediately imply the existence of an IT in G.

We now apply Theorem 2.2 with (I_0, T_0), to obtain a feasible pair (I, T). Then by Theorem 2.2(iii), I is an IMC. Moreover Theorem 2.2(i) implies that I contains e. $\square$

Putting the above results together gives the structural theorem of minimal counterexamples.

Theorem 3.7. *Let G be an r-partite graph with vertex partition $V_1 \cup \cdots \cup V_r$, where $r \geq 7$ and $|V_i| = n$ for each i. Suppose G has no independent transversal, but the deletion of any edge creates one. If $\Delta < \frac{r-1}{2r-4}n$, then G is a union of $r-1$ vertex-disjoint complete bipartite graphs.*

Proof. By Theorem 2.2(iii), there is an IMC I_0 in G. Fix one, say $I_0 = \{v_i, w_i : 1 \leq i \leq r-1\}$ where v_i is adjacent to w_i for each i. Then by Lemmas 3.4 and 3.5 we have that the vertex set of G is partitioned into partite sets $A_1^*, \ldots, A_{r-1}^*, B_1^*, \ldots, B_{r-1}^*$, where each $G[A_i^*, B_i^*]$ is a complete bipartite graph, and $A_{v_i}(I_0) \subseteq A_i^*$ and $A_{w_i}(I_0) \subseteq B_i^*$ for each i.

Suppose G has an edge xy that does not lie in any of the bipartite subgraphs $G[A_i^*, B_i^*]$. By Lemma 3.6 we know that xy lies in some IMC I. Let us assume without loss of generality that $x \in A_1^*$, then $y \notin B_1^*$.

Suppose first that $y \in A_1^*$. Then the whole class B_1^* is dominated more than once by the IMC I. Therefore by Lemma 3.1(iii) the number of vertices in B_1^* is at most $2(r-1)\Delta - rn$. On the other hand $|B_1^*| \geq |A_{w_1}(I_0)| \geq (5-4r)\Delta + 2rn$ by Lemma 3.1(iv), giving us $(2r-2)\Delta - rn \geq (5-4r)\Delta + 2rn$. This implies $\Delta \geq \frac{3r}{6r-7}n$, a contradiction for $r \geq 7$.

Suppose now that $y \in C$ for some $C \in \{A_2^*, \ldots, A_r^*, B_2^*, \ldots, B_r^*\}$, say without loss of generality $y \in A_2^*$. By Lemma 3.1(i) and (iv), the edge xy lies in a complete bipartite graph

J with a total of at least $(6-4r)\Delta + 2rn$ vertices, and with at least $(5-4r)\Delta + 2rn$ vertices in each class J_x and J_y. Here J_x and J_y denote the partite sets of J containing x and y respectively, and J_x contains all vertices whose only neighbour in I is y, and J_y contains all those whose only neighbour in I is x.

Suppose first that both $J_y \cap B_1^*$ and $J_x \cap B_2^*$ are non-empty and let $u \in J_y \cap B_1^*$ and $w \in J_x \cap B_2^*$ be arbitrary vertices. Then $\deg(w) + \deg(u) \geq (|A_2^*| + |J_y \cap B_1^*|) + (|A_1^*| + |J_x \cap B_2^*|) \geq |A_2^*| + |J_y| + |B_1^*| - |J_y \cup B_1^*| + |A_1^*| + |J_x| + |B_2^*| - |J_x \cup B_2^*|$. But since $|J_y \cup B_1^*| \leq \deg(x) \leq \Delta$ and $|J_x \cup B_2^*| \leq \deg(y) \leq \Delta$ we find by Lemma 3.1(iv)

$$2\Delta \geq \deg(w) + \deg(u) \geq (|J_y| + |J_x|) + (|A_1^*| + |A_2^*| + |B_1^*| + |B_2^*|) - 2\Delta$$
$$\geq (6-4r)\Delta + 2rn + (8-4r)\Delta + 2rn - 2\Delta = (12 - 8r)\Delta + 4rn.$$

This implies $\Delta \geq \frac{2r}{4r-5}$, a contradiction for $r \geq 5$.

Suppose now that $J_x \cap B_2^*$ is empty. (The case when $J_y \cap B_1^* = \emptyset$ is similar.) This means that all vertices in B_2^* are dominated more than once by I, since they are dominated by y and J_x contains all those dominated only by y. The number of such vertices is at most $2(r-1)\Delta - rn$ by Lemma 3.1(iii), while the size of one partite set is at least $(5-4r)\Delta + 2rn$ by Lemma 3.1(iv). This implies $(2r-2)\Delta - rn \geq (5-4r)\Delta + 2rn$, giving $\Delta \geq \frac{3r}{6r-7}n$, a contradiction for $r \geq 7$.

Therefore no such edge xy can exist, and G must be the union of vertex-disjoint complete bipartite graphs. $\square$

4. Proof of Theorem 1.1

In this section we focus on the case of odd number of parts. From the previous section we know that any minimal counterexample to our main theorem has to be the vertex-disjoint union of complete bipartite graphs. By proving the following theorem we finish the proof of Theorem 1.1.

Theorem 4.1. *Let $r = 2t+1$ be an odd integer. Let G be the union of $2t$ vertex-disjoint complete bipartite graphs, with a vertex r-partition $V(G) = V_1 \cup \cdots \cup V_r$ into classes of size n. If the maximum degree $\Delta(G) < \frac{t}{2t-1}n$ then G has an independent transversal of the classes $V_1, \ldots, V_r$.*

Proof. Suppose on the contrary that G has no independent transversal. By Theorem 2.2(iii) there is an IMC I that induces $r-1$ edges, which defines a tree structure $\mathscr{G}_I$ on the classes of G. Let us choose a root-class, for which all subtrees not containing the root have order at most t (for any tree of order $2t+1$ one can find such a vertex; this is basically the only time we use the fact that the number of parts is odd). We colour the vertices of I according to the tree structure: for an edge induced by I we colour the vertex in the parent class white and the vertex in the child class black. We call these $2t$ black and $2t$ white vertices *distinguished* and denote them by $b_1, \ldots, b_{2t}$ and $w_1, \ldots, w_{2t}$, respectively, where b_i is adjacent to w_i. The root contains only white vertices and all other classes contain exactly one black vertex. We call the white neighbour of the unique black vertex in class C, the *parent vertex* of C. By Lemma 2.1 the set T of black vertices is an almost complete independent transversal, which is only missing a vertex from the root class. For a white vertex w, we define the subtree $\mathcal{G}_w$ of w to be the subtree of the class-graph $\mathcal{G}_I$ that contains all the classes that are descendants of the class of the black neighbour of w (including the class of the black neighbour).

Our plan is to change the black vertices in some classes such that we still have a partial independent transversal, but now we are able to make a black/white switch on a path to the root and create a complete independent transversal.

Note that, since G is the vertex-disjoint union of $r - 1$ complete bipartite graphs, each edge induced by I lies in a distinct complete bipartite graph. Therefore every vertex of G has exactly one distinguished neighbour, and every component of G contains exactly one black vertex. This immediately implies the following.

Fact 1. Let Z and X be disjoint sets of black vertices. Then $N_G(Z) \cup X$ is an independent set.

We also note here two more technical facts that we will need in the proof.

Fact 2. The unique distinguished neighbour of a vertex v is either a white vertex on the path from the class of v to the root class, or a black vertex in a class that is not on the path from the class of v to the root.

Proof. For distinguished vertices the statement is true by definition. Let $v \in V_i$ be non-distinguished and suppose the statement is false. Although v is dominated by I, so Lemma 2.1 cannot be applied directly, by our assumption the lone distinguished neighbour of v is *not* in T_i^I (as it is defined in the proof of Lemma 2.1). So the transversal $T_i^I \cup \{v\}$ is independent, a contradiction. □

Fact 3. No set $\mathscr{L}$ of at most t classes is dominated by *fewer* than $2|\mathscr{L}|$ vertices. In particular, for any white vertex w_i there is a non-distinguished vertex $u \in \bigcup \{V_i : V_i \in V(\mathcal{G}_{w_i})\}$ in the classes of the subtree $\mathcal{G}_{w_i}$ of w_i, which has a distinguished neighbour $w \neq w_i$ in a class *outside* the classes of $\mathcal{G}_{w_i}$.

Proof. A set of $2|\mathscr{L}| - 1$ vertices can dominate at most $(2|\mathscr{L}| - 1)\Delta$ vertices, and $(2|\mathscr{L}| - 1)\Delta < (2|\mathscr{L}| - 1)\frac{t}{2t-1}n \leqslant (2|\mathscr{L}| - 1)\frac{|\mathscr{L}|}{2|\mathscr{L}|-1}n = |\mathscr{L}|n$.

For the second part, by Fact 2 the black vertex b_i cannot dominate any vertex in any class of $\mathcal{G}_{w_i}$. Thus the $|\mathcal{G}_{w_i}| \leqslant t$ classes in $\mathcal{G}_{w_i}$ cannot be completely dominated by the $2|\mathcal{G}_{w_i}| - 1$ distinguished vertices in $I \cap (\bigcup \{V_i : V_i \in V(\mathcal{G}_{w_i})\} \cup \{w_i\} - \{b_i\})$. □

In our proof we define a sequence of vertices $z_1, u_1, z_2, u_2, \ldots, z_q, u_q$ with the following properties.

- The z_i are distinct black vertices; the u_i are not black.
- z_i and u_i are in the same class. Note that this implies the u_i are all distinct.
- For $i = 1, \ldots, q - 1$, u_i is adjacent to z_{i+1}.

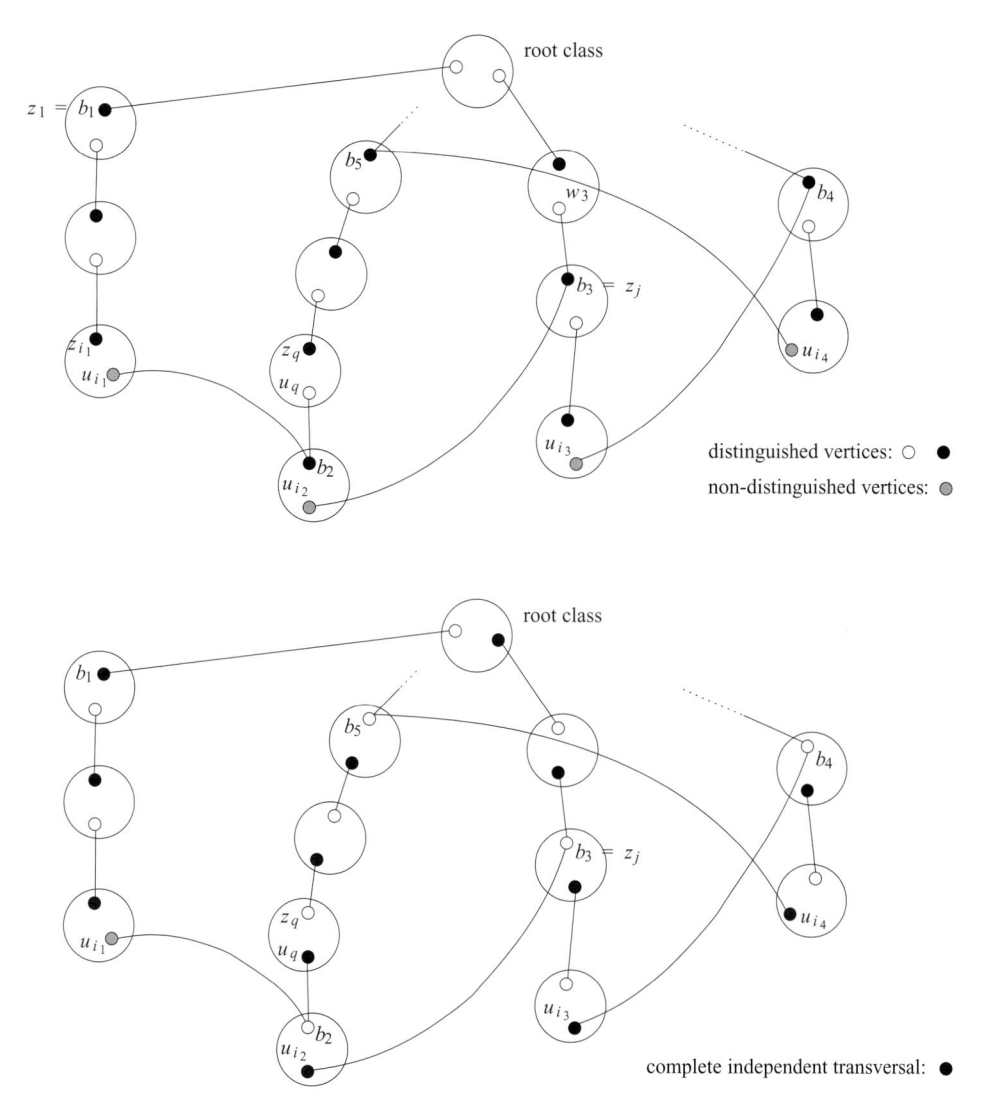

Figure 3. Case 1, before and after the switch.

The construction of the sequence goes as follows (see Figure 3). Let w_1 be an arbitrary white vertex in the root-class. (Here for convenience we may re-number the vertices b_i and w_i.) We define $z_1 = b_1$, the black neighbour of w_1. By Fact 3 there is a non-distinguished vertex g_1 in the classes of the subtree $\mathscr{G}_{w_1}$ that has a distinguished neighbour $w \ne w_1$ *outside* the subtree $\mathscr{G}_{w_1}$. By Fact 2, w must be black, say $w = b_2$. The initial segment $z_1, u_1, \ldots, z_{i_1}, u_{i_1}$ of our sequence is then defined by $u_{i_1} = g_1$, the z_k's for $k \le i_1$ are the black vertices in the classes on the path from the class of b_1 to the class of g_1 in $\mathscr{G}_I$ in the same order, while u_k for $1 \le k \le i_1 - 1$ is the white neighbour of z_{k+1}.

In general, if b_j is defined, we define g_j to be the (existing) non-distinguished vertex in a class in the subtree $\mathscr{G}_{w_j}$ of w_j, that has a distinguished neighbour $w \ne w_j$ outside $\mathscr{G}_{w_j}$.

(If more than one such vertex exists we just choose one arbitrarily.) We then define the next segment $z_{i_{j-1}+1}, u_{i_{j-1}+1}, \ldots, z_{i_j}, u_{i_j}$ of our sequence by $z_{i_{j-1}+1} = b_j$, $u_{i_j} = g_j$, the z_k's for $i_{j-1} < k \leq i_j$ are the black vertices in the classes that are on the path from the class of b_j to the class of g_j in $\mathcal{G}$, while u_k for $i_{j-1} + 1 \leq k \leq i_j - 1$ is the white neighbour of z_{k+1}.

If any of these new black vertices participated already in our sequence, we stop the sequence right before the repetition, so the last vertex is u_q, and the candidate for z_{q+1} is already some z_i in the sequence.

If the distinguished neighbour w of g_j is black, say $w = b_{j+1}$, then we go on and construct the next segment of our sequence.

We can build our sequence as long as we do not repeat a black vertex z_i and the distinguished neighbour of g_j outside the subtree $\mathcal{G}_{w_j}$ is not white. Since our graph is finite, so will be our sequence.

Case 1. Our sequence ends, because z_{q+1} would be equal to some z_i, $i \leq q$. We improve on the almost complete independent transversal T of black vertices by making switches (see Figure 3). Let z_t be the black vertex whose class V_t is closest to the root among all z_j, $i \leq j \leq q$, i.e., its path P to the root is shortest. Let $b(P)$ denote the set of black vertices in the classes of P, and let $w(P)$ denote their white neighbours. (Note then that $b(P) \cap \{z_i, \ldots, z_q\} = z_t$ and $w(P)$ contains a vertex in the root class.) We form the set T' by removing $Z = \{z_i, \ldots, z_q\} \cup b(P)$ from T and adding $U = \{u_i, \ldots, u_q\} \cup w(P)$. We claim that T' is a (complete) independent transversal of G. To see that T' is independent, apply Fact 1 to the sets Z and $X = T \setminus Z$ of black vertices, and observe that $U \subseteq N(Z)$ because u_q is adjacent to z_i. To check that T' is a transversal, recall that u_j and z_j were in the same class for each j, and note that $w(P)$ contains a vertex of each class of P including the root class, except for V_t. But V_t contains the vertex $u_t \in T'$. Therefore T' is an independent transversal as claimed.

Case 2. Our sequence stops, because the distinguished neighbour $w \neq w_k$ of $g_k = u_{i_k}$ outside the subtree $\mathcal{G}_{w_k}$ is white. Note then that $k \geq 2$.

By Fact 2, the class of w is above the class of $u_{i_k} = u_q$ in the class-graph tree $\mathcal{G}_1$. Since it is outside the subtree of w_k, w is also above b_k (but $w \neq w_k$!).

We identify the *last time* our sequence entered the subtree $\mathcal{G}_w$. By the property of the sequence, the last vertex in our sequence *not* contained in this subtree is not black, say u_j (note this vertex exists since $k \geq 2$). Then we claim $w \neq u_j$. To see this, note that if $w = u_j$ then for some index $l \leq k$, the vertex $b_l = z_{i_{l-1}+1}$ would be in a class above (or equal to) the class of w, while $g_l = u_{i_l}$ would be in a class below w. So $l < k$, since b_k is below w. But then by definition z_{i_l+1}, the distinguished neighbour of g_l, is outside the subtree of w_l, and thus outside the subtree of w as well. This is a contradiction, since then the entry of the sequence into $\mathcal{G}_w$ from u_j was not the last one.

We create a complete independent transversal T' as follows (see Figure 4). Let P denote the path from the class V_j containing u_j to the root class, then since V_j is not in $\mathcal{G}_w$ we know that none of the classes containing $\{u_{j+1}, \ldots, u_q\}$ are in P. Let $b(P)$ denote the set of black vertices in the classes of P, and let $w(P)$ denote their white neighbours (again $w(P)$ contains a vertex in the root class). We form the set T' by removing $Z = \{z_{j+1}, \ldots, z_q\} \cup b(P)$ from T and adding $U = \{u_j, \ldots, u_q\} \cup w(P)$. We claim

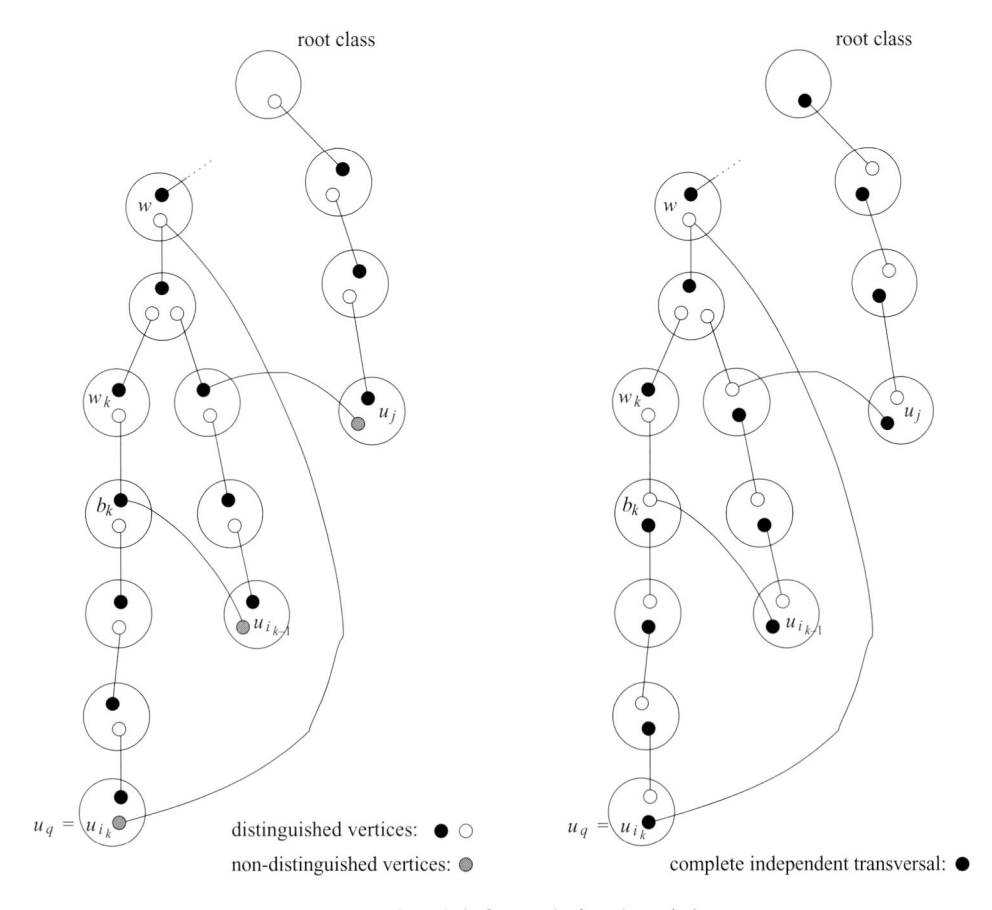

Figure 4. Case 2, before and after the switch.

that T' is a (complete) independent transversal of G. It is a transversal because u_i replaces z_i for each $j+1 \leq i \leq q$, and each class of P including the root class gets a vertex of $w(P)$, except for V_j. But V_j contains the vertex $u_j \in T'$, so every class of G contains an element of T'.

To check that T' is independent, first apply Fact 1 to the sets Z and $X = T \setminus Z$ of black vertices, and observe that $U \setminus \{u_q\} \subseteq N(Z)$. Thus it remains only to show that u_q is not adjacent to any vertex of $T' \setminus \{u_q\}$. Since u_q is adjacent to w, it is certainly independent of $X = T \setminus Z$ because it has exactly one distinguished neighbour. Let b denote the black neighbour of w. Suppose on the contrary that u_q is adjacent to some $x \in U$. Let z denote the black neighbour of x, then $z \in Z$, and x, z, w and b are all in the same component of G. But then we must have $z = b$ because this component contains only one black vertex, say $z = z_s$ where $j+1 \leq s \leq q$. Now $s = j+1$ is not possible, since otherwise by construction the next non-distinguished vertex in the sequence will be u_v for some $v \leq q$, and the distinguished neighbour z_{v+1} of u_v will be outside $\mathscr{G}_w$ and *not* equal to w. This would contradict the fact that this is the last time the sequence enters $\mathscr{G}_w$. Therefore $j+2 \leq s \leq q$. Then $x = u_{s-1} \in T'$ must be a non-distinguished vertex in our sequence, since w is the only distinguished neighbour of $z_s = b$. But all non-distinguished vertices in $U \setminus \{u_j\}$ that are candidates for u_{s-1} are in classes below $b = z_s$, so by Fact 2 they cannot be adjacent to z_s (which is black). Thus u_q cannot have any such neighbour $x \in U$, giving that T' is an independent transversal, and thus contradicting our assumption on G. This completes the proof of the theorem. □

5. Open problems

An intriguing problem remains unsolved regarding the number of independent transversals if the maximum degree is below the threshold $\Delta_r(n)$. In particular, Bollobás, Erdős and Szemerédi introduced the function $f_r(n)$ which is the largest number f such that every r-partite graph with parts of size n and maximum degree $\Delta_r(n) - 1$ has at least f independent transversals. Trivially $f_2(n) = n$. In [7] Bollobás, Erdős and Szemerédi determined $f_3(n)$ precisely and obtained that, quite surprisingly, $f_3(n) = 4$ for every $n \geqslant 4$. Jin [12] proved that $f_4(n) = \Theta(n^3)$, but the behaviour of the function $f_r(n)$ for $r \geqslant 5$ is a complete mystery. For r even we conjecture that $f_r(n) = \Theta(n^{r-1})$. For odd r the only thing we dare to predict is that $f_r(n) = O(n^{r-2})$; for the threshold $\Delta(r,n)$ is so 'unnaturally' high for odd r. At this point even $f_r(n) = \Theta(1)$ is a possibility. It would certainly be very interesting to gain more information; maybe the structural theorems of the present paper could be of use.

Another, less precise goal is related to the alternative proof of $\Delta_r \geqslant \frac{r}{2(r-1)}$ through labelled triangulations and Sperner's Lemma, a method developed by Aharoni and others in *e.g.*, [1] and [2]. It would be very desirable to understand the difference between the even and odd case by means of the topological properties of odd- and even-dimensional triangulated spheres.

Finally, we only stated the structural theorems of Section 3 for $r \geqslant 7$. It should certainly be possible to extend our methods and obtain results about the structure of slightly sub-optimal independent transversal-free examples for $r \leqslant 6$, that is, for $r = 4, 6$. Also, we did not investigate the stability of the optimal examples in the case of odd r. For $r = 3, 5$ the example is not unique; there are examples where the base graph is *not* the union of $r - 2$ bipartite graphs. We do not know what happens for odd $r \geqslant 7$.

Acknowledgement

The authors thank an anonymous referee for helpful suggestions that improved the presentation of this paper.

References

[1] Aharoni, R., Chudnovsky, M. and Kotlov, A. (2002) Triangulated spheres and colored cliques. *Discrete Comput. Geom.* **28** 223–229.
[2] Aharoni, R. and Haxell, P. (2000) Hall's Theorem for hypergraphs. *J. Graph Theory* **35** 83–88.
[3] Aharoni, R. and Haxell, P. Systems of disjoint representatives. Manuscript.
[4] Alon, N. (1988) The linear arboricity of graphs. *Israel J. Math.* **62** 311–325.
[5] Alon, N. (1992) The strong chromatic number of a graph. *Random Struct. Alg.* **3** 1–7.
[6] Alon, N. (2003) Problems and results in Extremal Combinatorics I. *Discrete Math.* **273** 31–53.

[7] Bollobás, B., Erdős, P. and Szemerédi, E. (1975) On complete subgraphs of r-chromatic graphs. *Discrete Math.* **13** 97–107.
[8] Haxell, P. (1995) A condition for matchability in hypergraphs. *Graphs Combin.* **11** 245–248.
[9] Haxell, P. (2001) A note on vertex list colouring. *Combin. Probab. Comput.* **10** 345–348.
[10] Haxell, P., Szabó, T. and Tardos, G. (2003) Bounded size components: partitions and transversals. *J. Combin. Theory Ser. B* **88** 281–297.
[11] Jin, G. (1992) Complete subgraphs of r-partite graphs, *Combin. Probab. Comput.* **1** 241–250.
[12] Jin, G. (1998) The number of complete subgraphs of equi-partite graphs. *Discrete Math.* **186** 157–165.
[13] Meshulam, R. (2001) The clique complex and hypergraph matchings. *Combinatorica* **21** 89–94.
[14] Szabó, T. and Tardos, G. Extremal problems for transversals in graphs with bounded degree. *Combinatorica*, to appear.
[15] Yuster, R. (1997) Independent transversals in r-partite graphs. *Discrete Math.* **176** 255–261.

The First Eigenvalue of Random Graphs

SVANTE JANSON

Department of Mathematics, Uppsala University,
PO Box 480, SE-751 06 Uppsala, Sweden
(e-mail: svante.janson@math.uu.se)
URL http://www.math.uu.se/~svante/)

Received 3 August 2003; revised 17 February 2005

For Béla Bollobás on his 60th birthday

We extend a result by Füredi and Komlós and show that the first eigenvalue of a random graph is asymptotically normal, both for $G_{n,p}$ and $G_{n,m}$, provided $np \geqslant n^\delta$ or $m/n \geqslant n^\delta$ for some $\delta > 0$. The asymptotic variance is of order p for $G_{n,p}$, and n^{-1} for $G_{n,m}$. This gives a (partial) solution to a problem raised by Krivelevich and Sudakov.
The formula for the asymptotic mean involves a mysterious power series.

1. Introduction

Füredi and Komlós [2] investigated the eigenvalues of random symmetric matrices. In particular, their result shows that for constant $p \in (0,1)$, the first eigenvalue λ_1 of the adjacency matrix of the random graph $G_{n,p}$ is asymptotically normal, with

$$\lambda_1(G_{n,p}) - (n-1)p - 1 + p \xrightarrow{d} N(0, 2p(1-p)) \qquad \text{as } n \to \infty. \qquad (1.1)$$

In fact, [2] showed that the random fluctuation of $\lambda_1(G_{n,p})$ asymptotically can be completely explained by the fluctuation of the number of edges in $G_{n,p}$. More precisely, they showed that if $e(G_{n,p}) \sim N(\binom{n}{2}, p)$ is the number of edges in $G_{n,p}$, then

$$\lambda_1(G_{n,p}) - \frac{2e(G_{n,p})}{n} - (1-p) = O_p(n^{-1/2}) \xrightarrow{p} 0,$$

which immediately implies (1.1) by the central limit theorem.

This suggests studying $\lambda_1(G_{n,p})$ conditioned on a given $e(G_{n,p})$, or, equivalently, $\lambda_1(G_{n,m})$, where m is a given function of n. Assume first, in analogy to the case studied by Füredi and Komlós, that $m/\binom{n}{2} \to p$, with $p \in (0,1)$ fixed. We will show that then $\lambda_1(G_{n,m})$ too is asymptotically normal, but with an asymptotic variance of order only n^{-1}.

We will also extend the results to $p \to 0$ and $m/\binom{n}{2} \to 0$, as long as $np \gg n^\delta$ and $m \gg n^{1+\delta}$ for some $\delta > 0$.

Krivelevich and Sudakov [6] have found the first-order asymptotics of $\lambda_1(G_{n,p})$ for all $p = p(n)$; in particular, for p in the range treated here, their result gives $\lambda_1(G_{n,p})/(np) \xrightarrow{p} 1$. They leave the question of the limit distribution as an open problem, which we thus (partially) answer. Note also the large deviation result by Alon, Krivelevich and Vu [1].

Our main results are the following. Here and elsewhere in this paper, $(a_i)_{i=0}^\infty$ is a certain sequence of integers, defined in Section 4. We have computed a_j for $j \leqslant 10$ by calculations with Pascal and Maple and found (unless we made a mistake)

$$A(z) := \sum_0^\infty a_j z^j = 1 + z + z^2 + z^5 + z^7 + 5z^8 + 2z^9 + 17z^{10} + \cdots.$$

No simple form is evident.

Theorem 1.1. *Suppose that $n \to \infty$, $p \to p_0 \in [0, 1)$ and $n^{1-\delta} p \to \infty$, for some fixed $\delta > 0$. Let, for some integer J with $2J + 1 \geqslant 1/\delta$,*

$$\alpha_{n,p} := (n-2)p + \sum_{j=1}^J a_j (np)^{1-j} = \sum_{j=0}^J a_j (np)^{1-j} - 2p.$$

Then

$$p^{-1/2}(\lambda_1(G_{n,p}) - \alpha_{n,p}) \xrightarrow{d} N(0, 2(1-p_0)).$$

Theorem 1.2. *Suppose that $n \to \infty$, $m/\binom{n}{2} \to p_0 \in [0, 1)$ and $n^{-1-\delta} m \to \infty$ for some fixed $\delta > 0$. Let, for some integer J with $2J \geqslant 1/\delta$,*

$$\alpha_{n,m} := \frac{2m}{n} + \sum_{j=1}^J a_j \left(\frac{2m}{n}\right)^{1-j} - \frac{2m}{n^2} = \frac{2m}{n} \left(\sum_{j=0}^J a_j \left(\frac{2m}{n}\right)^{-j} - \frac{1}{n} \right).$$

Then

$$n^{1/2}(\lambda_1(G_{n,m}) - \alpha_{n,m}) \xrightarrow{d} N(0, 2(1-p_0)^2).$$

Note that J is chosen such that terms $a_j(np)^{1-j}$ or $a_j(2m/n)^{1-j}$ with $j > J$ can be ignored.

The definition of $(a_i)_{i=0}^\infty$ in Section 4 is rather involved, and we find the numbers a_j quite mysterious. Lemma 3.1 exhibits the combinatorial significance of these numbers perhaps better than the theorems above. Nevertheless we are lacking a simple combinatorial interpretation of a_j, and leave it as an open problem to understand these numbers better.

Theorem 1.1 follows easily from Theorem 1.2. We will, however, prove both in parallel by the same method. Not surprisingly, the details are somewhat simpler for Theorem 1.1, but we will see that with our methods, the difference is not great.

Remark 1. Also for $p \to 0$, the random variation of $\lambda_1(G_{n,p})$ is explained by the variation of the number of edges $e(G_{n,p})$ in the sense of linear regression. In other words, we have $\lambda_1(G_{n,p}) = a(n,p)e(G_{n,p}) + b(n,p) + R$ for certain constants $a(n,p)$ and $b(n,p)$ and a random error term R such that $p^{-1/2}R \xrightarrow{p} 0$, while $p^{-1/2}a(n,p)(e(G_{n,p}) - \mathbb{E}\,e(G_{n,p}))$ converges in distribution.

For $G_{n,m}$, where the number of edges is constant and explains nothing, the proof shows that the variation is explained in this way by the number of paths of length 2 (or, equivalently, by the sum of the squares of the vertex degrees).

The proof uses the traditional method of computing the trace of a suitable power of the adjacency matrix as the number of closed walks of a given length in the graph. This number is closely related to subgraph counts, and we use methods from [3] to find the required asymptotics.

We consider the case $np \gg n^\delta$ ($m/n \gg n^\delta$) for some $\delta > 0$. It turns out that the smaller δ is, the higher matrix powers and the longer walks have to be employed (otherwise we cannot ignore the other eigenvalues); we also need more terms in the sums defining $\alpha_{n,p}$ and $\alpha_{n,m}$. We thus give general arguments treating arbitrarily long walks below. If we restricted ourselves to, say, $p \geqslant n^{-1/2}$, we would only have to consider a few small values of this length, and the general arguments could be replaced by explicit calculations, which would make the proof simpler but perhaps less interesting.

Remark 2. Note that we only study the case when p or m is so large that there is a large gap between the first and second eigenvalue. It seems that different methods are needed in the case of sparser graphs. Perhaps the methods of [6] could be useful.

Remark 3. Füredi and Komlós [2] studied more general random symmetric matrices where the entries are not restricted to 0 and 1. We leave it to the reader to extend the results of this paper to such matrices.

Remark 4. Note that $\lambda_1(G) \geqslant 2e(G)/n$ for every graph G with n vertices, since $\mathbf{v}A\mathbf{v}^t = 2e(G)/n$ if $\mathbf{v} = n^{-1/2}(1,\ldots,1)$ and A is the adjacency matrix of G. The results above show that, with high probability, we almost have equality for the random graphs studied here, which witnesses that the eigenvector for λ_1 is close to $\mathbf{v}$.

If X_n are random variables and c_n positive numbers, we write $X_n = o_p(c_n)$ if $X_n/c_n \xrightarrow{p} 0$, and $X_n = O_p(c_n)$ if the sequence X_n/c_n is stochastically bounded (tight).

If H is a graph, $v(H)$, $e(H)$ and $\text{aut}(H)$ denote the numbers of vertices, edges and automorphisms of H.

2. Matrices

We denote the eigenvalues (with multiplicities) of a real symmetric matrix M by $\lambda_1(M) \geqslant \lambda_2(M) \geqslant \ldots \geqslant \lambda_v(M)$. For a graph G, we similarly denote the eigenvalues of its adjacency matrix by $\lambda_1(G) \geqslant \cdots$.

The algebraic part of our proofs is the following lemma.

Lemma 2.1. *Let M be a real symmetric matrix and let $T_k := \mathrm{Tr}(M^k) = \sum_i \lambda_i(M)^k$. Suppose that, for some even $k \geqslant 2$ and $\mu > 0$,*

$$\lambda_1(M) \geqslant \mu, \tag{2.1}$$
$$T_k \leqslant \mu^k(1 + 2^{-k}). \tag{2.2}$$

Then

$$T_k\left(2 - \frac{T_{k-2}T_{k+2}}{T_k^2}\right) \leqslant \lambda_1(M)^k \leqslant T_k.$$

Proof. Let $\delta_i = \lambda_i/\lambda_1$, $1 \leqslant i \leqslant v$, where v is the size of M. First, by (2.1) and (2.2),

$$1 + \sum_{i=2}^{v} \delta_i^k = T_k/\lambda_1^k \leqslant 1 + 2^{-k}.$$

Hence $|\delta_i| \leqslant 1/2$ for $i \geqslant 2$. In particular,

$$(1 - \delta_i^2)^2 \geqslant \left(\tfrac{3}{4}\right)^2 > \tfrac{1}{2} \geqslant 2\delta_i^2.$$

Consequently,

$$T_{k-2}T_{k+2} - T_k^2 = \sum_{i,j=1}^{v}(\lambda_i^{k-2}\lambda_j^{k+2} - \lambda_i^k\lambda_j^k) = \sum_{i<j}(\lambda_i^{k-2}\lambda_j^{k+2} + \lambda_i^{k+2}\lambda_j^{k-2} - 2\lambda_i^k\lambda_j^k)$$

$$= \sum_{i<j}^{v} \lambda_i^{k-2}\lambda_j^{k-2}(\lambda_i^2 - \lambda_j^2)^2 \geqslant \sum_{j=2}^{v} \lambda_1^{k+2}\lambda_j^{k-2}(1 - \delta_j^2)^2$$

$$\geqslant \sum_{j=2}^{v} \lambda_1^{k+2}\lambda_j^{k-2} \cdot 2\delta_j^2 = 2\lambda_1^k \sum_{j=2}^{v} \lambda_j^k = 2\lambda_1^k(T_k - \lambda_1^k)$$

$$\geqslant T_k(T_k - \lambda_1^k).$$

The left inequality follows. The right one is immediate. □

Lemma 2.2. *Let M_n, $n \geqslant 1$, be random symmetric matrices (of arbitrary sizes), and let $T_{k,n} := \mathrm{Tr}(M_n^k)$. Suppose that $\mu_n > 0$ are real numbers such that for every $\eta > 0$*

$$\mathbb{P}(\lambda_1(M_n) \geqslant (1-\eta)\mu_n) \to 1 \quad \text{as } n \to \infty, \tag{2.3}$$

and that Y is a random variable and $\varepsilon_n \to 0$ are positive numbers such that

$$\varepsilon_n^{-1}\left(\frac{T_{k,n}}{\mu_n^k} - 1\right) \xrightarrow{d} kY \quad \text{as } n \to \infty, \tag{2.4}$$

jointly for three fixed consecutive even values of k. Then

$$\varepsilon_n^{-1}\left(\frac{\lambda_1(M_n)}{\mu_n} - 1\right) \xrightarrow{d} Y \quad \text{as } n \to \infty.$$

Proof. Write

$$T_{k,n} = \mu_n^k(1 + \varepsilon_n k Y_{k,n}). \tag{2.5}$$

Thus $Y_{k,n} \xrightarrow{d} Y$ jointly for three even values of k, say $k = m-2, m$ and $m+2$, and hence $(m-2)Y_{m-2,n} + (m+2)Y_{m+2,n} - 2mY_{m,n} \xrightarrow{p} 0$. Then

$$\begin{aligned} Q_n &:= \frac{T_{m-2,n} T_{m+2,n}}{T_{m,n}^2} = \frac{(1+\varepsilon_n(m-2)Y_{m-2,n})(1+\varepsilon_n(m+2)Y_{m+2,n})}{(1+\varepsilon_n m Y_{m,n})^2} \\ &= 1 + \varepsilon_n((m-2)Y_{m-2,n} + (m+2)Y_{m+2,n} - 2mY_{m,n}) + o_p(\varepsilon_n) \\ &= 1 + o_p(\varepsilon_n). \end{aligned} \tag{2.6}$$

(Some readers may prefer to use the Skorohod representation theorem [5, Theorem 4.30] and assume for simplicity that $Y_{k,n} \to Y$ a.s. for $k = m-2, m, m+2$; then o_p may be replaced by o.)

Moreover, with $\widetilde{\mu} := \mu_n(1-\eta)$, where $\eta > 0$ is so small that $(1-\eta)^{-k} < 1 + 2^{-k}$,

$$\mathbb{P}(T_{k,n} \leqslant \widetilde{\mu}^k(1+2^{-k})) = \mathbb{P}(1 + \varepsilon_n k Y_n \leqslant (1-\eta)^k(1+2^{-k})) \to 1 \quad \text{as } n \to \infty.$$

Since $\mathbb{P}(\lambda_1(M_n) \geqslant \widetilde{\mu}) \to 1$ as $n \to \infty$ by (2.3), we see that with probability tending to 1 as $n \to \infty$, M_n satisfies the assumptions of Lemma 2.1 (with m and $\widetilde{\mu}$) and thus

$$\mathbb{P}[T_{m,n}(2 - Q_n) \leqslant \lambda_1(M_n)^m \leqslant T_{m,n}] \to 1. \tag{2.7}$$

Combined with (2.6), this yields

$$\lambda_1(M_n)^m = T_{m,n}(1 + o_p(\varepsilon_n))$$

and thus

$$\begin{aligned} \lambda_1(M_n) &= T_{m,n}^{1/m}(1+o_p(\varepsilon_n)) = \mu_n(1+\varepsilon_n m Y_{m,n})^{1/m}(1+o_p(\varepsilon_n)) \\ &= \mu_n(1 + \varepsilon_n Y_{m,n} + o_p(\varepsilon_n)). \end{aligned}$$

The result follows. □

We apply Lemma 2.2 to $G_{n,p}$ and $G_{n,m}$, with $\mu_n = \alpha_{n,p}$ and $\mu_n = \alpha_{n,m}$, respectively. Note that $\alpha_{n,p} = np(1+o(1))$ and $\alpha_{n,m} = \frac{2m}{n}(1+o(1))$. By Remark 4, $\lambda_1(G_{n,m}) \geqslant 2m/n$, and (2.3) follows. For $G_{n,p}$, similarly, $\lambda_1 \geqslant 2e(G_{n,p})/n$ and $2e(G_{n,p})/(n^2p) \xrightarrow{p} 1$ by the law of large numbers; again (2.3) follows.

Note further that if M is the adjacency matrix of a graph G, then $\text{Tr}(M^k)$ equals the number of closed walks of length k in G; i.e., sequences $v_0, \ldots, v_k$ of vertices such that $v_0 = v_k$ and v_{i-1} and v_i are adjacent for $1 \leqslant i \leqslant k$; we denote this number by $W_k(G)$. Theorems 1.1 and 1.2 therefore follow by Lemma 2.2 from the following two lemmas. (The assumptions $k \geqslant 6/\delta$ are made for convenience and could be weakened. However, the results are not true for, say, $k = 2$ or $k = 4$, even for constant p.)

Lemma 2.3. *Under the hypotheses of Theorem 1.1, if* $Y \sim N(0, 2(1-p_0))$, *then for every* $k \geqslant 6/\delta$,

$$np^{1/2}\left(\frac{W_k(G_{n,p})}{\alpha_{n,p}^k} - 1\right) \xrightarrow{d} kY$$

and the convergence holds jointly for any set of such k.

Lemma 2.4. *Under the hypotheses of Theorem 1.2, if* $Y \sim N(0, 2(1-p_0)^2)$, *then for every* $k \geqslant 6/\delta$,

$$2mn^{-1/2}\left(\frac{W_k(G_{n,m})}{\alpha_{n,m}^k} - 1\right) \xrightarrow{d} kY$$

and the convergence holds jointly for any set of such k.

3. Random graphs

We prove Lemmas 2.3 and 2.4 using the orthogonal decomposition method of [3], summarized in [4, Section 6.4]. For convenience, we repeat the main definitions and results here, referring to [3] for proofs. We begin by defining an orthogonal family of functionals of $G_{n,p}$.

Let H be a graph. Consider the $(n)_{v_H}$ injective mappings from the vertex set of H into $\{1, \ldots, n\}$. Each such mapping φ maps H onto a copy $\varphi(H)$ of H in K_n, and we define

$$S_{n,p}(H) := \sum_{\varphi} \prod_{e \in \varphi(H)} (I_e - p), \qquad (3.1)$$

where $I_e = \mathbf{1}[e \in G_{n,p}]$ is the indicator that the edge e is present. In other words, we sum $\prod_{e \in H'} (I_e - p)$ over all copies of H in $G_{n,p}$, counted with multiplicities aut(H). Note that if $X_H(G)$ denotes the number of copies of H in G, each counted with multiplicity aut(H), we have the similar formula

$$X_H(G_{n,p}) = \sum_{\varphi} \prod_{e \in \varphi(H)} I_e, \qquad (3.2)$$

where, however, the terms in the sum are not orthogonal.

$S_{n,p}(H)$ depends on H only up to isomorphism. Hence we may regard H as an unlabelled graph.

Let $\mathcal{U}^0$ denote the set of unlabelled graphs without isolated vertices. Then the random variables $\{S_{n,p}(H)\}_{H \in \mathcal{U}^0}$ are orthogonal, and each functional of $G_{n,p}$ that depends only on the isomorphism type is a linear combination of these variables. In particular,

$$W_k(G_{n,p}) = \sum_{H \in \mathcal{U}^0} \hat{w}_k(n, p; H) S_{n,p}(H) \qquad (3.3)$$

for some coefficients $\hat{w}_k$.

Here we allow H to be the empty graph $\emptyset$ with $v(\emptyset) = e(\emptyset) = 0$; then $S_{n,p}(\emptyset) = 1$. Since $\mathbb{E} S_{n,p}(H) = 0$ when $H \neq \emptyset$, we have

$$\hat{w}_k(n, p; \emptyset) = \mathbb{E} W_k(G_{n,p}). \qquad (3.4)$$

We can find the decomposition (3.3) as follows. A closed walk of length k may have a finite number (depending on k) different shapes, since one or several vertices may be repeated. Hence W_k can be written as a linear combination of different subgraph counts X_H. For example, with $k = 4$ we can have a 4-cycle, a path of length 2 with each edge traversed twice, or a single edge traversed four times, and we find

$$W_4 = X_{C_4} + 2X_{P_2} + X_{K_2}.$$

(P_l denotes the path with l edges and thus $l + 1$ vertices.)

Next, substituting $I_e = (I_e - p) + p$ in (3.2) and expanding, each X_H becomes a linear combination of $S_{n,p}(K)$ for $K \subseteq H$. For example, straightforward calculations yield, with $(n)_k = n(n-1)\cdots(n-k+1)$,

$$X_{K_2}(G_{n,p}) = S_{n,p}(K_2) + (n)_2 p$$
$$X_{P_2}(G_{n,p}) = S_{n,p}(P_2) + 2(n-2)pS_{n,p}(K_2) + (n)_3 p^2$$
$$X_{C_4}(G_{n,p}) = S_{n,p}(C_4) + 4pS_{n,p}(P_3) + 4(n-3)p^2 S_{n,p}(P_2) + 2p^2 S_{n,p}(2K_2)$$
$$\qquad + 4(n-2)(n-3)p^3 S_{n,p}(K_2) + (n)_4 p^4.$$

In this way, we can obtain a decomposition (3.3) for any k explicitly (but the amount of work increases rapidly with k). Note that only terms with $e(H) \leqslant k$ appear.

For $H \in \mathcal{U}^0$,

$$S_{n,p}(H) = O_p\bigl(n^{v(H)/2} p^{e(H)/2}\bigr). \tag{3.5}$$

Hence we also define

$$S^*_{n,p}(H) := n^{-v(H)/2} p^{-e(H)/2} S_{n,p}(H), \tag{3.6}$$
$$\hat{w}^*_k(n,p;H) := n^{v(H)/2} p^{e(H)/2} \hat{w}_k(n,p;H); \tag{3.7}$$

thus (3.3) can be rewritten

$$W_k(G_{n,p}) = \sum_{H \in \mathcal{U}^0} \hat{w}^*_k(n,p;H) S^*_{n,p}(H), \tag{3.8}$$

where by (3.5), for every H,

$$S^*_{n,p}(H) = O_p(1). \tag{3.9}$$

If further $H \neq \emptyset$ and H is connected, we have the limit result ([3, Theorem 1], [4, Theorem 6.43]) that if $n \to \infty$, $p \to p_0 \in [0,1]$ and $np^{m(H)} \to \infty$, where $m(H) := \max\{e(F)/v(F) : F \subseteq H, v(F) > 0\}$, then, for some random variables $U(H)$,

$$S^*_{n,p}(H) \xrightarrow{d} U(H) \sim N\bigl(0, \mathrm{aut}(H)(1-p_0)^{e(H)}\bigr). \tag{3.10}$$

To prove Lemma 2.3, it is now sufficient to verify, for $k \geqslant 6/\delta$,

$$\mathbb{E}\, W_k(G_{n,p}) = \alpha^k_{n,p}\bigl(1 + o(n^{-1}p^{-1/2})\bigr) \tag{3.11}$$
$$\hat{w}^*_k(n,p;K_2) = \alpha^k_{n,p} n^{-1} p^{-1/2}(k + o(1)) \tag{3.12}$$
$$\hat{w}^*_k(n,p;H) = o\bigl(\alpha^k_{n,p} n^{-1} p^{-1/2}\bigr), \qquad H \in \mathcal{U}^0, v(H) \geqslant 3, \tag{3.13}$$

because then (3.8) yields by (3.4), (3.9)

$$\alpha_{n,p}^{-k} W_k(G_{n,p}) = \alpha_{n,p}^{-k} \mathbb{E} W_k(G_{n,p}) + \alpha_{n,p}^{-k} \hat{w}_k^*(n, p; K_2) S_{n,p}^*(K_2) + o_p(n^{-1} p^{-1/2})$$
$$= 1 + n^{-1} p^{-1/2} k S_{n,p}^*(K_2) + o_p(n^{-1} p^{-1/2}),$$

and Lemma 2.3 follows by (3.10), with $Y = U(K_2)$.

To prove Lemma 2.4, we define $p := m/\binom{n}{2}$ and note that

$$\alpha_{n,p} = \alpha_{n,m} + O(1/n) = \alpha_{n,m}(1 + O(n^{-2} p^{-1})) = \alpha_{n,m}(1 + o(n^{-3/2} p^{-1})).$$

For Lemma 2.4, we now need, for $k \geq 6/\delta$,

$$\mathbb{E} W_k(G_{n,p}) = \alpha_{n,p}^k \left(1 + o(n^{-3/2} p^{-1})\right) \tag{3.14}$$

$$\hat{w}_k^*(n, p; P_2) = \alpha_{n,p}^k n^{-3/2} p^{-1}(k + o(1)) \tag{3.15}$$

$$\hat{w}_k^*(n, p; H) = o(\alpha_{n,p}^k n^{-3/2} p^{-1}), \qquad H \in \mathcal{U}^0, H \neq \emptyset, K_2, P_2. \tag{3.16}$$

(No condition on $\hat{w}_k^*(n, p; K_2)$ is needed.) Indeed, using these estimates ([3, Theorem 7] or [4, Theorem 6.54]) with $\beta_n := n^{-3/2} p^{-1} \alpha_{n,m}^k$, shows that

$$n^{3/2} p \left(\frac{W_k(G_{n,m})}{\alpha_{n,m}^k} - 1 \right) \xrightarrow{d} k U(P_2)$$

(again jointly for different k), which yields Lemma 2.4 and thus Theorem 1.2.

It is important to note that we here draw a conclusion for $G_{n,m}$ from the estimates (3.14)–(3.16) for $G_{n,p}$. In the remainder of the paper, we thus consider $G_{n,p}$ only.

It remains to prove the estimates (3.11)–(3.13) and (3.14)–(3.16). Using $\alpha_{n,p} \sim np$ and changing J, we restate (and partly improve) them slightly as the following lemmas, which thus contain the combinatorial part of the proof of Theorems 1.1 and 1.2. (We treat $\hat{w}_k(n, p; \emptyset) = \mathbb{E} W_k(G_{n,p})$ separately because a much smaller relative error is required.)

Lemma 3.1. *Let $\delta > 0$ and let k and J be fixed integers with $J \geq 1/\delta$ and $k \geq 6/\delta$. If $n \to \infty$ and $np/n^\delta \to \infty$, then*

$$\mathbb{E} W_k(G_{n,p}) = (np)^k \left(1 - \frac{2}{n} + \sum_{j=1}^{J} a_j (np)^{-j} + O(n^{-2} p^{-1}) \right)^k.$$

Lemma 3.2. *Let $\delta > 0$ and $k \geq 4/\delta$ be fixed, and suppose that $H \in \mathcal{U}^0$ with $H \neq \emptyset$. If $n \to \infty$ and $np/n^\delta \to \infty$, then*

$$(np)^{-k} \hat{w}_k(n, p; H) = \begin{cases} k n^{-2} p^{-1} + o(n^{-2} p^{-1}), & H = K_2, \\ k n^{-3} p^{-2} + o(n^{-3} p^{-2}), & H = P_2, \\ o(n^{-v(H)/2 - 3/2} p^{-e(H)/2 - 1}), & H \neq \emptyset, K_2, P_2. \end{cases}$$

4. Proof of Lemma 3.1

We begin by giving an explicit, although rather opaque, definition of the numbers a_j in Theorems 1.1 and 1.2.

For a tree T, let $b_k(T)$ be the number of (not necessarily closed) walks of length k on T that traverse every edge at least twice. Let $\mathcal{T}_n$ be the set of the n^{n-2} trees on $\{1,\ldots,n\}$, and let $\mathcal{T} := \bigcup_{n=1}^{\infty} \mathcal{T}_n$, and define the formal power series

$$\Psi(\varepsilon, z) := \sum_{k=0}^{\infty} \sum_{n=1}^{\infty} \frac{1}{n!} \sum_{T \in \mathcal{T}_n} b_k(T) \varepsilon^{k-e(T)} z^k = \sum_{k=0}^{\infty} \sum_{T \in \mathcal{T}} \frac{1}{v(T)!} b_k(T) \varepsilon^{k-e(T)} z^k.$$

By symmetry, we can eliminate the factor $1/n!$ by only considering walks on $T \in \mathcal{T}_n$ such that the first visits to the vertices come in order $1, 2, \ldots, n$. Thus Ψ has integer coefficients.

If a term $\varepsilon^j z^k$ appears in $\Psi(\varepsilon, z)$ with nonzero coefficient, then $j = k - e(T)$ for some tree with a walk of length k that uses every edge at least twice. Thus $k \geqslant 2e(T)$, so $k/2 \leqslant j \leqslant k$. We can thus regard $\Psi(\varepsilon, z)$ as a power series in ε, with coefficients that are polynomials in z with integer coefficients. Note also that the constant term $\Psi(0, z) = 1$. It follows that there exists a unique power series $Z(\varepsilon)$ such that

$$Z(\varepsilon) \Psi(\varepsilon, Z(\varepsilon)) = 1. \tag{4.1}$$

Z has integer coefficients and $Z(0) = 1$. Finally, define the formal power series

$$A(\varepsilon) = \sum_{k=0}^{\infty} a_k \varepsilon^k := \frac{1}{Z(\varepsilon)}. \tag{4.2}$$

Note that each a_k is an integer and $a_0 = 1$.

Proof of Lemma 3.1. A closed walk with k steps defines a connected graph F consisting of all vertices and edges in the walk. Since $e(F) \leqslant k$, there is only a finite number of possible F (regarded as unlabelled graphs). The contribution to $(np)^{-k} \mathbb{E} W_k(G_{n,p})$ for a given unlabelled F is clearly

$$(np)^{-k} O\left(n^{v(F)} p^{e(F)}\right) = O\left(n^{v(F)-k} p^{e(F)-k}\right). \tag{4.3}$$

We consider three cases separately.

Case 1. F is a tree, $v(F) = e(F) + 1$.

Since a closed walk on a tree has to traverse each edge at least twice, we have $2e(F) \leqslant k$ and thus the contribution is, by (4.3),

$$O\left(n^{v(F)-k} p^{e(F)-k}\right) = O\left(n(np)^{e(F)-k}\right) = O\left(n(np)^{-k/2}\right) = O\left(n^{-2}\right) \tag{4.4}$$

because $(np)^{k/2} \geqslant (np)^{3/\delta} \gg n^3$.

Case 2. F has more than one cycle, $v(F) < e(F)$.

The contribution from F is by (4.3)

$$O\left(n^{-(e(F)-v(F))}(np)^{e(F)-k}\right)$$

which is $O(n^{-2} p^{-1})$ except when $v(F) = e(F) - 1$ and $e(F) = k$. The latter case means that the edges of the walk are distinct but one vertex is repeated. Labelling the vertices

$v_1, \ldots, v_k$, we thus have $v_i = v_j$ for two indices i and j, while the v_is otherwise are distinct. Moreover, $3 \leqslant |i-j| \leqslant k-3$ since each of the two cycles in F has at least 3 vertices. The indices i and j may thus be chosen in $k(k-5)/2$ ways, and thus the contribution from such walks is

$$(np)^{-k} \frac{k(k-5)}{2} (n)_{k-1} p^k = \frac{k(k-5)}{2} n^{-1} + O(n^{-2}).$$

The total contribution from F with $v(F) < e(F)$ is thus

$$\frac{k(k-5)}{2} n^{-1} + O(n^{-2} p^{-1}). \tag{4.5}$$

Case 3. F is unicyclic, $v(F) = e(F)$.

Then F consists of a cycle with attached trees. Given a closed walk on F traversing all edges, colour all edges of F that are traversed at least twice red and colour the remaining edges green. Each edge in the attached trees is red, while the edges in the cycle may be either red or green. Let $l \geqslant 0$ be the number of green edges.

If there are $l \geqslant 1$ green edges, the removal of them from F leaves l red components $T_1, \ldots, T_l$. Each T_i is a tree (possibly a single vertex only) and $v(F) = \sum_{i=1}^{l} v(T_i)$; moreover, the green edges join the red components into a cycle.

Fix $l \geqslant 3$ and trees $T_1, \ldots, T_l$ (regarded as disjoint subgraphs of K_n), and consider together all F that are obtained by joining the trees by l edges, one from each T_i to T_{i+1} (and from T_l to T_1). A closed walk on one of these F with red subtrees $T_1, \ldots, T_l$, that starts with the green edge leading from T_l to T_1, is called *special*. A special closed walk thus consists of a walk in each T_i that traverses each edge at least twice, together with single (green) steps linking the walks. The green links are determined by the walks in the trees, and thus the number of special walks with k_i steps inside T_i, $i = 1, \ldots, l$, is $\prod_{i=1}^{l} b_{k_i}(T_i)$; summing we find that the number of special walks with length k, for given $T_1, \ldots, T_l$, is, with $B(x; T) := \sum_{k=0}^{\infty} x^k b_k(T)$ and using $[x^j] f(x)$ to denote the coefficient of x^j in a power series $f(x)$,

$$\sum_{k_1 + \cdots + k_l = k - l} \prod_{i=1}^{l} b_{k_i}(T_i) = [x^{k-l}] B(x; T_1) \cdots B(x; T_l). \tag{4.6}$$

Each of these walks uses $\sum_1^l v(T_i)$ edges, so to get the contribution to $(np)^{-k} \mathbb{E} W_k(G_{n,p})$ we multiply by $(np)^{-k} p^{\sum v(T_i)}$.

Summing first over all choices of $T_1, \ldots, T_l$ with given vertex sets and then over all ways to choose these vertex sets in $\{1, \ldots, n\}$ we obtain

$$(np)^{-k} \sum_{n_1, \ldots, n_l \geqslant 1} \binom{n}{n_1, \ldots, n_l} \sum_{T_i \in \mathcal{T}_{n_i}} [x^{k-l}] B(x; T_1) \cdots B(x; T_l) p^{\sum_i v(T_i)}$$

$$= (np)^{-k} [x^{k-l}] \left(\sum_{T \in \mathcal{T}} \frac{B(x; T)}{v(T)!} (np)^{v(T)} \right)^l \left(1 + O\left(\frac{1}{n}\right) \right). \tag{4.7}$$

This is, for a given $l \geqslant 3$, the contribution from the walks that generate a unicyclic F with l red subgraphs, and that begin with a green edge. A walk generating such an F may be shifted (cyclically) in k ways by changing the starting point, and l of these shifts begin with a green edge; hence, the contribution from the walks that begin with a green edge is l/k times the total contribution for this F. Consequently, the contribution from all walks that generate a unicyclic F with l red subgraphs (for given $l \geqslant 3$) is k/l times the value in (4.7).

If $l < k$, then $v(T_i) > 1$ for some i so F contains a red edge. This means that $k > e(F) = v(F) = \sum_i v(T_i)$. Since each term in the sum in (4.7) then is

$$O\big((np)^{\sum_i v(T_i) - k}\big) = O\big((np)^{-1}\big),$$

the contribution of the term $O(1/n)$ in (4.7) then is $O\big((n^2p)^{-1}\big)$. Moreover,

$$\sum_{T \in \mathcal{T}} \frac{B(x; T)}{v(T)!} (np)^{v(T)} = \sum_{k=1}^{\infty} \sum_{T \in \mathcal{T}} \frac{b_k(T) x^k}{v(T)!} (np)^{e(T)+1} = np \Psi\left(\frac{1}{np}, npx\right).$$

Hence we find from (4.7) that the contribution to $(np)^{-k} \mathbb{E}\, W_k(G_{n,p})$ from all walks that generate a unicyclic F with l red subtrees is, for $3 \leqslant l < k$,

$$\frac{k}{l}(np)^{-k}[x^{k-l}](np)^l \Psi\left(\frac{1}{np}, npx\right)^l + O\big((n^2p)^{-1}\big) = \frac{k}{l}[x^{k-l}]\Psi\left(\frac{1}{np}, x\right)^l + O\big((n^2p)^{-1}\big). \quad (4.8)$$

For $l = k$ we are considering walks without repeated edges, i.e., cycles. Clearly, the contribution from them is

$$(np)^{-k}(n)_k p^k = 1 - \binom{k}{2}\frac{1}{n} + O(n^{-2}) = [x^0]\Psi\left(\frac{1}{np}, x\right)^k - \binom{k}{2}\frac{1}{n} + O(n^{-2}). \quad (4.9)$$

For $l \leqslant 2$, the formulas above are not quite correct. However, with $l \geqslant 0$ green edges and thus $e(F) - l$ red edges, we have $k \geqslant l + 2(e(F) - l)$ and thus $e(F) \leqslant (k+l)/2$. If $l \leqslant 2$ we thus have $e(F) \leqslant 1 + k/2$, and by (4.3), the contribution from such F is, since $k/2 \geqslant 3/\delta$,

$$O\big((np)^{e(F)-k}\big) = O\big((np)^{1-k/2}\big) = O\big(n(np)^{-3/\delta}\big) = O(n^{-2}). \quad (4.10)$$

Summing (4.4), (4.5), (4.8) for $3 \leqslant l < k$ and (4.9), (4.10) we find

$$(np)^{-k} \mathbb{E}\, W_k(G_{n,p}) = \sum_{l=3}^{k} \frac{k}{l}[x^{k-l}]\Psi\left(\frac{1}{np}, x\right)^l - 2\frac{k}{n} + O(n^{-2}p^{-1}).$$

Lemma 3.1 now follows from the next algebraic lemma. □

Lemma 4.1. *If $J \geqslant 0$ and $k \geqslant m \geqslant 2J$, then*

$$\sum_{l=k-m}^{k} \frac{k}{l}[z^{k-l}]\Psi(\varepsilon, z)^l = \left(\sum_{j=0}^{J} a_j \varepsilon^j\right)^k + O(\varepsilon^{J+1}).$$

Here and in the proof, $O(\varepsilon^a)$, with a real, denotes a polynomial or power series in ε containing only powers ε^j with $j \geqslant a$.

Proof. Define Φ_ε as the power series that solves the equation

$$\Psi(\varepsilon, z) = \Phi_\varepsilon(z\Psi(\varepsilon, z)). \tag{4.11}$$

Since $\Psi(\varepsilon, 0) = 1$, it is easily seen that Φ_ε exists and is unique; moreover, by an easy induction, each coefficient $[z^k]\Phi_\varepsilon(z)$ is a polynomial in ε with nonzero terms $c_j \varepsilon^j$ for $k/2 \leqslant j \leqslant k$ only, because Ψ is of this type. The same is then true for any power of $\Phi_\varepsilon(z)$.

By Lagrange's inversion formula [7, Theorem 5.4.2], for $1 \leqslant l \leqslant k$,

$$\frac{k}{l}[z^{k-l}]\Psi(\varepsilon, z)^l = \frac{k}{l}[z^k](z\Psi(\varepsilon, z))^l = [u^{k-l}]\Phi_\varepsilon(u)^k.$$

This is a polynomial in ε and is $O(\varepsilon^{(k-l)/2})$. Hence,

$$\sum_{l=k-m}^{k} \frac{k}{l}[z^{k-l}]\Psi(\varepsilon, z)^l = \sum_{j=0}^{m}[u^j]\Phi_\varepsilon(u)^k = \sum_{j=0}^{\infty}[u^j]\Phi_\varepsilon^k(u) + O(\varepsilon^{(m+1)/2}),$$

where the infinite sum is well defined as a power series in ε. This sum of all coefficients of Φ_ε^k is

$$\Phi_\varepsilon^k(1) = \Phi_\varepsilon(1)^k$$

and, substituting (4.1) in (4.11) and using (4.1) and (4.2),

$$\Phi_\varepsilon(1) = \Phi_\varepsilon(Z(\varepsilon)\Psi(\varepsilon, Z(\varepsilon))) = \Psi(\varepsilon, Z(\varepsilon)) = \frac{1}{Z(\varepsilon)} = A(\varepsilon).$$

(These manipulations are easily justified modulo ε^N for any fixed N.) The lemma follows. □

5. Proof of Lemma 3.2

It is easily seen from the discussion in Section 3 that $\hat{w}_k(n, p; H)$ can be computed as follows. Fix a copy H_0 of H in K_n and consider the set $\mathscr{W}$ of closed walks of length k in K_n that use every edge in H_0 at least once. If $\gamma \in \mathscr{W}$, let $\bar{\gamma}$ denote its trace, i.e., the subgraph of K_n consisting of the edges and vertices in γ. Then

$$\hat{w}_k(n, p; H) = \frac{1}{\mathrm{aut}(H)} \sum_{\gamma \in \mathscr{W}} p^{e(\bar{\gamma}) - e(H)}. \tag{5.1}$$

Let $c = c(H)$ be the number of components of H, and note that $v(H) \leqslant c + e(H)$.

Fix $j \geqslant 0$ and consider the closed walks γ in this sum that pass through j vertices outside H_0. Clearly, the number of such γ is $O(n^j)$.

Since $\bar{\gamma}$ connects the j vertices outside H_0 and the c components of H_0, it has at least $j + c - 1$ edges outside H_0, i.e.,

$$e(\bar{\gamma}) - e(H) \geqslant j + c - 1.$$

Case 1. $e(\bar{\gamma}) - e(H) = j + c - 1$.

In this case, if we collapse each component of H_0 to a single point, $\bar{\gamma}$ becomes a connected graph with $j + c$ vertices and $j + c - 1$ edges, i.e., a tree. The closed walk γ has to traverse each edge in this tree an even number of times, and thus

$$k \geqslant 2(j+c-1) + e(H).$$

The contribution to $(np)^{-k}\hat{w}_k(n,p;H)$ from all γ in Case 1 is thus, using (5.1),

$$O((np)^{-k}n^j p^{j+c-1}) = O\big((np)^{-k/2} n^{1-c-e(H)/2} p^{-e(H)/2}\big)$$
$$= o\big(n^{-k\delta/2+1-c/2-v(H)/2} p^{-e(H)/2}\big)$$
$$= o\big(n^{-3/2-v(H)/2} p^{-e(H)/2}\big).$$

This is covered by the o term in the lemma.

Case 2. $e(\bar{\gamma}) - e(H) \geqslant j + c$.

Then $k \geqslant e(\bar{\gamma}) \geqslant j + c + e(H)$. The contribution to $(np)^{-k}\hat{w}_k(n,p;H)$ is, using (5.1),

$$O((np)^{-k} n^j p^{j+c}) = O\big(n^{-c-e(H)} p^{-e(H)}\big) \qquad (5.2)$$
$$= O\big((np)^{-e(H)/2} n^{-c/2-v(H)/2} p^{-e(H)/2}\big).$$

If $e(H) \geqslant 3$, or if $e(H) = 2$ and $c > 1$, this is $o(n^{-3/2} p^{-1} n^{-v(H)/2} p^{-e(H)/2})$, which verifies the lemma for these H, i.e., all H except K_2 and P_2.

For $H = P_2$, the calculation in (5.2) yields $O(n^{-3} p^{-2})$, and $o(n^{-3} p^{-2})$ unless $k = e(\bar{\gamma}) = j + c + e(H) = j + 3$. We thus only have to consider γ that go through k different vertices, i.e., cycles of length k. The number of such cycles passing through H_0 is $2k(n)_k/(n)_3$, since there are $(n)_k/(n)_3$ choices of the cycle $\bar{\gamma}$, and for each $\bar{\gamma}$, γ may start at k places and in 2 directions. Thus, (5.1) yields

$$(np)^{-k}\hat{w}_k(n,p;P_2) = (np)^{-k}\frac{k(n)_k}{(n)_3} p^{k-2} + o(n^{-3}p^{-2}) = kn^{-3}p^{-2} + o(n^{-3}p^{-2}).$$

Finally, for $H = K_2$, (5.2) yields $O(n^{-2} p^{-1})$, and again we have o unless $k = e(\bar{\gamma}) = j + c + e(H) = j + v(H)$. Thus, again, we only have to consider γ that go through k different vertices, i.e., cycles of length k. Arguing as for P_2 we find that the number of such cycles passing through H_0 is $2k(n)_k/(n)_2$, and

$$(np)^{-k}\hat{w}_k(n,p;K_2) = (np)^{-k}\frac{k(n)_k}{(n)_2} p^{k-1} + o(n^{-2}p^{-1}) = kn^{-2}p^{-1} + o(n^{-2}p^{-1}). \qquad \square$$

Acknowledgement

Part of this research was done on a previous visit to Cambridge. I thank Béla Bollobás for interesting discussions.

References

[1] Alon, N., Krivelevich, M. and Vu, V. H. (2002) On the concentration of eigenvalues of random symmetric matrices. *Israel J. Math.* **131** 259–267.

[2] Füredi, Z. and Komlós, J. (1981) The eigenvalues of random symmetric matrices. *Combinatorica* **1** 233–241.

[3] Janson, S. (1994) *Orthogonal Decompositions and Functional Limit Theorems for Random Graph Statistics*, Mem. Amer. Math. Soc., Vol. 111, no. 534, AMS, Providence, RI.
[4] Janson, S., Łuczak, T. and Ruciński, A. (2000) *Random Graphs*, Wiley, New York.
[5] Kallenberg, O. (2002) *Foundations of Modern Probability*, 2nd edn, Springer, New York.
[6] Krivelevich, M. and Sudakov, B. (2003) The largest eigenvalue of sparse random graphs. *Combin. Probab. Comput.* **12** 61–72.
[7] Stanley, R. P. (1999) *Enumerative Combinatorics*, Vol. 2, Cambridge University Press.

On the Number of Monochromatic Solutions of $x + y = z^2$

AYMAN KHALFALAH[1] and ENDRE SZEMERÉDI[2]

[1] Faculty of Engineering, PO Box Alexandria 21544, Egypt
(e-mail: akhalfal@cs.rutgers.edu)
[2] Department of Computer Science, Rutgers, State University of NJ, New Brunswick, NJ 08903, USA
(e-mail: szemered@cs.rutgers.edu)

Received 4 February 2004; revised 30 June 2005

For Béla Bollobás on his 60th birthday

In the present work we prove the following conjecture of Erdős, Roth, Sárközy and T. Sós: Let f be a polynomial of integer coefficients such that $2|f(z)$ for some integer z. Then, for any k-colouring of the integers, the equation $x + y = f(z)$ has a solution in which x and y have the same colour. A well-known special case of this conjecture referred to the case $f(z) = z^2$.

1. Introduction, notation and definitions

Let $A_1, A_2, \ldots, A_k$ be subsets of $[1 \ldots N]$ such that

$$\bigcup_{i=1}^{k} A_i = [1, \ldots, N],$$

$$A_i \cap A_j = \phi \quad \text{for } i \neq j.$$

The family $\{A_1, A_2, \ldots, A_k\}$ is called a k-partition or k-colouring of $[1, \ldots, N]$ and the subsets $A_1, A_2, \ldots A_k$ are called colour classes. Let $S \subset [1, \ldots, 2N]$. For a fixed k-colouring we say that the equation

$$x + y = z, \quad z \in S \tag{1.1}$$

has a monochromatic solution if $\exists x, y \in A_i$ for some i, $x + y = z$ and $z \in S$.

Roth, Erdős, Sárközy and Sós [2] asked the following question. Let S be the set of perfect squares between 1 and $2N$, i.e., $S = \{z^2 : 1 \leqslant z^2 \leqslant 2N\}$; does equation (1.1) have a monochromatic solution for all possible k-colourings?

They showed that this true if $k \leqslant 3$, namely if you partition the natural numbers into three colour classes, then equation (1.1) has infinitely many monochromatic solutions. They also conjectured the following.

Conjecture 1.1. *If you partition the set of natural numbers into k colour classes, then there is a distinct monochromatic pair x and y such that x and y add up to a perfect square.*

In Section 2 we will prove the following theorem which settles the above conjecture positively.

Theorem 1.2. *Given k, there exists $N_0(k)$, depending only on k, such that, for all $N > N_0(k)$, any k-colouring of the first N positive integers must have a monochromatic pair x and y, $x \neq y$, such that they add up to a perfect square.*

Notice that the density version of this theorem is not true: consider, for example, the set of numbers congruent to 1 mod 3. Another conjecture from [2] that generalizes Conjecture 1.1 is as follows.

Conjecture 1.3. *If f is a non-constant polynomial of integer coefficients such that*

$$2 | f(z) \quad \text{for some integer } z, \tag{1.2}$$

then the equation

$$x + y = f(z)$$

has a solution for all possible k-colourings of the set of the integers, where x and y are of the same colour class and $x \neq y$.

Notice that $f(z) = z^2$ trivially satisfies condition (1.2), so Conjecture 1.3 is a generalization of Conjecture 1.1. In Section 3 we will prove the following theorem which settles the above conjecture positively.

Theorem 1.4. *Given a fixed k, and polynomial f of integer coefficients such that 2 is a prime divisor of f, there exists an absolute constant $N_0(k)$ such that: $\forall N > N_0(k)$, any k-colouring of the first N positive integers must have a monochromatic pair x and y, $x \neq y$, adding up to $f(z)$ for some z.*

2. Proof of Theorem 1.2

2.1. Notation
We use the abbreviation

$$e(\alpha) = e^{2\pi i \alpha}.$$

We denote the trigonometrical sum over a set X of integers as

$$f_X(\alpha) = \sum_{x \in X} e(\alpha x).$$

We will denote by S_Q the set of perfect squares that are less than $2N$:

$$S_Q = \{x^2 : 1 \leq x^2 \leq 2N\}.$$

Let P be a big prime. Define M as follows:

$$M = \prod_{p \text{ prime}, p \leq P, P \leq p^{\alpha_p} \leq P^2} p^{\alpha_p}.$$

Let ϵ and δ denote arbitrarily small real numbers greater than 0 such that $\delta > 6\epsilon$. For a set A, the set $A + d$ is defined as

$$A + d \stackrel{\text{def}}{=} \{x + d : x \in A\}. \tag{2.1}$$

We will denote by W the following:

$$W = \sum_{0 \leq i < b} e\left((Mi^2 + i)\frac{a}{b}\right).$$

We will denote by $|X|$ the cardinality of the set X.

2.2. Outline of proof

We are going to consider only a special type of monochromatic solution, namely:

$$(4Mx + 2) + (4My + 2) = (Mz + 2)^2, \quad \text{that is,} \tag{2.2}$$
$$4M(x + y) + 4 = M^2z^2 + 4Mz + 4, \quad \text{that is,}$$
$$x + y = M'z^2 + z,$$

where $M' = \frac{M}{4}$.

Define $B_l = \{x : 4Mx + 2 \in A_l\}$ for $1 \leq l \leq k$. Notice that B_l is an induced colouring of the first $\frac{N}{4M}$ numbers. Let

$$S = \{M'z^2 + z; M'z^2 + z \leq 2N\}.$$

From the discussion above it is clear that the existence of a monochromatic solution for equation (2.3) below implies the existence of a monochromatic solution to the original problem:

$$x + y = M'z^2 + z \quad x, y \in B_l \quad \text{for some } l. \tag{2.3}$$

We will try to show that equation (2.3) has a monochromatic solution, rather than equation (1.1), for the following reason: given $\frac{t}{2N}$ we will get $\frac{a}{b}$ such that $|\frac{t}{2N} - \frac{a}{b}| < \frac{1}{bQ}$ for $Q \approx n^{1-\delta}$. We will show that $f_S(\frac{a}{b})$ is small if $\frac{a}{b} \neq 0$, a statement not true about $f_{S_Q}(\frac{a}{b})$. Also the density version of the theorem is true for equation (2.3), whereas it is not true for the original equation.

From now on, without any loss of generality, we will write N instead of $\frac{N}{4M}$ and M instead of M'. We assume without loss of generality that $A = B_1$ has the maximum size among the B_l. Then $|A| \geq \frac{N}{k}$. We will look for a pair (x, y), $x, y \in A$ such that $x + y \in S$. Let N_S be the number of solutions of $x, y \in A$, such that $x + y \in S$. Then obviously

$$N_S = \frac{1}{2N} \sum_{t=0}^{2N-1} f_S\left(\frac{-t}{2N}\right) f_A\left(\frac{t}{2N}\right) f_A\left(\frac{t}{2N}\right). \tag{2.4}$$

We assume from now on that there do not exist distinct $x, y \in A$ with $x + y \in S$. This implies that $N_S \leq |S|$.

The proof will proceed as follows. For each t we will approximate $\frac{t}{2N}$ by $\frac{a(t)}{b(t)}$, such that $|\frac{a(t)}{b(t)} - \frac{t}{2N}| \leq \frac{1}{bQ}$, where $Q \approx N^{1-\delta}$. We will split the sum in equation (2.4) into two parts,

$$\sum{}^1 = \frac{1}{2N} \sum_{t=0:a(t)\neq 0}^{2N-1} f_S\left(\frac{-t}{2N}\right) f_A\left(\frac{t}{2N}\right) f_A\left(\frac{t}{2N}\right), \qquad (2.5)$$

and

$$\sum{}^2 = \frac{1}{2N} \sum_{t=0:a(t)=0}^{2N-1} f_S\left(\frac{-t}{2N}\right) f_A\left(\frac{t}{2N}\right) f_A\left(\frac{t}{2N}\right). \qquad (2.6)$$

We will prove that $|\sum^1|$ is small, and, since $N_S = \sum^1 + \sum^2 = 0$, $|\sum^2|$ must also be small. Next, we will examine the number of solutions (N_{S+d}) to the following problem:

$$x + y = z, \quad z \in S + d, \quad x \in A, \quad y \in A \quad \text{and} \quad d \leq N^{1-2\delta}. \qquad (2.7)$$

We will call that 'shifting'. Using the same reasoning,

$$N_{S+d} = \frac{1}{2N} \sum_{t=0}^{2N-1} f_{S+d}\left(\frac{-t}{2N}\right) f_A\left(\frac{t}{2N}\right) f_A\left(\frac{t}{2N}\right), \qquad (2.8)$$

We will show that $|N_{S+d} - N_S|$ is 'small' using estimations concerning exponential sums. Since there was no solution to the original problem then N_{S+d} is 'small'. The set S is 'nicely' distributed, i.e., each interval I larger than $N^{1/2+\epsilon}$ will contain at least $\frac{|I|}{4N} \cdot |S|$ of the elements of S; then using a counting argument $\sum_{d=1}^{N^{1-2\delta}} N_{S+d}$ must be much larger than the value obtained analytically, giving a contradiction and proving the theorem.

2.3. Lemmas
The following two lemmas are well known. We will state them without proof.

Lemmas 2.1 (Dirichlet). *For any real number α and any integer $Q \geq 1$, there exist integers a and q such that: $(a,b) = 1$, $1 \leq b \leq Q$, and*

$$\left|\alpha - \frac{a}{b}\right| \leq \frac{1}{bQ}.$$

Lemmas 2.2 (Weyl's Inequality [8], [9]). *Suppose that $(a,q) = 1$, $|\alpha - \frac{a}{q}| \leq q^{-2}$, $\epsilon > 0$,*

$$\phi(x) = \alpha x^k + \alpha_1 x^{k-1} + \cdots + \alpha_{k-1} x + \alpha_k,$$

and

$$T(\phi) = \sum_{x=1}^{Z} e(\phi(x)).$$

Then

$$|T(\phi)| < c(\epsilon) Z^{1+\epsilon} (q^{-1} + Z^{-1} + qZ^{-k})^{\frac{1}{K}},$$

where $K = 2^{k-1}$, ϵ is an arbitrarily small positive number, and $c(\epsilon)$ is a constant depending on ϵ.

In the next three lemmas, we will use Lemma 2.1 (Dirichlet) to approximate $\frac{t}{2N}$ by $\frac{a}{b}$ such that

$$\left|\frac{a}{b} - \frac{t}{2N}\right| < \frac{1}{bQ} \quad \text{with} \quad Q \approx N^{1-\delta}.$$

We split into three cases depending on b: b large ($> N^\delta$), b small and coprime to M, and b small and sharing a common factor with M.

Lemma 2.3. *Given $\frac{t}{2N}$, let $\frac{a}{b}$ be such that*

$$\left|\frac{a}{b} - \frac{t}{2N}\right| < \frac{1}{bQ},$$

$Q = N^{1-\delta}$, $(a,b) = 1$, $a > 0$, $b < Q$ and $b > N^\delta$, $\delta > 0$. *Then*

$$\left|f_S\left(\frac{t}{2N}\right)\right| < c(\epsilon)\sqrt{3}\frac{|S|}{N^{\delta/3}},$$

provided that $N > N_0(\delta)$.

Proof. Write $Z = |S|$. Then

$$f_S\left(\frac{t}{2N}\right) = \sum_{u \leqslant Z} e\left(\frac{t}{2N}(Mu^2 + u)\right).$$

In view of

$$\left|\frac{t}{2N} - \frac{a}{b}\right| < \frac{1}{bQ} < \frac{1}{b^2}, b > N^\delta, \tag{2.9}$$

and $\delta > 6\epsilon$, we can apply Lemma 2.2, with $k = 2$, $K = 2^{2-1} = 2$. We will divide the result into two cases. (Notice that $Z \approx (\sqrt{N/M})$.)

Case 1. $N^\delta < b < Z$. Then $N^{-\delta} > b^{-1} > Z^{-1} > bZ^{-2}$, so we will get

$$f_S\left(\frac{t}{2N}\right) < c(\epsilon)Z^{1+\epsilon}(b^{-1} + Z^{-1} + bZ^{-2})^{\frac{1}{2}} < c(\epsilon)Z^{1+\epsilon}(3N^{-\delta})^{\frac{1}{2}} < c(\epsilon)\sqrt{3}ZN^{-\delta/3}.$$

Case 2. $b \geqslant Z$. In this case we will get, using $b < Q = N^{1-\delta}$, $bZ^{-2} < MN^{-\delta}$, and $b^{-1}Z^{-1} < N^{-\frac{1}{2}}M^{\frac{1}{2}}$, that

$$f_S\left(\frac{t}{2N}\right) < c(\epsilon)Z^{1+\epsilon}(3MN^{-\delta})^{\frac{1}{2}} < c(\epsilon)\sqrt{3MZ}N^{-\delta/3}. \qquad \square$$

Lemma 2.4. *Given $\frac{t}{2N}$, let $\frac{a}{b}$ be such that*

$$\left|\frac{a}{b} - \frac{t}{2N}\right| < \frac{1}{bQ},$$

$Q = N^{1-\delta}$, $(a,b) = 1$, $a > 0$, and $b \leqslant N^\delta$. *Then*

$$\left|f_S\left(\frac{t}{2N}\right)\right| \leqslant \frac{|S|}{b}|W| + O(N^\delta).$$

Proof. Divide S into $\lfloor \frac{|S|}{b} \rfloor$ intervals each of size b. These intervals are

$$\{Mu^2 + u : db \leq u < (d+1)b\}, \quad 0 \leq d \leq \left\lfloor \frac{|S|}{b} \right\rfloor.$$

Let $\frac{t}{2N} = \frac{a}{b} + r$, where $|r| \leq \frac{1}{bQ}$. Then

$$f_S\left(\frac{t}{2N}\right) = \sum_d W_d + O(N^\delta), \quad \text{where} \tag{2.10}$$

$$W_d = \sum_{db \leq u < (d+1)b} e\left((Mu^2 + u)\frac{t}{2N}\right).$$

Here the $O(N^\delta)$ term is to compensate for the at most $b - 1$ terms each of magnitude not exceeding 1 that are missing from the summation. We can write $O(N^\delta)$ instead of $O(b)$ since $b \leq N^\delta$. We can write W_d as

$$W_d = \sum_{0 \leq i < b} e\left((M(db + i)^2 + (db + i))\frac{t}{2N}\right) \tag{2.11}$$

$$= \sum_{0 \leq i < b} e\left((M(d^2b^2 + 2dbi + i^2) + (db + i))\frac{t}{2N}\right)$$

$$= \sum_{0 \leq i < b} e\left(((M(d^2b^2 + 2dbi + i^2) + (db + i))\left(\frac{a}{b} + r\right)\right)$$

$$= \sum_{0 \leq i < b} e\left((Mi^2 + i)\frac{a}{b}\right) e(2Mdai) e(Md^2ab + da)$$

$$\times e(Mi^2r + ir + 2Mrdbi) e((Md^2b^2 + db)r)$$

$$= e((Md^2b^2 + db)r) e(Md^2ab + da)$$

$$\times \sum_{0 \leq i < b} e\left((Mi^2 + i)\frac{a}{b}\right) e(2Mdai) e(Mi^2r + ir + 2Mrdbi).$$

Notice that $|e((Md^2b^2 + db)r)| = 1$ and $e(Md^2ab + da) = 1 = e(2Mdai)$, and also $\sum_{0 \leq i < b} e((Mi^2 + i)\frac{a}{b}) = W$. We have

$$|W_d| < \left| \sum_{0 \leq i < b} e\left((Mi^2 + i)\frac{a}{b}\right) \right| \tag{2.12}$$

$$+ \left| \sum_{0 \leq i < b} \left(e\left((Mi^2 + i)\frac{a}{b}\right)\right) e(Mi^2r + ir + 2Mrdbi) - 1) \right|$$

$$< |W| + O\left(\sum_{0 \leq i < b} Mi^2r + ir + 2Mrdbi\right)$$

$$< |W| + O(Mb^3 r + b^2 r + 2Mrdb^3)$$
$$< |W| + O(rdb^3)$$
$$< |W| + O\left(\frac{|S|b}{Q}\right)$$
$$< |W| + O(bN^{-\frac{1}{2}+\delta}).$$

From equation (2.12) and equation (2.10) we conclude that

$$\left| f_S\left(\frac{t}{2N}\right) \right| < \frac{|S|}{b}(|W| + O(bN^{-\frac{1}{2}+\delta})) + O(N^\delta) \tag{2.13}$$

$$< \frac{|S|}{b}|W| + O(N^\delta). \qquad \square$$

Lemma 2.5. *Given $\frac{t}{2N}$, let $\frac{a}{b}$ be such that*

$$\left| \frac{a}{b} - \frac{t}{2N} \right| < \frac{1}{bQ},$$

$(a,b) = 1$, $a > 0$, $(M,b) = 1$, and $b < N^\delta$. *Then*

$$\left| f_S\left(\frac{t}{2N}\right) \right| \leqslant c(\epsilon)\sqrt{3}P^\epsilon \frac{|S|}{\sqrt{P}},$$

where P is the largest prime divisor of M.

Proof. Using Lemma 2.4 we can write

$$\left| f_S\left(\frac{t}{2N}\right) \right| = \frac{|S|}{b}|W| + O(N^\delta).$$

Let us evaluate

$$W = \sum_{0 \leqslant i < b} e\left((Mi^2 + i)\frac{a}{b}\right).$$

By Lemma 2.2 we can bound $|W|$ above by

$$c(\epsilon)b^{1+\epsilon}(b^{-1} + b^{-1} + b^{-1})^{\frac{1}{2}} < c(\epsilon)\sqrt{3}b^{\frac{1}{2}+\epsilon}$$

since z and q in the lemma are equal to b and k is equal to 2. Notice that since b is relatively prime to M then $b > P$ and the result follows. $\qquad \square$

Lemma 2.6. *Given $\frac{t}{2N}$, let $\frac{a}{b}$ be such that*

$$\left| \frac{a}{b} - \frac{t}{2N} \right| < \frac{1}{bQ},$$

$(a,b) = 1$, $a > 0$, $(M,b) > 1$, and $b < N^\delta$. *Then*

$$\left| f_S\left(\frac{t}{2N}\right) \right| \leqslant O(N^\delta).$$

Proof. Using Lemma 2.4, we can write

$$\left| f_S\left(\frac{t}{2N}\right) \right| = \frac{|S|}{b}|W| + O(N^\delta).$$

Let us evaluate

$$W = \sum_{0 \leqslant i < b} e\left((Mi^2 + i)\frac{a}{b}\right).$$

Let $b = g_1 g$, $(b > 1)$ where $g_1 = (M, b)$. Notice that $g_1 > 1$. Let $i = gu + v$, where $0 \leqslant u \leqslant g_1 - 1$ and $0 \leqslant v \leqslant g - 1$. Then

$$W = \sum_{i=0}^{b-1} e\left(\frac{a}{b}(Mi^2 + i)\right)$$

$$= \sum_{v=0}^{g-1} \sum_{u=0}^{g_1-1} e\left(\frac{a}{b}(M(g^2 u^2 + 2guv + v^2) + gu + v)\right).$$

Notice that $b | Mg$, so

$$W = \sum_{v=0}^{g-1} \sum_{u=0}^{g_1-1} e\left(\frac{a}{b}(Mv^2 + gu + v)\right) \quad (2.14)$$

$$= \sum_{v=0}^{g-1} e\left(\frac{a}{b}(Mv^2 + v)\right) \sum_{u=0}^{g_1-1} e\left(\frac{a}{g_1 g} gu\right)$$

$$= \sum_{v=0}^{g-1} e\left(\frac{a}{b}(Mv^2 + v)\right) \sum_{u=0}^{g_1-1} e\left(\frac{a}{g_1} u\right).$$

Notice that, since $(a, g_1) = 1$, we have

$$\sum_{u=0}^{g_1-1} e\left(\frac{a}{g_1} u\right) = 0.$$

Therefore $W = 0$, completing the proof. □

Lemma 2.7. *Fix $a(t)$ and $b(t)$ with $(a(t), b(t)) = 1$ such that*

$$\left| \frac{a(t)}{b(t)} - \frac{t}{2N} \right| < \frac{1}{bQ}$$

and $Q \approx N^{1-\delta}$. We have

$$\left| \frac{1}{2N} \sum_{\substack{t:a(t)=0}}^{2N-1} f_{S+d}\left(\frac{-t}{2N}\right) f_A\left(\frac{t}{2N}\right) f_A\left(\frac{t}{2N}\right) - \frac{1}{2N} \sum_{\substack{t:a(t)=0}}^{2N-1} f_S\left(\frac{-t}{2N}\right) f_A\left(\frac{t}{2N}\right) f_A\left(\frac{t}{2N}\right) \right|$$

$$= O(N^{-\delta}|S||A|),$$

for all $d \leqslant N^{1-2\delta}$.

Proof. When $a(t) = 0$, $\frac{t}{2N}d < N^{-\delta}$. Therefore

$$\left| f_{S+d}\left(\frac{t}{2N}\right) - f_S\left(\frac{t}{2N}\right) \right| = \left| \sum_{x \in S} e\left(\frac{t}{2N}(x+d)\right) - \sum_{x \in S} e\left(\frac{t}{2N}x\right) \right|$$

$$= \left| \sum_{x \in S} e\left(\frac{t}{2N}x\right) \left(e\left(\frac{t}{2N}d\right) - 1 \right) \right|$$

$$= O\left(\sum_{x \in S} \frac{t}{2N}d \right)$$

$$= O\left(\frac{t}{2N}d|S| \right)$$

$$= O(N^{-\delta}|S|).$$

Let us write

$$\sum^2 = \frac{1}{2N} \sum_{t:a(t)=0}^{2N-1} f_S\left(\frac{-t}{2N}\right) f_A\left(\frac{t}{2N}\right) f_A\left(\frac{t}{2N}\right)$$

$$\sum_d^2 = \frac{1}{2N} \sum_{t:a(t)=0}^{2N-1} f_{S+d}\left(\frac{-t}{2N}\right) f_A\left(\frac{t}{2N}\right) f_A\left(\frac{t}{2N}\right)$$

$$\left| \sum_d^2 - \sum^2 \right| \leq \frac{1}{2N} \sum_{t:a(t)=0}^{2N-1} \left| f_{S+d}\left(\frac{t}{2N}\right) - f_S\left(\frac{t}{2N}\right) \right| \left| f_A\left(\frac{t}{2N}\right) \right|^2$$

$$= O\left(N^{-\delta}|S| \frac{1}{2N} \sum_{t:a(t)=0}^{2N-1} \left| f_A\left(\frac{t}{2N}\right) \right|^2 \right)$$

$$= O\left(N^{-\delta}|S||A| \right).$$

In the last step, we use Parseval's identity. □

Lemma 2.8. *Fix $a(t)$ and $b(t)$ with $(a(t), b(t)) = 1$ such that*

$$\left| \frac{a(t)}{b(t)} - \frac{t}{2N} \right| < \frac{1}{bQ}$$

and $Q \approx N^{1-\delta}$. We have

$$\left| \frac{1}{2N} \sum_{t:a(t) \neq 0}^{2N-1} f_{S+d}\left(\frac{-t}{2N}\right) f_A\left(\frac{t}{2N}\right) f_A\left(\frac{t}{2N}\right) \right| \leq c(\epsilon) \sqrt{3} P^\epsilon \frac{|S||A|}{\sqrt{P}},$$

for all $d \leq N^{1-2\delta}$, where P is the largest prime divisor of M.

Proof. Let

$$F = \max_{t:a(t) \neq 0} \left| f_S\left(\frac{t}{2N}\right) \right|.$$

It follows from Lemmas 2.3, 2.5 and 2.6 that

$$F < c(\epsilon)\sqrt{3}P^\epsilon \frac{|S|}{\sqrt{P}}.$$

We have

$$\left|\frac{1}{2N}\sum_{t:a(t)\neq 0}^{2N-1} f_{S+d}\left(\frac{-t}{2N}\right)f_A\left(\frac{t}{2N}\right)f_A\left(\frac{t}{2N}\right)\right|$$

$$= \left|\frac{1}{2N}\sum_{t:a(t)\neq 0}^{2N-1} f_S\left(\frac{-t}{2N}\right)f_{A-d}\left(\frac{t}{2N}\right)f_A\left(\frac{t}{2N}\right)\right|$$

$$\leq \frac{1}{2N}\sum_{t:a(t)\neq 0}^{2N-1} \left|f_S\left(\frac{-t}{2N}\right)\right|\left|f_{A-d}\left(\frac{t}{2N}\right)f_A\left(\frac{t}{2N}\right)\right|$$

$$\leq F\frac{1}{2N}\sqrt{\left(\sum_{t:a(t)\neq 0}^{2N-1}\left|f_{A-d}\left(\frac{t}{2N}\right)\right|^2\right)\left(\sum_{t:a(t)\neq 0}^{2N-1}\left|f_A\left(\frac{t}{2N}\right)\right|^2\right)}$$

$$\leq \frac{1}{2N}F\sqrt{(2N|A|)(2N|A|)}$$

$$\leq FA < c(\epsilon)\sqrt{3}P^\epsilon \frac{|S||A|}{\sqrt{P}}.$$

We have used the Cauchy–Schwartz inequality and Parseval's identity. □

Lemma 2.9. *For any interval* $I \subset [0,\ldots,2N-1]$ *of length* H, *such that* $H > N^{1/2+\epsilon}$, *we have*

$$|S \cap I| > \frac{|S|H}{4N}.$$

Proof. Suppose the interval is $[r, r+1, r+2, \ldots, r+(H-1)]$.

So we want to count the number of integers l such that

$$\frac{-1+\sqrt{1+4Mr}}{2M} \leq l \leq \frac{-1+\sqrt{1+4M(r+(H-1))}}{2M}.$$

It follows that the number of such l that satisfy these inequalities is at least

$$\frac{\sqrt{1+4M(r+(H-1))}-\sqrt{1+4Mr}}{2M} - 2 \geq \frac{4(H-1)M}{4M(\sqrt{1+4M(r+(H-1))})} - 2$$

$$\geq \frac{H}{\sqrt{1+8MN}} - 3$$

$$> \frac{|S|H}{4N},$$

because $|S| < \sqrt{\frac{2N}{M}} - 1$. □

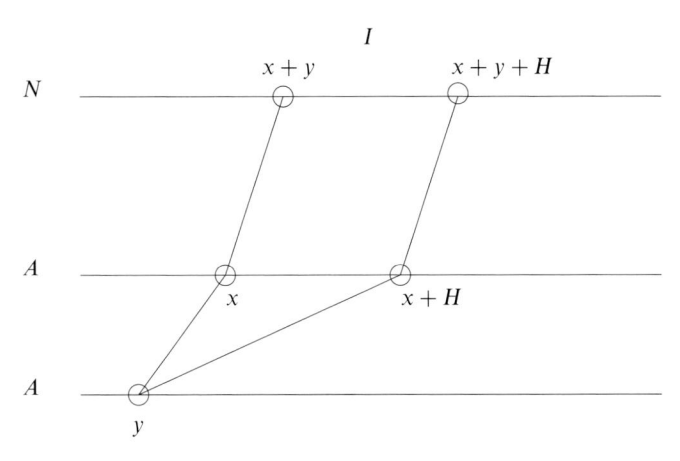

Figure 1. Shifting through an interval of length H

Lemma 2.10. *The total number of triples x, y, d satisfying $x + y \in S + d$, $x \in A$, $y \in B$ and $d < N^{1-2\delta}$ is at least*

$$\frac{N^{1-2\delta}|A|^2|S|}{4N}.$$

Proof. Pick $x \in A$ and $y \in A$ (as shown in Figure 1). Now let

$$I = \{x + y - d : 0 \leqslant d \leqslant N^{1-2\delta}\}.$$

Now clearly $|I| = H = N^{1-2\delta}$. So, using Lemma 2.9, there are at least

$$\frac{|S|H}{4N}$$

elements of S inside I. The total number of solutions is at least

$$\sum_{x \in A} \sum_{y \in A} \frac{|S|H}{4N} > \frac{|A||A||S||H|}{4N} > \frac{N^{1-2\delta}|A|^2|S|}{4N}. \qquad (2.15)$$

$\square$

2.4. Proof of main theorem

Let us write

$$N_S = \frac{1}{2N} \sum_{t=0}^{2N-1} f_S\left(-\frac{t}{2N}\right) f_A\left(\frac{t}{2N}\right) f_A\left(\frac{t}{2N}\right) \qquad (2.16)$$

$$= \frac{1}{2N} \sum_{t, a(t) \neq 0} f_S\left(-\frac{t}{2N}\right) f_A\left(\frac{t}{2N}\right) f_A\left(\frac{t}{2N}\right)$$

$$+ \frac{1}{2N} \sum_{t, a(t) = 0} f_S\left(-\frac{t}{2N}\right) f_A\left(\frac{t}{2N}\right) f_A\left(\frac{t}{2N}\right) \qquad (2.17)$$

$$= \sum\nolimits^1 + \sum\nolimits^2,$$

where $|\frac{t}{2N} - \frac{a(t)}{b(t)}| \leq \frac{1}{bQ}$ and $Q \leq N^{1-2\delta}$. Let

$$N_{S+d} = \frac{1}{2N} \sum_{t=0}^{2N-1} f_S\left(-\frac{t}{2N}\right) f_{A+d}\left(\frac{t}{2N}\right) f_A\left(\frac{t}{2N}\right) \quad (2.18)$$

$$= \frac{1}{2N} \sum_{t, a(t) \neq 0} f_S\left(-\frac{t}{2N}\right) f_{A+d}\left(\frac{t}{2N}\right) f_A\left(\frac{t}{2N}\right)$$

$$+ \frac{1}{2N} \sum_{t, a(t) = 0} f_S\left(-\frac{t}{2N}\right) f_{A+d}\left(\frac{t}{2N}\right) f_A\left(\frac{t}{2N}\right) \quad (2.19)$$

$$= \sum\nolimits_d^1 + \sum\nolimits_d^2,$$

and let us evaluate

$$|N_{S+d} - N_S| = \left| \left(\sum\nolimits_d^1 + \sum\nolimits_d^2 \right) - \left(\sum\nolimits^1 + \sum\nolimits^2 \right) \right| \quad (2.20)$$

$$\leq \left| \sum\nolimits_d^1 - \sum\nolimits^1 \right| + \left| \sum\nolimits_d^2 - \sum\nolimits^2 \right|$$

$$\leq \left| \sum\nolimits_d^2 - \sum\nolimits^2 \right| + \left| \sum\nolimits_d^1 \right| + \left| \sum\nolimits^1 \right|.$$

Lemma 2.7 showed that $|\sum_d^2 - \sum^2| = O(N^{-\delta}|S||A|)$ and Lemma 2.8 showed that

$$\left| \sum\nolimits_d^1 \right| < c(\epsilon)\sqrt{3}P^\epsilon \frac{|S||A|}{\sqrt{P}}.$$

Since we assumed that the problem had no solution, $N_S \leq |S|$, and thus $N_{S+d} \leq |N_{S+d} - N_S| + |S|$, so

$$\sum_{d \leq N^{1-2\delta}} N_{S+d} < N^{1-2\delta} \left(2c(\epsilon)\sqrt{3}P^\epsilon \frac{|S||A|}{\sqrt{P}} + O(N^{-\delta}|S||A|) \right) + O(N^{1-2\delta}|S|), \quad (2.21)$$

while Lemma 2.10 shows that, by combinatorial counting,

$$\sum_{d \leq N^{1-2\delta}} N_{S+d} > N^{1-2\delta} \frac{|S||A|^2}{4N}. \quad (2.22)$$

Notice that $k|A| > N$, so (2.21) and (2.22) give a contradiction if

$$k < \frac{\sqrt{P}}{8c(\epsilon)\sqrt{3}P^\epsilon}.$$

Finally we discuss the choice of the various constants we used. We fix $\delta = 1/10$ and $\epsilon = 1/60$. Then we fix P such that

$$k < \frac{\sqrt{P}}{8c(\epsilon)\sqrt{3}P^\epsilon}.$$

This defines M. Then we pick N large enough that all the asymptotic bounds are tight enough to give a contradiction. $\square$

3. Proof of Theorem 1.4

Let $\{A_1, \ldots, A_k\}$ be a k-colouring of $[1, \ldots, N]$. Let

$$f(x) = \sum_{i=0}^{n} c_i x^i$$

be a polynomial of integer coefficients c_i and $2 | f(z)$ for some z. Notice that $2 | f(z + 2l)$ for all l. Write $u = z + 2v$, for some v to be chosen later.

Let us evaluate $f(Mx + u)$:

$$f(Mx + u) = \sum_{i=0}^{n} c_i (Mx + u)^n \tag{3.1}$$

$$= \sum_{i=0}^{n} \sum_{j=0}^{i} c_i M^j x^j \binom{i}{j} u^{i-j} \quad \text{using the binomial theorem}$$

$$= \sum_{j=0}^{n} M^j x^j \sum_{i=j}^{n} c_i \binom{i}{j} u^{i-j}$$

$$= \sum_{j=0}^{n} E_j M^j x^j, \quad \text{where } E_j = \sum_{i=j}^{n} c_i \binom{i}{j} u^{i-j}.$$

Pick u such that $E_1 \neq 0$ and $E_2 \neq 0$. Notice that

$$E_0 = \sum_{i=0}^{n} c_i u^i = f(u) = 2l \quad \text{for some } l,$$

so we can write

$$f(Mx + u) = \sum_{j=3}^{n} E_j M^j x^j + E_2 M^2 x^2 + E_1 M x + 2l.$$

Consider the equation

$$(E_1 M y + l) + (E_1 M w + l) = f(Mx + u), \tag{3.2}$$

$$E_1 M(y + w) + 2l = \sum_{j=3}^{n} E_j M^j x^j + E_2 M^2 x^2 + E_1 M x + 2l,$$

$$y + w = \sum_{j=3}^{n} \frac{E_j M^j}{E_1 M} x^j + \frac{E_2 M}{E_1} x^2 + x.$$

Notice that, if we choose M large enough, this polynomial is integral since E depends only on the polynomial and not N. As before, if equation (3.2) has a monochromatic solution, i.e., a pair $E_1 M y + l$ and $E_1 M w + l \in A_i$ for some i, then Theorem 1.4 is true.

We will proceed as in the proof of Theorem 1.2. Let

$$S = \left\{ s : s = \sum_{j=3}^{n} \frac{E_j M^j}{E_1 M} x^j + \frac{E_2 M}{E_1} x^2 + x, s < 2N \right\}.$$

Let
$$B_i = \{y : E_1 My + l \in A_i\},$$
and $A = B_l$ be such that
$$|A| > \frac{|N|}{k}.$$
We ask whether the following equation has a solution:
$$y + w = z \quad y, w \in A, \quad z \in S.$$
Writing N_S for the number of solutions to this equation, we have
$$N_S = \frac{1}{2N} \sum_{t=0}^{2N} f_S\left(-\frac{t}{2N}\right) f_A\left(\frac{t}{2N}\right) f_A\left(\frac{t}{2N}\right).$$

Notice that Lemma 2.2 is for a general polynomial, so we can argue that if we approximate $\frac{t}{2N}$ by $\frac{a(t)}{b(t)}$ then $f_S(\frac{t}{2N})$ will be small for all t such that $a(t) \neq 0$. Indeed, all we used in the above lemmas were Lemma 2.2, the periodicity $e()$, and the fact that we have a linear term with coefficient one in our polynomial; all of these are true in this case also. So we can show that

$$f_S\left(\frac{t}{2N}\right) < c(\epsilon)\sqrt{3}P^\epsilon \frac{|S|}{P^{\frac{1}{K}}},$$

where $K = 2^{n-1}$ and P is the largest prime divisor of M. Also, let

$$N_{S+d} = \frac{1}{2N} \sum_{t=0}^{2N} f_S\left(-\frac{t}{2N}\right) f_{A+d}\left(\frac{t}{2N}\right) f_A\left(\frac{t}{2N}\right).$$

Again N_{S+d} can be shown to be approximately N_S analytically for all d such that $d < N^{1-2\delta}$, with error

$$2c(\epsilon)\sqrt{3}P^\epsilon \frac{N|A||S|}{P^{\frac{1}{K}}}.$$

On the other hand, it can be easily shown that, inside each interval of length $N^{1-2\delta}$, S has at least the expected number of points divided by a constant. Let this constant be F, so combinatorially

$$\sum_d NS_d > N^{1-2\delta} \frac{|S||A|^2}{2kNF},$$

which gives a contradiction if

$$k < \frac{P^{\frac{1}{K}}}{8Fc(\epsilon)\sqrt{3}P^\epsilon}. \qquad \square$$

References

[1] Erdős, P. and Graham, R. L. (1980) Old and new problems and results in combinatorial number theory. *Monogr. Enseignment Mathe.* **28**.

[2] Erdős, P. and Sárközy, A. (1977) On differences and sums of integers II. *Bull. Greek Math. Society* **18** 204–223.
[3] Khalfalah, A. A., Lodha, S. P. and Szemerédi, E. (2002) Tight bound for the density of sequence of integers the sum of no two of which is a perfect square. *Discrete Math.*, **256** 243–255.
[4] Lagarias, J. P., Odlyzko, A. M. and Shearer, J. B. (1982) On the density of sequences of integers the sum of no two of which is a square I: arithmetic progressions. *J. Combin. Theory Ser. A* **33** 167–185.
[5] Lagarias, J. P., Odlyzko, A. M. and Shearer, J. B. (1983) On the density of sequences of integers the sum of no two of which is a square II: general sequences. *J. Combin. Theory Ser. A* **34** 123–139.
[6] Massias, J. P. Sur les suites dont les sommes des terms 2 á 2 ne sont par des carrés.
[7] Sárközy, A. (1978) On difference sets of sequences of integers I. *Acta Math. Acad. Sci. Hungar.* **31** 125–149.
[8] Vaughan, R. C. (1997) *The Hardy–Littlewood Method*, 2nd edition, Cambridge University Press.
[9] Vinogradov, I. M. (1977) *The Method of Trigonometrical Sums in the Theory of Numbers*, Interscience Publishers. Translated from the Russian, revised and annotated by K. F. Roth and Anne Davenport.

Rapid Steiner Symmetrization of Most of a Convex Body and the Slicing Problem

B. KLARTAG[†] and V. MILMAN[‡]

School of Mathematical Sciences, Tel Aviv University, Tel Aviv 69978, Israel
(e-mail: klartag@math.ias.edu, milman@post.tau.ac.il)

Received 19 January 2004; revised 11 June 2004

For Béla Bollobás on his 60th birthday

For an arbitrary n-dimensional convex body, at least almost n Steiner symmetrizations are required in order to symmetrize the body into an isomorphic ellipsoid. We say that a body $T \subset \mathbb{R}^n$ is 'quickly symmetrizable with function $c(\varepsilon)$' if for any $\varepsilon > 0$ there exist only $\lfloor \varepsilon n \rfloor$ symmetrizations that transform T into a body which is $c(\varepsilon)$-isomorphic to an ellipsoid. In this note we ask, given a body $K \subset \mathbb{R}^n$, whether it is possible to remove a small portion of its volume and obtain a body $T \subset K$ which is quickly symmetrizable. We show that this question, for $c(\varepsilon)$ polynomially depending on $\frac{1}{\varepsilon}$, is equivalent to the slicing problem.

1. Introduction

We work in $\mathbb{R}^n$, endowed with the usual scalar product $\langle \cdot, \cdot \rangle$ and the Euclidean norm $|\cdot|$. Let $K \subset \mathbb{R}^n$ be a convex body, and let $H = \{x \in \mathbb{R}^n; \langle x, h \rangle = 0\}$ be a hyperplane through the origin in $\mathbb{R}^n$. For every $x \in \mathbb{R}^n$ there exists a unique decomposition $x = y + th$ where $y \in H, t \in \mathbb{R}$, so we can refer to (y, t) as coordinates in $\mathbb{R}^n$. The result of a 'Steiner symmetrization of K with respect to H' is the body:

$$S_H(K) = \left\{(x, t); \ K \cap (x + \mathbb{R}h) \neq \emptyset, \ |t| \leqslant \frac{1}{2}\mathrm{Meas}\{K \cap (x + \mathbb{R}h)\}\right\}$$

where Meas is the one-dimensional Lebesgue measure in the line $x + \mathbb{R}h$. Steiner symmetrization is a well-known operation in convexity. It preserves the volume of a body and transforms convex sets to convex sets (e.g., [3]). For a vector $h \in \mathbb{R}^n$ we often call the symmetrization with respect to the hypoplane $h^\perp$ also the symmetrization with respect to h. A suitably chosen finite sequence of Steiner symmetrizations may transform an arbitrary convex body into a body that is close to a Euclidean ball. Less expected is the

---
[†] Supported by the Israel Clore Foundation.
[‡] Partially supported by a grant from the Israel Science Foundation.

fact that relatively few symmetrizations suffice for obtaining a body that is close to a Euclidean ball. The following theorem, which improves a previous result of [7], appears in [11] ($D = \{x \in \mathbb{R}^n; |x| \leqslant 1\}$ is the standard Euclidean ball in $\mathbb{R}^n$).

Theorem 1.1. *For any $n \geqslant 2$ and any convex body $K \subset \mathbb{R}^n$ with $\mathrm{Vol}(K) = \mathrm{Vol}(D)$, there exist $3n$ Steiner symmetrizations that transform the body K into $\tilde{K}$ such that*

$$\frac{1}{c} D \subset \tilde{K} \subset cD,$$

where $c > 0$ is a numerical constant.

Given a convex body $K \subset \mathbb{R}^n$, we define its 'geometric distance' from a convex body $T \subset \mathbb{R}^n$ as

$$d_G(K, T) = \inf\left\{ab; \frac{1}{a} T \subset K \subset bT,\ a, b > 0\right\}$$

and we set $d_G(K) = d_G(K, D)$, the geometric distance of K from a Euclidean ball. The Banach–Mazur distance of K from a Euclidean ball is $d_{BM}(K) = \inf_T d_G(TK)$, where the infimum runs over all invertible linear transformations. The quantity $d_{BM}(K)$ measures the distance of K from an ellipsoid. Notice that we do not allow translations of the convex body when defining the distances.

The constant '3' in Theorem 1.1 is not optimal (see more accurate results in [11]). However, for bodies such as the cross-polytope $B_1^n = \{x \in \mathbb{R}^n; \sum |x_i| \leqslant 1\}$, at least $n - C \log n$ symmetrizations are required in order to symmetrize B_1^n into a body which is $\sqrt{C/2}$-close to an ellipsoid (see [11]). Therefore it is impossible to symmetrize a general convex body in $\mathbb{R}^n$ into an isomorphic ellipsoid, using significantly less than n symmetrizations. Let us consider another example: the cube $B_\infty^n = \{x \in \mathbb{R}^n; \forall i\ |x_i| \leqslant 1\}$ has a very short symmetrization process. For any $\varepsilon > 0$, there exist $\lfloor \varepsilon n \rfloor$ symmetrizations that transform B_∞^n into a body whose distance from a Euclidean ball is smaller than $c\sqrt{\frac{1}{\varepsilon} \log \frac{1}{\varepsilon}}$ for some numerical constant $c > 0$. Given a convex body $K \subset \mathbb{R}^n$, we say that 'K is $c(\varepsilon)$-symmetrizable' if, for any $\varepsilon > 0$, there exist $\lfloor \varepsilon n \rfloor$ symmetrizations that transform K into $\tilde{K}$ with $d_{BM}(\tilde{K}) < c(\varepsilon)$. Using this terminology, the cube is $c(\varepsilon)$-symmetrizable for $c(\varepsilon) = c\sqrt{\frac{1}{\varepsilon} \log \frac{1}{\varepsilon}}$. Note that here $c(\varepsilon)$ does not depend on the dimension n, and grows polynomially in $\frac{1}{\varepsilon}$ as ε tends to zero. Here we ask whether an arbitrary convex body $K \subset \mathbb{R}^n$ contains a large part which is $c(\varepsilon)$-symmetrizable, with $c(\varepsilon)$ being a polynomial in $\frac{1}{\varepsilon}$, whose coefficients do not depend on the dimension n.

Question 1.2. *Does there exist a function $c(\varepsilon)$, which is a polynomial in $\frac{1}{\varepsilon}$, such that, for any dimension n, and for any convex body $K \subset \mathbb{R}^n$, there exists a convex body $T \subset K$ with $\mathrm{Vol}(T) > \frac{9}{10} \mathrm{Vol}(K)$ such that T is $c(\varepsilon)$-symmetrizable?*

The number '$\frac{9}{10}$' has no special meaning, and may be replaced with any $\alpha < 1$. An *a priori* unrelated question is concerned with the isotropic constant. Let $K \subset \mathbb{R}^n$ be a convex body. K has an affine image $\tilde{K}$, which is unique up to orthogonal transformations, such that the barycentre of $\tilde{K}$ is at the origin, $\mathrm{Vol}(\tilde{K}) = 1$, and

$$\int_{\tilde{K}} \langle x, \theta \rangle^2 dx = L_K^2 |\theta|^2$$

for any $\theta \in \mathbb{R}^n$, where L_K does not depend on θ (see [15]). We say that L_K is the isotropic constant of K. A fundamental question in asymptotic convex geometry is the following.

Question 1.3. *Does there exist a constant $c > 0$ such that, for any integer n, and for any convex body $K \subset \mathbb{R}^n$, we have $L_K < c$?*

The main goal of this note is to show that Question 1.3 and Question 1.2 are equivalent.

Theorem 1.4. *Question 1.2 and Question 1.3 have the same answer.*

Theorem 1.4 connects two properties of the class of all convex bodies in all dimensions, yet formally it does not say anything about an individual body $K \subset \mathbb{R}^n$. We also obtain here results that are applicable to individual bodies. Proposition 4.4 states that given a body $K \subset \mathbb{R}^n$ that contains a large portion which is $c(\varepsilon)$-symmetrizable for some polynomial $c(\varepsilon)$, the isotropic constant of K may be bounded by a quantity that depends solely on the polynomial $c(\varepsilon)$. See also Proposition 3.2 for the opposite direction.

Before turning to the details of the proofs, let us shed some light on the concept of a $c(\varepsilon)$-symmetrizable body $T \subset \mathbb{R}^n$, for a polynomial $c(\varepsilon)$. Assume that T is such a body, for $c(\varepsilon) < c_1 \left(\frac{1}{\varepsilon}\right)^{c_2}$, where $c_1, c_2 > 0$ are universal constants. Then for any $\varepsilon > 0$ there exist $\lfloor \varepsilon n \rfloor$ symmetrizations of T with respect to special vectors $v_1, \ldots, v_{\lfloor \varepsilon n \rfloor}$, that transform T into $\tilde{T}$ that is $c(\varepsilon)$-close to an ellipsoid. Denote by E the subspace $\{v_1, \ldots, v_{\lfloor \varepsilon n \rfloor}\}^\perp$. By [11, Lemma 2.4],

$$\mathrm{Proj}_E(T) = \mathrm{Proj}_E(\tilde{T}) \implies d_{BM}(\mathrm{Proj}_E(T)) < c_1 \left(\frac{1}{\varepsilon}\right)^{c_2},$$

where Proj_E is the orthogonal projection onto E in $\mathbb{R}^n$. Therefore $T \subset \mathbb{R}^n$ has, for any $\varepsilon > 0$, projections to subspaces of dimension $\lceil (1-\varepsilon)n \rceil$ whose distance from an ellipsoid is smaller than some polynomial in $\frac{1}{\varepsilon}$. In fact, as will be explained later, by Theorem 1.1 a body is $c(\varepsilon)$-symmetrizable for a polynomial $c(\varepsilon)$ if and only if it has large projections which are polynomially close to an ellipsoid. Since the latter notion is clearly linearly invariant, then also $c(\varepsilon)$-symmetrizability with a polynomial $c(\varepsilon)$ is a linearly invariant property.

Throughout the paper we denote by $c, c', \tilde{c}, C$, etc., some positive universal constants whose value is not necessarily the same on different appearances. Whenever we write $A \approx B$, we mean that there exist universal constants $c, c' > 0$ such that $cA < B < c'A$. Also, $\mathrm{Vol}(T)$ denotes the volume of a set $T \subset \mathbb{R}^n$ relative to its affine hull. A random k-dimensional subspace in $\mathbb{R}^n$ is chosen according to the unique rotation invariant probability measure in the Grassman manifold $G_{n,k}$.

2. An M-position of order α and the isotropic position

For $K, T \subset \mathbb{R}^n$ denote the covering number of K by T by

$$N(K, T) = \min \left\{ N; \exists x_1, \ldots, x_N \in \mathbb{R}^n, K \subset \bigcup_{i=1}^{N} x_i + T \right\}.$$

Let $K \subset \mathbb{R}^n$ be a convex body. An ellipsoid $\mathcal{E} \subset \mathbb{R}^n$ is an M-ellipsoid of K with constant $c > 0$ if

$$\max\{N(K, \mathcal{E}), N(\mathcal{E}, K)\} < e^{cn}.$$

If $\mathcal{E} = D$, we say that K is in M-position with constant $c > 0$. A result by Milman states that any centrally symmetric (i.e., $K = -K$) convex body has a linear image in M-position with some absolute constant (see [13], or [18, Chapter 7]). Furthermore, we say that K is in M-position of order α with constants c_α, c'_α if for all $t > 1$

$$\max\{N(K, tc_\alpha D), N(D, tc_\alpha K)\} < e^{c'_\alpha \frac{n}{t^\alpha}}.$$

Another common term to describe this property is α-regular M-position with the appropriate constants. By a duality theorem [1], if K is centrally symmetric and is in M-position of order α, then also

$$\max\{N(K^\circ, c_\alpha t D), N(D, c_\alpha t K^\circ)\} < e^{\tilde{c}_\alpha \frac{n}{t^\alpha}}$$

where $K^\circ = \{y \in \mathbb{R}^n; \forall x \in K, \langle x, y \rangle \leq 1\}$. A theorem of Pisier [18] states that given a centrally symmetric $K \subset \mathbb{R}^n$, for any $\alpha < 2$, there exists a linear image of K which is in M-position of order α with some constants that depend solely on α.

The assumption of central symmetry in the above discussion is not crucial. In [14, 16] it is proved that any convex body whose barycentre lies at the origin has a linear image in M-position with some absolute constant. However, the literature seems to contain no discussion on the existence of regular M-positions for non-symmetric convex bodies. Next, we deduce that a regular M-position exists for any convex body.

Lemma 2.1. *Let $K, T \subset \mathbb{R}^n$ be convex bodies. Let $\alpha > 1$, and assume that for any $t > 1$,*

$$N(K, tT) < \exp\left(\frac{\alpha n}{t}\right).$$

Then, for any subspace $E \subset \mathbb{R}^n$ with $\dim(E) = k = \lambda n$,

$$\mathrm{Vol}(K \cap E)^{\frac{1}{k}} < \frac{c\alpha}{\lambda} \max_{x \in \mathbb{R}^n} \mathrm{Vol}(T \cap [E + x])^{\frac{1}{k}}$$

where $c > 0$ is a universal constant.

Proof. For any $t > 1$, set $N = N(t) = \exp\left(\frac{\alpha n}{t}\right)$. There exist points $x_1, \ldots, x_N \in \mathbb{R}^n$, such that

$$K \subset \bigcup_{i=1}^{N(t)} x_i + tT \Rightarrow K \cap E \subset \bigcup_{i=1}^{N(t)} (x_i + tT) \cap E.$$

Therefore, for any $t > 1$,

$$\text{Vol}(K \cap E) \leqslant N(t) \cdot \max_{x \in \mathbb{R}^n} \text{Vol}([x + tT] \cap E) = t^k \exp\left(\frac{\alpha n}{t}\right) \max_{x \in \mathbb{R}^n} \text{Vol}(T \cap [E + x]).$$

Selecting $t = \frac{\alpha}{\lambda} = \frac{\alpha n}{k}$ implies the conclusion of the lemma, with $c = e$. □

Proposition 2.2. *Let $K \subset \mathbb{R}^n$ be a convex body whose barycentre is at the origin. Then there exists a linear transformation L such that $\tilde{K} = L(K)$ satisfies, for any $t > 1$,*

$$\max\{N(\tilde{K}, tD), N(D, t\tilde{K}), N(\tilde{K}^\circ, tD), N(D, t\tilde{K}^\circ)\} < \exp\left(c\frac{n}{t^{1/5}}\right)$$

where $c > 0$ is a numerical constant.

Proof. Consider the centrally symmetric convex bodies $\underline{K} = K \cap (-K)$ and $\overline{K} = \text{conv}(K, -K)$ where 'conv' denotes convex hull. Then $\underline{K} \subset K \subset \overline{K}$ and also $\overline{K}^\circ = (\underline{K})^\circ$. Applying a linear transformation if needed, we may assume that $\underline{K}$ is in M-position of order 1. Let $L : \mathbb{R}^n \to \mathbb{R}^n$ be a linear mapping such that $L(\overline{K})$ is in M-position of order 1. Denote $\mathscr{E} = L^{-1}D$, i.e., $\mathscr{E}$ is an M-ellipsoid of order 1 of $\overline{K}$. Then, for any $t > 1$,

$$N(\mathscr{E}, t\overline{K}) < \exp\left(\frac{cn}{t}\right).$$

Let $E \subset \mathbb{R}^n$ be a subspace with $\dim(E) = k = \lambda n$. According to Lemma 2.1,

$$\text{Vol}(\mathscr{E} \cap E)^{\frac{1}{k}} < \frac{c'}{\lambda} \max_{x \in \mathbb{R}^n} \text{Vol}(\overline{K} \cap [E + x])^{\frac{1}{k}} = \frac{c'}{\lambda} \text{Vol}(\overline{K} \cap E)^{\frac{1}{k}} \quad (2.1)$$

since $\overline{K}$ is centrally symmetric. Note that the inequality of [20] and the inequality in [8, Theorem 1] imply that

$$\text{Vol}(\overline{K} \cap E)^{\frac{1}{k}} < \frac{c}{\lambda} \max_{x \in \mathbb{R}^n} \text{Vol}(K \cap [E + x])^{\frac{1}{k}} < \frac{C}{\lambda^2} \text{Vol}(K \cap E) \quad (2.2)$$

because the barycentre of K lies at the origin. Since $\underline{K}$ is in M-position of order 1, we have that $N(K, tD) < \exp\left(\frac{cn}{t}\right)$ for any $t > 1$. Lemma 2.1 implies that

$$\text{Vol}(K \cap E)^{\frac{1}{k}} < \frac{\tilde{c}}{\lambda} \text{Vol}(D \cap E)^{\frac{1}{k}}.$$

Combining this with (2.1) and (2.2), we get that

$$\text{Vol}(\mathscr{E} \cap E)^{\frac{1}{k}} < \frac{C}{\lambda^4} \text{Vol}(D \cap E)^{\frac{1}{k}}.$$

This is true for any λn-dimensional subspace E, for any $0 < \lambda < 1$ such that λn is an integer. By standard estimates for the covering number of an ellipsoid by Euclidean balls (e.g., [18, Remark 5.15]), we get that, for any $t > 1$,

$$N(\mathscr{E}, tD) < \exp\left(cnt^{-\frac{1}{4}}\right), \quad N(D, t\mathscr{E}^\circ) < \exp\left(cnt^{-\frac{1}{4}}\right)$$

and hence

$$N(K, tD) \leqslant N(\overline{K}, tD) \leqslant N\left(\overline{K}, t^{\frac{1}{5}}\mathscr{E}\right) N\left(\mathscr{E}, t^{\frac{4}{5}}D\right) < \exp\left(cnt^{-\frac{1}{5}}\right),$$

$$N(D, tK^\circ) \leqslant N(D, t(\overline{K})^\circ) \leqslant N\left(D, t^{\frac{1}{5}}\mathscr{E}^\circ\right) N\left(\mathscr{E}^\circ, t^{4/5}(\overline{K})^\circ\right) < \exp\left(cnt^{-\frac{1}{5}}\right).$$

Trivially $N(D, tK) < N(D, t\underline{K}) < \exp(c\frac{n}{t})$ and also $N(K°, tD) < N((\underline{K})°, tD) < \exp(c\frac{n}{t})$. We conclude that D is an M-ellipsoid of K of order $\frac{1}{5}$. □

Remark. The power '$\frac{1}{5}$' in Proposition 2.2 is clearly non-optimal and may be improved. We do not know what the optimal power is.

If K is in M-position, then proportional sections of K typically have a small diameter, and proportional projections of K typically contain a large Euclidean ball. If K is also in M-position of order α, then typical sections of dimension $\lfloor(1-\varepsilon)n\rfloor$ have a diameter which is smaller than some polynomial in $\frac{1}{\varepsilon}$, as follows from the next theorem (see, e.g., [9]).

Theorem 2.3. *Let $K \subset \mathbb{R}^n$ be a convex body in M-position of order α with constants c_α, c'_α. Let E be a random subspace of dimension $(1-\varepsilon)n$. Then, with probability larger than $1 - e^{-c'\varepsilon n}$,*

$$K \cap E \subset \left(\frac{c(c_\alpha, c'_\alpha)}{\varepsilon^{\frac{1}{2}+\frac{1}{\alpha}}}\right) D$$

$$\left(\frac{\varepsilon^{\frac{1}{2}+\frac{1}{\alpha}}}{c(c_\alpha, c'_\alpha)}\right) D \cap E \subset \text{Proj}_E(K)$$

where $c' > 0$ is a numerical constant, and $c(c_\alpha, c'_\alpha)$ depends neither on K nor on n, but solely on its arguments.

Assume that Question 1.3 has an affirmative answer. Our next proposition proves the existence of large projections that contain large Euclidean balls as in Theorem 2.3, for bodies in isotropic position (compare with [10, Proposition 5.4]).

Proposition 2.4. *Assume a positive answer to Question 1.3. Let $K \subset \mathbb{R}^n$ be a convex isotropic body with volume one whose barycentre is at the origin. Then, for any integer $k = (1-\varepsilon)n$ where $0 < \varepsilon < 1$, there exists a subspace E of dimension k with*

$$c\varepsilon^\beta \sqrt{n} D \cap E \subset \text{Proj}_E(K)$$

where $c > 0$ depends only on the constant in Question 1.3, and $\beta \leqslant 11$ is a numerical constant. If in addition K is centrally symmetric, then $\beta \leqslant 3$.

Proof. We shall use the following observation, which appears in [15, Proposition 3.11] and in [2]. Although it is stated there for centrally symmetric bodies, the generalization to the non-symmetric case is straightforward (a formulation appears in [6]). A positive answer to Question 1.3 yields that for any subspace F of dimension k,

$$c_1 < \text{Vol}(K \cap F)^{\frac{1}{n-k}} < c_2 \qquad (2.3)$$

where c_1, c_2 depend only on the constant in Question 1.3. Since the barycentre of K is at the origin, then $\text{Vol}(K \cap F)\text{Vol}(\text{Proj}_{F^\perp}(K)) \geq \text{Vol}(K) = 1$ for any subspace F (see [21]). By (2.3),

$$\text{Vol}(\text{Proj}_{F^\perp}(K))^{\frac{1}{n-k}} > \frac{1}{c_2}. \tag{2.4}$$

Assume for simplicity that K is centrally symmetric. Let $\mathscr{E}$ be an M-ellipsoid of order 1 of K, i.e.,

$$\max\{N(K, t\mathscr{E}), N(\mathscr{E}, tK)\} < e^{c\frac{n}{t}},$$

where $c > 0$ is a numerical constant. Let $0 < \lambda_1 \leq \cdots \leq \lambda_n$ be the axes of $\mathscr{E}$. Let $0 < \delta < 1$, and denote by F_1 the subspace spanned by the shortest $\lfloor \delta n \rfloor$ axes of $\mathscr{E}$. Since $N(\text{Proj}_{F_1}(K), t\,\text{Proj}_{F_1}(\mathscr{E})) < e^{c\frac{n}{t}}$, we obtain that

$$\text{Vol}(\text{Proj}_{F_1}(K)) < e^{c\frac{n}{t}} \left(t\lambda_{\lfloor \delta n \rfloor}\right)^{\lfloor \delta n \rfloor} \text{Vol}(D \cap E)$$

for any $t > 0$. Using (2.4) and the fact that $\text{Vol}(D \cap E)^{\frac{1}{\lfloor \delta n \rfloor}} \approx \frac{1}{\sqrt{\delta n}}$, when we set $t = \frac{1}{\delta}$ we get that $\lambda_{\lfloor \delta n \rfloor} > c'\sqrt{n}\delta^{3/2}$. Assume that $\lfloor \delta n \rfloor = \lfloor \frac{\varepsilon n}{2} \rfloor$ (hence $\delta \leq \frac{1}{2}$), and let F_2 denote the subspace of the longest $\lceil (1-\delta)n \rceil$ axes of $\mathscr{E}$. Since

$$N(\text{Proj}_{F_2}(K), t(\mathscr{E} \cap F_2)) \leq N(K, t\mathscr{E}) < e^{c\frac{n}{t}} < e^{2c\frac{(1-\delta)n}{t}}$$

and since a similar inequality holds for $N(\mathscr{E} \cap F_2, t\,\text{Proj}_{F_2}(K))$, then $\mathscr{E} \cap F_2$ is an M-ellipsoid of order 1 of $\text{Proj}_{F_2}(K)$. Also $c'\delta^{3/2}\sqrt{n}D \cap F_2 \subset \mathscr{E} \cap F_2$. By Theorem 2.3, there exists a subspace $E \subset F_2$ of dimension $(1-2\delta)n \geq (1-\varepsilon)n$, with

$$c\varepsilon^3 \sqrt{n}D \cap E \subset c'\varepsilon^{3/2}\mathscr{E} \cap E \subset \text{Proj}_E(\text{Proj}_{F_2}(K)) = \text{Proj}_E(K)$$

and the proposition is proved for centrally symmetric bodies. Regarding non-symmetric convex bodies, we may repeat the argument using an M-ellipsoid of order $\frac{1}{5}$, whose existence is guaranteed by Proposition 2.2. We obtain the same conclusion as in the symmetric case, but with a different power of ε. $\square$

Remark. Even in the centrally symmetric case, our bound $\beta \leq 3$ in Proposition 2.4 is not optimal, and may be improved by considering M-ellipsoids of higher order. We do not know what the best β is.

3. Slicing implies rapid symmetrization

The following lemma is standard. For completeness, we include its proof, which is trivial for centrally symmetric bodies. We would like to remind the reader that our definitions of distances forbid translations of the bodies.

Lemma 3.1. *Let $K \subset \mathbb{R}^n$ be a convex body and let $E \subset \mathbb{R}^n$ be a subspace such that:*
(1) $\operatorname{Proj}_E(K) = K \cap E$ *and* $d_{BM}(K \cap E) < A$ *for some* $A \geqslant 1$,
(2) $d_{BM}(K \cap E^\perp) < B$ *for some* $B \geqslant 1$.
Then $d_{BM}(K) < cAB$, where $c > 0$ is a numerical constant.

Proof. Applying a linear transformation inside E if necessary, we may assume that $D \subset K \cap E \subset AD$. Let $x \in \operatorname{Proj}_E(K)$ be any point. We claim that
$$-x + [K \cap (x + E^\perp)] \subset (A+1)K \cap E^\perp.$$
Indeed, since $|x| \leqslant A$ and since $-\frac{x}{A} \in K$, by convexity of K,
$$\frac{-x + [K \cap (x + E^\perp)]}{A+1} \subset \operatorname{conv}\left[-\frac{x}{A}, K \cap (x + E^\perp)\right] \cap E^\perp \subset K \cap E^\perp.$$
Therefore, $\operatorname{Proj}_{E^\perp}(K) \subset (A+1)K \cap E^\perp$. Now, let $\mathscr{E}$ be an ellipsoid, symmetric with respect to E, such that $\mathscr{E} \cap E \subset K \cap E \subset A\mathscr{E} \cap E$ and $\mathscr{E} \cap E^\perp \subset K \cap E^\perp \subset B\mathscr{E} \cap E^\perp$. Then,
$$\frac{1}{\sqrt{2}} \mathscr{E} \subset \operatorname{conv}(K \cap E, K \cap E^\perp) \subset K \subset \operatorname{Proj}_E(K) \times \operatorname{Proj}_{E^\perp}(K)$$
$$\subset K \cap E \times (A+1)K \cap E^\perp \subset (A\mathscr{E} \cap E) \times [(A+1)B\mathscr{E} \cap E^\perp] \subset \sqrt{2}(A+1)B\mathscr{E}$$
and the lemma is proved. □

Let $K \subset \mathbb{R}^n$ be a convex body, and assume that for any $\varepsilon > 0$ there exists a subspace E of dimension $\lfloor \varepsilon n \rfloor$ such that $d_{BM}(\operatorname{Proj}_{E^\perp}(K)) < c(\varepsilon)$, for some function $c(\varepsilon)$. Consider the body $K \cap E$. According to Theorem 1.1, after $\lfloor 3\varepsilon n \rfloor$ symmetrizations $K \cap E$ may be transformed into an isomorphic Euclidean ball. Apply the same symmetrizations to K, to obtain $\tilde{K}$. Since these symmetrizations include symmetrizations with respect to an orthogonal basis of E, elementary properties of the symmetrization (e.g., [11]) together with Lemma 3.1 imply that
$$d_{BM}(\tilde{K}) < c'c(\varepsilon).$$
We conclude, as was mentioned in the Introduction, that a convex body is $c(\varepsilon)$-symmetrizable with a $c(\varepsilon)$ which is polynomial in $\frac{1}{\varepsilon}$ if and only if it has projections to dimension $\lfloor (1-\varepsilon)n \rfloor$ whose distance from an ellipsoid is smaller than some polynomial in $\frac{1}{\varepsilon}$.

Before proving one direction of Theorem 1.4, which assumes a positive answer to Question 1.3, let us prove a weaker statement (with an exponential dependence, rather than a polynomial one), that is applicable to an individual body $K \subset \mathbb{R}^n$, and does not require uniform boundedness of the isotropic constant.

Proposition 3.2. *Let $\varepsilon > 0$, and let $K \subset \mathbb{R}^n$ be a convex body with $L_K < A$ for some $A > 0$. Then there exists a body $T \subset K$ with $\operatorname{Vol}(T) > \frac{9}{10}\operatorname{Vol}(K)$ and $\lfloor \varepsilon n \rfloor$ Steiner symmetrizations that transform T into $\tilde{T}$ such that*
$$d_{BM}(\tilde{T}, D) < (cA)^{\frac{1}{\varepsilon}},$$
where $c > 0$ is a universal constant.

Rapid Steiner Symmetrization of Convex Body and the Slicing Problem   437

Proof. Assume that the barycentre of K is at the origin. Let $\mathscr{E}$ be the isotropy ellipsoid of K normalized so that $\mathrm{Vol}(\mathscr{E}) = \mathrm{Vol}(K)$ (i.e., if $\tilde{K} = L(K)$ is isotropic for a linear operator L, then $\mathscr{E}$ is defined so that $L(\mathscr{E})$ is a Euclidean ball of volume one). Let $T = K \cap cA\mathscr{E}$. By the Borell lemma (e.g., [17, Theorem III.3]) $\mathrm{Vol}(T) > \frac{9}{10}\mathrm{Vol}(K)$, if $c > 0$ is suitably chosen. Note that $T \subset cA\mathscr{E}$, and

$$\left(\frac{\mathrm{Vol}(cA\mathscr{E})}{\mathrm{Vol}(T)}\right)^{1/n} < c'A.$$

By a theorem of Szarek and Tomczak-Jaegermann [22, 23] there exists a subspace E of dimension $\lceil(1-\varepsilon)n\rceil$ such that

$$d_G(\mathrm{Proj}_E(T), \mathrm{Proj}_E(\mathscr{E})) < (cA)^{\frac{1}{\varepsilon}}.$$

Let us apply the $3\varepsilon n$ symmetrizations that suit $T \cap E^\perp$ according to Theorem 1.1, to the body T, and obtain the body $\tilde{T}$. By [11, Lemma 2.1 and Lemma 2.4] (these $3\varepsilon n$ symmetrizations include symmetrizations with respect to an orthogonal basis),

$$\tilde{T} \cap E = \mathrm{Proj}_E(\tilde{T}) = \mathrm{Proj}_E(T)$$

and also $\tilde{T} \cap E^\perp$ has a universally bounded distance from a Euclidean ball. By Lemma 3.1,

$$d_{BM}(\tilde{T}, D) < (c'A)^{\frac{1}{\varepsilon}},$$

which completes the proof. □

The following proposition proves one part of Theorem 1.4.

Proposition 3.3. *Assume that Question 1.3 has a positive answer. Let $\varepsilon > 0$, and let $K \subset \mathbb{R}^n$ be a convex body. Then there exists a body $T \subset K$ with $\mathrm{Vol}(T) > \frac{9}{10}\mathrm{Vol}(K)$ and $\lfloor \varepsilon n \rfloor$ Steiner symmetrizations that transform T into $\tilde{T}$ such that*

$$d_{BM}(\tilde{T}) < c\frac{1}{\varepsilon^\beta}$$

where $c > 0$ is a constant that depends only on the constant in Question 1.3 and $0 < \beta < 11$ is a numerical constant.

Proof. Assume that $\mathrm{Vol}(K) = 1$ and that the barycentre of K is at the origin. Let $\mathscr{E}$ be the isotropy ellipsoid of K normalized so that $\mathrm{Vol}(\mathscr{E}) = \mathrm{Vol}(K)$, and denote $T = K \cap c\mathscr{E}$, where $c > 0$ depends linearly on the constant in Question 1.3. As before, by the Borell lemma, $\mathrm{Vol}(T) > \frac{9}{10}$. Also, if $c > 0$ is chosen properly, then the isotropy ellipsoid $\mathscr{F}$ of T satisfies $d_G(\mathscr{F}, \mathscr{E}) < c_1$ (e.g., [4]). By Proposition 2.4 there exists a subspace E of dimension $> (1-\varepsilon)n$ with

$$c\varepsilon^\beta \mathrm{Proj}_E(\mathscr{E}) \subset c'\varepsilon^\beta \mathrm{Proj}_E(\mathscr{F}) \subset \mathrm{Proj}_E(T) \subset c''\mathrm{Proj}_E(\mathscr{E})$$

where $c, c'' > 0$ depend only on the constant in Question 1.3. By Theorem 1.1 there exist some special $\lfloor 3\varepsilon n \rfloor$ symmetrizations designed specific to the body $T \cap E^\perp$. Apply these $\lfloor 3\varepsilon n \rfloor$ symmetrizations to T itself. Reasoning as in Proposition 3.2, we obtain a body $\tilde{T}$

with

$$d_{BM}(\tilde{T}) < c\frac{1}{\varepsilon^\beta}.$$

3.1. Dual symmetrization

Let $K \subset \mathbb{R}^n$ be a convex body and let H be a hyperplane in $\mathbb{R}^n$. For simplicity, assume that K is centrally symmetric. The result of a dual Steiner symmetrization of K is the body

$$S_H^\circ(K) = [S_H(K^\circ)]^\circ,$$

i.e., we symmetrize the dual body with respect to H. Next, we propose an alternative short symmetrization process for an arbitrary convex body $K \subset \mathbb{R}^n$. Rather than cutting a small portion of the volume, we combine symmetrizations of two kinds: Steiner symmetrization and dual Steiner symmetrization.

Theorem 3.4. *Let $K \subset \mathbb{R}^n$ be a centrally symmetric convex body. Then there exists $\tilde{K}$, a linear image of K, such that, for any $0 < \varepsilon < 1$, there exist $\lfloor \varepsilon n \rfloor$ Steiner symmetrizations that transform $\tilde{K}$ into K_1, and $\lfloor \varepsilon n \rfloor$ dual Steiner symmetrizations that transform K_1 into K_2 such that*

$$d_G(K_2) < \frac{c}{\varepsilon^3},$$

where $c > 0$ is a numerical constant.

Proof. Assume that $\varepsilon < \frac{1}{2}$. Let $\tilde{K}$ be a linear image of K which is in M-position of order 1. By Theorem 2.3 there exists a subspace E of dimension $\lfloor \varepsilon n \rfloor$ such that

$$c\varepsilon^{3/2} D \cap E^\perp \subset \text{Proj}_{E^\perp}(\tilde{K}). \tag{3.1}$$

Also, since $N(D \cap E, \frac{1}{\varepsilon}\text{Proj}_E(\tilde{K})) < \exp(c\varepsilon n)$, then

$$\left(\frac{\text{Vol}(\text{Proj}_E(\tilde{K}))}{\text{Vol}(\varepsilon D \cap E)}\right)^{\frac{1}{\dim(E)}} > C.$$

We apply $\lfloor 3\varepsilon n \rfloor$ symmetrization to $\tilde{K}$, all in the subspace E according to Theorem 1.1, to obtain the body K_1. The body K_1 satisfies

$$c\varepsilon D \cap E \subset K_1 \cap E.$$

In addition, $K_1 \cap E^\perp = \text{Proj}_{E^\perp}(K_1) = \text{Proj}_{E^\perp}(\tilde{K})$ (see, e.g., [11]). By (3.1) we conclude that

$$c\varepsilon^{3/2} D \subset K_1. \tag{3.2}$$

Note that (3.2) also remains true if we replace K_1 with a dual Steiner symmetrization of K_1. Next, as in the proof of Proposition 2.4, we have that $D \cap E^\perp$ is an M-ellipsoid of order 1 for $\text{Proj}_{E^\perp}\tilde{K} = \text{Proj}_{E^\perp}K_1 = K_1 \cap E^\perp$. By Theorem 2.3 there exists a subspace F of dimension $\lfloor 2\varepsilon n \rfloor$ that contains E such that

$$K_1 \cap F^\perp \subset \frac{c}{\varepsilon^{3/2}} D \cap F^\perp.$$

Note that all Steiner symmetrizations were carried out with respect to vectors inside F and hence the volume of $\tilde{K} \cap F$ is preserved. Reasoning as before, since $\tilde{K}$ is in M-position of order 1,

$$\left(\frac{\mathrm{Vol}(K_1 \cap F)}{\mathrm{Vol}(\frac{1}{\varepsilon}D \cap F)}\right)^{\frac{1}{\dim(F)}} = \left(\frac{\mathrm{Vol}(\tilde{K} \cap F)}{\mathrm{Vol}(\frac{1}{\varepsilon}D \cap F)}\right)^{\frac{1}{\dim(F)}} < C.$$

We apply $\lfloor 2\varepsilon n \rfloor$ dual Steiner symmetrizations to K_1, all in the subspace F according to Theorem 1.1, to obtain the body K_2. As before, we obtain that the body K_2 satisfies

$$\mathrm{Proj}_F K_2 \subset \frac{c}{\varepsilon} D \cap F, \qquad \mathrm{Proj}_{F^\perp} K_2 \subset \frac{c}{\varepsilon^{3/2}} D \cap F^\perp.$$

Combining this with (3.2) we get that

$$c\varepsilon^{3/2} D \subset K_2 \subset \frac{C}{\varepsilon^{3/2}} D$$

and the proof is complete. □

Remark. It is possible to avoid the use of a linear image in Theorem 3.4, at the cost of replacing the geometric distance with a Banach–Mazur distance. That is, for any centrally symmetric convex body $K \subset \mathbb{R}^n$ there exist $\lfloor \varepsilon n \rfloor$ Steiner symmetrizations followed by $\lfloor \varepsilon n \rfloor$ dual Steiner symmetrizations that transform K into a body which is $\frac{c}{\varepsilon^3}$ close to an ellipsoid.

4. Rapid symmetrization implies slicing

It remains to prove the second implication in Theorem 1.4, that a positive answer to Question 1.2 implies a positive answer to Question 1.3. We begin with a few lemmas, the first of which is standard and well known, and is proved here only for completeness.

Lemma 4.1. *Let $\mathscr{E}$ be an ellipsoid in $\mathbb{R}^n$. Then, among all k-dimensional sections of $\mathscr{E}$, the intersection of $\mathscr{E}$ with the subspace spanned by the shortest k axes of the ellipsoid has a minimal volume.*

Proof. Choose orthogonal coordinates such that $\mathscr{E} = TD$ for a diagonal matrix T. Let $0 < \lambda_1 \leqslant \lambda_2 \leqslant \cdots \leqslant \lambda_n$ be the numbers on the diagonal. Let V be a matrix of k rows and n columns such that its rows are orthonormal vectors in $\mathbb{R}^n$. Writing volumes as determinants, we need to show that

$$\sqrt{\det(VT^2V^t)} \geqslant \prod_{i=1}^{k} \lambda_i.$$

We will use the Cauchy–Binet formula. The sums in the next formula are over all subsets $A \subset \{1, \ldots, n\}$ with exactly k elements. For such A, we write V_A for the matrix obtained from V by taking the columns whose indices are in A. Then,

$$\det(VT^2V^t) = \sum_A \det(V_A T^2(V_A)^t) = \sum_A \left(\prod_{i \in A} \lambda_i^2\right) \det(V_A(V_A)^t)$$

$$\geq \left(\prod_{i=1}^k \lambda_i^2\right) \sum_A \det(V_A(V_A)^t) = \left(\prod_{i=1}^k \lambda_i^2\right) \det(VV^t) = \prod_{i=1}^k \lambda_i^2. \quad \square$$

Lemma 4.2. *Let $K \subset \mathbb{R}^n$ be a convex body of volume one whose barycentre is at the origin. Assume that K is in isotropic position, and denote $d = d_{BM}(K)$. Then, for any subspace E of dimension εn,*

$$\mathrm{Vol}(K \cap E)^{1/n} > \left(\frac{c}{d}\right)^\varepsilon,$$

where $c > 0$ is a numerical constant.

Proof. Let $\mathcal{E}$ be such that $\mathcal{E} \subset K \subset d\mathcal{E}$, and select an orthonormal basis $\{e_1, \ldots, e_n\}$ and $0 < \lambda_1 \leq \cdots \leq \lambda_n$ such that $\mathcal{E} = \{x \in \mathbb{R}^n; \sum \frac{\langle x, e_i \rangle^2}{\lambda_i^2} \leq 1\}$. Since $K \subset d\mathcal{E}$,

$$c \sum_{i=1}^n \frac{1}{\lambda_i^2} < L_K^2 \sum_{i=1}^n \frac{1}{\lambda_i^2} = \int_K \sum_{i=1}^n \frac{\langle x, e_i \rangle^2}{\lambda_i^2} dx \leq d^2.$$

Therefore, by the geometric–harmonic means inequality,

$$\left(\prod_{i=1}^{\varepsilon n} \lambda_i\right)^{\frac{1}{\varepsilon n}} \geq \sqrt{\frac{\varepsilon n}{\sum_{i=1}^{\varepsilon n} \frac{1}{\lambda_i^2}}} \geq \sqrt{\frac{c\varepsilon n}{d^2}} = c' \frac{\sqrt{\varepsilon n}}{d}.$$

Let E_ε denote the subspace spanned by the shortest εn axes, $e_1, \ldots, e_{\varepsilon n}$. By Lemma 4.1, $\mathrm{Vol}(\mathcal{E} \cap E) \geq \mathrm{Vol}(\mathcal{E} \cap E_\varepsilon)$ and

$$\mathrm{Vol}(K \cap E)^{1/n} \geq \mathrm{Vol}(\mathcal{E} \cap E)^{1/n} \geq \mathrm{Vol}(\mathcal{E} \cap E_\varepsilon)^{1/n}.$$

Since

$$\mathrm{Vol}(\mathcal{E} \cap E_\varepsilon)^{1/n} > \left(\prod_{i=1}^{\varepsilon n} \lambda_i\right)^{1/n} \frac{c}{(\sqrt{\varepsilon n})^\varepsilon} > \left(c' \frac{\sqrt{\varepsilon n}}{d\sqrt{\varepsilon n}}\right)^\varepsilon,$$

the lemma is proved. $\quad \square$

Let $K \subset \mathbb{R}^n$ be a convex body, and let $E \subset \mathbb{R}^n$ be a subspace of dimension k. We define the Schwartz symmetrization of K with respect to E, as the unique body $S_E(K)$ such that:

(i) for any $x \in E^\perp$, $\mathrm{Vol}(K \cap (x + E)) = \mathrm{Vol}(S_E(K) \cap (x + E))$,
(ii) for any $x \in E^\perp$, the body $S_E(K) \cap (x + E)$ is a Euclidean ball centred at $E^\perp$.

That is, we replace any section of K parallel to E with a Euclidean ball of the same volume. Schwartz symmetrization is a limit of a sequence of Steiner symmetrizations, and preserves volume and convexity. The following lemma is a reformulation of [6, Theorem 2.5]. For a convex body $K \subset \mathbb{R}^n$ of volume one whose barycentre is at the origin, denote by M_K the operator defined by

$$\forall u, v \in \mathbb{R}^n, \quad \langle u, M_K v \rangle = \int_K \langle x, u \rangle \langle x, v \rangle dx.$$

Define also $\text{Iso}(K) = L_K M_K^{-1/2} K$. Then $\text{Iso}(K)$ is the unique isotropic image of K under a positive definite linear transformation.

Lemma 4.3. *Let $K \subset \mathbb{R}^n$ be a convex body of volume one whose barycentre is at the origin, and let $E \subset \mathbb{R}^n$ be a k-dimensional invariant subspace of M_K. Then,*

$$\left(\frac{1}{c} \frac{k}{n} \right)^{\frac{k}{n}} < \frac{L_{S_E(K)}}{L_K^{1-\frac{k}{n}} \text{Vol}(\text{Iso}(K) \cap E)^{\frac{1}{n}}} < \left(c \frac{n}{k} \right)^{\frac{k}{n}}$$

where $c > 0$ is a numerical constant.

Proof. If K is isotropic, then the lemma is just a particular case of [6, Theorem 2.5]. Otherwise, since E is an invariant subspace of M_K,

$$\text{Iso}(S_E(\text{Iso}(K))) = \text{Iso}(S_E(K))$$

and hence the isotropic constant of $S_E(K)$ equals the isotropic constant of $S_E(\text{Iso}(K))$, and the lemma follows. □

The following proposition finishes the proof of Theorem 1.4.

Assume that there exist k Steiner symmetrizations that transform T into a body $\tilde{T}$ with $d_{BM}(\tilde{T}) < A$. Then also a Schwartz symmetrization of T with respect to a k-dimensional subspace that contains these k symmetrization vectors, transforms T into $\tilde{\tilde{T}}$ with $d_{BM}(\tilde{\tilde{T}}) < A$.

Proposition 4.4. *Let $K \subset \mathbb{R}^n$ be a convex body. Assume that there exists $T \subset K$ with $\text{Vol}(T) > \frac{9}{10} \text{Vol}(K)$, such that for any $\varepsilon > 0$ there exist $\lfloor \varepsilon n \rfloor$ symmetrizations, that transform T into $\tilde{T}$ with $d_{BM}(\tilde{T}) < c_1 \frac{1}{\varepsilon^{c_2}}$, where c_1, c_2 are independent of ε.*

Then $L_K < c(c_1, c_2)$ where $c(c_1, c_2)$ depends solely on its arguments.

Proof. By the discussion at the end of Section 1, we may assume that the barycentre of T is at the origin, that $\text{Vol}(T) = 1$ and that T is isotropic (symmetrizability is an affine invariant property). Also, for any $\varepsilon > 0$, there exists a subspace $E_{\varepsilon n} \subset \mathbb{R}^n$ of dimension $\lfloor \varepsilon n \rfloor$ such that the Schwartz symmetrization of T with respect to any subspace that contains $E_{\varepsilon n}$ is $\frac{c_1}{\varepsilon^{c_2}}$-close to an ellipsoid. Let us denote $\log^{(0)} n = n$ and $\log^{(i+1)} n = \log \max\{\log^{(i)} n, e\}$. Substitute $\delta_i = \frac{1}{(\log^{(i)} n)^2}$, and for i such that $\delta_i < \frac{1}{2}$ let

$$F_i = sp\{E_{\delta_1 n}, \ldots, E_{\delta_i n}\}$$

where sp denotes linear span. Denote $\varepsilon_i = \frac{1}{n}\dim(F_i)$. Then

$$\frac{1}{\left(\log^{(i)} n\right)^2} \leqslant \varepsilon_i \leqslant \sum_{j=1}^{i} \frac{1}{\left(\log^{(j)} n\right)^2} < \frac{2}{\left(\log^{(i)} n\right)^2}.$$

Let T_i denote the Schwartz symmetrization of T with respect to F_i. Since $F_{i-1} \subset F_i$ we can think of T_i as the Schwartz symmetrization of T_{i-1} with respect to F_i. According to our assumptions,

$$cL_{T_i} \leqslant d_{BM}(T_i) < c_1 \left(\frac{1}{\delta_i}\right)^{c_2} < c_1 \left(\log^{(i)} n\right)^{2c_2}$$

where the leftmost inequality appears in [15]. By Lemma 4.2, since $\varepsilon_{i+1} < \frac{2}{(\log^{(i+1)} n)^2}$,

$$\text{Vol}(\text{Iso}(T_i) \cap E_{i+1})^{1/n} > \left(\frac{c}{c_1 \left(\log^{(i)} n\right)^{2c_2}}\right)^{\frac{2}{(\log^{(i+1)} n)^2}} > C^{\frac{1}{\log^{(i+1)} n}}$$

and hence by Lemma 4.3, since F_{i+1} is an invariant subspace of M_{T_i} (recall that T is isotropic, and symmetrizations were applied only with respect to subspaces contained in F_{i+1}),

$$L_{T_{i+1}} > \left(\frac{c}{\left(\log^{(i+1)} n\right)^2}\right)^{\frac{2}{(\log^{(i+1)} n)^2}} L_{T_i}^{1 - \frac{2}{(\log^{(i+1)} n)^2}} C^{\frac{1}{\log^{(i+1)} n}}$$

and since $L_{T_i} < c(\log^{(i)} n)^{2c_2}$,

$$L_{T_{i+1}} > c^{\frac{1}{\log^{(i+1)} n}} L_{T_i} > \cdots > c^{\sum_{j=1}^{i+1} \frac{1}{\log^{(j)} n}} L_T.$$

Let i^* be the largest integer such that $\varepsilon_i < \frac{1}{2}$. Then T_{i^*} has a bounded distance from an ellipsoid, and $L_{T_{i^*}} < c(c_1, c_2)$ (see [15]). Therefore,

$$L_T < c^{\sum_{j=1}^{i^*} \frac{1}{\log^{(j)} n}} c(c_1, c_2) < c'(c_1, c_2)$$

and since $L_K \approx L_T$ (e.g., [4] or the Borell lemma), the proposition is proved. □

References

[1] Artstein, S., Milman, V. and Szarek, S. (2004) Duality of metric entropy. *Ann. Math.*, **159** 1313–1328.

[2] Ball, K. M. (1988) Logarithmically concave functions and sections of convex sets in $\mathbb{R}^n$. *Studia Math.* **88** 69–84.

[3] Bonnesen, T. and Fenchel, W. (1934) *Theorie der Konvexen Körper*, Springer, Berlin. English translation: *Theory of Convex Bodies*, BCS Associates (1987).

[4] Bourgain, J. (2002) On the isotropy-constant problem for 'Psi-2' bodies. In *Geometric Aspects of Functional Analysis 2001/02*, Vol. 1807 of *Lecture Notes in Mathematics*, Springer, Berlin, pp. 114–121.

[5] Bourgain, J. and Milman, V. D. (1987) New volume ratio properties for convex symmetric bodies in $\mathbb{R}^n$. *Invent. Math.* **88** 319–340.

[6] Bourgain, J., Klartag, B. and Milman, V. D. (2004) Symmetrization and isotropic constants of convex bodies. In *Geometric Aspects of Functional Analysis 2002/03*, Vol. 1850 of *Lecture Notes in Mathematics*, Springer, Berlin, pp. 101–116.

[7] Bourgain, J., Lindenstrauss, J. and Milman, V. D. (1989) Estimates related to Steiner symmetrizations. *Geometric Aspects of Functional Analysis 1987/88*, Vol. 1376 of *Lecture Notes in Mathematics*, Springer, pp. 264–273.

[8] Fradelizi, M. (1997) Sections of convex bodies through their centroid. *Arch. Math.* **69** 515–522.

[9] Giannopoulos, A. A. and Milman, V. D. (1998) Mean width and diameter of proportional sections of a symmetric convex body. *J. Reine Angew. Math.* **497** 113–139.

[10] Klartag, B. (2005) An isomorphic version of the slicing problem. *J. Funct. Anal.* **218** 372–394.

[11] Klartag, B. and Milman, V. (2003) Isomorphic Steiner symmetrization. *Invent. Math.* **153** 463–485.

[12] Meyer, M. and Pajor, A. (1990) On the Blaschke–Santaló inequality. *Arch. Math. (Basel)* **55** 82–93.

[13] Milman, V. D. (1986) Inégalité de Brunn–Minkowski inverse et applications à le théorie locale des espaces normés. *CR Acad. Sci. Paris, Ser. I* **302** 25–28.

[14] Milman, V. D. (1988) Isomorphic symmetrizations and geometric inequalities. In *Geometric Aspects of Functional Analysis 1986/87*, Vol. 1317 of *Lecture Notes in Mathematics*, Springer, Berlin, pp. 107–131.

[15] Milman, V. D. and Pajor, A. (1989) Isotropic position and inertia ellipsoids and zonoids of the unit ball of a normed n-dimensional space. In *Geometric Aspects of Functional Analysis 1987/88*, Vol. 1376 of *Lecture Notes in Mathematics*, Springer, Berlin, pp. 64–104.

[16] Milman, V. D. and Pajor, A. (2000) Entropy and asymptotic geometry of non-symmetric convex bodies. *Adv. Math.* **152** 314–335.

[17] Milman, V. D. and Schechtman, G. (1986) *Asymptotic Theory of Finite-Dimensional Normed Spaces*, Vol. 1200 of *Lecture Notes in Mathematics*, Springer, Berlin.

[18] Pisier, G. (1997) *The Volume of Convex Bodies and Banach Space Geometry*, Vol. 94 of *Cambridge Tracts in Math.*, Cambridge University Press.

[19] Rogers, C. A. and Shephard, G. C. (1957) The difference body of a convex body. *Arch. Math.* **8** 220–233.

[20] Rudelson, M. (2000) Sections of the difference body. *Discrete Comput. Geom.* **23** 137–146.

[21] Spingarn, J. E. (1993) An inequality for sections and projections of a convex set. *Proc. Amer. Math. Soc.* **118** 1219–1224.

[22] Szarek, S. (1978) On Kashin's almost Euclidean orthogonal decomposition of l_n^1. *Bull. Acad. Polon. Sci. Sér. Sci. Math. Astronom. Phys.* **26** 691–694.

[23] Szarek, S. and Tomczak-Jaegermann, N. (1980) On nearly Euclidean decomposition for some classes of Banach spaces. *Compositio Math.* **40** 367–385.

A Note on Bipartite Graphs Without 2k-Cycles

ASSAF NAOR[1] and JACQUES VERSTRAËTE[2]

[1]Microsoft Research, One Microsoft Way, Redmond, WA 98052-6399, USA
(e-mail: anaor@microsoft.com)

[2]Faculty of Mathematics, University of Waterloo,
200 University Avenue West, Waterloo, Canada N2L 3G1
(e-mail: jverstra@math.uwaterloo.ca)

Received 17 November 2003; revised 27 May 2005

For Béla Bollobás on his 60th birthday

The question of the maximum number $\mathrm{ex}(m,n,C_{2k})$ of edges in an m by n bipartite graph without a cycle of length $2k$ is addressed in this note. For each $k \geqslant 2$, it is shown that

$$\mathrm{ex}(m,n,C_{2k}) \leqslant \begin{cases} (2k-3)\bigl[(mn)^{\frac{k+1}{2k}} + m + n\bigr] & \text{if } k \text{ is odd,} \\ (2k-3)\bigl[m^{\frac{k+2}{2k}} n^{\frac{1}{2}} + m + n\bigr] & \text{if } k \text{ is even.} \end{cases}$$

1. Introduction

In this note, we study the maximum number of edges in an m by n bipartite graph containing no $2k$-cycles. This problem was studied in the papers by Erdős, Sós and Sárközy [3] and by Győri [4], in the context of a number-theoretic problem. It is also related to the size of subsets of points in projective planes, such as arcs and caps. A connection to a geometric problem involving points and lines in Euclidean space is described by de Caen and Székely [1].

Throughout the material to follow, $\gamma(m,n,g,k)$ denotes the maximum number of edges in an m by n bipartite graph of girth at least $2g$ containing no $2k$-cycle. In particular, we write $\gamma(m,n,2,k) = \mathrm{ex}(m,n,C_{2k})$, which is the maximum number of edges in a $2k$-cycle-free m by n bipartite graph. For $d, g \geqslant 2$, let $c(d,g)$ be the largest integer such that every bipartite graph of average degree at least $2d$ and girth at least $2g$ contains a cycle of length at least $c(d,g)$ with at least one chord. Our main result is as follows.

Theorem 1.1. Let $k, g \geq 2$ be integers. Then, for any d such that $c(d, g) \geq 2(k - g + 1)$,

$$\gamma(m, n, g, k) \leq \begin{cases} 2d\left[(mn)^{\frac{k+1}{2k}} + m + n\right] & \text{if } k \text{ is odd,} \\ 2d\left[m^{\frac{k+2}{2k}} n^{\frac{1}{2}} + m + n\right] & \text{if } k \text{ is even.} \end{cases}$$

Theorem 1.1 depends explicitly on the parameter $c(d, g)$. Let us make a few remarks about $c(d, g)$. First, a straightforward argument shows that $c(d, g) \geq 2d(g - 1) + 1$. Stronger results were obtained by Erdős, Faudree, Rousseau and Schelp [2]. We deduce from their paper that $c(d, g) \geq d^{g/4}$. In particular, by using these two bounds on $c(d, g)$ in Theorem 1.1, we respectively obtain the following two corollaries.

Corollary 1.2.

$$ex(m, n, C_{2k}) \leq \begin{cases} (2k - 3)\left[(mn)^{\frac{k+1}{2k}} + m + n\right] & \text{if } k \text{ is odd,} \\ (2k - 3)\left[m^{\frac{k+2}{2k}} n^{\frac{1}{2}} + m + n\right] & \text{if } k \text{ is even.} \end{cases}$$

Corollary 1.3. *For any number $\delta > 0$, there exists a constant $c(\delta)$ such that*

$$\gamma(m, n, \delta \log k, k) \leq \begin{cases} c(\delta)\left[(mn)^{\frac{k+1}{2k}} + m + n\right] & \text{if } k \text{ is odd,} \\ c(\delta)\left[m^{\frac{k+2}{2k}} n^{\frac{1}{2}} + m + n\right] & \text{if } k \text{ is even.} \end{cases}$$

Using the requirement on $c(d, g)$ from Theorem 1.1 and the inequality $c(d, g) \geq d^{g/4}$, a short computation shows that we may certainly take $c(\delta) = 4^{4/\delta}$ in Corollary 1.3. Recent results of Hoory [5] and Lam [6] show that $\gamma(m, n, k, k)$ satisfies the bounds given in Corollary 1.3, with $c(\delta) = 1$. In the case $k \in \{2, 3, 5\}$, the existence of rank two geometries known as generalized polygons (see [7] for constructions) show that the constant $c(\delta) = 1$ is best possible when $m = n$. The strength in Corollary 1.3 is that it shows that excluding cycles of length $O(\log k)$ has, to within an absolute constant factor, the same effect on the upper bounds as excluding all cycles of length at most $2k$. On the other hand, our next result gives an indication that $\gamma(m, n, k, k)$ and $\gamma(m, n, 2, k)$ may differ substantially.

Theorem 1.4. *For all integers m, n and $k \geq 3$,*

$$\gamma((k - 1)m, n, k, k) \geq (k - 1) \cdot \gamma(m, n, 2, k).$$

If we make the assumption that, for each even positive integer k, $\gamma(m, n, 2, k)$ is asymptotically $c_1(mn)^{1/2 + 1/(2k)} + c_2(m + n)$ as $m, n \to \infty$, for some constants c_1, c_2, then Theorem 1.4 gives

$$\liminf_{m, n \to \infty} \frac{\gamma(m, n, k, k)}{\gamma(m, n, 2, k)} \geq \sqrt{k - 1}.$$

Similar observations may be made for odd values of k, namely

$$\liminf_{m, n \to \infty} \frac{\gamma(m, n, k, k)}{\gamma(m, n, 2, k)} \geq (k - 1)^{1/2 - 1/(2k)}.$$

One can deduce from these observations and the constructions of generalized quadrangles and hexagons that

$$\liminf_{n\to\infty} \frac{\gamma(n,2n,3,3)}{\gamma(n,2n,2,3)} \geq 2^{1/3}, \tag{1.1}$$

$$\liminf_{n\to\infty} \frac{\gamma(n,2n,5,5)}{\gamma(n,2n,2,5)} \geq 4^{1/5}. \tag{1.2}$$

Inequality (1.1) comes from [8] and (1.2) is implicit in the work of Lazebnik, Ustimenko and Woldar [7]. The next section is devoted to proving Theorem 1.1, and the construction for Theorem 1.4 is presented in Section 3.

2. Proof of Theorem 1.1

The main tool in our proof will be the first theorem in [9]. Although the next proposition is not the statement of this theorem, it is straightforward to verify, from the proof appearing in [9].

Proposition 2.1. *Let G be a bipartite graph of average degree at least 4d and girth 2g. Then G contains cycles of $\frac{1}{2}c(2d,2g)$ consecutive even lengths, the shortest of which has length at most twice the radius of G.* □

Proof of Theorem 1.1. We proceed by induction on $m+n$. Suppose, for a contradiction, that G is a bipartite graph without cycles of length at most $2g-2$ and containing no $2k$-cycle, with parts A and B of sizes m and n, respectively, and with more edges than the corresponding upper bound in Theorem 1.1. By Proposition 2.1, and using the definition of d in the statement of Theorem 1.1, it is sufficient to show that G contains a subgraph of radius at most k and average degree at least d. Indeed, since G has even girth $2g$, Proposition 2.1 would then imply that there is an integer r such that $g \leq r \leq k$ and such that G contains the cycles $C_{2r}, C_{2r+2}, \ldots, C_{2r+2k-2g}$. Since $2k \in [2r, 2r+2k-2g]$ whenever $g \leq r \leq k$, one of these cycles has length $2k$, as required.

Case 1. k is odd.
In this case, we may assume that the minimum degree in A is at least

$$d_A = d\frac{m^{\frac{1}{2}+\frac{1}{2k}}}{n^{\frac{1}{2}-\frac{1}{2k}}} + d.$$

Indeed, if there were a vertex $v \in A$ with degree less than d_A then by deleting it we would arrive at an $m \times (n-1)$ bipartite graph G' with

$$e(G') > e(G) - d_A \geq 2dm^{\frac{1}{2}+\frac{1}{2k}}\left(n^{\frac{1}{2}+\frac{1}{2k}} - \frac{1}{2}n^{-\frac{1}{2}+\frac{1}{2k}}\right) + 2d(m+n) - d$$
$$> 2d\left([m(n-1)]^{\frac{1}{2}+\frac{1}{2k}} + (m+n-1)\right),$$

so that G' contains a $2k$-gon by the inductive hypothesis. Similarly, we may assume that the minimum degree in B is at least

$$d_B = d\frac{n^{\frac{1}{2}+\frac{1}{2k}}}{m^{\frac{1}{2}-\frac{1}{2k}}} + d.$$

Choose a vertex $v \in A$ and let H_r be the subgraph of G induced by vertices at distance at most r from v. Let us show that H_r has average degree at least d for some $r \leqslant k$. Suppose this cannot be done. Let D_r denote the set of vertices of H_r at distance exactly r from v. Then the average number of neighbours in D_{r-1} of a vertex in D_r is less than d. It follows that if $D_r \subset A$, then $|D_{r-1}|(d_B - d) < d|D_r|$ and if $D_r \subset B$, then $|D_{r-1}|(d_A - d) < d|D_r|$. Therefore,

$$|D_r| > \begin{cases} \left(\frac{d_A}{d} - 1\right)|D_{r-1}| = \frac{m^{\frac{1}{2}+\frac{1}{2k}}}{n^{\frac{1}{2}-\frac{1}{2k}}}|D_{r-1}| & \text{if } r \text{ is odd,} \\ \left(\frac{d_B}{d} - 1\right)|D_{r-1}| = \frac{n^{\frac{1}{2}+\frac{1}{2k}}}{m^{\frac{1}{2}-\frac{1}{2k}}}|D_{r-1}| & \text{if } r \text{ is even.} \end{cases}$$

Iterating these inequalities for $r = 1, 2, \ldots, k$ we get that since k is odd,

$$|D_k| > \left(\frac{m^{\frac{1}{2}+\frac{1}{2k}}}{n^{\frac{1}{2}-\frac{1}{2k}}}\right)^{\lceil \frac{k}{2} \rceil} \left(\frac{n^{\frac{1}{2}+\frac{1}{2k}}}{m^{\frac{1}{2}-\frac{1}{2k}}}\right)^{\lfloor \frac{k}{2} \rfloor} = m.$$

On the other hand, using the fact that k is odd once more, $D_k \subset B$, so that $|D_k| \leqslant m$, which is a contradiction. The proof of the second part is complete.

Case 2. k is even.

The proof here is similar, so we only indicate the necessary changes to the argument. The inductive hypothesis implies that the minimum degree in A is at least

$$d'_A = d\frac{m^{\frac{1}{2}+\frac{1}{k}}}{\sqrt{n}} + d,$$

and the minimum degree in B is at least

$$d'_B = d\frac{\sqrt{n}}{m^{\frac{1}{2}-\frac{1}{k}}} + d.$$

We now start with a vertex $v \in B$, and repeat the above argument. Since k is even, $D_k \subset B$, so that $|D_k| \leqslant m$, but

$$|D_k| > \left(\frac{m^{\frac{1}{2}+\frac{1}{k}}}{\sqrt{n}}\right)^{\frac{k}{2}} \left(\frac{\sqrt{n}}{m^{\frac{1}{2}-\frac{1}{k}}}\right)^{\frac{k}{2}} = m,$$

so we once more arrive at a contradiction. □

3. Proof of Theorem 1.4

Suppose we are given an m by n bipartite graph H, of girth at least $2k + 2$. From H, we construct a $(k-1)m$ by n bipartite graph containing no $2k$-cycles, and with $k-1$ times as many edges as H. Let A, B be the parts of H, let $A_1, A_2, \ldots, A_{k-1}$ be disjoint sets, and let $\phi : \bigcup_{i=1}^{k-1} A_i \to A$ be defined so that ϕ restricted to A_i is a bijection $A_i \leftrightarrow A$. Define a new graph G with parts $\bigcup_{i=1}^{k-1} A_i$ and B, with edge set

$$E = \{ab : \phi(a)b \in H\}.$$

In words, we are taking $(k-1)$ identical edge-disjoint copies of H which share B as one of their parts. We now show that G has no $2k$-cycles.

Suppose, for a contradiction, that G contains a $2k$-cycle $C = (a_1, b_1, a_2, a_2, \ldots, a_k, b_k, a_1)$ with $b_i \in B$ for all $i \in \{1, 2, \ldots, k\}$. Then

$$W = (\phi(a_1), b_1, \phi(a_2), b_2, \ldots, \phi(a_k), b_k, \phi(a_1))$$

is a closed walk of length $2k$ in H. As H has girth at least $2k + 2$, W takes place on a tree $T \subset H$ with at most k edges. On the other hand, the tree contains the k vertices in $V(C) \cap B$, and there are at least two vertices $a, a' \in V(C) \cap A_i$ for some $i \in \{1, 2, \ldots, k-1\}$, by the pigeonhole principle. Now $\phi(a)$ and $\phi(a')$ are distinct, since ϕ restricted to A_i is a bijection. Therefore the tree has at least $k + 2$ vertices, a contradiction. Therefore the graph G is $2k$-cycle-free.

References

[1] de Caen, D. and Székely, L. A. (1992) The maximum size of 4- and 6-cycle free bipartite graphs on m, n vertices. In *Sets, Graphs and Numbers* (Budapest, 1991), Vol. 60 of *Colloq. Math. Soc. János Bolyai*, North-Holland, Amsterdam, pp. 135–142.

[2] Erdős, P., Faudree, R. J., Rousseau, C. C. and Schelp, R. H. (1999) The number of cycle lengths in graphs of given minimum degree and girth. In *Paul Erdős Memorial Collection, Discrete Math.* **200** 55–60.

[3] Erdős, P., Sárközy, A. and Sós, V. T. (1995) On product representations of powers I. *European J. Combin.* **16** 567–588.

[4] Győri, E. (1997) C_6-free bipartite graphs and product representation of squares. In *Graphs and Combinatorics* (Marseille, 1995), *Discrete Math.* **165/166** 371–375.

[5] Hoory, S. (2002) The size of bipartite graphs with a given girth. *J. Combin. Theory Ser. B* **86** 215–220.

[6] Lam, T. (2001) Graphs without cycles of even length. *Bull. Austral. Math. Soc.* **63** 435–440.

[7] Lazebnik, F., Ustimenko, V. A. and Woldar, A. J. (1999) Polarities and $2k$-cycle-free graphs. In *16th British Combinatorial Conference* (London, 1997), *Discrete Math.* **197/198** 503–513.

[8] Naor, A. and Verstraëte, J. A. (2005) On the Turán number for the hexagon. To appear in *Adv. Math.*

[9] Verstraëte, J. A. (2000) Arithmetic progressions of cycle lengths in graphs. *Combin. Probab. Comput.* **9** 369–373.

Book Ramsey Numbers and Quasi-Randomness

V. NIKIFOROV, C. C. ROUSSEAU and R. H. SCHELP

Department of Mathematical Sciences, University of Memphis, Memphis, Tennessee, TN 38152, USA
(e-mail: vnikifrv@memphis.edu, ccrousse@memphis.edu, rschelp@memphis.edu)

Received 1 August 2003; revised 27 August 2004

For Béla Bollobás on his 60th birthday

A set of n triangles sharing a common edge is called a book with n pages and is denoted by B_n. It is known that the Ramsey number $r(B_n)$ satisfies $r(B_n) = (4 + o(1))n$. We show that every red–blue edge colouring of $K_{\lfloor(4-\varepsilon)n\rfloor}$ with no monochromatic B_n exhibits quasi-random properties when ε tends to 0. This implies that there is a constant $c > 0$ such that for every red–blue edge colouring of $K_{r(B_n)}$ there is a monochromatic B_n whose vertices span at least $\lfloor cn^2 \rfloor$ edges of the same colour as the book.

As an application we find the Ramsey number for a class of graphs.

1. Introduction

Our notation and terminology are standard (see, e.g., [1]). Thus, $G(n)$ is a graph of order n, and $G(n,m)$ is a graph of order n and size m. Given a graph G and a vertex $u \in V(G)$, we write $\Gamma(u)$ for the set of vertices adjacent to u; $d_G(u) = |\Gamma_G(u)|$ is the degree of u. Departing from common usage, we set $d_G(U) = |\bigcap_{x \in U} \Gamma_G(x)|$; in particular, $d_G(uv) = |\Gamma_G(u) \cap \Gamma_G(v)|$. We shall write $d(u)$ and $\Gamma(u)$ instead of $d_G(u)$ and $\Gamma_G(u)$ when the graph G is understood.

If the graphs G_1 and G_2 are vertex-disjoint then $G_1 + G_2$ denotes the union of G_1 and G_2 together with the edges joining vertices of G_1 to vertices of G_2. We write $k_3(G)$ for the number of triangles of G. Given a graph G, the Ramsey number $r(G)$ is the minimal n such that every red–blue colouring of K_n contains a monochromatic copy of G.

For $q \geq 1$, $k \geq 1$, a k-book $B_q^{(k)}$ of size q is a graph consisting of q distinct $(k+1)$-cliques, sharing a common k-clique. We denote by $bk^{(k)}(G)$ the size of the largest k-book in a graph G. We call 2-books simply books, and write $bk(G)$ for $bk^{(2)}(G)$.

Books have attracted considerable attention in Ramsey graph theory (see, e.g., [8], [4] and [6]). In particular, Erdős, Faudree, Rousseau, and Schelp [3] proved that, for every fixed k,

$$r(B_n^{(k)}) > (2^k + o(1))n.$$

On the other hand, Rousseau and Sheehan in [8] proved that

$$r(B_n) \leqslant 4n + 2,$$

and observed that the Paley graphs imply $r(B_n) = 4n + 2$ whenever $4n + 1$ is a prime power. Hence,

$$r(B_n) = (4 + o(1))n.$$

The main results of this note, stated and proved in Section 3, continue this line of investigation. In particular, we show that red–blue edge colourings of $K_{\lfloor(4-\varepsilon)n\rfloor}$ with no monochromatic B_n exhibit quasi-random properties when ε tends to 0.

We deduce the following Ramsey result: there is a constant $c > 0$ such that every red–blue edge colouring of $K_{r(B_n)}$ contains a monochromatic book B_n whose vertices span at least $\lfloor cn^2 \rfloor$ edges of the same colour as the book.

This, in turn, implies that, for n sufficiently large and every $0 < \beta < 1$, there exist

$$\alpha = \alpha(\beta) > 0, \qquad a \geqslant \lfloor \alpha \log n \rfloor, \qquad b \geqslant \lfloor n^{1-\beta} \rfloor$$

such that

$$r(H) = r(B_n),$$

where the graph H is obtained by adding $K_{a,b}$ to B_n, i.e.,

$$H = K_2 + \left(K_{a,b} \cup \overline{K}_{n-a-b}\right).$$

We show that this result is tight in the following sense: $r(H) > r(B_n)$ for any graph H defined as

$$H = K_2 + \left(F \cup \overline{K}_{n-k}\right),$$

where F is a nonbipartite graph of order $k \leqslant n$.

2. Some preliminary results

In this section we provide some necessary background for proofs of the aforementioned results in Ramsey theory. Much of this can be readily understood in the context of quasi-randomness and experts in this area will find no surprises. This material is included in order to make the paper more self-contained and accessible to readers with limited background in quasi-randomness.

Among the many results that Chung, Graham and Wilson state in [2] is the following one, attributed to Thomason [10, 11].

Theorem 2.1. *For every $\delta > 0$ there exists $\varepsilon = \varepsilon(\delta) > 0$ such that, if G is a graph of sufficiently large order n and*

$$\sum_{u,v \in V(G)} \left| d_G(uv) - \frac{1}{4}n \right| < \varepsilon n^3, \tag{2.1}$$

then
$$\left| e(G[X]) - \frac{1}{4}|X|^2 \right| < \delta n^2$$

for every $X \subset V(G)$.

It is not an easy task to extract a short proof of Theorem 2.1 from [2]; for the reader's benefit we present one at the end of this section.

We shall start with a simple condition for 'almost-regularity' of a graph.

Lemma 2.2. *For every ε and sufficiently large n, if $G = G(n)$ satisfies*

$$\sum_{uv \in E(G)} \left| d_G(uv) - \frac{n}{4} \right| + \sum_{uv \in E(\overline{G})} \left| d_{\overline{G}}(uv) - \frac{n}{4} \right| < \varepsilon n^3,$$

then

$$\sum_{u \in V(G)} \left| d_G(u) - \frac{n}{2} \right| < \sqrt{\varepsilon} n^2. \tag{2.2}$$

Proof. We calculate $k_3(G) + k_3(\overline{G})$ in two ways. First, by a well-known relation of Lorden [5],

$$k_3(G) + k_3(\overline{G}) = \binom{n}{3} - \frac{1}{2} \sum_{u \in V(G)} d_G(u)(n - 1 - d_G(u))$$

$$= \frac{n(n-1)(n-5)}{24} + \frac{1}{2} \sum_{u \in V(G)} \left(d_G(u) - \frac{n-1}{2} \right)^2 \tag{2.3}$$

and secondly,

$$k_3(G) + k_3(\overline{G}) = \frac{1}{3} \sum_{uv \in E(G)} d_G(uv) + \frac{1}{3} \sum_{uv \in E(\overline{G})} d_{\overline{G}}(uv).$$

Equating the two expressions for $k_3(G) + k_3(\overline{G})$, we obtain

$$\sum_{u \in V(G)} \left(d_G(u) - \frac{n}{2} \right)^2 = \frac{2}{3} \left(\sum_{uv \in E(G)} \left(d_G(uv) - \frac{n}{4} \right) + \sum_{uv \in E(\overline{G})} \left(d_{\overline{G}}(uv) - \frac{n}{4} \right) \right) + O(n^2).$$

Hence the hypothesis implies

$$\sum_{u \in V(G)} \left(d_G(u) - \frac{n}{2} \right)^2 < \varepsilon n^3,$$

and, using Cauchy's inequality, (2.2) follows for sufficiently large n. $\square$

Lemma 2.3. *For every $\delta > 0$ there exists $\varepsilon = \varepsilon(\delta) > 0$ such that if G is a graph of sufficiently large order n, and*

$$\sum_{uv \in E(G)} \left| d_G(uv) - \frac{n}{4} \right| + \sum_{uv \in E(\overline{G})} \left| d_{\overline{G}}(uv) - \frac{n}{4} \right| < \varepsilon n^3, \tag{2.4}$$

then

$$\sum_{u,v \in V(G)} \left| d_G(uv) - \frac{n}{4} \right| < \delta n^3.$$

Proof. Set $\varepsilon = \min\{(\delta/3)^2, 1\}$. Observe that for $uv \in E(\overline{G})$,

$$d_G(uv) = d_{\overline{G}}(uv) + d_G(u) + d_G(v) - (n-2).$$

Consequently,

$$\sum_{u,v \in V(G)} \left| d_G(uv) - \frac{n}{4} \right| = \sum_{uv \in E(G)} \left| d_G(uv) - \frac{n}{4} \right|$$

$$+ \sum_{uv \in E(\overline{G})} \left| d_{\overline{G}}(uv) - \frac{n}{4} + d_G(u) + d_G(v) - (n-2) \right|$$

$$\leqslant \sum_{uv \in E(G)} \left| d_G(uv) - \frac{n}{4} \right| + \sum_{uv \in E(\overline{G})} \left| d_{\overline{G}}(uv) - \frac{n}{4} \right|$$

$$+ \sum_{uv \in E(\overline{G})} \left| d_G(u) - \frac{n}{2} \right| + \sum_{uv \in E(\overline{G})} \left| d_G(v) - \frac{n}{2} \right| + 2\binom{n}{2}.$$

Applying Lemma 2.2 we find that

$$\sum_{uv \in E(\overline{G})} \left| d_G(u) - \frac{n}{2} \right| + \sum_{uv \in E(\overline{G})} \left| d_G(v) - \frac{n}{2} \right| \leqslant n \sum_{u \in V(G)} \left| d_G(u) - \frac{n}{2} \right|$$

$$< \sqrt{\varepsilon} n^3.$$

Hence,

$$\sum_{u,v \in V(G)} \left| d_G(uv) - \frac{n}{4} \right| < \varepsilon n^3 + \sqrt{\varepsilon} n^3 + 2\binom{n}{2} < 3\sqrt{\varepsilon} n^3 = \delta n^3. \qquad \square$$

Lemma 2.4. *For every $\delta > 0$ there exists $\varepsilon = \varepsilon(\delta) > 0$ such that if G is a graph of of sufficiently large order n, and*

$$bk(G) < \left(\frac{1}{4} + \varepsilon\right)n \quad \text{and} \quad bk(\overline{G}) < \left(\frac{1}{4} + \varepsilon\right)n,$$

then

$$\sum_{u,v \in V(G)} \left| d_G(uv) - \frac{n}{4} \right| < \delta n^3.$$

Proof. From (2.3) we have
$$3(k_3(G) + k_3(\overline{G})) \geqslant \frac{n(n-1)(n-5)}{8}.$$

Hence
$$\sum_{uv \in E(G)} \left(d_G(uv) - \frac{n}{4}\right) + \sum_{uv \in E(\overline{G})} \left(d_{\overline{G}}(uv) - \frac{n}{4}\right) \geqslant -\frac{5n(n-1)}{8},$$

and, by hypothesis, the terms in the above sums are each less than εn. It follows that
$$\sum_{uv \in E(G)} \left|d_G(uv) - \frac{n}{4}\right| + \sum_{uv \in E(\overline{G})} \left|d_{\overline{G}}(uv) - \frac{n}{4}\right| < \varepsilon n^3,$$

and the desired result follows from Lemma 2.3. □

Observe that it just as easy to show that (2.4) also implies a necessary condition for quasi-randomness as follows from the next lemma whose proof is omitted.

Lemma 2.5. *For every $\delta > 0$ there exists $\varepsilon = \varepsilon(\delta) > 0$ such that if G is a graph of sufficiently large order n, and*
$$\sum_{u,v \in V(G)} \left|d_G(uv) - \frac{1}{4}n\right| < \varepsilon n^3,$$

then
$$\sum_{uv \in E(G)} \left|d_G(uv) - \frac{1}{4}n\right| + \sum_{uv \in E(\overline{G})} \left|d_{\overline{G}}(uv) - \frac{1}{4}n\right| < \delta n^3.$$ □

From Lemma 2.4 and Theorem 2.1 we obtain the following.

Corollary 2.6. *For every $\delta > 0$ there exists $\varepsilon = \varepsilon(\delta) > 0$ such that if G is a graph of sufficiently large order n, and*
$$bk(G) < \left(\frac{1}{4} + \varepsilon\right)n, \qquad bk(\overline{G}) < \left(\frac{1}{4} + \varepsilon\right)n,$$

then
$$\left|e(G[X]) - \frac{1}{4}|X|^2\right| < \delta n^2.$$

for every $X \subset V(G)$. □

Applying a well-known inequality of Nordhaus and Stewart [7] we obtain the following simple lower bound on the booksize of a graph.

Lemma 2.7. *For every graph $G = G(n,m)$,*
$$bk(G) \geqslant \left(\frac{4m}{n^2} - 1\right)n.$$

Proof. For the sake of completeness we start with a brief proof of Nordhaus–Stewart inequality.

For every edge $uv \in E(G)$ we have

$$|\Gamma(u) \cap \Gamma(v)| = |\Gamma(u)| + |\Gamma(v)| - |\Gamma(u) \cup \Gamma(v)| \geq d(u) + d(v) - n.$$

Summing this inequality over all $uv \in E(G)$ we obtain

$$3k_3(G) \geq \sum_{uv \in E(G)} (d(u) + d(v) - n) = \sum_{u \in V(G)} d^2(u) - nm \geq \frac{4m^2}{n} - nm.$$

Hence, from

$$bk(G) \geq \frac{3k_3(G)}{m},$$

the desired result follows. □

2.1. Proof of Theorem 2.1
In our proof we borrow some ideas of [11], Theorem 1.

Proof. Set $\varepsilon = \min\{\delta^4, 1\}$ and choose $X \subset V(G)$, $X \neq \emptyset$, $X \neq V$.
For every $u \in V(G)$, let $x(u) = |\Gamma_G(u) \cap X|$, and set

$$|X| = k, \qquad e_1 = e(G[X]), \qquad e_2 = e(X, V(G)\setminus X).$$

We immediately see that

$$2e_1 = \sum_{u \in X} x(u), \qquad e_2 = \sum_{u \in V(G)\setminus X} x(u).$$

Also, from (2.1), we have

$$\sum_{u \in X} \binom{x(u)}{2} + \sum_{u \in V(G)\setminus X} \binom{x(u)}{2} = \sum_{u,v \in X} d_G(uv) < \frac{n}{4}\binom{k}{2} + \varepsilon n^3.$$

The convexity of $\binom{x}{2}$ implies

$$k \binom{2e_1/k}{2} + (n-k)\binom{e_2/(n-k)}{2} < \frac{n}{4}\binom{k}{2} + \varepsilon n^3,$$

and hence

$$\frac{(2e_1)^2}{k} + \frac{(e_2)^2}{n-k} < \frac{nk^2}{4} + 2\varepsilon n^3. \tag{2.5}$$

From (2.1) and Lemma 2.2 we obtain

$$\sum_{u \in X} \left(d_G(u) - \frac{n}{2}\right) \geq -\sum_{u \in V(G)} \left|d_G(u) - \frac{n}{2}\right| \geq -\sqrt{\varepsilon} n^2.$$

Hence

$$e_2 - \left(\frac{kn}{2} - 2e_1\right) = 2e_1 + e_2 - \frac{kn}{2} = \sum_{u \in X} \left(d_G(u) - \frac{n}{2}\right) \geq -\sqrt{\varepsilon} n^2,$$

and therefore

$$(e_2)^2 > \left(\frac{kn}{2} - 2e_1\right)^2 - \sqrt{\varepsilon}n^2\left(e_2 + \frac{kn}{2} - 2e_1\right) > \left(\frac{kn}{2} - 2e_1\right)^2 - \sqrt{\varepsilon}n^4.$$

This, in view of (2.5), implies

$$(n-k)(2e_1)^2 + k\left(\frac{kn}{2} - 2e_1\right)^2 < \frac{nk^3(n-k)}{4} + 2\varepsilon n^3 k(n-k) + \sqrt{\varepsilon}n^4 k$$

$$< \frac{nk^3(n-k)}{4} + 3\sqrt{\varepsilon}n^5.$$

Rearranging terms and dividing by $4/n$, we obtain

$$e_1^2 - \frac{k^2 e_1}{2} + \frac{k^4}{16} < \frac{3}{4}\sqrt{\varepsilon}n^4 < \sqrt{\varepsilon}n^4,$$

and hence

$$\left|e(G[X]) - \frac{1}{4}|X|^2\right| = \left|e_1 - \frac{1}{4}k^2\right| < \sqrt[4]{\varepsilon}n^2 = \delta n^2,$$

completing the proof. □

3. Some Ramsey-type results

In this section we shall prove some consequences for Ramsey theory.

Theorem 3.1. *For every $\delta > 0$ there exist $\varepsilon = \varepsilon(\delta) > 0$ and $n_0 = n_0(\delta)$ such that if $n > n_0$, $l \geq (4-\varepsilon)n$, and the graph $G = G(l)$ satisfies*

$$bk(G) < n, \qquad bk(\overline{G}) < n \qquad (3.1)$$

then

$$\left|e(G[X]) - \frac{1}{4}|X|^2\right| < \delta l^2.$$

for every $X \subset V(G)$.

Proof. Observe that

$$n \leq \frac{l}{(4-\varepsilon)} < \left(\frac{1}{4} + \varepsilon\right)l$$

for every sufficiently small $\varepsilon > 0$. Hence, by hypothesis, we obtain

$$bk(G) < \left(\frac{1}{4} + \varepsilon\right)l, \qquad bk(\overline{G}) < \left(\frac{1}{4} + \varepsilon\right)l,$$

and by Corollary 2.6 the proof is completed. □

Theorem 3.2. *There exists a constant $c > 0$ such that every red–blue edge colouring of $K_{r(B_n)}$ contains a monochromatic book B_n whose vertices span at least $\lfloor cn^2 \rfloor$ edges of the same colour as the book.*

Proof. Clearly we need to prove the theorem only for n sufficiently large, so assume n is as large as needed. Let $\varepsilon(\delta)$ be the function from Corollary 2.6 and set

$$c = \min\left\{\frac{1}{10}, \frac{1}{16}\varepsilon\left(\frac{1}{128}\right)\right\}. \tag{3.2}$$

Given a red–blue edge colouring of $K_{r(B_n)}$, let the graph G be on the same vertex set as $K_{r(B_n)}$, and let $E(G)$ be the set of red edges. The order of G is $r(B_n)$, so either $B_n \subset G$ or $B_n \subset \overline{G}$. Assume $B_n \subset G$ – say $uv \in E(G)$ – and $uv + X \subset G$ is a book such that $|X| = n$. Then, if

$$\left|e(G[X]) - \frac{1}{4}|X|^2\right| < \frac{1}{128}(r(B_n))^2 \tag{3.3}$$

holds, we have

$$e(G[X]) > \frac{1}{4}|X|^2 - \frac{1}{128}(r(B_n))^2 \geq \frac{n^2}{4} - \frac{(4n+2)^2}{128} > \frac{n^2}{10},$$

and in this case the desired result follows in view of (3.2).

Suppose now that (3.3) does not hold. Applying Corollary 2.6 with $\delta = 1/128$, we see that for $\varepsilon = \varepsilon(1/128)$ we have either

$$bk(G) \geq (1/4 + \varepsilon)r(B_n)$$

or

$$bk(\overline{G}) \geq (1/4 + \varepsilon)r(B_n).$$

Assume $bk(G) \geq (1/4 + \varepsilon)r(B_n)$ – say $uv \in E(G)$ – and $uv + X \subset G$ is a book such that $|X| \geq (1/4 + \varepsilon)r(B_n)$.

For n sufficiently large we may assume $r(B_n) > (4 - \varepsilon)n$, and so

$$|X| > \left(\frac{1}{4} + \varepsilon\right)(4 - \varepsilon)n > (1 + \varepsilon)n.$$

If $e(G[X]) \geq \varepsilon|X|^2/16$ then the proof is completed by (3.2), so assume $e(G[X]) < \varepsilon|X|^2/16$. Hence, setting $H = \overline{G}[X]$, we find that

$$e(H) > \binom{|X|}{2} - \frac{\varepsilon}{16}|X|^2.$$

From Lemma 2.7 applied to H, it follows that

$$bk(H) \geq \left(1 - \frac{2}{|X|} - \frac{\varepsilon}{4}\right)|X| > \left(1 - \frac{\varepsilon}{2}\right)|X|$$

$$> \left(1 - \frac{\varepsilon}{2}\right)(1 + \varepsilon)n > n.$$

Therefore, $B_{n+1} \subset H$, i.e., X induces a blue book of size $(n+1)$ – say $zw \in E(H)$ – and $zw + Y \subset H$ is a book such that $|Y| = n + 1$. Clearly Y induces at least $\binom{n+1}{2}/2 > n^2/4$ edges of the same colour, so taking either the red book $uv + Y$ or the blue book $zw + Y$, the proof is completed in view of $c < 1/4$. □

Recalling the well-known result (see, *e.g.*, [1, p. 113]) that for every $c > 0$ and $0 < \beta < 1$, there exist
$$\alpha = \alpha(c, \beta) > 0, \qquad a \geqslant \lfloor \alpha \log n \rfloor, \qquad b \geqslant \lfloor n^{1-\beta} \rfloor,$$
such that every $G(n, \lfloor cn^2 \rfloor)$ contains $K_{a,b}$, we obtain the result announced in the Introduction.

Theorem 3.3. *For n sufficiently large and every $0 < \beta < 1$ there exist*
$$\alpha = \alpha(\beta) > 0, \qquad a \geqslant \lfloor \alpha \log n \rfloor, \qquad b \geqslant \lfloor n^{1-\beta} \rfloor$$
such that
$$r(H) = r(B_n),$$
where the graph H is defined by
$$H = K_2 + \left(K_{a,b} \cup \overline{K}_{n-a-b} \right).$$
In particular, if $n = 4k + 1$ is a sufficiently large prime power then $r(H) = 4n + 2$. □

Observe that Theorem 3.3 is, in a sense, best possible, since for every nonbipartite graph F of order $k \leqslant n$, the graph H defined by
$$H = K_2 + \left(F \cup \overline{K}_{n-k} \right)$$
satisfies $r(H) > 4n + 4 \geqslant r(B_n)$. Indeed, consider the complete 4-partite graph $K_4(n+1)$. Its complement consists of 4 disjoint cliques of order $n + 1$ and consequently contains no B_n; for every book $uv + X \subset K_4(n+1)$, the set X induces a bipartite graph.

4. Concluding remarks

Despite its simplicity, Theorem 3.1 corroborates the expectation that for some sequences of graphs $\{H_n\}$, every sequence of graphs $\{G_n\}$ such that
$$|G_n| = (1 + o(1))r(H_n), \qquad H_n \not\subseteq G_n, \qquad H_n \not\subseteq \overline{G}_n,$$
is quasi-random.

It is also curious that techniques borrowed from the theory of quasi-random graphs could be useful in finding exact Ramsey numbers.

We shall conclude with three conjectures, the first one being due to Thomason [9].

Conjecture 4.1. *For every fixed integer $k > 2$,*
$$r\left(B_n^{(k)}\right) = (2^k + o(1))n.$$

Conjecture 4.2. *Let $k > 2$ be a fixed integer. For every $\delta > 0$, there exist $\varepsilon = \varepsilon(\delta) > 0$ and $n_0 = n_0(\delta)$ such that if $n > n_0$, $l \geqslant r\left(B_n^{(k)}\right) - \varepsilon n$, and $G = G(l)$ satisfies*
$$B_n^{(k)} \not\subseteq G, \qquad B_n^{(k)} \not\subseteq \overline{G},$$

then

$$\left| e(G[X]) - \frac{1}{4}|X|^2 \right| < \delta l^2,$$

for every $X \subset V(G)$.

Conjecture 4.3. *For every fixed integer* $k > 2$, *there exists a constant* $c = c(k) > 0$ *such that every red–blue edge colouring of* $K_{r(B_n^{(k)})}$ *contains a monochromatic* $B_n^{(k)}$ *whose vertices span at least* $\lfloor cn^2 \rfloor$ *edges of the same colour as the book.*

Acknowledgement

We are grateful to the referee for valuable suggestions.

References

[1] Bollobás, B. (1998) *Modern Graph Theory*, Vol. 184 of *Graduate Texts in Mathematics*, Springer, New York.
[2] Chung, F., Graham, R. and Wilson, R. M. (1989) Quasi-random graphs. *Combinatorica* **9** 345–362.
[3] Erdős, P., Faudree, R., Rousseau, C. C. and Schelp, R. H. (1978) The size Ramsey number. *Period. Math. Hungar.* **9** 145–161.
[4] Faudree, R., Rousseau, C. C. and Sheehan, J. (1982) Strongly regular graphs and finite Ramsey theory. *Linear Algebra Appl.* **46** 221–241.
[5] Lorden, G. (1962) Blue-empty chromatic graphs. *Amer. Math. Monthly* **69** 114–120.
[6] Nikiforov, V. and Rousseau, C. C. (2005) A note on Ramsey numbers for books. *J. Graph Theory* **49** 168–176.
[7] Nordhaus, E. and Stewart, B. (1963) Triangles in an ordinary graph. *Canad. J. Math.* **15** 33–41.
[8] Rousseau, C. C. and Sheehan, J. (1978) On Ramsey numbers for books. *J. Graph Theory* **2** 77–87.
[9] Thomason, A. (1982) On finite Ramsey numbers. *European J. Combin.* **3** 263–273.
[10] Thomason, A. (1987) Random graphs, strongly regular graphs and pseudorandom graphs. *Surveys in Combinatorics 1987*, Vol. 123 of *London Math. Soc. Lecture Note Ser.*, Cambridge University Press, Cambridge, pp. 173–195.
[11] Thomason, A. (1989) Dense expanders and pseudo-random bipartite graphs. *Discrete Math.* **75** 381–386.

Homomorphism and Dimension

PATRICE OSSONA de MENDEZ and PIERRE ROSENSTIEHL

Centre d'Analyse et de Mathématiques Sociales,
CNRS, UMR 8557, 54 Bd Raspail, 75006 Paris, France
(e-mail: pom@ehess.fr, pr@ehess.fr)

Received 28 October 2003; revised 28 November 2003

For Béla Bollobás on his 60th birthday

The dimension of a graph, that is, the dimension of its incidence poset, has become a major bridge between posets and graphs. Although allowing a nice characterization of planarity, this dimension behaves badly with respect to homomorphisms.

We introduce the *universal dimension* of a graph G as the maximum dimension of a graph having a homomorphism to G. The universal dimension, which is clearly homomorphism monotone, is related to the existence of some balanced bicolouration of the vertices with respect to some realizer.

Nontrivial new results related to the original graph dimension are subsequently deduced from our study of universal dimension, including chromatic properties, extremal properties and a disproof of two conjectures of Felsner and Trotter.

1. Introduction

Half a century ago, from the concept of dimension of a partial order [9], Dushnik started an investigation of the poset of one- and k-element subsets of $[n] = \{1,\ldots,n\}$ [8], and introduced a construction which has a natural interpretation in terms of abstract simplicial complexes, and which has been used or re-introduced in several contexts, such as mathematical economy [18], integer programming [2], and commutative algebra [3].

The dimension of the poset of one- and two-element subsets of $[n]$ (the complete graph incidence poset) has, in particular, been investigated by Spencer [20], Füredi, Hajnal, Rödl and Trotter [12], and more recently by Hoşten and Morris [13].

The study of the dimension of the incidence poset of graphs (other than complete ones) have attracted particular attention in the past few years, since W. Schnyder proved that the incidence poset of a graph has dimension at most 3 if and only if the graph is planar [19]. This theorem has been generalized by Brightwell and Trotter [6, 7] to a characterization of the dimension of the vertex, edge and face posets of planar graphs, and to a generalization to multigraphs.

Although the generalization of Schnyder's result to the problem of the geometric realization of an abstract simplicial complex is quite natural [16] and may be viewed as the existence of embeddings of posets in Euclidean space with a general separation property [17], the property of the class of planar graphs to be minor closed disappears in dimension at least 4. The incidence poset dimension is not homomorphism monotone (any nonplanar graph with a homomorphism to K_4 has dimension 4, although K_4 has dimension 3). In this paper we introduce a new graph dimension, the *universal dimension* of a graph, which will make this connection clear. We will prove that this universal dimension differs by at most two from the incidence poset dimension, while being (by its definition) homomorphism monotone.

We will relate the universal dimension of a graph to the existence of realizers admitting a special bicolouration and will give some applications of our results, such as the non-bounding of chromatic numbers of graphs of dimension 4, or extremal results on graphs of incidence poset dimension d, in the framework investigated by Bollobás and Thomason [5]. We also disprove two conjectures of Felsner and Trotter.

2. Definitions and notation

In the following we will only consider finite simple loopless graphs with at least one edge.

Let $P = (X, \leqslant)$ be a finite poset. A *realizer* of P is a nonempty family $\mathscr{R} = (<_1, \ldots, <_t)$ of linear orders on X whose intersection $\bigcap_{i=1}^{t} <_i$ is P. The *dimension* of P is the minimum cardinality of a realizer of P [9].

Homomorphisms of graphs are adjacency-preserving maps: a map $f : V(H) \to V(G)$ is a homomorphism of the graph H to the graph G if $\{f(x), f(y)\} \in E(G)$ whenever $\{x, y\} \in E(H)$. We will denote by $H \to G$ the existence of a homomorphism of H to G. The *chromatic number* $\chi(H)$ of a graph H is thus the smallest integer n, such that H has a homomorphism to K_n.

Following the definition of [14], the *multiplication* $G^{\cdot\cdot}(W_1, \ldots, W_n)$ of a graph G with vertex set $V(G) = \{v_1, \ldots, v_n\}$ is defined by

$$V(G^{\cdot\cdot}) = W_1 \cup \cdots \cup W_n,$$

$$\forall 1 \leqslant i \leqslant n, \quad |W_i| \geqslant 1$$

$$\forall 1 \leqslant i < j \leqslant n, \quad W_i \cap W_j = \emptyset,$$

$$\forall 1 \leqslant i \leqslant j \leqslant n, \forall u \in W_i, \forall v \in W_j, \quad \{u, v\} \in E(G^{\cdot\cdot}) \text{ if and only if } \{v_i, v_j\} \in E(G)$$

The sets $W_1, \ldots, W_n$ are called the *multivertices* corresponding to vertices $v_1, \ldots, v_n$, respectively.

A *property* of graphs $\mathscr{P}$ is an (infinite) class of labelled graphs closed under isomorphism. A property $\mathscr{P}$ is *monotone* if it is closed under taking subgraphs; it is *hereditary* if it is closed under taking induced subgraphs; it is *additive* if it is closed under disjoint unions of graphs. The *n-level* $\mathscr{P}^n$ of a property $\mathscr{P}$ is the set of graphs in $\mathscr{P}$ with vertex set $[n]$. The *speed* of $\mathscr{P}$, $|\mathscr{P}^n|$ is the number of graphs in the property on n vertices. The *size* of a monotone property $\mathscr{P}$ at level n is the maximum number of edges in a graph of $\mathscr{P}^n$, that is $e_{\mathscr{P}}(n) = \max\{|E(G)|, G \in \mathscr{P}^n\}$.

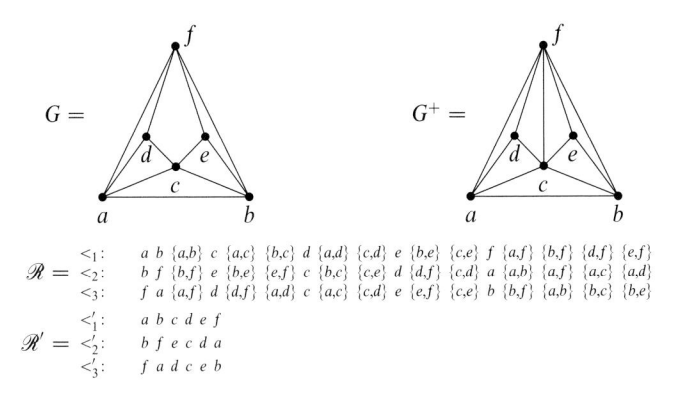

$$
\mathcal{R} = \begin{matrix} <_1: \\ <_2: \\ <_3: \end{matrix} \quad \begin{matrix} a\ b\ \{a,b\}\ c\ \{a,c\}\ \{b,c\}\ d\ \{a,d\}\ \{c,d\}\ e\ \{b,e\}\ \{c,e\}\ f\ \{a,f\}\ \{b,f\}\ \{d,f\}\ \{e,f\} \\ b\ f\ \{b,f\}\ e\ \{b,e\}\ \{e,f\}\ c\ \{b,c\}\ \{c,e\}\ d\ \{d,f\}\ \{c,d\}\ a\ \{a,b\}\ \{a,f\}\ \{a,c\}\ \{a,d\} \\ f\ a\ \{a,f\}\ d\ \{d,f\}\ \{a,d\}\ c\ \{a,c\}\ \{c,d\}\ e\ \{e,f\}\ \{c,e\}\ b\ \{b,f\}\ \{a,b\}\ \{b,c\}\ \{b,e\} \end{matrix}
$$

$$
\mathcal{R}' = \begin{matrix} <'_1: \\ <'_2: \\ <'_3: \end{matrix} \quad \begin{matrix} a\ b\ c\ d\ e\ f \\ b\ f\ e\ c\ d\ a \\ f\ a\ d\ c\ e\ b \end{matrix}
$$

Figure 1. A graph G, a realizer $\mathcal{R}$ of P_G, a graph realizer $\mathcal{R}'$ of both G and G^+.

Definition 1. Let G be a graph. The *incidence poset* P_G of G is the height 2 poset having $V(G) \cup E(G)$ as its ground set, such that $a < b$ in P_G if $a \in V(G)$, $b \in E(G)$ and b is incident to a.

Definition 2. Let G be a graph. A *graph realizer* of G of size t is a family $\mathcal{R} = (<_1, \ldots, <_t)$ of linear orders on $V(G)$, such that:[1]

$$\forall x \neq y \in V(G), \quad \exists i \in [t], \quad x <_i y$$
$$\forall \{x,y\} \in E(G), \forall z \in V(G) \setminus \{x,y\}, \quad \exists i \in [t], \quad x <_i z \text{ and } y <_i z$$

The *dimension* $\dim G$ of G is the smallest size of a graph realizer of G.

As already noticed in [8] (in a more general framework), a graph realizer of G may be obtained from a realizer of P_G by reducing the ground set to $V(G)$ and, conversely, a realizer of P_G may be obtained from a graph realizer by inserting the edges of G in lexicographic order. Hence $\dim G = \dim P_G$. Notice that a graph G cannot be reconstructed from one of its graph realizers, although it can be reconstructed from one realizer of P_G (see Figure 1).

Remark 1. Let G be a graph and let H be a subgraph of G. Then, the restriction of the linear orders of a graph realizer of G to $V(H)$ defines a graph realizer of H. Thus:

$$H \subseteq G \Longrightarrow \dim H \leqslant \dim G. \tag{2.1}$$

Hence, the property ($\dim \leqslant d$) of graphs with dimension at most d is monotone. It is also obviously additive.

---

[1] The first condition is missing in the definition of the dimension of a graph which appears in [11], causing the dimension of a graph to be possibly different from the dimension of its incidence poset.

3. The universal dimension of a graph

We shall introduce a new dimension associated to a graph, which will be defined in order to be homomorphism monotone.

Definition 3. Let G be a graph. The *universal dimension* $\operatorname{udim} G$ is the supremum of the dimensions of the graphs having a homomorphism to G:

$$\operatorname{udim} G = \sup_{H \to G} \dim H \tag{3.1}$$

Remark 2. Thus, we have this extension of (2.1) for universal dimension:

$$H \to G \implies \operatorname{udim}(H) \leqslant \operatorname{udim}(G). \tag{3.2}$$

Obviously, the property ($\operatorname{udim} \leqslant d$) of graphs with universal dimension at most d is monotone and additive.

Remark 3. Obviously, $\operatorname{udim} K_1 = 1$. However, for any graph G with at least one edge, $K_{3,3} \to G$. According to Schnyder's theorem, $\dim K_{3,3} > 3$ as $K_{3,3}$ is not planar. Hence no graph has universal dimension 2 or 3, and a graph has universal dimension at least 4 if and only if it has at least one edge.

We shall introduce the concept of bicolourable graph realizer, motivated by Theorem 3.3, that will assert that the universal dimension G will be the smallest size of a bicolourable graph realizer of G (similar to the definition of the dimension of G as the smallest size of a graph realizer of G).

Definition 4. Let $\mathscr{R} = \{<_1, \ldots, <_t\}$ be a graph realizer of a graph G. A *balanced bicolouration* of G with respect to $\mathscr{R}$ is a map $\Gamma : V(G) \times [t] \to \{0, 1\}$ so that, for every $x \in V(G)$ and for every vertex y adjacent to x (in G), there exists $i, j \in [t]$ so that

$$\begin{cases} y <_i x \text{ and } \Gamma(x, i) = 0, \\ y <_j x \text{ and } \Gamma(x, j) = 1. \end{cases} \tag{3.3}$$

When such a map exists, $\mathscr{R}$ is said to be a *bicolourable graph realizer* of G.

Remark 4. Notice that any graph G has a bicolourable graph realizer of size ($\dim G + 2$) obtained from a graph realizer of G by adding two dual linear orders.

Lemma 3.1. *Let G be a graph of order n, let $d > 3$ be an integer, let $a_1, \ldots, a_n$ be positive integers, and let $G^+ = G^{\cdot\cdot}(W_1, \ldots, W_n)$ be a multiplication of G, where (for $1 \leqslant p \leqslant n$)*

$$|W_p| > (a_p - 1) \left(\sum_{i=1}^{p-1} a_i + 1 \right)^d.$$

Then, if $\mathcal{R} = (<_1, \ldots, <_d)$ is a graph realizer of G^+, there exist $W'_1 \subseteq W_1, \ldots, W'_n \subseteq W_n$ such that

- for any $1 \leq i \leq n$, $|W'_i| = a_i$,
- for any $1 \leq i < j \leq n$, no two elements of W'_j are separated by an element of W'_i in some linear order in $\mathcal{R}$.

Proof. Consider the following assertion for $1 \leq p \leq n$.

Π_p: there exists $W'_1 \subseteq W_1, \ldots, W'_p \subseteq W_p$, such that

- for any $1 \leq i \leq p$, $|W'_i| = a_i$,
- for any $1 \leq i < j \leq p$, no two elements of W'_j are separated by an element of W'_i in some linear order in $\mathcal{R}$.

We shall prove by induction that Π_p holds for any $1 \leq p \leq n$. Obviously, Π_1 holds ($W'_1 = W_1$).

Assume Π_p holds for some $1 \leq p < n$. Then, the elements of $W'_1 \cup \cdots \cup W'_p$ determine at most $s_p = \left(\sum_{i=1}^{p} a_i\right) + 1$ intervals in each of the linear orders of $\mathcal{R}$. Colour any vertex x in W_{p+1} by the d-tuple corresponding to the indexes of the intervals in which x appear in $<_1, \ldots, <_d$. As $|W_{p+1}| > (a_{p+1} - 1)s_p^d$, a_{p+1} elements of W_{p+1} are coloured the same. Choosing these a_{p+1} elements for W'_{p+1}, the assertion Π_{p+1} clearly holds. $\square$

Lemma 3.2. *Let G be a graph of order n, let $d > 3$ be an integer, and let $G^+ = G^{::}(W_1, \ldots, W_n)$ be the multiplication of G where $|W_p| = (2n)^{d(d+2)}$ (for $1 \leq p \leq n$).*

If $\mathcal{R} = (<_1, \ldots, <_d)$ is a graph realizer of G^+, then there exists $W'_1 \subseteq W_1, \ldots, W'_n \subseteq W_n$ such that, for any $1 \leq i \leq n$:

- *$|W'_i| = 2$,*
- *the elements of W'_i are consecutive in each of the restrictions of the linear orders in $\mathcal{R}$ to $W'_1 \cup \cdots \cup W'_n$.*

Proof. According to Lemma 3.1, there exists $W''_1 \subseteq W_1, \ldots, W''_n \subseteq W_n$, such that

- for any $1 \leq i \leq n$, $|W''_i| = (2i-1)^d + 1$,
- for any $1 \leq i < j \leq n$, no two elements of W''_j are separated by an element of W''_i in some linear order in $\mathcal{R}$.

According to Lemma 3.1 again (considering the indexes of the W''_i in reverse order), there exists $W'_n \subseteq W''_n, \ldots, W'_1 \subseteq W''_1$, such that

- for any $1 \leq i \leq n$, $|W'_i| = 2$,
- for any $1 \leq i < j \leq n$, no two elements of W'_i are separated by an element of W'_j in some linear order in $\mathcal{R}$.

Altogether, the elements of W'_i form an interval of each of the restrictions of the linear order in $\mathcal{R}$ to $W'_1 \cup \cdots \cup W'_n$. $\square$

Theorem 3.3. *Let G be a graph of order n, let $d > 3$ be an integer, and let $G^+ = G^{\cdot\cdot}(W_1,\ldots,W_n)$ be the multiplication of G, where $|W_i| = (2n)^{d(d+2)}$. Then, the following assertions are equivalent:*

(1) *G has a bicolourable graph realizer of size d,*
(2) *$\operatorname{udim} G \leq d$,*
(3) *$\dim G^+ \leq d$,*

Thus, the universal dimension $\operatorname{udim} G$ of a graph G is the smallest size of a bicolourable graph realizer of G.

Proof. Assume $\mathscr{R} = (<_1,\ldots,<_d)$ is a bicolourable graph realizer of G, with balanced bicolouration Γ. If $H \to G$, then there exists a multiplication $H^+ = G^{\cdot\cdot}(W'_1,\ldots,W'_n)$ of G such that $H \subseteq H^+$. Let $V(G) = \{v_1,\ldots,v_n\}$ and let $W'_i = \{(v_i,1),\ldots,(v_i,k_i)\}$ for $1 \leq i \leq n$. In each linear order $<_j$ of $\mathscr{R}$, replace the vertex v_i by the sequence $(v_i,1),\ldots,(v_i,k_i)$ if $\Gamma(v_i,j) = 0$, or by the sequence $(v_i,k_i),\ldots,(v_1,1)$ if $\Gamma(v_i,j) = 1$. One easily checks that the so-obtained d linear orders realize the graph H^+ and thus they also realize H. Hence, (1) implies (2).

As $G^+ \to G$, we have $\dim G^+ \leq \operatorname{udim} G$. Thus, (2) implies (3).

Assume $\dim G^+ \leq d$. Let $\mathscr{R} = (<_1,\ldots,<_d)$ be a graph realizer of G^+. According to Lemma 3.2, there exists $W'_1 \subseteq W_1,\ldots,W'_n \subseteq W_n$, such that, for any $1 \leq i \leq n$, W'_i is a two-element set $\{x_i,y_i\}$, and such that x_i and y_i are consecutive in each of the restriction of the linear orders in $\mathscr{R}$ to $W'_1 \cup \cdots \cup W'_n$.

Let $\mathscr{R}' = (<'_1,\ldots,<'_d)$, where $<'_i$ is the restriction of $<_i$ to $\{x_1,\ldots,x_n\}$. According to Remark 1, $\mathscr{R}'$ is a graph realizer of G. Define $\Gamma(x_i,k)$ as 0 if $x_i <_k y_i$, and 1 otherwise. Let x_i be any vertex of G and let x_j be a neighbour of x_i in G. Then x_j is adjacent to both x_i and y_i in G^+. Hence, as x_i and y_i are always consecutive in the restriction of the linear orders to $W'_1 \cup \cdots \cup W'_n$, there exists $\alpha, \beta \in [d]$, such that $x_j <_\alpha x_i <_\alpha y_i$ and $x_j <_\beta y_i <_\beta x_i$. Thus, Γ is a balanced bicolouration of G with respect to $\mathscr{R}'$. Hence (3) implies (1). □

Corollary 3.4. *Let $1 < k < d$ be integers. Assume G has a graph realizer $(<_1,\ldots,<_d)$ such that two adjacent vertices are not comparable in $\bigcap_{i=1}^{k} <_i$ and not comparable in $\bigcap_{i=k+1}^{d} <_i$. Then $(<_1,\ldots,<_d)$ is bicolourable and thus $\operatorname{udim} G \leq d$.*

Proof. Define $\Gamma(x,i) = 0$ if $i \leq k$, and 1 otherwise. Then, Γ is a balanced bicolouration of G with respect to $(<_1,\ldots,<_d)$. □

Corollary 3.5. *For any graph G, $\dim G \leq \operatorname{udim} G \leq \dim G + 2$*

Proof. Let $(<_1,\ldots,<_d)$ be a graph realizer of G and $<$ any linear order on $V(G)$. Then $(<_1,\ldots,<_d,<,\overline{<})$ is a graph realizer of G to which Corollary 3.4 applies. □

We shall now bound the universal dimension of the complete graph of order t in terms of graph dimension of complete graphs.

Theorem 3.6. *Let t be a positive integer. Then,*

$$\dim K_{2t+\mathrm{udim}\,K_t} \leqslant \mathrm{udim}\,K_t \leqslant \dim K_{t+1} + 1. \tag{3.4}$$

Proof. Let $d = \mathrm{udim}\,K_t$. In order to prove $\dim K_{2t+d} \leqslant \mathrm{udim}\,K_t$, we may assume that t is maximal, namely that $\mathrm{udim}\,K_{t+1} > d$. Consider a graph realizer $(<_1, \ldots, <_d)$ of K_t with a balanced bicolouration Γ. Assume some vertex x of K_t is the minimum of at least two linear orders in $\mathscr{R}$, for instance $<_1$ and $<_2$. As the value of $\Gamma(x, 1)$ and $\Gamma(x, 2)$ is free, we may impose $\Gamma(x, 1) = 0$ and $\Gamma(x, 2) = 1$. Then, consider the complete graph on $t + 1$ vertices obtained by adding a vertex v. In the linear orders $<_i$ ($3 \leqslant i \leqslant d$), replace x by xv if $\Gamma(x, i) = 0$ and by vx, otherwise. In $<_1$, replace x by vx and in $<_2$ replace x by xv. Extend Γ by setting $\Gamma(v, i) = \Gamma(x, i)$ for $1 \leqslant i \leqslant d$. Then one easily checks that we have obtained a bicolourable realizer of K_{t+1} of size d, contradicting the maximality assumption on t.

Hence, no vertex of K_t is minimal in more than one linear order. In each of the $<_i$, replace each vertex x with the sequence

- $(x, 1)(x, 2)$, if $\Gamma(x, i) = 0$ and x is minimal in no linear order,
- $(x, 2)(x, 1)$, if $\Gamma(x, i) = 1$ and x is minimal in no linear order,
- $(x, 1)(x, 2)(x, 3)$, if $\Gamma(x, i) = 0$ and x is minimal in a linear order $<_j$ with $j \neq i$,
- $(x, 3)(x, 2)(x, 1)$, if $\Gamma(x, i) = 1$ and x is minimal in a linear order $<_j$ with $j \neq i$,
- $(x, 1)(x, 3)(x, 2)$, if x is minimal in $<_i$.

The so-obtained set of d linear orders is a graph realizer of K_{2t+d}. Thus, $\mathrm{udim}\,K_t \geqslant \dim K_{2t+d}$.

Let $<_1, \ldots, <_d$ be a graph realizer of K_{t+1}, and let z be the maximum of $<_1$. Let $<'_i$ be the restriction of $<_i$ to $V(K_{t+1}) - z$, and let $<'_{d+1}$ be the dual of $<'_1$. Colour the vertices 0 in $<'_1$ and $<'_{d+1}$, and 1 in all the other linear orders. Then, for any $x <'_1 y$, we have $y <'_{d+1} x$, so that each is less than the other coloured 0. Moreover, there exists $1 < i \leqslant d$ such that $y <'_i x$ (as $<_1 \cdots <_d$ is an antichain). So y is smaller than a x coloured 1. Assume x is not smaller than a y coloured 1. Then, y is never above $\{x, z\}$, contradicting the fact that $\{x, z\}$ is an edge. Thus, we have defined a bicolourable graph realizer of K_t of size $d + 1$. Hence, $\mathrm{udim}\,K_t \leqslant \dim K_{t+1} + 1$. □

Remark 5. Let $p \geqslant 3$. If $\dim K_{p+1} > \dim K_p$, then $\dim K_{2(p-1)} = \mathrm{udim}\,K_{p-1} = \dim K_p + 1$.

The following strengthening of the upper bound of $\mathrm{udim}\,K_t$ has been proposed by the referee of this paper.

Proposition 3.7. *Let t be a positive integer. Then*

$$\mathrm{udim}\,K_t \leqslant \min\{r : D(r - 2) \geqslant t + 2\}, \tag{3.5}$$

where $D(n)$ is the nth Dedekind number. □

Proof. In [11], the authors prove that K_t has a realizer of type $(<_1, \overline{<_1}, <_3, \ldots, <_d)$ if and only if $D(d - 2) \geqslant t$. Let $<_2 = \overline{<_1}$. Label the vertices of K_t with $1, \ldots, t$, according to the linear order $<_1$. Let $1 < x < y < t$. As $\{1, y\}$ is an edge of K_t, there exists $1 \leqslant i_x \leqslant d$, such that x is greater than 1 and y with respect to $<_{i_x}$. Hence $i_x > 2$. Similarly, as $\{x, n\}$ is an edge of K_t, there exists $2 < i_y \leqslant t$ such that y is greater than n and x with respect to $<_{i_y}$, and we have

$$\begin{cases} x <_1 y \text{ and } \Gamma(y, 1) = 0, \\ x <_{i_y} y \text{ and } \Gamma(y, i_y) = 1, \\ y <_2 x \text{ and } \Gamma(x, 2) = 0, \\ y <_{i_x} x \text{ and } \Gamma(x, i_x) = 1. \end{cases}$$

The restriction of $(<_1, \overline{<_1}, <_3, \ldots, <_d)$ to $\{2, \ldots, t-1\}$ thus defines with Γ a bicoloured realizer of K_{t-2}. □

The last part proof of Theorem 3.6 actually shows that every graph G has a vertex x such that $\text{udim}(G - x) \leqslant \dim G + 1$. One may ask if such a deletion is actually necessary.

Problem 1. Let G be a graph with dimension at least 3. Is it true that $\text{udim}\, G \leqslant \dim G + 1$?

Notice that the restriction $\dim G > 2$ is necessary, as $\dim K_2 = 2$ and $\text{udim}\, K_2 = 4$. Moreover, according to the Schnyder's theorem and the 4-colour theorem, any graph of dimension 3 is planar and has universal dimension at most 4 (as $\text{udim}\, K_4 = 4$). Thus, the answer is 'yes' for the special case where $\dim G = 3$.

4. Applications

4.1. Chromatic number

We shall now prove that graphs with universal dimension 4 have no bounded chromatic number. For this we will make use of the following Theorem of Kříž and Nešetřil.

Theorem 4.1 ([15]). *For each $k > 0$, there exists a finite set X and linear orderings $<_1, <_2$ on X such that the Hasse diagram H of $<_1<_2$ has chromatic number at least k.*

Lemma 4.2. *Let $k \geqslant 2$ be an integer, X a finite set and $<_1, \ldots, <_k$ linear orders on X. For $I \subseteq [k]$, denote P_I the partial order $(\bigcap_{i \in I} <_i) \cap (\bigcap_{i \in [k] \setminus I} \overline{<_i})$, and $H(P_I)$ its Hasse diagram. Then, the graph G with vertex set X and edge set $\bigcup_{I \subseteq [k]} E(H(P_I))$ has $(<_1, \ldots, <_k, \overline{<_k}, \ldots, \overline{<_1})$ as graph realizer, and thus has universal dimension at most $2k$.*

Proof. Define $<_{2k+1-i} = \overline{<_i}$, for $1 \leqslant i \leqslant k$ and $\mathcal{R} = (<_1, \ldots, <_{2k})$. Let $I \subseteq [k]$, let $\{x, y\}$ be an edge of $H(P_I)$ and let $z \notin \{x, y\}$. Without loss of generality we may assume that x is smaller than y with respect to P_I. Assume z is not greater than both x and y with respect to some $<_i$ ($1 \leqslant i \leqslant 2k$). Then, z is also never smaller than both x and y with respect to some $<_i$ ($1 \leqslant i \leqslant 2k$), as it would be greater than both x and y with respect to $<_{2k+1-i}$. Thus z lies between x and y in all the $<_i$. Hence $x < z < y$ with respect to P_I, contradicting $\{x,y\} \in H(P_I)$. We conclude that z is greater than x and y in some $<_i$, and thus that $\mathscr{R}$ is a graph realizer of G.

This graph realizer may be rewritten as $(<_1, \overline{<_1}, <_2, \ldots, <_k, \overline{<_2}, \ldots, \overline{<_k})$. As two vertices are not comparable in $<_1 \cap \overline{<_1}$ and not comparable in $(\bigcap_{i=2}^k <_i) \cap (\bigcap_{i=2}^k \overline{<_i})$, Corollary 3.4 applies and thus $\operatorname{udim} G \leqslant 2k$. □

Corollary 4.3. *Let $P = (X, \leqslant)$ be a poset of dimension at least 2 and let $H(P)$ be its Hasse diagram. Then $H(P)$ has a graph realizer of the form $(<_1, \ldots, <_{\dim P}, \overline{<_1}, \ldots, \overline{<_{\dim P}})$ and thus $\operatorname{udim}(H(P)) \leqslant 2 \dim P$.*

Theorem 4.4. *For any integer $k \geqslant 2$, there exists a graph G_k with chromatic number at least k, universal dimension 4, and having a realizer of the form $(<_1, <_2, \overline{<_1}, \overline{<_2})$.*

Proof. The existence of a graph G_k with chromatic number at least k, universal dimension at most 4, and having a realizer of the form $(<_1, <_2, \overline{<_1}, \overline{<_2})$ is a direct consequence of Theorem 4.1 and Corollary 4.3. According to Remark 3, G_k has universal dimension at least 4 and thus universal dimension exactly 4. □

Corollary 4.5. *There exists no integer k, such that any graph G of dimension 4 having a realizer of the form $(<_1, <_2, \overline{<_1}, \overline{<_2})$ has a vertex partition into at most k sets $V(G) = V_1 \cup \cdots \cup V_k$, each inducing a planar graph.*

Proof. Assume $V(G)$ has a partition in k sets, each inducing a planar graph. Then $\chi(G) \leqslant 4k$. As graphs of dimension 4 having a realizer of the form $(<_1, <_2, \overline{<_1}, \overline{<_2})$ have no bounded chromatic number, the result follows. □

Felsner and Trotter proposed the following two conjectures, which are obviously disproved by Corollary 4.5.

Conjecture 4.6 ([11]). *Let $G = (V, E)$ be a graph. Then $\dim G \leqslant 4$ if and only if the vertex set V can be partitioned into four parts so that each part induces an outerplanar graph.*

Conjecture 4.7 ([11]). *Let $G = (V, E)$ be a graph. Then $\dim G \leqslant 4$ and G has a realizer of the form $(<_1, <_2, <_3, \overline{<_3})$ if and only if the vertex set V can be partitioned into two parts so that each part induces an outerplanar graph.*

Problem 2. Does there exist a graph G of universal dimension 4 having no graph realizer of the form $(<_1, <_2, \overline{<_1}, \overline{<_2})$?

4.2. Extremal graphs

An (r,s)-*colouring* of a graph G is a colouring of the vertices of G such that each of the first s ones induces a complete graphs and each of the $r - s$ remaining ones induces an empty graph. The *colouring number* $r(\mathscr{P})$ is the maximal r for which there exists an s, $0 \leqslant s \leqslant r$, such that every (r,s)-colourable graph has property $\mathscr{P}$ (see [4]).

The speed $|(\dim \leqslant d)^n|$ of the property $(\dim \leqslant d)$ is the number of graphs of order n having dimension at most d. These numbers are related to the colouring number of the property $(\dim \leqslant d)$ by the following theorem.

Theorem 4.8 (Bollobás and Thomason [5]). *Let $\mathscr{P}$ be a hereditary property of graphs and let $\mathscr{P}^n$ be the set of graphs in $\mathscr{P}$ with vertex set $[n]$. Then*
$$|\mathscr{P}^n| = 2^{(1-1/r+o(1))n^2/2},$$
where $r = r(\mathscr{P})$ is the colouring number of $\mathscr{P}$.

The colouring number of the property $(\dim \leqslant d)$ (and hence its speed) is given by the following simple formula.

Theorem 4.9. *Let $d \geqslant 4$ be an integer, and let t be the greatest integer such that $\operatorname{udim} K_t = d$. Then the colouring number of the property $(\dim \leqslant d)$ is*
$$r(\dim \leqslant d) = t.$$
Thus, the number of labelled graphs of order n having dimension at most d is $2^{(1-1/t+o(1))n^2/2}$.

Proof. Any $(t, 0)$-colourable graph is t-chromatic and thus has dimension at most d (as $\operatorname{udim} K_t = d$). Hence, $r(\dim \leqslant d) \geqslant t$.

There exists a $(t+1)$-chromatic graph with dimension strictly greater than d, as $\operatorname{udim} K_{t+1} > d$. Thus, there exists a $(t+1, 0)$ colourable graph with universal dimension strictly greater than d. Moreover, as complete graphs are $(t+1, s)$-colourable for any $s \geqslant 1$ but have unbounded dimension, we deduce $r(\dim \leqslant d) < t + 1$. □

The size $e_{\dim \leqslant d}(n)$ (resp. $e_{\operatorname{udim} \leqslant d}(n)$) of the property $(\dim \leqslant d)$ (resp. $(\operatorname{udim} \leqslant d)$) is the maximum number of edges a graph G can have when it has n vertices and dimension at most d (resp. universal dimension at most d). It is proved in [1] that $\lim_{p\to\infty} e_{\dim \leqslant 4}(n)/n^2 = 3/8$.

We shall show that $e_{\dim \leqslant d}(n)$ is closely related to the universal dimension of complete graphs.

Theorem 4.10. *Let $d \geqslant 4$ be an integer, and let t be the greatest integer such that $\operatorname{udim} K_t = d$. Then:*
$$\left(1 - \frac{1}{t}\right)\binom{n}{2} + O(n) \leqslant e_{\operatorname{udim} \leqslant d}(n) \leqslant e_{\dim \leqslant d}(n) \leqslant \left(1 - \frac{1}{t}\right)\binom{n}{2} + o(n^2).$$

Proof. As $\operatorname{udim}(K_{t+1}) > d$, the graph $G = K_{t+1}^+$, defined in Theorem 3.3, is such that $\dim G > d$.

As G is $(t+1)$-chromatic, the maximum number of edges possessed by a graph of order n not having G as a subgraph is, according to the Erdős–Stone theorem [10],
$$\operatorname{ex}(n; G) = \left(1 - \frac{1}{t}\right)\frac{p^2}{2} + o(n^2).$$

Table 1. Order of complete graphs which are extremal for dimension and universal dimension, respectively.

t	$\max\{n : \dim K_n \leqslant t\}$	$\max\{n : \operatorname{udim} K_n \leqslant t\}$
4	12	4
5	81	between 30 and 38
6	2646	between 1088 and 1320
7	1422564	between 583507 and 711273

Hence, $e_{\dim \leqslant d}(n) \leqslant \operatorname{ex}(n; G)$. As $\dim G \leqslant \operatorname{udim} G$, $e_{\operatorname{udim} \leqslant d}(n) \leqslant e_{\dim \leqslant d}(n)$.

Moreover, the complete balanced t-partite graph is t-colourable and has universal dimension at most d. This graph has $\left(1 - \frac{1}{t}\right)\binom{n}{2} + O(n)$ edges, thus concluding the proof. □

The upper bounds appearing in Table 1 result from Theorem 3.6.

Using a technique similar to that of [13] and selecting pairs of consecutive antichains (in lexicographic order) which form a cover, we build a realizer of K_{2n}, where vertices $2k - 1$ and $2k$ ($1 \leqslant k \leqslant n$) are consecutive in each of the linear orders. This enumeration technique allows us to build a bicoloured graph realizer of K_n, leading to the lower bounds shown in Table 1 (notice that the validity of the computed bicoloured graph realizer has been checked by computer). For instance, as evidence for $\operatorname{udim} K_{30} = 5$, we reproduce here the computed 5-balanced bicolouration of a graph realizer of K_{30}:

```
<₁:  17 15 16  5  1  2  3  4  7  6 12  8  9 10 11 14 13 21 22 18 19 20 28 27 25 23 24 26 29 30
<₂:  20 18 19 21 22  6  7  2  1  4  3  5 13 14  9  8 11 10 12 15 16 17 30 29 26 24 23 25 27 28
<₃:  26 25 23 24 28 27 30 29  7  6  5  3  4  1  2 14 13 12 10 11  8  9 17 16 15 20 19 18 22 21
<₄:  29 30 27 28 23 24 25 26 21 22 18 19 20 15 16 17  8  9 10 11 12 13 14  1  2  3  4  5  6  7
<₅:  30 29 28 27 26 25 24 23 22 21 20 19 18 17 16 15 14 13 12 11 10  9  8  7  6  5  4  3  2  1
```

5. Conclusion

The universal dimension of a graph appears to be a nontrivial connection between poset dimensions and graph homomorphisms. In particular, for $d \geqslant 4$, no finite family $\mathscr{F}$ of finite graphs exists for a characterization like $G \to H \in \mathscr{F} \iff \operatorname{udim} G \leqslant d$. However, it is possible that such a characterization could exist for a fixed chromatic number.

Problem 3. For any $k \geqslant 2$ and any $d \geqslant 4$, does there exist a finite k-chromatic graph $U_{d,k}$ of universal dimension d such that any k-chromatic graph of universal dimension d has a homomorphism to $U_{d,k}$?

An additive hereditary property is *reducible* if there exists two hereditary properties $\mathscr{P}_1$ and $\mathscr{P}_2$, such that a graph G belongs to $\mathscr{P}$ if and only if there exists a partition (V_1, V_2) of G, so that $G[V_1] \in \mathscr{P}_1$ and $G[V_2] \in \mathscr{P}_2$ (see [14]). The question of determining whether or not ($\operatorname{udim} \leqslant d$) is reducible leads to the following question.

Problem 4. Let G be an extremal graph of universal dimension d (i.e., udim $G = d$, but the addition of any edge to G increases the universal dimension). Then, is the complementary graph of G disconnected?

Acknowledgements

The authors would like to thank the referee for his comments and for the improvement of the second inequality of Theorem 3.6, which is stated in Proposition 3.7.

References

[1] Agnarsson, G., Felsner, S. and Trotter, W. T. (1999) The maximum number of edges in a graph of bounded dimension, with application to ring theory. *Discrete Math.* **201** 5–19.

[2] Barany, L., Howe, R. and Scarf, H. (1994) The complex of maximal lattice free simplices. *Math. Program.* (Ser. A) **66** 273–281.

[3] Bayer, D., Peeva, I. and Sturmfels, B. (1998) Monomial resolutions. *Math. Res. Lett.* 31–46.

[4] Bollobás, B. (1995) *Extremal Graph Theory*, Handbook of Combinatorics, Vol. 2 (R. L. Graham, M. Grötschel, and L. Lovász, eds), Elsevier.

[5] Bollobás, B. and Thomason, A. (1997) Hereditary and monotone properties of graphs. In *The Mathematics of Paul Erdős* II (R. L. Graham and J. Nešetřil, eds), Vol. 14 of *Algorithms and Combinatorics*, Springer, Berlin, pp. 70–78.

[6] Brightwell, G. and Trotter, W. T. (1993) The order dimension of convex polytopes. *SIAM J. Discrete Math.* **6** 230–245.

[7] Brightwell, G. and Trotter, W. T. (1997) The order dimension of planar maps. *SIAM J. Discrete Math.* **10** 515–528.

[8] Dushnik, B. (1950) Concerning a certain set of arrangements. *Proc. Amer. Math. Soc.* **1** 788–796.

[9] Dushnik, B. and Miller, E. W. (1941) Partially ordered sets. *Amer. J. Math.* **63** 600–610.

[10] Erdős, P. and Stone, A. H. (1946) On the structure of linear graphs. *Bull. Amer. Math. Soc.* **52** 1087–1091.

[11] Felsner, S. and Trotter, W. T. (2005) Posets and planar graphs. *J. Graph Theory* **49** 273–284.

[12] Füredi, Z., Hajnal, P., Rödl, V. and Trotter, W. T. (1991) *Interval Orders and Shift Graphs*, Vol. 60 of *Colloq. Math. Soc. Janos Bolyai*, pp. 297–313.

[13] Hoşten, S. and Morris Jr., W. D. (1999) The order dimension of the complete graph. *Discrete Math.* (special issue on partial ordered sets) **201** 133–139.

[14] Kratochví l, J. and Mihók, P. (2000) Hom-properties are uniquely factorizable into irreducible factors. *Discrete Math.* **213** 189–194.

[15] Kříž, I. and Nešetřil, J. (1991) Chromatic number of Hasse diagrams, eyebrows and dimension. *Order* (8) 41–48.

[16] Ossona de Mendez, P. (1999) Geometric realization of simplicial complexes. In *Graph Drawing* (J. Kratochvil, ed.), Vol. 1731 of *Lecture Notes in Computer Science*, Springer, pp. 323–332.

[17] Ossona de Mendez, P. (2002) Realization of posets. *J. Graph Algorithms Appl.* **6** 149–153.

[18] Scarf, H. (1973) *The Computation of Economic Equilibria*, Vol. 24 of *Cowles Foundation Monographs*, Yale University Press.

[19] Schnyder, W. (1989) Planar graphs and poset dimension. *Order* **5** 323–343.

[20] Spencer, J. (1971) Minimal scrambling sets of simple orders. *Acta Math. Acad. Sci. Hung.* **22** 349–353.

© 2006 Richard Pinch
Printed in the United Kingdom

The distance of a permutation from a subgroup of S_n

RICHARD G.E. PINCH

2 Eldon Road, Cheltenham, Glos GL52 6TU, U.K.
(e-mail: rgep@chalcedon.demon.co.uk)

We show that the problem of computing the distance of a given permutation from a subgroup H of S_n is in general NP-complete, even under the restriction that H is elementary Abelian of exponent 2. The problem is shown to be polynomial-time equivalent to a problem related to finding a maximal partition of the edges of an Eulerian directed graph into cycles and this problem is in turn equivalent to the standard NP-complete problem of Boolean satisfiability.

1. Introduction

We show that the problem of computing the distance of a given permutation from a subgroup H of S_n is in general NP-complete, even under the restriction that H is elementary Abelian of exponent 2. The problem is polynomial-time equivalent to finding a maximal partition of the edges of an Eulerian directed graph into cycles and this is in turn equivalent to the standard NP-complete problem **3-SAT**.

2. Distance in the symmetric group

We define *Cayley distance* in a symmetric group as the minimum number of transpositions which are needed to change one permutation to another by post-multiplication

$$d(\rho, \pi) = \min \{ n \mid \rho \tau_1 \ldots \tau_n = \pi, \quad \tau_i \text{ transpositions} \}.$$

It is well-known that Cayley distance is a metric on S_n and that it is homogeneous, that is, $d(\rho, \pi) = d(I, \rho^{-1}\pi)$. Further, the distance of a permutation π from the identity in S_n is n minus the number of cycles in π.

If H is a subgroup of S_n, then we define the distance of a permutation π from H as

$$d(H, \pi) = \min_{\eta \in H} d(\eta, \pi).$$

We refer to Critchlow [2] and Diaconis [3] for background and further material on the uses of the Cayley and other metrics on S_n.

Problem 1 (Subgroup–Distance).
INSTANCE: *Symmetric group S_n, element $\pi \in S_n$, elements $\{h_1,\ldots,h_r\}$ of S_n, integer K.*
QUESTION: *Is there an element $\eta \in H = \langle h_1,\ldots,h_r \rangle$ such that $d(\eta,\pi) \leqslant K$?*

The natural measure of this problem is nr where r is the length of the list of generators. The following result shows that every subgroup of S_n has a set of generators of length at most n^2 and hence we are justified in taking n as the measure of the various problems derived from **Subgroup–Distance**.

Proposition 1. *Every subgroup of S_n can be generated by at most n^2 elements.*

Proof. Let H be a subgroup of S_n. It is clear that H is generated by the union of one Sylow subgroup for every prime p dividing the order $\#H$. The order of a Sylow p-subgroup of H is p^b where p^b divides $\#H$ and hence $n!$. It is well-known (e.g. Dickson [4] I, chap 9) that the power of p dividing $n!$ is at most $\frac{n}{p-1}$, so $b \leqslant n$. But consideration of the composition factors shows that a p-group of order p^b can be generated by a set of at most b elements: hence any Sylow subgroup of H can be generated by at most n elements. Further, the set of prime factors of the order of H forms a subset of the set of prime factors of $n!$, that is, of the primes up to n, and there are at most n such primes. Hence H can be generated by a set of at most n^2 elements. □

Although we do not need the stronger result, it can be shown that any subgroup of S_n can be generated by at most $3n - 2$ elements.

We define a subset of S_n to be *involutions with disjoint support (IDS)* to be a set of elements of the form $\gamma_j = \left(x_j^{(1)} y_j^{(1)}\right) \ldots \left(x_j^{(r_j)} y_j^{(r_j)}\right)$ where the $x_j^{(i)}, y_j^{(i)}$ are all distinct. The subgroup generated by an IDS is clearly elementary Abelian with exponent 2. Define the *width* of an IDS to be the maximum number of 2-cycles r_j in the generators γ_j. The problem **IDSw–Subgroup–Distance** is the problem **Subgroup–Distance** with the list of generators restricted to be an IDS of width at most w.

Theorem 2. *The problem* **IDS6–Subgroup–Distance** *is NP-complete.*

The Theorem will follow from combining Theorem 3 and Theorem 7. We deduce immediately that the more general problem **Subgroup–Distance** is also NP-complete.

By contrast, the problem of deciding whether the distance is zero, that is, testing for membership of a subgroup of S_n, has a polynomial-time solution, an algorithm first given by Sims [8] and shown to have a polynomial-time variant by Furst, Hopcroft and Luks [5]. See Babai, Luks and Seress [1] and Kantor and Luks [7] for a survey of related results.

3. Switching circuits

Let $G = (V, E)$ be a directed graph with vertex set V and edge set E. (We allow loops and multiple edges.) For each vertex v define $e_+(v)$ to be the set of edges out of v and $e_-(v)$ the set of edges into v. The in-valency $\partial_-(v) = \#e_-(v)$ and the out-valency $\partial_+(v) = \#e_+(v)$. We define a *switching circuit* to be a directed graph G for which $\partial_+(v) = \partial_-(v) = \partial(v)$, say, and for which there is a labelling $l_\pm(v)$ of each set $e_\pm(v)$ with the integers from 1 to $\partial(v)$. (The labels at each end of an edge are not related.) A *routing* ρ for a switching circuit is a choice of permutation $\rho(v) \in S_{\partial(v)}$ for each vertex v. Clearly there is a correspondence between routings for a switching circuit G and decompositions of the edge set of G into directed cycles. We define a *polarisation* T for a switching circuit G to be an equivalence relation on the set of vertices such that equivalent vertices have the same valency, and call (G, T) a *polarised switching circuit*. We say that a routing ρ *respects* the polarisation T if the permutations $\rho(x)$ and $\rho(y)$ are equal whenever x and y are equivalent vertices under T. We shall sometimes refer to a switching circuit without a polarisation, or with a polarisation for which all the classes are trivial, as *unpolarised*.

Problem 2 (Polarised–Switching–Circuit–Maximal–Routing).
INSTANCE: *Polarised switching circuit* (G, T), *positive integer* K.
QUESTION: *Is there a routing which respects T and has at least K cycles in the associated edge-set decomposition?*

We define the *width* of a polarisation to be the maximum number of vertices in an equivalence class of T. The problem **Width$w$–Valency$v$–Maximal–Routing** is the problem **Polarised–Switching–Circuit–Maximal–Routing** with the width of T constrained to be at most w and the in- and out-valency of each vertex in V constrained to be at most v.

Theorem 3. *Problem* **Width6–Valency2–Maximal–Routing** *is NP-complete.*

4. Proof of Theorem 3

We shall show that the problem **3-SAT**, [LO2] of Garey and Johnson [6], which is known to be NP-complete, can be reduced to the problem **Width6–Valency2–Maximal–Routing**.

We define a polarised switching circuit (G, T) to be *Boolean* (or *binary*) if every vertex has in- and out-valency 1 or 2. To each class C of the polarisation T we associate a Boolean variable $a(C)$. There is then a 1-1 correspondence between routings ρ which respect T and assignments of truth values to the variables $a(C), C \in T$ by specifying that $a(C)$ is 0 (false) if and only if the permutation $\rho(v)$ is the identity in S_2 for every v in C, and 1 (true) if and only if $\rho(v) = (1\ 2)$.

We denote a vertex in a polarisation class associated with the Boolean variable a as in Figure 1. Our convention for drawing the diagrams will be to assume the edges round each vertex labelled so that 1 is denoted by either "straight through" or "turn right".

We associate a vertex with the negated variable $\bar{a}$ by exchanging the input labels 1 and 2.

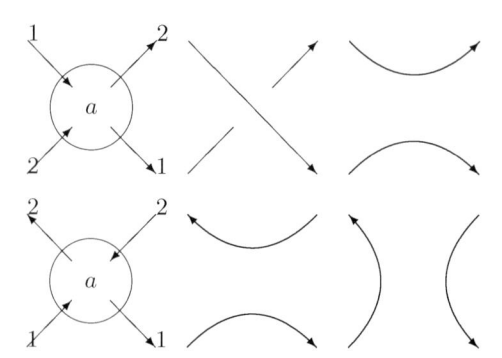

Figure 1. A vertex in a switching circuit associated with the Boolean variable a, and the routings with $a = 1$ and $a = 0$ respectively

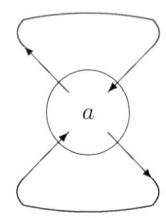

Figure 2. The switching circuit $I(a)$.

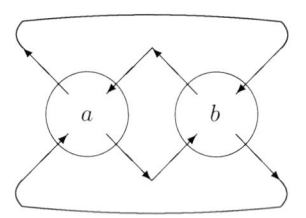

Figure 3. The switching circuit $E(a,b)$.

Our proof will proceed by finding polarised switching circuits for which the number of maximal cycles in a routing is a Boolean function of the variables.

For a single Boolean variable a define $I(a)$ to be the switching circuit in Figure 2.

For a pair of Boolean variables (a,b) define the polarised switching circuit $E(a,b)$ as in Figure 3.

Further define the polarised switching circuit $F(a,b)$ as in Figure 4.

Define $G(a,b)$ to be the disjoint union of $F(a,b)$ and $E(\bar{a},b)$.

Proposition 4.

1. The number of cycles in a routing for $I(a)$ is 2 if $a = 1$ and otherwise 1.
2. The number of cycles in a routing for $E(a,b)$ is 2 if $a = b$ and otherwise 1.

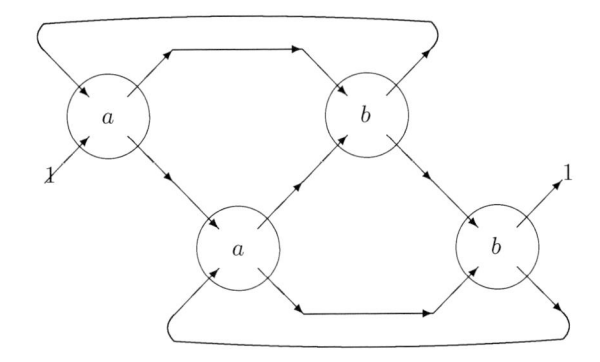

Figure 4. The switching circuit $F(a,b)$.

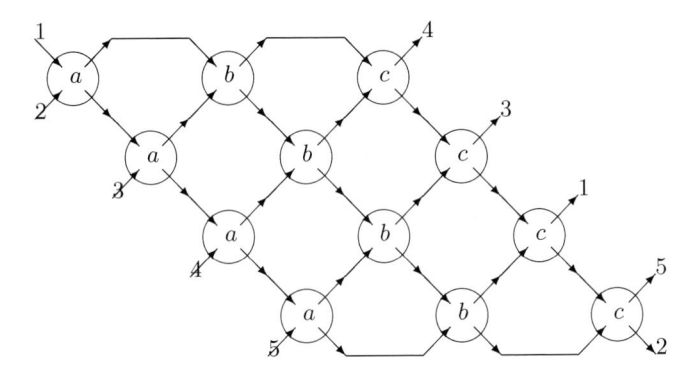

Figure 5. The switching circuit $A(a,b,c)$.

3 The number of cycles in a routing for $F(a,b)$ is 2 if $a \neq b$, 3 if $a = b = 1$ and 1 if $a = b = 0$.
4 The number of cycles in a routing for $G(a,b)$ is 2 if $a = b = 0$ and 4 otherwise.

Proof. In each case we simply enumerate the cases. □

For a triple of Boolean variables (a,b,c) define the polarised switching circuit $A(a,b,c)$ as in Figure 5.

Proposition 5. *The number of cycles in a routing for $A(a,b,c)$ is 1 if $a = b = c = 0$ and 3 otherwise.*

Proof. Again, in each case we simply enumerate the cases. □

Theorem 6. *There is a polynomial-time parsimonious transformation from the problem* **3-SAT** *to the problem* **Width6–Valency2–Maximal–Routing**.

Proof. Suppose we have an instance of **3-SAT**: that is, a Boolean formula Φ of length l in variables x_i which is a conjunct of k clauses each of which is a disjunct of at most three variables (possibly negated). We transform Φ into a formula Φ' in variables y_i^j by replacing the jth occurence of variable x_i by the variable y_i^j and conjoining clauses $\left(y_i^1 \equiv y_i^2\right) \wedge \ldots \wedge \left(y_i^{(r_i-1)} \equiv y_i^{(r_i)}\right)$ where the variable x_i occurs r_i times in Φ. Clearly Φ and Φ' represent the same Boolean function and have the same number of satisfying assignments. Every variable in Φ' occurs at most three times, and at most once in a disjunct deriving from a clause in Φ. Let n be the total number of variables in Φ'; certainly $n \leqslant l$.

We form a polarised switching circuit Ψ from Φ' as follows. Take a circuit $B(x, y, z)$ for every clause in Φ' of the form $(x \vee y \vee z)$; take a circuit $G(x, y)$ for every clause in Φ' of the form $(x \vee y)$; take a circuit $I(x)$ for every clause in Φ' of the form (x); take a circuit $E(x, y)$ for every clause in Φ' of the form $(x \equiv y)$. Let the number of circuits of types B, G, I and E taken to form Ψ be b, g, i, and e respectively. Put $M = 3b + 4g + 2i + 2e$. The resulting polarised switching circuit has n classes, and each class in the polarisation is involved in at most one circuit of the form B, G or I: hence each class contains at most $4 + 1 + 1 = 6$ vertices and the number of vertices in Ψ is thus at most $6n$. Furthermore, a routing for Ψ has M cycles if and only if the corresponding assignment of Boolean values gives Φ', and hence Φ, the value 1; otherwise a routing has less than M cycles. □

Since the problem **3-SAT** is known to be NP-complete, we immediately deduce that the problem **Width6–Valency2–Maximal–Routing** is NP-complete as well. This proves Theorem 3.

5. Switching circuits and IDS

In this section we obtain a polynomial-time equivalence between the problems **Width$w$–Valency2–Maximal–Routing** and **IDSw–Subgroup–Distance**.

Theorem 7. *There is a polynomial-time parsimonious equivalence between problems* **Width$w$–Valency2–Maximal–Routing** *and* **IDSw–Subgroup–Distance**.

Proof. Suppose we have an instance of **IDSw–Subgroup–Distance**, that is, an element π of S_n together with an IDS $\{\gamma_j\}$ of t generators $\gamma_j = \left(x_j^{(1)} y_j^{(1)}\right) \ldots \left(x_j^{(r_j)} y_j^{(r_j)}\right)$, the $x_j^{(i)}, y_j^{(i)}$ all distinct and all the $r_j \leqslant w$. We construct a polarised switching circuit on a graph, vertex set $V = \{P(1), \ldots, P(n)\} \cup \{Q(1,1), \ldots, Q(t, r_t)\}$. Each vertex $P(k)$ will be of in-valence and out-valence 1; each vertex $Q(j, i)$ will be of in-valence and out-valence 2. For each j up to t and i up to r_j we take edges from $P(x_j^i)$ and from $P(y_j^i)$ to $Q(j, i)$ labelled 1 and 2 respectively, and edges from $Q(j, i)$ to $P\left(\pi\left(x_j^i\right)\right)$ and to $P\left(\pi\left(y_j^i\right)\right)$ again labelled 1 and 2 respectively. We define a polarisation T on V by taking t classes $C_j = \{Q(j, i) \mid i = 1, \ldots, r_j\}$; clearly the width of T is at most w.

Conversely, suppose we have an instance of **Width$w$–Valency2–Maximal–Routing**, that is, a directed graph (V, E) with every vertex v having in- and out-valency two, a labelling

$l_{\pm}(v) : e_{\pm}(v) \to \{1, 2\}$ of edges into and out of each vertex v, and an equivalence relation T on V with t classes each of size at most w. Put $n = \#E$. We define a permutation π of E as follows. For an edge e into a vertex v, let $\pi(e)$ be the edge f out of v which has label $l_+(v)(f)$ equal to $l_-(v)(e)$. We further define an IDS by writing down a set of generators $\{\gamma_j\}$ as follows. For each class of vertices $C_j = \left\{ v_i^j \mid i = 1, \ldots, r_j \right\}$ in the polarisation T, let γ_j be the product of transpositions of the form $\left(f_i^j g_i^j \right)$ where f_i^j and g_i^j are the edges out of vertex v_i^j. Since each class in T has at most w elements, each generator γ_j is composed of at most w transpositions.

In each case there is a correspondence between routings ρ of the switching circuit which respect the polarisation T and permutations of the form $\pi\eta$ where η runs over the elements of the subgroup H of S_n generated by the γ_j: in this correspondence the number of cycles in the routing ρ is equal to the number of cycles in the permutation $\pi\eta$. Hence π is within distance d of the group generated by the γ_j if and only if there is a routing ρ with at least $n - d$ cycles. □

References

[1] L. Babai, E.M. Luks, and Seress Á. Fast management of permutation groups. *Proc. 29th annual symposium on foundation of computer science*, pages 272–282, 1988.

[2] D.E. Critchlow. *Metric methods for analysing partially ranked data*, volume 34 of *Lecture notes in statistics*. Springer Verlag, Berlin, 1985.

[3] P. Diaconis. *Group representations in probability and statistics*, volume 11 of *IMS lecture notes*. Institute of Mathematical Statistics, Hayward CA, 1988.

[4] Leonard E. Dickson. *History of the theory of numbers*. Chelsea, New York, 1971. original publication Carnegie Institute, Washington, 1919–23.

[5] M.L. Furst, J. Hopcroft, and E.M. Luks. Polynomial time algorithms for permutation groups. In *Proc. 21st IEEE FOCS*, pages 36–41, 1980.

[6] M.R. Garey and G.S. Johnson. *Computers and intractability: a guide to the theory of NP-completeness*. W.H. Freeman, San Francisco, CA, 1979.

[7] W.M. Kantor and E.M. Luks. Computing in quotient groups. In *Proceedings 22nd annual ACM symposium on theory of computing*, pages 524–534. ACM Press, 1990.

[8] C.C. Sims. Computational methods in the study of permutation groups. In J. Leech, editor, *Computational problems in abstract algebra*, pages 169–184, Oxford, 1970. Pergamon Press.

On Dimensions of a Random Solid Diagram

BORIS PITTEL[†]

Department of Mathematics, Ohio State University, 231 West 18th Avenue,
Columbus, Ohio, OH 43210-1174, USA
(e-mail: bgp@math.ohio-state.edu)

Received 1 August 2003; revised 8 November 2005

For Béla Bollobás on his 60th birthday

A solid diagram of volume n is a packing of n unit cubes into a corner so that the heights of vertical stacks of cubes do not increase in either of two horizontal directions away from the corner. An asymptotic distribution of the dimensions – heights, depths, and widths – of the diagram chosen uniformly at random among all such diagrams is studied. For each k, the planar base of k tallest stacks is shown to be Plancherel distributed in the limit $n \to \infty$.

1. Introduction

A solid diagram of volume n is defined as a set $\mathscr{P}$ of n positive integer lattice points $\mathbf{x} = (x_1, x_2, x_3) \in \mathbb{N}^3$, such that if $\mathbf{x} \in \mathscr{P}$, $\mathbf{x}' \leqslant \mathbf{x}$ then $\mathbf{x}' \in \mathscr{P}$ too. Such a diagram can also be interpreted as a descending plane partition of n, i.e., a set of nonnegative integers p_{ij}, $(i, j \geqslant 1)$, such that $\sum_{i,j} p_{ij} = n$, and p_{ij} decreases with i (j resp.) for each fixed j (i resp.) Indeed, p_{ij} is the height of the column of unit cubes stacked along the vertical line $x_1 = i, x_2 = j$, and the solid diagram is the union of all such columns. Given positive integers r, s, t, let $B(r, s, t)$ stand for the parallelepiped $\{\mathbf{x} \in \mathbb{N}^3 : x_1 \leqslant r, x_2 \leqslant s, x_3 \leqslant t\}$. Let $|\mathscr{P}|$ denote the volume of a diagram $\mathscr{P}$. According to MacMahon [19], for $|q| < 1$,

$$p_{r,s,t}(q) \stackrel{\text{def}}{=} \sum_{\mathscr{P} \subseteq B(r,s,t)} q^{|\mathscr{P}|} = \prod_{\mathbf{x} \in B(r,s,t)} \frac{1 - q^{|\mathbf{x}|-1}}{1 - q^{|\mathbf{x}|-2}}, \quad |\mathbf{x}| = x_1 + x_2 + x_3. \quad (1.1)$$

(We refer the reader to Andrews [1], Gessel and Vienott [11], Stanley [28] and Bressoud [3] for numerous references and eminently readable accounts of subsequent research on

---

[†] Supported by NSF grant DMS-0104104.

plane partitions spawned by MacMahon's studies.) Here is a more explicit version of (1.1):

$$p_{r,s,t}(q) = \prod_{\ell=1}^{\infty}(1-q^{\ell})^{-\ell}$$
$$\times \prod_{\ell>r}(1-q^{\ell})^{\ell-r} \cdot \prod_{\ell>s}(1-q^{\ell})^{\ell-s} \cdot \prod_{\ell>t}(1-q^{\ell})^{\ell-t}$$
$$\times \prod_{\ell>r+s}(1-q^{\ell})^{r+s-\ell} \cdot \prod_{\ell>r+t}(1-q^{\ell})^{r+t-\ell} \prod_{\ell>s+t}(1-q^{\ell})^{s+t-\ell}$$
$$\times \prod_{\ell>r+s+t}(1-q^{\ell})^{\ell-r-s-t}. \tag{1.2}$$

Sending r, s, t to infinity, we recover MacMahon's

$$p(q) = \prod_{\ell \geq 1}(1-q^{\ell})^{-\ell} \tag{1.3}$$

as the generating function of unrestricted diagrams. Likewise, fixing r and letting $s, t \to \infty$, we obtain that

$$p_r(q) = \prod_{\ell \geq 1}(1-q^{\ell})^{-\min(\ell,r)} \tag{1.4}$$

is the generating function of diagrams with the x_1-axis dimension at most r (see [19]). The formula (1.2) is ideally suited to our purposes. To prove it, we notice that, introducing

$$C_{\ell} = C_{\ell}(r,s,t) = |\{\mathbf{x} \in B(r,s,t) : |\mathbf{x}| = \ell\}|,$$

we transform (1.1) into

$$\sum_{\mathscr{P} \subseteq B(r,s,t)} q^{|\mathscr{P}|} = \prod_{\ell \geq 1}(1-q^{\ell})^{C_{\ell+1}-C_{\ell+2}}. \tag{1.5}$$

Further, as

$$|\{\mathbf{y} \in \mathbb{N}^3 : |\mathbf{y}| = \mu\}| = \binom{\mu-1}{2},$$

by the inclusion–exclusion principle we have

$$C_v = \binom{v-1}{2} - \binom{v-r-1}{2} - \binom{v-s-1}{2} - \binom{v-t-1}{2} \tag{1.6}$$
$$+ \binom{v-r-s-1}{2} + \binom{v-r-t-1}{2} + \binom{v-s-t-1}{2} - \binom{v-r-s-t}{2}.$$

Combining (1.5), (1.6) and

$$\binom{a}{2} - \binom{a-1}{2} = \binom{a-1}{1} = (a-1)^+,$$

we obtain (1.2).

Wright [30] used (1.3) to derive an asymptotic formula for $p_n = [q^n]p(q)$, the total number of plane partitions of n or the total number of solid diagrams of volume n:

$$p_n = an^{-25/36} e^{bn^{2/3}} \left(1 + \sum_{h=1}^{r} \delta_h n^{-2h/3} + O(n^{-2(r+1)/3})\right);$$

$$a = \frac{(2^{25}\zeta(3)^7)^{1/36}}{2\pi^{1/2}} e^c,$$

$$b = 3 \cdot 2^{-2/3} \zeta(3)^{1/3},$$

$$c = 2 \int_0^\infty \frac{y \log y}{e^{2\pi y} - 1} dy. \qquad (1.7)$$

Here $\zeta(\cdot)$ is the Riemann ζ-function, and δ_h are certain constants. Wright's powerful argument is quite technical, and adapting it for the more general case of restricted diagrams may well be quite problematic. However, a cruder formula – with the last factor of the form $1 + O(n^{-1/3})$ – is quite sufficient for our purposes. We will give a much simpler, and self-contained, proof of such a 'poor man's' asymptotic formula for both unrestricted and restricted diagrams. (We could also have used a rather general approach due to Meinardus – see [1] – but it is markedly more complex and yields a worse error bound.) In particular, we will determine an asymptotic formula for the total number of diagrams $\mathscr{P}$ that fit inside $B(r,s,t)$, with r,s,t all of order $n^{1/3} \log n$. Endowing the set of solid diagrams of volume n with uniform probability measure, we will then deduce the following probabilistic statement.

Theorem 1.1. *Let H_n, W_n, D_n denote the dimensions – height, width and depth – of the uniformly random diagram $\mathscr{P}$ of volume n. Define h_n, w_n, d_n by*

$$H_n = \frac{n^{1/3}}{\beta} \log \frac{\left(\frac{n^{1/3}}{\beta}\right)^2}{h_n}, \quad W_n = \frac{n^{1/3}}{\beta} \log \frac{\left(\frac{n^{1/3}}{\beta}\right)^2}{w_n}, \quad D_n = \frac{n^{1/3}}{\beta} \log \frac{\left(\frac{n^{1/3}}{\beta}\right)^2}{d_n}, \qquad (1.8)$$

where $\beta = (2\zeta(3))^{1/3}$. In the limit $n \to \infty$, h_n, w_n, d_n are asymptotically independent, and exponentially distributed, with parameter 1 each.

Note. For the random planar diagram of area n (linear partition of n) the corresponding result for the width W_n and the depth D_n, i.e., the lengths of the longest row and the longest column, was obtained by Erdős and Lehner [7], and Auluck, Chowla and Gupta [2]. It states that $\mathscr{W}_n, \mathscr{D}_n$ defined by

$$W_n = \frac{n^{1/2}}{\alpha}\left(\log \frac{n^{1/2}}{\alpha} + \mathscr{W}_n\right), \quad D_n = \frac{n^{1/2}}{\alpha}\left(\log \frac{n^{1/2}}{\alpha} + \mathscr{D}_n\right) \qquad (1.9)$$

($\alpha = \sqrt{\zeta(2)} = \frac{\pi}{6^{1/2}}$), are asymptotically independent, and *double*-exponentially distributed, i.e.,

$$P(\mathscr{W}_n \leq x), P(\mathscr{D}_n \leq x) \to e^{-e^{-x}}, \quad \forall x \in R.$$

If we define w_n, d_n by $\log 1/w_n = \mathscr{W}_n$, $\log 1/d_n = \mathscr{D}_n$, thus expressing W_n, D_n in the form of (1.8), then in the limit w_n, d_n are independent, exponentially distributed, with parameter 1.

The analogy is undeniably striking, and one is tempted to conjecture similar results for higher-dimensional diagrams. Any attempts to prove such an extension appear to be doomed at present, as even the basic problem of enumerating diagrams for four or more dimensions is wide open.

Using a conditioning device, Fristedt [10] considerably extended the results in [2, 7] and determined the limiting joint distribution of the first $k = o(n^{1/4})$ largest widths $W_n^{(i)}$ (row lengths) of the random planar diagram. The sequence of random variables $\{\mathscr{W}_n^{(i)}\}_{i \leqslant k}$, that are to $W_n^{(i)}$ what $\mathscr{W}_n = W_n^{(1)}$ is to W_n, was shown to be asymptotic to the Markov chain $\{Y_i\}_{i \leqslant k}$; here the density of Y_i conditioned on $Y_{i-1} = x$ is $e^{-y-e^{-y}+e^{-x}}$, $y \leqslant x$, and the density of Y_1 is $e^{-x-e^{-x}}$. By duality, the sequence $\{D_n^{(i)}\}$ of largest depths (column lengths) and $\{W_n^{(i)}\}$ are equidistributed, so the corresponding $\{\mathscr{D}_n^{(i)}\}_{i \leqslant k}$ is asymptotic to a copy of the above Markov chain. For the case $k = O(1)$, this result *plus* mutual independence of the two Markov chains are implicit in Erdős and Richmond [8]. In [24] it was found that $\{Y_i\}$ coincides, in distribution, with $\{\log(\sum_{j=1}^i E_i)^{-1}\}$, where $E_1, E_2, \ldots$ are independent copies of the exponential E. This means that, for k fixed, say, $\{w_n^{(i)}\}_{i \leqslant k}, \{d_n^{(i)}\}_{i \leqslant k}$ defined by

$$W_n^{(i)} = \frac{n^{1/2}}{\alpha} \log \frac{\frac{n^{1/2}}{\alpha}}{w_n^{(i)}}, \qquad D_n^{(i)} = \frac{n^{1/2}}{\alpha} \log \frac{\frac{n^{1/2}}{\alpha}}{d_n^{(i)}}$$

are asymptotic to $\{\sum_{j=1}^i E_i\}_{i \leqslant k}$ and $\{\sum_{j=1}^i E_i'\}_{i \leqslant k}$, where $E_1, E_1', E_2, E_2', \ldots$ are independent copies of E.

Extending the proof of Theorem 1.1, we will obtain the following 3-dimensional counterpart of the cited results for the random planar diagram.

Theorem 1.2. *Let $H_n^{(i)}$ denote the length of the ith-tallest vertical 'stack' of unit cubes in the random diagram of volume n. Let $W_n^{(i)}$ ($D_n^{(i)}$ resp.) denote the length of the ith-longest horizontal 'beam' of unit cubes which is parallel to the x_2-axis (to the x_1-axis resp.), $i = 1, 2, \ldots$. Define $h_n^{(i)}, w_n^{(i)}, d_n^{(i)}$ by formulas analogous to (1.8). Then, for each k, the sequences $\{h_n^{(i)} : i \leqslant k\}, \{w_n^{(i)} : i \leqslant k\}, \{d_n^{(i)} : i \leqslant k\}$ are asymptotic, in distribution, to $\{\sum_{j=1}^i E_j : i \leqslant k\}, \{\sum_{j=1}^i E_j' : i \leqslant k\}, \{\sum_{j=1}^i E_j'' : i \leqslant k\}$ respectively, where $\{E_i : i \leqslant k\}, \{E_i' : i \leqslant k\}, \{E_i'' : i \leqslant k\}$ are independent sequences, each formed by the independent copies of the random variable E.*

We could have stated and proved a stronger version of this theorem, covering $k = o(n^\sigma)$, for some $\sigma \in (0, 1/3)$, but decided to stick with k fixed to avoid clouding the basic ideas.

Our results imply that, with probability approaching 1, the k largest heights, widths, depths are all distinct. It's a dog-eat-dog world at each of the three 'extremities' of the diagram. Unlike linear partitions though, knowing the (limiting) distribution of, say, k tallest stacks of unit cubes gives us no information regarding the shape of the summit. The next theorem fills the gap.

First some definitions and facts. Given a planar (Young–Ferrers) diagram λ of area k, let $d(\lambda)$ denote the total number of ways to fill its k squares by numbers $1, \ldots, k$ so that the entries increase in every row and column in the direction away from the diagram's corner. The value of $d(\lambda)$ is given by a celebrated hook formula, due to Frame, Robinson and Thrall [9], namely

$$d(\lambda) = \frac{k!}{\prod_{\square \in \lambda} h(\square)}, \qquad (1.10)$$

where $h(\square)$ is the area of the hook that consists of the $\square$ itself, and of the squares right of and down from the $\square$. Remarkably, given k, $\{d^2(\lambda)/k!\}$ is a probability (Plancherel) distribution on the set Λ_k of all diagrams of area k, that is,

$$\sum_{\lambda \in \Lambda_k} \frac{d^2(\lambda)}{k!} = 1. \qquad (1.11)$$

This is a special case of a general theorem due to Burnside: the sum of squared dimensions of all irreducible representations of a group equals the group's order. (There is a bijection between the set of all irreducible representations of the symmetric group S_k (of order $k!$) and the set Λ_k, such that the degree of a representation corresponding to $\lambda \in \Lambda_k$ is $d(\lambda)$, Ledermann [17], Diaconis [6], Sagan [26].) Combinatorially, (1.11) is a consequence of a bijection, via the Robinson–Schensted algorithm, between elements of S_k and the set of pairs of Young tableaux of the same shape λ, $\lambda \in \Lambda_k$ (Knuth [15, Section 5.1.4], [28]).

Theorem 1.3. *Let $\mu_n^{(k)}$ denote the 'footprint' of the k tallest stacks, i.e., the planar diagram formed by the stacks bases. Then $\mu_n^{(k)}$ is, in the limit, Plancherel distributed on Λ_k, i.e.,*

$$\lim_{n \to \infty} P(\mu_n^{(k)} = \lambda) = \frac{d^2(\lambda)}{k!}, \quad \lambda \in \Lambda_k. \qquad (1.12)$$

Furthermore, conditioned on $\mu_n^{(k)} = \lambda$, in the limit $n \to \infty$ each of $d(\lambda)$ ways to allocate the bases of the k stacks among the k squares of λ is equally likely. Consequently, $\{\mu_n^{(i)}\}_{i \leqslant k}$ is asymptotic in distribution to the Markov chain $\{\mu^{(i)}\}_{i \leqslant k}$ such that, for every $\lambda \in \Lambda_i$ and $\lambda' \in \Lambda_{i+1}$ obtained from λ by adding an external square,

$$P(\mu^{(i+1)} = \lambda' \mid \mu^{(i)} = \lambda) = \frac{d(\lambda')}{(i+1)d(\lambda)}. \qquad (1.13)$$

Needless to say, analogous results hold for two other k-size projections, on the planes $x_1 = 0$ and $x_2 = 0$. In the spirit of Theorem 1.2, we conjecture that all three k-size projections are asymptotically independent.

Notes. (1) A new key ingredient of the proof is a formula, established independently by Krattenhaler [16, Section 6, Theorem 6.8] and Kadell [14, Section 6, Theorem 8] for the generating function of skew planar partitions $\mathscr{P}$ of shape $[1, r] \times [1, s] \setminus \mu$ where μ is a given (finite) diagram, such that the parts of $\mathscr{P}$ in the jth column are $t + \mu'_j$ at most. Here μ'_j is the number of squares in the jth column of μ. (Krattenhaler's result is slightly more general.) In the limit $r = s = \infty$, the formula becomes

$$\sum_{\mathscr{P}} q^{|\mathscr{P}|} = p_t(q) \prod_{\square \in \mu} (1 - q^{h(\square)})^{-1} \qquad (1.14)$$

(*cf.* [28, Chapter 7, Exercise 7.107(b)]); see (1.4) for $p_t(q)$. (Remarkably, the hook product factor in (1.14) had long been known as the generating function of reverse plane partitions of shape μ; see Stanley [27].) Ideally one would have preferred to have such a formula for the parts uniformly bounded by t, but the beautiful identity (1.14) will do the job perfectly anyway, as in our case $t \gg \max_j \mu'_j$.

(2) Greene, Nijenhuis and Wilf [12] discovered a random (hook) walk executed on a given (planar) diagram that provided a probabilistic proof for the hook formula. Subsequently [13] they showed how the hook walk can be used to generate the Markov chain $\{\mu^{(i)}\}$ that we have encountered in Theorem 1.3. Briefly, given the current diagram $\mu^{(i)}$, $\mu^{(i+1)}$ is obtained by executing the hook walk on the 'complementary' diagram, that starts at the corner, and adding the terminal square to $\mu^{(i)}$. Inspired by [12], and independently from [13], we found [22] a different hook-driven walk that builds $\mu^{(i+1)}$ from $\mu^{(i)}$. Unlike [13], this walk is executed on $\mu = \mu^{(i)}$ itself, starting with the corner square $(1,1)$. This is how it works. Call $\square \notin \mu$ 'external' if $\mu \cup \square$ is a diagram again, and $\square = (\alpha, \beta) \in \mu$ 'admissible' if both $(\alpha, \mu_\alpha + 1)$ and $(\mu'_\beta + 1, \beta)$ are external. The corner square is one of the admissible squares. Recursively, let an admissible (α, β) be the current state of the walk. Choose uniformly at random a square (γ, δ) belonging to the 'extended' hook, that is, the union of the hook of (α, β) in μ and two external squares, $(\alpha, \mu_\alpha + 1)$ and $(\mu'_\beta + 1, \beta)$. If (γ, δ) is external, then stop the walk and form a diagram $\mu \cup (\gamma, \delta)$. If (γ, δ) is admissible, then continue the walk from (γ, δ). Suppose (γ, δ) is not admissible, so that, say, $(\mu'_\delta + 1, \delta)$ is not external, in which case $\gamma = \alpha$. Then the next (admissible) state is set to be $(\mu'_\delta + 1, \beta)$, and the walk continues. Once an external $\square$ is hit, the walk stops, and we set $\mu^{(i+1)} = \mu \cup \square$. In light of Theorem 1.3, it is hard to resist interpreting the walk as a peculiar random descent from a mountain top.

(3) In a recent work McKay, Morse and Wilf [18] showed that the *uniformly* random Young tableau contains a subtableau of a shape λ with the limiting probability $d^2(\lambda)/|\lambda|!$. Then Stanley [29] derived an exact formula, which can be used, in principle, to obtain an asymptotic expansion of this probability.

We should mention that Cerf and Kenyon [4] were able to determine the limiting shape of the bulk of the random solid diagram (scaled by $n^{1/3}$), the deep analysis being based on connections with statistical physics, namely a 3-dimensional Ising model.

In [23] it was shown that the boundary of the random planar diagram undergoes, asymptotically, Gaussian fluctuations of order $n^{1/4}$ around its deterministic limiting shape. The corresponding problem for the solid diagram is open. It would seem plausible, by analogy with the planar partitions, that the fluctuations around the limiting shape should be of order $n^{1/6}$. However, Rick Kenyon (private communication) insisted that, on the basis of numerical experiments, the variance order is logarithmic, at most.

Cohn, Larsen and Propp [5] studied the limiting shape of a solid diagram chosen uniformly at random among all diagrams 'boxed' in $B(r,s,t)$, for large r,s,t, all of the same order of magnitude. At the core of their approach is the remarkable product formula for the number of Gelfand–Tsetlin patterns. Very recently Okounkov and Reshetikhin [21] studied the limiting shape and the asymptotic correlations in the bulk of the random solid diagram $\mathscr{P}$ with distribution proportional to $q^{|\mathscr{P}|}$, and $q \uparrow 1$. The analysis is based on a determinantal formula for the correlation functions of the Schur process; see Okounkov [20].

2. Auxiliary estimates

We will use the generating functions $p(q)$, $p_r(q)$, $p_{r,s,t}(q)$ (see (1.1)–(1.4)) to estimate the total numbers of unrestricted and restricted diagrams via the Cauchy integral formula. To this end we will need a sharp asymptotic formula for $p(q)$ for a complex-valued q close to 1.

Lemma 2.1. *Let $q = e^{-u}$. Suppose $\operatorname{Re} u > 0$ and $|\operatorname{Im} u| \leqslant \sigma \operatorname{Re} u$, where $\sigma > 0$ is fixed. Then*

$$\ln p(q) = \frac{\zeta(3)}{u^2} - \frac{1}{12} \ln \frac{1}{u} + c + O(|u|), \quad u \to 0, \tag{2.1}$$

where ln *stands for the principal value of the logarithm, the implied constant in $O(|u|)$ depends on σ only, and*

$$c = \int_0^\infty \left[\frac{e^z}{z(e^z - 1)^2} - \frac{1}{z^3} + \frac{e^{-z}}{12z} \right] dz. \tag{2.2}$$

Note. The integral in (2.2) is finite as the integrand is $O(z^{-3})$ for $z \to \infty$, and $-1/12 + O(z)$ for $z \to 0$.

Proof of Lemma 2.1. The argument is a considerably shortened variation of the proof of an asymptotic formula for $\prod_{\ell \geqslant 1}(1 - q^\ell)^{-1}$, the generating function of the linear partitions, due to Freiman; see Postnikov [25, Section 2.7].

Using (1.3),

$$\ln p(e^{-u}) = \sum_{m \geqslant 1} m \ln \frac{1}{1 - e^{-mu}} = \sum_{m \geqslant 1} m \sum_{j \geqslant 1} \frac{e^{-muj}}{j}$$

$$= \sum_{j \geqslant 1} \frac{1}{j} \sum_{m \geqslant 1} m e^{-muj} = \sum_{j \geqslant 1} \frac{1}{j} \frac{e^{-uj}}{(1 - e^{-uj})^2}$$

$$= u \sum_{j \geqslant 1} \frac{1}{uj} \frac{e^{uj}}{(e^{uj} - 1)^2}.$$

Now, for $z \to 0$,

$$\frac{e^z}{z(e^z - 1)^2} = \frac{1}{z^3} - \frac{1}{12z} + O(|z|) = \frac{1}{z^3} - \frac{e^{-z}}{12z} + O(1).$$

So, introducing

$$\phi(z) = \frac{e^z}{z(e^z - 1)^2} - \frac{1}{z^3} + \frac{e^{-z}}{12z},$$

we obtain

$$\ln p(e^{-u}) = u^{-2} \sum_{j \geq 1} \frac{1}{j^3} - \frac{1}{12} \sum_{j \geq 1} \frac{e^{-uj}}{j} + u \sum_{j \geq 1} \phi(uj)$$

$$= u^{-2}\zeta(3) - \frac{1}{12} \ln \frac{1}{1 - e^{-u}} + u \sum_{j \geq 1} \phi(uj)$$

$$= u^{-2}\zeta(3) - \frac{1}{12} \ln \frac{1}{u} + u \sum_{j \geq 1} \phi(uj) + O(|u|). \quad (2.3)$$

The key point of this identity is that the first two terms sharply capture the singular behaviour of $\ln p(e^{-u})$ for $u \to 0$, and the series term will turn out to be bounded for u in the wedge $\{u : \operatorname{Re} u > 0, |\operatorname{Im} u| \leq \sigma \operatorname{Re} u\}$.

Introduce $L = \{z = tu : t \in (0, \infty)\}$, the ray starting at $z = 0$ and passing through the point $z_0 = u$. Let L_n denote the part of L for t from $(n-1)$ to n. A standard argument shows that

$$\left| \int_L \phi(z)\, dz - u \sum_{j=1}^{\infty} \phi(uj) \right| \leq |u|^2 \sum_{j=1}^{\infty} \max_{z \in L_n} |\phi'(z)|. \quad (2.4)$$

Since $\phi(z)$ is regular for $|z| < 2\pi$, $|\phi'(z)| \leq K$ for some K and all z with $|z| \leq 1$. Let $z = \xi + i\eta$ be in the wedge; then $\xi \geq |z|(1 + \sigma^2)^{-1/2}$. Using

$$|e^z| = e^\xi, \quad |e^z - 1| \geq ||e^z| - 1| = e^\xi - 1,$$

we then have: for $|z| \geq 1$,

$$|\phi'(z)| \leq 2\frac{e^\xi}{(e^\xi - 1)^2} + 2\frac{e^{2\xi}}{(e^\xi - 1)^3} + \frac{3}{|z|^4} + |e^{-\xi}| = O(|z|^{-4}).$$

Thus, for all z in the wedge,

$$|\phi'(z)| \leq \frac{K_1}{1 + |z|^4},$$

for some absolute constant $K_1 = K_1(\sigma)$. Hence the right-hand expression in (2.4) is of order

$$|u|^2 \sum_{n=0}^{\infty} \frac{1}{1 + n^4|u|^4} \leq |u|^2 \left(1 + \int_0^\infty \frac{dx}{1 + x^4|u|^4} \right)$$

$$= |u|^2 + |u| \int_0^\infty \frac{dy}{1 + y^4}$$

$$= O(|u|), \quad u \to 0.$$

In addition,

$$\lim_{R \to \infty} \int_{\substack{z = Re^{i\theta} \\ \theta \in [0, \operatorname{Arg} u]}} \phi(z)\, dz = 0,$$

as $|\phi(z)| \leqslant K_2 R^{-3}$ on the integration arc, $K_2 = K_2(\sigma)$. Therefore

$$\int_L \phi(z)\,dz = \int_0^\infty \phi(z)\,dz.$$

This completes the proof. $\square$

Besides (2.1), we will need a bound for $|p(q)|$ when q is not necessarily close to 1.

Lemma 2.2. *Let* $q = re^{i\theta}$, $0 < r < 1$, $\theta \in (-\pi, \pi]$. *Then*

$$|p(q)| \leqslant p(r)\exp\left(-\frac{\alpha r \theta^2}{(1-r^2)^2(\alpha r\theta^2 + (1-r)^2)}\right), \quad \alpha = 4/\pi^2. \tag{2.5}$$

Note. This bound indicates that, on the circle of radius r, $|p(q)|$ attains its only maximum at $q = r$, and the maximum is sharply pronounced if r is close to 1.

Proof of Lemma 2.2. Using an inequality

$$\left|\frac{1}{1-z}\right| \leqslant \frac{1}{1-|z|}\exp(\operatorname{Re} z - |z|), \quad (|z| < 1),$$

(see [23]), we obtain

$$|p(re^{i\theta})| \leqslant p(r)\exp\left(\sum_{\ell \geqslant 1} \ell r^\ell (\cos(\theta\ell) - 1)\right).$$

Here

$$\sum_{\ell \geqslant 1} \ell r^\ell (\cos(\theta\ell) - 1) = -(1-r)^{-2} + \operatorname{Re}(1 - re^{i\theta})^{-2}$$

$$\leqslant -(1-r)^{-2} + (1 - 2r\cos\theta + r^2)^{-1}$$

$$\leqslant -\frac{2r(1-\cos\theta)}{(1-r^2)^2(2r(1-\cos\theta) + (1-r)^2)},$$

and it remains to notice that $1 - \cos\theta \geqslant 2\theta^2/\pi^2$. $\square$

3. Proof of Theorem 1.1

We begin with an asymptotic estimate of p_n, the total number of unrestricted diagrams of volume n.

Lemma 3.1. *Let* a, b *be defined by the first two relations in* (1.7), *with* c *defined in* (2.2). *Then*

$$p_n = an^{-25/36}e^{bn^{2/3}}(1 + O(n^{-1/3})). \tag{3.1}$$

Note. Comparing two formulas for c, (1.7) and (2.2), we thus obtain an identity

$$2\int_0^\infty \frac{y\ln y}{e^{2\pi y} - 1}\,dy = \int_0^\infty \left[\frac{e^z}{z(e^z-1)^2} - \frac{1}{z^3} + \frac{e^{-z}}{12z}\right]dz.$$

We have not tried to prove this identity directly. According to Maple, the approximate values of the integrals are -0.1654211437 and -0.1654211434, respectively.

Proof of Lemma 3.1. Given $0 < \rho < 1$, by Cauchy's integral formula,

$$p_n = (2\pi i)^{-1} \oint_{\substack{z=\rho e^{i\theta} \\ \theta \in (-\pi,\pi]}} z^{-(n+1)} p(z)\, dz. \tag{3.2}$$

Predictably, we want to choose ρ close to the root of $(\rho^{-n} p(\rho))' = 0$, or setting $\rho = e^{-\xi}$,

$$\sum_{\ell \geq 1} \frac{\ell^2}{e^{\ell \xi} - 1} = n \longrightarrow \xi^{-3}\left[\int_0^\infty \frac{y^2\, dy}{e^y - 1} + O(\xi)\right] = n.$$

Since the integral equals $2\zeta(3)$, we select $\xi = (2\zeta(3)/n)^{1/3}$. Break the interval $(-\pi, \pi]$ in two parts, $[-n^{-\delta}, n^{-\delta}]$ and $[-n^{-\delta}, n^{-\delta}]^c$, where $\delta \in (5/9, 2/3)$. By (2.1), and Lemma 2.2,

$$\left|\int_{|\theta| \geq n^{-\delta}} z^{-(n+1)} p(z)\, dz \right| \leq \rho^{-n} p(\rho) \int_{|\theta| \geq n^{-\delta}} e^{-\alpha_1 n^{4/3 - 2\delta}}\, d\theta \quad (\alpha_1 > 0)$$

$$= \exp(bn^{2/3} - \alpha_2 n^{4/3 - 2\delta}) \quad (\alpha_2 > 0). \tag{3.3}$$

Consider $|\theta| \leq n^{-\delta}$. Since $|\theta| = o(\xi)$, we apply (2.1) for $u = \xi - i\theta$ and after simple algebra obtain

$$\ln \frac{p(\rho e^{i\theta})}{(\rho e^{i\theta})^n} = bn^{2/3} - \frac{1}{36} \ln \frac{n}{2\zeta(3)} + c - \frac{1}{12\xi} i\theta - \theta^2 n^{4/3} \gamma_n$$
$$+ i\theta^3 n^{5/3} \gamma_n' + O(\theta^4 n^2) + O(n^{-1/3}), \tag{3.4}$$

where

$$\gamma_n = \zeta(3)^{-1/3} 2^{-4/3} n^{4/3} (1 + O(n^{-2(\delta - 1/3)})),$$

$\gamma_n' = O(1)$, and the last remainder term comes from $O(|u|)$ in (2.1). We notice that, by the definition of δ, $2(\delta - 1/3) > 1/3$ and $|\theta|^3 n^{5/3} = o(1)$, $\theta^4 n^2 = o(1)$, uniformly for θs in question. Using (3.4) and

$$\exp\left(-\frac{1}{12\xi} i\theta + i\theta^3 n^{5/3} \gamma_n' + O(\theta^4 n^2)\right) = 1 - \frac{i\theta}{12\xi} + i\theta^3 n^{5/3} \gamma_n' + O(\theta^4 n^2 + n^{-1/3}),$$

we obtain

$$\oint_{|\theta| \leq n^{-\delta}} z^{-(n+1)} p(z)\, dz$$

$$= \frac{e^{bn^{2/3} + c}}{2\pi \left(\frac{n}{2\zeta(3)}\right)^{1/36}} \times \int_{|\theta| \leq n^{-\delta}} e^{-\gamma_n \theta^2} \left(1 + i\theta^3 n^{5/3} \gamma_n' + O(\theta^4 n^2 + n^{-1/3})\right) d\theta. \tag{3.5}$$

Here

$$\int_{|\theta|\leqslant n^{-\delta}} e^{-\gamma_n \theta^2} d\theta = (\pi/\gamma_n)^{1/2}(1+o(n^{-2/3})),$$

$$\int_{|\theta|\leqslant n^{-\delta}} e^{-\gamma_n \theta^2} \left(-\frac{i\theta}{12\xi} + i\theta^3 n^{5/3}\gamma_n'\right) d\theta = 0,$$

$$\int_{|\theta|\leqslant n^{-\delta}} e^{-\gamma_n \theta^2}(\theta^4 n^2 + n^{-1/3}) d\theta = O(\gamma_n^{-1/2} n^{-1/3}).$$

Hence the integral in (3.5) equals, within a factor $1 + O(n^{-1/3})$,

$$\frac{e^{bn^{2/3}+c}}{2\pi\left(\frac{n}{2\zeta(3)}\right)^{1/36}} \cdot \frac{\pi^{1/2}}{\sqrt{\zeta(3)^{-1/3} 2^{-4/3} n^{4/3}}} = an^{-25/36} e^{bn^{2/3}}.$$

The proof of Lemma 3.1 is complete. $\square$

Turn now to $p_n(r,s,t)$, the total number of diagrams of volume n that fit inside the parallelepiped $B_{r,s,t}$.

Lemma 3.2. *Given $h, w, d > 0$, let r, s, t be the integer parts of*

$$\frac{n^{1/3}}{\beta} \ln\left[\frac{\left(\frac{n^{1/3}}{\beta}\right)^2}{h}\right], \quad \frac{n^{1/3}}{\beta} \ln\left[\frac{\left(\frac{n^{1/3}}{\beta}\right)^2}{w}\right], \quad \text{and} \quad \frac{n^{1/3}}{\beta} \ln\left[\frac{\left(\frac{n^{1/3}}{\beta}\right)^2}{d}\right],$$

respectively. Then

$$p_n(r,s,t) = an^{-25/36} e^{bn^{2/3}} e^{-(h+w+d)}(1 + O(n^{-1/3})). \tag{3.6}$$

Using Lemmas 3.1 and 3.2 we obtain immediately that the probability that the (uniformly) random diagram fits into the parallelepiped $B(r,s,t)$, with r,s,t defined in Lemma 3.2, is asymptotic to $e^{-h} e^{-w} e^{-d}$. This proves Theorem 1.1.

Proof of Lemma 3.2. The number $p_n(r,s,t)$ is given by the Cauchy integral formula, like (3.2), with $p_{r,s,t}(q)$ instead of $p(q)$. We choose again the circle of radius $\rho = e^{-\xi}$. On this circle, the total of the last four products in the formula (2.2) for $p_{r,s,t}(q), (q = \rho e^{i\theta})$, is (uniformly) asymptotic to

$$\exp\left(1 + O\left(\frac{\rho^{r+s} + \rho^{r+t} + \rho^{s+t}}{(1-\rho)^2}\right)\right) \cdot \exp\left(1 + O\left(\frac{\rho^{r+s+t}}{(1-\rho)^2}\right)\right) = 1 + O(n^{-2/3}). \tag{3.7}$$

Furthermore, the total of the three products in the second line of (1.2) is asymptotic to

$$\exp\left(-\frac{q^r + q^s + q^t}{(1-q)^2} + O\left(\frac{\rho^{2r} + \rho^{2s} + \rho^{2t}}{(1-\rho^2)^2}\right)\right) = \exp\left(-\frac{q^r + q^s + q^t}{(1-q)^2} + O(n^{-2/3})\right)$$

$$= \exp(O(1)). \tag{3.8}$$

Using (3.7), (3.8), we see that – analogously to (3.3) – the contribution of the set $\{\theta : |\theta| \geqslant n^{-\delta}\}$ to the value of the contour integral is of order $\exp(bn^{2/3} - \alpha_3 n^{4/3-2\delta})$, at most. Now, for $|\theta| \leq n^{-\delta}$,

$$\frac{q^r + q^s + q^t}{(1-q)^2} = \frac{e^{-\xi r} + e^{-\xi s} + e^{-\xi t}}{(1-e^{-\xi})^2} + i\theta(r + s + t + 6/\xi) + O(n^{-1/3})$$
$$= (h + w + d) + i\theta(r + s + t + 6/\xi) + O(n^{-1/3}). \tag{3.9}$$

From (3.7)–(3.9), it follows that the contribution of the interval $[-n^{-\delta}, n^{-\delta}]$ to the contour integral is obtained via inserting an extra factor $e^{-(h+w+d)}(1 + i\theta(r + s + t + 6/\xi) + O(n^{-1/3}))$ into the integrand on the right-hand side of (3.5). So, arguing analogously, we obtain that

$$p_{r,s,t}(n) = an^{-25/36} e^{bn^{2/3}} e^{-(h+w+d)}(1 + O(n^{-1/3})).$$

Lemma 3.2 is proved. □

4. Proof of Theorem 1.2

Consider first the joint distribution of the largest heights $H_n^{(i)}$ alone. (By symmetry, the distributions of the largest widths $W_n^{(i)}$ and the largest depths $D_n^{(i)}$ are exactly the same.)

Given $k \geq 1$, $\mathbf{d}^{(k)} = (d_1, \ldots, d_k)$ where $d_1 > \cdots > d_k \geq 1$, let $f(q; \mathbf{d}^{(k)})$ denote the generating function of the diagrams with the k largest heights equal to $d_1, \ldots, d_k$. For $k \geq 2$, introduce also $F(q; \mathbf{d}^{(k)})$, the generating function of the diagrams with the $(k-1)$ largest heights equal $d_1, \ldots, d_{k-1}$, and all other heights being d_k, at most. Define $F(q;d) = p_d(q)$, so $F(q;d)$ is the generating function of diagrams with all heights being d, at most. Clearly

$$f(q; \mathbf{d}^{(k)}) = F(q; \mathbf{d}^{(k)}) - F(q; \tilde{\mathbf{d}}^{(k)}), \qquad \tilde{\mathbf{d}}^{(k)} = (d_1, \ldots, d_{k-1}, d_k - 1). \tag{4.1}$$

Furthermore, given $d_1 > d_2 > \cdots > d_k > d_{k+1}$, the ratio of $f(q; d_1, \ldots, d_{k-1}, d_{k+1})$ to $q^{d_{k+1}}$ is the generating function of partitions with the k largest heights equal to $d_1, \ldots, d_{k-1}, d_{k+1}$, enumerated according to the sum of all heights, excluding the height d_{k+1}. (In all such partitions, the $(k+1)$th-largest height is d_{k+1} at most.) Multiplying this fraction by q^{d_k}, we get the generating function of all partitions with the k largest heights equal to $d_1, \ldots, d_{k-1}, d_k$, and all other heights being d_{k+1}, at most, which is $F(q; \mathbf{d}^{(k+1)})$. Thus

$$F(q; \mathbf{d}^{(k+1)}) = q^{d_k - d_{k+1}} f(q; \hat{\mathbf{d}}^{(k)}), \qquad \hat{\mathbf{d}}^{(k)} = (d_1, \ldots, d_{k-1}, d_{k+1}). \tag{4.2}$$

The equations (4.1), (4.2) allow us to express $f(q; d^{(k)})$ through $p_d(q)$ directly.

Lemma 4.1. *Let $d_1 > \cdots > d_k \geq k$. Then*

$$f(q; d^{(k)}) = q^{\sum_{t=1}^{k-1}(d_t - d_k)} \sum_{j=0}^{k} (-1)^j q^{\binom{j}{2}} \begin{bmatrix} k \\ j \end{bmatrix}_q p_{d_k - j}(q), \tag{4.3}$$

where $\begin{bmatrix} k \\ j \end{bmatrix}_q$ is the q-binomial coefficient,

$$\begin{bmatrix} k \\ j \end{bmatrix}_q = \frac{[k!]_q}{[j!]_q [(k-j)!]_q}, \qquad [m!]_q \stackrel{\text{def}}{=} (1-q)(1-q^2) \cdots (1-q^m).$$

Notes. **(1)** The coefficients of p_{d_k-j} in the (4.3) sum are exactly those in the q-binomial theorem, [1], namely

$$\prod_{j=0}^{k-1}(1-xq^j) = \sum_{j=0}^{k}(-1)^j q^{\binom{j}{2}} \begin{bmatrix} k \\ j \end{bmatrix}_q x^j, \qquad (4.4)$$

a fact of critical importance to us later.

(2) The derivation of (4.3) works without any changes for the ordinary, linear, partitions, in which case $p_d(q)$ is the generating function of partitions with all parts d, at most. Here

$$p_d(q) = \prod_{j=1}^{d}(1-q^j)^{-1},$$

and, without using (4.3),

$$f(q; d^{(k)}) = q^{d_1+\cdots+d_k} p_{d_k}(q) = q^{d_1+\cdots+d_k} \prod_{j=1}^{d_k}(1-q^j)^{-1}.$$

Plugging both relations into (4.3), after cancellations and setting $d_k = d$, we obtain an identity

$$\frac{q^{kd}}{[d!]_q} = \sum_{j=0}^{k}(-1)^j q^{\binom{j}{2}} \begin{bmatrix} k \\ j \end{bmatrix}_q \frac{1}{[(d-j)!]_q}, \quad d \geq k.$$

Has the reader seen this identity anywhere? (Christian Krattenhaler has informed me that this identity is a special case of the q-Chu–Vandermonde summation formula.)

Proof of Lemma 4.1. Since $\begin{bmatrix} 1 \\ 0 \end{bmatrix}_q = \begin{bmatrix} 1 \\ 1 \end{bmatrix}_q = 1$, (4.3) and (4.2) coincide for $k = 1$. Suppose (4.3) holds for some $k \geq 1$. Using (4.2) and (4.3), we then obtain

$$F(q; \mathbf{d}^{(k+1)}) = q^{\sum_{t=1}^{k}(d_t-d_{k+1})} \sum_{j=0}^{k}(-1)^j q^{\binom{j}{2}} \begin{bmatrix} k \\ j \end{bmatrix}_q p_{d_{k+1}-j}(q).$$

So, by (4.1) with $k + 1$ instead of k,

$$f(q; \mathbf{d}^{(k+1)}) = q^{\sum_{t=1}^{k}(d_t-d_{k+1})} \sum_{j=0}^{k}(-1)^j q^{\binom{j}{2}} \begin{bmatrix} k \\ j \end{bmatrix}_q p_{d_{k+1}-j}$$

$$- q^{\sum_{t=1}^{k}(d_t-(d_{k+1}-1))} \sum_{j=0}^{k}(-1)^j q^{\binom{j}{2}} \begin{bmatrix} k \\ j \end{bmatrix}_q p_{d_{k+1}-(1+j)}$$

$$= q^{\sum_{t=1}^{k}(d_t-d_{k+1})} \sum_{j=0}^{k+1}(-1)^j q^{\binom{j}{2}} \left(\begin{bmatrix} k \\ j \end{bmatrix}_q + q^{k-(j-1)} \begin{bmatrix} k \\ j-1 \end{bmatrix}_q \right) p_{d_{k+1}-j},$$

($\begin{bmatrix} k \\ -1 \end{bmatrix}_q = 0$ by definition). It remains to use the basic recurrence for q-binomial coefficients, namely

$$\begin{bmatrix} k+1 \\ j \end{bmatrix}_q = \begin{bmatrix} k \\ j \end{bmatrix}_q + q^{k+1-j} \begin{bmatrix} k \\ j-1 \end{bmatrix}_q. \qquad (4.5)$$

So (4.3) holds for $k + 1$, too. Lemma 4.1 is proved. □

While Lemma 4.1 holds for both linear and planar partitions, the next statement is specific for planar partitions.

Corollary 4.2.

$$f(q;\mathbf{d}^{(k)}) = q^{\sum_{t=1}^{k-1}(d_t-d_k)} p_{d_k}(q) F_k(q;d_k),$$

$$F_k(q;d) := \sum_{j=0}^{k} (-1)^j q^{\binom{j}{2}} \begin{bmatrix} k \\ j \end{bmatrix}_q \prod_{v=1}^{j} x(d-v),$$

$$x(\tau) := \prod_{\ell > \tau} (1 - q^\ell). \tag{4.6}$$

Proof of Corollary 4.2. In view of Lemma 4.1, and

$$\frac{p_{d-j}}{p_d(q)} = \prod_{v=1}^{j} \frac{p_{d-v}(q)}{p_{d-v+1}(q)},$$

it suffices to show that

$$\frac{p_{d-v}(q)}{p_{d-v+1}(q)} = x(d-v), \tag{4.7}$$

which in turn follows directly from the formula (1.4) for $p_r(q)$. The corollary is proved. □

However explicit, the formula (4.6) is unsuitable for use in the Cauchy integral formula. The reason is that the terms' signs alternate, and the absolute value of $F_k(q;d)$ for $q \to 1$ falls far below the absolute values of individual terms, and of their differences. By analogy with the q-binomial theorem, one would expect a closed, product-type formula for $F_k(q;d)$, which would tell one how small $F_k(q;d)$ actually is. No such formula is forthcoming though. Still, the q-binomial nature of the coefficients in the sum for F_k leads to the following *asymptotic* formula.

Lemma 4.3. Let $q \to 1$ and $d \to \infty$ in such a way that $1 - q = O(|1 - |q||)$, and $|q|^d = o(|1-q|)$. Then

$$F_k(q;d) = \left(\frac{q^d}{1-q}\right)^k + O\left(\frac{|q|^{(k+1)d}}{|1-q|^{k+1}}\right). \tag{4.8}$$

Proof of Lemma 4.3. For $k = 1$, we have

$$F_1(q;d) = 1 - x(d-1) = 1 - \prod_{\ell \geq d}(1-q^\ell)$$

$$= \sum_{t \geq 1}(-1)^{t-1} \sum_{\ell_t > \cdots > \ell_1 \geq d} q^{\ell_t} \cdots q^{\ell_1}$$

$$= \sum_{t \geq 1} (-1)^{t-1} \frac{q^{td}}{1-q^t} \prod_{v=1}^{t-1} \frac{q^v}{1-q^v}$$
$$= \sum_{t \geq 1} g_{1,t}(q) \left(\frac{q^d}{1-q}\right)^t, \qquad (4.9)$$

where

$$g_{1,t}(q) = (-1)^{t-1} q^{\binom{t}{2}} \prod_{v=1}^{t} (1 + \cdots + q^{v-1})^{-1}.$$

In particular, $g_{1,1}(q) = 1$. The series in (4.8) converges absolutely. Indeed, using $1-q = O(|1-|q||)$, we have

$$|1 + \cdots + q^{v-1}| = \frac{|1-q^v|}{|1-q|} \geq \omega^{-1} \left|\frac{1-|q|^v}{1-|q|}\right| \geq \omega^{-1},$$

for some constant $\omega > 0$, so that $|g_{1,t}(q)| \leq \omega^t$ and the tth term in (4.9) is of order $(o(1))^t$. Therefore (4.8) holds for $k = 1$. Suppose that, for some $k \geq 1$,

$$F_k(q; d) = \sum_{t \geq k} g_{k,t}(q) \left(\frac{q^d}{1-q}\right)^t, \qquad (4.10)$$

where $g_{k,t}(q) = O(\omega_k^t)$ for q in question; $\omega_k > 0$ is a constant. (We have just established (4.10) for $k = 1$, with $\omega_1 = \omega$.) Let us prove (4.10) for $F_{k+1}(q; d)$. To this end, first we use (4.5) and (4.6) for $F_{k+1}(q; d)$ to get

$$F_{k+1}(q; d) = F_k(q; d) + q^k \sum_{j=1}^{k+1} (-1)^j q^{\frac{(j-1)(j-2)}{2}} \begin{bmatrix} k \\ j-1 \end{bmatrix} \prod_{v=1}^{j} x(d-v)$$
$$= F_k(q; d) - q^k x(d-1) F_k(q; d-1).$$

So, using the inductive assumption (4.10) and the last equality in (4.9), we have

$$F_{k+1}(q; d) = \sum_{t \geq k} g_{k,t} \left(\frac{q^d}{1-q}\right)^t - q^k \left(1 - \sum_{t_1 \geq 1} g_{1,t_1} \left(\frac{q^d}{1-q}\right)^{t_1}\right) \sum_{t_2 \geq k} g_{k,t_2} \left(\frac{q^{d-1}}{1-q}\right)^{t_2}$$
$$= \sum_{t \geq k} g_{k,t}(1-q^{k-t}) \left(\frac{q^d}{1-q}\right)^t + \sum_{t \geq k+1} \left(\frac{q^d}{1-q}\right)^t \left(\sum_{\substack{t_1 \geq 1, t_2 \geq k \\ t_1 + t_2 = t}} q^{k-t} g_{1,t_1} g_{k,t_2}\right)$$
$$= \sum_{t \geq k+1} g_{k+1,t} \left(\frac{q^d}{1-q}\right)^t,$$

where

$$g_{k+1,t} = g_{k,t}(1-q^{k-t}) + \sum_{\substack{t_1 \geq 1, t_2 \geq k \\ t_1 + t_2 = t}} q^{k-t} g_{1,t_1} g_{k,t_2}, \quad t \geq k+1. \qquad (4.11)$$

Consequently, $g_{k+1,t} = O(\omega_{k+1}^t)$, for $\omega_{k+1} = 2 \max\{\omega_1, \omega_k\}$, say. So (4.10) holds for all $k \geq 1$.

Setting $t = k+1$ in (4.11), and remembering $g_{1,1} = 1$, we get

$$g_{k+1,k+1} = q^{-1}g_{k,k} + g_{k,k+1}(1 - q^{-1}).$$

Therefore, after simple algebra,

$$g_{k+1,k+1} = q^{-k} + q^{-k}(q-1)\sum_{t=0}^{k-1} q^t g_{t+1,t+2} = 1 + O(|1-q|), \qquad (4.12)$$

for every $k \geq 0$. The relations (4.10), (4.12) taken together complete the proof of (4.8). $\square$

Corollary 4.2 and Lemma 4.3 imply the following.

Corollary 4.4. *Let $q \to 1$ and $d_k \to \infty$ in such a way that $1 - q = O(|1 - |q||)$ and $|q|^{d_k} = o(|1-q|)$. Then*

$$f(q; \mathbf{d}^{(k)}) = \frac{q^{\sum_{t=1}^k d_t}}{(1-q)^k} p_{d_k}(q)\left[1 + O\left(\frac{|q|^{d_k}}{|1-q|}\right)\right]. \qquad (4.13)$$

Introduce $p_n(\mathbf{d}^{(k)})$, the total number of diagrams with the k largest heights equal $d_1, d_2, \ldots, d_k$. Of course, $p_n(\mathbf{d}^{(k)})$ is the coefficient by q^n in the generating function $f(q; \mathbf{d}^{(k)})$, expressed, as before, through the Cauchy integral

$$p_n(\mathbf{d}^{(k)}) = (2\pi i)^{-1} \oint_{\substack{z=e^{-\xi+i\theta} \\ \theta \in (-\pi,\pi]}} z^{-(n+1)} f(z; \mathbf{d}^{(k)}) \, dz. \qquad (4.14)$$

Let the integers $d_1 > \cdots > d_k$ be such that

$$d_j = \frac{n^{1/3}}{\beta}\left[\log\left(\frac{n^{1/3}}{\beta}\right)^2 + u_j\right],$$

where $u_j = u_j(n) = O(1)$. Consider $\theta \in [-n^{-\delta}, n^\delta]$, $\delta \in (5/9, 2/3)$. For these θs, using $\xi = \beta/n^{1/3}$,

$$q^{\sum_{t=1}^k d_t} = \exp\left((-\xi + i\theta)\sum_{t=1}^k \frac{n^{1/3}}{\beta}\left[\log\left(\frac{n^{1/3}}{\beta}\right)^2 + u_j\right]\right)$$
$$= \left(\frac{\beta}{n^{1/3}}\right)^{2k} \exp\left(-\sum_{t=1}^k u_t\right)\left(1 + i\theta k \frac{n^{1/3}}{\beta}\log\left(\frac{n^{1/3}}{\beta}\right)^2 + O(n^{-1/3})\right),$$

then

$$(1-q)^{-k} = (1 - e^{-\xi+i\theta})^{-k}$$
$$= (1-e^{-\xi})^{-k}\left(1 + \frac{1-e^{i\theta}}{e^\xi - 1}\right)^{-k}$$
$$= \left(\frac{n^{1/3}}{\beta}\right)^k\left(1 - ik\frac{\theta}{\xi} + O(n^{-1/3})\right),$$

and (see the proof of Lemma 3.2)

$$\begin{aligned}p_{d_k}(e^{-\xi+i\theta}) &= p(e^{-\xi+i\theta})\exp\left(-\frac{e^{-\xi d_k}}{(1-e^{-\xi})^2}+i\theta(d_k+2/\xi)+O(n^{-1/3})\right)\\&=e^{-e^{-u_k}}\left(1+i\theta(d_k+2/\xi)+O(n^{-1/3})\right).\end{aligned}$$

Therefore (4.13) becomes

$$\begin{aligned}f(e^{-\xi+i\theta};\mathbf{d}^{(k)}) &= \left(\frac{\beta}{n^{1/3}}\right)^k \exp\left(-\sum_{t=1}^k u_t - e^{-u_k}\right) p(e^{-\xi+i\theta})\\&\quad \times \left(1+i\theta\frac{n^{1/3}}{\beta}\left(2(k+1)\log\frac{n^{1/3}}{\beta}-(k-2)\right)+O(n^{-1/3})\right).\end{aligned}$$

So, as in the proof of Lemma 3.2, the contribution of the small θs to the contour integral (4.14) is, within a factor $1+O(n^{-1/3})$,

$$an^{-25/36}e^{bn^{2/3}} \cdot \left(\frac{\beta}{n^{1/3}}\right)^k \exp\left(-\sum_{t=1}^k u_t - e^{-u_k}\right). \quad (4.15)$$

As for the remaining θs, the exact formula (4.6) can be used to show – just as in the proofs of Lemmas 3.1 and 3.2 – that their overall contribution is by a factor $\exp(-\alpha^* n^{4/3-2\delta})$ smaller than the quantity in (4.15). Combining this with the asymptotic formula for p_n, the total number of solid diagrams, we have thus proved that the probability that the k largest heights are $(n^{1/3}/\beta)[\log(n^{1/3}/\beta)^2+u_j]$, $u_j = O(1)$, $1 \leqslant j \leqslant k$, is asymptotic to

$$\left(\frac{\beta}{n^{1/3}}\right)^k \exp\left(-\sum_{t=1}^k u_t - e^{-u_k}\right). \quad (4.16)$$

Therefore we have proved that the sequence $\{\mathcal{H}_n^{(i)}\}_{i\leqslant k}$, defined by

$$H_n^{(i)} = (n^{1/3}/\beta)(\log(n^{1/3}/\beta)+\mathcal{H}_n^{(i)}),$$

converges – in distribution – to $\{Y_i\}_{i\leqslant k}$ with the density

$$\exp\left(-\sum_{t\leqslant k} u_t - e^{-u_k}\right), \quad u_1 \geqslant \cdots \geqslant u_k.$$

As we know, it is the Markov chain, with the density of Y_1 equal to $e^{-u-e^{-u}}$, and the conditional density of Y_i, given $Y_{i-1}=u$, equal to $e^{-v-e^{-v}+e^{-u}}$, $v \leqslant u$. Since $\{Y_i\}_{i\leqslant k}$ is distributed as $\{\log(\sum_{j\leqslant i} E_j)^{-1}\}_{i\leqslant k}$, $E_1, E_2, \ldots$ being independent copies of E, this establishes the part of Theorem 1.2 concerning the *marginal* distribution of the largest heights, whence – by symmetry – the largest widths, and the largest depths. What is still lacking is the proof that the *joint* distribution of $\{h_n^{(i)}, w_n^{(i)}, d_n^{(i)}\}_{i\leqslant k}$ converges to that of $\{\sum_{j\leqslant i} E_j, \sum_{j\leqslant i} E'_j, \sum_{j\leqslant i} E''_j\}_{i\leqslant k}\}$, where the sequences $\{E_j\}$, $\{E'_j\}$, $\{E''_j\}$ are mutually independent. Somewhat surprisingly, we will prove it by using – three times in succession – the core of the '1-dimensional' argument above.

First, notation. For $\mathbf{k} = (k_1, k_2, k_3) \geqslant (1,1,1)$, denote

$$\mathbf{d}_i^{(k_i)} = (d_{i,1},\ldots,d_{i,k_i}), \quad (d_{i,1} > \cdots > d_{i,k_i} \geqslant 1), \quad i=1,2,3.$$

Set $\mathbf{d}^{(\mathbf{k})} = (\mathbf{d}_1^{(k_1)}, \mathbf{d}_2^{(k_2)}, \mathbf{d}_3^{(k_3)})$. Let $f(q; \mathbf{d}^{(\mathbf{k})})$ denote the generating function of the diagrams with k_1 largest heights, k_2 largest widths, and k_3 largest depths comprising the tuple $\mathbf{d}^{(\mathbf{k})}$. Let $F_1(q; \mathbf{d}^{(\mathbf{k})})$ denote the generating function of the diagrams with $(k_1 - 1)$ largest heights, k_2 largest widths, and k_3 largest depths comprising the $((k_1 - 1) + k_2 + k_3)$-tuple $(\mathbf{d}_1^{(k_1-1)}, \mathbf{d}_2^{(k_2)}, \mathbf{d}_3^{(k_3)})$, and the remaining heights being at most d_{1,k_1}. Define $F_2(q; \mathbf{d}^{(\mathbf{k})})$, $F_3(q, \mathbf{d}^{(\mathbf{k})})$ analogously.

As in (4.1),

$$\begin{aligned} f(q; \mathbf{d}^{(\mathbf{k})}) &= F_1(q; \mathbf{d}^{(\mathbf{k})}) - F_1(q; \mathbf{d}^{(\mathbf{k})} - \mathbf{e}_{k_1}) \\ &= F_2(q; \mathbf{d}^{(\mathbf{k})}) - F_2(q; \mathbf{d}^{(\mathbf{k})} - \mathbf{e}_{k_1+k_2}) \\ &= F_3(q; \mathbf{d}^{(\mathbf{k})}) - F_3(q; \mathbf{d}^{(\mathbf{k})} - \mathbf{e}_{k_1+k_2+k_3}), \end{aligned} \quad (4.17)$$

where $\mathbf{e}_j = (\overbrace{0, \ldots, 0}^{j-1}, 1, \overbrace{0, \ldots, 0}^{|\mathbf{k}|-j})$, $|\mathbf{k}| = k_1 + k_2 + k_3$. Furthermore, as in (4.2),

$$\begin{aligned} F_1(q; \mathbf{d}^{(\mathbf{k}+\mathbf{e}_1)}) &= q^{d_{1,k_1} - d_{1,k_1+1}} f(q; ((\mathbf{d}_1^{(k_1-1)}, d_{1,k_1+1}), \mathbf{d}_2^{(k_2)}, \mathbf{d}_3^{(k_3)})), \\ F_2(q; \mathbf{d}^{(\mathbf{k}+\mathbf{e}_2)}) &= q^{d_{2,k_2} - d_{2,k_2+1}} f(q; (\mathbf{d}_1^{(k_1)}, (\mathbf{d}_2^{(k_2-1)}, d_{2,k_2+1}), \mathbf{d}_3^{(k_3)})), \\ F_3(q; \mathbf{d}^{(\mathbf{k}+\mathbf{e}_3)}) &= q^{d_{3,k_3} - d_{3,k_3+1}} f(q; (\mathbf{d}_1^{(k_1)}, \mathbf{d}_2^{(k_2)}, (\mathbf{d}_3^{(k_3-1)}, d_{3,k_3+1}))) \end{aligned} \quad (4.18)$$

($\mathbf{e}_i$ being ith unit vector in R^3), provided that $d_{k_i+1} \geqslant \max\{k_j : j \neq i\}$ for the ith identity, $i = 1, 2, 3$. Let us check, say, the first identity. Consider a diagram $\mathscr{P}$ with the parameter $((\mathbf{d}_1^{(k_1-1)}, d_{1,k_1+1}), \mathbf{d}_2^{(k_2)}, \mathbf{d}_3^{(k_3)})$. If we replace the k_1-tallest vertical stack of unit cubes (whose height is d_{1,k_1+1}) by the taller stack of height d_{1,k_1} we get another diagram $\mathscr{P}'$. Its heights with rank exceeding (strictly) k_1 are all d_{1,k_1+1} at most. Also, the depths and the widths in $\mathscr{P}$ which are affected by this operation are among those having rank exceeding d_{1,k_1+1}. Since $d_{1,k_1+1} \geqslant \max\{k_2, k_3\}$, the diagram $\mathscr{P}'$ inherits the parameters $\mathbf{d}_2^{(k_2)}, \mathbf{d}_3^{(k_3)}$ from $\mathscr{P}$ fully intact. Therefore the right-hand expression in (4.18) is indeed the generating function of all diagrams with k_1 largest heights, k_2 largest widths, and k_3 largest depths comprising $\mathbf{d}^{(\mathbf{k})}$, and all other heights being d_{1,k_1+1} at most, that is $F_1(q; (\mathbf{d}_1^{(k_1+1)}, \mathbf{d}_2^{(k_2)}, \mathbf{d}_3^{(k_3)})) = F_1(q; \mathbf{d}^{(\mathbf{k}+\mathbf{e}_1)})$.

Using (4.17), (4.18), we will prove the following counterpart of Corollary 4.4.

Lemma 4.5. *Let $\mathbf{k}$ be fixed. Let $q \to 1$ and $d_{\min} = \min\{d_{i,k_i} : i = 1, 2, 3\} \to \infty$ in such a way that $1 - q = O(|1 - q|)$ and $|q|^{d_{\min}} = O(|1 - q|^2)$. Then*

$$f(q; \mathbf{d}^{(\mathbf{k})}) = \prod_{i=1}^{3} \frac{q^{\sum_{t=1}^{k_i} d_{i,t}}}{(1-q)^{k_i}} p_{d_{1,k_1}, d_{2,k_2}, d_{3,k_3}}(q)(1 + O(|1-q|)). \quad (4.19)$$

Proof of Lemma 4.5. As in the proofs of Lemma 4.3 and Corollary 4.4, first we need to show that

$$f(q; \mathbf{d}^{(\mathbf{k})}) = p_{d_{1,k_1}, d_{2,k_2}, d_{3,k_3}}(q) \prod_{i=1}^{3} q^{\sum_{t=1}^{k_i-1}(d_{i,t} - d_{i,k_i})}$$

$$\times \sum_{\mathbf{t} \geqslant (\mathbf{k},0)} h_{\mathbf{k},\mathbf{t}}(1-q)^{t_4} \prod_{i=1}^{3} \left(\frac{q^{d_{i,k_i}}}{1-q}\right)^{k_i}, \quad (4.20)$$

where $\mathbf{t} = (t_1, t_2, t_3, t_4)$, $|h_{\mathbf{k},\mathbf{t}}| = O(\omega_\mathbf{k}^{|\mathbf{t}|})$, ($|\mathbf{t}| = \sum_i t_i$), for some constant $\omega_\mathbf{k} > 0$ and all $\mathbf{t}$.
For $\mathbf{k} = (1,1,1)$, (4.20) follows from the obvious inclusion–exclusion formula

$$f(q; \mathbf{d}_1) = \sum_{\varepsilon \in \{0,1\}^3} (-1)^{|\varepsilon|} p_{\mathbf{d}_1 - \varepsilon}(q), \quad \mathbf{d}_1 = (d_{1,1}, d_{1,2}, d_{1,3}),$$

and the relations

$$\frac{p_{r-1,s,t}(q)}{p_{r,s,t}(q)} = \frac{x(r-1)x(r+s+t-1)}{x(r+s-1)x(r+t-1)},$$
$$\frac{p_{r,s-1,t}(q)}{p_{r,s,t}(q)} = \frac{x(s-1)x(r+s+t-1)}{x(r+s-1)x(s+t-1)},$$
$$\frac{p_{r,s,t-1}(q)}{p_{r,s,t}(q)} = \frac{x(t-1)x(r+s+t-1)}{x(r+t-1)x(s+t-1)}, \quad (4.21)$$

implied by (1.2). (See (4.6) for the definition of $x(\cdot)$.) Omitting somewhat tedious computations, we just mention that the leading coefficient $h_{\mathbf{k},(\mathbf{k},0)}$ is $1 + O(|1-q|)$ in this case because, for $(r,s,t) = (d_{1,1}, d_{2,1}, d_{3,1})$, the ratios in (4.21) are $1 - (q^{d_{i,1}}/(1-q))[1 + O(|1-q|)]$, $i = 1,2,3$.

Now consider $k_1 \geq 1$, but keep $k_2 = k_3 = 1$. Using (4.17), (4.18), we can use induction on k_1 to show that, analogously to (4.3),

$$f(q; \mathbf{d}^{(\mathbf{k})}) = q^{\sum_{t=1}^{k_1-1}(d_{1,t}-d_{1,k_1})}$$
$$\times \sum_{j=0}^{k_1-1}(-1)^j \begin{bmatrix} k_1-1 \\ j \end{bmatrix}_q q^{\binom{j+1}{2}} f(q; (d_{1,k_1}-j, d_{2,1}, d_{3,1})). \quad (4.22)$$

Combining (4.22) and (4.20) ($\mathbf{k} = (1,1,1)$), and arguing as in the proof of Corollary 4.2, we obtain

$$f(q; \mathbf{d}^{(\mathbf{k})}) = p_{d_{1,k_1}, d_{2,1}, d_{3,1}}(q) \cdot q^{\sum_{t=1}^{k_1-1}(d_{1,t}-d_{1,k_1})} \phi_{k_1}(q; (d_{1,k_1}, d_{2,1}, d_{3,1}));$$
$$\phi_{k_1}(q; (d_1, d_2, d_3)) = \sum_{j=0}^{k_1}(-1)^j \begin{bmatrix} k_1 \\ j \end{bmatrix}_q \prod_{v=1}^{j} \xi(d_1 - v, d_2, d_3)$$
$$\times \sum_{\mathbf{t} \geq (1,1,1,0)} h_{(1,1,1),\mathbf{t}}(1-q)^{t_4} \frac{q^{d-j}}{1-q} \prod_{i=2}^{3} \frac{q^{d_2}}{1-q}, \quad (4.23)$$

where $\xi(r,s,t)$ is the first ratio in (4.21). In view of the first relation in (4.23), and (4.20), our task is to show that for $|q|^{\min_i d_i} = O(|1-q|^2)$,

$$\phi_{k_1}(q; (d_1, d_2, d_3)) = \sum_{\mathbf{t} \geq (\mathbf{k},0)} h_{\mathbf{k},\mathbf{t}}(1-q)^{t_4} \prod_{i=1}^{3}\left(\frac{q^{d_i}}{1-q}\right)^{k_i}. \quad (4.24)$$

For k_1 this definitely holds by the second relation in (4.23). Using (4.5) for $k = k_1 - 1$, by the definition of $\phi_\cdot(\cdot)$, we obtain

$$\phi_{k_1+1}(q; (d_1, d_2, d_3)) = \phi_{k_1}(q; (d_1, d_2, d_3)) - q^{k_1} \xi(d_1-1, d_2, d_3) \phi_{k_1}(q; (d_1-1, d_2, d_3)).$$

Suppose that (4.24) holds for some $k_1 \geq 1$. Then, by the last recurrence,

$$\phi_{k_1+1}(q;(d_1,d_2,d_3)) = \sum_{\mathbf{t} \geq (\mathbf{k},0)} h_{\mathbf{k},\mathbf{t}}(1-q)^{t_4} \prod_{i=1}^{3} \left(\frac{q^{d_i}}{1-q}\right)^{k_i}$$

$$- q^k \xi(d_1-1,d_2,d_3) \sum_{\mathbf{t} \geq (\mathbf{k},0)} h_{\mathbf{k},\mathbf{t}}(1-q)^{t_4} \prod_{i=1}^{3} \left(\frac{q^{d_i-\delta(i,1)}}{1-q}\right)^{t_i}. \quad (4.25)$$

Here, by the definition of $\xi(\cdot)$ (see the comment following (4.21)),

$$\xi(d_1,d_2,d_3) = 1 - \frac{q^{d_1}}{1-q} + \sum_{t_1 \geq 1, |\mathbf{t}| \geq 2} a_{\mathbf{t}}(1-q)^{t_4} \prod_{i=1}^{3} \left(\frac{q^{d_i}}{1-q}\right)^{t_i}. \quad (4.26)$$

So every coefficient with subindex $\mathbf{t} \geq (\mathbf{k},0)$ such that $t_1 = k_1$ in the resulting expansion of $\phi_{k_1+1}(q;(d_1,d_2,d_3))$ equals $1-1 = 0$. Therefore the expansion contains only the terms with $t_1 \geq k_1 + 1$. That there exists $\omega_{\mathbf{k}+\mathbf{e}_1}$ such that the $\mathbf{t}$th coefficient in ϕ_{k_1+1} is $O(\omega_{\mathbf{k}+\mathbf{e}_1}^{|\mathbf{t}|})$ is done as in the proof of Lemma 4.3.

In addition, let $c_{k_1,\mathbf{t}}$ denote the coefficient of the $\mathbf{t}$th term in the expansion of ϕ_{k_1}, $\mathbf{t} \geq (\mathbf{k},0)$, and let c_{k_1} be the first (potentially) nonzero coefficient, that is, $c_{k_1} = c_{k_1,(\mathbf{k},0)}$. From (4.25), (4.26) it follows that

$$c_{k_1+1} = q^{-1} c_{k_1} + (1-q^{-1}) c_{k_1,(\mathbf{k}+\mathbf{e}_1,0)}.$$

Using this recurrence and $c_1 = 1 + O(|1-q|)$, we can easily show that $c_{k_1} = 1 + O(|1-q|)$. Thus we have proved (4.20) and $h_{\mathbf{k},(\mathbf{k},0)} = 1 + O(|1-q|)$ for $\mathbf{k} = (k_1,1,1)$.

Now, only notational changes are needed to use this result as a basis for the inductive proof of (4.20) for $\mathbf{k} = (k_1,k_2,1)$, together with the estimate of the first nonzero coefficient $h_{\mathbf{k},(\mathbf{k},0)}$. Of course, this time we use the second relations in (4.17), (4.18), and (4.21). The final step is the third inductive proof, with the $\mathbf{k} = (k_1,k_2,1)$ as the basis and the third relations in (4.17), (4.18), and (4.21) coming into play, that establishes (4.20) for all $\mathbf{k} = (k_1,k_2,k_3) \geq (1,1,1)$, and the formula

$$h_{\mathbf{k},(\mathbf{k},0)} = 1 + O(|1-q|).$$

Lemma 4.5 is proved completely. □

Now let $\mathbf{d}^{(\mathbf{k})}$ be such that

$$d_{i,t} = \frac{n^{1/3}}{\beta} \left[\log\left(\frac{n^{1/3}}{\beta}\right)^2 + u_{i,t} \right],$$

where $u_{i,t} = u_{i,t}(n) = O(1)$, $t \leq k_i$, $i = 1,2,3$. Closely following the lines of the proof of (4.16), with Lemma 4.5 replacing Corollary 4.4, we obtain that the probability that k_1 largest heights, k_2 largest widths, and k_3 largest depths comprise the tuple $\mathbf{d}^{(\mathbf{k})} = (\mathbf{d}_1^{(k_1)}, \mathbf{d}_2^{(k_2)}, \mathbf{d}_3^{(k_3)})$ is asymptotic to

$$\prod_{i=1}^{3} \left(\frac{\beta}{n^{1/3}}\right)^{k_i} \exp\left(-\sum_{t=1}^{k_i} u_{i,t} - e^{-u_{i,k_i}}\right).$$

This, of course, is equivalent to what Theorem 1.2 states. □

5. Proof of Theorem 1.3

Let $d_1 > \cdots > d_k$ and λ be a planar diagram of area k. Let T be a Young tableau of shape λ with entries $1,\ldots,k$. Introduce $f(q;\mathbf{d^{(k)}}, T)$, the generating function of the plane partitions with k largest heights $d_1,\ldots,d_k$, such that the base of the ith-tallest stack is labelled i in $T^{(k)}$. Let $p_n(\mathbf{d^{(k)}}, T)$ be the number of such solid diagrams of volume n, i.e., $p_n(\mathbf{d^{(k)}}, T) = [q^n] f(q;\mathbf{d^{(k)}}, T)$. From Kadell's formula (1.14) it follows that

$$p_n(\mathbf{d^{(k)}}, T) \leqslant p_n^*(\mathbf{d^{(k)}}, T), \qquad p_n^*(\mathbf{d^{(k)}}, T) \stackrel{\text{def}}{=} [q^n] \frac{q^{\sum_{t=1}^k d_t}}{\prod_{\square \in \lambda^{(k)}} (1 - q^{h(\square)})} p_{d_k}(q). \tag{5.1}$$

Observe that

$$\prod_{\square \in \lambda^{(k)}} (1 - q^{h(\square)}) = (1-q)^k \left(\prod_{\square \in \lambda^{(k)}} h(\square) \right) (1 + O(|1-q|)), \quad q \to 1.$$

So, by (4.13), for $|q|^{d_k} = O(|1-q|^2)$, $p_n^*(\mathbf{d^{(k)}}, T)$ differs from $f(q;\mathbf{d^{(k)}})$ by the factor $\prod_{\square \in \lambda^{(k)}} h^{-1}(\square) + O(|1-q|)$. Consequently, as in the derivation of (4.15), we obtain the following: let $u_1 > \cdots > u_k$, $u_j = O(1)$, be such that $d_j = (n^{1/3}/\beta)\left[\log(n^{1/3}/\beta)^2 + u_j\right]$ are integers; then, within a factor $1 + O(n^{-1/3})$,

$$p_n^*(\mathbf{d^{(k)}}, T) = a n^{-25/36} e^{bn^{2/3}} \cdot \left(\frac{\beta}{n^{1/3}}\right)^k$$

$$\times \exp\left(-\sum_{t=1}^k u_t - e^{-u_k}\right) \prod_{\square \in \lambda^{(k)}} h^{-1}(\square). \tag{5.2}$$

Therefore, the probability that the k tallest stacks have heights $d_1,\ldots,d_k$ and their bases form the tableau T is asymptotically

$$\left(\frac{\beta}{n^{1/3}}\right)^k \exp\left(-\sum_{t=1}^k u_t - e^{-u_k}\right) \prod_{\square \in \lambda^{(k)}} h^{-1}(\square), \tag{5.3}$$

at most. To bound this probability from below, we use (1.14) again to obtain

$$p_n(\mathbf{d^{(k)}}, T) \geqslant p_n^{**}(\mathbf{d^{(k)}}, T), \qquad p_n^{**}(\mathbf{d^{(k)}}, T) \stackrel{\text{def}}{=} [q^n] \frac{q^{\sum_{t=1}^{k-1} d_t + \tilde{d}_k}}{\prod_{\square \in \lambda} (1 - q^{h(\square)})} p_{\tilde{d}_k}(q), \tag{5.4}$$

where $\tilde{d}_k = d_k - \max_j \lambda'_j$. Like $p_n^*(\mathbf{d^{(k)}}, T)$, $p_n^{**}(\mathbf{d^{(k)}}, T)$ is asymptotic to the right-hand side expression in (5.2). Therefore the probability in question is asymptotically *at least* as large as the expression in (5.3), whence it is equivalent to (5.3).

We know that $\{\mathcal{H}_n^{(i)}\}_{i \leqslant k}$ defined by $H_n^{(i)} = (n^{1/3}/\beta)\left[\log(n^{1/3}/\beta)^2 + \mathcal{H}_n^{(i)}\right]$ has the limiting density equal to the exponential factor in (5.3). So, summing over $u_1,\ldots,u_k$, we obtain that

$$\lim_{n \to \infty} P(\mathcal{T}_n^{(k)} = T) = \prod_{\square \in \lambda} h^{-1}(\square). \tag{5.5}$$

Here $\mathcal{T}_n^{(k)}$ is the random Young tableau formed by the bases of the k tallest stacks. (To be sure, $\mathcal{T}_n^{(k)}$ is defined for all partitions $\mathcal{P}$ except a set of partitions whose probability approaches zero as $n \to \infty$.)

Denoting the shape of $\mathcal{T}_n^{(k)}$ by $\mu_n^{(k)}$ and using the hook formula (1.10), we then arrive at

$$\lim_{n\to\infty} \mathrm{P}(\mu_n^{(k)} = \lambda) = d(\lambda) \prod_{\square \in \lambda} h^{-1}(\square) = \frac{d^2(\lambda)}{k!}.$$

In particular, denoting the set of all Young diagrams of area k by Λ_k, we re-establish the identity

$$\sum_{\lambda \in \Lambda_k} \frac{d^2(\lambda)}{k!} = 1.$$

The proof of Theorem 1.3 is complete. □

Note. It seems certain that the condition 'k is fixed' can be relaxed considerably, especially if one is content with an integral version of Theorem 1.3. We conjecture that for $k = o(n^{1/3})$ and $\mathcal{M}_k \subseteq \Lambda_k$,

$$\mathrm{P}(\mu_n^{(k)} \in \mathcal{M}_k) \sim \sum_{\lambda \in \mathcal{M}_k} \frac{d^2(\lambda)}{k!},$$

as long as the sum on the right is bounded away from zero.

Acknowledgement

I am grateful to Sergey Fomin, Christian Krattenhaler, and an anonymous referee for their interest in this paper and expert comments.

References

[1] Andrews, G. E. (1976) *The Theory of Partitions*, Vol. 2 of *Encyclopedia of Mathematics and its Applications*, Addison-Wesley, Reading, MA.

[2] Auluck, F. C., Chowla, S. and Gupta, H. (1942) On the maximum value of partitions of n into k parts. *J. Indian Math. Soc.* **6** 105–112.

[3] Bressoud, D. M. (1999) *Proofs and Confirmations: The Story of the Alternating Sign Matrix Conjecture*. Spectrum Series, The Mathematical Association of America, Cambridge University Press.

[4] Cerf, R. and Kenyon, R. (2001) The low temperature expansion of the Wulff crystal in the 3D Ising model. *Comm. Math. Phys.* **222** 147–179.

[5] Cohn, H., Larsen, M. and Propp, J. (1998) The shape of a typical boxed plane partition. *New York J. Math.* **4** 137–165.

[6] Diaconis, P. (1988) *Group Representations in Probability and Statistics*, Vol. 11 of *Lecture Notes Monogr. Ser.*, Institute of Mathematical Statistics, Hayward, CA.

[7] Erdős, P. and Lehner, J. (1941) The distribution of the number of summands in the partition of a positive integer. *Duke Math. J.* **8** 335–345.

[8] Erdős, P. and Richmond, L. B. (1993) On graphical partitions. *Combinatorica* **13** 57–63.

[9] Frame, J. S., Robinson, G. de B. and Thrall, R. M. (1954) The hook graphs of the symmetric group. *Canad. J. Math.* **6** 316–324.
[10] Fristedt, B. (1993) The structure of random partitions of large integers. *Trans. Amer. Math. Soc.* **337** 703–735.
[11] Gessel, I. M. and Vienott, X. G. (1989) Determinants, paths, and plane partitions. Preprint, available at http://www.cs.brandeis.edu/~ira/
[12] Greene, C., Nijenhuis, A. and Wilf, H. S. (1979) A probabilistic proof of a formula for the number of Young tableaux of a given shape. *Adv. Math.* **31** 104–109.
[13] Greene, C., Nijenhuis, A. and Wilf, H. S. (1984) Another probabilistic method in the theory of Young tableaux. *J. Combin. Theory Ser. A* **37** 127–135.
[14] Kadell, K. W. J. (1997) Schützenberger's Jeu de Taquin and plane partitions *J. Combin. Theory Ser. A* **77** 110–133.
[15] Knuth, D. E. (1998) *The Art of Computer Programming*, Vol. 3, *Sorting and Searching*, second edition, Addison-Wesley, Reading, MA.
[16] Krattenhaler, C. (1990) Generating functions for plane partitions of a given shape. *Mathematica* pp. 173–201.
[17] Ledermann, W. (1987) *Introduction to Group Characters*, second edition, Cambridge University Press, New York.
[18] McKay, B. D., Morse, J. and Wilf, H. S. (2002) The distributions of the entries of Young tableaux *J. Combin. Theory Ser. A* **97** 117–128.
[19] MacMahon, P. A. (1960) *Combinatory Analysis*, Chelsea, New York. Originally published in two volumes by Cambridge University Press (1915–1916),
[20] Okounkov, A. (2001) Infinite wedge and random partitions. *Selecta Math., New Ser.* **7** 57–81.
[21] Okounkov, A. and Reshetikhin, N. (2003) Correlation function of Schur process with applications to local geometry of a random 3-dimensional Young diagram. *J. Amer. Math. Soc.* **16** 581–603.
[22] Pittel, B. (1986) On growing a random Young tableau. *J. Combin. Theory Ser. A* **41** 278–285.
[23] Pittel, B. (1997) On a likely shape of the random Ferrers Diagram. *Adv. Appl. Math.* **18** 432–488.
[24] Pittel, B. (1999) Confirming two conjectures about the integer partitions. *J. Combin. Theory Ser. A* **88** 123–135.
[25] Postnikov, A. G. (1988) *Introduction to Analytic Number Theory*, Vol. 68 of *Translations of Mathematical Monographs*, AMS, Providence, RI.
[26] Sagan, B. E. (1991) *The Symmetric Group*, Wadsworth and Brooks, Pacific Grove, CA.
[27] Stanley, R. (1971) Theory and applications of plane partitions, Part 2. In *Studies in Applied Mathematics*, **50** 259–279.
[28] Stanley, R. (1999) *Enumerative Combinatorics 2*, Vol. 62 of *Cambridge Studies in Advanced Mathematics*, Cambridge University Press.
[29] Stanley, R. (2003) On the enumeration of skew Young tableaux. *Adv. Appl. Math.* **30** 283–294.
[30] Wright, E. M. (1931) Asymptotic partition formulae I: Plane partitions. *Quart. J. Math.* (Oxford Series) **2** 177–189.

The Small Giant Component in Scale-Free Random Graphs

OLIVER RIORDAN[†]

Department of Pure Mathematics and Mathematical Statistics,
University of Cambridge, Cambridge CB3 0WB, UK
and
Trinity College, Cambridge CB2 1TQ, UK
(e-mail: o.m.riordan@dpmms.cam.ac.uk)

Received 12 March 2004; revised 29 October 2004

For Béla Bollobás on his 60th birthday

Building on the methods developed in joint work with Béla Bollobás and Svante Janson, we study the phase transition in four 'scale-free' random graph models, obtaining upper and lower bounds on the size of the giant component when there is one. In particular, we determine the extremely slow rate of growth of the giant component just above the phase transition. We greatly reduce the significant gaps between the existing upper and lower bounds, giving bounds that match to within a factor $1 + o(1)$ in the exponent.

In all cases the method used is to couple the neighbourhood expansion process in the graph on n vertices with a continuous-type branching process that is independent of n. It can be shown (requiring some separate argument for each case) that with probability tending to 1 as $n \to \infty$ the size of the giant component divided by n is within $o(1)$ of the survival probability σ of the branching process. This survival probability is given in terms of the maximal solution ϕ to certain non-linear integral equations, which can be written in the form $\phi = \mathbf{F}(\phi)$ for a certain operator $\mathbf{F}$. Upper and lower bounds are found by constructing trial functions ϕ_0, ϕ_1 with $\mathbf{F}(\phi_0) \leqslant \phi_0$ and $\mathbf{F}(\phi_1) \geqslant \phi_1$ holding pointwise; basic properties of branching processes then imply that $\phi_1 \leqslant \phi \leqslant \phi_0$, giving upper and lower bounds on σ.

1. Introduction

In the last few years there has been much interest in new random graph models motivated by the empirical and heuristic study of large-scale real-world graphs such as the world-wide web. For extensive surveys of this area, see [2, 15]. To date, most of the work on these models has been experimental or heuristic, but there is an increasing body of precise mathematical work. For a partial survey of this see [7]. One of the questions frequently asked about random graphs in general, and these new models in particular, is the following. How robust is the graph under random failures? More precisely, if vertices

[†] Royal Society Research Fellow.

or edges of the n-vertex graph are deleted independently with a certain probability $1-p$, how large is the largest component in what remains? In particular, for which p is there a *giant component*, i.e., a component of order $\Theta(n)$? If there is a critical probability p_0 above which such a giant component appears, how large is the giant component when $p = p_0 + \varepsilon$? These are the standard finite versions of the basic questions in percolation theory; of course, here the medium in which percolation is being studied is itself random. Our aim is to answer the last question very precisely for certain models, greatly improving the existing bounds.

Probably the most studied of the new random graph models mentioned above is the Barabási–Albert or BA 'model' of [4]. Given an integer $m \geq 1$, the BA model describes a graph which grows by adding one vertex at a time, the new vertex sending out m edges to earlier vertices, chosen according to the 'preferential attachment' rule: the probability that an old vertex is chosen to receive an edge from the new vertex is proportional to its degree. Unfortunately, the description given by Barabási and Albert in [4] (repeated in [2]) is incomplete, and also inconsistent; see [9] for the details, or [7] for a fuller discussion. In fact, it is not obvious how to define a simple graph that has exactly the desired properties. In order to prove results, one must first have a well-defined model, and this is why the *linearized chord diagram* or LCD model was introduced in [9]. This is a precise way of defining a multi-graph (which will have loops and multiple edges, but only very few) that fits the BA description. The LCD model also has the very useful property that it has a static description: there is a way of constructing the n-vertex graph directly, rather than by adding one vertex at a time.

Typical graphs generated by the vague BA model or precise LCD model are 'scale-free' in the sense that their degree sequences follow a power-law (see [4, 11]); this perhaps unfortunate terminology is widely used in this area. The robustness of these graphs has been studied experimentally in [3], heuristically in [13, 14] and rigorously in [8]. In the last paper it is proved that for $m \geq 2$, if a constant fraction $p > 0$ of the edges or vertices is retained, then there is always a giant component, but for small p it is very small; the fraction of the vertices that it contains is between $\exp(-\Theta(p^{-2}))$ and $\exp(-\Theta(p^{-1}))$ as $p \to 0$. Here we show that the larger of these two bounds is correct, and find the constant in the exponent in terms of m.

In this way, and in many other ways, the LCD model behaves very differently from a classical random graph $G(n, c/n)$ with a comparable number of edges, which has a giant component if and only if c exceeds a critical value, $c_0 = 1$. One can ask whether this is due to the growth in time, or to preferential attachment. To answer this it makes sense to study graphs which grow in time, with each new vertex joining to a set of m, say, earlier vertices chosen uniformly at random. This model, the *uniformly grown m-out random graph*, studied in [10], is extremely natural in its own right as a model of graphs which grow in time, but where *a priori* we know nothing else about the graph than that it has a number of edges linear in the number of vertices. Even more natural is a slight modification where edges are present independently, the *uniformly grown random graph* or c/j-graph, studied in [6], which was introduced much earlier by Dubins in 1984. See the next section for precise definitions and results, as well as further background on the specific models. One can argue that these models are 'scale-free' in a very different and more natural sense than the BA/LCD model. In any case, we shall see in the next section that the models considered here are related closely enough to justify studying them together.

In [6] and [10] it was shown that the uniformly grown graphs described above are in one sense more like $G(n, c/n)$ than the LCD model: there is a threshold in terms of the parameter c (or the edge-retention probability p) such that a giant component appears only above this threshold. (The result for the c/j-graph essentially dates back to 1989; see [19, 21] for the related earlier results, and [17, 6] for a discussion of their relevance.) However, in [6, 10] it is shown that, above the threshold, the giant component emerges very slowly indeed: for $c = c_0 + \varepsilon$, where c_0 is the critical value of the parameter, upper and lower bounds are given on the size of the giant component which are of the form $\exp(-\Theta(\varepsilon^{-1/2}))$. Here we shall give much more precise bounds for a common generalization of these two models, showing that the lower bounds in [6, 10] are essentially correct, and finding the constant in the exponent exactly.

The questions studied here are all of the form: What is the size of the giant component (if any) in a certain inhomogeneous random graph? Such questions have been studied previously in various cases, generally requiring independence between the edges, which we do not have for two of the models studied here. For example, Söderberg [22, 23, 24] considers a general model with finitely many types of vertices, where the edge probabilities depend on the vertex types, giving formulae for the critical point and giant component size without proof. This model does not cover cases such as that studied here, where the space of types is infinite and the ratios between edge probabilities are unbounded. The case of a rather different growing network model (that nevertheless essentially includes the c/j-graph) has been studied heuristically by Dorogovtsev, Mendes and Samukhin [16]; we return to this in the next section.

Of the models we study, perhaps the most interesting and important is the LCD model. This is the prime example of a properly defined model fitting the vague Barabási–Albert description; this BA 'model' is the model in which there is most current interest. On the other hand, the percolation behaviour near the phase transition is most interesting for the uniformly grown models; the very slow emergence of a giant component above some nonzero threshold (an 'infinite order phase transition') is very striking, and it is surprising that accurate bounds on the size of this giant component can be obtained.

2. Models and results

Throughout this paper we shall write $C_1(G)$ for the maximal order (number of vertices) of a component of a graph G. We say that an event holds *with high probability* (w.h.p.) if it holds with probability tending to 1 as $n \to \infty$, with all other parameters fixed.

2.1. The LCD model

The first model we shall consider is the LCD model of [9], a precise version of the vague Barabási–Albert 'growth with preferential attachment' model [4]. The LCD (multi-)graph $G_m^{(n)}$ on n vertices, where $m \geqslant 1$ is a fixed integer parameter, may be defined as follows. Start with $G_m^{(0)}$ the empty graph with no vertices, or with $G_m^{(1)}$ the graph with one vertex, 1, and m loops. Given $G_m^{(n-1)}$, construct $G_m^{(n)}$ by adding a single new vertex n, and then successively adding m edges $e_1,\ldots,e_m$ incident with n. The other end of each e_i is a vertex chosen from $\{1,2,\ldots,n\}$ with probability proportional to its degree, where we count $e_1,\ldots,e_{i-1}$ as already contributing to the degrees of their endpoints, and furthermore count e_i as contributing one to the degree of n. $G_m^{(n)}$ contains some loops and multiple edges, but not very many; their presence will not affect our results.

At first sight it is not clear why one should take exactly this definition for $G_m^{(n)}$, which is expanded on in [9]. The answer is that $G_m^{(n)}$ defined in this way has a static description derived via random pairings, or linearized chord diagrams; see [9]. This static description seems to be needed to prove all but the simplest results about $G_m^{(n)}$.

As mentioned in the introduction, the robustness of $G_m^{(n)}$ was considered in [8]. Given $0 < p < 1$, let $G_m^{(n)}(p)$ be the graph formed from $G_m^{(n)}$ by keeping each vertex with probability p, independently of the other vertices and of $G_m^{(n)}$. In [8] it was shown that if $m \geq 2$ is fixed and p is small, then w.h.p.

$$\exp(-\Theta(p^{-2})) \leq C_1(G_m^{(n)}(p))/n \leq \exp(-\Theta(p^{-1})).$$

Here we show that the upper bound is correct, and give a much more precise form of the bound.

Theorem 2.1. *Let $m \geq 2$ and $0 < p < 1$ be fixed, and let $G_m^{(n)}(p)$ be the graph formed from $G_m^{(n)}$ by keeping vertices independently with probability p. There is a constant $\lambda_m(p)$ such that*

$$C_1(G_m^{(n)}(p)) = (\lambda_m(p) + o(1))n$$

holds w.h.p. as $n \to \infty$. Furthermore, as $p \to 0$ with m fixed, we have

$$\Omega\left(p^2\left(\frac{m-1}{m+1}\right)^{\frac{1}{2p}}\right) = \lambda_m(p) = O\left(\left(\frac{m-1}{m+1}\right)^{\frac{1}{2p}}\right). \tag{2.1}$$

In particular, as $p \to 0$ with m fixed,

$$\lambda_m(p) = \exp\left(-\frac{c_m}{p} + O(\log(1/p))\right) = \exp\left(-\frac{c_m + o(1)}{p}\right), \tag{2.2}$$

where

$$c_m = \frac{1}{2}\log\left(\frac{m+1}{m-1}\right) = \frac{1}{m} + \frac{1}{3m^3} + O(m^{-5}). \tag{2.3}$$

There will be essentially no difference if we delete edges rather than vertices: the only change is that our bounds on the size of the largest component increase by a factor $1/p$. This affects the bounds (2.1) on $\lambda_m(p)$, but not the asymptotic form (2.2). A similar comment applies to all our results.

The restriction to $m \geq 2$ above is essential (a fact missed by some heuristic approaches). For $m = 1$ the model gives a forest with loops, and, as noted in [8], for *any* $p < 1$ w.h.p. there is no giant component.

2.2. The $c/\sqrt{ij}$-graph

It is easy to check that in the j-vertex LCD graph, the expected degree of vertex $i < j$ is $m\sqrt{j/i}(1 + O(1/i))$. It follows from the preferential attachment rule that in $G_m^{(n)}$ the probability of the edge ij, $i < j$, is $\frac{m}{2\sqrt{ij}}(1 + O(1/i))$. This suggests that we should compare $G_m^{(n)}$ with the $c/\sqrt{ij}$-graph, i.e., the graph on $\{1, 2, \ldots, n\}$ in which each edge ij is present independently, with probability $c/\sqrt{ij}$. Here $c > 0$ is a parameter, and if $c > \sqrt{2}$ we should write $\min\{1, c/\sqrt{ij}\}$ for the edge probabilities. We shall not do so because our interest is in c tending to zero, and in any case all our arguments do not depend on the probabilities of the first $O(1)$ edges.

The $c/\sqrt{ij}$-graph $G_{1/\sqrt{ij}}^{(n)}(c)$ is very much easier to study that $G_m^{(n)}$, due to the independence between edges, and because the form of the 'kernel' $1/\sqrt{ij}$ makes summing over paths very simple. We shall prove the following rather precise result concerning the largest component in $G_{1/\sqrt{ij}}^{(n)}(c)$.

Theorem 2.2. *There is a function $f(c)$ such that for $c > 0$ fixed, w.h.p. the largest component of $G_{1/\sqrt{ij}}^{(n)}(c)$ has order $(f(c) + o(1))n$ as $n \to \infty$. Furthermore,*

$$f(c) = 2e^{1-\gamma} \exp\left(-\frac{1}{2c}\right)\left(1 + O\left(c^{-1} e^{-1/(2c)}\right)\right), \tag{2.4}$$

as $c \to 0$. In particular,

$$f(c) \sim 2e^{1-\gamma} \exp(-1/(2c)),$$

where $\gamma = 0.57721566\ldots$ is Euler's constant.

This result shows that the LCD graph does not behave as if the edges were independent, but that in a certain numerical sense, the difference is fairly small when m is large. For comparison, the correct normalization is to take $c = pm/2$, so that $G_m^{(n)}(p)$ and $G_{1/\sqrt{ij}}^{(n)}(c)$ have (approximately) the same individual edge probabilities. If edges were independent in the LCD graph, Theorem 2.2 shows that the constant c_m given by (2.3) would be equal to $1/m$. As we can see this is not true, but the difference in the constant in the exponent shrinks surprisingly quickly as m becomes large. Note, however, that small m is the case most often considered. This kind of behaviour has been seen before, in a much simpler context: it was shown in [7] that the number of triangles (or l-cycles) in the LCD graph is almost the same as if edges were independent – there is an extra factor $1 + O(1/m)$. Such results are not so surprising in the light of the equivalent description of the LCD graph given in [9]. There it is shown that, roughly speaking, we may construct the LCD graph by taking certain random variables $X(i)$ for each vertex i, whose distribution is that of a sum of m exponentials, and then putting in each edge ij, $i < j$, with probability $X(i)/\sqrt{ij}$. (There is still some dependence, due to the fixed out degree.) Typically, the $X(i)$ do not vary by more than a factor $1 + O(1/m)$.

2.3. The uniformly grown m-out random graph

In the light of the very substantial differences between the LCD/BA graph and a classical random graph (for example, $G(n, c/n)$, or an m-out random graph), it makes sense to ask whether it is growth or preferential attachment that causes these differences. This has lead several people to consider the following model $H_m^{(n)}$, the *uniformly grown m-out random graph*. This is the graph on $\{1, 2, \ldots, n\}$ in which each vertex $i \geq m + 1$ is joined to a set of

m vertices from $\{1, 2, \ldots, i-1\}$, chosen uniformly at random from all such sets, the choices being independent over different i. Such a model was considered by Barabási and Albert in their original paper [4], where they noted that it does not have a power-law degree sequence. In fact, the degree sequence of $H_m^{(n)}$ is not that far from a classical random graph with constant average degree. For example, the maximum degree is $\Theta(\log n)$. It is thus not surprising that $H_m^{(n)}$ does display a phase transition: let $H_m^{(n)}(p)$ be the random subgraph of $H_m^{(n)}$ obtained by keeping each edge (or vertex) independently with probability p. It was shown in [8] that for $p < p_l$ there is no giant component, while for $p > p_u$ there is, with $p_l, p_u = \frac{1}{4m}(1 + O(1/m))$. The critical probability $p_c = p_u$ was found in [10], building on the work in [6], and upper and lower bounds on the size of the giant component were given for $p = p_c + \varepsilon$, matching up to a factor π in the exponent. Here we shall show that the lower bound from [10] is correct, and give a rather more precise form.

Theorem 2.3. *Let $m \geqslant 2$ and $0 < p < 1$ be fixed, and set*

$$p_c = p_c(m) = \frac{1}{2}\left(1 - \sqrt{\frac{m-1}{m}}\right) = \frac{1}{4m} + \frac{1}{16m^2} + O(m^{-3}). \tag{2.5}$$

Let $H_m^{(n)}(p)$ be the random subgraph of the uniformly grown m-out random graph $H_m^{(n)}$ obtained by retaining each edge (or each vertex) independently with probability p. There exists a function $f_m(p)$ of m and p, not depending on n, such that $C_1(H_m^{(n)}(p)) = (f_m(p) + o(1))n$ holds w.h.p. as $n \to \infty$, with $f_m(p) = 0$ for $p \leqslant p_c$, and

$$f_m(p_c + \varepsilon) = \exp\left(-\frac{\pi}{2(m(m-1))^{1/4}}\frac{1}{\sqrt{\varepsilon}} + O(\log(1/\varepsilon))\right) \tag{2.6}$$

as $\varepsilon \to 0$ from above with m fixed.

Again, it does not matter whether it is edges or vertices that we retain independently with probability p; the precise value of $f_m(p)$ changes by a factor of p, but this does not affect (2.6). Also, the restriction to $m \geqslant 2$ is again essential; otherwise there is never a giant component. Although we have written the result in a slightly different way, the constant in the exponent is the same as that given by the lower bound in [10].

2.4. The uniformly grown or c/j-graph

The model described in the previous subsection represents a graph which grows in time, each new vertex sending edges to old vertices chosen uniformly at random. While always sending out the same number m of edges is one natural choice, it is perhaps not the most natural. One could consider instead sending out (essentially) a Poisson number of edges with a certain mean c. This has the advantage that different edges are then present independently, leading to the *uniformly grown random graph* or c/j-graph $G_{1/j}^{(n)}(c)$, the graph on $\{1, 2, \ldots n\}$ in which edges are present independently, and each edge ij, $i < j$, has probability c/j. (Or $\min\{1, c/j\}$ if $c > 2$; again this will not be relevant here.) It might be more natural to use $j - 1$ instead of j so that the expected out-degrees are all exactly c; such minor changes to the model will not matter here, as can be seen from our general result in Section 8.

The claim that such a growing random graph model is very natural is supported by its independent introduction in several places. Callaway, Hopcroft, Kleinberg, Newman and Strogatz [12] introduced a (more or less) equivalent model, the CHKNS model, in 2001, giving heuristic arguments that there is a phase transition at a parameter equivalent to $c = 1/4$, and heuristic and numerical evidence for an infinite order phase transition. In fact, the c/j-graph is much older; it was proposed by Dubins in 1984 (see [19, 21]) who asked (in an essentially equivalent infinite form) the question: What is the critical value of c for percolation? The answer, $c = 1/4$, was proved by Kalikow and Weiss [19] who gave the lower bound, and Shepp [21] who gave the upper bound. A more general result was proved by Durrett and Kesten [18]. As pointed out in [6], and also independently by Durrett [17], these results carry over to the CHKNS model. The question of the rate of emergence of the giant component in the c/j-graph was treated rigorously in [6]; as in [10], upper and lower bounds matching up to a factor π in the exponent were proved, showing that the transition is indeed infinite order. This work predates that in [10] on the more complicated m-out version, which builds very much on the work in [6].

Here we again close the gap, proving the following result.

Theorem 2.4. *There is a function $f(c)$ such that for $c > 0$ fixed, $C_1(G_{1/j}^{(n)}(c)) = (f(c) + o(1))n$ holds w.h.p. as $n \to \infty$. Furthermore, $f(c) = 0$ for $c \leqslant 1/4$, and*

$$f(1/4 + \varepsilon) = \exp\left(-\frac{\pi}{2}\frac{1}{\sqrt{\varepsilon}} + O(\log(1/\varepsilon))\right) \tag{2.7}$$

as $\varepsilon \to 0$ from above.

Again this result carries over to the CHKNS model, taking $c = 2\delta$, where δ is the edge-density parameter in [12]. The argument is exactly the same as in [6], so we do not repeat it here.

For comparison between Theorems 2.3 and 2.4, note that the individual edge probabilities in $H_m^{(n)}(p)$ and $G_{1/j}^{(n)}(mp)$ are (essentially) the same. Thus, if edges in $H_m^{(n)}(p)$ behaved exactly as if independent, we would have $p_c(m) = 1/(4m)$ in place of (2.5), and the constant in the exponent in (2.6) would increase by a factor $((m-1)/m)^{1/4}$, noting that $p = p_c + \varepsilon$ in Theorem 2.3 corresponds to $c = 1/4 + m\varepsilon$ in Theorem 2.4. Hence, we see that the dependence between edges going out from a vertex to earlier vertices, due to the fixed out-degree, affects the critical probability and rate of emergence of the giant component by a factor $1 + O(m^{-1})$. Again this is in some sense a small change; however, one might expect dependence that is large enough to shift the critical probability to radically change the size of the extremely small giant component just above that probability. This does not happen here.

Dorogovtsev, Mendes and Samukhin [16] found by 'rigorous' methods the critical probability and size of the giant component in a family of growing scale-free random graphs. One special case is exactly the CHKNS model. The results of [16] in this case are consistent with, and much more precise than, Theorem 2.4. However, their arguments are not a proof; various possible problems to do with limits are not addressed, for example, including the possible existence of $\Theta(n)$ vertices in 'intermediate' sized components. It may well be, however, that their (difficult) arguments can be made rigorous with further work, perhaps using as a starting point the branching process approximation proved here (see Section 3).

The range of growing models considered in [16] includes graphs with power-law degree distributions, with variable exponents. However, the model is very different from $G_m^{(n)}(p)$; not only are edges added at each step between two random old vertices, but, more importantly, preferential attachment is based on degree *in the final graph*. In contrast, percolation in a preferential attachment graph means growth with preferential attachment, followed by the deletion of some edges; in [16] a random subgraph of a graph generated in some way is never considered. This makes a huge difference to the behaviour of the giant component (except in one case, namely uniform attachment). Setting the parameters of their model to produce a sparse graph with the same degree exponent as $G_m^{(n)}$ or $G_m^{(n)}(p)$ (namely, -3), Dorogovtsev, Mendes and Samukhin [16] find a component containing a constant fraction of the edges, rather than the fraction $\exp(-\Theta(1/p))$ obtained for $G_m^{(n)}(p)$. The heuristics of [16] are not obviously applicable to $G_m^{(n)}(p)$ and related graphs.

2.5. Further models

Our results about the uniformly grown random graph and uniformly grown m-out random graph will be proved by considering a common generalization we call the *uniformly grown Z-out random graph*. Roughly speaking, this is the graph in which each vertex i chooses independently a number Z_i of edges to send to earlier vertices, which are then chosen uniformly at random. Each Z_i has (roughly) the distribution of Z. Precise definitions and a result generalizing Theorems 2.3 and 2.4 are given in Section 8. The case where $Z = m$ is constant is $H_m^{(n)}$, the case where Z is Poisson with mean c is $G_{1/j}^{(n)}(c)$. This graph (called X-out rather than Z-out) was introduced in [10] but with stronger assumptions than we shall make here. Again our result (Theorem 8.1) gives much better bounds. As in [10], it turns out that it is the first two moments of Z that matter.

Having noted that the uniformly grown Z-out graph is much easier to handle when Z has a Poisson distribution, so edges are independent, it is natural to ask whether the LCD model can be modified in this way. The answer is, of course, yes; taking the definition for the LCD model given at the start of this section, one can certainly add a random number of edges for each new vertex rather than just m. Such a model could be analysed using the methods of [9]; just as $G_m^{(n)}$ is constructed by first constructing $G_1^{(n)}$ (via random pairings of random points in $[0, 1]$, to achieve some independence), and then merging vertices m at a time, a generalization could be constructed from $G_1^{(n)}$ by merging vertices in groups of random sizes. However, the gain is not so great as for the uniformly grown case – we will not achieve independence between edges. The construction outlined above shows that the positive correlation between edges ij, ik, $i < j < k$ will remain (and is likely to be of the form $1 + O(1/\mathbf{E}(Z))$). In fact, it seems likely that the very close (second-order) agreement between the constant c_m appearing in (2.3) for the LCD graph $G_m^{(n)}$ and the value $1/m$ for the related graph with independence arises from the near cancellation of two effects: in $G_m^{(n)}$ there is positive correlation by a factor of roughly $1 + 1/m$ between edges ij, ik, $i < j < k$, arising from preferential attachment, and negative correlation with correlation factor of roughly $1 - 1/m$ between edges ik, jk, $i < j < k$, resulting from the out-degrees being fixed at m. It seems likely that in the Z-out version of the LCD graph, where Z is Poisson with mean m, as only the positive correlation effect would remain, the equivalent of (2.3) would be $c'_m = 1/m - \Theta(1/m^2)$. More precisely, perhaps

$$c'_m = \frac{1}{m} - \frac{1}{3m^2} + O(m^{-3})$$

would hold. (This is based on solving (7.3) with m in place of $m - 1$ and writing $(p'/2)\log(\rho)$ as a series in m.) It is possible that the methods used in [9], [8] and here could be used to prove such a result, but any proof is not likely to be short!

Finally, in [18], Durrett and Kesten consider an inhomogeneous random graph generalizing the c/j graph: edges are present independently, and the probability that the edge ij, $i < j$, is present is given by $f(i/j)/j$ for some function f. Working in a slightly different context (asking whether the infinite graph is almost surely connected), they give a criterion in terms of f that determines whether or not there is a giant component. It would be interesting to see whether the corresponding phase transition always has infinite order, perhaps with growth of the form $\exp(-\Theta(\varepsilon^{-1/2}))$, where the constant depends on f.

3. The method

We shall use a single basic method to analyse all the models we consider. This is to compare (in fact couple) the neighbourhood expansion process in the random graph G with a certain branching process. Here the neighbourhood expansion process is the random process $(\Gamma_k) = (\Gamma_0, \Gamma_1, \Gamma_2, \ldots)$, where Γ_0 consists of a single uniformly chosen random vertex v_0 of G, and Γ_k is the set of vertices at graph distance exactly k from v_0. In all cases, it is not just the number of vertices in each Γ_k that matters, but also where in the graph they are, and often how they were reached. Thus the branching process $(X_k) = (X_0, X_1, \ldots)$ we compare with will consist of sets of points x each of which has a certain *type* $\tau(x)$ associated to it. In the simplest case these types will be real numbers in $(0, 1]$. The distribution of the number and types of the children of a point in the branching process will depend on its type.

This coupling result needs to some extent separate arguments in each case. It has already been proved in [8] for the LCD model. After proving a suitable coupling result, there is still much work to be done: having checked that almost all vertices not in small components are in a single giant component, we obtain the relative size of the giant component as the survival probability of the branching process, which in turn is given by the solution to certain non-linear integral equations. At first sight it is not clear how to accurately bound the solution to these equations. It turns out that splitting off early vertices, which are likely to have large degrees, and solving exactly the linearized version of the equations for the resulting process (with a slightly modified value of the edge-probability parameter) gives good results.

A consequence of the method used here is that any potential improvement of the bounds in any of the results requires no further combinatorial work. For fixed parameters, the limit of the normalized size of the giant component is given exactly by the survival probability of a certain branching process, and hence by the solution to certain equations. The task of obtaining better bounds is exactly the task of obtaining better estimates on the solutions to these equations.

Another consequence is that it does not matter whether it is edges or vertices that we retain independently with probability p; in the branching process approximation to the neighbourhood expansion process we only consider components that are trees. For these, it makes no difference whether it is each 'outward' edge or the vertex at its 'far end' that is retained with probability p. The expected number of vertices in small components, and hence the size of the giant component, will (in the $n \to \infty$ limit) be exactly a factor of p smaller for vertex deletion, due to the reduced number of starting vertices for the neighbourhood expansion. Of course we could also consider deletion of both edges and vertices.

The rest of the paper is organized as follows. In the next section we consider the simplest model, the $c/\sqrt{ij}$-graph. We shall show that for this model good bounds on the giant component can be obtained directly, by counting paths, rather than by coupling with a branching process. This method will give some intuition into the arguments used later: the way the path counting is done corresponds to the separate consideration of early vertices described above.

In Section 5 we describe precisely the type of branching process we consider, and state and prove some very simple results about such branching processes that we will use to estimate the survival probability. In Section 6 we return to the $c/\sqrt{ij}$-graph and apply the branching process method to prove Theorem 2.2, obtaining much more precise results than by path counting. The analysis of the branching process in this case is very simple and does not require any tricks. The coupling argument is also relatively simple; although this can be viewed as a greatly simplified version of the argument in [8] we give a complete proof. In Section 7 we turn to the LCD model, proving Theorem 2.1; here the relevant coupling result has already been proved, but the analysis of the branching process is more difficult. Finally, in Section 8 we turn to the uniformly grown Z-out model, proving a common generalization of Theorems 2.3 and 2.4.

In all cases our aim is to find the normalized size of the giant component. For this, the branching process method works very well. One might want more information, such as the approximate size of the second largest component above the critical probability, and of the largest component below. It is not clear whether good bounds can easily be obtained by the methods used here. One might guess that the answer to both questions would always be $O(\log n)$, on general grounds based on the behaviour of branching processes. But this is not true: it was shown in [6] by a direct method that the subcritical c/j-graph always has a component of order $n^{\Omega(1)}$.

Throughout the rest of the paper we shall always assume that n, the number of vertices, is at least some sufficiently large constant n_0, which may depend on the parameters (e.g., m, c) of the result we are trying to prove. Thus we shall use inequalities such as $m \log n \leqslant n^{1/10}$ without further comment. We also ignore probability zero events. In particular, we shall

4. The $c/\sqrt{ij}$-graph: direct analysis

We start by considering a simple model with edges present independently, motivated by the BA or LCD growth with preferential attachment model. This model is much easier to analyse than the LCD model, due to the independence between edges. In fact, even among models with independence, this model is one of the easiest to analyse, due to the form of the kernel, $c/\sqrt{ij}$. This makes counting paths and finding the eigenfunction of the relevant branching process much simpler than for the uniformly grown random graph considered in [6], with kernel $c/\max\{i,j\}$.

Let $G = G^{(n)}_{1/\sqrt{ij}}(c)$ be the graph on $\{1, 2, \ldots, n\}$ in which each pair $\{i, j\}$ of vertices is joined independently with probability $c/\sqrt{ij}$. For now, we shall prove a much weaker result than Theorem 2.2, namely, that

$$f_1(c)n \leq C_1(G^{(n)}_{1/\sqrt{ij}}(c)) \leq f_2(c)n \tag{4.1}$$

holds w.h.p. for functions $f_1(c)$, $f_2(c)$ satisfying

$$f_1(c), f_2(c) = \exp\left(-\frac{1+o(1)}{2c}\right)$$

as $c \to 0$. The proof of this weaker result will be much simpler.

Proof of (4.1). We start with the upper bound. Fix $\eta > 0$ and set a 'cutoff' $\rho = \exp(-(1-\eta)/c)$. We call a vertex i of G *early* if $i \leq \rho n$, and *late* otherwise. It is easy to check that any component of order $\Theta(n)$ is very likely to contain an early vertex: otherwise, the subgraph of G induced by the late vertices contains a component of order $\Theta(n)$. But each individual early vertex has positive probability of being joined to such a component, and the probability that no early vertex joins to it is exponentially small.

We bound the expected number of paths from late to early vertices. More precisely, let us call a path $x_0 x_1 \cdots x_l$ a *late-early path* if $x_0, \ldots, x_{l-1}$ are late vertices and x_l is an early vertex. The expected number N_l of late-early paths of length l is at most

$$\sum_{\rho n < x_0 \leq n} \cdots \sum_{\rho n < x_{l-1} \leq n} \sum_{0 < x_l \leq \rho n} \prod_{j=0}^{l-1} \frac{c}{\sqrt{x_j x_{j+1}}}.$$

In fact, N_l is equal to the sum above when the summation variables are restricted to be distinct. Assuming that ρn is an integer (which we can ensure by adjusting η slightly as n varies), we obtain

$$N_l \leq \int_{\rho n < x_0 \leq n} \cdots \int_{\rho n < x_{l-1} \leq n} \int_{0 < x_l \leq \rho n} \prod_{j=0}^{l-1} \frac{c}{\sqrt{x_j x_{j+1}}} \, dx_l \, dx_{l-1} \cdots dx_0$$

$$= c^l n \int_{\rho < x_0 \leq 1} \cdots \int_{\rho < x_{l-1} \leq 1} \int_{0 < x_l \leq \rho} \frac{1}{\sqrt{x_0 x_l}} \prod_{j=1}^{l-1} \frac{1}{x_j} \, dx_l \, dx_{l-1} \cdots dx_0$$

$$= c^l n \left(\int_{x_0=\rho}^1 x_0^{-1/2} \, dx_0 \right) \left(\int_{x_l=0}^\rho x_l^{-1/2} \, dx_l \right) \left(\int_{x=\rho}^1 x^{-1} \, dx \right)^{l-1}$$

$$= 4c^l n (1-\rho^{1/2}) \rho^{1/2} (\log(1/\rho))^{l-1}$$

$$\leqslant 4cn\rho^{1/2} (c \log(1/\rho))^{l-1}.$$

By our choice of ρ, we have $c \log(1/\rho) = 1 - \eta < 1$. It follows that the expectation of the total number of late-early paths is at most

$$\sum_{l \geqslant 1} N_l \leqslant 4c\eta^{-1} \rho^{1/2} n \leqslant \exp(-(1/2-\eta)/c) n \qquad (4.2)$$

if η is fixed and c is small enough. In fact, arguing as at the end of Section 6 of [6], we can obtain the same formula (4.2) as an upper bound on the number of late-early paths holding w.h.p., rather than just in expectation. Since w.h.p. any giant component contains an early vertex, w.h.p. the size of the giant component is at most the number ρn of early vertices plus the number of late-early paths, so for small enough c, w.h.p.

$$C_1(G) \leqslant \rho n + \exp(-(1/2-\eta)/c) n \leqslant \exp(-(1/2-2\eta)/c) n.$$

As $\eta > 0$ was arbitrary, this proves the upper bound in (4.1).

We remark that a more complicated version of the argument above appeared in Section 6 of [6]; the argument for the apparently rather distantly related uniformly grown random graph considered there is more complicated, but proceeded by comparison to a path counting argument in G.

The lower bound is an easy application of Theorem 10 of [6]. Again we take a cutoff, this time $\rho = \exp(-(1+\eta)/c)$, and consider the integral operator on $C[\rho, 1]$ with kernel $\kappa(x, y) = c/\sqrt{xy}$. Since

$$\int_{y=\rho}^1 \frac{1}{\sqrt{xy}} \frac{1}{\sqrt{y}} \, dy = \log(1/\rho) \frac{1}{\sqrt{x}},$$

this has an eigenfunction $\phi(x) = x^{-1/2}$ on $[\rho, 1]$, with eigenvalue $c \log(1/\rho) = (1 + \eta)$. Viewing κ as the kernel corresponding to an induced subgraph of G, we may thus apply Theorem 10 of [6], with $c = 1/(1+\eta)$ and $\gamma = \eta$. As in [6], there is a very minor technical point concerning the scaling, as the theorem we apply concerns a continuous kernel defined on $[0, 1]^2$ rather than $[\rho, 1]^2$; it turns out that rescaling does not introduce an extra factor. We conclude that w.h.p. the induced subgraph of G on the vertices $\{\rho n, \ldots, n\}$ has a component of order at least

$$\frac{\eta}{1+\eta} n \int_\rho^1 \phi(x) dx / \sup_{\rho \leqslant x \leqslant 1} \phi(x).$$

Since $\int_\rho^1 \phi(x) dx \to 2$ as $c \to 0$, while

$$\sup_{\rho \leqslant x \leqslant 1} \phi(x) = \phi(\rho) = \rho^{-1/2} = \exp((1+\eta)/(2c)),$$

this shows that w.h.p. we have

$$C_1(G) \geqslant (2 + o(1)) \eta \exp(-(1+\eta)/(2c)) n \geqslant \exp(-(1+2\eta)/(2c)) n$$

if c is small enough. Since $\eta > 0$ was arbitrary, this completes the proof. $\square$

We shall return to this model after introducing the basic method we use to analyse all the models considered here. However, the result and proof above illustrate a fact that we shall use later in the analysis of the branching process itself: starting from a supercritical random graph, or branching process, good bounds on the size of the giant component can often be obtained by dealing separately with the few 'early' vertices of highest degree. This turns out to apply to all the models considered here.

5. Branching process background

In this section we state and prove some simple results about the survival probability of certain continuous-type branching processes, in which each point of the process has a *type*, here a real number (or a real number and an integer) associated to it, which affects the distribution of the number and type of its offspring.

Almost certainly, all results we need are special cases of standard results in the theory of branching processes. However, because our particular processes are not of a very standard type, it is likely that any result of sufficient generality to include the cases we need would actually have much more generality than we would like to go into here. Also, it would be misleading to give the impression that we need anything deep from the theory of branching processes: all we need are some very basic first results, which are intuitive and easy to prove. Thus the presentation here will be self-contained. Our aim is to prove only what we need, so we shall solve all problems of analysis (existence of integrals/limits, *etc.*) in the easiest way; rather than aiming to keep generality, we shall impose much stronger conditions than those likely to be really needed whenever it is convenient to do so.

Our branching processes will be associated with a certain *type space* $\mathcal{T}$. It will be convenient to think of $\mathcal{T}$ as the half-open interval $(0,1]$; in all the examples $\mathcal{T}$ will be either $(0,1]$, or $(0,1] \times \{1,2,\ldots r\}$ for $r=2$ or $r=3$. We shall write our branching process as $(X_k) = (X_0, X_1, X_2, \ldots)$. Each generation X_k consists of a set of *points* x, each of which has a *type* $\tau(x) \in \mathcal{T}$ associated to it. The first generation X_0 will consist of a single point x_0 whose type is chosen according to some distribution. Each $x \in X_k$ with type $\tau(x) = \alpha$ will give rise independently to a set of children $y_1, \ldots, y_n$, $0 \leqslant n < \infty$, with types $\beta_i = \tau(y_i)$, where n has a certain distribution depending on α, and, given n, the vector $(\beta_1, \ldots, \beta_n)$ has a certain continuous distribution, also depending on α. The next generation X_{k+1} will be the disjoint union of the sets of children of all $x \in X_k$.

In general there is a probability density

$$f_n(\alpha, \beta_1, \ldots, \beta_n) \, d\beta_1 \cdots d\beta_n$$

that a point x with $\tau(x) = \alpha$ has exactly n children and that their types lie in small intervals around $\beta_1, \ldots, \beta_n$. We shall write $X_1(\alpha)$ for the distribution of the number and types of the children of a point x with type $\tau(x) = \alpha$. The simplest example is the Poisson case, where x gives rise to children whose types β are distributed as a Poisson process with a certain density $g(\alpha, \beta)$. In this case

$$f_n(\alpha, \beta_1, \ldots, \beta_n) = \exp\left(-\int_{\mathcal{T}} g(\alpha, \beta) \, d\beta\right) \frac{1}{n!} \prod_{i=1}^n g(\alpha, \beta_i). \tag{5.1}$$

(Here we treat $(\beta_1,\ldots,\beta_n)$ as an n-tuple; of course, it is only the set $\{\beta_1,\ldots,\beta_n\}$ that matters in the end.)

Another example is the binomial case, where the number of children of x has a binomial distribution whose parameters depend on α, and the type of each child is independently distributed according to some density depending on α but not the number of children. Two of our examples will involve a mixture of the Poisson and binomial cases.

We shall make the following assumptions about $X_1(\alpha)$. Firstly, we shall assume that $X_1(\alpha)$ is continuous in the following strong sense: for every α and every $\varepsilon > 0$, there is a $\delta > 0$ such that if $|\alpha' - \alpha| < \delta$ then $X_1(\alpha)$ and $X_1(\alpha')$ can be coupled so as to agree exactly (same number of children, and exactly the same types) with probability at least $1 - \varepsilon$. (In the more general case $\mathcal{T} = (0,1] \times \{1,2,\ldots r\}$, the condition $|\alpha' - \alpha| < \delta$ of course means that the discrete parts of α, α' coincide, and the continuous parts are close.)

Secondly, we shall assume that the branching process is connected, in the sense that there is some $k > 0$ such that, for any $\alpha, \beta \in \mathcal{T}$ and any neighbourhood N_β of β, starting from $X_0 = \{x_0\}$ with $\tau(x_0) = \alpha$ there is a positive probability that X_k contains some y with $\tau(y) \in N_\beta$. Roughly speaking, this condition says that a point of any type can have descendants of any other type. (The weaker assumption where k is allowed to depend on α and β should suffice.)

Finally, we assume that there is some $\alpha \in \mathcal{T}$ such that a point of type α has positive probability of having at least two children.

Let us say that an $x \in X_k$ *survives* if x has descendants in all later generations. Let $S_\infty(\alpha)$ be the probability that x survives, given that $\tau(x) = \alpha$. Consider the following operator $\mathbf{F}$ which takes as input a (measurable) function $\phi: \mathcal{T} \to [0,1]$ and returns another such function:

$$\mathbf{F}(\phi)(\alpha) = \sum_{n\geq 0} \int_{\beta_1,\ldots,\beta_n \in \mathcal{T}} f_n(\alpha, \beta_1, \ldots, \beta_n) \left(1 - \prod_{i=1}^n (1 - \phi(\beta_i))\right) d\beta_n \cdots d\beta_1.$$

This has the following interpretation. Start with a point x_0 with $\tau(x_0) = \alpha$, and generate its children. To each child with type β, assign 'success' independently with probability $\phi(\beta)$. Then x_0 has at least one successful child with probability $\mathbf{F}(\phi)(\alpha)$. In particular, we may take $\phi = S_\infty$, assigning 'success' to each point that survives. In this case, the probability that x_0 has at least one successful child is exactly the probability that x_0 survives, so

$$S_\infty = \mathbf{F}(S_\infty).$$

Let us note the following properties of $\mathbf{F}$. Whatever the function $\phi: \mathcal{T} \to [0,1]$, the function $\mathbf{F}(\phi)$ is continuous on $\mathcal{T}$. This follows from our strong continuity assumption: for α and α' close, we can couple the two sets of children to agree with high probability; when they disagree, we use only the fact that ϕ takes values in $[0,1]$ to bound the difference between $\mathbf{F}(\phi)(\alpha)$ and $\mathbf{F}(\phi)(\alpha')$. Also, $\mathbf{F}(\phi)$ clearly takes values in $[0,1]$. Furthermore, $\mathbf{F}$ is monotone: if $\phi_1 \geq \phi_2$ holds pointwise, then so does $\mathbf{F}(\phi_1) \geq \mathbf{F}(\phi_2)$.

Finally, $\mathbf{F}$ is *superlinear* in the sense that given $\phi: \mathcal{T} \to [0,1]$ and a real number λ, $0 < \lambda < 1$, we have

$$\mathbf{F}(\lambda \phi) \geq \lambda \mathbf{F}(\phi). \tag{5.2}$$

(Perhaps sublinear would be the more normal term, but working on $[0,1]$, superlinear seems more natural.) To see this, note that $\mathbf{F}(\lambda\phi)(\alpha)$ has the following interpretation. Start with x_0 with $\tau(x_0) = \alpha$. Generate the children of x_0. Assign to each child having type β 'initial success' with probability $\phi(\beta)$, independently of the other children. Then pick a subset of the initially successful children to be 'successful', selecting each independently with probability λ. $\mathbf{F}(\lambda\phi)$ is the probability that at least one child is successful. In contrast, $\lambda\mathbf{F}(\phi)$ is just λ times the probability that at least one child is initially successful. If p_k is the probability that exactly k children are initially successful, then we have

$$\mathbf{F}(\lambda\phi)(\alpha) = \sum_{k \geq 1}(1 - (1-\lambda)^k)p_k \geq \sum_{k \geq 1} \lambda p_k = \lambda \mathbf{F}(\phi)(\alpha),$$

proving (5.2). Note also that if $\phi(\beta) > 0$ for all β, then we have strict inequality in (5.2) for some α, in particular for any α for which x_0 has positive probability of having at least two children.

We have noted already that the survival probability $S_\infty(\alpha)$ solves $S_\infty = \mathbf{F}(S_\infty)$. In fact, it is easy to see that S_∞ is the pointwise supremum of all solutions S to $S = \mathbf{F}(S)$. Taking S_k to be the probability of having descendants in at least the next k generations, we see from the interpretation of $\mathbf{F}$ that $S_{k+1} = \mathbf{F}(S_k)$. In particular, $S_k = \mathbf{F}^k(1)$, the result of applying $\mathbf{F}$ k times to the constant function taking value one. But $S_\infty = \inf_k S_k$, so we have

$$S_\infty(\alpha) = \lim_{k \to \infty} \mathbf{F}^k(1)(\alpha),$$

where the limit is of a decreasing sequence. If S is any function satisfying $S = \mathbf{F}(S)$, then $S \leq 1$. From monotonicity of $\mathbf{F}$ we thus have $S = \mathbf{F}^k(S) \leq \mathbf{F}^k(1)$ holding pointwise for all k, and $S \leq S_\infty$ follows.

Actually, there is at most one nonzero solution to $S = \mathbf{F}(S)$; as pointed out by James Norris [20], this can be shown by looking at the ratio of two solutions. Suppose that S is a solution of $S = \mathbf{F}(S)$ which is not identically zero or equal to S_∞. We may assume that S_∞ is not identically zero, as $S \leq S_\infty$. Note that S is continuous, as $S = \mathbf{F}(S)$, and $\mathbf{F}(\phi)$ is always continuous. From connectedness, using $S = \mathbf{F}^k(S)$ for all k, we see that $S(\alpha) > 0$ holds pointwise. (S is nonzero at some β, and therefore in a neighbourhood of that β. Now apply connectedness to show that $S(\alpha) = \mathbf{F}^k(S)(\alpha)$ is positive.) Let $\lambda = \inf S(\alpha)/S_\infty(\alpha)$. Note that $\lambda < 1$ as $S \neq S_\infty$, and $S_\infty \geq S$ holds pointwise.

We claim that the infimum above is attained at some α_0; this will require an extra assumption not so far stated. If $\mathcal{T}$ were compact there would be no problem, as S and S_∞ are continuous. We shall use the following additional property of $X_1(\alpha)$: there is some $k_0 \geq 1$ such that, given any neighbourhood N_β of some $\beta \in \mathcal{T}$ and any $M > 0$, as $\alpha \to 0$ (or $\alpha \to (0, i)$ in the more general case), the probability that x_0 with $\tau(x_0) = \alpha$ has at least M k_0th generation offspring with types in N_β tends to one. Roughly speaking, this just says that points with very small α have very many children, the expected number (in any interval) tending to infinity. This assumption will hold with $k_0 = 2$ in all cases we consider. This assumption implies that if $\phi : \mathcal{T} \to [0,1]$ is any continuous function not identically zero, then

$$\mathbf{F}^{k_0}(\phi)(\alpha) \to 1 \qquad (5.3)$$

as $\alpha \to 0$. (Or as $\alpha \to (0, i)$.) Using (5.3), and the fact that $S = \mathbf{F}^{k_0}(S)$ and $S_\infty = \mathbf{F}^{k_0}(S_\infty)$, we see that $S(\alpha)/S_\infty(\alpha)$ tends to $1/1 = 1$ as $\alpha \to 0$. From this and continuity it follows that the infimum $\lambda = \inf_\alpha S(\alpha)/S_\infty(\alpha) < 1$ must be attained at some $\alpha_0 \in \mathcal{T}$. But now as $S(\alpha_0) > 0$ we have $\lambda > 0$. We have $S(\alpha) \geqslant \lambda S_\infty(\alpha)$ for every α by definition of λ, so from (5.2) and monotonicity of $\mathbf{F}$,

$$S(\alpha_0) = \mathbf{F}(S)(\alpha_0) \geqslant \mathbf{F}(\lambda S_\infty)(\alpha_0) \geqslant \lambda \mathbf{F}(S_\infty)(\alpha_0) = \lambda S_\infty(\alpha_0). \tag{5.4}$$

As we have equality between the first and last terms, equality holds throughout. In particular, $\mathbf{F}(S)(\alpha_0) = \mathbf{F}(\lambda S_\infty)(\alpha_0)$. Since $\mathbf{F}$ is monotone, $S \geqslant \lambda S_\infty$, and S and S_∞ are continuous, it follows that $S(\alpha_1) = \lambda S_\infty(\alpha_1)$ for any α_1 that can be reached from α_0. (That is, for any α_1 such that for any neighbourhood N of α_1 the probability given that $\tau(x_0) = \alpha_0$ that x_0 has a child with type in N is positive.) But then the argument giving (5.4) holds with α_1 in place of α_0. Repeating, and using our connectivity assumption, we see that equality holds in (5.4) if α_0 is replaced by any value α. But this contradicts the fact that strict inequality holds in (5.2) for some α, proving the claim that any solution to $S = \mathbf{F}(S)$ is either S_∞ or identically zero. (Of course, S_∞ may itself be identically zero, if the branching process is subcritical.)

In general, we cannot calculate S_∞; the form of the equation $S = \mathbf{F}(S)$ will be too complicated. What we shall actually use is two simple facts giving bounds on S_∞ in terms of trial functions ϕ satisfying certain conditions.

Fact 1. If $\phi : \mathcal{T} \to [0, 1]$ satisfies $\mathbf{F}(\phi) \geqslant \phi$ pointwise, then $S_\infty \geqslant \phi$ holds pointwise.

This is very easy to see: set $\phi_k = \mathbf{F}^k(\phi)$. Then by assumption $\phi_1 \geqslant \phi_0$. By induction and monotonicity of $\mathbf{F}$, $\phi_{k+1} \geqslant \phi_k$. Thus the pointwise limit $\phi_\infty = \lim_{k \to \infty} \phi_k$ exists, and by monotone convergence of the integral involved,

$$\mathbf{F}(\phi_\infty) = \mathbf{F}(\lim_{k \to \infty} \phi_k) = \lim_{k \to \infty} \mathbf{F}(\phi_k) = \lim_{k \to \infty} \phi_{k+1} = \phi_\infty$$

holds pointwise. Thus ϕ_∞ solves $\mathbf{F}(\phi_\infty) = \phi_\infty$, and as S_∞ is the supremum of all such solutions, $S_\infty \geqslant \phi_\infty \geqslant \phi$.

Fact 2. Suppose that $\phi : \mathcal{T} \to [0, 1]$ satisfies $\mathbf{F}(\phi) \leqslant \phi$ pointwise, and that $\mathbf{F}(\phi)$ is not identically zero. Then $S_\infty \leqslant \phi$ holds pointwise.

The assumption that $\mathbf{F}(\phi)$ is not identically zero is only required to avoid trivialities such as ϕ almost everywhere zero. Recall that $\mathbf{F}(\phi)$ is continuous. As $\mathbf{F}(\phi)$ is not identically zero, it follows from the connectivity assumption that for some k, $\mathbf{F}^k(\phi)$ is nowhere zero. We may and shall assume that $k \geqslant k_0$, the quantity appearing in (5.3) and the preceding assumption, which we use again here. Note that $\mathbf{F}^k(\phi) \leqslant \phi$ holds pointwise, applying monotonicity of $\mathbf{F}$ $k - 1$ times to $\mathbf{F}(\phi) \leqslant \phi$. Hence it suffices to show that $S_\infty \leqslant \mathbf{F}^k(\phi)$. Suppose not, and set $\lambda = \inf_{\alpha \in \mathcal{T}} \mathbf{F}^k(\phi)(\alpha)/S_\infty(\alpha)$, so $\lambda < 1$. This infimum is attained at some $\alpha_0 \in \mathcal{T}$. (Again we use continuity of $\mathbf{F}^k(\phi)$ and S_∞, together with (5.3).) Since $\mathbf{F}^k(\phi)$ is nowhere zero, $\lambda > 0$. Now $\mathbf{F}^k(\phi) \geqslant \lambda S_\infty$ holds pointwise by definition of λ. Thus by monotonicity of $\mathbf{F}$, by the assumption $\phi \geqslant \mathbf{F}(\phi)$, and by (5.2),

$$\mathbf{F}^k(\phi) \geqslant \mathbf{F}^{k+1}(\phi) \geqslant \mathbf{F}(\lambda S_\infty) \geqslant \lambda \mathbf{F}(S_\infty) = \lambda S_\infty.$$

At α_0 we have equality between the first and last terms, and hence throughout. Again it follows that $\mathbf{F}^k(\phi) = \lambda S_\infty$ holds pointwise, and hence that we have equality throughout for all α, not just at α_0. But this contradicts the fact that strict inequality in (5.2) always occurs for some α.

6. The $c/\sqrt{ij}$ model revisited

Let us return to the model considered in Section 4, this time using the branching process for more precise analysis. For this section, let $G = G^{(n)}_{1/\sqrt{ij}}(c)$ be the graph on $[n] = \{1, 2, \ldots, n\}$ in which each pair of vertices $\{i, j\}$ is joined with probability $c/\sqrt{ij}$, independently of all other pairs.

Our aim is to prove Theorem 2.2, showing that w.h.p. the largest component of G has order $(f(c) + o(1))n$, where $f(c)$ satisfies (2.4). This is a much more precise result than that proved in Section 4.

Let $(\Gamma_k) = (\Gamma_0, \Gamma_1, \Gamma_2, \ldots)$ be the neighbourhood expansion process in the graph G, defined as follows: $\Gamma_0 = \{v_0\}$, where v_0 is a vertex of G chosen uniformly at random. For $k \geq 0$, set $N_k = \bigcup_{r \leq k} \Gamma_r$, and let Γ_{k+1} be the set of vertices of $G \setminus N_k$ adjacent in G to some vertex of Γ_k. In other words, for all k, let Γ_k be the set of vertices at graph distance exactly k from v_0.

6.1. Coupling

We aim to couple (Γ_k) with a branching process $(X_k) = (X_0, X_1, \ldots)$ of the general type described in Section 5. Here each generation X_k consists of a set of points x each having a type $\alpha(x) \in (0, 1]$ associated with it. Each such point x gives rise to children y in the next generation whose types $\alpha(y)$ are distributed as a Poisson process on $(0, 1]$ with density $c/\sqrt{\alpha(x)\alpha(y)}$. In particular, the number of children is Poisson with mean $\int_0^1 c/\sqrt{\alpha(x)\beta}\,d\beta = 2c/\sqrt{\alpha(x)}$. The first generation X_0 consists of a single point x_0 with $\alpha(x_0)$ uniformly distributed on $(0, 1]$.

For a vertex v of G, set $\alpha(v) = v/n$. The aim is to couple the two processes so that in each generation the number of vertices/points coincides, and so do their α values. The intuition is very simple: the chance that a vertex i of G with $\alpha(i) = i/n = \alpha$ has a neighbour j with $\alpha(j) = j/n$ in a certain small interval $[\beta, \beta + d\beta]$ is roughly

$$\sum_{j \in [n\beta, n(\beta+d\beta)]} \frac{c}{\sqrt{ij}} \sim \sum_{j \in [n\beta, n(\beta+d\beta)]} \frac{c}{n\sqrt{\alpha\beta}} \sim \frac{c}{\sqrt{\alpha\beta}}\,d\beta,$$

which is the probability that a point of type α has a child with type in $[\beta, \beta + d\beta]$.

In this case, the proof of a suitable coupling result is relatively straightforward and perhaps rather tedious, but we shall write it out in full as this is the basic method underlying all the results of this paper, and this model is much the simplest case.

Lemma 6.1. *Let $c > 0$ be fixed. For each n the processes (Γ_k) and (X_k) can be coupled so that with probability $1 - o(n^{-1/100})$ we have*

$$|\Gamma_k| = |X_k|$$

for $0 \leqslant k \leqslant K$, and either $|\Gamma_K| = |X_K| = 0$ or

$$\sum_{k=0}^{K+1} |\Gamma_k|, \ \sum_{k=0}^{K+1} |X_k| \geqslant n^{1/100}. \tag{6.1}$$

In other words, unless a certain event of small probability holds, as far as size is concerned the two processes behave in exactly the same way in the two models until their total size reaches at least $n^{1/100}$ or they die out. A corresponding comparison result is proved in [8] for the much more complicated LCD model, and an appropriate branching process. There almost all the difficulties are due to the lack of independence in the model.

Proof. We shall write ε for $1/100$ throughout, emphasizing that ε is a small positive constant appearing as the exponent of n in the result, whose value we have not attempted to optimize. We shall construct our coupling so that the α values of corresponding points/vertices also agree. Exact agreement is not possible, so we shall ask for agreement only of rounded α values, $\tilde{\alpha}(v) = n^{-1/2} \lceil n^{1/2} \alpha(v) \rceil$, where v is either a vertex of G or a point of some X_k.

We shall inductively couple Γ_k and X_k, $k = 0, 1, 2, \ldots$, until both processes die out, both processes become large (as in the statement of the lemma), or some bad event of small probability occurs. Our coupling will be such that there is a bijection from Γ_k to X_k so that if v maps to x then $\tilde{\alpha}(v) = \tilde{\alpha}(x)$. There is no problem coupling $\Gamma_0 = \{v_0\}$ and $X_0 = \{x_0\}$ in this way, as $\alpha(x_0)$ is by definition uniform on $(0, 1]$, while $\alpha(v_0)$ is uniform on $\{1/n, 2/n, \ldots, n/n\}$, so the rounded values can be coupled to agree with probability $1 - O(n^{-1/2})$.

Let us suppose that we have constructed $\Gamma_0, \ldots, \Gamma_k$ and $X_0, \ldots, X_k$ coupled as above. We aim to extend the coupling to Γ_{k+1} and X_{k+1}. Note that so far we have only 'looked at' potential edges of the graph G that are induced by $N_k = \Gamma_0 \cup \cdots \cup \Gamma_k$, or are incident with N_{k-1}. Hence, conditional on $\Gamma_0, \ldots, \Gamma_k$, each edge from Γ_k to $G \setminus N_k$ is present independently with its original unconditioned probability. Note that from the stopping rule we do not need to consider too many vertices of G. In particular, there is no problem allowing a small ($o(n^{-2\varepsilon})$) failure probability in our coupling for every vertex $v \in N_k$ considered.

We claim that we may assume that $\tilde{\alpha}(v) \geqslant n^{-1/4}$ holds for every $v \in \Gamma_k$, and thus that $\tilde{\alpha}(x) \geqslant n^{-1/4}$ holds for every $x \in X_k$. Suppose not. Then there is a $v \in \Gamma_k$ with $v < n^{3/4} + O(n^{1/2})$. (The $O(n^{1/2})$ allows for the rounding.) Such a vertex v is joined to each $j \in V(G) \setminus N_k$ with $j \geqslant n/2$ independently, with probability $c/\sqrt{vj} \geqslant cn^{-7/8}/2$. But, from our stopping rule, $|N_k| \leqslant n^{\varepsilon}$, so there are at least $n/3$ candidates for the vertex j, and the number of neighbours of v in Γ_{k+1} stochastically dominates a Binomial $\text{Bi}(n/3, cn^{-7/8}/2)$ distribution with mean $cn^{1/8}/6$. As this mean is much larger than n^{ε}, with very high probability, certainly at least $1 - o(n^{-2\varepsilon})$, say, v has at least n^{ε} neighbours in Γ_{k+1}, and in particular, $|\Gamma_{k+1}| \geqslant n^{\varepsilon}$. Similarly, but more simply, with very high probability the corresponding point $x \in X_k$ with $\tilde{\alpha}(x) \geqslant n^{-1/4}$ has at least n^{ε} children, so $|X_{k+1}| \geqslant n^{\varepsilon}$. Thus, we may abandon the coupling at this stage: our stopping rule (6.1) is satisfied for $K = k$ with sufficiently high probability.

We assume from now on that the $\tilde{\alpha}(.)$ values for $v \in \Gamma_k$ or $x \in X_k$ are all at least $n^{-1/4}$. Let us consider the vertices $v \in \Gamma_k$ one by one, coupling their contributions to Γ_{k+1} with the children of the corresponding $x \in X_k$. When we come to a particular vertex v, let S be the set of neighbours in Γ_{k+1} of vertices $v' \in \Gamma_k$ previously considered, noting that conditional on everything so far, each vertex w of $V(G) \setminus (N_k \cup S)$ is joined to v independently with probability $c/\sqrt{vw}$. We may assume that $|S| \leqslant n^\varepsilon$; otherwise, unless our coupling has already broken down, both Γ_{k+1} and X_{k+1} have cardinality at least n^ε, and our stopping rule (6.1) allows us to abandon the coupling. Consider any possible value β of $\tilde{\alpha}(.)$, with $\beta \geqslant n^{-3/8}$. In $V(G)$, there are $n^{1/2} + O(1)$ vertices j with $\tilde{\alpha}(j) = \beta$. Hence, in $V(G) \setminus (N_k \cup S)$ there are $n^{1/2} + O(n^\varepsilon)$ such vertices. Let us write B for the set of these vertices. Each $j \in B$ is joined to v independently with probability

$$c/\sqrt{vj} = \frac{c}{n\sqrt{\alpha(v)\alpha(j)}} = \frac{c}{n\sqrt{\tilde{\alpha}(v)\beta}}\left(1 + O\left(n^{-1/8}\right)\right).$$

For the last step note that for any w the absolute difference between $\alpha(w)$ and $\tilde{\alpha}(w)$ is at most $n^{-1/2}$, and we are assuming that $\tilde{\alpha}(v) > n^{-1/4}$ and $\beta > n^{-3/8}$. Since the individual edge probabilities above are very small, the number of $j \in B$ joined to v is very close in distribution to a Poisson distribution with mean

$$\mu = \sum_{j \in B} \frac{c}{\sqrt{vj}} = \mu_0\left(1 + O\left(n^{-1/8}\right)\right),$$

where

$$\mu_0 = \frac{c}{n^{1/2}\sqrt{\tilde{\alpha}(v)\beta}}.$$

In fact, more simply, a direct calculation shows that the probability that v has at least two neighbours in B is at most $\mu^2/2 = O(\mu_0^2)$, while the probability that v has exactly one neighbour in B is $\mu(1 - O(\mu)) = \mu_0 + O(\mu_0^2)$. Hence, the probability that v has no neighbours in B is $1 - \mu_0 + O(\mu_0^2)$. Note that from our bounds on $\tilde{\alpha}(v)$ and β, we have $\mu_0 \leqslant 1/\sqrt{n^{1-1/4-3/8}} = n^{-3/16}$. Also, in the continuous branching process, the number of children y of the point x corresponding to v having $\tilde{\alpha}(y) = \beta$ has a Poisson distribution with mean $\mu' = \mu_0(1 + O(n^{-1/2}))$. We can couple this number with the number of neighbours of v in G so that with probability $1 - O(\mu_0^2 + \mu_0 n^{-1/2}) = 1 - O(\mu_0 n^{-3/16})$ the two numbers agree (and are either 0 or 1). Using independence in both models, we may construct this coupling simultaneously for all possible values $\beta > n^{-3/8}$, with total error probability $O(n^{-3/16} \sum_\beta \mu_0(\beta))$. Since

$$\sum_\beta \mu_0(\beta) \sim \int_{\beta = n^{-3/8}}^1 \frac{c}{\sqrt{\tilde{\alpha}(v)\beta}}\, d\beta < \frac{2c}{\sqrt{\tilde{\alpha}(v)}} \leqslant 2cn^{1/8},$$

this total error probability is $O(n^{-1/16})$ and thus $o(n^{-2\varepsilon})$. As noted earlier, such an error probability per vertex considered is acceptable.

At this point we have coupled the neighbours j of v with $\tilde{\alpha}(j) > n^{-3/8}$ with the children y of x with $\tilde{\alpha}(y) > n^{-3/8}$ so that (with high enough probability) there is a bijection between them preserving $\tilde{\alpha}(.)$ values. To complete the proof we must consider neighbours with $\tilde{\alpha}(j) \leqslant n^{-3/8}$. But the expected number of such neighbours is at most

$\sum_{j \leqslant n^{5/8}} c/\sqrt{vj} \leqslant 2cn^{5/16}/\sqrt{v}$. As $\tilde{\alpha}(v) \geqslant n^{-1/4}$, so $v \geqslant n^{3/4} - O(n^{1/2})$, this expectation is at most $3cn^{-1/16} = o(n^{-2\varepsilon})$, so the probability that v has such a neighbour is $o(n^{-2\varepsilon})$. A similar argument applies to x, so we see that (with high enough probability) we have already coupled all neighbours of v in the next generation with all children of x.

Following the argument above for each $v \in \Gamma_k$ in turn, we see that we can construct the required coupling between Γ_{k+1} and X_{k+1} unless either some bad event with probability $o(|\Gamma_k|n^{-2\varepsilon})$ occurs, or both Γ_{k+1} and X_{k+1} have cardinality at least $n^{1/\varepsilon}$, in which case (6.1) holds. Since we always stop before considering more than n^ε vertices, we can construct the required coupling with probability $1 - o(n^\varepsilon)$, completing the proof of the lemma. □

As a simple corollary, Lemma 6.1 gives good bounds on the number of vertices of G in small components. Let us write $\#_r(G)$ for the number of vertices of G in components of order r, and let us write s_r for the probability that the branching process (X_k) has total size $|(X_k)| = \sum_{k \geqslant 0} |X_k|$ equal to r. Note that s_r is independent of n, and that $\sum_{r \geqslant 0} s_r = 1 - \sigma$, where σ is the survival probability of the branching process, i.e., $\sigma = \Pr(|(X_k)| = \infty)$.

Corollary 6.2. *Let $c > 0$ be fixed. There is a positive constant ε (independent of c) such that with probability $1 - o(n^{-1})$ we have*

$$|\#_r(G) - s_r n| \leqslant n^{1-2\varepsilon} \tag{6.2}$$

for all r in the range $0 \leqslant r \leqslant n^\varepsilon$.

Proof. We shall take $\varepsilon = 1/1000$, and use Lemma 6.1. Note that $\mathbf{E}(\#_r(G))$ is exactly n times the probability q_r that a random vertex of G lies in a component of order r. From Lemma 6.1, for $r \leqslant n^\varepsilon$ we have $|s_r - q_r| = o(n^{-10\varepsilon})$. (Note that we are using $\varepsilon = 1/1000$ here, compared to $1/100$ in the (statement and) proof of Lemma 6.1.) Hence,

$$|\mathbf{E}(\#_r(G)) - s_r n| \leqslant n^{1-3\varepsilon} \tag{6.3}$$

certainly holds for $0 \leqslant r \leqslant n^\varepsilon$.

It remains to show concentration of $\#_r(G)$ about its mean. We can do this using standard martingale methods. There is a slight annoyance due to the unbounded (and potentially very large) degrees in G. Perhaps the simplest way to deal with this is to avoid large degrees altogether, as follows. Let the *down-degree* of a vertex i be the number of $j < i$ to which i is adjacent. Note that the down-degree of i in G has expectation $\sum_{j<i} c/\sqrt{ij} \leqslant 2c$. From the independence of the edges, it is easy to check that the chance that a given vertex i has down-degree in G at least $(\log n)^2$, say, is $o(n^{-2})$. Thus with probability $1 - o(n^{-1})$ every vertex of G has down-degree at most $(\log n)^2$.

Let H be the graph obtained from G by removing all edges ij where $j < i$ and i has down-degree at least $(\log n)^2$ in G. Then by construction every vertex of H has down-degree at most $(\log n)^2$. Also, G and H coincide exactly with probability $1 - o(n^{-1})$. Hence $\#_r(G)$ and $\#_r(H)$ coincide with probability $1 - o(n^{-1})$, and $\mathbf{E}(\#_r(G)) = \mathbf{E}(\#_r(H)) + o(1)$. Thus it suffices to prove that $\#_r(H)$ is concentrated about its mean.

Fix $0 \leqslant r \leqslant n^\varepsilon$. The graph H can be constructed in n independent steps, at each step deciding the down-neighbourhood of a vertex i. Changing the random choice made at one such step modifies the final graph H in at most $2(\log n)^2$ edges, and thus changes $\#_r(H)$ by, crudely, at most $M = 4r(\log n)^2 = o(n^{2\varepsilon})$. (Adding or deleting a single edge in any graph changes the number of vertices in components of order r by at most $2r$.) It follows from standard martingale inequalities (for example, the Hoeffding–Azuma inequality) that for any $t > 0$ the probability that $\#_r(H)$ deviates from its mean by more than t is at most $2\exp(-t^2/(2nM^2))$. Taking $t = n^{3/4}$ and recalling that $M = o(n^{2\varepsilon})$ this probability is exponentially small in some power of n, and certainly at most n^{-2}, say. Summing over r, we see that with probability $1 - o(n^{-1})$ we have $|\#_r(H) - \mathbf{E}(\#_r(H))| \leqslant n^{3/4}$ for all r, $0 \leqslant r \leqslant n^{\varepsilon}$. Combined with (6.3) and the very close agreement between $\#_r(H)$ and $\#_r(G)$ this proves the corollary. □

Having obtained good bounds on the number of vertices in small components, we would now like to study the order of the giant component. Let $\sigma = \sigma(c)$ be the survival probability of the branching process (X_k), i.e., the probability that $|(X_k)| = \infty$.

Lemma 6.3. *Let $c > 0$ be fixed. Then*
$$C_1(G) = (\sigma(c) + o(1))n$$
holds w.h.p. as $n \to \infty$.

Proof. It only remains to show that w.h.p. almost all vertices not in small components are in the giant component. We could prove this as in [9, 8], by running the neighbourhood expansion argument further. However, there is a much easier way, namely adding vertices to the graph, as in [10]. This method will be applicable to the other models we consider as well.

Let $\varepsilon = 1/1000$ be the constant appearing in Corollary 6.2, and set $L = (\log n)^2/2$. (Here we could take $L = n^{\varepsilon}/2$, but later we shall need to work with smaller L.)

From Corollary 6.2, w.h.p. the number of vertices of G in *small* components, i.e., components of order at most L, is $(\sum_{r \leqslant L} s_r + o(1))n$. Here we use the fact that (6.2) holds with probability $o(L^{-1})$ for each $r \leqslant L$. Now s_r is independent of n, and $\sum_{r \geqslant 0} s_r = 1 - \sigma(c)$. Thus $\sum_{r \leqslant L} s_r = 1 - \sigma(c) + o(1)$, using only $L \to \infty$. Hence, w.h.p. G has $(\sigma(c) + o(1))n$ vertices in *large* components, i.e., components of order greater than L. We shall assume that $\sigma(c) > 0$, as otherwise there is nothing left to prove.

Let $\eta = \eta(n) = (\log n)^{-1/2}$. Let G_t be the subgraph of G induced by the first t vertices, where t runs from $n' = \lfloor(1 - \eta)n\rfloor$ to n. Note that $G' = G_{n'}$ has exactly the distribution of $G_{1/\sqrt{ij}}^{(n')}(c)$. Hence, using Corollary 6.2 as above, noting that $L \leqslant (n')^{\varepsilon}$, the number of vertices of G' in large components is w.h.p. $(\sigma(c) + o(1))n' = (\sigma(c) + o(1))n$. Let $C_1, \ldots, C_r$ be the large components of G'. It suffices to show that in G, w.h.p. all the C_i lie within a single component; this large component then contains $(\sigma(c) + o(1))n$ vertices, as required.

We shall condition on G', assuming, as we may, that G' contains at least $\sigma(c)/2$ vertices in large components, i.e., in $\bigcup C_r$. We shall use only one property of the process $(G' = G_{n'}, G_{n'+1}, \ldots, G_{n-1}, G_n = G)$: we claim that there is a $\gamma > 0$ (depending on c but not on n) such that, conditioning on G_{t-1} and fixing two disjoint sets A, B of vertices of G_{t-1}, the probability that the new vertex t of G_t is adjacent to some $a \in A$ and some $b \in B$

is at least $\gamma|A||B|/n^2$. For the model we consider here this is very easy to see: the edges between vertex t and G_{t-1} are independent of G_{t-1}, and each $a \in A$ has independently probability $c/\sqrt{at} \geq c/n$ of being adjacent to t. It follows that t sends an edge to A with probability at least $1 - (1-c/n)^{|A|} = \Theta(|A|/n)$, using $c = \Theta(1)$ and $|A| \leq n$. Since t sends an edge independently to B with probability $\Theta(|B|/n)$ this proves the claim.

It remains to prove that w.h.p. the C_i all lie in a single component of G. Suppose not. Then there is a partition of $[r] = \{1, 2, \ldots, r\}$ into (D, D') such that there is no path from $C_D = \bigcup_{i \in D} C_i$ to $C_{D'}$ in G. In particular, no vertex t, $n' < t \leq n$, sends edges to both C_D and $C_{D'}$. Let us call such a partition bad.

Fix a partition D, D'. Without loss of generality we may take $d = |D| \leq |D'|$. Note that as $|D|, |D'| \geq d$ and the C_i are all large we have $|C_D|, |C_{D'}| \geq dL$. Also, $C_D \cup C_{D'}$ contains all $\geq \sigma(c)n/2$ vertices of G' in large components, so at least one of C_D and $C_{D'}$ has cardinality at least $\sigma(c)n/4$. It follows that $|C_D||C_{D'}| \geq dL\sigma(c)n/4$. Hence, for each t, $n' < t \leq n$, the probability that the new vertex t joins to both C_D and $C_{D'}$ is at least $\gamma dL\sigma(c)/(4n)$. But, as this lower bound on the probability holds for each t conditional on G_{t-1}, the probability that no t, $n' < t \leq n$, joins to both C_D and $C_{D'}$ is at most

$$(1 - \gamma dL\sigma(c)/(4n))^{n-n'} \leq \exp(-\gamma dL\sigma(c)(n-n')/(4n)) \leq \exp(-\gamma L\sigma(c)\eta/4)^d.$$

But the number of partitions with $|D| = d$ is at most $\binom{r}{d} \leq r^d$. Hence the probability that there is some bad partition is at most

$$\sum_{d=1}^{r/2} (r \exp(-\gamma L\sigma(c)\eta/4))^d.$$

Now $L = \Theta((\log n)^2)$, while γ and $\sigma(c)$ are constant, and $\eta = (\log n)^{-1/2}$. Hence,

$$r \exp(-\gamma L\sigma(c)\eta/4) = r \exp(-\Theta((\log n)^{3/2})) = o(rn^{-100}) = o(1),$$

and the sum above is $o(1)$, completing the proof of the lemma. □

The argument above is similar to that used in [10] for a different model. We have written the argument in a cleaner and more general form so that we can apply it to all the models we consider.

6.2. Analysis of the branching process

We have now found the size of the giant component of the $c/\sqrt{ij}$-graph G in terms of $\sigma = \sigma(c)$, the survival probability of a certain branching process. It remains to calculate this survival probability. Of all the branching processes we consider, this is the only one where this can be done exactly (in terms of the inverse of a standard function). In this case we shall not need to use the two facts given at the end of Section 5.

Let us recall that we are considering the continuous type branching process $(X_k) = (X_0, X_1, \ldots)$ where each $x \in X_k$ has a type $\alpha(x) \in (0, 1]$. The initial condition is that $X_0 = \{x_0\}$ where $\alpha(x_0)$ is uniformly distributed on $(0, 1]$. Each $x \in X_k$ gives rise independently to children $y \in X_{k+1}$ according to a Poisson process, so that the values $\alpha(y)$ associated to these children form a Poisson process on $(0, 1]$ with density $c/\sqrt{\alpha(x)\alpha(y)}$.

As in Section 5, we say that an $x \in X_k$ survives if x has descendants in all later generations. Note that x_0 survives if and only if at least one of its offspring does. Let $S_\infty(\alpha) = S_\infty(\alpha, c)$ be the probability that x_0 survives given that $\alpha(x_0) = \alpha$. Then, given that $\alpha(x_0) = \alpha$, the expected number of surviving children of x_0 is exactly

$$\mu = \int_{\beta=0}^1 \frac{c}{\sqrt{\alpha\beta}} S_\infty(\beta) \, d\beta,$$

since each child y with $\alpha(y) = \beta$ survives with probability $S_\infty(\beta)$. Now, from the Poisson nature of the process, given $\alpha(x_0)$, the number of surviving children of x_0 has a Poisson distribution. Since this number has expectation μ, the conditional probability that x_0 survives is thus $1 - \exp(-\mu)$. In other words, for $0 < \alpha \leq 1$ we have

$$S_\infty(\alpha) = 1 - \exp\left(-\int_{\beta=0}^1 \frac{cS_\infty(\beta)}{\sqrt{\alpha\beta}} \, d\beta\right).$$

This equation is exactly the equation $S_\infty = \mathbf{F}(S_\infty)$ of Section 5, and we could have derived it using the Poisson density (5.1) with $g(\alpha, \beta) = c/\sqrt{\alpha\beta}$. However, the direct argument given above is probably clearer.

Let us write $A = A(c)$ for $c \int_{\beta=0}^1 S_\infty(\beta)/\sqrt{\beta} \, d\beta$. Note that as $S_\infty \leq 1$ we have $A \leq 2c$. Now

$$S_\infty(\alpha) = 1 - \exp(-A/\sqrt{\alpha}),$$

and it only remains to determine A. However, by definition of A we have

$$A = c \int_{\beta=0}^1 \frac{S_\infty(\beta)}{\sqrt{\beta}} \, d\beta = c \int_{\beta=0}^1 \frac{1}{\sqrt{\beta}} \left(1 - e^{-A/\sqrt{\beta}}\right) d\beta.$$

Let us write $g(x)$ for the function

$$g(x) = \int_{\beta=0}^1 \frac{1}{\sqrt{\beta}} \left(1 - e^{-x/\sqrt{\beta}}\right) d\beta.$$

Then it is easy to see that $g(x) \sim 2x \log(1/x)$ as $x \to 0$. Now the relationship between A and c is exactly that $A = cg(A)$, i.e., that $g(A)/A = 1/c$. As $c \to 0$ we certainly have $A \to 0$ (recall that $A \leq 2c$). Thus it follows that $2 \log(1/A) \sim 1/c$, i.e., that

$$A = \exp\left(-\frac{1 + o(1)}{2c}\right).$$

Actually, we can obtain much more precise results, by writing $g(x)$ in terms of the *exponential integral*

$$E_1(x) = \int_{t=1}^\infty \exp(-xt)/t \, dt.$$

Here we use the notation of [1, equation 5.1.4, p. 228]. Substituting $\beta = t^{-2}$ and integrating by parts shows that

$$g(x) = 2 - 2e^{-x} + 2xE_1(x).$$

Now from [1, equation 5.1.11], we have

$$E_1(x) = -\gamma - \log(x) - \sum_{k=1}^{\infty}(-1)^k \frac{x^k}{k.k!}$$

where $\gamma = 0.57721566\ldots$ is Euler's constant. It follows that

$$g(x)/x = 2\log(1/x) + 2(1-\gamma) + x - x^2/6 + x^3/36 + O(x^4).$$

Finally, using $g(A)/A = 1/c$ we see that

$$A(c) = e^{1-\gamma} \exp\left(-\frac{1}{2c}\right)\left(1 + \Theta(e^{-1/(2c)})\right)$$

as $c \to 0$, where we could continue the expansion in terms of $\exp(-1/(2c))$ if we liked.
It remains to calculate $\sigma(c)$. By the definition of the process this is given by

$$\int_{\alpha=0}^{1} S_\infty(\alpha)\,d\alpha = \int_{\alpha=0}^{1}(1 - \exp(-A/\sqrt{\alpha}))\,d\alpha.$$

Another reduction to $E_1(x)$ shows that this is $2A + O(A^2 \log(1/A))$. (We could obtain further terms in the series, but there seems little point.) Putting this all together we see that as $c \to 0$,

$$\sigma(c) = 2e^{1-\gamma} \exp\left(-\frac{1}{2c}\right)\left(1 + O(c^{-1}e^{-1/(2c)})\right).$$

Combined with Lemma 6.3, this gives the size of the giant component in the $c/\sqrt{ij}$-graph, completing the proof of Theorem 2.2.

7. The LCD model

In this section we turn to one of the most basic and most studied scale-free graph models, the 'growth with preferential attachment' model introduced in vague terms by Barabási and Albert in [4] and made precise as the LCD model $G_m^{(n)}$ in [9]. A precise definition of the model was given in Section 2. This model is the hardest to analyse of all those considered here, owing to the very important effect of dependence between edges. After all, the preferential attachment rule is what gives rise to the unbalanced power-law degree sequence, and this rule is based exactly on dependence between edges. Fortunately, most of the work needed has already been done in [8]; a nontrivial amount of calculation remains, however.

Our aim now is to prove Theorem 2.1. For convenience, we recall the statement: let $m \geqslant 2$ and $0 < p < 1$ be fixed, and let $G_m^{(n)}(p)$ be the graph formed from $G_m^{(n)}$ by keeping vertices independently with probability p. Then

$$C_1(G_m^{(n)}(p)) = (\lambda_m(p) + o(1))n$$

holds w.h.p. as $n \to \infty$ for some constant $\lambda_m(p) > 0$. Furthermore, as $p \to 0$ with m fixed, $\lambda_m(p)$ satisfies the bounds (2.1).

At first sight it is not at all clear how to compare the neighbourhood expansion process in $G_m^{(n)}(p)$ with a branching process. In the case of the $c/\sqrt{ij}$-graph, for example, the independence between edges meant that in the neighbourhood expansion process different vertices $v \in \Gamma_k$ at distance exactly k from the random starting vertex v_0 give rise to neighbours at distance $k+1$ essentially independently of each other. (The only dependence is due to the initially small chance that some w can be reached from v_0 by more than one path of length $k+1$.) Hence it is clear that the neighbourhood expansion process should behave like a branching process, at least initially. Here, however, the presence of an edge ij, $i < j$, makes it likely that i has relatively large degree (as if it has large degree, the edge is more likely), and hence makes it more likely that any given other possible edge ik, $k > i$, is present. (The correlation factor between a pair of edges of this form turns out to be approximately $(m+1)/m$; see [7].) It turns out, however, that partial independence can be re-introduced by introducing auxiliary variables associated to each vertex. This is shown in [9], and relies heavily on the static alternative description of $G_m^{(n)}$. In conclusion, after much work it is shown in [8], using the methods of [9], that the neighbourhood expansion process in $G_m^{(n)}(p)$ can be coupled with a certain branching process, although in a rather indirect way.

Let $0 < p < 1$ and $0 \leqslant \rho < 1$ be fixed; we shall consider a branching process corresponding to the neighbourhood expansion process in the subgraph of $G_m^{(n)}(p)$ induced by the vertices $\{\lceil \rho n \rceil, \ldots, n-1, n\}$. The coupling result we shall quote from [8] applies without a cutoff, i.e., with $\rho = 0$; this is all we need to obtain the size of the giant component. However, in working with the $\rho = 0$ branching process, it will be convenient to also consider a $\rho > 0$ branching process.

Let $L(\alpha), R(\alpha) : [\rho, 1] \to [0, \infty)$ be the maximal solution to

$$L(\alpha) = \frac{p}{2\sqrt{\alpha}} \int_{\beta=\rho}^{\alpha} \frac{1}{\sqrt{\beta}} \left(1 - \frac{(1-L(\beta))^m}{(1+R(\beta))^{m+1}}\right) d\beta,$$

$$R(\alpha) = \frac{p}{2\sqrt{\alpha}} \int_{\beta=\alpha}^{1} \frac{1}{\sqrt{\beta}} \left(1 - \frac{(1-L(\beta))^{m-1}}{(1+R(\beta))^m}\right) d\beta,$$

(7.1)

and let

$$\sigma(p,\rho) = p \int_{\alpha=\rho}^{1} \left(1 - \frac{(1-L(\alpha))^m}{(1+R(\alpha))^m}\right) d\alpha.$$

(7.2)

When $\rho = 0$, $R(0)$ is not defined, and $R(\alpha) \to \infty$ as $\alpha \to 0$. In this case we work with functions L, R on $(0,1]$. Theorem 3.1 of [8] states that in the graph $G_m^{(n)}(p)$ obtained from the LCD graph $G_m^{(n)}$ by keeping vertices independently with probability p, there is w.h.p. a unique giant component of size $(\lambda_m(p) + o(1))n$. Examining the proof in [8], the value of $\lambda_m(p)$ is given by $\sigma(p,0)$. For edge deletion rather than vertex deletion the proof and result are the same, except that the initial factor of p in (7.2) is absent. Upper and lower bounds on $\lambda_m(p) = \sigma(p,0)$ are given in [8]; here we shall use the exact solution of the linearized form of (7.1), with $\rho > 0$, to obtain much better bounds on $\sigma(p,0)$, matching up to a factor $1 + o(1)$ in the exponent.

The description of the branching process, proof of the required coupling, and derivation of the survival probability equations (7.1) are all given in [8], so there is no need to repeat them here. However, we would like to use the two facts proved at the end of Section 5. We shall write (7.1) in the form $(L, R) = \mathbf{F}_\rho((L, R))$, where $\mathbf{F}_\rho$ is the operator defined by the right-hand side of (7.1). Unfortunately, $\mathbf{F}_\rho$ does not have quite the right form for the operator $\mathbf{F}$ of Section 5. Indeed, L and R are not exactly the survival probabilities of particular vertices. However, L and R can be viewed as transformed survival probabilities in a certain modified branching process that survives if and only if the (branching process associated naturally to the) neighbourhood expansion process does. This will involve a distinction between vertices *reached from the left*, i.e., $v \in \Gamma_k$ adjacent to $w \in \Gamma_{k-1}$ with $w < v$, or *reached from the right* ($w > v$). This distinction occurs again in Section 8; in $G_m^{(n)}$ there there are additional complications.

The transformation described above is simple, but cannot be written concisely without reference to the details of the branching process in [8]. In outline, we consider a triple of survival probabilities $(S_1(\alpha), S_2(\alpha), S_3(\alpha))$; these will represent respectively the probabilities of: (1) the survival of the vertex at the end of an outgoing leftwards edge from a vertex x with $\alpha(x) = \alpha$, and (2)/(3) the survival of *some* vertex to the right of x joined directly to x, in the cases that x itself was reached from the left (S_2) and x reached from the right (S_3). There is an operator $\mathbf{F}'$ of the form in Section 5 such that $(S_1, S_2, S_3) = \mathbf{F}'((S_1, S_2, S_3))$. It is shown in [8] that, from the form of the process, if $(\psi_1, \psi_2, \psi_3) = \mathbf{F}'((\phi_1, \phi_2, \phi_3))$ then $\log((1 - \psi_3)/(1 - \psi_2)) = (m+1)/m$. In the arguments of Section 5 we need only work with functions of the form $\mathbf{F}'(.)$, so we can consider instead the corresponding $\mathbf{F}''((S_1, S_2))$. The monotone transformation $S_1 = L$, $S_2 = 1 - (1 + R)^{-m}$ transforms $\mathbf{F}''$ into $\mathbf{F}_\rho$.

The key point is that, applying the results of Section 5 to $\mathbf{F}''$, it follows that trial functions (L', R') mapping under $\mathbf{F}_\rho$ to functions above/below themselves give lower/upper bounds on the maximal solution (L, R) to (7.1).

Proof of Theorem 2.1. Consider the following equations, obtained by linearizing (7.1):

$$\tilde{L}(\alpha) = \frac{p'}{2\sqrt{\alpha}} \int_{\beta=\rho}^{\alpha} \frac{1}{\sqrt{\beta}} \left(m\tilde{L}(\beta) + (m+1)\tilde{R}(\beta)\right) d\beta,$$

$$\tilde{R}(\alpha) = \frac{p'}{2\sqrt{\alpha}} \int_{\beta=\alpha}^{1} \frac{1}{\sqrt{\beta}} \left((m-1)\tilde{L}(\beta) + m\tilde{R}(\beta)\right) d\beta. \quad (7.3)$$

We have replaced p by p' since, given p, we shall solve the linearized equations for a slightly different value p' to obtain bounds on the solutions to the non-linear equations. The equations (7.3) can be solved by multiplying through by $\sqrt{\alpha}$ and differentiating. We claim that the functions

$$\tilde{L}(\alpha) = \alpha^{-1/2}\left(\frac{m+1}{m-1}\alpha^{p'/2} - \alpha^{-p'/2}\right)$$

$$\tilde{R}(\alpha) = \alpha^{-1/2}\left(\alpha^{-p'/2} - \alpha^{p'/2}\right)$$

solve (7.3), provided

$$\rho = \left(\frac{m-1}{m+1}\right)^{1/p'}. \quad (7.4)$$

This is easy to verify by substituting back. We shall take (7.4) as the definition of ρ from now on. Note that $\tilde{L}$ and $\tilde{R}$ are nonnegative on $[\rho, 1]$, and are pointwise bounded above by $\frac{10}{m}\alpha^{-1/2}$, say. Indeed, $\alpha^{1/2}\tilde{L}(\alpha)$ is increasing, and hence bounded by $\tilde{L}(1) = 2/(m-1)$, while $\alpha^{1/2}\tilde{R}(\alpha)$ is decreasing, and hence bounded by $\rho^{1/2}\tilde{R}(\rho) = \sqrt{\frac{m+1}{m-1}} - \sqrt{\frac{m-1}{m+1}}$. Note also that for m fixed and p' not too large,

$$\int_{\alpha=\rho}^{1} \tilde{L}(\alpha)\,d\alpha = \Omega(1), \tag{7.5}$$

due to the contribution from α between $1/2$ and 1, say.

Throughout we shall consider a function $\varepsilon = \varepsilon(p)$ tending to zero at some suitable rate as $p \to 0$. In the end it turns out that we shall take $\varepsilon = p$, but the roles played by ε and p are rather different, so we preserve the separate notation.

To obtain a lower bound on $\lambda_m(p) = \sigma(p, 0)$, set $p' = p(1 - \varepsilon)$. Set

$$L_0(\alpha) = \varepsilon \rho^{1/2}\tilde{L}(\alpha)/100, \qquad R_0(\alpha) = \varepsilon \rho^{1/2}\tilde{R}(\alpha)/100.$$

Then L_0 and R_0 solve (7.3) and are bounded above by $\varepsilon/(10m)$. Let us write (7.1) in the form $(L, R) = \mathbf{F}_\rho((L, R))$, where $\mathbf{F}_\rho$ is an operator taking a pair of functions on $[\rho, 1]$ to another such pair. When ε is small enough, since L_0, R_0 are bounded by $\varepsilon/(10m)$, when evaluating $\mathbf{F}_\rho((L_0, R_0))$ the linear approximation of the integrand is correct to within a factor $1 \pm \varepsilon/2$, say. Hence, since (L_0, R_0) solves (7.3) and $p = p'/(1 - \varepsilon)$, for p small enough the two functions given by $\mathbf{F}_\rho((L_0, R_0))$ are pointwise at least L_0 and R_0. It follows from the first fact proved at the end of Section 5, or simply by iterating $\mathbf{F}_\rho$, starting from (L_0, R_0), to obtain a solution to $(L, R) = \mathbf{F}_\rho((L, R))$, that L_0, R_0 are pointwise lower bounds for the maximal solution L, R to $(L, R) = \mathbf{F}_\rho((L, R))$. Hence,

$$\lambda_m(p) = \sigma(p, 0) \geqslant \sigma(p, \rho) \geqslant p \int_{\alpha=\rho}^{1} \left(1 - \frac{(1 - L_0(\alpha))^m}{(1 + R_0(\alpha))^m}\right)d\alpha.$$

Again, the linearized form of the above integral is a good approximation. Using (7.5) we see that this integral is $\Omega(\varepsilon \rho^{1/2})$, so $\lambda_m(p) = \Omega(p\varepsilon \rho^{1/2}) = \Omega(p^2 \rho^{1/2})$, recalling that $\varepsilon = p$. Now, as $p' = p(1 - \varepsilon) = p(1 - p)$,

$$\rho^{1/2} = \exp\left(\frac{\log((m-1)/(m+1))}{2p}(1-p)^{-1}\right) = \Theta\left(\left(\frac{m-1}{m+1}\right)^{\frac{1}{2p}}\right)$$

as $p \to 0$, so our bound $\lambda_m(p) = \Omega(p^2 \rho^{1/2})$ gives exactly the lower bound claimed in (2.1).

For the upper bound, we consider $p' = p(1 + \varepsilon)$, defining $\tilde{L}, \tilde{R}$ and ρ as above. Again we shall take $\varepsilon = p$. Set $\tau = \rho(1 + \varepsilon)^{2/p'}$, noting that $\tau = \Theta(\rho)$. Recall from (7.4) that $\rho^{p'/2} = \sqrt{\frac{m-1}{m+1}}$, so

$$\tau^{p'/2} = (1 + \varepsilon)\sqrt{\frac{m-1}{m+1}},$$

and

$$\left(\frac{m+1}{m-1}\tau^{p'/2} - \tau^{-p'/2}\right) = \left((1+\varepsilon) - \frac{1}{1+\varepsilon}\right)\sqrt{\frac{m+1}{m-1}} = \Theta(\varepsilon).$$

Hence, for $\alpha \geqslant \tau$ we have

$$\alpha^{1/2}\tilde{L}(\alpha) \geqslant \tau^{1/2}\tilde{L}(\tau) = \Omega(\varepsilon). \tag{7.6}$$

Set

$$L_1(\alpha) = \begin{cases} f\tilde{L}(\alpha) & \alpha \geq \tau \\ p & \alpha < \tau, \end{cases}$$

$$R_1(\alpha) = \begin{cases} f\tilde{R}(\alpha) & \alpha \geq \tau \\ p/\sqrt{\alpha} & \alpha < \tau, \end{cases}$$

where $f > 0$ is a scale-factor to be chosen later, and define L_2, R_2 by $(L_2, R_2) = \mathbf{F}_0((L_1, R_1))$. Thus,

$$L_2(\alpha) = \frac{p}{2\sqrt{\alpha}} \int_{\beta=0}^{\alpha} \frac{1}{\sqrt{\beta}} \left(1 - \frac{(1-L_1(\beta))^m}{(1+R_1(\beta))^{m+1}}\right) d\beta, \qquad (7.7)$$

and

$$R_2(\alpha) = \frac{p}{2\sqrt{\alpha}} \int_{\beta=\alpha}^{1} \frac{1}{\sqrt{\beta}} \left(1 - \frac{(1-L_1(\beta))^{m-1}}{(1+R_1(\beta))^m}\right) d\beta.$$

We claim that $L_2(\alpha) \leq L_1(\alpha)$ and $R_2(\alpha) \leq R_1(\alpha)$ hold for every α, provided p is small enough.

For $\alpha < \tau$ the claim is trivial: the large brackets inside the integrals above are each at most 1, so

$$L_2(\alpha) \leq \frac{p}{2\sqrt{\alpha}} \int_{\beta=0}^{\alpha} \frac{1}{\sqrt{\beta}} d\beta = p = L_1(\alpha),$$

and

$$R_2(\alpha) \leq \frac{p}{2\sqrt{\alpha}} \int_{\beta=\alpha}^{1} \frac{1}{\sqrt{\beta}} d\beta = p \frac{2-2\sqrt{\alpha}}{2\sqrt{\alpha}} \leq \frac{p}{\sqrt{\alpha}} = R_1(\alpha).$$

Suppose now that $\alpha \geq \tau$. The integral defining $R_2(\alpha)$ only involves values $L_1(\beta), R_1(\beta)$ for which $\beta \geq \alpha \geq \tau$. In this range $L_1 = f\tilde{L}$ and $R_1 = f\tilde{R}$. Now the linearized form of the right-hand side of (7.1) is always an upper bound on the original form, as $1 - (1-x)^i(1+y)^{-j} < ix + jy$ for positive real numbers x, y and for $i, j \geq 1$. Hence,

$$R_2(\alpha) \leq \frac{p}{2\sqrt{\alpha}} \int_{\beta=\alpha}^{1} \frac{1}{\sqrt{\beta}} ((m-1)f\tilde{L}(\beta) + mf\tilde{R}(\beta)) d\beta.$$

Since $\tilde{L}$ and $\tilde{R}$ solve (7.3), it follows that

$$R_2(\alpha) \leq (p/p')f\tilde{R}(\alpha) \leq f\tilde{R}(\alpha) = R_1(\alpha).$$

Finally, we must estimate $L_2(\alpha)$ for $\alpha \geq \tau$. We shall split the integral on the right-hand side of (7.7) into two parts, bounding the contribution from $\beta < \tau$ using only that the large bracket involved is at most 1, and the contribution from $\beta \geq \tau$ as above:

$$L_2(\alpha) = \frac{p}{2\sqrt{\alpha}} \int_{\beta=0}^{\tau} \frac{1}{\sqrt{\beta}} \left(1 - \frac{(1-L_1(\beta))^m}{(1+R_1(\beta))^{m+1}}\right) d\beta$$

$$+ \frac{p}{2\sqrt{\alpha}} \int_{\beta=\tau}^{\alpha} \frac{1}{\sqrt{\beta}} \left(1 - \frac{(1-L_1(\beta))^m}{(1+R_1(\beta))^{m+1}}\right) d\beta$$

$$\leqslant \frac{p}{2\sqrt{\alpha}} 2\sqrt{\tau} + \frac{p}{2\sqrt{\alpha}} \int_{\beta=\tau}^{\alpha} \frac{1}{\sqrt{\beta}} (mL_1(\beta) + (m+1)R_1(\beta)) \, d\beta$$

$$\leqslant p\sqrt{\frac{\tau}{\alpha}} + \frac{p}{p'} f\tilde{L}(\alpha)$$

$$\leqslant f\tilde{L}(\alpha) + p\sqrt{\frac{\tau}{\alpha}} - \frac{\varepsilon f}{2} \tilde{L}(\alpha),$$

using $p' = p(1+\varepsilon)$. The final bound will be at most $f\tilde{L}(\alpha) = L_1(\alpha)$ as long as we choose f large enough so that $2p\sqrt{\tau} < \varepsilon f\tilde{L}(\alpha)\sqrt{\alpha}$ holds for all $\alpha > \tau$. Using (7.6) and recalling that $\varepsilon = p$, we see that it suffices to take $f = A\sqrt{\tau}/p$ for some constant A.

For this value of f we have shown that our trial functions L_1, R_1 iterate under $\mathbf{F}_0$ to functions below themselves. Using the second fact proved at the end of Section 5, which applies by the remarks made before the start of this proof, it follows that L and R, the maximal solutions to $(L, R) = \mathbf{F}_0((L, R))$, are bounded above by L_1 and R_1. Thus from (7.2) we have

$$\lambda_m(p) = \sigma(p, 0) \leqslant p \int_{\alpha=0}^{1} \left(1 - \frac{(1 - L_1(\alpha))^m}{(1 + R_1(\alpha))^m}\right) d\alpha$$

$$\leqslant p\tau + p \int_{\alpha=\tau}^{1} (mL_1(\alpha) + mR_1(\alpha)) \, d\alpha.$$

Here the last inequality is obtained by splitting the integral at τ, using 1 as an upper bound on the integrand for $\alpha < \tau$, and the linear form as an upper bound for $\alpha > \tau$. As $\tilde{L}$ and $\tilde{R}$ are bounded by $C\alpha^{-1/2}$ for some universal constant, the final integral is bounded by $2mCf$ times $\int_0^1 \alpha^{-1/2} \, d\alpha = 2$, so

$$\lambda_m(p) \leqslant p\tau + O(pf) = p\tau + O(\sqrt{\tau}).$$

The second term dominates, and, as before, we have

$$\sqrt{\tau} = \Theta(\sqrt{\rho}) = \Theta\left(\left(\frac{m-1}{m+1}\right)^{\frac{1}{2p}}\right),$$

giving the upper bound stated in (2.1) and completing the proof of Theorem 2.1. □

The proof above consists of one way of bounding the solution to certain rather unpleasant equations (7.1). It may be that there is a better way of bounding the solution, eliminating the gap of a factor p^2 between the upper and lower bounds. Although in one sense rather large, in another sense this gap is rather small: not only have we shown that $\lambda_m(p) = \exp(-\Theta(1/p))$, improving the lower bound of $\exp(-\Theta(1/p^2))$ given in [8], but we have found the constant in the exponent to within $o(1)$.

The calculation used in the proof is in some sense a version of the direct argument given for the $c/\sqrt{ij}$ model in Section 4. Indeed, our method of obtaining an upper bound on the size of the giant component involved removing a small number of 'early' vertices (with $i \leqslant \rho n$), so that the graph on the remaining 'late' vertices was subcritical. The final bound on $L_2(\alpha)$ above is really a way of estimating the total number of paths from late to early vertices, using the fact that the branching process restricted to late vertices is subcritical, and that there are very few early vertices. This is analogous to the upper bound on the giant component of the $c/\sqrt{ij}$ graph proved by directly estimating the number of late-early paths in Section 4. For the lower bound, we again removed some early vertices, even though this decreases the size of the giant component. The gain was that the resulting branching process, which is still supercritical, has an eigenfunction that is not too unbalanced (does not become too large as α becomes small). This means that a not too small multiple of the eigenfunction gives a lower bound on the survival probability. Again this method is analogous to, though much more precise than, the method used in Section 4, or in [6] for a different model.

8. The uniformly grown Z-out random graph

We now turn to graphs that grow by adding one vertex at a time, each new vertex sending edges to some number of old vertices that are chosen uniformly at random, considering a common generalization of the uniformly grown m-out random graph and the c/j-graph, as in [10]. The argument for the c/j-graph would be considerably simpler than for this general case, but it turns out that the actual calculations come out more or less the same in the general case.

8.1. Precise definitions and conditions

In Section 2 we introduced the uniformly grown Z-out graph in a rather vague way, leaving precise details and the conditions we shall need for our theorem to this section. The definition will be essentially as in [10] (where X was used instead of Z – here that would clash with our branching process notation), except that we allow rather greater generality.

Let Z be a nonnegative integer-valued random variable; the main examples to bear in mind are the cases where Z is constant, and where Z has a Poisson distribution. Let Z_i, $i = 1, 2, \ldots$, be independent random variables with Z_i taking values in $\{0, 1, \ldots, i-1\}$, and Z_i converging to Z in distribution as $i \to \infty$. (We shall make additional assumptions later.) We say that the random graph process $(G_Z^{(n)})_{n \geq 0}$ is a *uniformly growing Z-out process* if $G_Z^{(0)}$ is the empty graph with no vertices, and $G_Z^{(n)}$ is obtained from $G_Z^{(n-1)}$ by adding a new vertex n and joining it to a uniformly chosen set of Z_n vertices from $\{1, 2, \ldots, n-1\}$, this choice being independent of $G_Z^{(n-1)}$. We say that the graph $G_Z^{(n)}$ defined in this way is a *uniformly grown Z-out graph*. Note that two uniformly grown Z-out graphs need not have exactly the same distribution, as the only condition we have so far imposed on the Z_i is that they converge to Z.

Given a $G_Z^{(n)}$ as above, we wish to study the size of the largest component of $G_Z^{(n)}(p)$, the random subgraph of $G_Z^{(n)}$ obtained by retaining each edge independently with probability p. (We could instead keep vertices independently, but as usual it makes no difference in the end; the argument for the vertex-deleted subgraph is notationally more complex but the final bounds differ by exactly a factor of p.) Here we may take $p = 1$, as $G_Z^{(n)}(p)$ has the same distribution as $G_{Z(p)}^{(n)}$, where $Z(p)$ is the random variable defined by the number of successes in Z trials, each of which succeeds with probability p, independently of Z and of all other trials. More precisely, we must replace each Z_i with a corresponding $Z_i(p)$.

Throughout we assume that $\Pr(Z \geq 2) > 0$, and that the first two moments of Z are finite. We write c for $\mathbf{E}(Z)$ for consistency with earlier work. We also write $r = r(Z)$ for $\mathbf{E}(Z(Z-1))/\mathbf{E}(Z)^2$, noting that $r > 0$, and that $r = 1$ if Z has a Poisson distribution. If $Z(p)$ is defined from Z as above, then $\mathbf{E}(Z(p)) = p\mathbf{E}(Z)$ and $r(Z(p)) = r(Z)$. Indeed, the kth factorial moment of $Z(p)$ is p^k times the corresponding moment of Z; compare the expected number of k-element subsets of a set of size Z and one of size $Z(p)$.

We impose the following additional conditions on Z and the Z_i, which may or may not be needed, but are used in our proof. First, we assume that $\mathbf{E}(Z^3)$ is finite. In fact, as we shall consider asymptotics as the distribution of Z varies, we assume that $\mathbf{E}(Z^3) = O(1)$, i.e., that as we vary Z, the third moment is bounded by some absolute constant. This is only used at one point in the proof; see (8.13). Secondly, we assume that

$$d_{TV}(Z_i, Z) = O((\log i)^{-8}) \tag{8.1}$$

as $i \to \infty$. Here d_{TV} is the total variation distance, namely half the sum of the absolute value of the difference of the probabilities of taking certain values. In (8.1), an exponent of $-2 + \varepsilon$ would perhaps suffice if we were more careful in the proof, but this would lead to additional complications. Finally, we assume that $\mathbf{E}(Z_i^2)$ is bounded above by some constant independent of i.

Theorem 8.1. *Let Z and Z_i, $i = 1, 2, \ldots$, be random variables satisfying the conditions above, and set $c = \mathbf{E}(Z)$ and*

$$r = r(Z) = \frac{\mathbf{E}(Z(Z-1))}{\mathbf{E}(Z)^2}.$$

Let $G_Z^{(n)}$ be the uniformly grown Z-out graph defined by the Z_i. Then there is a constant $\lambda(Z)$ such that

$$C_1(G_Z^{(n)}) = (\lambda(Z) + o(1))n$$

holds w.h.p. as $n \to \infty$. Furthermore, writing

$$c_0 = \frac{1}{2}\frac{1 - \sqrt{r}}{1 - r}$$

if $r \neq 1$ and $c_0 = 1/4$ if $r = 1$, we have $\lambda(Z) = 0$ if $c \leq c_0$, and

$$\lambda(Z) = \Omega\left(\varepsilon^{3/2} \exp\left(-\frac{\pi}{2r^{1/4}}\varepsilon^{-1/2}\right)\right) \tag{8.2}$$

and

$$\lambda(Z) = O\left(\varepsilon^{-3/2} \exp\left(-\frac{\pi}{2r^{1/4}}\varepsilon^{-1/2}\right)\right) \tag{8.3}$$

whenever $c = c_0 + \varepsilon$ with $\varepsilon \to 0$ from above and $r(Z) = r$ is fixed.

Thus, while the exact normalized size of the giant component does depend on the whole distribution Z, the existence of such a giant component, and its asymptotic size when small, are determined by the first two moments of Z only.

Taking Z to be Poisson with mean c, and Z_i to be Binomial with parameters $i-1$ and $c/(i-1)$, Theorem 8.1 immediately implies Theorem 2.4.

For Theorem 2.3, note that all our conditions on Z and the Z_i are preserved if we pass to $Z(p)$ and $Z_i(p)$ for $0 < p < 1$. In particular, starting from a Z with mean μ and $r(Z) = r$ satisfying the conditions above, applying Theorem 8.1 with Z replaced by $Z(p)$ shows that in the random subgraph $G_Z^{(n)}(p)$ of $G_Z^{(n)}$ there is a giant component if and only if $p > p_0$, where

$$p_0 = \mu^{-1} c_0 = \frac{1}{2\mu} \frac{1 - \sqrt{r}}{1 - r}$$

for $r \neq 1$ and $p_0 = \frac{1}{4\mu}$ if $r = 1$, and that if $p = p_0 + \varepsilon$ then the size of the giant component is w.h.p.

$$n \exp\left(-\left(\frac{\pi}{2r^{1/4}\mu^{1/2}} + o(1)\right)\varepsilon^{-1/2}\right)$$

as $\varepsilon \to 0$ from above. (This time we have written the upper and lower bounds together in a weaker but simpler form.) In particular, taking $Z = m$ to be constant, so $\mu = m$ and $r = (m-1)/m$, and taking $Z_i = m$ for all $i > m$ (and $Z_i = 0$ for $i \leqslant m$), we obtain Theorem 2.3.

8.2. Comparison with a branching process

As in Section 7, or in [8], we shall prove results about the largest component in $G_Z^{(n)}$ by comparison with a branching process. This method was used in [10], but with more restrictive conditions. As in [10], each point in the branching process has a type $(\alpha(x), T(x))$, where $T(x)$ has two possible values, which we shall write here as L for points 'reached from the left' and R for points 'reached from the right'. Each point will have a random number of left- and right-children in the next generation, where the α value for a left/right-child is smaller/larger than the α value of the parent, and all left/right-children are reached from the right/left.

The precise definition is that we start from $X_0 = \{x_0\}$, a single point of type (α, R), where α is uniformly distributed on $(0,1)$. Each point $x \in X_k$ gives rise to children in the next generation as follows: x has a random number N of left-children, of types (β_i, R), $i = 1, \ldots, N$, where the β_i are independently uniformly distributed on $(0, \alpha(x))$. Here N has the distribution of Z if $T(x) = R$, and of $Y - 1$ if $T(x) = L$, where Y is the size-biased version of Z, satisfying $\Pr(Y = t) = t \Pr(Z = t)/E(Z)$. Note that $E(Y - 1) = E(Z(Z-1))/E(Z) = rE(Z) = rc$. The point x also has a Poisson number of right-children of types (β_i, L), where the β_i values are distributed according to a Poisson process on $(\alpha(x), 1)$ with density $c\beta^{-1} d\beta$. Let (X_k) be the branching process defined in this way.

As before, given a (large enough) value of n, let (Γ_k) be the neighbourhood expansion process in $G_Z^{(n)}$, so $\Gamma_0 = \{v_0\}$, where v_0 is chosen uniformly at random from $\{1, 2, \ldots, n\}$, and Γ_k is the set of vertices at graph distance exactly k from v_0 in $G_Z^{(n)}$. We claim that (X_k) and (Γ_k) can be coupled so that with high probability the sizes of corresponding generations agree until they become 'large' or die out. This time, we shall use a smaller meaning of 'large', namely larger than $M = (\log n)^2$. (Perhaps any function $M(n)$ with $M(n)/\log n \to \infty$ would do if we were more careful in the proof.)

Lemma 8.2. *Let Z and $Z_1, Z_2, \ldots$ be given, satisfying the conditions of Theorem 8.1. For each n the processes (Γ_k) and (X_k) can be coupled so that with probability $1 - O(M^{-2})$ we have*

$$|\Gamma_k| = |X_k|$$

for $0 \leqslant k \leqslant K$, and either $|\Gamma_K| = |X_K| = 0$ or

$$\sum_{k=0}^{K+2} |\Gamma_k|, \sum_{k=0}^{K+2} |X_k| \geqslant M, \tag{8.4}$$

where $M = (\log n)^2$.

Outline proof. The basic method is the same as the proof of Lemma 6.1. Also, a very similar result is proved in [10] though under stronger assumptions. Thus we shall only give an outline proof, concentrating on the difficulties, and explaining why with the smaller choice of M here our assumptions are sufficient.

As usual, we inductively couple the branching and neighbourhood expansion processes so that the number and types of points agree. Agreement in type means two things: firstly, agreement as to whether the point/vertex was reached from the left or the right, where a vertex $v \in \Gamma_k$ is reach from the left/right according to whether its (with high enough probability unique) neighbour $w \in \Gamma_{k-1}$ satisfies $w < v$ or $w > v$. Secondly, agreement between α values after rounding to multiples of $n^{-1/2}$, say, where the α value of a vertex v is v/n.

As before, there are two key points. Firstly, we may ignore any bad event that has probability $O(M^{-3})$ per point/vertex reached in X_k or Γ_k, since we abandon the coupling before considering more than M points/vertices. Secondly, we can ignore points/vertices with small α values, say smaller than $n^{-1/4}$; we shall prove the second claim later. As a consequence we may assume that Z is bounded by M^3, and that the Z_i and Z are identically distributed. To see this, note that as Z has finite mean and variance, the probability that Z exceeds M^3 is $O(M^{-6})$. As Y, the size-biased version of Z, has finite mean, the probability that Y exceeds M^3 is $O(M^{-3})$. Defining Z' to have the distribution of Z conditioned on $Z \leqslant M^3$, and Y' to be the size-biased version of Z', we see that $d_{TV}(Z, Z') = O(M^{-6})$ and $d_{TV}(Y, Y') = O(M^{-3})$. Thus Z and Z', and also Y and Y', may be coupled to agree with probability $1 - O(M^{-3})$. It follows that we can couple the branching processes defined using the distributions Z and Z' so that with probability $1 - O(M^{-2})$ they agree until both become large. Similarly, we can couple $G_Z^{(n)}$ with a graph where the down-degree $Z'(i)$ of each vertex $i \geqslant n^{3/4}$ has distribution Z', and these down-degrees are independent. As $d_{TV}(Z, Z') = O(M^{-6})$, while, from (8.1), $d_{TV}(Z_i, Z) = O(M^{-4})$ for all $i \geqslant n^{3/4}$, we can do this coupling so that the down-neighbourhoods of all but $O(nM^{-4})$ vertices agree; call the remaining vertices *bad*. We claim that it is unlikely that the neighbourhood expansion process reaches a bad vertex before becoming large. If we reach a bad vertex of large degree (larger than M), there is no problem, as the next generation is large by definition, so the stopping condition (8.4) in the lemma applies. (Here it helps that we continue the sums in (8.4) to $k = K + 2$ rather than $k = K + 1$.) For bad vertices of degree at most M it is easy to see that the relative chance of reaching some bad vertex to reaching some good vertex is $O(M)$ times the density of bad vertices, and hence $O(M^{-3})$. Such a probability per vertex reached can safely be ignored.

We now turn to the claim that we can ignore points/vertices with α values less than $n^{-1/4}$. We assume here and from now on, as we may, that Z is bounded by M^3, and that all Z_i, $i \geqslant n^{3/4}$, have exactly the distribution of Z. Now a vertex v, $v \geqslant n^{3/4}$, is very unlikely to have a neighbour w with $w < v/M^6$, say. Indeed, as Z_v is bounded by M^3, and from the uniform distribution of neighbours, this probability is $O(M^{-3})$. The same bound holds conditional on Z_v, so we can apply this to a vertex $v \in \Gamma_k$ whether v was reached from the left or the right. It follows that we may assume as we construct the coupling that no $v \in \Gamma_k$ has a left-neighbour $w < v/M^6$. We may also assume that $v_0 \geqslant n/M^3$. Suppose now that some Γ_k, reached before the stopping condition (8.4) holds, contains a vertex v with $\alpha(v) < n^{-1/4}$, i.e., $v < n^{3/4}$. Then we must have $k \geqslant 3$, and Γ_{k-3} must contain a w with $w < M^{18} n^{3/4} < n^{0.76}$. But any such w is likely to have many right-neighbours in Γ_{k-2}: its expected number of neighbours w' with $n^{0.76} < w' < n^{0.77}$ is $\Omega(\log n)$, and from independence and exponential concentration of the Binomial distribution, it follows that with probability $1 - O(M^{-100})$ the number of actual neighbours w' in this range is at least $\Omega(\log n)$ with a reduced implicit constant. Similarly, with very high probability each of these w' has $\Omega(\log n)$ neighbours w'' in Γ_{k-1} with $n^{0.77} < w'' < n^{0.78}$. After one more step we find with high enough probability $\Omega((\log n)^3) > M$ vertices in Γ_k. Thus the probability that some Γ_k contains a v with $\alpha(v) < n^{-1/4}$ but that $|\bigcup_{t \leqslant k} \Gamma_k| < M$ holds is $O(M^{-2})$, and we may abandon our coupling if this event occurs. A similar but simpler argument applies to the branching process.

From this point the rest of the argument is easy, though not easy to write down precisely. We give only an outline; a more complete argument in a special case is given in [10]. The basic method is as in Section 7, but we must deal with dependence between edges introduced via the Z_is. There are two key observations. Firstly, suppose that $i < j$. Then the conditional probability that $ij \in E(G_Z^{(n)})$ given Z_j is proportional to Z_j. Hence, the conditional distribution of Z_j given that $ij \in E(G_Z^{(n)})$ is exactly the size-biased version Y of Z. Thus, ignoring other complications, when we reach a vertex from the left we expect its number of other left-neighbours to have the distribution of $Y - 1$. The second point is that we stop before reaching too many vertices; when we come to consider some $v \in \Gamma_k$, asking what are its neighbours in Γ_{k+1}, a certain set U of vertices have been 'used up'. This set consists of $\bigcup_{t \leqslant k} \Gamma_k$ together with any vertices of Γ_{k+1} found as neighbours of $w \in \Gamma_k$ considered before v. By the stopping rule, we may assume that $|U| < M$. Conditioning on the part of the graph uncovered so far, all we know about the neighbourhood of v is that it includes a certain $u \in \Gamma_{k-1}$, and otherwise avoids a certain subset of U. Similarly, for any $w \in [n] \setminus U$, all we know about the neighbourhood of w is that it avoids a certain subset of U. Now when we come to consider whether some edge vw is present, with $w < v$, say, we need to know the conditional distribution of Z_v. The key point is that since U is so small, Z_v is never too large, and v is large enough, given any possible value of Z_v the probability that v sends an edge to U is extremely small ($o(n^{-1/2})$, say). Hence the conditional distribution of Z_v is very close to Z or to Y, depending on how v was reached. Similarly, the chance that some $w > v$ sends an edge to v is proportional to the conditional expectation of Z_w, which is very close to its unconditional value. □

As before, this coupling result enables us to compare the size of the giant component (if any) in $G_Z^{(n)}$ with the survival probability $\sigma(Z)$ of the branching process (X_k).

Lemma 8.3. *Let Z and $Z_1, Z_2, \ldots$ be given, satisfying the conditions of Theorem 8.1. Then*
$$C_1(G_Z^{(n)}) = (\sigma(Z) + o(1))n$$
holds w.h.p. as $n \to \infty$.

Proof. Let us write $\#_k(G)$ for the number of vertices of a graph G in components of order k. Writing $M = (\log n)^2$ as before, Lemma 8.2 implies immediately that
$$\mathbf{E}(\#_k(G_Z^{(n)})) = (s_k(Z) + O(M^{-2}))n$$
for $0 \leq k \leq M$, where s_k is the probability that the branching process has total size k, so $\sigma(Z) = 1 - \sum_{k=0}^{\infty} s_k$. We shall in fact apply the lemma to a slightly modified graph.

As usual, the next step is to show concentration. Again there is a potential problem with large degrees, which can be dealt with in the following global way. Let Z' be Z capped at $B = \lceil M^3 \rceil$, say, so Z' is the smaller of Z and B. Define Z'_i from Z_i similarly. Let $G_{Z'}^{(n)}$ be the uniformly grown Z'-out graph defined using the Z'_i. By construction of Z'_i, we have $G_{Z'}^{(n)} \subset G_Z^{(n)}$. Furthermore, if ij, $i < j$, is an edge present in $G_Z^{(n)}$ but not $G_{Z'}^{(n)}$, then j has large degree (at least B) in both graphs. It follows that the only way a vertex v can lie in a component of order $k < B$ in one graph and not in the other is if v is in a component of order k in $G_{Z'}^{(n)}$, and that component meets one of the extra edges of $G_Z^{(n)}$. Hence we have
$$|\#_k(G_Z^{(n)}) - \#_k(G_{Z'}^{(n)})| \leq k|E(G_Z^{(n)} \setminus G_{Z'}^{(n)})|.$$
Now the right-hand side above is just $k \sum_i (Z_i - Z'_i)$. For each i we have
$$\mathbf{E}(Z_i - Z'_i) = \sum_{t \geq B} \Pr(Z_i = t)(t - B) \leq \sum_{t \geq B} \Pr(Z_i = t)t^2/B \leq \mathbf{E}(Z_i^2)/B.$$
From our assumptions, the $\mathbf{E}(Z_i^2)$ are uniformly bounded above. Hence,
$$\mathbf{E}(|\#_k(G_Z^{(n)}) - \#_k(G_{Z'}^{(n)})|) = O(nk/B) = O(nk/M^3),$$
and the probability that $\sum_{k \leq M} |\#_k(G_Z^{(n)}) - \#_k(G_{Z'}^{(n)})|$ exceeds $n/\sqrt{M}$ is $o(1)$.

We know from the proof of Lemma 8.2 that truncating Z at $B \geq M^3$ has little effect on the branching process, and in particular that $|s_k(Z') - s_k(Z)| = O(M^{-2})$. Since $G_{Z'}^{(n)}$ satisfies the conditions of Lemma 8.2 we thus have
$$\mathbf{E}(\#_k(G_{Z'}^{(n)})) = (s_k(Z') + O(M^{-2}))n = (s_k(Z) + O(M^{-2}))n$$

for $0 \leq k \leq M$. Finally, $\#_k(G_{Z'}^{(n)})$ is concentrated about its mean: in constructing the graph $G_{Z'}^{(n)}$ we have n independent steps, the construction of the down-neighbourhoods of the vertices. Changing one such step adds/deletes at most B edges to the final graph, and hence changes $\#_k$ by at most $4kB \leq n^{1/10}$. It follows that with probability $1 - O(M^{-2})$ we have $\#_k(G_{Z'}^{(n)})$ within $n^{3/4}$ of its mean, say. Combined with the observations above this shows that

$$\Pr\left(\sum_{k \leq M} |\#_k(G_Z^{(n)}) - ns_k(Z)| \geq 2n/\sqrt{M}\right) = o(1).$$

It follows that w.h.p. the graph $G_Z^{(n)}$ has $\sigma(Z)n + o(n)$ vertices in 'large' components, where we may take large to mean greater than L for any $L \to \infty$ with $L \leq M$.

It remains to show that most (all but $o(n)$) vertices in large components are in a single giant component. The argument is exactly as in the proof of Lemma 6.3: recall that $G_Z^{(n)}$ was defined as one graph in a graph process $(G_Z^{(t)})_{t \geq 0}$. We consider the graph process from $t = n'$ to $t = n$, where $n' = \lfloor n(1-\eta) \rfloor$ for $\eta = (\log n)^{-1/2}$, say. Each time we add a new vertex, passing from $G_Z^{(t-1)}$ to $G_Z^{(t)}$, there is a positive probability that we add at least two edges. Thus, given disjoint sets of vertices A and B, there is a probability $\Omega(|A||B|/n^2)$ that the new vertex sends edges to both A and B. This is exactly the single condition needed in the proof of Lemma 6.3 to show that almost all vertices in large components are in a single giant component. □

8.3. Analysis of the branching process

We have shown that the size of the giant component in $G_Z^{(n)}$ is w.h.p. close to n times the survival probability $\sigma(Z)$ of a certain branching process defined above; it remains to estimate $\sigma(Z)$. As before, it will be convenient to modify the branching process, by deleting all points with $\alpha(.)$ values less than ρ. The survival probability $\sigma(Z, \rho)$ of the modified process is of course at most $\sigma(Z) = \sigma(Z, 0)$.

We could write down directly the equation giving the survival probabilities of points in our branching process, but as in Section 7 and in [10] it is more convenient to work with a transformation of these probabilities. In the (possibly) truncated process ($\rho \geq 0$), let $L(\alpha)$ be the probability that a given left-child of a point x with $\alpha(x) = \alpha$ survives, and let us write the probability that no right-children of x survive as $\exp(-cR(\alpha))$. (This is natural due to the Poisson form of the distribution of right-children.) Then, as noted in [10] (see equations (21), (22) and the modifications in Section 9), it is easy to check that

$$L(\alpha) = \frac{1}{\alpha} \int_{\beta=\rho}^{\alpha} \left(1 - e^{-cR(\beta)} \sum_{t \geq 0} p_t (1 - L(\beta))^t\right) d\beta, \tag{8.5}$$

and

$$R(\alpha) = \int_{\beta=\alpha}^{1} \frac{1}{\beta} \left(1 - e^{-cR(\beta)} \sum_{t \geq 0} q_t (1 - L(\beta))^t\right) d\beta, \tag{8.6}$$

where $p_t = \Pr(Z = t)$ and $q_t = \Pr(Y - 1 = t)$, where Y is the size-biased version of Z, so

$$q_t = (t+1)p_{t+1} / \sum_i i p_i.$$

Also, recalling that the branching process is started from a single point x_0 with $\alpha(x_0)$ uniformly chosen from $[0,1]$ (and then killed immediately if $\alpha(x_0) < \rho$), the survival probability $\sigma(Z,\rho)$ of the process is given by

$$\sigma(Z,\rho) = \int_{\beta=\rho}^{1} \left(1 - e^{-cR(\beta)} \sum_{t\geq 0} p_t(1-L(\beta))^t\right) d\beta = L(1). \tag{8.7}$$

For the comparison with [10] note that the factor p present there is absent here, as we have absorbed the edge deletion into the distribution of Z. Note also that although all bounds we shall prove depend only on c and r, $\sigma(Z) = \sigma(Z,0)$ itself depends on the whole distribution of Z.

Proof of Theorem 8.1. To complete the proof of Theorem 8.1 it suffices to prove that the bounds (8.2) and (8.3) claimed for $\lambda(Z)$ hold for $\sigma(Z)$. Note that $\sigma(Z)$ is monotonic in Z, in the sense that if Z_1 stochastically dominates Z_2, then $\sigma(Z_1) \geq \sigma(Z_2)$. In fact, corresponding random graphs, and hence the branching processes, can be coupled so that one contains the other. It follows that we may restrict our attention to the case $c > c_0$. Indeed, given Z with $c \leq c_0$, and given any $c' = c_0 + \varepsilon > c_0$ with $c' < 1$, it is easy to check that we can construct a distribution Z' with $\mathbf{E}(Z') = c'$ and $r(Z') = r(Z)$. (The idea is to shift some weight from value zero to larger values, using $c' < 1$. If we shift weight only to value 1, we can achieve the required mean without increasing the second factorial moment (SFM) at all. Shifting less weight to very large values, we can achieve a large increase in the SFM. Something in between gives exactly the increase of a factor $(c'/c)^2$ in the SFM required.) Now if we prove the upper bound on $\sigma(Z')$ corresponding to (8.3), then it follows that $\sigma(Z) = 0$, since (8.3) tends to 0 as $\varepsilon \to 0$. Similarly, but more simply, given any $c_2 > c_0$ it suffices to consider $c \leq c_2$; for larger values of c we may simply use 1 as an upper bound on $\sigma(Z)$, and consider $Z' = Z(c_2/c)$ to obtain a lower bound.

As before, let us write $\mathbf{F}_\rho$ for the operator defined by the right-hand sides of equations (8.5) and (8.6), so we are looking for the maximal solution to $(L,R) = \mathbf{F}_\rho((L,R))$. To obtain upper and lower bounds, it suffices to find pairs of functions which are mapped under $\mathbf{F}_\rho$ to functions above/below themselves. This is because L and the monotonic transformation $1 - \exp(-cR)$ of R are exactly the survival probabilities of a certain branching process. This process is not that considered in the previous subsection, but is equivalent to it, in that one can be transformed into the other. The transformed process has one point of type 1 for each left-edge from a vertex (or left-child of the untransformed process), and one point of type 2 corresponding to the (possibly empty) set of all right-edges (right-children).

We consider the linearized form of (8.5), (8.6). Note that $\sum t p_t = \mathbf{E}(Z) = c$, while $\sum t q_t = \mathbf{E}(Y-1) = \mathbf{E}(Z(Z-1))/\mathbf{E}(Z) = rc$. Thus the linearized form of these equations is

$$\tilde{L}(\alpha) = \frac{1}{\alpha} \int_{\beta=\rho}^{\alpha} (c'\tilde{L}(\beta) + c'\tilde{R}(\beta))\,d\beta,$$

$$\tilde{R}(\alpha) = \int_{\beta=\alpha}^{1} \frac{1}{\beta}(rc'\tilde{L}(\beta) + c'\tilde{R}(\beta))\,d\beta. \tag{8.8}$$

Here we have again replaced c by c' as we shall solve the linearized equations with c' slightly different from c to obtain bounds on the solutions to the real equations (8.5), (8.6).

It will be convenient to think of r as fixed and c as variable. But we shall only consider values of c, c' just above the critical value c_0, which lies strictly between 0 and 1/2. Thus we have $c, c' = \Theta(1)$, and $1/2 - c' = \Theta(1)$.

As in [10], let

$$\theta = \frac{1}{2}\sqrt{-1 + 4c' - 4(c')^2(1-r)},$$

set

$$A = \frac{\theta}{c'r}, \qquad B = \frac{1/2 - c'}{c'r},$$

and let

$$\tilde{L}(\alpha) = \alpha^{-1/2}(A\cos(\theta \log \alpha) + B\sin(-\theta \log \alpha)),$$

$$\tilde{R}(\alpha) = \alpha^{-1/2}\sin(-\theta \log \alpha).$$

Then, as shown in [10], and as may be verified by substituting back, $\tilde{L}$ and $\tilde{R}$ solve (8.8), provided

$$\rho = \exp\left(-\frac{\pi - \tan^{-1}(A/B)}{\theta}\right).$$

From now on we shall define ρ by this condition.

Let us note the following properties of θ, ρ, A, B, $\tilde{L}$ and $\tilde{R}$. As c' tends to c_0 from above, $\theta \to 0$. (In fact, c_0 is defined by $\theta(c_0) = 0$.) In particular, if $c' = c_0 + \delta^2$, then

$$\theta = r^{1/4}\delta + O(\delta^3). \tag{8.9}$$

This may be checked by differentiating the square of $\theta(c)$ with respect to c. We shall always consider $c' = c_0 + o(1)$, so we have $\theta = o(1)$.

Now $A = \Theta(\theta) = o(1)$ while $B = \Theta(1)$. Thus $\pi - \tan^{-1}(A/B) = \pi - \Theta(\theta)$, and

$$\rho = \Theta(\exp(-\pi/\theta)). \tag{8.10}$$

Turning to $\tilde{L}$ and $\tilde{R}$, note that they are nonnegative on $[\rho, 1]$; indeed, ρ is defined as the largest zero of $\tilde{L}$ in $(0, 1)$. Secondly, $\tilde{R}$ is decreasing; this is easy to see from the fact that $\tilde{R}$ solves (8.8). Thirdly,

$$\tilde{L}(\alpha), \tilde{R}(\alpha) = O\left(\theta \rho^{-1/2}\right) \tag{8.11}$$

uniformly for $\alpha \in [\rho, 1]$. This is clear for $\tilde{R}$, as $\tilde{R}$ is decreasing and

$$\tilde{R}(\rho) = \rho^{-1/2}\sin(-\theta \log \rho) = \rho^{-1/2}\sin(\pi - \tan^{-1}(A/B)) = \rho^{-1/2}\sin(\tan^{-1}(A/B)),$$

and the argument of the arctangent is $\Theta(\theta)$. To see that $\tilde{L}(\alpha) = O(\theta \rho^{-1/2})$, note that the sin term contributing to $\tilde{L}(\alpha)$ is $\Theta(\tilde{R}(\alpha))$. For the cos term note that $A = \Theta(\theta)$ and $|\cos(x)| \leq 1$. Finally, we claim that

$$\tilde{L}(e^2 \rho) = \Theta\left(\theta \rho^{-1/2}\right). \tag{8.12}$$

This is easy to check, as the derivative of $\alpha^{1/2}\tilde{L}(\alpha)$ with respect to $\log \alpha$ is $\theta(-A\sin(\theta \log \alpha) - B\cos(-\theta \log \alpha))$, which is asymptotically $B\theta$ for $\alpha = e^x \rho$ and x bounded, because $-\theta \log \alpha \sim \pi$. Hence, noting that $\tilde{L}(\rho) = 0$ and taking $x = 2$, at $\alpha = e^2 \rho$ we have $\alpha^{1/2}\tilde{L}(\alpha) \sim 2B\theta$.

We are now in a position to prove good upper and lower bounds on $\sigma(Z)$, arguing along the lines used in Section 7.

For the lower bound, given Z with $r(Z) = r$ and $c = \mathbf{E}(Z) = c_0 + \varepsilon$, set $c' = c - \varepsilon^{3/2}$, noting that $c' > c_0$ if ε is small enough. Thus we may write $c' = c_0 + \delta^2$, noting that $\delta = \varepsilon^{1/2}(1 - O(\varepsilon^{1/2}))$. Let θ, ρ, $\tilde{L}$ and $\tilde{R}$ be defined as above, using this value of c', and set

$$L_0(\alpha) = f\tilde{L}(\alpha), \qquad R_0(\alpha) = f\tilde{R}(\alpha)$$

for $\rho \leq \alpha \leq 1$. We shall take f to be a very small constant η times $\theta^{-1}\varepsilon^{3/2}\rho^{1/2}$. Thus, from (8.11) we have that L_0 and R_0 are bounded on $[\rho, 1]$ by $O(\eta\varepsilon^{3/2})$. It follows that if we consider $\mathbf{F}_\rho((L_0, R_0))$, the linearized forms of the integrands in (8.5), (8.6) are pointwise good approximations to the original forms, and in particular, that the non-linear forms are at least $1 - O(\eta\varepsilon^{3/2})$ times the linearized versions. To see this, note that for $x > 0$ and $0 < y < 1$ we have

$$1 - e^{-x}(1-y)^t \geq 1 - e^{-x}e^{-ty}$$
$$= 1 - e^{-(x+ty)}$$
$$\geq x + ty - (x+ty)^2/2.$$

Thus, recalling that $\sum_t p_t = 1$ and $\sum_t t p_t = \mathbf{E}(Z) = c$,

$$1 - e^{-cR_0(\beta)}\sum_{t \geq 0} p_t(1 - L_0(\beta))^t = \sum_t p_t\left(1 - e^{-cR_0(\beta)}(1 - L_0(\beta))^t\right)$$
$$\geq cR_0(\beta) + cL_0(\beta) - O((R_0(\beta) + L_0(\beta))^2).$$

In the last step we used $\sum t^2 p_t = \mathbf{E}(Z^2) = rc^2 + c = O(1)$. Setting $(L_0', R_0') = \mathbf{F}_\rho(L_0, R_0)$, and recalling that $L_0, R_0 = O(\eta\varepsilon^{3/2})$, it follows that

$$L_0'(\alpha) \geq \left(1 - O\left(\eta\varepsilon^{3/2}\right)\right)\frac{1}{\alpha}\int_{\beta=\rho}^{\alpha}(cL_0(\beta) + cR_0(\beta))\,d\beta$$
$$= \left(1 - O\left(\eta\varepsilon^{3/2}\right)\right)(c/c')L_0(\alpha),$$

since L_0 and R_0 solve (8.8). A similar argument proves a corresponding inequality for R_0, except that we require the additional condition

$$\sum t^2 q_t = O(1). \tag{8.13}$$

This is ensured by our condition on $\mathbf{E}(Z^3)$, and is the only place this condition is used. Finally, choosing η small enough, the factor $(1 - O(\eta\varepsilon^{3/2}))(c/c')$ is larger than 1, so $\mathbf{F}_\rho((L_0, R_0)) \geq (L_0, R_0)$ holds pointwise. Thus L_0 and R_0 are pointwise lower bounds on the maximal solution to $\mathbf{F}_\rho((L, R)) = (L, R)$, and from (8.7),

$$\sigma(Z, \rho) \geq \int_{\beta=\rho}^{1}\left(1 - e^{-cR_0(\beta)}\sum_{t \geq 0}p_t(1 - L_0(\beta))^t\right)d\beta = L_0'(1) \geq L_0(1).$$

As $L_0(1) = f\tilde{L}(1) = fA = \Theta(\theta f)$, and we have chosen f to be a small constant times $\theta^{-1}\varepsilon^{3/2}\rho^{1/2}$, we have

$$\sigma(Z) \geqslant \sigma(Z,\rho) = \Omega(\theta f) = \Omega(\varepsilon^{3/2}\rho^{1/2}) = \Omega\left(\varepsilon^{3/2}\exp\left(-\frac{\pi}{2\theta}\right)\right),$$

where the last step is from (8.10). Now from (8.9), $\theta = r^{1/4}\delta + O(\delta^3) = r^{1/4}\varepsilon^{1/2} + O(\varepsilon)$. It follows that $\exp(-1/\theta) = \Theta(\exp(-1/(r^{1/4}\varepsilon^{1/2})))$, so we obtain the bound on $\sigma(Z)$ required for (8.2).

The argument for the upper bound is similar; this time there is no problem with the linear approximation, as the integrands in (8.5), (8.6) are always at most their linearized forms. However, we must put back the vertices we deleted. The method is exactly analogous to that used in Section 7.

Given Z with $r(Z) = r$ and $c = \mathbf{E}(Z) = c_0 + \varepsilon$, set $c' = c + \varepsilon^{3/2}$, and define δ, θ, ρ, $\tilde{L}$ and $\tilde{R}$ as above, using this value of c'. Let $\tau = e^2\rho$, set

$$L_1(\alpha) = \begin{cases} f\tilde{L}(\alpha) & \alpha \geqslant \tau \\ 1 & \alpha < \tau, \end{cases}$$

$$R_1(\alpha) = \begin{cases} f\tilde{R}(\alpha) & \alpha \geqslant \tau \\ \log(1/\alpha) & \alpha < \tau, \end{cases}$$

where $f > 0$ is a scale-factor to be chosen later, and define L_2, R_2 by $(L_2, R_2) = \mathbf{F}_0((L_1, R_1))$. We claim that L_2, R_2 are pointwise at most L_1, R_1. This is immediate for $\alpha < \tau$, using only the fact that the integrands in (8.5), (8.6) are at most one and $1/\beta$ respectively. Suppose that $\alpha \geqslant \tau$. As in Section 7, $R_2(\alpha) \leqslant R_1(\alpha)$ is almost immediate: the integral defining $R_2(\alpha)$ only involves values of $L_1(\beta)$ and $R_1(\beta)$ for $\beta \geqslant \alpha$, where $L_1 = f\tilde{L}$ and $R_1 = f\tilde{R}$. The integral is bounded above by its linearized form, which is exactly $f(c/c')\tilde{R}(\alpha)$, as $\tilde{L}$, $\tilde{R}$ solve (8.8). This in turn is at most $R_1(\alpha)$ as $c' > c$.

Turning to L_2, bounding the integrand by 1 for $\beta \leqslant \tau$ and by its linearized form for $\beta \geqslant \tau$, and noting that $\rho < \tau$, for $\alpha \geqslant \tau$ we have

$$L_2(\alpha) \leqslant \frac{\tau}{\alpha} + \frac{1}{\alpha}\int_{\beta=\tau}^{\alpha}(cL_1(\beta) + cR_1(\beta))\,d\beta$$

$$\leqslant \frac{\tau}{\alpha} + \frac{c}{c'}f\tilde{L}(\alpha)$$

$$= f\tilde{L}(\alpha) + \frac{\tau}{\alpha} - \Omega(\varepsilon^{3/2}f\tilde{L}(\alpha)),$$

using $c' = c + \varepsilon^{3/2}$. This will be at most $f\tilde{L}(\alpha) = L_1(\alpha)$ provided we choose f large enough that

$$C\varepsilon^{3/2}f\tilde{L}(\alpha)\alpha \geqslant \tau \tag{8.14}$$

holds for all $\alpha \geqslant \tau$, where C is the constant implied by the $\Omega(.)$ above. Now $\alpha\tilde{L}(\alpha)$ is increasing in α, as $\tilde{L}$ is given by (8.8), so it suffices to ensure that $C\varepsilon^{3/2}f\tilde{L}(\tau) \geqslant 1$. But $\tau = e^2\rho$, so from (8.12) we have $\tilde{L}(\tau) = \Theta(\theta\rho^{-1/2})$, and we may choose f so that (8.14) is satisfied and at the same time

$$f = O(\varepsilon^{-3/2}\theta^{-1}\rho^{1/2}).$$

With this choice of f we have $\mathbf{F}_0((L_1, R_1)) = (L_2, R_2) \leqslant (L_1, R_1)$ holding pointwise, so by the second fact proved at the end of Section 5, L_1 and R_1 are pointwise upper bounds on the maximal solution (L, R) to $\mathbf{F}_0((L, R)) = (L, R)$. It follows that $\sigma(Z)$ is at most $L_1(1) = f\tilde{L}(1) = \Theta(\theta f)$. Hence

$$\sigma(Z) = O(\varepsilon^{-3/2} \rho^{1/2}).$$

The same calculation as used at the end of the proof of the lower bound shows that

$$\rho^{1/2} = \Theta\left(\exp\left(-\frac{\pi}{2\theta}\right)\right) = \Theta\left(\exp\left(-\frac{\pi}{2r^{1/4}\varepsilon^{1/2}}\right)\right),$$

completing the proof of the bound (8.3) and hence of Theorem 8.1. □

As noted after the statement of Theorem 8.1, this perhaps rather unwieldy result immediately implies specific results for the uniformly grown m-out random graph and the uniformly grown (or c/j) random graph, namely Theorems 2.3 and 2.4. In addition, Theorem 8.1 gives an explanation of the difference between these two specific results: as well as its mean, which controls the edge density, the second moment of Z is important. This is because the second moment (but no higher moments) appear when we count the expected numbers of paths between certain vertices in the Z-out random graph.

References

[1] Abramowitz, M. and Stegun, I. A., eds (1965) *The Handbook of Mathematical Functions*, Dover, New York. (Ninth printing, 1970.)
[2] Albert, R. and Barabási, A.-L. (2002) Statistical mechanics of complex networks. *Rev. Mod. Phys.* **74** 47–97.
[3] Albert, R., Jeong, H. and Barabási, A.-L. (2000) Error and attack tolerance of complex networks. *Nature* **406** 378–382.
[4] Barabási, A.-L. and Albert, R. (1999) Emergence of scaling in random networks. *Science* **286** 509–512.
[5] Bollobás, B. (2001) *Random Graphs*, second edn, Vol. 73 of *Cambridge Studies in Advanced Mathematics*, Cambridge University Press, Cambridge.
[6] Bollobás, B., Janson, S. and Riordan, O. (2005) The phase transition in the uniformly grown random graph has infinite order. *Random Struct. Alg.* **26** 1–36.
[7] Bollobás, B. and Riordan, O. (2002) Mathematical results on scale-free random graphs. In *Handbook of Graphs and Networks* (S. Bornholdt and H. G. Schuster, eds), Wiley-VCH, Weinheim, pp. 1–34.
[8] Bollobás, B. and Riordan, O. (2003) Robustness and vulnerability of scale-free random graphs. *Internet Mathematics* **1** 1–35.
[9] Bollobás, B. and Riordan, O. (2004) The diameter of a scale-free random graph. *Combinatorica* **24** 5–34.
[10] Bollobás, B. and Riordan, O. (2005) Slow emergence of the giant component in the growing m-out graph. *Random Struct. Alg.* **27** 1–24.
[11] Bollobás, B., Riordan, O., Spencer, J. and Tusnády, G. (2001) The degree sequence of a scale-free random graph process. *Random Struct. Alg.* **18** 279–290.
[12] Callaway, D. S., Hopcroft, J. E., Kleinberg, J. M., Newman, M. E. J. and Strogatz, S. H. (2001) Are randomly grown graphs really random? *Phys. Rev. E* **64** 041902.
[13] Callaway, D. S., Newman, M. E. J., Strogatz, S. H. and Watts, D. J. (2000) Network robustness and fragility: percolation on random graphs. *Phys. Rev. Lett.* **85** 5468–5471.

[14] Cohen, R., Erez, K., ben-Avraham, D. and Havlin, S. (2000) Resilience of the Internet to random breakdowns. *Phys. Rev. Lett.* **85** 4626–4628.
[15] Dorogovtsev, S. N. and Mendes, J. F. F. (2002) Evolution of networks. *Adv. Phys.* **51** 1079.
[16] Dorogovtsev, S. N., Mendes, J. F. F. and Samukhin, A. N. (2001) Anomalous percolation properties of growing networks. *Phys. Rev. E* **64** 066110.
[17] Durrett, R. (2003) Rigorous result for the CHKNS random graph model. In *Proc. Discrete Random Walks 2003* (C. Banderier and C. Krattenthaler, eds), *Discrete Mathematics and Theoretical Computer Science* **AC** 95–104. http://dmtcs.loria.fr/proceedings/
[18] Durrett, R. and Kesten, H. (1990) The critical parameter for connectedness of some random graphs. In *A Tribute to Paul Erdős* (A. Baker, B. Bollobás and A. Hajnal, eds), Cambridge University Press, Cambridge, pp. 161–176.
[19] Kalikow, S. and Weiss, B. (1988) When are random graphs connected? *Israel J. Math.* **62** 257–268.
[20] Norris, J. private communication.
[21] Shepp, L. A. (1989) Connectedness of certain random graphs. *Israel J. Math.* **67** 23–33.
[22] Söderberg, B. (2002) General formalism for inhomogeneous random graphs. *Phys. Rev. E* **66** 066121.
[23] Söderberg, B. (2003) Random graphs with hidden colour. *Phys. Rev. E* **68** 015102(R).
[24] Söderberg, B. (2003) Properties of random graphs with hidden colour. *Phys. Rev. E* **68** 026107.

A Dirac-Type Theorem for 3-Uniform Hypergraphs

VOJTĚCH RÖDL,[1][†] ANDRZEJ RUCIŃSKI[2][‡]
and ENDRE SZEMERÉDI[3]

[1] Emory University, Atlanta, GA, USA
(e-mail: rodl@mathcs.emory.edu)

[2] A. Mickiewicz University, Poznań, Poland
(e-mail: rucinski@amu.edu.pl)

[3] Rutgers University, New Brunswick, USA
(e-mail: szemered@cs.rutgers.edu)

Received 13 May 2004; revised 2 February 2005

For Béla Bollobás on his 60th birthday

A Hamiltonian cycle in a 3-uniform hypergraph is a cyclic ordering of the vertices in which every three consecutive vertices form an edge. In this paper we prove an approximate and asymptotic version of an analogue of Dirac's celebrated theorem for graphs: for each $\gamma > 0$ there exists n_0 such that every 3-uniform hypergraph on $n \geqslant n_0$ vertices, in which each pair of vertices belongs to at least $(1/2 + \gamma)n$ edges, contains a Hamiltonian cycle.

1. Introduction

A substantial amount of research in graph theory continues to concentrate on the existence of Hamiltonian cycles. A classic theorem of Dirac states that a sufficient condition for an n-vertex graph to be Hamiltonian is that the minimum degree is at least $n/2$, and there are obvious counterexamples showing that this is best possible.

The study of Hamiltonian cycles in hypergraphs was initiated in [2] where, however, a different definition from the one considered here was introduced. From now on, by a hypergraph we will always mean a 3-uniform hypergraph, that is, a hypergraph where every edge is of size three. Given a set $U \subset V(H)$, we denote by $H[U]$ the sub-hypergraph of H induced by U.

[†] Research supported by NSF grant DMS-0300529.
[‡] Research supported by KBN grant 2 P03A 015 23. Part of the research was performed at Emory University, Atlanta.

Definition 1. *A cycle of order k* is a hypergraph C on k vertices and k edges, whose vertices can be labelled $v_1, \ldots, v_k$ in such a way that for each $i = 1, \ldots, k-2$, $\{v_i, v_{i+1}, v_{i+2}\} \in C$ as well as $\{v_{k-1}, v_k, v_1\} \in C$ and $\{v_k, v_1, v_2\} \in C$ (there are $2k$ such labellings). By a *Hamiltonian cycle* in an n-vertex hypergraph we mean a subhypergraph which is a cycle of order n. In other words, we say that a hypergraph H with $|V(H)| = n$ is *Hamiltonian* if its vertices can be labelled $v_1, \ldots, v_n$ in such a way that for each $i = 1, \ldots, n-2$, $\{v_i, v_{i+1}, v_{i+2}\} \in H$ as well as $\{v_{n-1}, v_n, v_1\} \in H$ and $\{v_n, v_1, v_2\} \in H$.

This notion and its generalizations have potential to be applicable in many contexts which still need to be explored. An application in the relational database theory can be found in [5]. As observed in [9], the square of a (graph) Hamiltonian cycle naturally coincides with a Hamiltonian cycle in a hypergraph built on top of the triangles of the graph. More precisely, given a graph G, let $\text{Tr}(G)$ be the set of triangles in G. Define a hypergraph $H^{\text{Tr}}(G) = (V(G), \text{Tr}(G))$. Then there is a one-to-one correspondence between Hamiltonian cycles in $H^{\text{Tr}}(G)$ and the squares of Hamiltonian cycles in G. For results on the existence of squares of Hamiltonian cycles see, *e.g.*, [10].

Example 1. Consider a robot walking through a tough terrain with the task of visiting n designated locations and return to the base (one may view these locations as fuel providers). In order for the robot to move from one location to another, after reaching any one of them it has to be able to 'see' the next one. To optimize, we do not want the robot to visit a location more than once. So far, this is just the standard travelling salesman problem, but suppose that in order to speed up the motion, or to smooth out the trajectory, we request that the robot 'sees' the next two locations. Then our problem becomes that of finding a Hamiltonian cycle in $H^{\text{Tr}}(G)$, where G is the graph of those pairs of n locations which can 'see' each other.

Of course, a reader with strong imaginary skills can replace the robotics terminology with something else, like mountain hiking or the travelling salesman problem with an option of skipping a town.

Our next example cannot be formulated in terms of $H^{\text{Tr}}(G)$ for any graph G.

Example 2. Consider a patient taking 24 different pills on a daily basis, one at a time every hour. Certain combinations of three pills can be deadly if taken within 2.5 hours. Let D be the set of deadly triplets of pills. Then any safe schedule corresponds to a Hamiltonian cycle in the hypergraph which is precisely the complement of D.

In [9] the authors gave a sufficient condition for a hypergraph to have a Hamiltonian cycle. They proved that if every pair of vertices belongs to more than $\frac{5}{6}(n-1) + 1$ edges, then the hypergraph contains a Hamiltonian cycle. They also conjectured that, in fact, a much stronger result is true, namely that $\frac{5}{6}(n-1) + 1$ can be replaced by $n/2$. If true this would be in close analogy with Dirac's degree condition for graphs. Some support for this conjecture stems from a construction of an edge-maximal hypergraph with each pair degree at least $\lfloor n/2 \rfloor - 1$, but not containing a Hamiltonian cycle (see [9, Theorem 3]). In this paper we prove an approximate and asymptotic version of this conjecture.

We say that a hypergraph H is an (n, γ)-graph if H has n vertices and every pair of vertices belongs to at least $(1/2 + \gamma)n$ edges.

Theorem 1.1. *For each $\gamma > 0$ there exists n_0 such that every (n, γ)-graph with $n \geqslant n_0$ is Hamiltonian.*

Remark 1. Note that an (n, γ)-graph is also an (n, γ')-graph for all $\gamma' < \gamma$. Therefore it is enough to prove Theorem 1.1 only for sufficiently small γ.

2. Preliminary lemmas

All statements in this section assume that $0 < \gamma < 1$ is sufficiently small (see Remark 1), n is sufficiently large and H is an (n, γ)-graph on a vertex set V.

Definition 2. A *k-path* is a hypergraph P on k vertices and $k - 2$ edges, whose vertices can be labelled $v_1, \ldots, v_k$ in such a way that for each $i = 1, \ldots, k - 2$, $\{v_i, v_{i+1}, v_{i+2}\} \in P$ (there are two such labellings). We say that P connects the (ordered) pairs $v_1 v_2$ and $v_k v_{k-1}$, which will be referred to as *the endpairs of* P. Note that by saying that ab is an endpair of a hyperpath, we always mean that a is the first (or the last) vertex on the path, while b is the second (or penultimate). We will often call a hyperpath, simply, a path.

For two paths P and Q, let ab be an endpair of P and ba be an endpair of Q, and assume further that $V(P) \cap V(Q) = \{a, b\}$. By $P \circ Q$ we denote the path obtained (in a unique way) as a concatenation of P and Q. This definition extends naturally to more than two paths.

Lemma 2.1. (Connecting Lemma) *For every two disjoint and ordered pairs of vertices xy and cd there is a k-path in H, $k \leqslant 4/\gamma$, which connects xy and cd.*

Proof. We construct sets $A_0, A_1, A_2, \ldots$ and bipartite graphs $G_1, G_2, \ldots$, where $V(G_i) = A_{i-1} \cup A_i$, as follows. Let $A_0 = \{y\}$ and $A_1 = \{z : xyz \in H\}$, and let G_1 be the star with y as the centre and A_1 as the set of its leaves. Note that $|A_1| \geqslant (1/2 + \gamma)n$. Further, let

$$A_2' = \{w : \exists z \in A_1 \text{ such that } yzw \in H\} \text{ and } G_2' = \{zw : z \in A_1, w \in A_2', yzw \in H\}.$$

Observe that for every edge $zw \in G_2'$ with $w \neq x$ the vertices $xyzw$ form a 4-path in H. Also, for each $z \in A_1$, we have $\deg_{G_2'}(z) \geqslant (1/2 + \gamma)n$.

Let $A_2^0 = \{w \in A_2' : \deg_{G_2'}(w) < \sqrt{n}\}$, $A_2 = A_2' \setminus A_2^0$ and $G_2 = G_2'[A_1 \cup A_2]$. Note that

$$|A_1|(1/2 + \gamma)n \leqslant |G_2'| \leqslant n^{3/2} + |A_2||A_1|$$

which implies that $|A_2| > n/2$.

Having constructed $A_0, A_1, \ldots, A_j$ and $G_1, \ldots G_j$, $j \geqslant 2$, consider, for every $w \in A_j$, an auxiliary bipartite graph B_w^j between the neighbours of w in G_j and all vertices in V,

where a pair zu, $z \in N_{G_j}(w)$, $u \in V$, is an edge of B_w^j if $zwu \in H$. Define
$$A'_{j+1} = \{u : \exists w \in A_j \text{ such that } \deg_{B_w^j}(u) \geqslant n^{1/4}\}$$
and
$$G'_{j+1} = \{wu : \text{ such that } w \in A_j \text{ and } \deg_{B_w^j}(u) \geqslant n^{1/4}\}.$$

Finally, let
$$A^0_{j+1} = \{w \in A'_{j+1} : \deg_{G'_{j+1}}(w) < \sqrt{n}\},$$
$$A_{j+1} = A'_{j+1} \setminus A^0_{j+1}$$
and
$$G_{j+1} = G'_{j+1}[A_j \cup A_{j+1}].$$

Notice that some sets A_j may intersect or even coincide (in fact, at some point the construction starts to repeat itself forever). Nevertheless, for the sake of our construction, we treat them as disjoint, cloning the vertices as much as necessary. Let us call the entire structure, consisting of the sets $A_0, A_1, A_2 \ldots$ and the graphs $G_1, G_2 \ldots$, an *xy-cascade*.

We had to alter our construction for $j \geqslant 3$ and require $\deg_{B_w^j}(u) \geqslant n^{1/4}$ and not just $\deg_{B_w^j}(u) \geqslant 1$, in order to be able to return from any edge of G_j back to xy by a legitimate hyperpath, on which all vertices must be distinct. With the above definition, in any xy-cascade, there is always a hyperpath from any edge of G_j going backward all the way down to xy as long as $j < n^{1/4}$. Indeed, when choosing a next (backward) vertex, we can avoid any given set of vertices of size less than $n^{1/4}$. In particular, we can avoid all vertices which are already on the path, as well as x and y. (In fact, we will need this property only to avoid sets of size $O(1)$.)

A vertex $u \in A_j$ is called *heavy* if $\deg_{G_j}(u) \geqslant (1/2 + \gamma/2)n$.

Claim 2.2. *There exists an index $j \leqslant j_0 = \lceil 1/\gamma \rceil + 1$ such that A_j contains at least one heavy vertex.*

Proof. We will first show that for $j \geqslant 2$ every vertex $w \in A_j$ has in G'_{j+1} degree at least $(1/2 + \gamma)n - n^{3/4}$. Indeed, let s be the number of vertices $u \in V$ with $\deg_{B_w^j}(u) < n^{1/4}$. Then
$$sn^{1/4} + (n-s)|N_{G_j}(w)| \geqslant |B_w^j| \geqslant |N_{G_j}(w)|(1/2 + \gamma)n$$
which yields, using $|N_{G_j}(w)| = \deg_{G_j}(w) \geqslant \sqrt{n}$ and $s \leqslant n$,
$$n - s \geqslant (1/2 + \gamma)n - \frac{sn^{1/4}}{|N_{G_j}(w)|} \geqslant (1/2 + \gamma)n - n^{3/4}.$$

Note also that the total number of edges of G'_{j+1} incident to the vertices of A^0_{j+1} is smaller than $n^{3/2}$.

Now suppose that the claim is not true. Then, using the above estimates and remembering that the set A_j contains no heavy vertices, for each $j = 2, \ldots, j_0$,
$$|A_{j-1}|(1/2 + \gamma)n - n^{7/4} - n^{3/2} \leqslant |G_j| \leqslant |A_j|(1/2 + \gamma/2)n$$

and, consequently, since $|A_1| \geq (1/2 + \gamma)n$, we have

$$|A_{j_0}| > \frac{1+2\gamma}{1+\gamma}|A_{j_0-1}| - O(n^{3/4}) > \left(\frac{1+2\gamma}{1+\gamma}\right)^{\lceil 1/\gamma \rceil} \frac{n}{2} > n,$$

a contradiction. (For the last inequality we used the fact that $(1-x)e^x \leq 1$ with $x = \gamma/(1+2\gamma)$ and assumed that $(1+2\gamma)\ln 2 \leq 2$.) $\square$

Given two disjoint, ordered pairs of vertices xy and cd, consider the xy-cascade $(A_j^{(1)}, G_j^{(1)})$ and the cd-cascade $(A_j^{(2)}, G_j^{(2)})$. For $i = 1, 2$, let $b^{(i)} \in A_{j^{(i)}}^{(i)}$ be a heavy vertex in the corresponding cascade, where $j^{(i)} \leq j_0$.

Assume first that $b^{(1)} = b^{(2)} := b$. Then, by the definitions of a heavy vertex and of an (n, γ)-graph, for each $i = 1, 2$ there exists $a^{(i)} \in A_{j^{(i)}-1}^{(i)}$ such that $a^{(i)}b \in G_{j^{(i)}}^{(i)}$ and $a^{(1)}ba^{(2)}$ is an edge of H. Similarly, when $b^{(1)} \neq b^{(2)}$, for each $i = 1, 2$ there exists $a^{(i)} \in A_{j^{(i)}-1}^{(i)}$ such that $a^{(i)}b^{(i)} \in G_{j^{(i)}}^{(i)}$ and both $a^{(1)}b^{(1)}b^{(2)}$ and $a^{(2)}b^{(2)}b^{(1)}$ are edges of H.

Moreover, by the definition of the xy-cascade, there is a $(j^{(1)} + 2)$-path $P^{(1)}$ connecting xy and $b^{(1)}a^{(1)}$ and, by the definition of the cd-cascade, there is a $(j^{(2)} + 2)$-path $P^{(2)}$, disjoint from $P^{(1)}$, connecting cd and $b^{(2)}a^{(2)}$. Hence, for some

$$k = (j^{(1)} + 2) + (j^{(2)} + 2) \leq 2(j_0 + 2) \leq 4/\gamma,$$

there is a k-path in H which connects xy and cd (for the last inequality we have assumed that $\gamma \leq 1/4$). $\square$

Lemma 2.3. (Absorbing Lemma) *There is an l-path A in H with $l = |V(A)| \leq 20\gamma^2 n$, such that for every subset $U \subset V \setminus V(A)$ of size at most $\gamma^5 n$ there is a path A_U in H with $V(A_U) = V(A) \cup U$ and such that A_U has the same endpairs as A.*

In other words, this lemma asserts that there is *one* not too long path such that *every* not too large subset can be 'absorbed' into this path by creating a longer path with the same endpairs. Consequently, if this path happens to be a segment of a cycle C of order at least $(1 - \gamma^5)n$ then, setting $U = V \setminus V(C)$, the path A_U together with the path $C \setminus A$ form a Hamiltonian cycle. We will use this observation at the end of our proof of Theorem 1.1.

Proof. An ordered set (or a sequence) of four vertices will be called a 4-*tuple*. Given a vertex v we say that a 4-tuple of vertices x, y, z, w absorbs v if $xyz, yzw, xyv, yvz, vzw \in H$. A 4-tuple is called *absorbing* if it absorbs a vertex. This terminology reflects the fact that the path $xyzw$ can be extended by inserting (or absorbing) vertex v to create the path $xyvzw$. Note that both paths have the same set of endpairs.

Claim 2.4. *For every $v \in V$ there are at least $2\gamma^2 n^4$ 4-tuples absorbing v.*

Proof. Because H is an (n, γ)-graph, there are at least $(n-1)(1/2 + \gamma)n$ ordered pairs yz such that $vyz \in H$. For each such pair there are at least $2\gamma n$ common neighbours x of vy and yz, and at least $2\gamma n - 1$ common neighbours w of vz and yz, yielding together

at least
$$(n-1)(1/2+\gamma)n2\gamma n(2\gamma n-1) > 2\gamma^2 n^4$$
4-tuples absorbing v. □

For each $v \in V$, let $\mathscr{A}_v$ be the family of all 4-tuples absorbing v. The next claim is obtained by the probabilistic method.

Claim 2.5. *There exists a family $\mathscr{F}$ of at most $2\gamma^3 n$ disjoint, absorbing 4-tuples of vertices of H such that for every $v \in V$, $|\mathscr{A}_v \cap \mathscr{F}| > \gamma^5 n$.*

Proof. We first select a family $\mathscr{F}'$ of 4-tuples at random by including each of $n(n-1)(n-2)(n-3) \sim n^4$ of them independently with probability $\gamma^3 n^{-3}$ (some of the selected 4-tuples may not be absorbing at all). By Chernoff's inequality (see, e.g., [8]), with probability $1 - o(1)$, as $n \to \infty$,

- $|\mathscr{F}'| < 2\gamma^3 n$, and
- for each $v \in V$, $|\mathscr{A}_v \cap \mathscr{F}'| > \frac{3}{2}\gamma^5 n$.

Moreover, the expected number of intersecting pairs of 4-tuples in $\mathscr{F}'$ is at most
$$n^4 \times 4 \times 4 \times n^3 \times (\gamma^3 n^{-3})^2 = 16\gamma^6 n,$$
and so, by Markov's inequality, with probability at least $1/17$,

- there are at most $17\gamma^6 n$ pairs of intersecting 4-tuples in $\mathscr{F}'$.

Thus, with positive probability, a random family $\mathscr{F}'$ possesses all three properties marked by the bullets above, and hence there exists at least one such family which, with a little abuse of notation, we also denote by $\mathscr{F}'$. After deleting from $\mathscr{F}'$ all 4-tuples intersecting other 4-tuples in $\mathscr{F}'$, as well as those which do not absorb any vertex, we obtain a subfamily $\mathscr{F}$ of $\mathscr{F}'$ consisting of disjoint and absorbing 4-tuples and such that for each $v \in V$,
$$|\mathscr{A}_v \cap \mathscr{F}| > \frac{3}{2}\gamma^5 n - 34\gamma^6 n > \gamma^5 n. \qquad \square$$

Set $f = |\mathscr{F}|$ and let $F_1, \ldots, F_f$ be the elements of $\mathscr{F}$. For each $i = 1, \ldots, f$, F_i is absorbing and thus spans a 4-path in H. We will further denote these paths also by F_i and set $F = \bigcup_{i=1}^{f} F_i$.

Our next task is to connect all these 4-paths into one, not too long path A. To this end, we will repeatedly apply Lemma 2.1 and, for each $i = 1, \ldots, f-1$, connect the endpairs of F_i and F_{i+1} by a short path. Recall that the operation $P \circ Q$ has been defined at the beginning of this section.

Claim 2.6. *There exists a path A in H of the form*
$$A = F_1 \circ C_1 \cdots \circ F_{f-1} \circ C_{f-1} \circ F_f$$
where the paths $C_1, \ldots, C_{f-1}$ have each at most $8/\gamma$ vertices.

Proof. We will prove by induction on i that for each $i = 1, \ldots f$, there exists a path A_i in H of the form $A_1 = F_1$ and, for $i \geq 2$,

$$A_i = F_1 \circ C_1 \cdots \circ F_{i-1} \circ C_{i-1} \circ F_i,$$

where the paths $C_1, \ldots, C_{i-1}$ have each at most $8/\gamma$ vertices. Then $A = A_f$.

There is nothing to prove for $i = 1$. Assume the statement is true for some $1 \leq i \leq f - 1$. Let ab be an endpair of A_i and let cd be an endpair of F_{i+1}. Denote by H_i the subhypergraph induced in H by the set of vertices $V_i = (V \setminus V(F \cup A_i)) \cup \{a, b, c, d\}$. Since

$$|V(F \cup A_i)| < |\mathscr{F}|(4 + 8/\gamma) < 10f/\gamma < 20\gamma^2 n,$$

H_i is a $(|V_i|, \gamma/2)$-graph, where $0 < n - |V_i| < 20\gamma^2 n$. By Lemma 2.1 applied to H_i and the pairs ba and dc, there is a path $C_i \subset H_i$ of length at most $4/(\gamma/2) = 8/\gamma$, connecting these pairs. Note that $V(C_i) \setminus \{a, b, c, d\}$ is disjoint from $V(F \cup A_i)$, and thus

$$A_{i+1} = A_i \circ C_i \circ F_{i+1}$$

is the desired path. □

Claim 2.6 states that we may connect all 4-paths in $\mathscr{F}$ into one path A of length at most $f(4 + 8/\gamma) < 20\gamma^2 n$. It remains to show that A has the absorbing property. Let $U \subset V \setminus V(A)$, $|U| \leq \gamma^5 n$. Because for every $v \in U$ we have $|\mathscr{A}_v \cap \mathscr{F}| > \gamma^5 n$, that is, there are at least $\gamma^5 n$ disjoint, v-absorbing 4-tuples in A, we can insert all vertices of U into A one by one, each time using a fresh absorbing 4-tuple. □

Given $U \subseteq V$ and $x, y \in V$, let

$$\deg_H(xy, U) = |\{z \in U : xyz \in H\}|,$$

and, in particular, $\deg_H(xy) = \deg_H(xy, V)$. Note that in an (n, γ)-graph H we have $\deg_H(xy) \geq (1/2 + \gamma)n$ for all pairs of vertices $x, y \in V$.

Lemma 2.7. (Reservoir Lemma) For every subset $W \subset V$, $|W| < \gamma n/4$, there exists a subset $R \subset V \setminus W$ (a reservoir) such that $|R| = \lceil \gamma^5 n/2 \rceil$ and for every pair of vertices $x, y \in V$

$$\deg_H(xy, R) \geq (1/2 + \gamma/2)(|R| + 4).$$

In particular, for every $S \subset V \setminus R$, $|S| = 4$, the induced sub-hypergraph $H[R \cup S]$ is an $(|R| + 4, \gamma/2)$-graph.

Proof. Set $r = \gamma^5 n/2$ (to avoid irrelevant complications, we assume that r is an integer). We choose R randomly out of all $\binom{n-|W|}{r}$ possibilities and apply the probabilistic method. For each pair of vertices x, y, the random variable $X = X_{xy}$, counting the vertices of R which are neighbours of x, y, has the hypergeometric distribution with expectation $\mathbb{E}X$ satisfying

$$r \geq \mathbb{E}X \geq \frac{\deg_H(xy) - |W|}{n - |W|} r \geq \left(\frac{1}{2} + \frac{3}{4}\gamma\right) r.$$

Hence, by Chernoff's bound ([8], (2.6) on p. 26 and Theorem 2.10 on p. 29)

$$\mathbb{P}\left(X < \left(\frac{1}{2} + \frac{1}{2}\gamma\right)(r+4)\right) \leqslant \mathbb{P}\left(X \leqslant \mathbb{E}X - \frac{1}{4}\gamma r + 2\right) \leqslant \exp\left\{-\frac{\gamma^2 r}{33}\right\},$$

and consequently,

$$\mathbb{P}\left(\exists x, y : X_{xy} < \left(\frac{1}{2} + \frac{1}{2}\gamma\right)(r+4)\right) \leqslant \binom{n}{2}\exp\left\{-\frac{\gamma^7 n}{66}\right\} = o(1). \qquad \square$$

3. Proof of Theorem 1.1

We first outline the forthcoming proof. Let H be an (n, γ)-graph.
- By Lemma 2.3 fix an absorbing l-path A, $l = |V(A)| \leqslant 20\gamma^2 n$.
- By Lemma 2.7 with $W = V(A)$, fix a reservoir set $R \subset V \setminus V(A)$, $|R| = \lceil \frac{1}{2}\gamma^5 n \rceil$.
- Set $H_1 = H[V \setminus (V(A) \cup R)]$ and cover all but at most $\frac{1}{2}\gamma^5 n$ vertices of H_1 by disjoint paths $P_1, \ldots, P_p$, where $p \leqslant \gamma^8 n$. Denote the set of uncovered vertices by T.
- By $p+1$ applications of Lemma 2.1 and by the property of R, connect all paths $P_1, \ldots, P_p$, as well as A, into one cycle C in H, leaving only a leftover subset R' of R and the trash set T outside C. Note that $|R' \cup T| \leqslant \gamma^5 n$.
- Using the absorbing property of A insert $R' \cup T$ into C, obtaining a Hamiltonian cycle in H.

It remains to explain the third task on the above list: how to cover almost all vertices of H_1 by disjoint paths $P_1, \ldots, P_p$. This will be taken care of by Lemma 3 stated below. However, as this lemma relies heavily on a regularity lemma for hypergraphs and some related results, we devote two separate sections, Section 4 and Section 5, to its proof.

Lemma 3.1. (Path-Cover Lemma) *For every $\gamma > 0$ there exists n_0 such that every (n, γ)-graph, $n > n_0$, contains a family of at most $\gamma^8 n$ vertex-disjoint paths, covering all but at most $\gamma^5 n/2$ vertices.*

Proof of Theorem 1.1. Let us assume that $\gamma < \frac{1}{400}$ (see Remark 1) and let A be an absorbing l-path in H, $l = |V(A)| \leqslant 20\gamma^2 n$, whose existence is guaranteed by Lemma 2.3. By Lemma 2.7 applied to H with $W = V(A)$, there exists a reservoir set $R \subset V \setminus V(A)$ of size $|R| = \frac{1}{2}\gamma^5 n$ (for simplicity we are assuming that this is an integer) with the property described in that lemma. Set $H_1 = H[V \setminus (V(A) \cup R)]$ and note that H_1 is an (n_1, γ_1)-graph, where

$$n - 20\gamma^2 n - \frac{1}{2}\gamma^5 n \leqslant n_1 \leqslant n$$

and

$$0 < \gamma - 20\gamma^2 - \frac{1}{2}\gamma^5 < \gamma_1 < \gamma.$$

We apply Lemma 3.1 to H_1, obtaining a family of $p \leqslant \gamma_1^8 n_1 \leqslant \gamma^8 n$ vertex-disjoint paths $P_1, \ldots, P_p$, and the set $T \subset V(H_1)$ of vertices not covered by these paths of size $|T| \leqslant \gamma_1^5 n_1 / 2 \leqslant \gamma^5 n / 2$.

To connect all these paths as well as the path A into one cycle C, we successively apply Lemma 2.1 to ever shrinking sub-hypergraphs of the form $H[R_i \cup S]$, where a subset $R_i \subseteq R$ will be defined below in (3.1), while S consists of all four vertices from the endpairs of the two paths to be connected at the current stage. Thus, new vertices of the connecting paths will be entirely contained in the set R. The next claim, very similar to Claim 2.6, describes the procedure of connecting together all path $P_1, \ldots, P_p$ into one path L.

Claim 3.2. *There exists a path L in H of the form*
$$L = P_1 \circ C_1 \cdots \circ P_p \circ C_p \circ A,$$
where the paths $C_1, \ldots, C_p$ have each at most $20/\gamma$ vertices and are such that
$$V(C_1 \cup \cdots \cup C_p) \setminus V(P_1 \cup \cdots \cup P_p \cup A) \subset R.$$

Proof. Set $P_{p+1} = A$ to unify notation. We will prove by induction on i that for every $i = 1, \ldots, p+1$, there exists a path L_i in H of the form
$$L_i = P_1 \circ C_1 \cdots \circ P_{i-1} \circ C_{i-1} \circ P_i, \tag{3.1}$$
where the paths $C_1, \ldots, C_{i-1}$ have each at most $20/\gamma$ vertices and are such that
$$V(C_1 \cup \cdots \cup C_{i-1}) \setminus V(P_1 \cup \cdots \cup P_i) \subset R.$$
Then L_{p+1} is the desired path L.

There is nothing to prove for $i = 1$. Assume that the statement is true for some $1 \leqslant i \leqslant p$. Let ab be an endpair of L_i and cd be an endpair of P_{i+1}. Since
$$|V(C_1 \cup \cdots \cup C_{i-1})| \leqslant (i-1)(20/\gamma) < 20p/\gamma < 20\gamma^7 n,$$
the remaining subset
$$R_i = R \setminus V(C_1 \cup \cdots \cup C_{i-1})$$
of R still maintains the property described in Lemma 2.7 but with $\frac{2}{5}\gamma$ instead of $\gamma/2$. Indeed, since $\gamma < \frac{1}{400}$, for every pair $x, y \in V$
$$\deg_H(xy, R_i) \geqslant \left(\frac{1}{2} + \frac{1}{2}\gamma\right)(|R| + 4) - 20\gamma^7 n \leqslant \left(\frac{1}{2} + \frac{2}{5}\gamma\right)(|R_i| + 4).$$
Hence, the sub-hypergraph $H[R_i \cup \{a, b, c, d\}]$ is an $(|R_i| + 4, \frac{2}{5}\gamma)$-graph. By Lemma 2.1 applied to $H[R_i \cup \{a, b, c, d\}]$ and the pairs ba and dc there is a path C_i in $H[R_i \cup \{a, b, c, d\}]$ of length at most $20/\gamma$ connecting these pairs. Note that $V(C_i) \setminus \{a, b, c, d\}$ is disjoint from $V(C_1 \cup \cdots \cup C_{i-1}) \cup V(P_1 \cup \cdots \cup P_{p+1})$ and thus
$$L_{i+1} = L_i \circ C_i \circ P_{i+1}$$
is as in (3.1). □

To obtain the cycle C, let ab and cd be the two endpairs of $L = L_{p+1}$ (we follow the notation from the proof of Claim 3.2). Again, applying Lemma 2.1 to $H[R_{p+1} \cup \{a, b, c, d\}]$ and the pairs ba and dc we obtain a path C_{p+1} of length at most $20/\gamma$ connecting these pairs (and thus forming the desired cycle C) and such that $V(C_{p+1}) \setminus \{a,b,c,d\} \subseteq R_{p+1}$. Set $R_{p+2} = R_{p+1} - V(C_{p+1})$. There are at most $\gamma^5 n$ vertices left outside C. Indeed, we have

$$n - |V(C)| = |T| + |R_{p+2}| < |T| + |R| < \frac{1}{2}\gamma^5 n + \frac{1}{2}\gamma^5 n = \gamma^5 n.$$

Finally, we absorb these remaining vertices into the path A, which is now part of the cycle C. Set $U = V - V(C)$. By Lemma 2.3 there is a path A_U with the same endpairs as A and such that $V(A_U) = V(A) \cup U$. Then $A_U \cup C$ contains a Hamiltonian cycle of H. □

4. Regularity of hypergraphs

In the previous section we stated Lemma 3.1, so crucial for the proof of our main result. Here we make thorough preparations towards its proof which is contained in Section 5. Our proof will be based on a simplified version of the regularity lemma for hypergraphs from [6].

4.1. Regularity of graphs

We say that a bipartite graph G with bipartition $V(G) = X \cup Y$ is (d, ε)-*regular* if for all $A \subseteq X$ and $B \subseteq Y$ with $|A| > \varepsilon |X|$ and $|B| > \varepsilon |Y|$, we have

$$|d_G(A, B) - d| < \varepsilon,$$

where

$$d_G(A, B) = \frac{e_G(A, B)}{|A||B|}$$

is the *density* of the pair (A, B) and $e_G(A, B)$ is the number of edges in G with one endpoint in A and the other in B. We will write d_G or $d(G)$ for $d_G(X, Y)$. We say that G is ε-*regular* if it is (d, ε)-regular for some d.

Note that the (bipartite) complement of a (d, ε)-regular graph is itself $(1 - d, \varepsilon)$-regular. Also, if G_i is (d_i, ε_i)-regular, $i = 1, 2$, and G_1 and G_2 have the same vertex set (and the same bipartition), but are edge-disjoint, then their union $G_1 \cup G_2$ is $(d_1 + d_2, \varepsilon_1 + \varepsilon_2)$-regular.

A triple $\mathcal{T} = (P^{12}, P^{13}, P^{23})$ of bipartite graphs with vertex sets $V_1 \cup V_2$, $V_1 \cup V_3$ and $V_2 \cup V_3$ will be referred to as a *triad*. Let $\text{tr}(\mathcal{T})$ stand for the number of triangles in $P = P^{12} \cup P^{13} \cup P^{23}$. It is easy to estimate the number of triangles in a triad consisting of ε-regular graphs (see, e.g., Fact A in [6]). Here we will need a slight extension of that result, assuming that only two out of the three bipartite graphs are ε-regular. We include a simple proof for completeness.

Fact 4.1. *Let $\mathcal{T} = (P^{12}, P^{13}, P^{23})$ be a triad, where for some $0 \leqslant d_{13}, d_{23}, \varepsilon \leqslant 1$, the graphs P^{13} and P^{23} are, respectively, (d_{13}, ε)-regular and (d_{23}, ε)-regular. Then*

$$d(P^{12})d_{13}d_{23} - 4\varepsilon < \frac{\text{tr}(\mathcal{T})}{|V_1||V_2||V_3|} < d(P^{12})d_{13}d_{23} + 6\varepsilon.$$

In particular, if all three graphs are (d, ε)-regular, then

$$d^3 - 5\varepsilon < \frac{\text{tr}(\mathcal{T})}{|V_1||V_2||V_3|} < d^3 + 7\varepsilon.$$

Proof. Assume for simplicity that $|V_i| = n$ for each $i = 1, 2, 3$. For $v \in V_1$, let $N(v)$ be the set of neighbours of v in P^{13}, and for $v \in V_1$ and $u \in V_2$, let $N(v, u)$ be the subset of $N(v)$ consisting of the neighbours of u in P^{23} (which are thus also neighbours of v in P^{13}).

Let U^+ and U^ε be the sets of those vertices $v \in V_1$ for which, respectively, $|N(v)| > (d_{13} + \varepsilon)n$ and $|N(v)| \leqslant \varepsilon n$. By the ε-regularity of P^{13} we have $|U^+| \leqslant \varepsilon n$. If $v \in U^\varepsilon$, then, clearly, $|N(v, u)| \leqslant \varepsilon n$ for every $u \in V_2$. For each $v \in U = V_1 \setminus (U^+ \cup U^\varepsilon)$, let U_v^+ be the set of those vertices $u \in V_2$ for which $|N(v, u)| > (d_{23} + \varepsilon)|N(v)|$. Then, by the ε-regularity of P^{23} we have $|U_v^+| \leqslant \varepsilon n$.

Let us express P^{12} as a union of four edge-disjoint subgraphs, $P^{12} = F_1 \cup F_2 \cup F_3 \cup F_4$, where F_1 consists of all edges vu with $v \in U^+$, F_2 – with $v \in U^\varepsilon$, F_3 – with $v \in U$ and $u \in U_v^+$, and, finally, F_4 consists of all edges vu with $v \in U$ and $u \notin U_v^+$.

By the above estimates, $\sum_{vu \in F_i} |N(v, u)| \leqslant \varepsilon n^3$, $i = 1, 2, 3$, while

$$\sum_{vu \in F_4} |N(v, u)| \leqslant |P^{12}|(d_{13} + \varepsilon)(d_{23} + \varepsilon)n.$$

Altogether,

$$\mathrm{tr}(\mathcal{T}) = \sum_{vu \in P^{12}} |N(v,u)| \leqslant 3\varepsilon n^3 + d(P^{12})(d_{13} + \varepsilon)(d_{23} + \varepsilon)n^3 < (d(P^{12})d_{13}d_{23} + 6\varepsilon)n^3.$$

For the lower bound, we may assume that $\min(d(P^{12}), d_{13}, d_{23}) > 4\varepsilon$ and consider the set U^- of all $v \in V_1$ for which $|N(v)| < (d_{13} - \varepsilon)n$, and for each $v \in V_1 \setminus U^-$, the set U_v^- of those vertices $u \in V_2$ for which $|N(v, u)| < (d_{23} - \varepsilon)|N(v)|$. We have $|U^-| \leqslant \varepsilon n$ and, for all $v \in V_1 \setminus U^-$, $|U_v^-| \leqslant \varepsilon n$, because $|N(v)| \geqslant (d_{13} - \varepsilon)n > \varepsilon n$. Thus, for all but at most $2\varepsilon n^2$ pairs v, u, we have $|N(v, u)| > (d_{13} - \varepsilon)(d_{23} - \varepsilon)n$. Consequently, as for the upper bound,

$$\mathrm{tr}(\mathcal{T}) \geqslant (d(P^{12}) - 2\varepsilon)(d_{13} - \varepsilon)(d_{23} - \varepsilon)n^3 > (d(P^{12})d_{13}d_{23} - 4\varepsilon)n^3. \qquad \square$$

4.2. Regularity of hypergraphs

For a triad $\mathcal{T} = (P^{12}, P^{13}, P^{23})$ with $\mathrm{tr}(\mathcal{T}) > 0$ and a 3-uniform, 3-partite hypergraph H with vertex set $V(H) = V_1 \cup V_2 \cup V_3$ we define the *density of H over $\mathcal{T}$* as

$$d_H(\mathcal{T}) = \frac{|H \cap \mathrm{Tr}(\mathcal{T})|}{\mathrm{tr}(\mathcal{T})},$$

where $\mathrm{Tr}(\mathcal{T})$ is the set of triplets formed by the vertex sets of all triangles in P. (If $\mathrm{tr}(\mathcal{T}) = |\mathrm{Tr}(\mathcal{T})| = 0$ then we set $d_H(\mathcal{T}) = 0$.)

Definition 3. Let $\delta > 0$. We will say that a hypergraph H is δ-*regular* with respect to the triad $\mathcal{T} = (P^{12}, P^{13}, P^{23})$ if for every triad $\mathcal{S} = (Q^{12}, Q^{13}, Q^{23})$ such that $Q^{ij} \subseteq P^{ij}$, $1 \leqslant i < j \leqslant 3$, and $\mathrm{tr}(\mathcal{S}) > \delta \mathrm{tr}(\mathcal{T})$, we have $|d_H(\mathcal{S}) - d_H(\mathcal{T})| < \delta$. A triad with respect to which a hypergraph is not δ-regular will be called δ-*irregular*.

The hereditary nature of regularity is captured by the following, simple fact.

Fact 4.2. *Let $\mathcal{T} = (P^{12}, P^{13}, P^{23})$ be a triad with vertex sets V_1, V_2, V_3, all of equal size n, and let H be a hypergraph, $V(H) = V_1 \cup V_2 \cup V_3$. Furthermore, for $0 < \eta < 1$, let $U_i \subseteq V_i$, $|U_i| > \eta n$, $i = 1, 2, 3$, and $Q^{ij} = P^{ij}[U_i, U_j]$, $1 \leqslant i < j \leqslant 3$.*

(a) *If P^{12} is (d,ε)-regular with $d > \varepsilon$ and $\varepsilon < \eta < 1$ then Q^{12} is $(d, \varepsilon/\eta)$-regular and, moreover, $d - \varepsilon < d_{Q^{12}}(U_1, U_2) < d + \varepsilon$.*
(b) *If all graphs P^{ij}, $1 \leqslant i < j \leqslant 3$, are (d, ε)-regular, $d^3 > 11\varepsilon/\eta$, and H is δ-regular with respect to $\mathcal{T}$, where $\delta < \eta^3/3$, then H is $3\delta/\eta^3$- regular with respect to the triad $\mathcal{S} = (Q^{12}, Q^{13}, Q^{23})$ and has density $d_H(\mathcal{S})$ satisfying $|d_H(\mathcal{S}) - d_H(\mathcal{T})| < \delta$.*

Proof. Part (a) is obvious. For part (b), note that by Fact 4.1, $\text{tr}(\mathcal{T}) < n^3(d^3 + 7\varepsilon)$ and, similarly,

$$\text{tr}(\mathcal{S}) > (\eta n)^3(d^3 - 5\varepsilon/\eta) > \frac{1}{3}\eta^3(d^3 + 7\varepsilon)n^3 > \frac{1}{3}\eta^3\text{tr}(\mathcal{T}),$$

where for the middle inequality we used the assumption that $d^3 > 11\varepsilon/\eta$. Thus, if $\mathcal{R}$ is a subtriad of $\mathcal{S}$ with $\text{tr}(\mathcal{R}) \geqslant (3\delta/\eta^3)\text{tr}(\mathcal{S})$ then, since $\eta^3/3 > \delta$, we have $\text{tr}(\mathcal{R}) > \delta\text{tr}(\mathcal{T})$ and by the δ-regularity of H with respect to $\mathcal{T}$,

$$|d_H(\mathcal{R}) - d_H(\mathcal{T})| < \delta.$$

Since the above applies in particular to $\mathcal{R} = \mathcal{S}$, we conclude that $|d_H(\mathcal{R}) - d_H(\mathcal{S})| < 2\delta < 3\delta/\eta^3$, which proves that H is $3\delta/\eta^3$-regular with respect to $\mathcal{S}$. □

We now state the regularity lemma for 3-uniform hypergraphs from [6] in a simplified form, suitable for our needs. It is a special case of a regularity lemma for k-uniform hypergraphs proved by M. Schacht and one of the authors in [16]. Set $K(U, W)$ for the complete bipartite graph with vertex sets U and W.

Lemma 4.3. (Regularity Lemma for Hypergraphs) *For every $\delta > 0$, an integer t_0 and for all decreasing sequences $0 < \varepsilon(l) < 1$, there exist constants T_0, L_0 and N_0 such that every 3-uniform hypergraph H with at least N_0 vertices admits a partition Π of $\binom{V(H)}{2}$ consisting of an auxiliary vertex set partition $V(H) = V_0 \cup V_1 \cup \cdots \cup V_t$, where $t_0 \leqslant t < T_0$, $|V_0| < t$ and $|V_1| = |V_2| = \cdots = |V_t|$, and, for each pair i, j, $1 \leqslant i < j \leqslant t$, of a partition $K(V_i, V_j) = \bigcup_{a=1}^{l} P_a^{ij}$, where $1 \leqslant l < L_0$, satisfying the following conditions:*

(i) *all graphs P_a^{ij} are $(1/l, \varepsilon(l))$-regular,*
(ii) *for all but at most $\delta l^3 t^3$ triads $\mathcal{T}_{abc}^{hij} = (P_a^{hi}, P_b^{hj}, P_c^{ij})$, the hypergraph H is δ-regular with respect to $\mathcal{T}_{abc}^{hij}$.*

Remark 2. There are three essential differences between the original version in [6] and the one stated above. Firstly, instead of counting triangles contained in irregular triads, in (ii) we count the irregular triads themselves, which is essentially equivalent (*cf.* Proposition 4.6 in [13]). Secondly, we have no exceptional graphs P_0^{ij} whatsoever, and all graphs P_a^{ij}, $a \geqslant 1$, are $(1/l, \varepsilon(l))$-regular. Moreover, the number of graphs P_a^{ij} between each pair (V_i, V_j) of clusters, which varied in the original setting, is now, conveniently, precisely l. This makes the statement somewhat clearer and easier to apply.

Finally, in contrast with Definition 3, in [6] a more general concept of (δ, r)-regularity is considered. In this paper we use only the case when $r = r(t, l) = 1$. However, the proof of

Lemma 4.3 given below extends easily to a more general version with r being an arbitrary function of t and l as in [6].

Now we show how the version we use follows from the original statement, Theorem 3.5 in [6] (see [16] for a more general case of k-uniform hypergraphs).

Proof of Lemma 4.3. Given $\delta > 0$, t_0 and $0 < \varepsilon(l) < 1$, apply Theorem 3.5 in [6] to H with $\hat{\delta} = \delta/5$, $\varepsilon_1 = 2(\delta/5)^4$, t_0, $l_0 = \lceil 2/\delta \rceil$, $r(l, t) \equiv 1$ and $\varepsilon_2(l) = \varepsilon(l)/(l+1)$ (which yield constants $\hat{T}_0, \hat{L}_0, \hat{N}_0$) obtaining a partition $\hat{\Pi}$ as in Theorem 3.5 of [6], consisting of a vertex partition $V(H) = V_0 \cup V_1 \cup \cdots \cup V_t$, $t_0 \leq t < \hat{T}_0$, $|V_0| < t$, and $|V_1| = |V_2| = \cdots = |V_t| = m$, and, for each pair i, j, $1 \leq i < j \leq t$, of a partition $K(V_i, V_j) = \bigcup_{a=0}^{l_{ij}} \hat{P}_a^{ij}$, $l_{ij} \leq l$, $1 \leq l < \hat{L}_0$, such that

(a) all but at most $\varepsilon_1 \binom{t}{2} m^2$ edges of $\bigcup_{1 \leq i < j \leq t} K(V_i, V_j)$ belong to ε_2-regular graphs $\hat{P}_a^{ij}$;
(b) for all but at most $\varepsilon_1 \binom{t}{2}$ exceptional pairs $\{i, j\}$, $1 \leq i < j \leq t$, we have $|\hat{P}_0^{ij}| \leq \varepsilon_1 m^2$ and $||\hat{P}_a^{ij}| - m^2/l| \leq \varepsilon_2 m^2$, $a \geq 1$;
(c) all but at most $\hat{\delta} n^3$ triangles of $\bigcup_{1 \leq i < j < k \leq t} K(V_i, V_j, V_k)$ belong to $\hat{\delta}$-irregular triads.

We will prove Lemma 4.3 with $T_0 = \hat{T}_0$, $L_0 = \hat{L}_0$ and $N_0 = \max(\hat{N}_0, N_1)$, where N_1 is chosen so that the probability estimates below hold true. In our proof we will use properties (a) and (c), and only the first statement in (b) saying that the graphs $\hat{P}_0^{ij}$ are sparse.

Once Theorem 3.5 from [6] has been applied, and thus a value of l revealed, we set

$$\varepsilon = \varepsilon(l) \quad \text{and} \quad \varepsilon_2 = \varepsilon_2(l).$$

For each pair i, j, $1 \leq i < j \leq t$, call a graph $\hat{P}_a^{ij}$ good if it is $(1/l, \varepsilon_2)$-regular. Let $s = s_{ij} \leq l$ be the number of good graphs $\hat{P}_a^{ij}$ and assume that these are the graphs $\hat{P}_a^{ij}$, $a = 1, \ldots, s$. Define

$$R^{ij} = K(V_i, V_j) - \bigcup_{a=1}^{s} \hat{P}_a^{ij}.$$

Note that R^{ij}, as a complement (in $K(V_i, V_j)$) of a union of s $(1/l, \varepsilon_2)$-regular graphs, is $(1 - s/l, s\varepsilon_2)$-regular. Note also that $R^{ij} \supseteq \hat{P}_0^{ij}$.

Partition each R^{ij} with $s = s_{ij} < l$ into $l - s$ $(1/l, \varepsilon)$-regular graphs P_a^{ij}, $a = s+1, \ldots, l$. Guided by the proof of Lemma 3.8 in [6], this can be done as follows. For $s = l - 1$ we do nothing, since $s\varepsilon_2 < \varepsilon$. For $s \leq l - 2$, apply a random partition: with probability $1/(l-s)$ assign, independently, each edge of R^{ij} to one of the graphs P_a^{ij}, $a = s+1, \ldots, l$. Then, by Chernoff's inequality, with probability approaching 1, for all $a = s+1, \ldots, l$ and all $A \subseteq V_i, B \subseteq V_j$, $|A|, |B| > \varepsilon m > s\varepsilon_2 m$, we have

$$d_{P_a^{ij}}(A, B) \in \left(\frac{1}{l} - \frac{s\varepsilon_2}{l-s} + o(1), \frac{1}{l} + \frac{s\varepsilon_2}{l-s} + o(1) \right) \subset \left(\frac{1}{l} - \varepsilon, \frac{1}{l} + \varepsilon \right),$$

by our choice of ε. Let N_1 be such that the above probability is positive, and so the required partition exists.

Set $P_a^{ij} = \widehat{P}_a^{ij}$ for $a = 1,\ldots,s$. Since $\varepsilon \geqslant \varepsilon_2$, each graph P_a^{ij} is $(1/l,\varepsilon)$-regular. We will call the graphs P_a^{ij}, $a = s+1,\ldots,l$, *newborn*.

If $s_{ij} = l$, then $R^{ij} = \widehat{P}_0^{ij}$. Set $P_1^{ij} = \widehat{P}_1^{ij} \cup \widehat{P}_0^{ij}$ and $P_a^{ij} = \widehat{P}_a^{ij}$ for $a = 2,\ldots,l$. We will call the graph P_1^{ij} *swollen*. Note that, in this case, $\widehat{P}_0^{ij}$, as a complement of a union of l $(1/l,\varepsilon_2)$-regular graphs, is $(\varepsilon_2, l\varepsilon_2)$-regular, and consequently, P_1^{ij} is $(1/l,(l+1)\varepsilon_2)$-regular. Since $(l+1)\varepsilon_2 \leqslant \varepsilon$, again, all graphs P_a^{ij}, $a = 1,\ldots,l$ are $(1/l,\varepsilon)$-regular. As we also have $K(V_i,V_j) = \bigcup_{a=1}^{l} P_a^{ij}$, for all $1 \leqslant i < j \leqslant t$, this is the required partition Π.

It remains to estimate the number of δ-irregular triads in Π. First, note that $\hat{\delta}$-regularity implies δ-regularity. Second, having at most $\hat{\delta}n^3$ triangles in $\hat{\delta}$-irregular triads (*cf.* property (c) above), one can easily show that there are at most

$$(2\hat{\delta} + \varepsilon_1)l^3t^3 < \frac{1}{2}\delta l^3 t^3$$

$\hat{\delta}$-irregular triads in $\widehat{\Pi}$ (see Proposition 4.6(ii) in [13] with $r = 1$).

Third, the changes that lead from partition $\widehat{\Pi}$ to the new partition Π could possibly create new δ-irregular triads. These changes, however, affected only the edges of $\bigcup R^{ij}$, although in two different ways.

Recall that for $s = s_{ij} < l$ the edges of R^{ij} were divided into *newborn* graphs P_a^{ij}, $a = 1,\ldots,l-s$, while for $s_{ij} = l$, the graph $R^{ij} = \widehat{P}_0^{ij}$ was added to $\widehat{P}_1^{ij}$ to create a *swollen* graph P_1^{ij}. Due to the choice of l_0, there are at most

$$\binom{t}{2}tl^2 < \frac{1}{4}\delta l^3 t^3$$

triads in Π consisting of at least one swollen graph. Below, we will show that

(*) there are at most $\frac{1}{5}\delta l^3 t^3$ triads in Π which contain a newborn graph.

This will imply Lemma 4.3, since even if all the triads containing newborn or swollen graphs were δ-irregular, they together with the at most $\frac{1}{2}\delta l^3 t^3$ $\hat{\delta}$-irregular triads of $\widehat{\Pi}$, would sum up to at most

$$\left(\frac{1}{2} + \frac{1}{4} + \frac{1}{5}\right)\delta l^3 t^3 < \delta l^3 t^3$$

δ-irregular triads in Π.

To prove statement (*), we will estimate the size of the graph $R_{<l} = \bigcup_{s_{ij}<l} R^{ij}$. An edge belongs to $R_{<l}$ if it belongs to one of the following graphs:

(i) an ε_2-irregular graph of $\widehat{\Pi}$,
(ii) the graph $\widehat{P}_0^{ij}$, for each pair $\{i,j\}$ which is *not* exceptional (*cf.* property (b) above)
(iii) the complete bipartite graph $K(V_i,V_j)$, for each exceptional pair $\{i,j\}$.

We will now bound the number of edges belonging to the union of graphs for each category (i)–(iii) above. By property (a) above, there are at most $\varepsilon_1\binom{t}{2}m^2 < \varepsilon_1 n^2/2$ edges belonging to ε_2-irregular graphs of $\widehat{\Pi}$. Concerning (ii), by property (b) above, summing over all non-exceptional pairs $\{i,j\}$, we infer that

$$\sum_{i,j}|\widehat{P}_0^{ij}| < \binom{t}{2}\varepsilon_1 m^2 < \varepsilon_1 n^2/2.$$

Finally, by property (b) again, there are at most $\varepsilon_1 \binom{t}{2}$ exceptional pairs $\{i,j\}$, and, consequently, summing over all such pairs, we have

$$\sum_{i,j} |V_i||V_j| < \varepsilon_1 \binom{t}{2}(n/t)^2 < \varepsilon_1 n^2/2.$$

Altogether,

$$|R_{<l}| \leq \frac{3}{2}\varepsilon_1 n^2.$$

Since for every pair i,j with $s = s_{ij} < l$, the edges of R^{ij} are divided into $(1/l,\varepsilon)$-regular graphs, there are at most

$$\frac{\frac{3}{2}\varepsilon_1 n^2}{\frac{1}{2l}m^2} < 2\varepsilon_1 t^2 l$$

such graphs. Hence, there are at most $2\varepsilon_1 t^2 l \times l^2 t < \hat{\delta} l^3 t^3 < \frac{1}{5}\delta l^3 t^3$ triads in Π containing at least one of the newborn graphs. $\square$

4.3. Regularity and hyperpaths

In order to cover a δ-regular hypergraph by many disjoint paths, it suffices to construct one path, remove it, and use the heredity (see Fact 4.2). The lemma below is a very special case of Theorem 3.1.1 in [11], where all we want is just one path. In its full version, Theorem 3.1.1 in [11], under a stronger assumption of so called (δ,r)-regularity of H, guarantees the right number of copies of any fixed hypergraph, but has, therefore, a much more complicated proof which requires the strongest form of Lemma 4.3 (see [6]).

Lemma 4.4. *For all integers $k \geq 3$ and $l \geq 1$, real numbers $\alpha > 0$, and all*

$$\delta < \frac{1}{2}\min(k^{-1},\alpha) \quad \text{and} \quad \varepsilon \leq \delta/(15l^3),$$

the following holds. Suppose that

- *H is a k-partite, 3-uniform hypergraph with vertex partition $(V_1,\ldots,V_k)$, $|V_1| = \cdots = |V_k| := m$,*
- *P^i is a bipartite graph between V_i and V_{i+1}, $1 \leq i \leq k-1$, and Q^i is a bipartite graph between V_i and V_{i+2}, $1 \leq i \leq k-2$,*
- *all graphs P^i and Q^i are $(1/l,\varepsilon)$-regular, and*
- *for all $1 \leq i \leq k-2$, H is δ-regular with respect to the triad $\mathcal{T}^i = (P^i, P^{i+1}, Q^i)$ and has density $d_H(\mathcal{T}^i) \geq \alpha$.*

Then, H contains a path on k vertices, one from each set V_i.

Proof. Define graphs $B^{k-1},\ldots,B^1$ recursively. Set $B^{k-1} = \emptyset$ and for each $i = k-2,\ldots,1$, let B^i be the set of those edges $uv \in P^i$, $u \in V_i$, $v \in V_{i+1}$, for which there is no $w \in V_{i+2}$ such that $uvw \in H$ and $vw \in P^{i+1} - B^{i+1}$.

Claim 4.5. For each $i = 1, \ldots, k-1$, we have $|B^i| < 2(k-i)\delta|P^i|$.

Observe that once we have proved this claim, we are done. Indeed, we may create a required path as follows. Select any edge $v_1v_2 \in P^1 - B^1$, where $v_1 \in V_1$ and $v_2 \in V_2$. This is possible, because $|P^1| > |B^1|$. Then choose $v_3 \in V_3$ so that $v_1v_2v_3 \in H$ and $v_2v_3 \in P^2 - B^2$. Continue until an entire path on k vertices is found.

Proof of Claim 4.5. We prove the claim by induction on $i = k-1, \ldots, 1$. It is clearly true for $i = k-1$ (recall that $B^{k-1} = \emptyset$). Assume that for some $1 \leqslant i < k-1$ we have

$$|B^{i+1}| < 2(k-i-1)\delta|P^{i+1}| \quad \text{but} \quad |B^i| \geqslant 2(k-i)\delta|P^i|.$$

Set

$$\mathscr{T} = \mathscr{T}^i = (P^i, P^{i+1}, Q^i) \quad \text{and} \quad \mathscr{T}_0 = (B^i, P^{i+1} - B^{i+1}, Q^i).$$

Our goal is to prove that

$$\operatorname{tr}(\mathscr{T}_0) > \delta \operatorname{tr}(\mathscr{T}), \tag{4.1}$$

and thus, by the δ-regularity of H with respect to $\mathscr{T}$, that

$$d_H(\mathscr{T}_0) > d_H(\mathscr{T}) - \delta > \alpha - 2\delta > 0.$$

As $d_H(\mathscr{T}_0) = |H \cap \operatorname{Tr}(\mathscr{T}_0)|/\operatorname{tr}(\mathscr{T}_0)$, this would imply that there is an edge $uvw \in H$ with $uv \in B^i$ and $vw \in P^{i+1} - B^{i+1}$, a contradiction to the definition of B^i.

To prove (4.1), note first that, by Fact 4.1,

$$\operatorname{tr}(\mathscr{T}) < \left(\frac{1}{l^3} + 7\varepsilon\right)m^3 \tag{4.2}$$

and, trivially,

$$\operatorname{tr}(\mathscr{T}_0) \geqslant \operatorname{tr}(B^i, P^{i+1}, Q^i) - \operatorname{tr}(P^i, B^{i+1}, Q^i).$$

Now, apply Fact 4.1 two more times, first the lower bound to $\operatorname{tr}(B^i, P^{i+1}, Q^i)$, then the upper bound to $\operatorname{tr}(P^i, B^{i+1}, Q^i)$, obtaining

$$\operatorname{tr}(B^i, P^{i+1}, Q^i) > \left(\frac{2(k-i)\delta|P^i|}{l^2 m^2} - 4\varepsilon\right)m^3,$$

$$\operatorname{tr}(P^i, B^{i+1}, Q^i) < \left(\frac{2(k-i-1)\delta|P^{i+1}|}{l^2 m^2} + 6\varepsilon\right)m^3,$$

and, finally,

$$\operatorname{tr}(\mathscr{T}_0) > \left(\frac{2\delta}{l^2}\left(\frac{1}{l} - (2k-2i-1)\varepsilon\right) - 10\varepsilon\right)m^3 = \left(\frac{2\delta}{l^3} - \frac{2\delta(2k-2i-1)\varepsilon}{l^3} - 10\varepsilon\right)m^3$$

$$> \left(\frac{2\delta}{l^3} - 10\varepsilon - \frac{2\varepsilon}{l^2}\right)m^3 > \left(\frac{1}{l^3} + 7\varepsilon\right)\delta m^3 > \delta \operatorname{tr}(\mathscr{T}),$$

by our assumptions on δ and ε, and by (4.2). □

It is now relatively easy to show that a δ-regular hypergraph can be almost covered by disjoint paths of a given length.

Corollary 4.6. *For all $k \geq 1$, $l \geq 1$, $0 < \alpha < 1$,*

$$\delta \leq \frac{\alpha^4}{k^{16}18^4} \quad \text{and} \quad \varepsilon \leq \frac{\sqrt{\delta}}{11kl^3}$$

the following is true. If $\mathcal{T} = (P, Q, R)$ is a triad of $(1/l, \varepsilon)$-regular graphs and a hypergraph H is δ-regular with respect to $\mathcal{T}$ and has density $d_H(\mathcal{T}) \geq \alpha$, then at least $(1 - \delta^{1/4})|V(H)|$ vertices of H can be covered by vertex-disjoint $3k$-paths.

Proof. For clarity of exposition let the three vertex sets of $\mathcal{T}$ be U, V, W and $|U| = |V| = |W| = m$, where k divides m. We break each set U, V, W arbitrarily into k subsets of equal size: $U = U_1 \cup U_2 \cup \cdots \cup U_k$, etc., and claim that there are at least $(1 - \delta^{1/4})m/k$ vertex-disjoint $3k$-paths, each containing precisely one vertex from each set $U_1, V_1, W_1, U_2, V_2, W_2, \ldots, U_k, V_k, W_k$ (we will call such paths *transversal*).

Indeed, consider any family $\mathcal{Q}$ of less than $(1 - \delta^{1/4})m/k$ vertex-disjoint transversal $3k$-paths. Let subsets $U'_i \subset U_i$, $V'_i \subset V_i$, and $W'_i \subset W_i$, $i = 1, \ldots, k$, consist of all vertices not covered by these paths. Since for all $h, i, j = 1, \ldots, k$

$$|U'_h| = |V'_i| = |W'_j| > \delta^{1/4}m/k,$$

by Fact 4.2 with $\eta = \delta^{1/4}/k$ (note that $(1/l)^3 > 11\varepsilon/\eta$ and $\delta < \eta^3/3$), the subtriad $\mathcal{T}'$ induced by sets U'_h, V'_i, W'_j is such that the three bipartite subgraphs,

$$P[U'_h, V'_i], \quad Q[U'_h, W'_j], \quad R[V'_i, W'_j],$$

are all $(1/l, \varepsilon')$-regular with $\varepsilon' = \varepsilon/\eta$, and H is δ'-regular with respect to $\mathcal{T}'$ with $\delta' = 3\delta^{1/4}k^3$ and has density $d_H(\mathcal{T}') \geq \alpha - \delta \geq \alpha/2$. Since $\varepsilon' \leq \delta'/(15l^3)$ and $\delta' < \frac{1}{2}\min((3k)^{-1}, \alpha/2)$, by Lemma 4.4 applied to the sub-hypergraph

$$H' = H[U'_1 \cup V'_1 \cup W'_1 \cup U'_2 \cup V'_2 \cup W'_2 \cdots \cup U'_k \cup V'_k \cup W'_k]$$

with $3k$ and $\alpha/2$ in place of k and α, there is a transversal $3k$-path in H'. This path can be added to $\mathcal{Q}$. □

5. Proof of the Path-Cover Lemma

In this final section we prove Lemma 3.1, a crucial ingredient of the proof of our main Theorem 1.1. For the proof of Lemma 3.1, we need one more result, ensuring a large matching in a suitably structured hypergraph. It will be applied to the cluster graph resulting from the Regularity Lemma (Lemma 4.3).

A *matching* in a hypergraph is a set (sub-hypergraph) of disjoint edges. Our last lemma in this section guarantees an almost perfect matching in a special class of hypergraphs. Given a hypergraph K, let G_K be the graph of all pairs xy of vertices which belong to less than $|V(K)|/2$ edges of K, that is, for which $\deg_K(xy) < |V(K)|/2$.

Lemma 5.1. *If a hypergraph K on $t \geq 24$ vertices satisfies the inequality*

$$\Delta(G_K) \leq \frac{t}{12},$$

then K contains a matching M covering all but at most $\max(2, \Delta(G_K)) + 1$ vertices.

Proof. Let $M = e_1, \ldots e_g$, $g < t/3$, be a largest matching in K and suppose that a set U of more than $u = \max(2, \Delta(G_K)) + 1$ vertices remains uncovered. Note that

$$t - 3g = |U| \geq u + 1 \geq 4. \tag{5.1}$$

Call a pair of vertices of K *big* if it belongs to at least $t/2$ edges of K, that is, it is an edge of the complement G_K^c of the graph G_K. Let $v_1, v_2, v_3, v_4 \in U$ be four vertices of U, where $v_3 v_4$ forms a big pair. (The existence of a big pair in U is guaranteed, because for any vertex $v \in U$ its degree in G_K is at most $u - 1$, while there are at least u vertices in U besides v.)

The set $N(v_3, v_4)$ of neighbours of the pair $v_3 v_4$ in K intersects at least $t/6$ edges of M (since it is disjoint from U). Then, because $2(u - 1) < t/6$, there exists an edge of the matching M, say $e_g = \{x_1, x_2, x_3\}$, such that $v_3 v_4 x_3 \in K$ and both $v_1 x_1$ and $v_2 x_2$ are big pairs in K.

Note that if for some $s = 1, 2$, there is a vertex $v \in U \setminus \{v_3, v_4\}$ such that $v v_s x_s \in K$, then the edges $e_1, \ldots, e_{g-1}, v v_s x_s$ and $v_3 v_4 x_3$ would form a matching of K larger than M – a contradiction. Hence, for each $s = 1, 2$ we have $N(v_s x_s) \cap U \subseteq \{v_3, v_4\}$. Similarly, $v_1 x_1 x_2 \notin K$ and $v_2 x_2 x_1 \notin K$. Hence, for each $i = 1, 2$, all but at most three neighbours of $v_s x_s$ belong to $M - e_g$.

Let a_q^s, $s = 1, 2$, $q = 1, \ldots, g - 1$, be the number of vertices of e_q which together with $v_s x_s$ form an edge of K. Then

$$\sum_{q=1}^{g-1} (a_q^1 + a_q^2) \geq 2\left(\frac{t}{2} - 3\right) = t - 6.$$

By averaging, there must be an index q, $1 \leq q \leq g - 1$, such that

$$a_q^1 + a_q^2 \geq \frac{t-6}{g-1} > 3,$$

where the last inequality follows from (5.1). This means, however, that there are $y, z \in e_q$, $y \neq z$, such that $y v_1 x_1 \in K$ and also $z v_2 x_2 \in K$. But then

$$e_1, \ldots, e_{q-1}, e_{q+1}, \ldots, e_{g-1}, v_3 v_4 x_3, y v_1 x_1, z v_2 x_2$$

is a larger matching than M – a contradiction. $\square$

The *outline* of the proof of Lemma 3.1 goes as follows. With a suitable choice of δ, t_0, and $\varepsilon(l)$,

- obtain a partition Π with respect to H as in Lemma 4.3;
- select from Π a system of bipartite graphs $\mathscr{P} = \{P^{i,j}, 1 < i \leq j < t\}$, such that the corresponding cluster hypergraph $K = K(\mathscr{P})$ of dense and δ-regular triads preserves essentially the property of H (see Claim 5.3 below);

- relying on Lemma 5.1, choose from K a subsystem M of vertex-disjoint triads which cover most of the clusters;
- to each of the triads of M apply Corollary 4.6 and cover most of its vertices by disjoint paths.

To achieve the second task above, we will randomly choose one graph P_a^{ij} between each pair (V_i, V_j) and benefit from the assumption that H is an (n, γ)-graph (see Claim 5.3 below).

Proof of the Path-Cover Lemma (Lemma 3.1). Let $0 < \gamma < 1/2$, n be sufficiently large, and let H be an (n, γ)-graph. Let

$$\delta = \frac{\gamma^{132}}{(12 \times 13)^4} \quad \text{and} \quad \varepsilon(l) = \frac{\gamma^{74}}{11 \times 6^4 l^3}$$

and $t_0 > 200/\gamma$ such that for all $t \geq t_0$ we have

$$t^2 \exp\left\{-\frac{1}{32}\gamma^2 t\right\} < \frac{1}{2}.$$

Assuming $n > N_0(\delta, t_0, \varepsilon(l))$, apply Lemma 4.3 to H with the above choice of δ, t_0 and $\varepsilon(l)$, to obtain a partition Π of $\binom{V(H)}{2}$ satisfying conditions (i) and (ii) of that lemma. Set

$$\varepsilon = \varepsilon(l) \quad \text{and} \quad |V_1| = \cdots = |V_t| = m,$$

and recall that $|V_0| < t$.

A triad $\mathcal{T}$ in Π is said to be *dense* if $d_H(\mathcal{T}) \geq \gamma/2$. Using the assumption that every pair of vertices in H belongs to at least $(1/2 + \gamma)n$ hyperedges, we will now show that every graph P_a^{ij} belongs to nearly as large fraction of dense triads.

Claim 5.2. *In the partition Π every graph P_a^{ij} belongs to at least $(1/2 + \gamma/3)tl^2$ dense triads.*

Proof. For clarity of exposition, we assume that $V_0 = \emptyset$, or, equivalently, that t divides n. Suppose to the contrary that a graph $P = P_a^{ij}$ belongs to less than $(1/2 + \gamma/3)tl^2$ dense triads of Π. Let S be the set of hyperedges of H which contain an edge of P and a third vertex outside $V_i \cup V_j$. First note that, because every pair of vertices of H belongs to at least $(1/2 + \gamma)n$ hyperedges of H, we have

$$|S| \geq \left(\frac{1}{l} - \varepsilon\right) m^2 \left(\frac{1}{2}n + \gamma n - 2m\right).$$

Since $m = n/t \leq n/t_0$, $t_0 > 200/\gamma$ and $0.99\gamma < 1/2$, this leads to the bound

$$|S| > \left(\frac{1}{2} + 0.99\gamma - \varepsilon l\right)\frac{n^3}{t^2 l}.$$

We will find an upper bound on $|S|$ contradicting the above lower bound. Let us split the hyperedges of S into two classes. Let S_1 consist of those hyperedges of H which contain an edge of P, an edge of P_b^{ih} and an edge of P_c^{jh}, for some $h \in [t] \setminus \{i, j\}$ and $1 \leq b, c \leq l$, such that the triad $\mathcal{T}^{h,b,c} = (P, P_b^{ih}, P_c^{jh})$ is dense. For each such triad we will estimate the number of hyperedges of H by the number $\mathrm{tr}(\mathcal{T}^{h,b,c})$ of triangles in (P, P_b^{ih}, P_c^{jh}), which, by Fact 4.1, is at most $(1/l^3 + 7\varepsilon)m^3$. Since we assumed that there are at most $(1/2 + \gamma/3)tl^2$ such triads, we infer that

$$|S_1| < \left(\frac{1}{2} + \frac{1}{3}\gamma\right)tl^2\left(\frac{1}{l^3} + 7\varepsilon\right)m^3 < \left(\frac{1}{2} + \frac{1}{3}\gamma + 7\varepsilon l^3\right)\frac{n^3}{t^2 l}.$$

Let $S_2 = S \setminus S$ consist of those hyperedges of H which contain an edge of P, an edge of P_b^{ih} and an edge of P_c^{jh}, for some $h \in [t] \setminus \{i, j\}$ and $1 \leqslant b, c \leqslant l$, such that the triad $\mathcal{T}^{h,b,c} = (P, P_b^{ih}, P_c^{jh})$ is not dense. In this case, conversely, we estimate their number by tl^2 – the total number of triads containing P, while the number of hyperedges of H is now at most $\frac{1}{2}\gamma\mathrm{tr}(\mathcal{T}^{h,b,c})$. Hence,

$$|S_2| < \frac{1}{2}\gamma tl^2\left(\frac{1}{l^3} + 7\varepsilon\right)m^3 < \left(\frac{1}{2}\gamma + 7\varepsilon l^3\right)\frac{n^3}{t^2 l}$$

and

$$|S| = |S_1| + |S_2| < \left(\frac{1}{2} + \frac{5}{6}\gamma + 14\varepsilon l^3\right)\frac{n^3}{t^2 l}.$$

As $14\varepsilon l^3 + \varepsilon l < \gamma/7 < (0.99 - 5/6)\gamma$, this is a contradiction to the previously established lower bound on $|S|$. □

Using the probabilistic method we will now select one graph from each set $\{P_a^{ij} : a = 1,\ldots,l\}$, $1 \leqslant i < j \leqslant t$, which (almost) maintains the property established in Claim 5.2, and in which most triads are δ-regular.

Claim 5.3. *There exists a family $\mathcal{P}$ of bipartite graphs $P^{ij} = P_{a_{ij}}^{ij}$ between pairs (V_i, V_j), where $1 \leqslant i < j \leqslant t$, such that*

(a) *every graph of $\mathcal{P}$ belongs to at least $(1/2 + \gamma/12)t$ dense triads in $\mathcal{P}$, and*

(b) *all but at most $2\delta t^3$ triads of $\mathcal{P}$ are δ-regular.*

Proof. We apply the probabilistic method and Chernoff's and Markov's inequalities. For all $1 \leqslant i < j \leqslant t$, choose an index $a_{ij} \in \{1, 2, \ldots, l\}$ independently and uniformly at random. The selected indices determine a (random) family $\mathcal{P}$ of $\binom{t}{2}$ bipartite graphs.

For each $a = 1, 2, \ldots, l$, let $I_a^{ij} = 1$ if $a_{ij} = a$ and 0 otherwise. For convenience, we will abbreviate $P_{a_{ij}}^{ij} = P^{ij}$. Further, let X_a^{ij} be the number of indices $h \in [t] - \{i, j\}$ such that $(P_a^{ij}, P^{ih}, P^{jh})$ is a dense triad. Note that I_a^{ij} and X_a^{ij} are independent random variables and that X_a^{ij} is a sum of independent $0-1$ random variables with

$$\mathbb{E}(X_a^{ij}) \geqslant \left(\frac{1}{2} + \frac{1}{3}\gamma\right)tl^2\left(\frac{1}{l}\right)^2 = \left(\frac{1}{2} + \frac{1}{3}\gamma\right)t.$$

Thus, by Chernoff's inequality ([8], (2.6) on p. 26 and Theorem 2.8 on p. 29)

$$\mathbb{P}\left\{X_a^{ij} < \left(\frac{1}{2} + \frac{1}{12}\gamma\right)t\right\} \leqslant \mathbb{P}\left\{X_a^{ij} < \mathbb{E}X_a^{ij} - \frac{1}{4}\gamma t\right\} \leqslant \exp\left\{-\frac{1}{32}\gamma^2 t\right\}.$$

Further, let $Z_a^{ij}=1$ if both $I_a^{ij}=1$ and $X_a^{ij}<(1/2+\gamma/12)t$. Then

$$\mathbb{P}\left\{\sum_{a,i,j}Z_a^{ij}>0\right\} \leq \sum_{a,i,j}\mathbb{E}(Z_a^{ij}) = \sum_{a,i,j}\mathbb{E}(I_a^{ij})\mathbb{P}\left\{X_a^{ij}<\left(\frac{1}{2}+\frac{1}{12}\gamma\right)t\right\}$$

$$\leq t^2 l \frac{1}{l}\exp\left\{-\frac{1}{32}\gamma^2 t\right\} = t^2\exp\left\{-\frac{1}{32}\gamma^2 t\right\} < \frac{1}{2}$$

by our assumption on t_0.

On the other hand, by condition (ii) of Lemma 4.3, the expected number of δ-irregular triads of $\mathscr{P}$ is at most $\delta t^3 l^3 (1/l)^3 = \delta t^3$, and hence, by Markov's inequality, the probability that there are more than $2\delta t^3$ such triads is less than $1/2$. Thus there exists a selection $\mathscr{P}$ which satisfies both (a) and (b). □

To proceed, we define the *cluster hypergraph* $K = K(\mathscr{P})$ as the 3-uniform hypergraph consisting of all triplets ijh, $1 \leq i < j < h \leq t$, corresponding to dense and δ-regular triads $\mathscr{T}^{ijh} = (P^{ij}, P^{ih}, P^{jh})$, recalling that 'dense' means that $d_H(\mathscr{T}^{ijh}) \geq \gamma/2$.

It is convenient to consider also the auxiliary hypergraphs $D = D(\mathscr{P})$ and $IR = IR(\mathscr{P})$ of all triplets ijh for which the triad $\mathscr{T}^{ijh} = (P^{ij}, P^{ih}, P^{jh})$ is, respectively, dense and δ-irregular. Thus

$$K = D - IR.$$

Observe that for $\mathscr{P}$ satisfying the conclusion of Claim 5.3, we have that D is a $(t, \gamma/12)$-graph, while $|IR| \leq 2\delta t^3$. However, note that K does not necessarily satisfy the assumption of Lemma 5.1. To remedy this, we will select carefully a large sub-hypergraph K' of K as follows.

Call a pair i, j of vertices of K *malicious* if it belongs to more than $\sqrt{\delta}t$ triplets of IR, that is, if $\deg_{IR}(ij) > \sqrt{\delta}t$. Then at most $6\sqrt{\delta}t^2$ pairs are malicious, since otherwise the sum of pair degrees would be larger than $3|IR|$, a contradiction. Let B be the graph of malicious pairs. In turn, call a vertex i *malicious* if $\deg_B(i) > \delta^{1/4}t$. At most $12\delta^{1/4}t$ vertices are malicious, since otherwise the sum of vertex degrees in B would be larger than $2|B|$, a contradiction, again. Remove all malicious vertices obtaining a sub-hypergraph D' and a subgraph B', both on the same set of at least $t - 12\delta^{1/4}t$ vertices. Note that $\Delta(B') \leq \delta^{1/4}t$.

In the hypergraph $K' = D' - IR$, every pair i, j which is not an edge of B' has degree

$$\deg_{B'}(ij) \geq \left(\frac{1}{2}+\frac{1}{12}\gamma - 12\delta^{1/4} - \sqrt{\delta}\right)t \geq \frac{t}{2},$$

where the last inequality follows by our choice of δ. Hence, the graph $G_{K'}$, consisting of those pairs i, j for which $\deg_{K'}(ij) < t/2$, is a subgraph of B' and thus $\Delta(G_{K'}) \leq \Delta(B') \leq \delta^{1/4}t$. By Lemma 5.1, there is in K' a matching M covering all but at most $\Delta(G_{K'}) + 1 \leq 2\delta^{1/4}t$ vertices of K'.

Set $k = \lfloor \gamma^{-8} \rfloor$ and $\alpha = \gamma/2$, and note that each triad corresponding to a triplet of K' (and thus also of M) satisfies the assumptions of Corollary 4.6. Apply Corollary 4.6 to each triad corresponding to a triplet of M to conclude that all but $2\delta^{1/4}$-fraction of the vertices of each such triad can be covered by vertex-disjoint $3k$-paths. Hence, by our choice of δ, altogether there are only at most

$$|V_0| + (12 + 2 + 1)\delta^{1/4}n < 16\delta^{1/4}n < \frac{1}{2}\gamma^{33}n < \frac{1}{2}\gamma^5 n$$

vertices not covered by this system of disjoint $3k$-paths. Clearly, as the paths in the cover have each $3k > \gamma^{-8}$ vertices, the total number of these paths does not exceed $n/3k < \gamma^8 n$. □

6. Concluding remarks

Remark 3. Using a recent result from [7] and standard derandomization techniques our proof can be turned into a polynomial time algorithm constructing a Hamiltonian cycle in every (n, γ)-graph with sufficiently many vertices.

Remark 4. It is possible to generalize Theorem 1.1 to k-uniform hypergraphs, $k \geqslant 4$, in which every set of $k - 1$ vertices belongs to at least $(1/2 + \gamma)n$ edges (see [14].) Moreover, a much more refined and complicated argument allows us to strengthen Theorem 1.1 to the case $\gamma = 0$, which was originally conjectured in [9]. (This is work in progress: see [15].)

Remark 5. After completing their proof, the authors realized that the Frankl–Rödl Regularity Lemma for 3-uniform hypergraphs from [6] could be replaced by an application of a weaker form of the lemma, which is a straightforward generalization of the Szemerédi Regularity Lemma for graphs in [17] (see, *e.g.*, [3], [12], or [4] for a precise statement).

The method presented in this paper, however, will be more suitable for approaching problems of this kind which require more structure than just a path, such as hypergraph extensions of the results in [1] and [10]. This is because, unlike the weak regularity lemma, the Frankl–Rödl Regularity Lemma provides a partition into δ-regular blocks which contain many copies of any fixed sub-hypergraph (see the comment prior to Lemma 4.4).

Acknowledgements

We would like to thank Mathias Schacht and an anonymous referee for a thorough reading of the manuscript.

References

[1] Alon, N. and Yuster, R. (1996) H-factors in dense graphs. *J. Combin. Theory Ser. B* **66** 269–282.
[2] Bermond, J. C. *et al.* (1976) Hypergraphes hamiltoniens. *Prob. Comb. Theorie Graph Orsay* **260** 39–43.
[3] Chung, F. R. K. (1991) Regularity lemma for hypergraphs and quasi-randomness. *Random Struct. Alg.* **2** 241–252.
[4] Czygrinow, A. and Rödl, V. (2000) An algorithmic regularity lemma for hypergraphs. *SIAM J. Comput.* **30** 1041–1066.
[5] Demetrovics, J., Katona, G. O. H. and Sali, A. (1998) Design type problems motivated by database theory. *J. Statist. Planning and Inference* **72** 149–164.

[6] Frankl, P. and Rödl, V. (2002) Extremal problems on set systems. *Random Struct. Alg.* **20** 131–164.
[7] Haxell, P., Nagle, B. and Rödl, V. An algorithmic version of the hypergraph regularity method. To appear in *Proc. 46th Annual IEEE Symposium on Foundations of Computer Science* (FOCS 2005).
[8] Janson, S., Łuczak, T. and Ruciński, A. (2000) *Random Graphs*, Wiley, New York.
[9] Katona, G. Y. and Kierstead, H. A. (1999) Hamiltonian chains in hypergraphs. *J. Graph Theory* **30** 205–212.
[10] Komlós, J., Sárközy, G. N. and Szemerédi, E. (1998) On the Pósa–Seymour conjecture. *J. Graph Theory* **29** 167–176.
[11] Nagle, B. and Rödl, V. (2003) Regularity properties for triple systems. *Random Struct. Alg.* **23** 264–332.
[12] Prömel, H. J. and Steger, A. (1992) Excluding induced subgraphs III: A general asymptotic. *Random Struct. Alg.* **3** 19–31.
[13] Rödl, V. and Ruciński, A. (1998) Ramsey properties of random hypergraphs. *J. Combin. Theory Ser. A* **81** 1–33.
[14] Rödl, V., Ruciński, A. and Szemerédi, E. A Dirac-type theorem for k-uniform hypergraphs. Submitted.
[15] Rödl, V., Ruciński, A. and Szemerédi, E. Dirac theorem for 3-uniform hypergraphs. In preparation.
[16] Rödl, V. and Schacht, M. Regular partitions of hypergraphs. In preparation.
[17] Szemerédi, E. (1978) Regular partitions of graphs. In *Problèmes Combinatoires et Theorie des Graphes* (Colloq. Internat. CNRS, Univ. Orsay, Orsay, 1976), Vol. 260 of *Colloq. Internat. CNRS*, CNRS, Paris, pp. 399–401.

On Dependency Graphs and the Lattice Gas

ALEXANDER D. SCOTT[1] and ALAN D. SOKAL[2]

[1]Department of Mathematics, University College London, London WC1E 6BT, England
(e-mail: scott@math.ucl.ac.uk)
[2]Department of Physics, New York University, 4 Washington Place, New York, NY 10003 USA
(e-mail: sokal@nyu.edu)

Received 8 October 2004; revised 24 June 2005

For Béla Bollobás on his 60th birthday

We elucidate the close connection between the repulsive lattice gas in equilibrium statistical mechanics and the Lovász Local Lemma in probabilistic combinatorics. We show that the conclusion of the Lovász Local Lemma holds for dependency graph G and probabilities $\{p_x\}$ if and only if the independent-set polynomial for G is nonvanishing in the polydisc of radii $\{p_x\}$. Furthermore, we show that the usual proof of the Lovász Local Lemma – which provides a sufficient condition for this to occur – corresponds to a simple inductive argument for the nonvanishing of the independent-set polynomial in a polydisc, which was discovered implicitly by Shearer [28] and explicitly by Dobrushin [12, 13]. We also present a generalization of the Lovász Local Lemma that allows for 'soft' dependencies. The paper aims to provide an accessible discussion of these results, which are drawn from a longer paper [26] that has appeared elsewhere.

1. Introduction

In probabilistic combinatorics, one is often faced with a collection of events $(A_x)_{x \in X}$ in some probability space, for which one wishes to prove that $\mathbb{P}(\bigcap_{x \in X} \overline{A}_x) > 0$, i.e., that with positive probability none of the events happen. If the events are independent, then this is easily done. However, in practice there are usually some dependencies present, and so we must find some way to deal with them. One approach is to control the dependencies by a dependency graph: we say that a graph G with vertex set X is a *dependency graph* for the events $(A_x)_{x \in X}$ if, for each $x \in X$, the event A_x is independent from (the σ-algebra generated by) the collection $\{A_y : y \notin \Gamma^*(x)\}$, where $\Gamma^*(x) = \Gamma(x) \cup \{x\}$ is the closed neighbourhood of x.

Since we are interested in proving that $\mathbb{P}(\bigcap_{x \in X} \overline{A}_x) > 0$, the question is then how large the probabilities of the A_x can be while still being able to guarantee that $\mathbb{P}(\bigcap_{x \in X} \overline{A}_x) > 0$. More precisely, we have the following problem.

Problem 1.1. *Fix a graph G with vertex set X. For which sequences $\mathbf{p} = (p_x)_{x \in X} \in [0, 1]^X$ is it true that, for every collection $(A_x)_{x \in X}$ of events with dependency graph G such that $\mathbb{P}(A_x) \leq p_x$ for all $x \in X$, we have $\mathbb{P}(\bigcap_{x \in X} \overline{A}_x) > 0$?*

We shall say that a sequence $\mathbf{p}$ with this property is *good* for the dependency graph G.

An ostensibly unrelated problem arises in statistical mechanics, in the context of the repulsive lattice gas. In its simplest form (we discuss more general versions later), the 'lattice-gas partition function' associated with the graph G (on vertex set X) is simply the multivariate generating polynomial for independent subsets of vertices, *i.e.*, the polynomial

$$Z_G(\mathbf{w}) = \sum_{\substack{X' \subseteq X \\ X' \text{ independent}}} \prod_{x \in X'} w_x, \qquad (1.1)$$

where we associate a separate variable w_x (usually interpreted as a complex number) to each vertex $x \in X$. This polynomial is familiar in combinatorics, though usually in its single-variable form. (One of our contentions in this paper is that the multivariable form is quite natural, and often easier to analyse.) Since the empty set is trivially independent, we have $Z_G(\mathbf{0}) = 1$, so that Z_G is nonzero in some (complex) neighbourhood of the origin.

Mathematical physicists have devoted considerable effort to locating the zeros of the lattice-gas partition function, and in particular to finding complex polydiscs in which Z_G is nonvanishing.[1] For a sequence of radii $\mathbf{R} = (R_x)_{x \in X}$, let us define the closed polydisc $\overline{D}_{\mathbf{R}} = \{\mathbf{w} \in \mathbb{C}^X : |w_x| \leq R_x \,\forall x\}$. We then have the following problem.

Problem 1.2. *Fix a graph G with vertex set X. For which sequences $\mathbf{R} = (R_x)_{x \in X} \in [0, \infty)^X$ does the closed polydisc $\overline{D}_{\mathbf{R}}$ contain no zeros of Z_G?*

Although the two problems we have stated seem at first sight to be completely unrelated, they turn out to be closely connected. The main result that we will discuss in this paper is the following.

---

[1] Since the physically realizable values of w_x are *positive real*, the reader may be wondering why physicists (of all people!) would want to study the *complex* zeros. One reason is that the complex zeros of $Z_G(\mathbf{w})$ closest to the origin determine the radius of convergence of the Mayer expansion (2.14). A further reason is connected with the Yang–Lee [38] approach to studying phase transitions. Briefly: A *phase transition* is any point where one or more physical quantities depend nonanalytically on one or more control parameters. In any reasonable statistical-mechanical model (such as the polynomial (1.1)), such nonanalyticity is clearly impossible for finite graphs G; rather, phase transitions occur only in the *infinite-volume limit*. That is, we consider a countably infinite graph G_∞ (usually a regular lattice) and an increasing sequence of subgraphs $(G_n)_{n \geq 1}$ converging to G_∞, and study the *limiting free energy per unit volume* $f_{G_\infty}(\mathbf{w}) = \lim_{n \to \infty} |G_n|^{-1} \log Z_{G_n}(\mathbf{w})$. The nonanalyticities of $f_{G_\infty}(\mathbf{w})$ for real $\mathbf{w}$ arise from the singularities of $\log Z_{G_n}(\mathbf{w})$ for *complex* $\mathbf{w}$ that approach the real axis in the limit $n \to \infty$. But the singularities of $\log Z_{G_n}(\mathbf{w})$ are precisely the zeros of $Z_{G_n}(\mathbf{w})$. See, *e.g.*, [30, Section 1] or [31, Section 5] for a slightly more detailed explanation.

Theorem 1.3 (The equivalence theorem). *Let G be a finite graph with vertex set X, and let* $\mathbf{p} = (p_x)_{x \in X} \in [0,1]^X$. *Then the following two statements are equivalent:*

(a) $\mathbf{p}$ *is good for the dependency graph G,*
(b) *the closed polydisc* $\overline{D}_\mathbf{p}$ *contains no zeros of* Z_G.

This will follow from Theorem 3.1 below.

Physicists and mathematicians have each given sufficient conditions for a sequence of positive real numbers to have the properties specified in one (and hence both) of the problems above. For the repulsive lattice gas, Dobrushin [12, 13] gave a sufficient (but not necessary) condition on **R** for Z_G to be nonvanishing in $\overline{D}_\mathbf{R}$. (A slight generalization of Dobrushin's result is given below, as Theorem 4.3.) Likewise, the Lovász Local Lemma [14, 15] – which is an important tool in probabilistic combinatorics – gives a sufficient (but not necessary) condition for a sequence $\mathbf{p}$ to be good for a dependency graph G.

The equivalence between Problems 1.1 and 1.2 means that it is possible to compare the Lovász Local Lemma with Dobrushin's Theorem: both results give sufficient but not necessary conditions, and it is natural to wonder whether one of them might give stronger results than the other. Surprisingly, it turns out that the two theorems, proved in different fields and two decades apart (Dobrushin's Theorem in the 1990s and the Local Lemma in the 1970s), give *identical* criteria! Indeed, close examination of the (inductive) proofs shows that the two proofs are substantially isomorphic.

1.1. Outline of the paper

The main aim of this paper is to present the connections between dependency graphs and the lattice gas discovered in [26]. Although there are no new results in this paper, we hope that this selection of material from the (rather long) paper [26] will give a shorter and more accessible account of the topic aimed at a combinatorial audience. In order to simplify the presentation, we have omitted some of the proofs; detailed proofs of all the results in this paper, together with much more extensive discussion and many further results, can be found in [26], to which we refer the reader for further information.

In Section 2, we examine the lattice gas. After giving some basic properties and examples, we review the Mayer expansion, which is the expansion of $\log Z_G$ in a Taylor series around $\mathbf{w} = 0$. The crucial property that we shall exploit is the fact that the coefficients in the Mayer expansion have alternating signs (Proposition 2.1). We use this to prove (Theorem 2.2) that the closest zeros to the origin lie in the negative real quadrant $(-\infty, 0]^X$, along with some related facts.

In Section 3, we analyse the connection between dependency graphs and lattice gases, and prove the equivalence of Problems 1.1 and 1.2. (In fact, Theorem 3.1 asserts somewhat more than Theorem 1.3.) The proof of Theorem 3.1 has two key ingredients: first we use ideas of Shearer [28] to relate good sequences for a dependency graph G to the negative real zeros of the corresponding multivariate independent-set polynomial Z_G; then we use the results of Section 2 to relate the latter to the *complex* zeros of Z_G. These arguments can in fact be extended to the lattice gas with 'soft interactions', which are connected to dependency graphs with a suitably defined notion of 'weak' dependence; we conclude Section 3 by explaining this connection.

In Section 4, we turn to results giving *sufficient* conditions for a sequence **p** or **R** to have the properties specified in Problems 1.1 and 1.2. After reviewing the Lovász Local Lemma and Dobrushin's Theorem, we prove the equivalence of the criteria given by these two results, and discuss some consequences, including a 'softened' version of the Lovász Local Lemma. We also discuss an improved bound, inspired by the work of Shearer [28].

We conclude the paper, in Section 5, with some comments on the Lovász/Dobrushin bounds.

2. The repulsive lattice gas

2.1. Definition

In statistical mechanics, a 'grand-canonical gas' is defined by a *single-particle state space* X (here a nonempty finite set), a *fugacity vector* $\mathbf{w} = \{w_x\}_{x \in X} \in \mathbb{C}^X$, and a *two-particle Boltzmann factor* $W : X \times X \to \mathbb{C}$ with $W(x,y) = W(y,x)$. The (*grand*) *partition function* $Z_W(\mathbf{w})$ is then defined to be the sum over ways of placing $n \geq 0$ 'particles' on 'sites' $x_1, \ldots, x_n \in X$, with each configuration assigned a 'Boltzmann weight' given by the product of the corresponding factors w_{x_i} and $W(x_i, x_j)$:

$$Z_W(\mathbf{w}) = \sum_{n=0}^{\infty} \frac{1}{n!} \sum_{x_1, \ldots, x_n \in X} \left(\prod_{i=1}^n w_{x_i} \right) \left(\prod_{1 \leq i < j \leq n} W(x_i, x_j) \right) \quad (2.1\text{a})$$

$$= \sum_{\mathbf{n}} \left(\prod_{x \in X} \frac{w_x^{n_x} W(x,x)^{n_x(n_x-1)/2}}{n_x!} \right) \left(\prod_{\{x,y\} \subseteq X} W(x,y)^{n_x n_y} \right) \quad (2.1\text{b})$$

where in (2.1b) the sum runs over all multi-indices $\mathbf{n} = \{n_x\}_{x \in X}$ of nonnegative integers, and the product runs over all two-element subsets $\{x, y\} \subseteq X$ ($x \neq y$).[2] We shall use the notation $\mathbf{w}^{\mathbf{n}} = \prod_{x \in X} w_x^{n_x}$ and $|\mathbf{w}| = \{|w_x|\}_{x \in X}$ (although, abusing notation, we shall also write $|\mathbf{n}| = \sum_{x \in X} |n_x|$). We will also write $\mathbf{w} \geq \mathbf{0}$ to indicate that $\mathbf{w}$ is a vector of real numbers such that $w_x \geq 0$ for all $x \in X$.

In this paper we shall limit attention to the *repulsive* lattice gas in which $0 \leq W(x,y) \leq 1$ for all x, y. From this assumption it follows immediately that $Z_W(\mathbf{w})$ is an entire analytic function of $\mathbf{w}$ satisfying $|Z_W(\mathbf{w})| \leq \exp(\sum_{x \in X} |w_x|)$.

If $W(x,x) = 0$ for all $x \in X$ – in statistical mechanics this is called a *hard-core self-repulsion* – then the only nonvanishing terms in (2.1b) have $n_x = 0$ or 1 for all x (i.e., each site can be occupied by at most one particle), so that $Z_W(\mathbf{w})$ can be written as a sum over

---

[2] More precisely, $Z_W(\mathbf{w})$ sums $1/n!$ times the Boltzmann weight over ways of placing n *distinguishable* particles onto sites. In the case of hard-core self-repulsion (which is the case of principal interest to us), this is equivalent to summing the Boltzmann weight over ways of placing n *indistinguishable* particles onto sites. (For the two interpretations to be equivalent when multiple occupation is permitted, there would have to be an additional factor $\prod_{x \in X} n_x!$ in (2.1a) and (2.1b).)

subsets:

$$Z_W(\mathbf{w}) = \sum_{X' \subseteq X} \left(\prod_{x \in X'} w_x \right) \left(\prod_{\{x,y\} \subseteq X'} W(x,y) \right). \qquad (2.2)$$

In this case $Z_W(\mathbf{w})$ is a multiaffine polynomial, *i.e.*, of degree 1 in each w_x separately. Combinatorially, $Z_W(\mathbf{w})$ is the generating polynomial for *induced subgraphs* of the complete graph, in which each vertex x gets weight w_x and each edge xy gets weight $W(x,y)$.

If, in addition to hard-core self-repulsion, we have $W(x,y) = 0$ or 1 for each pair $x \neq y$ – in statistical mechanics this is called a *hard-core pair interaction* – then we can define a (simple loopless) graph $G = (X, E)$ by setting $xy \in E$ whenever $W(x,y) = 0$ and $x \neq y$, so that $Z_W(\mathbf{w})$ is precisely the *independent-set polynomial* for G:

$$Z_G(\mathbf{w}) = \sum_{\substack{X' \subseteq X \\ X' \text{ independent}}} \prod_{x \in X'} w_x. \qquad (2.3)$$

Traditionally the independent-set polynomial is defined as a univariate polynomial $Z_G(w)$ in which w_x is set equal to the same value w for all vertices x. But one of our main contentions in this paper is that Z_G is more naturally understood as a multivariate polynomial; this allows us, in particular, to exploit the fact that Z_G is multiaffine.

More generally, given any W satisfying $0 \leqslant W(x,y) \leqslant 1$ for all x, y, let us define a simple loopless graph $G = G_W$ (the *support graph* of W) by setting $xy \in E(G)$ if and only if $W(x,y) \neq 1$ and $x \neq y$. The partition function $Z_W(\mathbf{w})$ can be thought of as a 'soft' version of the independent-set polynomial for G, in which an edge $xy \in E(G)$ has 'strength' $1 - W(x,y) \in (0,1]$.

In the situation of hard-core self-repulsion (2.2), it is convenient to define, for each subset $\Lambda \subseteq X$, the restricted partition function

$$Z_\Lambda(\mathbf{w}) = \sum_{X' \subseteq \Lambda} \prod_{x \in X'} w_x \prod_{\{x,y\} \subseteq X'} W(x,y). \qquad (2.4)$$

Of course this notation is redundant, since the same effect can be obtained by setting $w_x = 0$ for $x \in X \setminus \Lambda$, but it is useful for the purpose of inductive computations and proofs. We have, for any $x \in \Lambda$, the *fundamental identity*

$$Z_\Lambda(\mathbf{w}) = Z_{\Lambda \setminus x}(\mathbf{w}) + w_x Z_{\Lambda \setminus x}(W(x, \cdot)\mathbf{w}) \qquad (2.5)$$

where

$$[W(x, \cdot)\mathbf{w}]_y = W(x,y) w_y; \qquad (2.6)$$

here the first term on the right-hand side of (2.5) covers the summands in (2.4) with $X' \not\ni x$, while the second covers $X' \ni x$. In the special case of a hard-core interaction (= independent-set polynomial) for a graph G, (2.5) reduces to

$$Z_\Lambda(\mathbf{w}) = Z_{\Lambda \setminus x}(\mathbf{w}) + w_x Z_{\Lambda \setminus \Gamma^*(x)}(\mathbf{w}), \qquad (2.7)$$

where we have used the notation $\Gamma^*(x) = \Gamma(x) \cup \{x\}$. The fundamental identity (2.5)/(2.7) plays an important role both in the inductive proof of the Lovász Local Lemma and in the Dobrushin–Shearer inductive argument for the nonvanishing of Z_W in a polydisc (Section 4).

Remarks. (1) It is worth remarking that the hard-core lattice gas (2.3) is not merely *one* interesting statistical-mechanical model; it turns out to be the *universal* statistical-mechanical model in the sense that any statistical-mechanical model living on a vertex set V_0 can be mapped onto a gas of nonoverlapping 'polymers' on V_0, *i.e.*, a hard-core lattice gas on the intersection graph of V_0 [29, Section 5.7].[3] This construction, which is termed the 'polymer expansion' or 'cluster expansion', is an important tool in mathematical statistical mechanics [27, 7, 17, 9, 6]; it is widely employed to prove the absence of phase transition at high temperature, low temperature, large magnetic field, low density, or weak nonlinear coupling. Further information on expansion methods in statistical mechanics (and related combinatorial problems) can be found in the excellent recent survey by Borgs [6].

(2) Repeated use of (2.5) obviously gives an algorithm to compute $Z_W(\mathbf{w})$. But this algorithm takes in general exponential time. In fact, calculating $Z_G(\mathbf{w})$ for general graphs G (or even for cubic planar graphs) is NP-hard (as noted by Shearer [28]), since even calculating the *degree* of $Z_G(\mathbf{w})$ – that is, the maximum size of an independent set – is NP-hard [16, pp. 194–195]. Therefore, if $P \neq NP$ it is impossible to calculate $Z_G(\mathbf{w})$ for general graphs in polynomial time.

In this section we give some general results concerning the partition function of a lattice gas; additional related results can be found in [26]. Most of these results are valid for an arbitrary repulsive lattice gas (2.1), in which multiple occupation of a site is permitted. A few of the results are restricted to the case of a hard-core self-repulsion (2.2), in which multiple occupation of a site is forbidden.

2.2. An example: The complete r-ary rooted tree [25, 28]

Before continuing with the general theory, we pause to compute an important example. Let $T_n^{(r)}$ be the complete rooted tree with branching factor r and depth n. We limit attention to the univariate independent-set polynomial. Fix $r \geq 1$; and to lighten the notation, let us write Z_n as a shorthand for $Z_{T_n^{(r)}}$. Applying the fundamental identity (2.7) to the root vertex, we obtain the nonlinear recursion

$$Z_n(w) = Z_{n-1}(w)^r + w Z_{n-2}(w)^{r^2}, \qquad (2.8)$$

which is valid for all $n \geq 0$ if we set $Z_{-1} \equiv Z_{-2} \equiv 1$. By defining

$$Y_n(w) = \frac{Z_n(w)}{Z_{n-1}(w)^r}, \qquad (2.9)$$

---

[3] The *intersection graph* of a finite set S is the graph whose vertices are the nonempty subsets of S, and whose edges are the pairs with nonempty intersection.

we can convert the second-order recursion (2.8) to a first-order recursion

$$Y_n(w) = 1 + \frac{w}{Y_{n-1}(w)^r} \tag{2.10}$$

with initial condition $Y_{-1} \equiv 1$. The polynomials $Z_n(w)$ can be reconstructed from the rational functions $Y_n(w)$ by

$$Z_n(w) = \prod_{k=0}^{n} Y_k(w)^{r^{n-k}}. \tag{2.11}$$

Let $w_n < 0$ be the negative real root of Z_n of smallest magnitude (set $w_n = -\infty$ if Z_n has no negative real root). Note that $w_{-1} = -\infty$ and $w_0 = -1$. Since $Z_n(0) = 1$, we have $Z_n(w) > 0$ for all $w \in (w_n, 0]$. Let us prove by induction that $w_{n-1} < w_n$ for $n \geq 0$. It is true for $n = 0$. For $n \geq 1$ we have

$$Z_n(w_{n-1}) = Z_{n-1}(w_{n-1})^r + w_{n-1} Z_{n-2}(w_{n-1})^{r^2} < 0 \tag{2.12}$$

since $Z_{n-1}(w_{n-1}) = 0$, $w_{n-1} < 0$ and $Z_{n-2}(w_{n-1}) > 0$ by the inductive hypothesis. Therefore Z_n vanishes somewhere between w_{n-1} and 0.

It follows that the w_n increase to a limit $w_\infty \leq 0$ as $n \to \infty$. Let us show, following Shearer [28], that

$$w_\infty = -\frac{r^r}{(r+1)^{r+1}} \tag{2.13}$$

by proving the two inequalities.

Proof of $\geq$. If $w \in [w_\infty, 0)$, we have $Z_n(w) > 0$ for all n and hence also $Y_n(w) > 0$ for all n. Since $Y_{-1} > Y_0$, it follows from the monotonicity of (2.10) that $\{Y_n(w)\}_{n \geq 0}$ is a strictly decreasing sequence of positive numbers, hence converges to a limit $y_* \geq 0$ satisfying the fixed-point equation $y_* = 1 + w/y_*^r$, or equivalently $w = y_*^{r+1} - y_*^r$. Elementary calculus then shows that $w \geq -r^r/(r+1)^{r+1}$; taking $w = w_\infty$ we obtain $w_\infty \geq -r^r/(r+1)^{r+1}$. □

Proof of $\leq$. If $-r^r/(r+1)^{r+1} \leq w < 0$, the equation $w = y_*^{r+1} - y_*^r$ has a unique solution $y_* \in [r/(r+1), 1)$. It then follows by induction (using (2.10) and the initial condition $Y_{-1} = 1$) that $1 = Y_{-1}(w) > Y_0(w) > \cdots > Y_{n-1}(w) > Y_n(w) > \cdots > y_*$ for all $n \geq 0$. In particular, $Y_n(w) > 0$ for all n, so that $w_n < w$ for all n. This shows that $w_\infty \leq -r^r/(r+1)^{r+1}$. □

Let us conclude by observing that (2.10) defines a degree-r rational map $R_w : y \mapsto 1 + w/y^r$ parametrized by $w \in \mathbb{C} \setminus 0$. Moreover, the zeros of $Z_n(w)$ correspond to those values w for which R_w has a (superattractive) orbit $0 \mapsto \infty \mapsto 1 \mapsto 1 + w \mapsto \cdots \mapsto 0$ of period $n + 3$ (or some divisor of $n + 3$). As $n \to \infty$, these points accumulate on a 'Mandelbrot-like' set in the complex w-plane [22].

2.3. The Mayer expansion

Let us now return to the general case of a repulsive lattice gas (2.1). Since $Z_W(\mathbf{w})$ is an entire function of $\mathbf{w}$ satisfying $Z_W(\mathbf{0}) = 1$, its logarithm is analytic in some neighbourhood of $\mathbf{w} = \mathbf{0}$, and so can be expanded in a convergent Taylor series:

$$\log Z_W(\mathbf{w}) = \sum_{\mathbf{n}} c_{\mathbf{n}}(W) \mathbf{w}^{\mathbf{n}}, \qquad (2.14)$$

where we have used the notation $\mathbf{w}^{\mathbf{n}} = \prod_{x \in X} w_x^{n_x}$, and of course $c_{\mathbf{0}} = 0$. In statistical mechanics, (2.14) is called the *Mayer expansion* [36, 6]; there is a beautiful combinatorial formula for the Mayer coefficients $c_{\mathbf{n}}(W)$, which we shall not need here (see [36, 6, 26]).

For our purposes, the crucial property of the Mayer expansion is the following.

Proposition 2.1 (Signs of Mayer coefficients). *Suppose that the lattice gas is repulsive, i.e., $0 \leq W(x,y) \leq 1$ for all $x, y \in X$. Then, for all $\mathbf{n} \geq \mathbf{0}$, the Mayer coefficients $c_{\mathbf{n}}(W)$ satisfy*

$$(-1)^{|\mathbf{n}|-1} c_{\mathbf{n}}(W) \geq 0. \qquad (2.15)$$

The alternating-sign property (2.15) for the Mayer coefficients of a repulsive gas has been known in the physics literature for over 40 years: see Groeneveld [18] for a brief sketch of one proof. Our own proof [26, Section 2.2], which is based on the partitionability of a matroid complex, also controls the signs of the first two derivatives of $c_{\mathbf{n}}(W)$ with respect to W. We think that the Mayer coefficients $c_{\mathbf{n}}(W)$ merit further study from a combinatorial point of view; we would not be surprised if new identities or inequalities were waiting to be discovered.

2.4. The fundamental theorem

Let us now state the principal result of this section. We use the notation $|\mathbf{w}| = \{|w_x|\}_{x \in X}$.

Theorem 2.2 (The fundamental theorem). *Consider any repulsive lattice gas, and let $\mathbf{R} = \{R_x\}_{x \in X} \geq \mathbf{0}$. Then the following are equivalent.*

(a) *There exists a connected set $C \subseteq (-\infty, 0]^X$ that contains both $\mathbf{0}$ and $-\mathbf{R}$, such that $Z_W(\mathbf{w}) > 0$ for all $\mathbf{w} \in C$. (Equivalently, $-\mathbf{R}$ belongs to the connected component of $Z_W^{-1}(0, \infty) \cap (-\infty, 0]^X$ containing $\mathbf{0}$.)*
(b) $Z_W(\mathbf{w}) > 0$ *for all $\mathbf{w} \in \mathbb{R}^X$ satisfying $-\mathbf{R} \leq \mathbf{w} \leq \mathbf{0}$.*
(c) $Z_W(\mathbf{w}) \neq 0$ *for all $\mathbf{w} \in \mathbb{C}^X$ satisfying $|\mathbf{w}| \leq \mathbf{R}$.*
(d) *The Taylor series for $\log Z_W(\mathbf{w})$ around $\mathbf{0}$ is convergent at $\mathbf{w} = -\mathbf{R}$.*
(e) *The Taylor series for $\log Z_W(\mathbf{w})$ around $\mathbf{0}$ is absolutely convergent for $|\mathbf{w}| \leq \mathbf{R}$.*

Moreover, when these conditions hold, we have $|Z_W(\mathbf{w})| \geq Z_W(-\mathbf{R}) > 0$ for all $\mathbf{w} \in \mathbb{C}^X$ satisfying $|\mathbf{w}| \leq \mathbf{R}$.

In the case of hard-core self-repulsion, (a)–(e) are also equivalent to the following.

(b′) $Z_W(-\mathbf{R}\, \mathbf{1}_S) > 0$ *for all $S \subseteq X$, where*

$$(\mathbf{R}\, \mathbf{1}_S)_x = \begin{cases} R_x & \text{if } x \in S, \\ 0 & \text{otherwise.} \end{cases} \qquad (2.16)$$

(f) $Z_W(-\mathbf{R}) > 0$, and $(-1)^{|S|} Z_W(-\mathbf{R}; S) \geqslant 0$ for all $S \subseteq X$, where

$$Z_W(\mathbf{w}; S) = \sum_{S \subseteq X' \subseteq X} \left(\prod_{x \in X'} w_x \right) \left(\prod_{\{x,y\} \subseteq X'} W(x, y) \right). \tag{2.17}$$

(g) There exists a probability measure P on 2^X satisfying $P(\varnothing) > 0$ and

$$\sum_{T \supseteq S} P(T) = \left(\prod_{x \in S} R_x \right) \left(\prod_{\{x,y\} \subseteq S} W(x, y) \right) \tag{2.18}$$

for all $S \subseteq X$. (This probability measure is unique and is given by

$$P(S) = (-1)^{|S|} Z_W(-\mathbf{R}; S).$$

In particular, $P(\varnothing) = Z_W(-\mathbf{R}) > 0$.)

Remarks. (1) The conditions (b'), (f) and (g) are inspired in part by Shearer [28, Theorem 1].

(2) Suppose that the univariate entire function $Z_W(w)$, defined by setting $w_x = w$ for all x, is strictly positive whenever $-R \leqslant w \leqslant 0$. Then in fact $Z_W(\mathbf{w}) > 0$ whenever $-R \leqslant w_x \leqslant 0$ for all x: this follows from (a) $\Longrightarrow$ (b) by taking C to be the segment $[-R, 0]$ of the diagonal.

Our proof of Theorem 2.2 hinges on the alternating-sign property (2.15) for the Taylor coefficients of $\log Z_W$. In preparation for this proof, let us recall the Vivanti–Pringsheim theorem in the theory of analytic functions of a single complex variable [19, Theorem 5.7.1]: if a power series $f(z) = \sum_{n=0}^{\infty} a_n z^n$ with *nonnegative* coefficients has a finite nonzero radius of convergence, then the point of the circle of convergence lying on the positive real axis is a singular point of the function f. Otherwise put, if f is a function whose Taylor series at 0 has all nonnegative coefficients and f is analytic on some complex neighbourhood of the real interval $[0, R)$, then f is in fact analytic on the open disc of radius R centred at the origin and its Taylor series is absolutely convergent there. Here we will need the following multidimensional generalization [26] of the Vivanti–Pringsheim theorem.

Proposition 2.3 (Multidimensional Vivanti–Pringsheim theorem). *Let C be a connected subset of $[0, \infty)^n$ containing $\mathbf{0}$, let U be an open neighbourhood of C in $\mathbb{C}^n$, and let f be a function analytic on U whose Taylor series around $\mathbf{0}$ has all nonnegative coefficients. Then the Taylor series of f around $\mathbf{0}$ converges absolutely on the set $\mathrm{hull}(C) \equiv \bigcup_{\mathbf{R} \in C} \bar{D}_\mathbf{R}$, where $\bar{D}_\mathbf{R}$ denotes the closed polydisc $\{\mathbf{w} \in \mathbb{C}^n : |w_i| \leqslant R_i \text{ for all } i\}$, and it defines a function that is continuous on $\mathrm{hull}(C)$ and analytic on its interior.*

We shall also make use of the following elementary result.

Lemma 2.4. *Let F be a function on 2^X, and define*

$$F_-(S) = \sum_{X' \subseteq S} F(X'), \qquad (2.19)$$

$$F_+(S) = \sum_{X' \supseteq S} F(X'). \qquad (2.20)$$

Then

$$F_-(S) = \sum_{Y \subseteq S^c} (-1)^{|Y|} F_+(Y) \qquad (2.21)$$

where $S^c \equiv X \setminus S$.

Finally, we shall need that fact that if $F : \mathbb{C}^X \to \mathbb{C}$ is multiaffine on a rectangle, then its value at any point inside the rectangle is a convex combination of the values at extreme points of the rectangle. For instance, if F is multiaffine on the rectangle $[0, 1]^X$, then

$$F(\mathbf{z}) = \sum_{Y \subseteq X} \left(\prod_{i \in Y} z_i \right) \left(\prod_{i \in X \setminus Y} (1 - z_i) \right) F(\mathbf{1}_Y). \qquad (2.22)$$

This is clear when $\mathbf{z} \in \{0, 1\}^X$, and equality for all $\mathbf{z} \in [0, 1]^X$ then follows because both sides are multiaffine. A similar statement holds for general rectangles.

We are now ready to prove Theorem 2.2.

Proof of Theorem 2.2. (c) $\Longrightarrow$ (b) $\Longrightarrow$ (a) is trivial (note that $Z_W(\mathbf{0}) = 1$).

(e) $\Longrightarrow$ (d) is trivial. The alternating sign property (2.15) implies that all terms $c_\mathbf{n} \mathbf{w}^\mathbf{n}$ in the Mayer expansion (2.14) are nonpositive when $\mathbf{w} \leq \mathbf{0}$, and so we get (d) $\Longrightarrow$ (e).

(e) implies that the sum of the Taylor series for $\log Z_W(\mathbf{w})$ defines an analytic function on the open polydisc $D_\mathbf{R}$ and a continuous function on the closed polydisc $\bar{D}_\mathbf{R}$. Its exponential equals $Z_W(\mathbf{w})$ on $D_\mathbf{R}$ and hence by continuity also on $\bar{D}_\mathbf{R}$. Therefore (e) $\Longrightarrow$ (c).

Finally, assume (a). Since Z_W is continuous on $\mathbb{C}^X$ (and has real coefficients), we can find an open connected neighbourhood C' of C in $(-\infty, 0]^X$ on which $Z_W > 0$, and an open neighbourhood U of C' in $\mathbb{C}^X \simeq \mathbb{R}^{2|X|}$ on which $Z_W \neq 0$. It is fairly easy to show that we can find a finite polygonal path $P \subset C'$ from $\mathbf{0}$ to $-\mathbf{R}$ (consider the set of points in C' that can be reached by a finite polygonal path from $\mathbf{0}$); taking a suitably small neighbourhood of P gives a simply connected open set U' in $\mathbb{C}^X$ with $P \subset U' \subset U$ (see [26] for details). Then $\log Z_W$ is a well-defined single-valued analytic function on U', once we specify $\log Z_W(\mathbf{0}) = 0$. Applying Proposition 2.3 to $\log Z_W$ on P and U' (using the alternating-sign property (2.15)), we conclude that the Taylor series for $\log Z_W$ around $\mathbf{0}$ is absolutely convergent on $\bar{D}_\mathbf{R}$. Therefore (a) $\Longrightarrow$ (e).

The bound $|Z_W(\mathbf{w})| \geq Z_W(-\mathbf{R})$ for $|\mathbf{w}| \leq \mathbf{R}$, which is equivalent to $\mathrm{Re}\log Z_W(\mathbf{w}) \geq \log Z_W(-\mathbf{R})$, is an immediate consequence of the alternating-sign property (2.15).

Now consider the special case of a hard-core self-repulsion. (b) $\Longrightarrow$ (b') is trivial, and (b') $\Longrightarrow$ (b) follows from the fact that Z_W is multiaffine (*i.e.*, of degree ≤ 1 in each w_x separately) because the value of Z_W at any point $\mathbf{w}$ in the rectangle $-\mathbf{R} \leq \mathbf{w} \leq \mathbf{0}$ is a convex combination of the values at extreme points of the rectangle, which correspond to possible choices of $S \subset X$ in (2.16).

To show that (b) $\Longrightarrow$ (f), note that

$$Z_W(\mathbf{w}; S) = \left(\prod_{x \in S} w_x\right) \left(\prod_{\{x,y\} \subseteq S} W(x, y)\right) Z_W(W(S, \cdot)\mathbf{w}), \tag{2.23}$$

where we have defined

$$[W(S, \cdot)\mathbf{w}]_y = \left(\prod_{x \in S} W(x, y)\right) w_y \tag{2.24}$$

(note in particular that this vanishes whenever $y \in S$). Hence

$$(-1)^{|S|} Z_W(-\mathbf{R}; S) = \left(\prod_{x \in S} R_x\right) \left(\prod_{\{x,y\} \subseteq S} W(x, y)\right) Z_W(-W(S, \cdot)\mathbf{R}) \geq 0 \tag{2.25}$$

since $-\mathbf{R} \leq -W(S, \cdot)\mathbf{R} \leq \mathbf{0}$, with strict inequality when $|S| = 0$ or 1 (since the product over $W(x, y)$ is in that case empty).

To show that (f) $\Longrightarrow$ (b'), use Lemma 2.4 applied to the set function

$$F(S) = \left(\prod_{x \in S} -R_x\right) \left(\prod_{\{x,y\} \subseteq S} W(x, y)\right). \tag{2.26}$$

We have

$$F_-(S) = Z_W(-\mathbf{R}\,\mathbf{1}_S), \tag{2.27}$$

$$F_+(S) = Z_W(-\mathbf{R}; S), \tag{2.28}$$

so that Lemma 2.4 asserts the identity

$$Z_W(-\mathbf{R}\,\mathbf{1}_S) = \sum_{Y \subseteq S^c} (-1)^{|Y|} Z_W(-\mathbf{R}; Y). \tag{2.29}$$

By (f), the $Y = \varnothing$ term is > 0 and the other terms are ≥ 0, so $Z_W(-\mathbf{R}\,\mathbf{1}_S) > 0$ for all S.

Finally, let us show that (f) $\Longleftrightarrow$ (g). By inclusion-exclusion, there are unique numbers $P(T)$ satisfying (2.18), namely $P(T) = (-1)^{|T|} Z_W(-\mathbf{R}; T)$. Moreover, taking $S = \varnothing$ in (2.18) we see that $\sum_T P(T) = 1$. Therefore, P is a probability measure if and only if $(-1)^{|T|} Z_W(-\mathbf{R}; T) \geq 0$ for all T; and $P(\varnothing) > 0$ if and only if $Z_W(-\mathbf{R}; \varnothing) = Z_W(-\mathbf{R}) > 0$. □

3. Dependency graphs and the lattice gas: The equivalence theorem

In this section, we begin our investigation of the relationship between dependency graphs and the lattice gas. In Section 3.1, we work with the lattice gas with hard-core pair interactions, which has partition function given by (2.3); in Section 3.2, we extend the results to the lattice gas with soft-core pair interactions, which has partition function given by (2.2).

3.1. Hard-core version

We begin by recalling the definition of a dependency graph. Let $(A_x)_{x \in X}$ be a finite family of events on some probability space, and let G be a graph with vertex set X. We say that G is a *dependency graph* for the family $(A_x)_{x \in X}$ if, for each $x \in X$, the event A_x is independent from the σ-algebra $\sigma(A_y : y \in X \setminus \Gamma^*(x))$. Note that this is much stronger than requiring merely that A_x be independent of each such A_y separately.

A family of events typically has many possible dependency graphs: for instance, if G is a dependency graph for events $(A_x)_{x \in X}$, then any graph obtained by adding edges to G is also a dependency graph. In particular, if the events A_x are independent, then any graph on X is a dependency graph. Nor must there be a unique minimal dependency graph. Consider, for instance, the set of binary strings of length n with odd digit sum (giving each such string equal probability), and let A_i be the event that the ith digit is 1. Any graph without isolated vertices is a dependency graph for this collection of events.

There is also a stronger notion of a dependency graph G for a collection of events $(A_x)_{x \in X}$, where we demand that if Y and Z are disjoint subsets of X such that G contains no edges between Y and Z, then the σ-algebras $\sigma(A_y : y \in Y)$ and $\sigma(A_z : z \in Z)$ are independent. In this case we shall refer to G as a *strong dependency graph* for the events $(A_x)_{x \in X}$. (For instance, this situation arises in any statistical-mechanical model with variables living on the set X and pair interactions only on the edges of G, where each A_x depends only on the variable at x.) Alternatively, the dependency-graph hypothesis can be replaced by a *weaker* hypothesis concerning conditional probabilities, as in the lopsided Lovász Local Lemma (Theorem 4.2). It will follow from Theorem 3.1 below that all three hypotheses lead to the same lower bound on $\mathbb{P}(\bigcap_{x \in X} \overline{A}_x)$.

Our aim is to relate dependency graphs to lattice gases. The following result (which is a development of Shearer [28, Theorem 1]) gives the connection. For a graph G, we define

$$\mathscr{R}(G) = \{\mathbf{R} \in [0, \infty)^X : Z_G(\mathbf{w}) \neq 0 \ \forall \mathbf{w} \in \overline{D}_\mathbf{R}\}. \tag{3.1}$$

Theorem 3.1 (The equivalence theorem, hard-core case). *Let $(A_x)_{x \in X}$ be a family of events on some probability space, and let G be a graph with vertex set X. Suppose that $(p_x)_{x \in X}$ are real numbers in $[0, 1]$ such that, for each x and each $Y \subseteq X \setminus \Gamma^*(x)$, we have*

$$\mathbb{P}\left(A_x \Big| \bigcap_{y \in Y} \overline{A}_y\right) \leq p_x. \tag{3.2}$$

(a) If $\mathbf{p} \in \mathscr{R}(G)$, then

$$\mathbb{P}\left(\bigcap_{x \in X} \overline{A}_x\right) \geq Z_G(-\mathbf{p}) > 0 \tag{3.3}$$

and more generally

$$\mathbb{P}\left(\bigcap_{x \in Y} \overline{A}_x \Big| \bigcap_{x \in Z} \overline{A}_x\right) \geq \frac{Z_G(-\mathbf{p}\,\mathbf{1}_{Y \cup Z})}{Z_G(-\mathbf{p}\,\mathbf{1}_Z)} > 0 \tag{3.4}$$

for any subsets $Y, Z \subseteq X$. Moreover, this lower bound is best possible in the sense that there exists a probability space on which there can be constructed a family of events $(B_x)_{x \in X}$ with probabilities $\mathbb{P}(B_x) = p_x$ and strong dependency graph G, such that $\mathbb{P}(\bigcap_{x \in X} \overline{B}_x) = Z_G(-\mathbf{p})$.

(b) *If $\mathbf{p} \notin \mathcal{R}(G)$, then there exists a probability space on which there can be constructed:*

  (i) *a family of events $(B_x)_{x \in X}$ with probabilities $\mathbb{P}(B_x) = p_x$ and strong dependency graph G, satisfying $\mathbb{P}(\bigcap_{x \in X} \overline{B}_x) = 0$; and*
  (ii) *a family of events $(B'_x)_{x \in X}$ with probabilities $\mathbb{P}(B'_x) = p'_x \leqslant p_x$ and strong dependency graph G, satisfying $\mathbb{P}(B'_x \cap B'_y) = 0$ for all $xy \in E(G)$ and $\mathbb{P}(\bigcap_{x \in X} \overline{B}'_x) = 0$.*

Remarks. (1) Please note that G is here an *arbitrary* graph with vertex set X; it need not be a dependency graph for the events $(A_x)_{x \in X}$. Rather, given G, we can regard $\mathbf{p}$ as *defined* by

$$p_x = \max_{Y \subseteq X \setminus \Gamma^*(x)} \mathbb{P}\left(A_x \Big| \bigcap_{y \in Y} \overline{A}_y\right) \tag{3.5}$$

(this is clearly the minimal choice). There is then a tradeoff in the choice of G: adding more edges reduces p_x (since there are fewer conditional probabilities to control) but also shrinks the set $\mathcal{R}(G)$ (see [26]).

(2) Though (3.2) is the *weak* hypothesis of the lopsided Lovász Local Lemma (Theorem 4.2), we will prove in (a) and (b) that the extremal families $(B_x)_{x \in X}$ and $(B'_x)_{x \in X}$ have G as a *strong* dependency graph. Therefore, all three dependency hypotheses lead to the same optimal lower bound on $\mathbb{P}(\bigcap_{x \in X} \overline{A}_x)$.

(3) The proofs given here of Theorems 3.1 and 3.2 are logically independent of nearly all of Theorem 2.2. More precisely, if we were to define $\mathcal{R}(G)$ by condition (b) of Theorem 2.2, then the only part of Theorem 2.2 that is used in the proofs of Theorems 3.1 and 3.2 is the (relatively easy) implication (b) $\Longrightarrow$ (f). But we have chosen to define $\mathcal{R}(G)$ instead by condition (c), in order to emphasize the connection with the *complex* zeros of the partition function.

Proof. For $\mathbf{p} \in \mathcal{R}(G)$, we wish to define a family of events $(B_x)_{x \in X}$ (on a new probability space) such that the hypotheses of the theorem are satisfied and $\mathbb{P}(\bigcap_{x \in X} \overline{B}_x)$ is as small as possible. An intuitively reasonable way to do this is to make the events B_x as disjoint as possible, consistent with the condition (3.2) (or with either of the two stronger notions of dependency graph). With this in mind, for $\Lambda \subseteq X$ let us define

$$\mathbb{P}\left(\bigcap_{x \in \Lambda} B_x\right) = \begin{cases} \prod_{x \in \Lambda} p_x & \text{if } \Lambda \text{ is independent in } G, \\ 0 & \text{otherwise.} \end{cases} \tag{3.6}$$

This defines a signed measure on the σ-algebra generated by $(B_x)_{x \in X}$; indeed, inclusion-exclusion gives

$$\mathbb{P}\left(\bigcap_{x \in \Lambda} B_x \cap \bigcap_{x \notin \Lambda} \overline{B}_x\right) = \sum_{I \supseteq \Lambda} (-1)^{|I|-|\Lambda|} \mathbb{P}\left(\bigcap_{x \in I} B_x\right) \quad (3.7\text{a})$$

$$= \sum_{I \supseteq \Lambda, I \text{ independent}} (-1)^{|I|-|\Lambda|} \prod_{x \in I} p_x \quad (3.7\text{b})$$

$$= (-1)^{|\Lambda|} Z_G(-\mathbf{p}; \Lambda), \quad (3.7\text{c})$$

where $Z_G(-\mathbf{p}; \Lambda)$ is defined as in (2.17). In particular, taking $\Lambda = \emptyset$, we have $\mathbb{P}(\bigcap_{x \in X} \overline{B}_x) = Z_G(-\mathbf{p})$. Now since $\mathbf{p} \in \mathscr{R}(G)$, condition (c) (and hence all the conditions) of Theorem 2.2 is satisfied. Thus Theorem 2.2(f) implies that (3.7c) is nonnegative for all Λ, so that (3.6) defines a probability measure on $\sigma(B_x : x \in X)$. (This is the probability measure defined in Theorem 2.2(g).)

If Y and Z are disjoint subsets of X such that G contains no edges between Y and Z, it follows from (3.6) that for $Y_0 \subseteq Y$ and $Z_0 \subseteq Z$ the events $\bigcap_{x \in Y_0} B_x$ and $\bigcap_{x \in Z_0} B_x$ are independent. This implies (see, for instance, [37, Theorem 4.2] or [4, Theorem 4.2]) that $\sigma(B_x : x \in Y)$ and $\sigma(B_x : x \in Z)$ are independent, and so G is a strong dependency graph.

We next show that $(B_x)_{x \in X}$ is a family minimizing $\mathbb{P}(\bigcap_{x \in X} \overline{B}_x)$. For $\Lambda \subseteq X$, we define

$$P_\Lambda = \mathbb{P}\left(\bigcap_{x \in \Lambda} \overline{A}_x\right), \quad (3.8)$$

$$Q_\Lambda = \mathbb{P}\left(\bigcap_{x \in \Lambda} \overline{B}_x\right). \quad (3.9)$$

Let us now prove by induction on $|\Lambda|$ that P_Λ/Q_Λ is monotone increasing in Λ. Note first that by inclusion-exclusion,

$$Q_\Lambda = \sum_{I \subseteq \Lambda} (-1)^{|I|} \mathbb{P}\left(\bigcap_{x \in I} B_x\right) \quad (3.10\text{a})$$

$$= \sum_{I \subseteq \Lambda, I \text{ independent}} (-1)^{|I|} \prod_{x \in I} p_x \quad (3.10\text{b})$$

$$= Z_G(-\mathbf{p}\, \mathbf{1}_\Lambda). \quad (3.10\text{c})$$

Thus $Q_\Lambda > 0$ for all Λ, since $\mathbf{p} \in \mathscr{R}(G)$ and $\mathscr{R}(G)$ is a down-set. Furthermore, for $y \notin \Lambda$,

$$Q_{\Lambda \cup \{y\}} = \sum_{I \subseteq \Lambda \cup \{y\}, I \text{ independent}} (-1)^{|I|} \prod_{x \in I} p_x \quad (3.11\text{a})$$

$$= Q_\Lambda - p_y \sum_{I \subseteq \Lambda \setminus \Gamma(y), I \text{ independent}} (-1)^{|I|} \prod_{x \in I} p_x \quad (3.11\text{b})$$

$$= Q_\Lambda - p_y Q_{\Lambda \setminus \Gamma(y)}. \quad (3.11\text{c})$$

(Note that this is just the fundamental identity (2.7) applied to $Z_G(-\mathbf{p}\mathbf{1}_\Lambda)$.) On the other hand,

$$P_{\Lambda \cup \{y\}} = P_\Lambda - \mathbb{P}\left(A_y \cap \bigcap_{x \in \Lambda} \overline{A}_x\right) \tag{3.12a}$$

$$\geq P_\Lambda - \mathbb{P}\left(A_y \cap \bigcap_{x \in \Lambda \setminus \Gamma(y)} \overline{A}_x\right) \tag{3.12b}$$

$$\geq P_\Lambda - p_y P_{\Lambda \setminus \Gamma(y)} \tag{3.12c}$$

by the hypothesis (3.2). Now we want to show that $P_{\Lambda \cup \{y\}}/Q_{\Lambda \cup \{y\}} \geq P_\Lambda/Q_\Lambda$, or equivalently that $P_{\Lambda \cup \{y\}} Q_\Lambda - Q_{\Lambda \cup \{y\}} P_\Lambda \geq 0$. By (3.11) and (3.12) we have

$$P_{\Lambda \cup \{y\}} Q_\Lambda - Q_{\Lambda \cup \{y\}} P_\Lambda \geq [P_\Lambda - p_y P_{\Lambda \setminus \Gamma(y)}] Q_\Lambda - [Q_\Lambda - p_y Q_{\Lambda \setminus \Gamma(y)}] P_\Lambda \tag{3.13a}$$

$$= p_y [P_\Lambda Q_{\Lambda \setminus \Gamma(y)} - Q_\Lambda P_{\Lambda \setminus \Gamma(y)}] \tag{3.13b}$$

$$\geq 0 \tag{3.13c}$$

since

$$\frac{P_\Lambda}{Q_\Lambda} \geq \frac{P_{\Lambda \setminus \Gamma(y)}}{Q_{\Lambda \setminus \Gamma(y)}} \tag{3.14}$$

by the inductive hypothesis.

Since P_Λ/Q_Λ is monotone increasing in Λ, we have $P_X/Q_X \geq P_\varnothing/Q_\varnothing = 1$, which proves (3.3). More generally, for any subsets $Y, Z \subseteq X$, we have $P_{Y \cup Z}/Q_{Y \cup Z} \geq P_Z/Q_Z$ and hence $P_{Y \cup Z}/P_Z \geq Q_{Y \cup Z}/Q_Z$, which gives (3.4).

For $\mathbf{p} \notin \mathscr{R}(G)$, choose a minimal vector $\mathbf{p}' \leq \mathbf{p}$ such that $\mathbf{p}' \geq 0$ and $Z_G(-\mathbf{p}') = 0$ (such a $\mathbf{p}'$ is in general non-unique). Then the family of events $(B'_x)_{x \in X}$ defined by (3.6) with p_x replaced by p'_x satisfies $\mathbb{P}(\bigcap_{x \in X} \overline{B}'_x) = Z_G(-\mathbf{p}') = 0$ (by (3.7c) with $\Lambda = \varnothing$). Since $\mathbf{p}'$ is in the closure of $\mathscr{R}(G)$, it follows by the minimality of $\mathbf{p}'$ and the continuity of Z_G that this is a well-defined probability measure; note that if x and y are adjacent then $\mathbb{P}(B'_x \cap B'_y) = 0$ by (3.6). Thus we have constructed a collection of events satisfying part (b)(ii) of the theorem.

To construct a collection of events satisfying part (b)(i), let $(C_x)_{x \in X}$ be an (independent) collection of independent events satisfying

$$[1 - \mathbb{P}(B'_x)][1 - \mathbb{P}(C_x)] = 1 - p_x. \tag{3.15}$$

Then the events $B_x = B'_x \cup C_x$ satisfy $\mathbb{P}(B_x) = p_x$ and $\mathbb{P}(\bigcap \overline{B}_x) \leq \mathbb{P}(\bigcap \overline{B}'_x) = 0$. $\square$

Remarks. (1) If $(A_x)_{x \in X}$ is a family of events satisfying (3.3) with equality, then we have $P_X = Q_X$ in the foregoing proof; and since $P_\varnothing = Q_\varnothing = 1$, the monotonicity of P_Λ/Q_Λ implies that we have $P_\Lambda = Q_\Lambda$ for every $\Lambda \subseteq X$. Thus, if $(A_x)_{x \in X}$ is an extremal family, the probabilities of all events in $\sigma(A_x : x \in X)$ are completely determined and are given by (3.6)/(3.7).

(2) More generally, dependencies between events can be expressed in terms of a dependency *digraph*: each event A_x is independent from the σ-algebra $\sigma(A_y : y \in X \setminus \Gamma^*_+(x))$,

where $\Gamma_+^*(x) = \Gamma_+(x) \cup \{x\}$ and $\Gamma_+(x)$ is the out-neighbourhood of x. See, *e.g.*, [2, Lemma 5.1.1] or [5, Theorem 1.17]. It would be interesting to have a digraph analogue of Theorem 3.1, but we do not know how to do this.

3.2. Soft-core version

Let us now consider how to extend Theorem 3.1 to the more general case of a soft-core pair interaction, *i.e.*, to allow 'soft edges' xy of strength $1 - W(x, y) \in [0, 1]$. The first step here is to replace the hard-core dependency condition (3.2) by an appropriate soft-core version.

Let $W: X \times X \to [0, 1]$ be symmetric and satisfy $W(x, x) = 0$ for all $x \in X$; and let $(A_x)_{x \in X}$ be a collection of events in some probability space. For each $x \in X$, let S_x be a random subset of X, independent of the σ-algebra $\sigma(A_x : x \in X)$, defined by the probabilities

$$\mathbb{P}(y \in S_x) = W(x, y) \qquad (3.16)$$

independently for each $y \in X$. (Thus in the case of a hard-core pair interaction, we have $S_x = X \setminus \Gamma^*(x)$ with probability 1.) Let $(p_x)_{x \in X}$ be real numbers in $[0, 1]$. We say that $(A_x)_{x \in X}$ satisfies the *weak dependency conditions with interaction W and probabilities* $(p_x)_{x \in X}$ if, for each $x \in X$ and each $Y \subseteq X \setminus x$ we have

$$\mathbb{E}\left(\mathbb{P}\left(A_x \cap \bigcap_{y \in Y \cap S_x} \overline{A}_y\right)\right) \leqslant p_x \mathbb{E}\left(\mathbb{P}\left(\bigcap_{y \in Y \cap S_x} \overline{A}_y\right)\right), \qquad (3.17)$$

where the expectations are taken over the random choice of subset S_x. (Note that in the special case of a hard-core pair interaction, we have $Y \cap S_x = Y \setminus \Gamma^*(x)$ with probability 1, so that (3.17) reduces to (3.2).) Of course, the reference here to a random subset S_x can be replaced by an explicit expression for the probabilities $\mathbb{P}(Y \cap S_x = Y')$, so that (3.17) is equivalent to

$$\sum_{Y' \subseteq Y} \left(\prod_{y \in Y'} W(x, y)\right) \left(\prod_{y \in Y \setminus Y'} [1 - W(x, y)]\right) \mathbb{P}\left(A_x \cap \bigcap_{y \in Y'} \overline{A}_y\right) \leqslant$$
$$p_x \sum_{Y' \subseteq Y} \left(\prod_{y \in Y'} W(x, y)\right) \left(\prod_{y \in Y \setminus Y'} [1 - W(x, y)]\right) \mathbb{P}\left(\bigcap_{y \in Y'} \overline{A}_y\right). \qquad (3.18)$$

We also replace (3.1) by the definition

$$\mathscr{R}(W) = \{\mathbf{R} \in [0, \infty)^X : Z_W(\mathbf{w}) \neq 0 \; \forall \mathbf{w} \in \overline{D}_\mathbf{R}\}. \qquad (3.19)$$

We can now state a soft-core version of Theorem 3.1.

Theorem 3.2 (The equivalence theorem, soft-core case). *Let $(A_x)_{x \in X}$ be a family of events in some probability space, and let $W : X \times X \to [0, 1]$ be symmetric and satisfy $W(x, x) = 0$ for all $x \in X$. Suppose that $(A_x)_{x \in X}$ satisfies the weak dependency conditions (3.17)/(3.18) with interaction W and probabilities $(p_x)_{x \in X}$.*

(a) *If* $\mathbf{p} \in \mathcal{R}(W)$, *then*

$$\mathbb{P}\left(\bigcap_{x \in X} \overline{A}_x\right) \geq Z_W(-\mathbf{p}) > 0 \tag{3.20}$$

and more generally

$$\mathbb{P}\left(\bigcap_{x \in Y} \overline{A}_x \Big| \bigcap_{x \in Z} \overline{A}_x\right) \geq \frac{Z_W(-\mathbf{p} \mathbf{1}_{Y \cup Z})}{Z_W(-\mathbf{p} \mathbf{1}_Z)} > 0 \tag{3.21}$$

for any subsets $Y, Z \subseteq X$. *Furthermore, this bound is best possible in the sense that there exists a family* $(B_x)_{x \in X}$ *with probabilities* $\mathbb{P}(B_x) = p_x$ *that satisfies the weak dependency conditions* (3.17)/(3.18) *with interaction* W *and probabilities* $(p_x)_{x \in X}$, *has strong dependency graph* G_W, *and has* $\mathbb{P}(\bigcap_{x \in X} \overline{B}_x) = Z_W(-\mathbf{p})$.

(b) *If* $\mathbf{p} \notin \mathcal{R}(W)$, *then there exists a probability space on which there can be constructed:*

(i) *a family of events* $(B_x)_{x \in X}$ *with probabilities* $\mathbb{P}(B_x) = p_x$ *and satisfying the weak dependency conditions* (3.17)/(3.18) *with interaction* W, *such that* $\mathbb{P}(\bigcap_{x \in X} \overline{B}_x) = 0$; *and*

(ii) *a family of events* $(B'_x)_{x \in X}$ *with probabilities* $\mathbb{P}(B'_x) = p'_x \leq p_x$ *and satisfying the weak dependency conditions* (3.17)/(3.18) *with interaction* W, *such that* $\mathbb{P}(B'_x \cap B'_y) = W(x, y)\mathbb{P}(B'_x)\mathbb{P}(B'_y)$ *for all* x, y *and* $\mathbb{P}(\bigcap_{x \in X} \overline{B}'_x) = 0$.

The proof of Theorem 3.2 is similar to that of Theorem 3.1; details can be found in [26].

4. Dependency graphs and the lattice gas: Sufficient conditions

In this section we shall consider sufficient conditions on a set of radii $\mathbf{R} = \{R_x\}_{x \in X}$ so that the partition function $Z_W(\mathbf{w})$ is nonvanishing in the closed polydisc $|\mathbf{w}| \leq \mathbf{R}$. Our main tool will be the fundamental identity (2.5), applied inductively.

4.1. The Lovász Local Lemma

Let G be a graph with vertex set X. Recall that G is a *dependency graph* for the family $(A_x)_{x \in X}$ if, for each $x \in X$, the event A_x is independent from the σ-algebra generated by the events $\{A_y : y \in X \setminus \Gamma^*(x)\}$. Erdős and Lovász [14] proved the following fundamental result.

Theorem 4.1 (Lovász Local Lemma). *Let G be a dependency graph for the family of events* $(A_x)_{x \in X}$, *and suppose that* $(r_x)_{x \in X}$ *are real numbers in* $[0, 1)$ *such that, for each x,*

$$\mathbb{P}(A_x) \leq r_x \prod_{y \in \Gamma(x)} (1 - r_y). \tag{4.1}$$

Then $\mathbb{P}(\bigcap_{x \in X} \overline{A}_x) \geq \prod_{x \in X}(1 - r_x) > 0$.

Erdős and Spencer [15] (see also [2, 23]) later noted that the same conclusion holds even if A_x and $\sigma(A_y : y \in X \setminus \Gamma^*(x))$ are not independent, provided that the 'harmful' conditional probabilities are suitably bounded. More precisely:

Theorem 4.2 (Lopsided Lovász Local Lemma). *Let $(A_x)_{x \in X}$ be a family of events on some probability space, and let G be a graph with vertex set X. Suppose that $(r_x)_{x \in X}$ are real numbers in $[0, 1)$ such that, for each x and each $Y \subseteq X \setminus \Gamma^*(x)$, we have*

$$\mathbb{P}\left(A_x \Big| \bigcap_{y \in Y} \overline{A}_y\right) \leq r_x \prod_{y \in \Gamma(x)} (1 - r_y). \qquad (4.2)$$

Then $\mathbb{P}(\bigcap_{x \in X} \overline{A}_x) \geq \prod_{x \in X} (1 - r_x) > 0$.

In fact, the arguments of [14, 15] (see also [32, 33]) show that in Theorems 4.1 and 4.2 a slightly stronger conclusion holds: for all pairs Y, Z of subsets of X we have

$$\mathbb{P}\left(\bigcap_{x \in Y} \overline{A}_x \Big| \bigcap_{x \in Z} \overline{A}_x\right) \geq \prod_{x \in Y \setminus Z} (1 - r_x). \qquad (4.3)$$

The Lovász Local Lemma has proved incredibly useful in probabilistic combinatorics. However, one limitation of the result is that it does not take into account the 'strength' of dependence. Our aim in this section is to relate the Lovász Local Lemma to Dobrushin's Theorem (presented in Section 4.2) and to discuss some consequences. In particular, we present an extension of the Lovász Local Lemma to the context of 'weak' dependence (Theorem 4.6). Here the precise definition of 'weak dependence' is essentially forced upon us by Theorem 3.2, and it may be a little difficult to use in practice. It would be very interesting to see some concrete applications of Theorem 4.6.

4.2. Basic bound

In this section, we will provide some sufficient conditions for the nonvanishing of Z_W in a closed polydisc $\bar{D}_\mathbf{R}$, based on 'local' properties of the interaction W (or of the graph G). Results of this type have traditionally been proved [24, 11, 27, 7, 8, 10, 29, 9] by explicitly bounding the terms in the Mayer expansion (2.14); this requires some rather nontrivial combinatorics (for example, facts about partitionability together with the counting of trees). Once this is done, an immediate consequence is that Z_W is nonvanishing in any polydisc where the series for $\log Z_W$ is convergent. Dobrushin's brilliant idea [12, 13] was to prove these two results in the opposite order. First one proves, by an elementary induction on the cardinality of the state space, that Z_W is nonvanishing in a suitable polydisc (Theorem 4.3); it then follows immediately that $\log Z_W$ is analytic in that polydisc, and hence that its Taylor series (2.14) is convergent there. Let us remark that the Dobrushin–Shearer inductive method employed in Section 4 is limited, at present, to models with hard-core self-repulsion (2.2), for which Z_W is a multiaffine polynomial. It is an interesting open question to know whether this approach can be made to work without the assumption of hard-core self-repulsion.[4]

---

[4] See also Kotecký and Preiss [21] for a third approach to proving the convergence of the Mayer expansion.

Our first (and most basic) bound is due to Dobrushin [12, 13] in the case of a hard-core interaction; the generalization to a soft repulsive interaction was proved a few years ago by one of us [30]. The method of proof is, however, already implicit (in more powerful form) in Shearer [28, Theorem 2].

Theorem 4.3 (Dobrushin [12, 13], Sokal [30]). *Let X be a finite set, and let W satisfy*

(a) $0 \leqslant W(x, y) \leqslant 1$ for all $x, y \in X$,
(b) $W(x, x) = 0$ for all $x \in X$.

Let $\mathbf{R} = \{R_x\}_{x \in X} \geqslant 0$. Suppose that there exist constants $\{K_x\}_{x \in X}$ satisfying $0 \leqslant K_x < 1/R_x$ and

$$K_x \geqslant \prod_{y \neq x} \frac{1 - W(x,y)K_y R_y}{1 - K_y R_y} \tag{4.4}$$

for all $x \in X$. Then, for each subset $\Lambda \subseteq X$, $Z_\Lambda(\mathbf{w})$ is nonvanishing in the closed polydisc $\bar{D}_{\mathbf{R}} = \{\mathbf{w} \in \mathbb{C}^X : |w_x| \leqslant R_x \text{ for all } x\}$ and satisfies there

$$\left| \frac{\partial \log Z_\Lambda(\mathbf{w})}{\partial w_x} \right| \leqslant \begin{cases} \dfrac{K_x}{1 - K_x |w_x|} & \text{for all } x \in \Lambda, \\ 0 & \text{for all } x \in X \setminus \Lambda. \end{cases} \tag{4.5}$$

Moreover, if $\mathbf{w}, \mathbf{w}' \in \bar{D}_{\mathbf{R}}$ and $w'_x/w_x \in [0, +\infty]$ for each $x \in \Lambda$, then

$$\left| \log \frac{Z_\Lambda(\mathbf{w}')}{Z_\Lambda(\mathbf{w})} \right| \leqslant \sum_{x \in \Lambda} \left| \log \frac{1 - K_x |w'_x|}{1 - K_x |w_x|} \right|, \tag{4.6}$$

where on the left-hand side we take the standard branch of the log, i.e., $|\operatorname{Im} \log \cdots| \leqslant \pi$.

Remark 1. It follows from (4.4) that $K_x \geqslant 1$ and hence that $R_x < 1$.

It is convenient to rewrite Theorem 4.3 in terms of the new variables $r_x = K_x R_x$.

Corollary 4.4. *Let X be a finite set, and let W satisfy $0 \leqslant W(x, y) \leqslant 1$ for all $x, y \in X$ and $W(x, x) = 0$ for all $x \in X$. Let $\mathbf{R} = \{R_x\}_{x \in X} \geqslant 0$. Suppose that there exist constants $0 \leqslant r_x < 1$ satisfying*

$$R_x \leqslant r_x \prod_{y \neq x} \frac{1 - r_y}{1 - W(x,y) r_y} \tag{4.7}$$

for all $x \in X$. Then, for all $\mathbf{w}$ satisfying $|\mathbf{w}| \leqslant \mathbf{R}$, the partition function Z_W satisfies

$$|Z_W(\mathbf{w})| \geqslant Z_W(-\mathbf{R}) \geqslant \prod_{x \in X} (1 - r_x) > 0 \tag{4.8}$$

and more generally

$$\left| \frac{Z_W(\mathbf{w} \mathbf{1}_{Y \cup Z})}{Z_W(\mathbf{w} \mathbf{1}_Z)} \right| \geqslant \prod_{x \in Y} (1 - r_x) > 0. \tag{4.9}$$

In particular, if we define the 'maximum weighted degree'

$$\Delta_W = \max_{x \in X} \sum_{y \neq x} [1 - W(x, y)] \tag{4.10}$$

and write

$$F(\Delta_W) = \frac{2 + \Delta_W - \sqrt{\Delta_W^2 + 4\Delta_W}}{2}, \tag{4.11}$$

$$R(\Delta_W) = F(\Delta_W) e^{-[1 - F(\Delta_W)]} \tag{4.12}$$

we have

$$|Z_W(\mathbf{w})| \geq [1 - F(\Delta_W)]^{|X|} > 0 \tag{4.13}$$

whenever $|w_x| \leq R(\Delta_W)$ for all $x \in X$.

Proof. Setting $r_x = K_x R_x$, we find that (4.4) becomes (4.7), and (4.6) with $\Lambda = X$ and $w' = 0$ becomes (4.8).

To obtain the last claim, note first that

$$\frac{1-r}{1-Wr} = \frac{1-r}{1-r+(1-W)r} = \frac{1}{1+(1-W)\frac{r}{1-r}} \geq e^{-(1-W)r/(1-r)} \tag{4.14}$$

whenever $0 \leq W \leq 1$ and $0 \leq r \leq 1$. Therefore, if we set $r_x = r$ for all $x \in X$, we have

$$r_x \prod_{y \neq x} \frac{1 - r_y}{1 - W(x, y) r_y} \geq r e^{-\Delta_W r/(1-r)}. \tag{4.15}$$

We then choose r to maximize the right-hand side of (4.15); simple calculus yields $\Delta_W r = (1-r)^2$ and $r = F(\Delta_W)$, so that the right-hand side of (4.15) is bounded below by $R(\Delta_W)$. It follows that if we define $R_x = R(\Delta_W)$ and $r_x = F(\Delta_W)$ for all $x \in X$ then (4.7) and so (4.8) are satisfied. □

Remarks. (1) The radius $R(\Delta_W)$ behaves as

$$R(\Delta_W) = \begin{cases} 1 - 2\Delta_W^{1/2} + \frac{5}{2}\Delta_W + O(\Delta_W^{3/2}) & \text{as } \Delta_W \to 0, \\ \frac{1}{e\Delta_W}[1 - \frac{1}{\Delta_W} + \frac{3}{2\Delta_W} + O(\Delta_W^{-3})] & \text{as } \Delta_W \to \infty. \end{cases} \tag{4.16}$$

Example 4.6 (the r-ary rooted tree) shows that this bound is sharp (to leading order) as $\Delta_W \to \infty$. At the other extreme, the $1 - \text{const} \times \Delta_W^{1/2}$ behaviour at small Δ_W is also best possible, since the two-site lattice gas with $W(x, x) = W(y, y) = 0$ and $W(x, y) = 1 - \epsilon$ has $Z_W(w) = 1 + 2w + (1-\epsilon)w^2$ and hence has a root at $w = -1/(1 + \sqrt{\epsilon})$. (However, the coefficient 2 rather than 1 in the $\Delta_W^{1/2}$ term of (4.16) may not be best possible.)

Specializing Corollary 4.4 to the case of a hard-core pair interaction for a graph G,

$$W(x, y) = \begin{cases} 0 & \text{if } x = y \text{ or } xy \in E(G), \\ 1 & \text{if } x \neq y \text{ and } xy \notin E(G), \end{cases} \tag{4.17}$$

we have the following result.

Corollary 4.5. *Let G be a finite graph with vertex set X, and let* $\mathbf{R} = \{R_x\}_{x \in X} \geqslant 0$. *Suppose that there exist constants* $0 \leqslant r_x < 1$ *satisfying*

$$R_x \leqslant r_x \prod_{y \in \Gamma(x)} (1 - r_y) \qquad (4.18)$$

for all $x \in X$. *Then, for all* $\mathbf{w}$ *satisfying* $|\mathbf{w}| \leqslant \mathbf{R}$, *the independent-set polynomial* Z_G *satisfies*

$$|Z_G(\mathbf{w})| \geqslant Z_G(-\mathbf{R}) \geqslant \prod_{x \in X}(1 - r_x) > 0 \qquad (4.19)$$

and more generally

$$\left| \frac{Z_G(\mathbf{w}\,\mathbf{1}_{Y \cup Z})}{Z_G(\mathbf{w}\,\mathbf{1}_Z)} \right| \geqslant \prod_{x \in Y}(1 - r_x) > 0. \qquad (4.20)$$

In particular, if G has maximum degree Δ, *then* $|Z_G(\mathbf{w})| \geqslant [\Delta/(\Delta+1)]^{|X|} > 0$ *whenever* $|w_x| \leqslant \Delta^\Delta/(\Delta+1)^{\Delta+1}$ *for all* $x \in X$.

Proof. The last claim is obtained by setting $r_x = 1/(\Delta + 1)$ for all $x \in X$. □

Remark. The radius $\Delta^\Delta/(\Delta+1)^{\Delta+1}$ behaves for large Δ as

$$\frac{\Delta^\Delta}{(\Delta+1)^{\Delta+1}} = \frac{1}{e\Delta}\left[1 - \frac{1}{2\Delta} + \frac{7}{24\Delta^2} - \frac{3}{16\Delta^3} + O(\Delta^{-4})\right], \qquad (4.21)$$

which agrees with (4.16) to leading order in $1/\Delta$ but is slightly larger (hence better) at order $1/\Delta^2$.

Combining Corollary 4.5 with Theorem 3.1, we immediately obtain the lopsided Lovász Local Lemma (Theorem 4.2). It is equally possible to go in the opposite direction, and deduce the Dobrushin bounds from the Lovász Local Lemma.

Let us remark that we have been able to relate the Lovász Local Lemma to a combinatorial polynomial (namely, the independent-set polynomial) only in the case of an *undirected* dependency graph G. Although the Local Lemma can be formulated quite naturally for a dependency *digraph* [2, 5, 23], we do not know whether the digraph Lovász problem can be related to any combinatorial polynomial. (Clearly the independent-set polynomial cannot be the right object in the digraph context, since exclusion of simultaneous occupation is manifestly a symmetric condition.)

The results also allow us to deduce a 'soft-core' version of the lopsided Lovász Local Lemma. Combining Corollary 4.4 with Theorem 3.2, we obtain the following result.

Theorem 4.6. *Let* $(A_x)_{x \in X}$ *be a family of events in some probability space, and let* $W: X \times X \to [0, 1]$ *be symmetric and satisfy* $W(x, x) = 0$ *for all* $x \in X$. *Suppose that* $(A_x)_{x \in X}$ *satisfies the weak dependency conditions* (3.17)/(3.18) *with interaction W and probabilities* $(p_x)_{x \in X}$. *Suppose further that* $(r_x)_{x \in X}$ *are real numbers in* $[0, 1)$ *satisfying*

$$p_x \leqslant r_x \prod_{y \in \Gamma(x)} (1 - r_y). \qquad (4.22)$$

Then

$$\mathbb{P}\left(\bigcap_{x \in X} \overline{A}_x\right) \geq \prod_{x \in X}(1-r_x) > 0, \tag{4.23}$$

and more generally for sets $Y, Z \subseteq X$, we have

$$\mathbb{P}\left(\bigcap_{x \in Y} \overline{A}_x \Big| \bigcap_{x \in Z} \overline{A}_x\right) \geq \prod_{x \in Y \setminus Z}(1-r_x) > 0. \tag{4.24}$$

Defining the weighted degree Δ_W as in (4.10), we obtain the following.

Lemma 4.7. *Let $(A_x)_{x \in X}$ satisfy the weak dependency conditions (3.17)/(3.18) with interaction W and probabilities $(p_x)_{x \in X}$. If $p_x < \Delta_W^{\Delta_W}/(\Delta_W + 1)^{\Delta_W + 1}$ for every $x \in X$, then $\mathbb{P}(\bigcap_{x \in X} \overline{A}_x) > 0$.*

Proof. As in the proof of Corollary 4.5, set $r_x = r \equiv 1/(\Delta_W + 1)$ for all $x \in X$. Then check (4.7):

$$r_x \prod_{y \neq x} \frac{1-r_y}{1-W(x,y)r_y} \leq r_x \prod_{y \neq x}(1-r_y)^{1-W(x,y)}$$

$$\leq r(1-r)^{\Delta_W}$$

$$\leq \frac{\Delta_W^{\Delta_W}}{(\Delta_W + 1)^{\Delta_W + 1}}. \tag{4.25}$$

In the first inequality we have used the fact that $1 - W(x,y)r_y \leq (1-r_y)^{W(x,y)}$ for $0 \leq W(x,y) \leq 1$. $\square$

As noted above, it would be interesting to see applications of Theorem 4.6 and Lemma 4.7.

4.3. Improved bound

Finally in this section, we note that Theorem 4.3 can be slightly sharpened. Note, first of all, that we need not insist that the bound (4.5) hold with the *same* constant K_x for all $\Lambda \ni x$; rather, we can use constants $K_{x,\Lambda}$ that depend on Λ.

Let us define the constants $K_{x,\Lambda} \in [0, +\infty]$ as a function of the family $\{R_x\}$ by the recursion

$$K_{x,\Lambda} = \prod_{\substack{y \in \Lambda \setminus x \\ W(x,y) \neq 1 \\ R_y > 0}} \frac{1 - W(x,y) K_{y,\Lambda \setminus x} R_y}{1 - K_{y,\Lambda \setminus x} R_y} \tag{4.26}$$

if $K_{y,\Lambda \setminus x} R_y < 1$ for all terms in the product, and $K_{x,\Lambda} = +\infty$ otherwise.

We define a graph G with vertex set $V = \{x \in X : R_x > 0\}$ and edge set $E = \{x, y \in V : W(x,y) \neq 1\}$; and for each $\Lambda \subseteq X$, let G_Λ be the subgraph of G induced by $\Lambda \cap V$. Then only the connected component of G_Λ containing x plays any role in the definition of $K_{x,\Lambda}$: that is, if G_Λ has several connected components with vertex sets $\Lambda_1, \ldots, \Lambda_k$ and $x \in \Lambda_i$, then $K_{x,\Lambda} = K_{x,\Lambda_i}$.

Let us now call a pair (x, Λ) 'good' if $K_{x,\Lambda} < \infty$ and $K_{x,\Lambda} R_x < 1$. It is easily shown that if (x, Λ) is good, then $(y, \Lambda \setminus x)$ is also good whenever $y \in \Lambda \setminus x$ with $W(x, y) \neq 1$ and $R_y > 0$, i.e., whenever y is a neighbour of x in G_Λ. (Indeed, this follows under the weaker hypothesis that $K_{x,\Lambda} < \infty$.)

The following result can then be proved (see [26] for a proof).

Theorem 4.8 (Improved Dobrushin–Shearer bound). *Let X be a finite set, and let W satisfy*

(a) $0 \leqslant W(x, y) \leqslant 1$ for all $x, y \in X$,
(b) $W(x, x) = 0$ for all $x \in X$.

Let $\mathbf{R} = \{R_x\}_{x \in X} \geqslant 0$. Define the constants $K_{x,\Lambda} \in [0, +\infty]$ as above. Suppose that in each connected component of G_Λ there exists at least one vertex x for which the pair (x, Λ) is good. Then $Z_\Lambda(\mathbf{w})$ is nonvanishing in the closed polydisc $\bar{D}_\mathbf{R}$; and for every good pair (x, Λ) and every $\mathbf{w} \in \bar{D}_\mathbf{R}$, we have

$$\left| \frac{\partial \log Z_\Lambda(\mathbf{w})}{\partial w_x} \right| \leqslant \frac{K_{x,\Lambda}}{1 - K_{x,\Lambda} |w_x|}. \tag{4.27}$$

Moreover, if $\mathbf{w}, \mathbf{w}' \in \bar{D}_\mathbf{R}$ and $w'_x / w_x \in [0, +\infty]$ for each $x \in \Lambda$, and in addition the pair (x, Λ) is good whenever $w'_x \neq w_x$, then

$$\left| \log \frac{Z_\Lambda(\mathbf{w}')}{Z_\Lambda(\mathbf{w})} \right| \leqslant \sum_{\substack{x \in \Lambda \\ w'_x \neq w_x}} \left| \log \frac{1 - K_{x,\Lambda} |w'_x|}{1 - K_{x,\Lambda} |w_x|} \right|, \tag{4.28}$$

where on the left-hand side we take the standard branch of the log, i.e., $|\operatorname{Im} \log \cdots| \leqslant \pi$.

As a corollary of Theorem 4.8, we can deduce a bound due originally (in the Lovász context) to Shearer [28, Theorem 2], which improves the last sentence of Corollary 4.5 by replacing Δ by $\Delta - 1$. Indeed, we can very slightly improve Shearer's bound by allowing one vertex x_0 to have a larger radius R_{x_0}.

Corollary 4.9. *Let $G = (X, E)$ be a finite graph of maximum degree $\Delta \geqslant 2$, and fix one vertex $x_0 \in X$. Suppose that $|w_{x_0}| \leqslant (\Delta - 1)^\Delta / \Delta^\Delta$ and that $|w_x| \leqslant (\Delta - 1)^{\Delta-1} / \Delta^\Delta$ for all $x \neq x_0$. Then $Z_G(\mathbf{w}) \neq 0$.*

Proof. Since Z_G factorizes over connected components, we can assume without loss of generality that G is connected. (Indeed, if G is disconnected, then we can allow one 'x_0-like' vertex in *each* connected component.) Set $R_{x_0} = (\Delta - 1)^\Delta / \Delta^\Delta$ and $R_x = (\Delta - 1)^{\Delta-1} / \Delta^\Delta$ for all $x \neq x_0$.

We first claim that if $x_0 \notin \Lambda$, and $x \in \Lambda$ is a vertex with at least one neighbour in $X \setminus \Lambda$, then

$$K_{x,\Lambda} < \left(\frac{\Delta}{\Delta - 1} \right)^{\Delta - 1} \tag{4.29}$$

(note the strict inequality). The proof is by induction on $|\Lambda|$, using the definition (4.26). It certainly holds if $\Lambda = \{x\}$. For general Λ, note first that since every y appearing in the product on the right-hand side of (4.26) has at least one neighbour outside of $\Lambda \setminus x$ (namely, x itself), $K_{y,\Lambda \setminus x}$ satisfies (4.29) by the inductive hypothesis and so $K_{y,\Lambda \setminus x} R_y < 1/\Delta$. Also, since x has at least one neighbour outside Λ, there are at most $\Delta - 1$ factors in the product. Thus

$$K_{x,\Lambda} < \left(\frac{1}{1 - 1/\Delta}\right)^{\Delta - 1} = \left(\frac{\Delta}{\Delta - 1}\right)^{\Delta - 1}. \tag{4.30}$$

It then follows that

$$K_{x_0, X} < \left(\frac{\Delta}{\Delta - 1}\right)^{\Delta}, \tag{4.31}$$

since the bound (4.29) applies to all the terms $K_{y,\Lambda \setminus x_0}$ appearing on the right-hand side of (4.26). We therefore have $K_{x_0, X} R_{x_0} < 1$, and so the pair (x_0, X) is good. The claim then follows from Theorem 4.8. □

Replacing $\Delta^\Delta / (\Delta + 1)^{\Delta + 1}$ by $(\Delta - 1)^{\Delta - 1} / \Delta^\Delta$ may seem to be a negligible improvement, since both quantities have the same leading behaviour $\approx 1/(e\Delta)$ as $\Delta \to \infty$, and differ only at higher order:

$$\frac{(\Delta - 1)^{\Delta - 1}}{\Delta^\Delta} = \frac{1}{e\Delta}\left[1 + \frac{1}{2\Delta} + \frac{7}{24\Delta^2} + \frac{3}{16\Delta^3} + O(\Delta^{-4})\right] \tag{4.32}$$

(cf. (4.21)).[5] But Shearer's bound $(\Delta - 1)^{\Delta - 1}/\Delta^\Delta$ has the great merit of being *best possible*: for, as he showed [28], if G is the complete rooted tree with branching factor $r = \Delta - 1$ and depth n, then $Z_G(w)$ has negative real zeros that tend to $w = -(\Delta - 1)^{\Delta - 1}/\Delta^\Delta$ as $n \to \infty$ (see Section 2.2 above).

We remark that Corollary 4.9 does not appear to extend naturally to the soft-core case (note that having one neighbour outside Λ in the argument around (4.29) need not reduce the weighted degree of a vertex in Λ by 1).

For additional detail and further results, as well as a discussion of the *optimal* bounds, we refer the reader to [26].

5. Conclusion

How good are the bounds given by the results in Section 4? We shall consider the diagonal case, where all radii are the same. Let us define, for a finite graph G,

$$\lambda_c(G) = \sup\{\lambda : \lambda \mathbf{1} \in \mathcal{R}(G)\}. \tag{5.1}$$

For a countably infinite graph G, we define $\lambda_c(G)$ to be the infimum of $\lambda_c(H)$ over finite induced subgraphs H of G.

For graphs G with maximum degree Δ, we know from Corollary 4.9 that $\lambda_c(G) \leq (\Delta - 1)^{\Delta - 1}/\Delta^\Delta$, and by Shearer's result this is optimal for the infinite Δ-regular tree.

---

[5] The amusing similarity of (4.21) and (4.32) arises from the fact that $-(-\Delta)^{-\Delta}/(-\Delta + 1)^{-\Delta + 1} = (\Delta - 1)^{\Delta - 1}/\Delta^\Delta$.

The bound is close to optimal for Δ-regular graphs of large girth (as can be seen from monotonicity results from [26]). However, the value of λ_c is less easy to determine for graphs with short cycles. For instance, consider the square lattice $\mathbb{Z}^2$. This is 4-regular, and so $\lambda_c \geqslant 3^3/4^4 = 0.105$, but the correct value of λ_c is not so clear. Fortunately, this problem has also been considered by physicists. Indeed, Todo [35] has given the extraordinarily precise (but nonrigorous) numerical estimate

$$\lambda_c(\mathbb{Z}^2) = 0.119\,338\,881\,88(1), \tag{5.2}$$

obtained by using transfer matrices and the phenomenological-renormalization method (a variant of finite-size scaling). It would be interesting to gain good rigorous estimates (see [26] for further discussion).

There are other probabilistic inequalities that are expressed in terms of a dependency graph (see for instance Suen [34] or Janson [20]); it would be interesting to know if any of these can be related to a combinatorial polynomial (= statistical-mechanical partition function) in a manner analogous to Theorem 3.1. However, even without such a result, there may be scope for proving further inequalities in the presence of weak dependency conditions of the form discussed in Section 3.2 above.

Finally, we note that combinatorics and statistical physics have seen a very extensive and fruitful interaction in recent years; we hope that the results in this paper indicate that much remains to be discovered.

Acknowledgements

We wish to thank Keith Ball for informing us of the important work of Shearer [28]; Keith Ball, Pierre Leroux and Joel Spencer for helpful discussions; and Synge Todo for communicating to us some of his unpublished numerical results. We would also like to thank the referees for their helpful comments.

This research was supported in part by US National Science Foundation grants PHY–9900769 and PHY–0099393 and UK Engineering and Physical Sciences Research Council grant GR/S26323/01.

References

[1] Alon, N. (1991) A parallel algorithmic version of the local lemma. *Random Struct. Alg.* **2** 367–378.

[2] Alon, N. and Spencer, J. (2000) *The Probabilistic Method*, 2nd edn, Wiley, New York.

[3] Beck, J. (1991) An algorithmic approach to the Lovász local lemma I. *Random Struct. Alg.* **2** 343–365.

[4] Billingsley, P. (1986) *Probability and Measure*, 2nd edn, Wiley, New York.

[5] Bollobás, B. (2001) *Random Graphs*, 2nd edn, Cambridge University Press, Cambridge.

[6] Borgs, C. (2003) *Expansion Methods in Combinatorics*. Preprint: to be published in the *Conference Board of the Mathematical Sciences* book series.

[7] Brydges, D. C. (1986) A short course on cluster expansions. In *Phénomènes Critiques, Systèmes Aléatoires, Théories de Jauge / Critical Phenomena, Random Systems, Gauge Theories*, Les Houches summer school, Session XLIII, 1984 (K. Osterwalder and R. Stora, eds), Elsevier/North-Holland, Amsterdam, pp. 129–183.

[8] Brydges, D. C. and Kennedy, T. (1987) Mayer expansions and the Hamilton–Jacobi equation. *J. Statist. Phys.* **48** 19–49.

[9] Brydges, D. C. and Martin, Ph. A. (1999) Coulomb systems at low density: A review. *J. Statist. Phys.* **96** 1163–1330, cond-mat/9904122 at arXiv.org.

[10] Brydges, D. C. and Wright, J. D. (1988) Mayer expansions and the Hamilton–Jacobi equation II: Fermions, dimensional reduction formulas. *J. Statist. Phys.* **51** 435–456. Erratum **97** (1999) 1027.

[11] Cammarota, C. (1982) Decay of correlations for infinite range interactions in unbounded spin systems. *Commun. Math. Phys.* **85** 517–528.

[12] Dobrushin, R. L. (1996) Estimates of semi-invariants for the Ising model at low temperatures. In *Topics in Statistical and Theoretical Physics*, Vol. 177 of *American Mathematical Society Translations*, Ser. 2, pp. 59–81.

[13] Dobrushin, R. L. (1996) Perturbation methods of the theory of Gibbsian fields. In *Lectures on Probability Theory and Statistics* Ecole d'Eté de Probabilités de Saint-Flour XXIV, 1994 (P. Bernard, ed.), Vol. 1648 of *Lecture Notes in Mathematics*, Springer, Berlin, pp. 1–66.

[14] Erdős, P. and Lovász, L. (1975) Problems and results on 3-chromatic hypergraphs and some related questions. In *Infinite and Finite Sets*, Vol. II, *Colloq. Math. Soc. Janos Bolyai*, Vol. 10, North-Holland, Amsterdam, pp. 609–627.

[15] Erdős, P. and Spencer, J. (1991) Lopsided Lovász local lemma and Latin transversals. *Discrete Appl. Math.* **30** 151–154.

[16] Garey, M. R. and Johnson, D. S. (1979) *Computers and Intractability: A Guide to the Theory of NP-Completeness*, Freeman, San Francisco.

[17] Glimm, J. and Jaffe, A. (1987) *Quantum Physics: A Functional Integral Point of View*, 2nd edn, Springer, New York.

[18] Groeneveld, J. (1962) Two theorems on classical many-particle systems. *Phys. Lett.* **3** 50–51.

[19] Hille, E. (1973) *Analytic Function Theory*, 2nd edn, Chelsea, New York.

[20] Janson, S. (1998) New versions of Suen's correlation inequality. *Random Struct. Alg.* **13** 467–483.

[21] Kotecký, R. and Preiss, D. (1986) Cluster expansion for abstract polymer models. *Commun. Math. Phys.* **103** 491–498.

[22] Milnor, J. (2000) On rational maps with two critical points. *Experiment. Math.* **9** 481–522.

[23] Molloy, M. and Reed, B. (2002) *Graph Colouring and the Probabilistic Method*, Springer, Berlin/New York.

[24] Penrose, O. (1967) Convergence of fugacity expansions for classical systems. In *Statistical Mechanics: Foundations and Applications* (T. A. Bak, ed.), Benjamin, New York/Amsterdam, pp. 101–109.

[25] Runnels, L. K. (1967) Phase transition of a Bethe lattice gas of hard molecules. *J. Math. Phys.* **10** 2081–2087.

[26] Scott, A. D. and Sokal, A. D. (2005) The repulsive lattice gas, the independent-set polynomial, and the Lovász Local Lemma. *J. Statist. Phys.* **118** 1151–1261, cond-mat/0309352 at arXiv.org.

[27] Seiler, E. G. (1982)) *Gauge Theories as a Problem of Constructive Quantum Field Theory and Statistical Mechanics*, Vol. 159 of *Lecture Notes in Physics*, Springer, Berlin/New York.

[28] Shearer, J. B. (1985) On a problem of Spencer. *Combinatorica* **5** 241–245.

[29] Simon, B. (1993) *The Statistical Mechanics of Lattice Gases*, Princeton University Press, Princeton, NJ.

[30] Sokal, A. D. (2001) Bounds on the complex zeros of (di)chromatic polynomials and Potts-model partition functions. *Combin. Probab. Comput.* **10** 41–77, cond-mat/9904146 at arXiv.org.

[31] Sokal, A. D. (2005) The multivariate Tutte polynomial (alias Potts model) for graphs and matroids. In *Surveys in Combinatorics, 2005*, (Bridget S. Webb, ed.), Cambridge University Press, pp. 173–226.

[32] Spencer, J. (1975) Ramsey's Theorem: a new lower bound. *J. Combin. Theory Ser. A* **18** 108–115.

[33] Spencer, J. (1977/78) Asymptotic lower bounds for Ramsey functions. *Discrete Math.* **20** 69–76.

[34] Suen, W. C. (1990) A correlation inequality and a Poisson limit theorem for nonoverlapping balanced subgraphs of a random graph. *Random Struct. Alg.* **1** 231–242.
[35] Todo, S. (1999) Transfer-matrix study of negative-fugacity singularity of hard-core lattice gas. *Int. J. Mod. Phys. C* **10** 517–529. cond-mat/9703176 at arXiv.org.
[36] Uhlenbeck, G. E. and Ford, G. W. (1962) The theory of linear graphs with applications to the theory of the virial development of the properties of gases. In *Studies in Statistical Mechanics*, Vol. I (J. de Boer and G. E. Uhlenbeck, eds), North-Holland, Amsterdam, pp. 119–211.
[37] Williams, D. (1991) *Probability with Martingales*, Cambridge University Press, Cambridge.
[38] Yang, C. N. and Lee, T. D. (1952) Statistical theory of equations of state and phase transitions I: Theory of condensation. *Phys. Rev.* **87** 404–409.

Solving Sparse Random Instances of Max Cut and Max 2-CSP in Linear Expected Time

ALEXANDER D. SCOTT[1] and GREGORY B. SORKIN[2]

[1]Department of Mathematics, University College London, London WC1E 6BT, UK
(e-mail: scott@math.ucl.ac.uk)
[2]Department of Mathematical Sciences, IBM T.J. Watson Research Center,
Yorktown Heights NY 10598, USA
(e-mail: sorkin@watson.ibm.com)

Received 19 September 2004; revised 7 August 2005

For Béla Bollobás on his 60th birthday

We show that a maximum cut of a random graph below the giant-component threshold can be found in linear space and linear expected time by a simple algorithm. In fact, the algorithm solves a more general class of problems, namely binary 2-variable constraint satisfaction problems. In addition to Max Cut, such Max 2-CSPs encompass Max Dicut, Max 2-Lin, Max 2-Sat, Max-Ones-2-Sat, maximum independent set, and minimum vertex cover. We show that if a Max 2-CSP instance has an 'underlying' graph which is a random graph $G \in \mathcal{G}(n,c/n)$, then the instance is solved in linear expected time if $c \leqslant 1$. Moreover, for arbitrary values (or functions) $c > 1$ an instance is solved in expected time $n \exp(O(1 + (c-1)^3 n))$; in the 'scaling window' $c = 1 + \lambda n^{-1/3}$ with λ fixed, this expected time remains linear.

Our method is to show, first, that if a Max 2-CSP has a connected underlying graph with n vertices and m edges, then $O(n 2^{(m-n)/2})$ is a deterministic upper bound on the solution time. Then, analysing the tails of the distribution of this quantity for a component of a random graph yields our result. Towards this end we derive some useful properties of binomial distributions and simple random walks.

1. Introduction

In this paper we prove that a maximum cut of a random graph below the giant-component threshold can be found in linear expected time.

Theorem 1.1. *For any $c \leqslant 1$, a maximum cut of a random graph $G \in \mathcal{G}(n,c/n)$ can be found in time whose expectation is $O(n)$, using space $O(m+n)$, where m is the size of the graph.*

We should point out the significance of requiring linear time 'in expectation' rather than just 'almost always'. With high probability, a random graph below the giant-component threshold consists solely of trees and unicyclic components, and a maximum cut in such a graph is easy to find. (It cuts all edges except for one edge in each odd cycle.) However, exponential time can be spent on finding optimal cuts in the rare multicyclic graphs, which is what makes the proof of Theorem 1.1 difficult.

Our approach is, first, to give a deterministic algorithm and an upper bound on its running time as a function of the input graph's 'excess' of edges over vertices, $m - n$.

Theorem 1.2. *Let G be a connected graph with n vertices and m edges. A maximum cut of G can be found in time $O(n 2^{(m-n)/2})$, using space $O(m + n)$.*

We then bound the distribution of the excess in a component of a sparse random graph. This enables us to bound the *expected* running time of our algorithm, and hence prove Theorem 1.1.

In fact, Theorems 1.1 and 1.2 are special cases of more general results (Theorems 7.1 and 3.3). Our algorithm employs local reductions that take us outside the class of Max Cut problems, forcing us to work with the larger class Max 2-CSP (defined and discussed in Section 2). Working in this broader class both simplifies our methods and means that our results apply not just to Max Cut but also to problems including Mix Dicut, Max 2-Lin, Max 2-Sat, Max-Ones-2-Sat, maximum independent set, minimum vertex cover, and weighted versions of these problems.

Throughout the paper, n and m are reserved for the number of vertices and edges of a graph G. By $G \in \mathcal{G}(n, p)$ as usual we denote a random graph with n vertices, where each potential edge is present with probability p, independently; we also write $G(n, p)$ as shorthand for such a graph.

1.1. Context

Our results are particularly interesting in the context of phase transitions for various maximum constraint-satisfaction problems. Since we are just situating our results, we will be informal. It is well known that a random 2-Sat formula with 'density' $c < 1$ (where the number of clauses is c times the number of variables) is satisfiable with probability tending to 1 as the number n of variables tends to infinity, while for $c > 1$, the probability of satisfiability tends to 0 as $n \to \infty$; see Chvátal and Reed [8], Goerdt [16], and Fernandez de la Vega [14]. Indeed there is now a detailed picture of the scaling window; see Bollobás, Borgs, Chayes, Kim and Wilson [7]. Max 2-Sat has since been shown to exhibit similar behaviour, so for $c < 1$, only an expected $\Theta(1/n)$ clauses go unsatisfied, while for $c > 1$, an expected $\Theta(n)$ clauses must go unsatisfied; see Coppersmith, Gamarnik, Hajiaghayi and Sorkin [11].

For a random graph $G(n, c/n)$, with $c < 1$ the graph almost surely consists solely of small trees and unicyclic components, while for $c > 1$, it almost surely contains a 'giant', complex component, of order $\Theta(n)$ (see for example Bollobás [6]). Again, [11] proves the related facts that in a maximum cut of such a graph, for $c < 1$ only an expected $\Theta(1)$ edges fail to be cut, while for $c > 1$ it is $\Theta(n)$.

For both Max 2-Sat and Max Cut, it seems likely that the mostly satisfiable (or mostly cuttable) sparse instances are algorithmically easy, while the not-so-satisfiable dense instances are algorithmically hard. While, as far as we are aware, little is known about the hardness of dense instances, our results here confirm that not only are typical sparse Max Cut instances easy, but even the atypical ones can be accommodated in linear expected time; see Section 8, Conclusions, for further discussion.

More generally, our interest here is in solving random instances of hard problems in polynomial expected time, and of course there is a substantial body of literature on this subject. For example, results on colouring random graphs in polynomial expected time can be found in Krivelevich and Vu [23], Coja-Oghlan, Moore and Sanwalini [9], and Coja-Oghlan and Taraz [10].

As already remarked, our expected-linear-time result comes from analysing an algorithm which, for arbitrary connected graphs, runs in time $O(n2^{(m-n)/2})$, deterministically. This parametrization in terms of $m-n$ is efficient for random graphs up to the giant-component threshold or even slightly beyond, because a random graph $G(n,(1+n^{-1/3})/n)$ typically has a giant component with $\Theta(n^{2/3})$ vertices and a similar number of edges, but excess only $\Theta(1)$. In a paper with a preliminary version of the present result [28], we also showed that, for any Max 2-CSP instance, the same Max 2-CSP algorithm has running time $\tilde{O}(2^{m/5})$, where the $\tilde{O}$ notation hides polynomial factors. Previously, Niedermeier and Rossmanith showed that Max 2-Sat could be solved in time $\tilde{O}(2^{0.347m})$ [27]; Hirsch improved this to $\tilde{O}(2^{m/4})$ [18]; Gramm, Hirsch, Niedermeier and Rossmanith improved it to $\tilde{O}(2^{m/5})$ and adapted it to solve Max Cut in time $\tilde{O}(2^{m/3})$ [17]; Fedin and Kulikov improved the Max Cut result to $\tilde{O}(2^{m/5})$ [24]; and in a forthcoming paper we refine our present algorithm and analysis to obtain time $\tilde{O}(2^{19m/100})$ for any Max 2-CSP [29].

1.2. Outline of proof

Our main result will be Theorem 1.1, generalized from $c \leqslant 1$ to a larger range, and from Max Cut to the class Max 2-CSP (to be defined in Section 2). Its proof has a few principal components. Since the maximum cut of a graph is the combination of maximum cuts of each of its connected components (and the same is true for any Max 2-CSP), it suffices to bound the expected time the algorithm spends on the component containing a fixed vertex.

In order to bound the expected running time of 'Algorithm A' (introduced in Section 3) we must control the distribution of the excess of a component of a random graph. This is done by 'exploring' the component as a branching process, dominating it with a similar process, and analysing the latter as a random walk. We obtain stochastic bounds on the component order u and, conditioned upon u, the 'width' w (defined later), and finally the excess, which is dominated by a binomial random variable $B(uw, c/n)$.

Finally, we combine the running times, which are exponentially large in the excess, with the exponentially small large-deviation bounds on the excess, to show that Algorithm A runs in linear expected time.

We could also have tried to bound the expected running time of Algorithm A by characterizing the excess using estimates for $C(n, n+k)$, the number of connected graphs on n vertices with $n+k$ edges. Such estimates are given by Bollobás [5], Łuczak [25] and Bender, Canfield and McKay [3], but they seem not to be immediately suitable for our purposes; we discuss this more extensively in Section 6.

2. Max 2-CSP

The problem Max Cut is to partition the vertices of a given graph into two classes so as to maximize the number of edges 'cut' by the partition. Think of each edge as being a function on the classes or 'colours' of its endpoints, with value 1 if the endpoints are of different colours, 0 if they are the same: Max Cut is equivalent to finding a 2-colouring of the vertices which maximizes the sum of these edge functions. This view naturally suggests a generalization.

An *instance* (G, S) of Max 2-CSP is given by an 'underlying' graph $G = (V, E)$ and a set S of 'score' functions. Writing $\{R, B\}$ for the colours Red and Blue, for each edge $e \in E$ there is a 'dyadic' score function $s_e : \{R, B\}^2 \to \mathbb{R}$, for each vertex $v \in V$ there is a 'monadic' score function $s_v : \{R, B\} \to \mathbb{R}$, and finally there is a single 'niladic' score function $s_0 : \{R, B\}^0 \to \mathbb{R}$ which takes no arguments and is just a constant convenient for bookkeeping. We allow an instance to have parallel edges (and further such edges may be generated while the algorithm runs).

A potential *solution* is a 'colouring' of the vertices, *i.e.*, a function $\phi : V \to \{R, B\}$, and an optimum solution is one which maximizes

$$s(\phi) := s_0 + \sum_{v \in V} s_v(\phi(v)) + \sum_{uv \in E} s_{uv}(\phi(u), \phi(v)). \tag{2.1}$$

Without belabouring the notation for edges, we wish to take each edge just once, and (since s_{uv} need not be a symmetric function) with a fixed notion of which endpoint is 'u' and which is 'v'. We will typically assume that $V = [n]$ and any edge uv is really an ordered pair (u, v) with $1 \leqslant u < v \leqslant n$. We remark that the '2' in the name Max 2-CSP refers to the fact that the score functions take 2 or fewer arguments (3-Sat, for example, is out of scope); replacing 2 by a larger value would mean replacing the underlying graph with a hypergraph.

Our assumption of an undirected underlying graph is sound even for a problem such as Max Dicut (maximum directed cut). Here one normally thinks of a directed edge as cut only if its head has colour 0 and its tail has colour 1, but for a directed edge (v, u) with $v > u$ this may be expressed by the undirected (or, equivalently, canonically directed) edge (u, v) with score 1 if $(\phi(u), \phi(v)) = (1, 0)$, and score 0 otherwise. That is, instead of directing the *edges* we incorporate the direction into the score functions. (In cases like this we do not mention the monadic and niladic score functions; they are 'unused', *i.e.*, taken to be identically 0.)

An obvious computational-complexity issue is raised by allowing scores to be arbitrary *real* values. Our algorithm will add, subtract, and compare these values (never introducing an absolute value larger than the sum of those in the input) and we assume that each such operation can be done in time $O(1)$ and its result represented in space $O(1)$. If desired, scores may be limited to integers, and the length of the integers factored in to the algorithm's complexity, but this seems uninteresting and we will not remark on it further.

We can solve minimization problems by replacing each score function with its negation (there is no assumption of positivity) and solving the resulting maximization problem. Max 2-CSP also models *weighted* problems: assigning a weight to a constraint just means multiplying the score function by the weight. Generalization to problems like

Max-Ones-2-Sat can be achieved by adding, to the usual 2-Sat formulation, small monadic cost functions $s_v(\phi(v)) = \epsilon\phi(v)$ (for instance, $\epsilon = 1/2n$ will do); this rewards setting variables to 1, but not at the expense of satisfying even a single clause.

Max 2-CSP further includes problems that are not obviously structured around pairwise constraints. Our original example of Max Cut may fall into this category, as do maximum independent set and minimum vertex cover (a minimum set of vertices dominating all edges). To model the problem of finding a maximum independent set in a graph as a Max 2-CSP, let $\phi(v) = 1$ if vertex is to be included in the set and 0 otherwise, define vertex scores $s_v(\phi(v)) = \phi(v)$ (rewarding a vertex for being included in the set), and define edge scores $s_{uv}(\phi(u),\phi(v)) = -2$ if $\phi(u) = \phi(v) = 1$, and 0 otherwise (penalizing violations of independence, and outweighing the reward for inclusion). Similarly, for minimum dominating set we penalize vertices for inclusion, but more heavily penalize edges neither of whose endpoints is included.

3. Solving a maximum constraint-satisfaction instance

In this section we describe our algorithm, Algorithm A, and analyse its performance on an arbitrary Max 2-CSP instance. The implications for random instances are taken up in subsequent sections.

The algorithm uses three types of reductions and an additional, trivial 'pseudo-reduction'. We begin by defining these reductions. We then show how the algorithm fixes a sequence in which to apply the reductions by looking at the underlying graph of the instance. This sequence defines a tree of instances, which can be solved bottom-up to solve the original one. Finally, we bound the algorithm's time and space requirements.

3.1. Reductions

Our first two reductions are 'transformations', each producing an equivalent problem with fewer vertices; the third is a 'splitting rule' producing a pair of problems, both with fewer vertices, one of which is equivalent to the original problem.

Reduction I. Let y be a vertex of degree 1, with neighbour x. Reducing (V, E, S) on y results in a new problem (V', E', S') with $V' = V \setminus y$ and $E' = E \setminus xy$. S' is the restriction of S to V' and E', except that for $C, D \in \{R, B\}$ we set

$$s'_x(C) = s_x(C) + \max_D \{s_{xy}(C,D) + s_y(D)\},$$

i.e., we set

$$s'_x(R) = s_x(R) + \max\{s_{xy}(R,R) + s_y(R), s_{xy}(R,B) + s_y(B)\},$$
$$s'_x(B) = s_x(B) + \max\{s_{xy}(B,R) + s_y(R), s_{xy}(B,B) + s_y(B)\}.$$

Note that any colouring ϕ' of V' can be extended to a colouring of V in two ways, namely ϕ_R and ϕ_B (corresponding to the two colourings of y), and the defining property of the reduction is that $S'(\phi') = \max\{S(\phi_R), S(\phi_B)\}$. In particular, $\max_{\phi'} S'(\phi') = \max_\phi S(\phi)$, and an optimal colouring ϕ' for the instance (V', E', S') can be extended to an optimal colouring ϕ for (V, E, S).

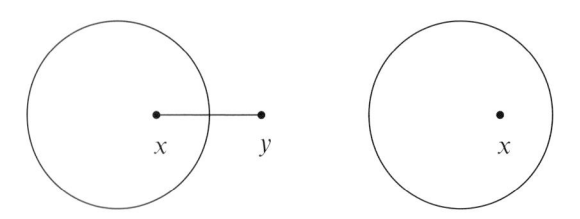

Reduction II. Let y be a vertex of degree 2, with neighbours x and z. If $x = z$ we have a pair of parallel edges: we combine the two edges and perform a type I reduction. Otherwise, reducing (V, E, S) on y results in a new problem (V', E', S') with $V' = V \setminus y$ and $E' = (E \setminus \{xy, yz\}) \cup \{xz\}$. S' is the restriction of S to V' and E', except that for $C, D, E \in \{R, B\}$ we set

$$s'_{xz}(C, D) = \max_{E}\{s_{xy}(C, E) + s_{yz}(E, D) + s_y(E)\},$$

i.e., we set

$$s'_{xz}(R, R) = \max\{s_{xy}(R, R) + s_{yz}(R, R) + s_y(R), s_{xy}(R, B) + s_{yz}(B, R) + s_y(B)\},$$
$$s'_{xz}(R, B) = \max\{s_{xy}(R, R) + s_{yz}(R, B) + s_y(R), s_{xy}(R, B) + s_{yz}(B, B) + s_y(B)\},$$
$$s'_{xz}(B, R) = \max\{s_{xy}(B, R) + s_{yz}(R, R) + s_y(R), s_{xy}(B, B) + s_{yz}(B, R) + s_y(B)\},$$
$$s'_{xz}(B, B) = \max\{s_{xy}(B, R) + s_{yz}(R, B) + s_y(R), s_{xy}(B, B) + s_{yz}(B, B) + s_y(B)\}.$$

This reduction creates a new edge xy, which may be parallel to one or more existing edges, each such edge having its associated score function. (The only reason we do not immediately merge parallel edges is that, working within our linear-time constraint, there is not time to identify them! Unfortunately our notation fails to distinguish among parallel edges and their scores, but this is only to keep the notation manageable; there is no deeper issue.) As in Reduction I, any colouring ϕ' of V' can be extended to V in two ways, ϕ_R and ϕ_B, and S' picks out the larger of the two scores. Also as in Reduction I, $\max_{\phi'} S'(\phi') = \max_\phi S(\phi)$, and an optimal colouring ϕ' for the instance (V', E', S') can be extended to an optimal colouring ϕ for (V, E, S).

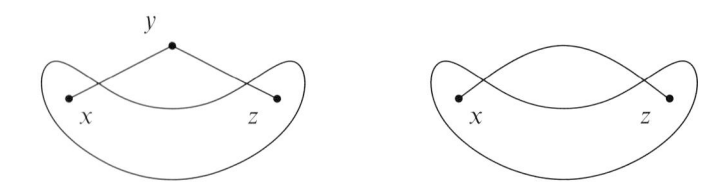

Reduction III. Let y be a vertex of degree 3 or higher. Where reductions I and II each had a single reduction of (V, E, S) to (V', E', S'), here we define a pair of reductions of (V, E, S), to (V', E', S^R) and (V', E', S^B), corresponding to assigning the colour R or B to y.

We define $V' = V \setminus y$, and E' as the restriction of E to V'. For $C, D \in \{R, B\}$, S^C is the restriction of S to $V \setminus y$, except that we set

$$(s^C)_0 = s_0 + s_y(C),$$

and, for every neighbour x of y,

$$(s^C)_x(D) = s_x(D) + s_{xy}(D, C).$$

In other words, S^R is the restriction of S to $V \setminus y$, except that we set $(s^R)_0 = s_0 + s_y(R)$ and, for every neighbour x of y,

$$(s^R)_x(R) = s_x(R) + s_{xy}(R, R)$$
$$(s^R)_x(B) = s_x(B) + s_{xy}(B, R).$$

Similarly S^B is given by $(s^B)_0 = s_0 + s_y(B)$ and, for every neighbour x of y,

$$(s^B)_x(R) = s_x(R) + s_{xy}(R, B)$$
$$(s^B)_x(B) = s_x(B) + s_{xy}(B, B).$$

As in the previous reductions, any colouring ϕ' of $V \setminus y$ can be extended to V in two ways, ϕ_R and ϕ_B, corresponding to the colour given to y, and now (this is different!) $S^R(\phi') = S(\phi_R)$ and $S^B(\phi') = S(\phi_B)$. Furthermore,

$$\max\{\max_{\phi'} S^R(\phi'), \max_{\phi'} S^B(\phi')\} = \max_\phi S(\phi),$$

and an optimal colouring on the left can be extended to an optimal colouring on the right.

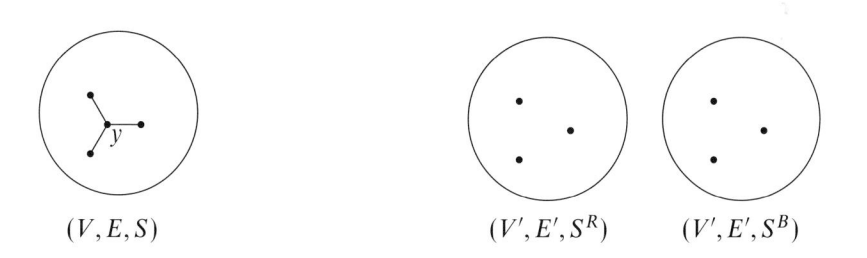

(V, E, S) $\qquad\qquad (V', E', S^R) \qquad (V', E', S^B)$

Reduction 0. We define one more 'pseudo-reduction'. If a vertex y has degree 0 (so it has no dyadic constraints), we simply delete it from the instance and incorporate its cost into the niladic score s_0. Specifically, reducing (V, E, S) on y results in a new problem (V', E', S') with $V' = V \setminus y$ and $E' = E$. S' is the restriction of S to V' and E', except that for $C \in \{R, B\}$ we set

$$s'_0 = s_0 + \max_C s_y(C).$$

As usual, $\max_{\phi'} S'(\phi') = \max_\phi S(\phi)$, and an optimal colouring ϕ' for (V', E', S') can be extended to an optimal colouring ϕ for (V, E, S).

3.2. Algorithm idea

A recursive algorithm for solving an input instance works as follows. Begin with the input problem instance. Given an instance $\mathcal{M} = (G, S)$:

(1) If any reduction of type 0, I or II is possible, apply it to reduce $\mathcal{M}$ to $\mathcal{M}'$, record certain information about the reduction, solve $\mathcal{M}'$ recursively, and use the recorded information to reverse the reduction and extend the solution to one for $\mathcal{M}$.
(2) If only a type III reduction is possible, reduce on a vertex of degree ≥ 4 if one exists, a vertex of degree 3 otherwise. In either case, first recursively solve $\mathcal{M}^R$ (the 'red' version of the reduction), then solve $\mathcal{M}^B$ (the 'blue' version), select the solution with the larger score, and use the recorded information to reverse the reduction and extend the solution to one for $\mathcal{M}$.
(3) If no reduction is possible then the graph has no vertices, there is a unique colouring (the empty colouring), and the score is s_0 (from the niladic score function).

The algorithm makes two runs as above. In the first run, in step (3) save the score if it sets a new record; this returns the optimal score but not the corresponding colouring. In the second run, when the score matches the record value, stop and return the colouring, by using the information stored at the ancestor nodes of the current leaf.

That the algorithm returns an optimal solution, *i.e.*, that it is correct, follows from the definitions of the reductions. The run time is analysed in the following sections, dealing with implementation details and data structures. We shall speak in terms of a *computation tree* implicitly defined by the recursive computation. For clarity, we will speak of 'nodes' of this tree, as opposed to 'vertices' of the instance's underlying graph. The tree's root node is the original problem instance, and each node's children are the subinstances derived from reducing it; III-reducing a node produces two children and the other reductions give a single child. We will also use a *reduced tree* which collapses nodes with exactly one child, so that a series of reductions starting with any number of 0, I and II-reductions and ending with a III-reduction (or when the instance is empty) is represented by a single edge. The depth r of the reduced tree is a key parameter: Corollary 3.2 shows that an appropriate algorithm implementation runs in time $O(n2^r)$ (and space $O(m + n)$).

3.3. Implementation details and data structures

Limiting the algorithm to linear space excludes saving copies of the instance as we descend a branch of the tree; rather, when the algorithm is processing a node of the tree, the corresponding instance should be the only one explicitly maintained, while the ancestor instances should be reconstructible by compact information stored at the ancestor nodes.

In this subsection we establish the following claim.

Claim 3.1. *After linear-time preprocessing, we can do the following.*

(1) *Identify the next reduction to perform, in time $O(1)$.*
(2) *Perform a III-reduction on a vertex y in time $O(n)$, creating an $O(\deg(y))$-space annotation enabling the reduction to be reversed and its colouring optimally extended.*
(3) *Perform a series of 0-, I- or II-reductions corresponding to an edge of the reduced tree in time $O(n)$, creating an $O(1)$-space annotation for each individual reduction.*

The implementation details are unrelated to the main direction of the paper, the characterization of a random process. However, they are important since the linear-time result depends on these tight space and time bounds.

3.3.1. Data structure. We assume a RAM model, so that a given memory location can be accessed in constant time.

We presume that the input graph is given in a sparse representation, consisting of a vector of vertices, each with its monadic score function (a 2-element table) and a doubly linked list of incident edges, each edge with its dyadic score function (a 4-element table) and a pointer to the edge's twin copy indexed by the other endpoint. We also assume that there is a doubly linked list of all the vertices. As vertices are removed from an instance to create a subinstance, they are bridged over in the linked list, so that there is always a linked list of just the vertices in the subinstance. We maintain an indication of whether each vertex is still unset or has been set to Red or Blue.

In time $O(m+n)$ and space $O(n)$, we transform the input instance into an equivalent instance without multiple edges. The simple procedure relies on a pointer array of length n, initially empty. For each vertex u, we iterate through the incident edges. For an edge to vertex v, if the vth entry of the pointer array is empty, we put a pointer to the edge uv. If the vth entry is not empty, this is not the first uv edge, and we coalesce it with the original one. That is, using the pointer to the original edge, we add the redundant edge's score function to that of the original one. We use the link from the redundant uv edge to its 'vu' twin copy to delete the twin and bridge over it, then delete and bridge over the redundant uv edge itself. After processing the last edge for vertex u we run through its edges again, clearing the pointer array. The time to process a vertex u is of order the number of its incident edges (or $O(1)$ if it is isolated), so the total time is $O(m+n)$ as claimed. Henceforth we assume without loss of generality that the input instance has no multiple edges.

3.3.2. Vertex degrees. One of the trickier points is to maintain information about the degree of each vertex. The algorithm may introduce multiple edges, and by a vertex's 'degree' we mean the number of *distinct* neighbours. Rather than keeping the precise degree of each vertex, we maintain a 'low-degree stack' containing all vertices of degrees 0, 1, and 2; a stack of degree-3 vertices; and a 'high-degree stack' of vertices of degree $\geqslant 4$. The stacks themselves are maintained as doubly linked lists, and from each vertex v we keep a pointer to its 'marker' in the stack. The stacks can be created in linear time from the input, and can also be maintained efficiently. The only difficulty comes from the possibility of multiple edges.

The key subroutine is a *degree-checking* procedure for a vertex x. Iterate through x's incident edges, keeping track of the number of distinct neighbouring vertices seen, stopping when we run out of edges or find 4 distinct neighbours. If a neighbour is repeated, coalesce the two edges. The time spent on x is $O(1)$ plus the number of edge coalescences. Once the degree of x is determined as 3, less, or more, x's marker is removed from its old stack (using the link from x to delete the marker, and links from the marker to its predecessor and successor to bridge over it), and a marker to x is pushed onto the appropriate new stack.

Table 1. Example of a II-reduction replacing score functions for yx and yz with a score function for xz and the associated optimal values of y.

(a)			(b)			(c)			
y	x	score	y	z	score	x	z	y	score
R	R	0	R	R	0	R	R	B	3
R	B	2	R	B	1	R	B	B	2
B	R	2	B	R	1	B	R	R	2
B	B	0	B	B	0	B	B	R	3

When reducing on vertex y, run the degree-checking procedure on each neighbour x of y. Of each neighbour's degree-checking time of $O(1)$ plus the number of edge coalescences, we charge the $O(1)$ to y, and subsume it into the time for the reduction on y which anyway is $\Theta(\deg(y))$. We account for the edge coalescences separately, claiming that in any sequence of reductions (i.e., within any branch of the recursion tree) there are at most n coalescences. It suffices to show that in any sequence of reductions, at most n duplicate edges are created; this is true because only a II-reduction can create such an edge, and even then it creates at most one.

Existence of the degree stacks assures the first part of Claim 3.1, that we can select a next reduction in time $O(1)$: simply pop a vertex, in preference order, from the low-degree stack, the high-degree stack, or the degree-3 stack. We have also shown that the stack maintenance can be performed within the times specified by the second and third parts of the claim. Thus we turn our attention to the remaining, more mathematical aspects of the reductions.

3.3.3. 0-, I- and II-reductions. We omit discussion of type 0 and I reductions and start with the slightly more complicated type II reductions. Suppose that the popped vertex y has two neighbours x and z. First we construct the score function s_{xz} replacing s_{xy} and s_{yz}. At the same time, we make a note of how to set y as a function of x and z. For example if xy is a 'cut' constraint of weight 2, and yz is a cut constraint of weight 1, these are replaced by a single anti-cut constraint on xz, with associated optimal values of y: Table (a) gives the score function for xy, Table (b) gives that for yz, and Table (c), gives the score function and the optimal value of y for xz. Table (c), mapping the colouring of xz to a score and an optimal colour for y, is associated with the instance $\mathcal{M}$ being reduced. The new instance $\mathcal{M}'$ is formed by deleting edge yx and its twin xy (bridging over them in the linked lists), deleting yz and its twin zy, and adding a new edge xz and twin zx with score function taken from Table (c). Finally, vertex y is deleted. Disregarding the degree-checking time accounted for in the previous subsection, the reduction takes time and space $O(1)$. Reversing the reduction is equally straightforward, if we associate x, y, z, and Table (c) with this step in the recursive calculation; this takes space $O(1)$ per node of the implicit computation tree, at most one root-to-leaf branch of which exists at any time.

3.3.4. III-reductions. A 'Red' III-reduction on y is performed by making a first sweep through the incident edges, for edge yx adding the score function $s_{yx}(R,.)$ to the monadic score function $s_x()$, then making a second sweep and deleting each edge yx and twin xy. (As each edge is deleted, we save a pointer to it and its predecessor and successor; we also save a pointer to each neighbour x of y.) For each edge xy deleted, we run a degree check on x and place it on the appropriate stack. This defines an instance $\mathcal{M}^R$. The same procedure is of course applied for a reduction to $\mathcal{M}^B$. Reversing a reduction is straightforward, if before performing it we record y's neighbours and the corresponding dyadic score functions, as well as y's monadic score function. This takes space $O(\deg(y))$.

3.3.5. Backtracking. To undo a reduction of any type, we use the saved pointers to reconstruct the deleted edges, 'un-bridging' the pointer bridges we built around them, correcting the vertex degrees, and undoing the changes to the score functions.

This establishes Claim 3.1.

3.4. Algorithm implementation and analysis

Having established Claim 3.1, it is easy to analyse the algorithm's complexity first in terms of the depth of the reduced computation tree, and then in terms of m and n.

Corollary 3.2. *An n-vertex, m-constraint Max 2-CSP instance whose computation tree has at most r type III reductions in any root-to-leaf path can be solved in time $O(n2^r)$ using space $O(m+n)$.*

Proof. If the computation tree has at most r III-reductions in any root-to-leaf path, then by definition the reduced tree is a binary tree of depth at most r, with $O(2^r)$ nodes.

On its first pass, the recursive algorithm performs a depth-first search of the reduced tree; this takes time linear in the tree's size, multiplied by the time for an elementary step. Moving from a node to its child means performing a series of 0-, I- and II-reductions and a single III-reduction, which by Claim 3.1 takes time $O(n)$. Reversing a reduction is equally easy, so returning from a child to its parent also takes time $O(n)$. Thus the total run time for the first pass is $O(n2^r)$.

On its second pass, the algorithm repeats a portion of the depth-first search until it reaches a leaf with optimal score. It then reconstructs the corresponding colouring by using the information stored at ancestors of that leaf; since there are at most r III-reductions to reverse, this pass takes time at most $O(n2^r) + O(rn) + O(n) = O(n2^r)$.

At any stage of the recursion, the sub-problem being solved corresponds to a node of the reduction tree, and at each ancestor node is recorded information to reconstruct that instance. By Claim 3.1 the annotation for reducing any vertex y is of size $O(1 + \deg(y))$, and so the space needed by the algorithm is $O(m+n)$. □

We can now bound the running time of Algorithm A in terms of the excess of the graph underlying the CSP. Note that Theorem 1.2 follows as a special case of Theorem 3.3.

Theorem 3.3. *Given a weighted Max 2-CSP whose underlying graph G is connected, has order n, size m, and excess $\kappa = m - n$, Algorithm A returns an optimal solution in time $O(n2^{\min(\kappa/2, m/4)})$, using space $O(m+n)$.*

The 'm/4' bound is used here only for Case 4 of the proof of Theorem 7.1, and we include a short proof to keep the argument self-contained. In fact a bound of $m/5$ holds for the same algorithm. A proof sketch was given in [28]; a more careful proof is given in [29], which also gives a more sophisticated algorithm and analysis resulting in a bound of $19m/100$.

Proof. In light of Corollary 3.2, it suffices to prove that the number of type III reduction steps $r(G)$ is bounded by both $\max\{0, \kappa/2\}$ and $m/4$.

We begin with the $\kappa/2$ bound. The proof is by induction on the order of G.[1] The base cases, order 0 and 1, is trivial. If the first reduction is of type I or II (it cannot be of type 0 as the graph is connected) then the induction is trivial.

Otherwise, the first type III reduction, from G to G', reduces the number of edges by at least 3 and the number of vertices by exactly 1, thus reducing the excess to $\kappa' \leqslant \kappa - 2$. If G' has components $G'_1, \ldots, G'_l$, then $r(G) = 1 + \sum_i r(G'_i)$. Given that we applied a type III reduction, G had minimum degree $\geqslant 3$ (that is, per Section 3.3.2, at least 3 distinct neighbours, independent of edge multiplicities), so G' has minimum degree $\geqslant 2$. Thus each component G'_i has minimum degree $\geqslant 2$, and so excess $\kappa'_i \geqslant 0$. Then, by induction, $r(G) = 1 + \sum_i r(G'_i) \leqslant 1 + \sum_i \max\{0, \kappa'_i/2\} = 1 + \sum_i \kappa'_i/2 = 1 + \kappa'/2 \leqslant 1 + (\kappa - 2)/2 = \kappa/2$.

For the $m/4$ bound, we simply argue that each III-reduction results in the destruction of at least 4 edges. A III-reduction on a high-degree vertex destroys at least 4 edges instantly. If the III-reduction is instead on a degree-3 vertex then its neighbours were also of degree 3. (If any were of higher degree we would have III-reduced on it instead; if any were of lower degree we would have 0-, I- or II-reduced on it.) The III-reduction converts these 3 neighbours to degree 2, they get pushed onto the (previously empty) low-degree stack, and the next step will be to II-reduce on the first of them, destroying a fourth edge following the 3 from the III-reduction. □

4. The binomial distribution and staircase random walks

Our analysis in the next section will centre on characterizing the order and excess of a component of a random graph, which we will do by showing how these quantities are dominated by parameters of a random walk. Characterization of the random walk itself requires a certain amount of work, and since it is independent of our Max CSP context we take it up in this separate section.

We would have expected the facts given here already to be known, but we have searched the literature and spoken to colleagues without turning up anything. Even if the main points of this section are not new (Definition 4.2, Theorem 4.5, Corollary 4.6, and Theorem 4.7) they seem not to be well known, and may be of independent interest.

---

[1] There is a simpler 'proof' from the fact that I- and II-reductions preserve excess, and III-reductions decrease it by at least 2. Unfortunately this overlooks 0-reductions, which increase excess, or equivalently, overlooks the fact that components consisting of an isolated vertex have negative excess.

Our aim parallels a well-known result for Brownian motion. Since we took both that result and its proof as our model, let us state it. Let $X : [0, 1] \to \mathbb{R}$ be a standard Brownian motion with $X(0) = 0$ and $X(1) = s$. Then, where ϕ denotes the density of the standard normal $N(0, 1)$, the following theorem is a classical result on the 'standard Brownian bridge' $X(t) - ts$.

Theorem 4.1. *For any $b \geq 0$, $\Pr(\max_t(X(t) - ts) \geq b) = \phi(2b)/\phi(0) = \exp(-2b^2)$.*

For a standard Brownian motion, an increment $X(t + \tau) - X(t)$ has Gaussian distribution $N(0, \tau)$, and the proof of the theorem applies the reflection principle to Brownian motion, using the symmetry $\phi(x) = \phi(-x)$ of the Gaussian density.

We require an analogous result for a simple random walk $X(t)$, by which we mean a walk which at each step increases by 1 with probability p, and stays the same with probability $1 - p$; we will condition the walk on $X(0) = 0$ and $X(n) = s$. The increments $X(t + \tau) - X(t)$ for the (unconditioned) random walk have binomial distribution $B(\tau, p)$, and our proof of Theorem 4.7 (analogous to the Brownian-motion theorem above) applies the reflection principle to the random walk, using the *asymmetry* of the binomial distribution as in Theorem 4.5 and Corollary 4.6.

We begin by defining and characterizing a continuous extension of the binomial density function.

Definition 4.2. For any real values $n > 0$, $0 \leq p \leq 1$, and any real k, we define

$$B_{n,p}(k) := \frac{\Gamma(n+1)}{\Gamma(k+1)\Gamma(n-k+1)} p^k (1-p)^{n-k} \tag{4.1}$$

if $0 \leq k \leq n$, and $B_{n,p}(k) := 0$ otherwise.

For integers n and $k \in \{0, \ldots, n\}$ this is of course just the binomial density $\binom{n}{k} p^k (1 - p)^{n-k}$. Although the continuous extension need not integrate to 1, we will still call it the 'continuous binomial density'.

It is a simple and well-known fact that the usual binomial density function (on integers) is unimodal with with a unique maximum lying at $k = \lfloor (n + 1)p \rfloor$, or two maxima if, for that k, $B_{n,p}(k) = B_{n,p}(k - 1)$. We first prove that the continuous extension is unimodal.

Theorem 4.3. *The continuous binomial density defined by (4.1) is unimodal; $B_{n,p}((n+1)p - 1) = B_{n,p}((n+1)p)$; every value of $B_{n,p}$ on the interval $[(n+1)p - 1, (n+1)p] = [np - (1 - p), np + p]$ exceeds every value outside it; and thus the maximum lies in this interval.*

Note that the maximum need not occur at np, as shown for instance by $n = 3$, $p = 1/3$, where the maximum occurs at around 0.82 rather than at 1.

Proof. We use Gauss's representation $\Gamma(x) = x^{-1} \prod_{i=1}^{\infty} \left[(1 + 1/i)^x (1 + x/i)^{-1} \right]$ [15, p. 450]. First, $B_{n,p}(k)$ is log-concave:

$$\ln B_{n,p}(k) = k \ln p + (n-k) \ln(1-p) + \ln(k+1) + \ln(n-k+1) + \ln \Gamma(n+1)$$
$$- \sum_{i \geq 1} [(k+1) \ln(1 + 1/i) - \ln(1 + (k+1)/i)]$$
$$- \sum_{i \geq 1} [(n-k+1) \ln(1 + 1/i) - \ln(1 + (n-k+1)/i)],$$

so

$$\frac{d}{dk} \ln B_{n,p}(k) = \ln p - \ln(1-p) + \sum_{i \geq 0} \left[\frac{1}{i+k+1} - \frac{1}{i+n-k+1} \right], \tag{4.2}$$

and

$$\frac{d^2}{dk^2} \ln B_{n,p}(k) = 0 + \sum_{i \geq 0} [-1/(i+k+1)^2 - 1/(i+n-k+1)^2] < 0.$$

Thus $B_{n,p}(k)$ is unimodal. Also,

$$B_{n,p}(k)/B_{n,p}(k-1) = \frac{n-k+1}{k} \frac{p}{1-p}, \tag{4.3}$$

so for $k = (n+1)p$, $B_{n,p}(k)/B_{n,p}(k-1) = 1$. Thus the maximum of $B_{n,p}(k)$ occurs for some k in the range $[(n+1)p - 1, (n+1)p]$, and moreover every value of $B_{n,p}(k)$ in this range is at least as large as every value outside it. □

We will need the following simple fact.

Remark 4.4. If a real-valued function f is convex on $[a - \lambda, b + \lambda]$, with $a < b$ and $\lambda \geq 0$, then

$$\frac{1}{b-a} \int_a^b f(x) dx \leq \frac{f(a-\lambda) + f(b+\lambda)}{2}.$$

Proof. Let $\bar{f}(x)$ be the linear interpolation at x from $f(a - \lambda)$ and $f(b + \lambda)$. By convexity, for all $x \in [a - \lambda, b + \lambda]$, $f(x) \leq \bar{f}(x)$. Integrating, $\frac{1}{b-a} \int_a^b f(x) dx \leq \frac{1}{b-a} \int_a^b \bar{f}(x) dx$. As the average value of a linear function, the latter quantity is $\frac{1}{2}(\bar{f}(a) + \bar{f}(b)) = \frac{1}{2}(\bar{f}(a - \lambda) + \bar{f}(b + \lambda)) = \frac{1}{2}(f(a - \lambda) + f(b + \lambda))$, concluding the proof. □

For a Gaussian distribution, the right and left tails are of course symmetric to one another. For fixed p and large n, a binomial distribution $B_{n,p}$ is approximately Gaussian and the two tails are nearly but not exactly symmetric. We use Claim 4.3 to show that, for $p \leq 1/2$, a binomial's right tail (slightly) dominates its left tail. (For $p > 1/2$ the opposite is true, by symmetry.)

Theorem 4.5. *For $p \in (0, 1/2)$, the continuous binomial density function $B_{n,p}(k)$ defined by (4.1) has the property that for all deviations $\delta \geqslant 0$,*

$$B_{n,p}((n+1)p - 1 - \delta) \leqslant B_{n,p}((n+1)p + \delta).$$

Proof. For notational convenience, let $N = n + 1$. The truth of the theorem for $\delta = 0$ is immediate from Claim 4.3's assertion that $B_{n,p}(Np - 1) = B_{n,p}(Np)$. It suffices, then, to prove the nonnegativity of

$$\frac{d}{d\delta} \ln\left(\frac{B_{n,p}(Np + \delta)}{B_{n,p}(Np - \delta - 1)}\right) = \frac{d}{d\delta} \ln B_{n,p}(Np + \delta) - \frac{d}{d\delta} \ln B_{n,p}(Np - \delta - 1).$$

(That is, the slope going forwards from the point $Np + \delta$ should 'outweigh' the slope going backwards from $Np - \delta - 1$.) Taking the derivatives from (4.2), then, we wish to show nonnegativity of

$$2(\ln(p) - \ln(1-p)) + \sum_{i \geqslant 0} \left[\frac{1}{i + Np + \delta + 1} - \frac{1}{i + N(1-p) - \delta} \right. \quad (4.4)$$

$$\left. + \frac{1}{i + Np - \delta} - \frac{1}{i + N(1-p) + \delta + 1} \right].$$

Before proving this, we note that if δ is fixed and we let $N \to \infty$, it can be seen (by approximating the sum by an integral) that (4.4) tends to 0; showing (4.4) to be positive will require a little care.

Let $f(x) = 1/[(p + x)(1 - p + x)]$. Note that $(1 - 2p) \int_0^\infty f(x) dx = \ln(1-p) - \ln(p)$, so f will be used to address the first summand in (4.4). Also $f''(x) = 2/[(1 - p + x)(p + x)^3] + 2/[(1 - p + x)^2(p + x)^2] + 2/[(1 - p + x)^3(p + x)]$, which is positive for $x > -p$, so f is convex on $(-p, \infty)$. Returning to the quantities in (4.4), then

$$\sum_{i \geqslant 0} \left[\frac{1}{i + Np + \delta + 1} - \frac{1}{i + N(1-p) - \delta} + \frac{1}{i + Np - \delta} - \frac{1}{i + N(1-p) + \delta + 1} \right]$$

$$= \sum_{i \geqslant 0} \left[\frac{N(1 - 2p)}{(i + Np - \delta)(i + N(1-p) - \delta)} + \frac{N(1 - 2p)}{(i + Np + \delta + 1)(i + N(1-p) + \delta + 1)} \right]$$

$$= \sum_{i \geqslant 0} \left[\frac{\frac{1}{N}(1 - 2p)}{(p + (i - \delta)/N)(1 - p + (i - \delta)/N)} \right.$$

$$\left. + \frac{\frac{1}{N}(1 - 2p)}{(p + (i + \delta + 1)/N)((1-p) + (i + \delta + 1)/N)} \right]$$

$$= \frac{1}{N}(1 - 2p) \sum_{i \geqslant 0} [f((i - \delta)/N) + f((i + \delta + 1)/N)]$$

$$\geqslant (1 - 2p) \sum_{i \geqslant 0} 2 \int_{i/N}^{(i+1)/N} f(x) dx \quad \text{(as explained below)}$$

$$= 2(1 - 2p) \int_0^\infty f(x) dx$$

$$= -2(\ln(p) - \ln(1-p)),$$

proving the nonnegativity of (4.4). The inequality follows from Remark 4.4, with $a = i/N$, $b = (i+1)/N$, and $\lambda = \delta/N$: f is convex on $(-p, \infty)$, which contains the relevant range because $a - \lambda = (i - \delta)/N \geq -\delta/N > -p$ as long as $\delta \leq np < Np$, while if $\delta > np$ then the theorem is true trivially, as $B_{n,p}((n+1)p - 1 - \delta)$ is 0 while $B_{n,p}((n+1)p + \delta)$ is positive. $\square$

We note that this has the following corollary for binomial random variables.

Corollary 4.6. *For a binomially distributed random variable $X \sim B(n, p)$, $p \leq 1/2$, for any $\delta \geq 0$, $\Pr(X = \lfloor (n+1)p - 1 - \delta \rfloor) \leq \Pr(X = \lfloor (n+1)p + \delta \rfloor)$.*

Proof. From Claim 4.3, $\Pr(X = \lfloor (n+1)p - 1 - \delta \rfloor) = B_{n,p}(\lfloor (n+1)p - 1 - \delta \rfloor) \leq B_{n,p}((n+1)p - 1 - \delta)$. Also, $\Pr(X = \lfloor (n+1)p + \delta \rfloor) = B_{n,p}(\lfloor (n+1)p + \delta \rfloor) \geq B_{n,p}((n+1)p + \delta)$: if $\lfloor (n+1)p + \delta \rfloor \geq (n+1)p$ the last inequality follows from the fact that $B_{n,p}$ decreases above $(n+1)p$, while if $\lfloor (n+1)p + \delta \rfloor < (n+1)p$ it follows from the fact that every value of $B_{n,p}$ in the interval $[(n+1)p - 1, (n+1)p]$ is larger than any value outside it. By Theorem 4.5, $B_{n,p}((n+1)p - 1 - \delta) \leq B_{n,p}((n+1)p + \delta)$. Putting the three inequalities together proves the claim. $\square$

Next we consider the deviation of a 'staircase' random walk above its linear interpolate. Our bound on the tail of this parameter is roughly the square of what would be obtained from a naive application of Hoeffding's inequality for sampling with replacement [19, Section 6].

As noted earlier, our result and proof are modelled on a classical equality (Theorem 4.1) for the Brownian bridge. Since the 'long-run' behaviour of a simple random walk converges to Brownian motion (in a sense we do not need to make precise), it is not surprising that we should be able to obtain a similar result.

Theorem 4.7. *Fix any positive integers n and $S \leq n/2$, and any integer discrepancy $b \geq 2$. Let $X_1, \ldots, X_n$ be a 0–1 sequence chosen uniformly at random from among all such sequences having sum $X_1 + \cdots + X_n = S$. Then*

$$\Pr\left(\max_i \left\{ X_1 + \cdots + X_i - \frac{i}{n} S \right\} \geq b \right) \leq \frac{B_{n, S/n}(S + 2b - 1)}{B_{n, S/n}(S)}.$$

Proof. First observe that a random 0–1 sequence with $X_1 + \cdots + X_n = S$ as above has precisely the same distribution as a sequence of n i.i.d. Bernoulli random variables conditioned on having sum S. For the remainder of the proof we adopt this view, in particular choosing to give each random variable the distribution $X_i \sim B(p)$ with $p = S/n \leq 1/2$. For notational convenience, let $X^\tau = \sum_{i=1}^\tau X_i$. Noting that $\mathbb{E}X^\tau = \tau p$, then, we are asking for the conditional probability that there is a time τ such that $X^\tau \geq \tau p + b$. If so, define the 'first crossing time' τ_b to be $\min\{\tau \leq n : X^\tau \geq \tau p + b\}$, and otherwise let $\tau_b = n + 1$. Because X^τ increases by at most 1 in a step, if $\tau_b \leq n$ then

$X^{\tau_b} = \lceil \tau_b p + b \rceil$. The event we are interested in is precisely that $\tau_b \leq n$, conditioned on $X^n = np$:

$$\Pr(\tau_b \leq n \mid X^n = np) = \frac{\Pr(\tau_b \leq n, X^n = np)}{\Pr(X^n = np)}. \quad (4.5)$$

The numerator of this expression is

$$\text{num} = \sum_{\tau=1}^{n} \Pr(\tau_b = \tau, X^n = np)$$

$$= \sum_{\tau=1}^{n} \Pr(\tau_b = \tau) \Pr(X^n = np \mid \tau_b = \tau)$$

which by the Markovian nature of the process

$$= \sum_{\tau=1}^{n} \Pr(\tau_b = \tau) \Pr(X^n = np \mid X^\tau = \lceil \tau p + b \rceil)$$

$$= \sum_{\tau=1}^{n} \Pr(\tau_b = \tau) B_{n-\tau, p}(np - \lceil \tau p + b \rceil)$$

$$= \sum_{\tau=1}^{n} \Pr(\tau_b = \tau) B_{n-\tau, p}((n - \tau + 1)p - 1 - \delta)$$

where, using b's integrality, $\delta = ((n - \tau + 1)p - 1) - (np - \lceil \tau p + b \rceil) = b - 1 + \lceil \tau p \rceil - \tau p + p \geq 0$, and thus we may apply the inequality of Theorem 4.5:

$$\leq \sum_{\tau=1}^{n} \Pr(\tau_b = \tau) B_{n-\tau, p}((n - \tau + 1)p + \delta)$$

$$= \sum_{\tau=1}^{n} \Pr(\tau_b = \tau) B_{n-\tau, p}(np + b - 1 - 2\tau p + \lceil \tau p \rceil + 2p).$$

For reasons that will shortly become clear, we wish to replace the binomial's argument by $np + 2b - 1 - \lceil \tau p + b \rceil$. We observe that the original argument is larger than $(n - \tau + 1)p$ (because $\delta \geq 0$); the new argument is smaller than the original one (because b is integral, and $-\lceil \tau p \rceil \leq -2\tau p + \lceil \tau p \rceil$); and the new argument is larger than $(n - \tau + 1)p - 1$ (because the difference is $b - \lceil \tau p \rceil + \tau p - p > b - 1 - p > 0$). Thus by Theorem 4.3, the binomial's value can only increase:

$$\leq \sum_{\tau=1}^{n} \Pr(\tau_b = \tau) B_{n-\tau, p}(np + 2b - 1 - \lceil \tau p + b \rceil)$$

$$= \sum_{\tau=1}^{n} \Pr(\tau_b = \tau) \Pr(X^n = np + 2b - 1 \mid X^\tau = \lceil \tau p + b \rceil),$$

which again by the Markovian nature of the process

$$= \sum_{\tau=1}^{n} \Pr(\tau_b = \tau) \Pr(X^n = np + 2b - 1 \mid \tau_b = \tau)$$

$$= \sum_{\tau=1}^{n} \Pr(\tau_b = \tau, X^n = np + 2b - 1)$$

$$= \Pr(X^n = np + 2b - 1),$$

where the last equality holds because $X^n = np + 2b - 1$ means that X^n exceeds its expectation np by $2b - 1 \geqslant b$ and thus implies that $\tau_b \leqslant n$. Returning to (4.5) and substituting $S = np$ yields the claim. □

Remark 4.8. For positive integers b and np with $b \leqslant 2np$,

$$\frac{B_{n,p}(np+b)}{B_{n,p}(np)} \leqslant \exp\bigl(-(3\ln(3) - 2)/4 \cdot b^2/np\bigr).$$

Proof. The proof is by simple calculation:

$$\frac{B_{n,p}(np+b)}{B_{n,p}(np)} = \frac{(n - (np+b-1)) \cdots (n - np)}{(np+b) \cdots (np+1)} \left(\frac{p}{1-p}\right)^b$$

$$= \prod_{i=1}^{b} \frac{1 - (i-1)/n(1-p)}{1 + i/np}$$

$$\leqslant \prod_{i=1}^{b} \frac{1}{1 + i/np}$$

$$\leqslant \exp\left(-\int_0^b \ln(1 + i/np) \, di\right)$$

$$= \exp\bigl(-np(1 + b/np)\ln(1 + b/np) + b\bigr). \tag{4.6}$$

If $b = x \cdot np$, then the value of c for which (4.6) equals $\exp(-cb^2/np)$ is $c = \frac{(x+1)\ln(x+1) - x}{x^2}$. Since this is decreasing in x, the worst-case (smallest) value of c occurs for the largest allowed value of $x = b/np$. By hypothesis, this is $x = 2$, where $c = (3\ln(3) - 2)/4$. For smaller values of $x = b/np$, then, (4.6) is smaller than $\exp\bigl(-(3\ln(3) - 2)/4 \cdot b^2/np\bigr)$, completing the proof. □

5. Stochastic size and excess of a random graph

We stochastically bound the excess $\kappa = m - n$ of a component of a random graph G via the branching-process approach pioneered by Karp [21] (see also Kendall [22], von Bahr and Martin-Löf [2], and Martin-Löf [26]). Given a graph G and a vertex x_1 in G, together with a linear order on the vertices of G, the branching process finds a spanning tree of the component G_1 of G that contains x_1 and, in addition, counts the number of non-tree edges of G_1 (i.e., calculates the excess plus 1).

At each step of the process, vertices are classified as 'living', 'dead', or 'unexplored', beginning with just x_1 living, and all other vertices unexplored. At the ith step, the process takes the earliest living vertex x_i. All edges from x_i to unexplored vertices are added to the spanning tree, and the number of non-tree edges is increased by 1 for each edge from x_i to a living vertex. Unexplored vertices adjacent to x_i are then reclassified as living, and x_i is made dead. The process terminates when there are no living vertices.

Now suppose G is a random graph in $\mathcal{G}(n, c/n)$, with the vertices ordered at random. Let $w(i)$ be the number of live vertices after the ith step and define the *width* $w = \max w(i)$. (Note that $w(i)$ and w are functions of the random process, not just the component, since they depend on the order in which the vertices are taken. Despite this, for convenience we will refer to the 'width of a component'.)

Let $u = |G_1|$, so that $w(0) = 1$ and $w(u) = 0$. The number of non-tree edges uncovered in the ith step is binomially distributed as $B(w(i) - 1, c/n)$, and so, conditioning on u and $w(1), \ldots, w(u)$, the number of excess edges is distributed as

$$B\left(\sum_{i=1}^{u}(w(i) - 1), c/n\right). \tag{5.1}$$

Since $\sum_{i=1}^{u}(w(i) - 1) \leq uw$, and also $\sum_{i=1}^{u}(w(i) - 1) \leq \sum_{i=1}^{u}(i - 1) = \binom{u}{2}$, the number of excess edges is dominated by the random variable $B(\min\{uw, \binom{u}{2}\}, c/n)$. At the ith stage of the process, there are at most $n - i$ unexplored vertices, and so the number of new live vertices is dominated by $B(n - i, c/n)$. This allows us to define a simpler random walk which dominates the graph edge-exposure branching process.

Definition 5.1. Given a constant $c > 0$ and integer $n > 0$, define the random walk RW by $X(0) = 1$ and, for $i > 0$, $X(i) = X(i - 1) + B(n - i, c/n)$, where the binomial increments are independent. Parametrize its time-i width by $W'(i) = X(i) - i$, its order by $U' = \min\{n, \min\{i: W'(i) = 0\}\}$, and its (maximum) width by $W' = \max_{i \leq U'} W'(i)$.

Claim 5.2. *The order U and width W of the component G_1 on vertex 1 of a random graph $G(n, c/n)$ are stochastically dominated by U' and W' of the random walk RW.*

Proof. Consider a variant of the branching process on the random graph in which at each step we add enough new special 'red' vertices to bring the number of unexplored vertices to $n - i$. This is equivalent to the random walk RW. It also dominates the original branching process: in the implicit coupling between the two, the variant has width at least as large at every step, and thus also has maximum width and order which are at least as large as those of the original process. □

Thus the excess κ_1 of G_1 is stochastically dominated by the same quantity for RW:

$$\kappa_1 \preceq B\left(\min\{U'W', \binom{U'}{2}\}, c/n\right). \tag{5.2}$$

Let $t(G_1)$ be the time spent by Algorithm A on G_1. We shall analyse the total running time by 'charging' $t(G_1)/|G_1|$ to each vertex of G_1; the running time is then the sum of these charges.

Claim 5.3. *The amortized running time $t(G_1)/|G_1|$ of Algorithm A on G_1, the component on vertex 1 of a random graph $G(n, c/n)$, satisfies*

$$\mathbb{E}(t(G_1)/|G_1|) = O(1)\,\mathbb{E}\exp\bigl(c(\sqrt{2}-1)\min\{U'W'/n,\,U'/2\}\bigr), \quad (5.3)$$

with U' and W' given by the random walk RW.

Proof. The running time of Algorithm A on a connected graph with n_1 vertices, m_1 edges and excess $\kappa_1 = m_1 - n_1$ is

$$O(n_1 2^{\kappa_1/2}). \quad (5.4)$$

The exponential moments of binomial random variables are simple and well known: if a random variable U has distribution $B(N, p)$, then

$$\mathbb{E}z^U = \sum_{i=0}^{N}\binom{N}{i}z^i p^i(1-p)^{N-i} = (pz+(1-p))^N = (1+p(z-1))^N \leqslant \exp(p(z-1)N),$$

and in particular,

$$\mathbb{E}\sqrt{2}^{U} \leqslant \exp((\sqrt{2}-1)Np). \quad (5.5)$$

Setting $N = \min\{U'W', \binom{U'}{2}\}$ and combining (5.2), (5.4), and (5.5) gives

$$\mathbb{E}(t(G_1)/|G_1|) = O(1)\,\mathbb{E}(2^{\kappa/2}) \leqslant O(1)\,\mathbb{E}\exp((\sqrt{2}-1)N\,c/n). \quad (5.6)$$

Noting that $U' \leqslant n$, and so $\binom{U'}{2}/n \leqslant U'/2$, yields (5.3). □

In the following, we therefore focus on finding bounds on the probability $\Pr(U, W)$ that the 'first' component of a random graph has order U and width W, or, since $(U, W) \leq (U', W')$ per Claim 5.2, the corresponding probability $\Pr(U', W')$ for the random walk RW.

We use a version of Chernoff's inequality (see Janson, Łuczak and Ruciński [20, Theorems 2.1 and 2.8]), which states that for a sum Z of independent 0–1 Bernoulli random variables with parameters $p_1,\ldots,p_n$ and expectation $\mu = \sum_{i=1}^{n} p_i$:

$$\Pr(Z \geqslant \mu + t) \leqslant \exp\bigl(-t^2/(2\mu + 2t/3)\bigr), \quad (5.7)$$
$$\Pr(Z \leqslant \mu - t) \leqslant \exp\bigl(-t^2/(2\mu)\bigr). \quad (5.8)$$

The next lemma describes the probability that U' is large, and has a corollary for the probability that $|G_1|$ is large. (We will not use the corollary, but we state it because it is natural and potentially useful.) Although the proof is framed in terms of the binomial increments $B(n-i, c/n)$ for RW (corresponding to vertex exposures in the random graph), it may also be helpful to think in terms of subdividing such an increment into $n-i$ Bernoulli increments $\text{Be}(c/n)$ (corresponding to edge exposures in the random graph).[2]

---

[2] The edge-exposure model was previously used by Spencer in an elegant short paper [30]. Spencer was studying a related problem, calculating $C(n, n+k)$ (the number of connected graphs with k vertices and $n+k$ edges) for k fixed and $n \to \infty$. We will discuss $C(n, n+k)$ in Section 6.

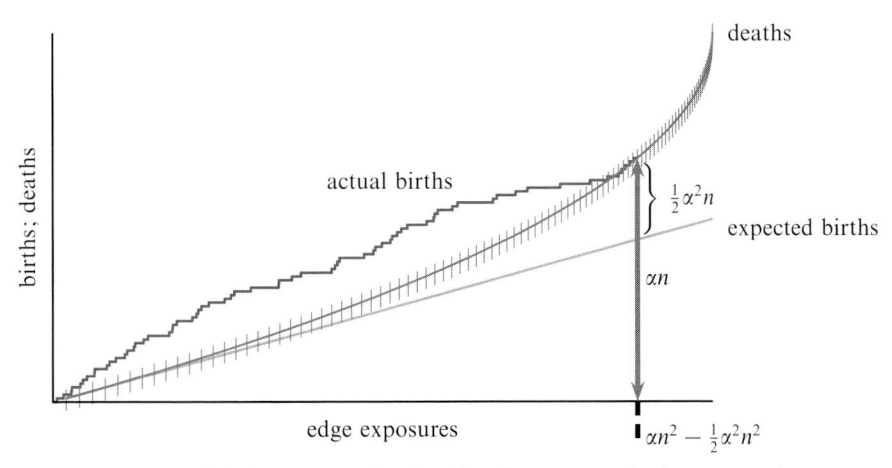

Figure 1. Birth-death process described by edge exposures in the augmented graph process, or equivalently by Bernoulli exposures in RW.

This view will be essential in proving Lemma 5.7, and is illustrated in Figure 1. The X axis indicates edge exposures j in the augmented graph model, or equivalently the number of Bernoulli random variables exposed in the random walk (a 'finer sampling' of the same RW). Since the number of edge exposures between successive deaths shrinks from $n-1$ to $n-2$ etc., the cumulative number of deaths (call it $d(j)$) grows super-linearly. (As it happens, $d(j)$ is a parabola rotated sideways.) The expected cumulative number of births grows linearly, as $(c/n)j$, and (for $c=1$) is tangent to the 'death curve' at the origin. The event that the actual number of births equals deaths equals αn means that a corresponding sum of Bernoulli random variables $\text{Be}(c/n)$ equals αn; the individual births comprising this sum describe a random walk, a sample of which is shown in the figure.

Lemma 5.4. *Let $n > 0$ be an integer, $\Lambda > 0$ a real, and $c > 0$ a real with $c \leqslant 1 + \Lambda$. For any integer $i > 0$, setting $\alpha = i/n$, the time-i widths $W'(i)$ of the random walk RW with parameters c, n satisfy*

$$\Pr(W'(\alpha n) > 0) \leqslant \exp\left(\frac{-3\alpha^3 n(1 - 6\Lambda/\alpha)}{24 - 8\alpha}\right). \tag{5.9}$$

Proof. We assume $\Lambda/\alpha < 1/6$ since otherwise the inequality is trivial. We also assume without loss of generality that $c = 1 + \Lambda$, since the corresponding process dominates that for any smaller c. Note that $W'(i)$ has distribution

$$W'(i) \sim B\left((n-1) + \cdots + (n-i), \frac{1+\Lambda}{n}\right) - i + 1$$

$$= B\left(ni - \binom{i+1}{2}, \frac{1+\Lambda}{n}\right) - i + 1$$

and so $W'(i) > 0$ means that

$$B\left(ni - \binom{i+1}{2}, \frac{1+\Lambda}{n}\right) \geq i = \alpha n. \tag{5.10}$$

This binomial r.v. has expectation

$$\left(\alpha n^2 - \binom{\alpha n + 1}{2}\right)\frac{1+\Lambda}{n} \leq (1+\Lambda)(\alpha - \alpha^2/2)n. \tag{5.11}$$

For convenience, define $q = \Lambda/\alpha < 1/6$. Thus if (5.10) holds, the r.v. exceeds its expectation by at least

$$\alpha^2 n/2 - \Lambda(\alpha - \alpha^2/2)n = \frac{\alpha^2 n}{2}(1 - 2q + \alpha q) \geq 0. \tag{5.12}$$

Together with (5.11) and (5.12), (5.7) implies that (5.10) has probability at most

$$\exp\left(\frac{-\alpha^4 n^2(1 - 2q + \alpha q)^2/4}{2(1 + q\alpha)(\alpha - \alpha^2/2)n + \alpha^2 n(1 - 2q + \alpha q)/3}\right). \tag{5.13}$$

A short calculation shows that this is at most (5.9). (More careful calculation can decrease the '6' in 6Λ a bit, but this is immaterial for our purposes.) □

We will use the lemma in the form of the following corollary.

Corollary 5.5. *Let $n > 0$ be an integer, $\Lambda > 0$ a real, and $c > 0$ a real with $c \leq 1 + \Lambda$. For any integers $i, j > 0$ with $i < j \leq 2i$, setting $\alpha = i/n$, we have*

$$\Pr(\exists k \in (i, j] : W'(k) = 0) \leq K \exp\left(\frac{-3\alpha^3 n(1 - 6\Lambda/\alpha)}{24 - 8\alpha}\right), \tag{5.14}$$

for some absolute constant K.

Proof. Assume without loss of generality that $j \leq n$. Let A be the event $(W'(i) > 0)$, let Z be the event $(\exists k \in (i, j] : W'(k) = 0)$, and for $k \in (i, j]$, define the event $B_k = (W'(k) = 0) \cap (W'(l) \neq 0 \, \forall l \in [k+1, j])$ (i.e., the last zero of $W'(\cdot)$ in $[i, j]$ is at k). Then the events B_k are disjoint and $Z = \bigcup_{k=i+1}^{j} B_k$.

By Lemma 5.4, it is sufficient to show that $\Pr(A \mid Z) \geq 1/K$, since then $\Pr(A) \geq \Pr(A \cap Z) = \Pr(Z)\Pr(A \mid Z) \geq \Pr(Z)/K$, and the inequality follows from (5.9). As the B_k are disjoint, $\Pr(A \mid Z) = \sum_{k=i+1}^{j} \Pr(A \mid B_k)\Pr(B_k \mid Z)$. Since $\sum_{k=i+1}^{j} \Pr(B_k \mid Z) = 1$, it is enough to show that, for $k \in (i, j]$, $\Pr(A \mid B_k) \geq 1/K$.

Now consider the random walk $W'(\cdot)$. For $k > 0$, let $e(k) = nk - \binom{k+1}{2}$ (the number of edge exposures by time k). Then $W'(k) \sim B(e(k), (1+\Lambda)/n) - k + 1$. Thus B_k holds only if we have exactly $k - 1$ successes in $e(k) = nk - \binom{k+1}{2}$ Bernoulli trials. Conditioning on B_k, $A \cap B_k$ is the event that $X \geq i$, where X is the number of these $k - 1$ successes which occur within the first $e(i)$ trials. The $k - 1$ successes are uniformly random over the $e(k)$ trials, so X is a hypergeometric random variable parametrized by $e(k)$ trials, $k - 1$ successes, and $e(i)$ samples. We wish to show that $\Pr(X \geq i) \geq 1/K$.

The times of the $k-1$ successes may be simulated by drawing them randomly without replacement from $\{1,\ldots,n\}$, the rth success falling among the first $e(i)$ trials with probability at least $(e(i)-(r-1))/(e(k)-(r-1))$ (even if all the previous $r-1$ successes fell among the first $e(i)$ trials). Since $(e(i)-(r-1))/(e(k)-(r-1)) \geq e(i)/e(k) - (k-1)/e(k) \geq i/k - 2/n \geq (i-2)/k$, X dominates a random variable $Y \sim B(k-1, (i-2)/k)$, and it suffices to show that $\Pr(Y \geq i) \geq 1/K$.

For any bounded k, $\Pr(Y \geq i) \geq \Pr(Y = k-1)$ is bounded away from 0, so it suffices to consider $k \geq 1001$. Since $i \geq k/2$, the binomial's probability parameter $(i-2)/k$ is at least 0.49. We now consider two cases. If $1-(i-2)/k \leq 100/(k-1)$ then $\Pr(Y \geq i) \geq \Pr(Y = k-1) \geq (1-100/(k-1))^{k-1} \geq 10^{-50}$. Otherwise, Y has variance $\sigma^2 \geq (k-1) \cdot 0.49 \cdot 100/(k-1) = 49$. By the Berry–Esseen theorem (see for example [4]), the error in estimating Y from the corresponding Gaussian is at most $1.88/\sigma$. Since the Gaussian itself exceeds its mean by 3 with probability at least $1/2 - 3/(\sqrt{2\pi}\,\sigma)$, $\Pr(Y \geq i) \geq \Pr(Y \geq \mathbb{E}Y + 3) \geq 1/2 - 3/(\sqrt{2\pi}\,\sigma) - 1.88/\sigma > 0.05$. $\square$

While we will not use it, we note that Lemma 5.4 has an immediate consequence for a component of a random graph.

Corollary 5.6. *Let $n > 0$ be an integer, $\Lambda > 0$ a real, and $c > 0$ a real with $c \leq 1 + \Lambda$. For any integer $i > 0$, setting $\alpha = i/n$, the order of the component G_1 containing vertex 1 in a random graph $G(n, c/n)$ satisfies*

$$\Pr(|G_1| > \alpha n) \leq \exp\left(\frac{-3\alpha^3 n(1 - 6\Lambda/\alpha)}{24 - 8\alpha}\right).$$

Lemma 5.7. *Given an integer $n \geq 4$ and any value $c > 0$, RW has the property that, for any $\alpha \in (0,1]$ and any $\beta \geq \frac{\alpha^2}{8-4\alpha} + \frac{4}{n}$,*

$$\Pr(W' \geq \beta n \mid W'(\alpha n) = 0) \leq \exp\left(-(3\ln(3) - 2)\left(\beta - \frac{\alpha^2}{8-4\alpha} - \frac{4}{n}\right)^2 \frac{n}{\alpha}\right). \qquad (5.15)$$

Proof. If $W'(\alpha n) = 0$ then $W' \leq \max_{t \leq \alpha n} W'(t)$. Since $W'(t)$ decreases by at most 1 per step, and is 0 at αn, we have $W' \leq \alpha n$ and thus we may assume $\beta \leq \alpha$ (or else the result is trivial).

Switching from the vertex-exposure to the edge-exposure view for the random walk, up to and including the tth step of RW, the total number of edge exposures is

$$e(t) := (n-1) + \cdots + (n-t) = tn - \binom{t+1}{2},$$

and the total number of births is $Z_1 + \cdots + Z_{e(t)}$, where the Zs are i.i.d. Bernoulli $\text{Be}(c/n)$ random variables.

For the remainder of the proof, set $i := \alpha n$. Since the number of deaths by the tth vertex exposure is t, and the number of births is $Z_1 + \cdots + Z_{e(t)}$, and we start with one live vertex, the condition $W'(i) = 0$ means that the sequence $Z_1, Z_2, \ldots$ is conditioned by $Z_1 + \cdots + Z_{e(i)} = i - 1$.

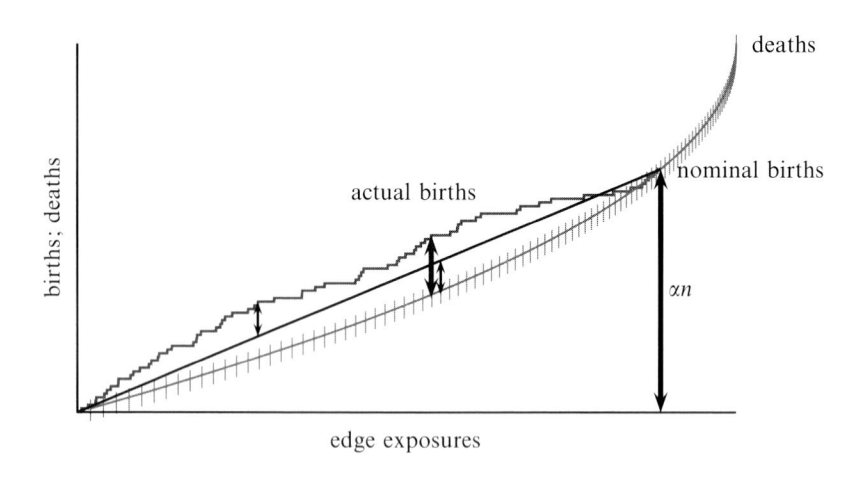

Figure 2. Birth/death process

For any time $t \leq i$, the number of live vertices is given by

$$W'(t) = Z_1 + \cdots + Z_{e(t)} - (t-1) \tag{5.16}$$

$$= \left[Z_1 + \cdots + Z_{e(t)} - \frac{e(t)}{e(i)}(i-1) \right] + \left[\frac{e(t)}{e(i)}(i-1) - (t-1) \right]; \tag{5.17}$$

that is, the gap between the actual number born and its expectation (conditioned on $Z_1 + \cdots + Z_{e(i)} = i - 1$), plus the gap between the expectation and the number of deaths.

This view may be more easily apprehended with reference to Figure 2. The maximum of $W'(t)$, the number of live vertices at any time, is the largest gap between this random walk and the death curve, which can be bounded as the maximum gap between the random walk and a linear interpolation of it, plus the maximum gap between the linear interpolation and the death curve. The first of these two gaps is a random quantity governed by Claim 4.7, while the second is a deterministic function of α and n.

The second term in (5.17) may be bounded as

$$\max_{t \leq i} \left[\frac{e(t)}{e(i)}(i-1) - (t-1) \right] \leq 1 + \max_{t \leq i} \left[\frac{e(t)}{e(i)} i - t \right]$$

$$= 1 + i^2/[8n - 4i - 4]$$

(the maximum occurs at $t = i/2$), and with $i = \alpha n$ and $n \geq 2$, it is easily checked that this is

$$\leq \alpha^2 n/(8 - 4\alpha) + 2.$$

Thus, defining a 'discrepancy' δ by $\delta n = \lfloor \beta n - \alpha^2 n/(8 - 4\alpha) - 2 \rfloor \geq \beta n - \alpha^2 n/(8 - 4\alpha) - 3$, we may rewrite (5.17) as $W'(t) \leq \left[Z_1 + \cdots + Z_{e(t)} - \frac{e(t)}{e(i)}(i-1) \right] + \beta n - \delta n$

to obtain

$$\Pr(\max_{t \leqslant \alpha n} W'(t) \geqslant \beta n \mid W'(\alpha n) = 0)$$
$$\leqslant \Pr\left(\max_{t \leqslant \alpha n}\left[Z_1 + \cdots + Z_{e(t)} - \frac{e(t)}{e(i)}(i-1)\right] \geqslant \delta n \mid W'(\alpha n) = 0\right)$$
$$\leqslant \Pr\left(\max_{j \leqslant e(i)}\left[Z_1 + \cdots + Z_j - \frac{j}{e(i)}(i-1)\right] \geqslant \delta n \mid W'(\alpha n) = 0\right).$$

Recalling that we have conditioned upon $Z_1 + \cdots + Z_{e(i)} = i - 1 = \alpha n - 1$, we apply Theorem 4.7, with $S = i - 1 = \alpha n - 1$ and $b = \delta n$. (The theorem's 'n' is $e(i)$, so its '$p = S/n$' is $\frac{i-1}{e(i)} = \frac{i-1}{in - \binom{i+1}{2}}$, and its hypothesis '$p \leqslant 1/2$' is guaranteed by $n \geqslant 4$.) Using the notation $B(n, p; k)$ for $B_{n,p}(k)$, this shows the probability to be

$$\leqslant \frac{B\big(e(i), (i-1)/e(i); (i-1) + (2\delta n - 1)\big)}{B\big(e(i), (i-1)/e(i); (i-1)\big)}. \tag{5.18}$$

By its definition, $\delta n \leqslant \beta n - 2$, and we have already argued that the probability is 0 unless $\beta \leqslant \alpha$, so we may assume $\delta n \leqslant \beta n - 2 \leqslant \alpha n - 2 = i - 2$, and in particular, $2\delta n - 1 < 2(i-1)$: the deviation of $2\delta n - 1$ in (5.18) is less than twice the mean. Thus we may apply Remark 4.8, showing the probability to be

$$\leqslant \exp\big(-(3\ln(3) - 2)/4 \cdot (2\delta n - 1)^2/(\alpha n - 1)\big)$$
$$\leqslant \exp\left(-(3\ln(3) - 2)\left(\beta - \frac{\alpha^2}{8 - 4\alpha} - \frac{4}{n}\right)^2 \frac{n}{\alpha}\right). \qquad \square$$

We will use the lemma in the form of the following corollary.

Corollary 5.8. *Given an integer $n \geqslant 4$ and any value $c > 0$, RW has the property that, for any $0 < \alpha_0 \leqslant \alpha_1 \leqslant 1$ and any $\beta \geqslant \frac{\alpha_0^2}{8 - 4\alpha_0} + \frac{4}{n}$,*

$$\Pr(W' \geqslant \beta n \mid \exists \alpha \in [\alpha_0, \alpha_1] : W'(\alpha n) = 0)$$
$$\leqslant \exp\left(-(3\ln(3) - 2)\left(\beta - \frac{\alpha_1^2}{8 - 4\alpha_1} - \frac{4}{n}\right)^2 \frac{n}{\alpha_1}\right). \tag{5.19}$$

Proof. Let B_k and Z be the events defined in the proof of Corollary 5.5, where we take $i = \alpha_0 n$ and $j = \alpha_1 n$, and let C be the event $(W' \geqslant \beta n)$. Then,

$$\Pr(C \mid Z) = \sum_{k=i}^{j} \Pr(C \mid B_k) \Pr(B_k \mid Z)$$
$$= \sum_{k=i}^{j} \Pr(C \mid W'(k) = 0) \Pr(B_k \mid Z) \qquad (W'(\cdot) \text{ is Markovian})$$
$$\leqslant \max_{k \in [i,j]} \Pr(C \mid W'(k) = 0).$$

We are now done, by Lemma 5.7. $\square$

6. Remarks on the bounds

Recall that we are aiming to bound κ, the excess of the component of a random graph containing the fixed vertex 1. That the tail bounds on κ must be done carefully – that the constants count – is illustrated by considering the probability that a large (linear-sized) excess arises simply because the random graph $G(n, 1/n)$ has many more edges than expected. Such a graph has $(n + \epsilon n)/2$ edges (rather than the expected $n/2$) with probability $\exp(-\Theta(\epsilon^2 n))$, and in this event the expected value of κ is $\Theta(\epsilon^3 n)$. Thus for $\epsilon = \Theta(1)$, the probability of excess $\kappa = \epsilon^3 n$ is at least $\exp(-\Theta(n))$, the running time is $\exp(\Theta(n))$, and it becomes critical which of the two constants hidden in the respective Thetas is larger. In particular, it is quite conceivable that it really might be essential to have our running-time bound of $2^{\kappa/2}$ rather than the more naive 2^κ that would be obtained by III-reducing on all vertices of degree 3 or more. This example also shows that good tail bounds on κ will be required even in the fantastically improbable regime $\kappa = \Theta(n)$.[3]

As remarked earlier, tail bounds on κ could also be computed by first-moment methods. Writing $C(u, u + \kappa)$ for the number of connected graphs with u vertices and $u + \kappa$ edges, and applying our algorithm to a random graph $G(n, p)$, the expected time spent on components of order u and excess κ is at most $2^{\kappa/2}$ times the expected number of them, namely

$$2^{\kappa/2} \binom{n}{u} C(u, u + \kappa) p^{u+\kappa} (1-p)^{u(n-u) + \binom{u}{2} - u - \kappa}. \tag{6.1}$$

Motivated by the preceding paragraph, we will consider the value of this expression at $p = 1/n$, $u = \epsilon n$ and $\kappa = \epsilon^3 n$, with ϵ constant; we fix these values for the remainder of this section.

Wright [32] shows that $C(u, u + \kappa)$ is asymptotically equal to some explicit constant times

$$(A/\kappa)^{\kappa/2} u^{u+(3\kappa-1)/2} \tag{6.2}$$

where $A = e/12$ and the formula is valid for $1 \ll \kappa \ll u^{1/3}$, i.e., for $\kappa \to \infty$ but $\kappa = o(u^{1/3})$. Since the algorithm's expected time is at most (6.1) summed over all u and κ, and Wright's formula for $C(u, u + \kappa)$ sacrifices only a constant factor, nothing is given up (outside of the $2^{\kappa/2}$ term in (6.1)), and this method of calculation must yield a suitable result (a value linear in n) *in the range* where Wright's formula is applicable. Unfortunately, this range does not include our $u = \epsilon n$, $\kappa = \epsilon^3 n$.

Similarly, Łuczak [25] shows that for $0 \ll \kappa \ll u$,

$$1/(e^8 \sqrt{\kappa})(A/\kappa)^{\kappa/2} u^{u+(3\kappa-1)/2} \leqslant C(u, u+\kappa) \leqslant \sqrt{u^3/\kappa}(A/\kappa)^{\kappa/2} u^{u+(3\kappa-1)/2}$$

where $A = e/12 + O(\kappa/u)$. With ϵ fixed, $u = \epsilon n$ and $\kappa = \epsilon^3 n$, this does not give the needed explicit bound on A. It might be possible to extract such a bound, and also to work with all parameter pairs (u, κ) (not just the demonstrative values we chose here), but even then we

---

[3] Were it not for the need to go up into the tails, we could capitalize on results such as those of Aldous [1], that the joint distribution of the largest component's order and excess is asymptotically normal. In Aldous's analysis, Brownian motion plays the same role as our random walk RW.

would be left with the polynomial leading factors: unless these can be banished, we would be unable to prove that the expected running time is linear rather than merely polynomial.

Bollobás [5] (see also Bollobás [6, V.4]) shows that (6.2) is a universal upper bound on $C(u, u + \kappa)$ for some universal constant A. Substituting (6.2) into (6.1) and evaluating at $p = 1/n$, $u = \epsilon n$ and $\kappa = \epsilon^3 n$ gives (up to small polynomial factors of n and ϵ) $[c(\epsilon)A]^{\epsilon^3 n/2}$, where $c(\epsilon)$ is an easily calculated explicit function. For this method to show that the expected time (for this u and κ) is polynomial in n, we would need a reasonably small upper bound on the constant A.

Bender, Canfield and McKay [3] (in the same journal issue as Łuczak's [25]) give extremely accurate estimates for $C(u, u + \kappa)$, whose substitution into expression (6.1) must in principle satisfy our needs. Their formula, though, is rather complex, involves an implicitly defined function, and appears difficult to work with.

In short, while suitable tail bounds could presumably be proved by a first-moment calculation, there are complications. We therefore chose to adopt the branching-process approach.

The branching-process approach also provides some intuition. It is a classical Erdős–Rényi result [13, 12] that a random graph at the critical density $1/n$ typically has a giant component whose size is $\Theta(n^{2/3})$. Since our component sizes are given as αn, the giant component (which is likely to be the most difficult component for our algorithm to solve, and thus the component we should focus on) would have $\alpha n = \Theta(n^{2/3})$, $\alpha = \Theta(n^{-1/3})$. The inequality (5.9) 'allows' such a component, giving its probability as $\exp(-\Theta(1))$.

Just above the scaling window, in a random graph $G(n, p)$ with $p = (1 + \epsilon)/n$ where $n^{-1/3} \ll \epsilon \ll 1$ (or equivalently $p = (1 + \lambda n^{-1/3})/n$ where $1 \ll \lambda \ll n^{1/3}$), up to constant factors the giant component's order is typically $\epsilon n = \lambda n^{2/3}$ and its excess typically $\epsilon^3 n = \lambda^3$, suggesting we consider $\alpha = \lambda n^{-1/3}$ and $\kappa = \lambda^3$. (See Bollobás [6, Chapter VI] and particularly Janson, Łuczak and Ruciński [20, Section 5.4] for this and related results.) Since our bound suggests that κ might be about $(\alpha n)(\beta n)(c/n)$, this would suggest that the typical 'width' (maximum number of live vertices) βn of the giant component would have $\beta = \lambda^2 n^{-2/3}$. Notice that (5.15) imposes no penalty on β until $\beta = \Theta(\alpha^2)$, and the above values $\alpha = \lambda n^{-1/3}$, $\beta = \lambda^2 n^{-2/3}$ fall just at this point.

In summary, we think that the branching-process analysis offers insight into the likelihood or unlikelihood of observing graph components of given order and excess. Especially if one takes as given the properties of binomial distributions and staircase random walks proved in Section 4, the branching process analysis is not unduly complicated, and is entirely self-contained.

7. Assembly

In this section we state and prove the main result, Theorem 7.1; note that Theorem 1.1 follows as a special case.

Theorem 7.1. *For any $\lambda = \lambda(n) > 0$ and $c \leqslant 1 + \lambda n^{-1/3}$, let $G \in \mathcal{G}(n, c/n)$ be a random graph, and let (G, S) be any weighted Max 2-CSP instance over this graph. Then (G, S) can be solved exactly in expected time $O(n) \exp(O(1 + \lambda^3))$, and in space $O(m + n)$.*

Proof. Consider Algorithm A applied to the graph G. We calculate the running time as follows: for each component G' of G, let $t(G')$ be the time taken by Algorithm A to find an optimal assignment for G'. For each vertex v of G', we define $t(v) = t(G')/|G'|$. Then the total running time of Algorithm A is $O(m+n) + \sum_{v \in V(G)} t(v)$. Choosing any fixed vertex $v \in V(G)$ (say, $v = 1$), we see that the expected running time is at most

$$O((1+c)n) + n\mathbb{E}t(v). \tag{7.1}$$

It therefore suffices to prove that

$$\mathbb{E}\,t(v) \leqslant \exp(O(1+\lambda^3)) = \exp(O(1+\Lambda^3 n)), \tag{7.2}$$

where $\Lambda = \lambda n^{-1/3}$.

Recall the random walk RW of Definition 5.1, with order U' and maximum width W'. We examine the contribution of each possible pair (U', W') to the expectation. Specifically, for integers αn and βn, define $E(\alpha, \beta)$ to be the expected time that the algorithm spends on cases with $U' = \alpha n$ and $W' = \beta n$. Recalling that if $U' < n$ then $W' \leqslant U'$, we compute the contribution separately for $0 \leqslant \beta \leqslant \alpha < 1$ (this forms the bulk of the argument to come, including Cases 1–4 below), and for $\alpha = 1$ (Case 5).

As we are computing asymptotics in n, we may assume that n is large ($n \geqslant 10\,000$). It is enough to prove that

$$\sum_{\alpha n, \beta n} E(\alpha, \beta) \leqslant \exp(O(1+\lambda^3)). \tag{7.3}$$

To do so, we break $[0,1]^2$ into rectangular regions and bound the sum of $E(\alpha, \beta)$ separately over each region. Assuming $\alpha < 1$, the amortized running time of Algorithm A is bounded by Claim 5.3. Thus for any rectangle $R = [\alpha_0, \alpha_1] \times [\beta_0, \beta_1] \subseteq [0,1) \times [0,1]$,

$$\sum_{(\alpha,\beta) \in R} E(\alpha, \beta) \leqslant O(1) \exp\left\{ \max_R [(1+\Lambda)(\sqrt{2}-1)\alpha\beta]n \right\} \Pr((U'/n, W'/n) \in R)$$

$$\leqslant O(1) \exp\left\{ \max_R [(1+\Lambda)(\sqrt{2}-1)\alpha\beta]n \right\}$$

$$\cdot \Pr(\exists \alpha \in [\alpha_0, \alpha_1] : W'(\alpha n) = 0)$$

$$\cdot \Pr(W'/n \in [\beta_0, \beta_1] \mid \exists \alpha \in [\alpha_0, \alpha_1] : W'(\alpha n) = 0).$$

By Corollaries 5.5 and 5.8,

$$\sum_{(\alpha,\beta) \in R} E(\alpha, \beta) \leqslant O(1) \exp\left\{ \left(\max_R [(1+\Lambda)(\sqrt{2}-1)\alpha\beta] - \left[\frac{3\tilde{\alpha}^3(1 - 6\Lambda/\tilde{\alpha})}{24 - 8\tilde{\alpha}} I(\tilde{\alpha}) \right] \right. \right. \tag{7.4}$$

$$\left. \left. - \left[(3\ln(3) - 2) \left(\beta_0 - \frac{\alpha_1^2}{8 - 4\alpha_1} - \frac{4}{n} \right)^2 \frac{1}{\alpha_1} J\left(\alpha_1, \beta_0 - \frac{4}{n}\right) \right] \right) n \right\},$$

where $\tilde{\alpha}$ is $\alpha_0 - 1/n$ (adjusting for the open interval of Corollary 5.5), $I(\tilde{\alpha})$ is the indicator function for $\tilde{\alpha} > 6\Lambda$, and $J(\alpha, \beta)$ is the indicator function for $\beta \geqslant \alpha^2/(8 - 4\alpha)$. We reiterate that (7.4) bounds the *sum* of $E(\alpha, \beta)$ over R, not just the maximum.

Without loss of generality we restrict Λ to $\Lambda > n^{-1/3}$, since for smaller Λ the quantity (7.4) decreases monotonically while the target bound $n \exp(O(1 + \Lambda^3 n))$ does not decrease below $n\exp(O(1))$. We may also restrict Λ to $\Lambda < 0.01$, since by then the target bound $n\exp(O(0.01^3 n))$ allows time for the algorithm to III-reduce on every vertex.

Given that $\Lambda < 0.01$ and $\alpha \leqslant 1$, in (7.4) we may replace each $\tilde{\alpha}$ by α at the expense of an $O(1)$ factor outside the whole expression: if $I(\tilde{\alpha}) = I(\alpha) = 1$ this is easy to check, while if $I(\tilde{\alpha}) = 0$ but $I(\alpha) = 1$ then $6\Lambda < \alpha \leqslant 6\Lambda + 1/n$, and a short calculation shows that the exponent (not forgetting the factor of n) changes by at most $O(1)$ and so the expression is multiplied by $\exp(O(1)) = O(1)$.

Also, given that $\Lambda < 0.01$ and $\alpha \leqslant 1$, on substituting $\beta = \beta' + \frac{4}{n}$ into (7.4), the first term (again taking into account the n inside the exponent) is $\leqslant \exp\{(1+\Lambda)(\sqrt{2}-1)\alpha\beta' n + 4\}$, and we may simply move the $\exp(4)$ outside and subsume it into the leading $O(1)$. That is, we may simply ignore the $\frac{4}{n}$ in (7.4), extending the range of summation to $\beta \in [-\frac{4}{n}, 1]$, and so summing, over rectangles $R = [\alpha_0, \alpha_1] \times [\beta_0, \beta_1]$, $O(1)$ times

$$\exp\left\{\left(\max_R[(1+\Lambda)(\sqrt{2}-1)\alpha\beta] - \left[\frac{3\alpha_0^3(1-6\Lambda/\alpha_0)}{24-8\alpha_0}I(\alpha_0)\right]\right. \\ \left. - \left[(3\ln(3)-2)\left(\beta_0 - \frac{\alpha_1^2}{8-4\alpha_1}\right)^2 \frac{1}{\alpha_1} J(\alpha_1, \beta_0)\right]\right)n\right\}. \quad (7.5)$$

We now consider five regimes of values (α, β), which together cover the space $[0,1] \times [-\frac{4}{n}, 1]$:

(1) α and β both small: $(\alpha, \beta) \in [0, 1000\Lambda] \times [0, 1000000\Lambda^2]$,
(2) $\beta \leqslant 0$,
(3) $\alpha < 1$ and $0 \leqslant \beta \leqslant 1.03\alpha^2/(8-4\alpha)$ (and excluding Case 1),
(4) $\alpha < 1$ and $\beta \geqslant 1.03\alpha^2/(8-4\alpha)$ (and excluding Case 1),
(5) $\alpha = 1$.

Only the first case contains likely and significant pairs (α, β), and it defines the bound in (7.3); the second case trivially makes a negligible contribution; the remaining cases also contribute negligibly but are a little trickier to analyse. Note that the $\alpha = 1$ case must be treated differently: if $U' = \alpha n = n$ we may have $W'(\alpha n) > 0$, Corollary 5.8 (with $\alpha_0 = \alpha_1 = 1$) does not apply, and thus neither does (7.5).

Case 1. $R = [0, 1000\Lambda] \times [0, 1000000\Lambda^2]$. This rectangle R contains likely pairs (α, β), and we simply estimate R's probability as 1. Thus (7.5) becomes

$$\sum_R E(\alpha, \beta) \leqslant \exp\{(1+\Lambda)(\sqrt{2}-1) \cdot 1000\Lambda \cdot 1000000\Lambda^2 \cdot n\} = \exp(O(\Lambda^3 n)).$$

Case 2. $R = [0,1] \times [-\frac{4}{n}, 1]$. Again we simply estimate R's probability as 1; since $\beta \leqslant 0$, the exponential's value is at most 1.

Case 3. Instead of considering only $\beta \leqslant 1.03\alpha^2/(8-4\alpha)$, we will treat a larger domain, $\beta \leqslant 1.03\alpha^2/4$. If $\alpha \leqslant 1000\Lambda$ this falls under Case 1, so we need only consider $\alpha \geqslant 1000\Lambda$. We cover the space $1000\Lambda \leqslant \alpha \leqslant 1$, $\beta \leqslant 1.03\alpha^2/4$ with rectangles

$$R_i = [\alpha_i^\star, 1.01\alpha_i^\star] \times [0, 1.03(1.01\alpha_i^\star)^2/4],$$

with $\alpha_i^\star = 1000\Lambda \cdot 1.01^i$, $i = 0, 1, \ldots, I - 1$, where $\alpha_I^\star$ is the first term larger than 1. Within any such rectangle R, writing $\alpha^\star$ for $\alpha_i^\star$ (and noting that $\alpha^\star \leqslant 1$), (7.5) is at most

$$\exp\left\{\left([(1+\Lambda)(\sqrt{2}-1) \cdot 1.01^3 \cdot 1.03\,\alpha^{\star 3}/4] - \left[\frac{3\alpha^{\star 3}(1 - 6\Lambda/\alpha^\star)}{24 - 8\alpha^\star}\right]\right)n\right\}$$

$$\leqslant \exp\left\{\left([1.01\,(\sqrt{2}-1) \cdot 1.01^3 \cdot 1.03/4] - \left[\frac{3(1-6/1000)}{24}\right]\right)\alpha^{\star 3} n\right\}$$

$$\leqslant \exp(-0.013\alpha^{\star 3} n).$$

Recalling that $\alpha_i^\star = 1000\Lambda \cdot 1.01^i$ and that $\Lambda \geqslant n^{-1/3}$, the contribution to the overall sum (7.3) is at most

$$\sum_{i=0}^{\infty} \exp(-0.013 \cdot 1000^3 \Lambda^3 \cdot 1.01^{3i} n) \leqslant \sum_{i=0}^{\infty} \exp(-13\,000\,000 \cdot 1.03^i) = O(1).$$

Case 4. We divide this case into two sub-cases, $\alpha \leqslant 100\Lambda$ and $\alpha > 100\Lambda$.

Sub-case: $\alpha \leqslant 100\Lambda$. If $\beta \leqslant 1000000\Lambda^2$ then Case 1 applies, so we need only consider $\beta > 1000000\Lambda^2$. Break this domain into rectangles,

$$R_j = [0, 100\Lambda] \times [\beta_j^\star, 1.01\beta_j^\star],$$

with $\beta_j^\star = 1000000\Lambda^2 \cdot 1.01^j$, $j = 0, 1, \ldots$; it does no harm to let j run to infinity. For any such rectangle R, writing $\beta^\star$ for $\beta_j^\star$, a crude upper bound on (7.5) is given by

$$\exp\left\{\left([(1+\Lambda)(\sqrt{2}-1)100\Lambda \cdot 1.01\beta^\star] - \left[(3\ln(3) - 2)(\beta^\star - (100\Lambda)^2)^2 \frac{1}{100\Lambda}\right]\right)n\right\}.$$

Since $\Lambda \leqslant 0.01$ and, by definition of $\beta_0^\star$, $(100\Lambda)^2 \leqslant 0.01\beta^\star$, this is

$$\leqslant \exp\left\{\left([(\sqrt{2}-1)1.01^2(100\Lambda)\beta^\star] - [(3\ln(3)-2)(0.99\beta^\star)^2/100\Lambda]\right)n\right\}.$$

Noting that $\beta^\star/100\Lambda \geqslant 10000\Lambda$, and factoring out $\Lambda\beta^\star$, this is

$$\leqslant \exp\left\{\left([(\sqrt{2}-1)1.01^2 \cdot 100] - [(3\ln(3)-2)\,0.99^2 \cdot 10000]\right)\Lambda\beta^\star n\right\}$$

$$\leqslant \exp(-12000\Lambda\beta^\star n).$$

Summing over the rectangles R_j with $\beta_j^\star = 1000000\Lambda^2 \cdot 1.01^j$ gives a contribution to (7.3) of at most

$$\sum_{j=0}^{\infty} \exp(-12000\Lambda^3 1.01^j n) \leqslant \sum_{j=0}^{\infty} \exp(-12000 \cdot 1.01^j) = O(1).$$

Sub-case: $\alpha > 100\Lambda$. This case, with $\alpha > 100\Lambda$ and $\beta > 1.03\alpha^2/(8 - 4\alpha)$, is the most delicate. Observations of such values α, and of β conditioned upon α, are both unlikely, and we need to keep all three terms in (7.5).

In this case, we break down (a superset of) the domain into rectangles

$$R_{ij} = [\alpha_i^\star, 1.001\alpha_i^\star] \times [\beta_{ij}^\star, 1.01\beta_{ij}^\star],$$

with $\alpha_i^\star = 100\Lambda \cdot 1.001^i$ and $\beta_{ij}^\star = 1.03(\alpha_i^\star)^2/(8-4\alpha_i^\star) \cdot 1.01^j$, over $i = 0,\ldots,I-1$ and $j = 0,\ldots,J(i)-1$, where I is the first value for which $\alpha_I > 1$, and $J(i)$ the first value for which $\beta_{i,J(i)} > 1$. Within any such rectangle R given by $\alpha^\star = \alpha_i^\star$ and $\beta^\star = \beta_{ij}^\star$, we have $J(1.001\alpha^\star, \beta^\star) = 1$ (from $1.03\alpha^{\star 2}/(8-4\alpha) \geqslant 1.001^2 \alpha^{\star 2}/(8 - 4 \cdot 1.001\alpha^\star)$), and thus (7.5) is at most

$$\exp\left\{\left([(1.01)(\sqrt{2}-1) \cdot 1.001\alpha^\star \cdot 1.01\beta^\star] - \left[\frac{3\alpha^{\star 3}(1-6/100)}{24-8\alpha^\star}\right]\right.\right.$$
$$\left.\left. - \left[(3\ln(3)-2)\left(\beta^\star - \frac{1.001^2 \alpha^{\star 2}}{8-4 \cdot 1.001\alpha^\star}\right)^2 \frac{1}{1.001\alpha^\star}\right]\right)n\right\}$$
$$=: \exp(-f(\alpha^\star, \beta^\star)n), \tag{7.6}$$

and we claim that $f(\alpha, \beta) \geqslant 0.07\alpha\beta$ for all $0 \leqslant \alpha \leqslant 1$, $1.03\alpha^2/(8-4\alpha) \leqslant \beta \leqslant 1$.

Verifying this is fairly straightforward. $f(\alpha, \beta)/(\alpha\beta)$ is, for a given α, extremized either by an extreme value of β – namely $\beta = 1$ or $\beta = 1.03\alpha^2/(8-4\alpha)$ – or by the point where its derivative $\frac{\partial}{\partial \beta}$ vanishes. The derivative can be computed in closed form (a computer-algebra package helps), and is a rational expression whose numerator is a quadratic expression (with no linear term) in β, whose positive root may thus be written down as an explicit function $\beta = \beta(\alpha)$. Substituting $\beta = 1$, $\beta = 1.03\alpha^2/(8-4\alpha)$, or $\beta = \beta(\alpha)$ into $f(\alpha, \beta)/(\alpha\beta)$ yields in each case a function which is tractable over $\alpha \in [0, 1]$. When the three are graphed, it is easily seen that the smallest value is achieved at $\alpha = 1$, $\beta = \beta(\alpha)$, where $f(\alpha, \beta)/(\alpha\beta) > 0.07$ (it is about 0.0704). This observation can be confirmed straightforwardly if tediously.

Now, summing the contributions of the rectangles R_{ij} to (7.3) is done as in the previous cases. It is at most

$$\sum_{i=0}^{I}\sum_{j=0}^{J(i)} \exp(-f(\alpha_i^\star, \beta_{ij}^\star)n)$$
$$\leqslant \sum_{i=0}^{\infty}\sum_{j=0}^{\infty} \exp(-0.07\alpha_i^\star \beta_{ij}^\star n)$$
$$\leqslant \sum_{i=0}^{\infty}\sum_{j=0}^{\infty} \exp(-0.07 \cdot (100\Lambda \cdot 1.001^i) \cdot (1.03 \cdot [100\Lambda \cdot 1.001^i]^2)/8 \cdot 1.01^j \cdot n)$$
$$\leqslant \sum_{i=0}^{\infty}\sum_{j=0}^{\infty} \exp(-9000\,\Lambda^3 1.003^i\, 1.01^j\, n)$$
$$\leqslant \sum_{i=0}^{\infty}\sum_{j=0}^{\infty} \exp(-9000 \cdot 1.003^i\, 1.01^j)$$
$$= O(1).$$

Case 5. If $U' = n$ then $W'(n) \geqslant 0$, and so we must have had at least $n-1$ births among the $\binom{n}{2}$ Bernoulli random variables comprising RW. Denote this (random) number of births by $M' \sim B\bigl(\binom{n}{2}, c/n\bigr)$. Let $M \sim B\bigl(\binom{n}{2}, c/n\bigr)$ be the number of edges in the whole graph. (M' and M have the same distribution but represent different quantities.) From Theorem 3.3, the amortized running time for v is at most $2^{M/4}$. It suffices to show that $\mathbb{E}(2^{M/4} \cdot \mathbf{1}_{M' \geqslant n-1}) = o(1)$, where $\mathbf{1}$ denotes the indicator function of an event.

Recall that without loss of generality we are assuming $\Lambda \leqslant 0.01$ and $n \geqslant 10000$ (so $n - 1 \geqslant 0.9999n$). Thus it suffices to compute an upper bound on

$$\mathbb{E}(2^{M/4} \cdot \mathbf{1}_{M' \geqslant 0.9999n})$$
$$= \mathbb{E}(2^{M/4} \cdot \mathbf{1}_{M < 1.053n} \mathbf{1}_{M' \geqslant 0.9999n}) + \mathbb{E}(2^{M/4} \cdot \mathbf{1}_{M \geqslant 1.053n} \mathbf{1}_{M' \geqslant 0.9999n})$$
$$\leqslant \mathbb{E}(2^{1.053n/4} \cdot \mathbf{1}_{M' \geqslant 0.9999n}) + \mathbb{E}(2^{M/4} \cdot \mathbf{1}_{M \geqslant 1.053n}).$$

We apply Chernoff's inequality (5.7) to both terms. Note that $\mathbb{E}M' = \mathbb{E}M = \binom{n}{2} \cdot c/n \leqslant \frac{1}{2} \cdot 1.01n = 0.505n$.

For the first term, $M' \geqslant 0.9999n$ means that $M' - \mathbb{E}M' \geqslant 0.4949n$, so

$$\mathbb{E}(2^{1.053n/4} \cdot \mathbf{1}_{M' \geqslant 0.9999n}) = 2^{1.053n/4} \Pr(M' \geqslant 0.9999n)$$
$$\leqslant 2^{1.053n/4} \exp\left(-\frac{(0.4949n)^2}{2 \cdot 0.505n + 2/3 \cdot 0.4949n}\right)$$
$$\leqslant \exp(1.053 \ln(2) n/4) \cdot \exp(-0.1827n)$$
$$\leqslant \exp(-0.0002n)$$
$$= o(1).$$

Noting that $\mathbb{E}M - 0.505n \leqslant 0$, the second term is at most

$$\sum_{m \geqslant 1.053n} 2^{m/4} \Pr(M \geqslant \mathbb{E}M + m - 0.505n)$$
$$= 2^{0.505n/4} \sum_{t \geqslant 0.548n} 2^{t/4} \Pr(M \geqslant \mathbb{E}M + t)$$
$$\leqslant 2^{0.505n/4} \sum_{t \geqslant 0.548n} 2^{t/4} \exp\left(\frac{-t^2}{2 \cdot 0.505n + 2t/3}\right). \quad (7.7)$$

Writing $t = \gamma n$, we see that the logarithm of the summand is

$$\frac{t}{4} \ln 2 - \frac{t^2}{1.01n + 2t/3} = \left(\frac{\gamma \ln 2}{4} - \frac{\gamma^2}{1.01 + 2\gamma/3}\right) n \leqslant -0.123n$$

for $\gamma \geqslant 0.548$. Since $\Pr(M \geqslant \mathbb{E}M + t) = 0$ for $t > \binom{n}{2}$, expression (7.7) is at most

$$\binom{n}{2} 2^{0.505n/4} \exp(-0.123n) \leqslant \tfrac{1}{2}n^2 \exp\bigl((0.505/4 \ln 2 - 0.123)n\bigr)$$
$$\leqslant \tfrac{1}{2}n^2 \exp(-0.03n)$$
$$= o(1).$$

Thus, the case $U' = n$ contributes $o(1)$ to the amortized time.

It follows that, in each of the five cases above, the contributions of the corresponding $U' = \alpha n$, $W' = \beta n$ sum to $\exp(O(1 + \lambda^3))$, and the result is proved. □

8. Conclusions

In the present paper we focus on Max Cut. Our result for random 'sparse' instances is strong in that it applies not only right up to $c = 1$, but through the scaling window $c = 1 + \lambda n^{-1/3}$.

We believe that our methods can be extended to Max 2-Sat, but the analysis is certainly more complicated. In fact our results already apply to any Max 2-CSP, and in particular to Max 2-Sat, but only in the regime where there are about $n/2$ clauses on n variables; since it is likely that random Max 2-Sat instances with up to about n clauses can be solved efficiently on average (the Max 2-Sat phase transition occurs around n clauses), our present result for Max 2-Sat is relatively weak. As was the case with Max Cut, it is easy to see that with $m = cn$ and $c < 1$ it is almost always easy to solve Max 2-Sat for a random formula with m clauses on n variables; this follows immediately from the facts that decision 2-Sat is easy and that with high probability such a formula is completely satisfiable. The hard part is to show that it is easy not just with high probability but in expectation.

Since Max Cut is in general NP-hard (and even NP-hard to approximate to better than a 16/17 factor; see Trevisan, Sorkin, Sudan and Williamson [31]), it would be interesting to resolve whether random instances *above* the giant-component threshold can be solved in polynomial expected time (as we have shown for those below the threshold); we conjecture that they cannot. A related question is whether it is possible to solve Max Cut for a random cubic graph in polynomial expected time; again, we conjecture that it is not. It would also be interesting to know if there is a 'mildly exponential' algorithm for such graphs (*e.g.*, one with running time $\exp(O(\sqrt{n}))$). If not, because the 'kernel' of a random graph just above the giant-component threshold is close to being a random cubic graph (see Janson, Łuzcak and Ruciński [20, Section 5.4]), our running time of $O(n)\exp(O(1 + \lambda^3))$ may be the best possible. Questions about average-case hardness are of broad interest, but no quick resolution is in sight.

Acknowledgements

We are grateful to Don Coppersmith for helpful and enjoyable conversations. We also thank the referee for a thorough reading and many constructive comments.

References

[1] Aldous, D. (1997) Brownian excursions, critical random graphs and the multiplicative coalescent. *Ann. Probab.* **25** 812–854.

[2] von Bahr, B. and Martin-Löf, A. (1980) Threshold limit theorems for some epidemic processes. *Adv. Appl. Probab.* **12** 319–349.

[3] Bender, E. A., Canfield, E. R. and McKay, B. D. (1990) The asymptotic number of labeled connected graphs with a given number of vertices and edges. *Random Struct. Alg.* **1** 127–169.

[4] Berry, A. C. (1941) The accuracy of the Gaussian approximation to the sum of independent variates. *Trans. Amer. Math. Soc.* **49** 122–136.

[5] Bollobás, B. (1984) The evolution of sparse graphs. In *Graph Theory and Combinatorics* (Cambridge 1983), Academic Press, London, pp. 35–57.

[6] Bollobás, B. (2001) *Random Graphs*, Vol. 73 of *Cambridge Studies in Advanced Mathematics*, Cambridge University Press, Cambridge.

[7] Bollobás, B., Borgs, C., Chayes, J. T., Kim, J. H. and Wilson, D. B. (2001) The scaling window of the 2-SAT transition. *Random Struct. Alg.* **18** 201–256.

[8] Chvátal, V. and Reed, B. (1992) Mick gets some (the odds are on his side). In *33th Annual Symposium on Foundations of Computer Science* (Pittsburgh, PA, 1992), IEEE Comput. Soc. Press, Los Alamitos, CA, pp. 620–627.

[9] Coja-Oghlan, A., Moore, C. and Sanwalani, V. Max k-cut and approximating the chromatic number of random graphs. To appear in *Random Struct. Alg.*

[10] Coja-Oghlan, A. and Taraz, A. (2003) Colouring random graphs in expected polynomial time. In *Proc. STACS 2003*, Vol. 2607 of *Lecture Notes in Computer Science*, Springer, pp. 487–498.

[11] Coppersmith, D., Gamarnik, D., Hajiaghayi, M. T. and Sorkin, G. B. (2004) Random MAX SAT, random MAX CUT, and their phase transitions. *Random Struct. Alg.* **24** 502–545.

[12] Erdős, P. and Rényi, A. (1960) On the evolution of random graphs. *Magyar Tud. Akad. Mat. Kutató Int. Közl.* **5** 17–61.

[13] Erdős, P. and Rényi, A. (1961) On the evolution of random graphs. *Bull. Inst. Internat. Statist.* **38** 343–347.

[14] Fernandez de la Vega, W. (1992) On random 2-SAT. Manuscript.

[15] Gellert, W., Kästner, H., Küstner, H. and Hellwich, M., eds (1977) *The VNR Concise Encyclopedia of Mathematics*, Van Nostrand Reinhold Co., New York.

[16] Goerdt, A. (1996) A threshold for unsatisfiability. *J. Comput. System Sci.* **53** 469–486.

[17] Gramm, J., Hirsch, E. A., Niedermeier, R. and Rossmanith, P. (2003) New worst-case upper bounds for MAX-2-SAT with an application to MAX-CUT. *Discrete Appl. Math.* **130** 139–155.

[18] Hirsch, E. A. (2000) A new algorithm for MAX-2-SAT. In *Proc. STACS 2000* (Lille), Vol. 1770 of *Lecture Notes in Computer Science*, Springer, Berlin, pp. 65–73.

[19] Hoeffding, W. (1963) Probability inequalities for sums of bounded random variables. *J. Amer. Statist. Assoc.* **58** 13–30.

[20] Janson, S., Łuczak, T. and Ruciński, A. (2000) *Random Graphs*, Wiley-Interscience Series in Discrete Mathematics and Optimization, Wiley-Interscience, New York.

[21] Karp, R. M. (1990) The transitive closure of a random digraph. *Random Struct. Alg.* **1** 73–93.

[22] Kendall, D. G. (1956) Deterministic and stochastic epidemics in closed populations. In *Proc. Third Berkeley Symposium on Mathematical Statistics and Probability, 1954–1955, Vol. IV* (Berkeley and Los Angeles), University of California Press, pp. 149–165.

[23] Krivelevich, M. and Vu, V. H. (2002) Approximating the independence number and the chromatic number in expected polynomial time. *J. Combin. Optim.* **6** 143–155.

[24] Kulikov, A. S. and Fedin, S. S. (2002) Solution of the maximum cut problem in time $2^{|E|/4}$. *Zap. Nauchn. Sem. S.-Peterburg. Otdel. Mat. Inst. Steklov.* (POMI) **293** 129–138, 183.

[25] Łuczak, T. (1990) On the number of sparse connected graphs. *Random Struct. Alg.* **1** 171–173.

[26] Martin-Löf, A. (1986) Symmetric sampling procedures, general epidemic processes and their threshold limit theorems. *J. Appl. Probab.* **23** 265–282.

[27] Niedermeier, R. and Rossmanith, P. (1999) New upper bounds for MaxSat. In *Automata, Languages and Programming* (Prague, 1999), Vol. 1644 of *Lecture Notes in Computer Science*, Springer, Berlin, pp. 575–584.

[28] Scott, A. D. and Sorkin, G. B. (2003) Faster algorithms for MAX CUT and MAX CSP, with polynomial expected time for sparse instances. In *Proc. Random 2003* (Baltimore, MD, 2003), Springer, pp. 382–395.

[29] Scott, A. D. and Sorkin, G. B. Faster exponential-time algorithms for Max 2-Sat, Max Cut, and Max k-Cut. In preparation.
[30] Spencer, J. (1997) Enumerating graphs and Brownian motion. *Comm. Pure Appl. Math.* **50** 291–294.
[31] Trevisan, L., Sorkin, G. B., Sudan, M. and Williamson, D. P. (2000) Gadgets, approximation, and linear programming. *SIAM J. Comput.* **29** 2074–2097.
[32] Wright, E. M. (1980) The number of connected sparsely edged graphs III: Asymptotic results. *J. Graph Theory* **4** 393–407.